D1624674

LINCOLN'S MENTORS

THE EDUCATION OF A LEADER

MICHAEL J. GERHARDT

ch.
CUSTOM
HOUSE

HarperCollins books may be purchased for educational, business, or sales promotional use. For information, please e-mail the Special Markets Department at SPsales@harpercollins.com.

FIRST EDITION

Designed by Leah Carlson-Stanisic
Photograph by Everett Collection/Shutterstock, Inc.

Library of Congress Cataloging-in-Publication Data

Names: Gerhardt, Michael J., 1956– author.
Title: Lincoln's mentors : the education of a leader / Michael J. Gerhardt.
Description: First edition. | New York, NY : Custom House, [2021] | Includes bibliographical references and index.
Identifiers: LCCN 2020038379 (print) | LCCN 2020038380 (ebook) | ISBN 9780062877192 (hardcover) | ISBN 9780062877185 (trade paperback) | ISBN 9780062877208 (ebook)
Subjects: LCSH: Lincoln, Abraham, 1809-1865—Friends and associates. | Lincoln, Abraham, 1809–1865—Political career before 1861. | United States—Politics and government— 19th century.
Classification: LCC E457.2 .G44 2021 (print) | LCC E457.2 (ebook) | DDC 973.7092—dc23
LC record available at https://lccn.loc.gov/2020038379
LC ebook record available at https://lccn.loc.gov/2020038380

ISBN 978-0-06-287719-2

21 22 23 24 25   LSC   10 9 8 7 6 5 4 3 2 1

IN MEMORY OF MY MOTHER, SHIVIA LEE GERHARDT
(1929–2020). HER SELFLESS DEVOTION, LOVE,
PATIENCE, COURAGE, AND SUPPORT ARE ENDURING
BLESSINGS AND MODELS TO US ALL.

Upon the subject of education, not presuming to dictate any plan or system respecting it, I can only say that I view it as the most important subject we as a people can be engaged in.

—ABRAHAM LINCOLN, First Campaign Speech, March 9, 1832

CONTENTS

LINCOLN'S MENTORS

THE SEARCH FOR LINCOLN'S TEACHERS

In times of crisis, Americans look to Abraham Lincoln. That impulse has been especially strong in the year 2020, as the nation has simultaneously grappled with a pandemic costing more than a quarter of a million American lives, a recession causing unemployment exceeding Depression-level numbers, mass protests against racism and police brutality, waves of violence across the land, and an impending presidential election bound to inflame divisions among America's public. For such a fraught time, Lincoln's eloquence, steady hand, and determination in leading an embattled nation to overcome secession, brutal civil war, and a severely weakened economy have become touchstones for Americans yearning for unifying, calming leadership.

Yet, at the heart of Abraham Lincoln's successes and story is a mystery that has intrigued historians and admirers. How did a man with no executive experience and only a single term in Congress become America's greatest president? That is the question nearly everyone asks about Lincoln. This book suggests an answer. The most common view of Lincoln as a political genius does not give Lincoln his due. To be certain, his political acumen and soaring rhetoric are matched by few others in American history. But Lincoln had a handful of men to whom he turned for guidance and inspiration throughout his life. Even as a young man, Lincoln knew enough to know he needed mentors. He could not learn in isolation all the skills he needed to become a great leader.

Consider, for example, the popular depiction of Abraham Lincoln in the years 1849–1856 as lost, alone, and in desperate need

of inspiration. True, when Lincoln returned home to Springfield, Illinois, after his seemingly lackluster two years in the U.S.
House of Representatives, his prospects were bleak. After years of
struggling to make a mark in national politics, he worried that he
had failed and that his political career was over. The Whig Party
opposed nominating him for a second term, in spite of his loyal
service to the party for more than a decade. He did not want to
leave office, but the strong stance he had taken in opposition to the
Mexican War had eroded his local support, given that Illinois had
been among the states with the largest numbers of volunteers for
the conflict. Many of the people back home, including his fellow
Whigs, mocked him as "Spotty Lincoln," a nickname they coined
after he failed to persuade the House to approve a resolution criticizing President James Polk for lying to Congress about the exact spot where Mexicans had fired the first shots that started the
Mexican War. Lincoln had campaigned for Zachary Taylor in 1848,
but, once in office, Taylor denied his application to become the
commissioner of the Land Office. Worse still, Taylor died in 1850,
elevating to the presidency his vice president, Millard Fillmore, a
Whig who did not take Lincoln seriously.

By the time Fillmore left office in 1853, Lincoln had to confront
the fact that the Whig Party was dying. Its demise left him without a party apparatus to sustain his political future. Dispirited,
he turned his attention to reviving his law practice with William
Herndon, building it into a highly respected but ultimately not
very lucrative firm. His dreams of becoming a great lawyer were
crushed when the nationally renowned Edwin Stanton of Ohio
dismissed him from the biggest case of his career in 1855.

As Lincoln's fortunes dimmed in the 1850s, his longtime Democratic rival, Stephen Douglas, had become a senator for Lincoln's
home state of Illinois and a rapidly rising star in national politics.
In 1854, Douglas took center stage nationally in drafting and securing passage of the controversial Kansas-Nebraska Act, a law that
allowed voters in the territories of Kansas and Nebraska to decide
for themselves whether or not to permit slavery within their borders. As violence erupted in Kansas and President Franklin Pierce

ordered the federal government to support proslavery forces there, Lincoln did little. Home was no respite. His sagging prospects exacerbated tensions with his wife, Mary Todd, who had great ambitions for him. Seemingly adrift, he read Euclid to sharpen his mind. He worried that he would never make a lasting mark on the world. His friends worried about his sanity.

Yet Lincoln's vexation was of a deliberative sort, a pause—albeit forced by circumstance—and not a surrender. His aspirations had not eroded, nor had much of his support for high office. In fact, his development as a serious presidential contender had begun many years earlier, and in 1849–1856, Lincoln was not so much reinventing himself as adapting the lessons he had learned over more than two decades in politics and law. In every critical phase of his life, including those seven years, Lincoln followed the same strategy, and it was not to turn completely inward. Besides reading voraciously to learn more about political philosophy, issues, and history of concern to him, he looked to others for guidance on the skills, vision, and strategies he needed in order to achieve his ambitions.

Certainly, the books, newspapers, plays, and poetry he read offered him a foundation that went beyond the accumulation of facts and phrases; they filled the vacuum left by the traumatizing absence of his father and the death of his mother when he was nine. An older, wiser Lincoln, who had benefited from years of self-education, noted that "a capacity, and taste, for reading, gives access to what has already been discovered by others. It is the key, or one of the keys, to the already solved problems. And not only so. It gives a relish, and a facility, for successfully pursuing the unsolved ones."[1] Books, unlike people, would never let him down. Yet he also realized that the words he read on the page could tell him only so much. To ascend in the tumultuous world of politics required a different sort of study.

Many people who knew Lincoln in his middle to late years spoke of how he read purely for utilitarian purposes, forgoing the pleasures of fiction. Even his study of poetry and drama was in pursuit of mastering the cadence of public speaking and better understanding human nature. Both as a boy and later in his life, Lincoln was intrigued with the founding of the republic, particularly with stories of the great men who were responsible for it, giants who still tread the earth in his lifetime. The books and speeches he read and reread sparked a lifelong fascination with politics, rhetoric, and the Constitution, and the debates in Illinois and throughout the nation were extensions of those the Founders had had in framing and ratifying the Constitution. Those men were hardly a distant memory for Lincoln and his generation. The nation had elected its first president only twenty years before Lincoln was born.

Lincoln's ambition to make an enduring mark on the world led him to five men, whose experiences, political insights, vision of the Constitution, example, and guidance helped him navigate the path to the presidency. From these five, Lincoln learned valuable lessons on how to master party politics, to campaign for office, to understand and use executive power, to negotiate, to manage a Cabinet, to craft a speech, and to develop policies and a constitutional vision that fit the times and became his most enduring legacy. In the nineteenth century, it was common, particularly for enterprising young men, to find flesh-and-blood mentors, men who would serve not only as father figures but also as teachers of the skills they needed to succeed. Lincoln's mentors were something different. For him, mentors were not just the men he actually interacted with but also sources of inspiration and instruction— men to be emulated, men whose mistakes he was determined to avoid, and men with whom he could argue or take issue without fear of alienation or retaliation. There is no reliable evidence that he ever met two of the men who so profoundly helped him steer his course to the White House. Yet Lincoln referred to these two men as mentors, seeing them as far more than inspirational figures. I have followed Lincoln's lead.

It is common for those who adore Lincoln to dismiss Andrew Jackson as a mentor because Lincoln campaigned so vigorously against Jackson and his political heirs, Martin Van Buren, James Polk, and Franklin Pierce, and for Jackson's most notable foe,

Henry Clay. Yet Jackson bracketed Lincoln's life: he was the president during Lincoln's formative years, and he was the first president to appoint Lincoln to federal office. Jackson was the president who initiated the Black Hawk War, in which Lincoln became captain of a company of volunteers, and Jackson was the only other president who formulated a coherent and compelling case against secession and the only one whose portrait hung in Lincoln's office throughout his presidency. Furthermore, Lincoln agreed with Jackson's declaration of the unique position of the president as the only federal official elected by and therefore representative of all the people of the Union. He adapted Jackson's understanding of democracy to the circumstances the nation confronted in a civil war, and he emulated Jackson's suspension of habeas corpus and Supreme Court–defying strategy.

Throughout his life, Lincoln was surrounded by Democrats who revered Jackson and by Whigs who cheered Jackson's great rival Henry Clay. Navigating this world sharpened Lincoln's ability to maintain lasting friendships with men who opposed him politically.

In 1832, Lincoln cast his first vote in a presidential election for Clay, not Jackson. For the rest of his life, he proudly proclaimed himself a Clay Whig and praised Clay as his "teacher," "mentor," and "beau ideal of a statesman." Lincoln followed both Clay and Jackson into the law, as a means for earning a living and making political and social connections. He considered himself to be a "self-made man" like Clay.

Clay was the source that Lincoln cited most often, not only in his debates with Douglas, but also later as president. Clay's oratory was more than a model that Lincoln studied and emulated. Lincoln also followed Clay's lead and that of Thomas Jefferson, whom Clay followed, in developing his vision, as a Senate candidate and later as president, of a connection between the promise of the Declaration of Independence that "all men are created equal" and the Constitution as the implementation of that pledge.

Lincoln's third mentor was Zachary Taylor. Lincoln followed Taylor closely when he led his army to important victories in the

Mexican War, and Lincoln was among the first national politicians to back Taylor for the presidency in 1848. Lincoln admired Taylor as another "self-made man" and lauded Taylor's devotion, like that of Clay and Jackson, to standing firmly against secession and against rebels threatening the Union.

Taylor bookends the years 1849 to 1861 for Lincoln, who eulogized him in 1850 and then, as he traveled to his own inauguration as president, singled Taylor out to a friendly audience in Pittsburgh as the man responsible for "my political education." The phrase *political education* meant something special to Lincoln and his generation; it referred to what people learned from real political experiences rather than books. Lincoln was not so idealistic that he perpetually ignored the chilly pragmatism of several of his mentors, especially Clay and Jackson. Recognizing the political advantages of such hardheaded expediency, he not only changed his mind about issues but also changed allegiances—even when it meant undermining those who had guided his ascent. He'd backed Taylor for president at the expense of Clay's candidacy, despite the fact that there may have been no politician who shaped Lincoln's beliefs more than Clay. Lincoln cited many of Taylor's actions as precedent for his own, including making record numbers of recess appointments and treating Southern forces as rebels and traitors. Lincoln followed Taylor in believing (as Jackson did) that a president must sometimes lead Congress, not the other way around, in fashioning national policy—in some ways a rejection of his earlier Whig conviction that legislative bodies should be paramount. On the other hand, Lincoln was not blinded by his mentors' virtues. The strategy he used to form his Cabinet was a barbed rejection of Taylor's choice not to use appointments to unify his party and administration.

A fourth mentor was Mary Todd's cousin John Todd Stuart. Lincoln met Stuart when Lincoln was assisting Sangamon County surveyor John Calhoun, a staunch Jacksonian (and no relation to the famous South Carolina senator of the same name). Stuart and Lincoln crossed paths again during the Black Hawk War. Jackson had appointed as leader of the federal troops Zachary Taylor, a personal friend who had served with him in the Seminole Wars more than a decade before. Stuart introduced Lincoln to law and politics in Illinois. He was a model for Lincoln in debating political opponents (they both faced off against Stephen Douglas, but Stuart did so much earlier) and in courtroom appearances, particularly in jury trials. Stuart made it possible for Lincoln to meet many of the people who would help him rise in Illinois politics. Lincoln sought Stuart's counsel and respected him even after Stuart joined the Democratic Party in 1856.

Orville Browning is the least likely of any of these men to have been a mentor to Lincoln. Whereas Browning, like Stuart, was a friend of Lincoln's, Browning, unlike Stuart, was a sometime rival. Yet Lincoln repeatedly turned to him for advice throughout his life. Their alliance helped move Whig policies through their respective chambers of the Illinois state legislature. Sometimes Browning (and Stuart, too) counseled Lincoln about his personal life. Browning and his wife tried to advise Lincoln through several amorous relationships, including the one they (as well as Stuart) thought the most troublesome, Lincoln's marriage to Mary Todd. Browning helped to found the national Republican Party as well as the Republican Party in Illinois. He constructed the Illinois Republican Party platform in 1856, on which Lincoln ran against Douglas. And Browning's advice after Lincoln's election to the presidency foreshadowed the strategy that Lincoln followed to keep the two promises he made in his First Inaugural Address—not to initiate hostilities with the South and to protect the Union.

The impact of these mentors on Lincoln's life is evident in his rise to power and his years as president. Only in the last few months of his presidency, with his reelection secured and the war won, did he begin to forgo consulting them. By then, his confidence in his political acumen, vision, and rhetoric was at its peak.

The giant figure of Lincoln seated in the iconic monument honoring him in the nation's capital captures the myth of Lincoln sitting alone, nearly godlike, contemplating the great issues of his day, head and shoulders above the fray. That is not how Lincoln learned to lead. This book is the story of how he did.

CHAPTER ONE

FINDING HIS MENTORS
(1809–1834)

In 1832, the United States was at a crossroads. It was hurtling toward its twelfth presidential election. The young nation, slightly more than four decades old, was sharply divided—politically, economically, regionally, and racially. No conflict generated more controversy and division than the legitimacy and maintenance of the institution of slavery. In 1832, there were twenty-four states, nearly equally divided over the future of slavery. The newest state, Missouri, was a slave state. Neighboring Illinois, admitted into the Union in 1818, was relatively small and had a peculiarly mixed record on slavery—with a state constitution adopted in 1819 forbidding it but a harsh Black Code adopted by the state legislature that restricted the presence and activities of African Americans.

The deep political divisions in the nation were reflected in the two fierce rivals vying for the nation's highest office. The contest pitted the stubborn, combative incumbent and champion of ordinary citizens and states' rights to regulate commercial interests and slavery without federal oversight, Andrew Jackson, against his most hated rival, Henry Clay, America's best-known orator and legislator, who championed a strong national government devoted to economic development and compromise, even on the question of the future of slavery in America.

As flatboats and other vessels navigated up and down the Mississippi and between these two westernmost states, their crews and passengers could not help but see their stark differences. Among those pilots was a young man who reputedly expressed his revulsion when he saw African Americans in chains auctioned along his route. Abraham Lincoln was not alone in such sentiments, but in his case his thoughts may well have gone beyond mere observation. Perhaps he might be the one to change things, to eradicate such demonic commerce. Enamored with the exploits of great men, he yearned to become one of them, though he did not come from a prosperous or powerful family and regularly had no money in his pocket. He had to start somewhere, and he chose to settle in the small town of New Salem, Illinois, home to no more than a few hundred people. Many of its residents came from Kentucky, where he had been born, but although the frontier could be merciless, young Abraham Lincoln believed there were no immutable limits on what a determined man might become. The only things he needed were the opportunities to educate himself and to find the right men who could help him earn his place in America's story.

I

He was born with a thirst for learning but had no teacher.

Lincoln never met his paternal grandfather, Abraham. Twenty-eight years before he was born, his namesake, then only forty-one, was killed by a Native American, while his three sons, including the younger Abraham's father, Thomas, watched in horror.[1] Abe's uncle Mordecai, aged fifteen, shot the native, giving rise to a story the family embellished over the years. Lincoln himself told and retold the story throughout his life, never missing the chance to imbue it with the character of a legend that underscored the harshness of the frontier, where he was born in a one-room cabin on February 9, 1809.

The realities of childhood are difficult to reconstruct, even more so when the man was a child in the early 1800s and later grows up to become president. In Lincoln's case, contemporaneous records of his childhood are hard to come by; most people who interacted with him were illiterate, and most who wrote about his youth did so later with the knowledge of what he became almost certainly shaping their memories. Even so, we can separate some clues of the boy who became Lincoln from the legends of his origins.

Neighbors recalled young Abe's mother, Nancy, as "superior"[2] to her husband because she knew how to read. To young Abe's delight, she regularly read aloud to him from the King James Bible. At seven, he was sent, with his sister, Sarah, to a one-room school a mile away from their home in Kentucky, where he learned the fundamentals of reading and ciphering. When Abe was nine, Nancy died after consuming tainted milk, writhing in excruciating agony in front of him and the rest of her family.

Dennis Hanks, who was ten years older than Abe, was an illegitimate cousin who had lived with Nancy's family and later with the Lincolns after she died. Dennis claimed to have given young Abe "his first lesson in spelling—reading and writing,"[3] explaining that he had "taught Abe to write with a buzzards [sic] quill, which I killed with a rifle and having made a pen—put Abes [sic] hand in mine and moving his fingers by my hand to give him the idea of how to write."[4] It is possible there is some truth to his recollections, for he was reputed to be "the best-educated" member of the Hanks family, though what passed for best might not have been all that good. It is equally if not more possible that his recollections were hyperbole, given that they were recorded later, after Lincoln's death, and Dennis may have surrendered to the temptation to overstate his importance, as many later did, in the education of a man who became president.

A more likely story, historian David Herbert Donald suggested, was that "Dilworth's Spelling-Book," which Abe and Sarah had begun to use in school in Kentucky, provided Abe's first serious introduction to the basics of grammar and spelling. He more likely credited his stepmother rather than his biological mother with the other essential help he received in learning how to read and write, declaring in the 1850s, "All that I am, or hope to be, I owe to my angel mother."

Beginning with the alphabet and Arabic and Roman numerals,
Dilworth's

> *proceeded to words of two letters, then three letters, and finally four
> letters. From these, students learned to construct sentences, like
> the one Abe constructed in school, "No man may put off the law
> of God." Dilworth's then went on to more advanced subjects, and
> the final section included the prose and verse selections, which were
> accompanied by crude woodcuts, which may have been the first pic-
> tures [Abe] had ever seen.[5]*

Other texts used in school, "like *The Columbian Class Book* and *The
Kentucky Preceptor,* expanded and reinforced what [Abe] learned
from Dilworth's."[6] Most contemporaries, particularly after Lincoln
died, remembered him as a prodigy, while John Hanks, another
cousin who lived with the Lincolns for a short while, remembered
young Abe as "somewhat dull, not a brilliant boy—but worked his
way by toil: to learn was hard for him, but he worked slowly, but
surely."[7]

When Abe was ten, his father, Thomas, abandoned both him
and his sister, Sarah, for several months. He left the children with
Dennis Hanks while he searched for a new wife. Thomas found
her in Kentucky, where they married in late 1819. He returned
with his new wife, Sarah, along with her two children, to Indiana.
She found Abe, Sarah, and Dennis not just emaciated and "wild,"
but Abe, she thought, was "the ugliest chap that ever obstructed
her view."[8] No matter her initial apprehension, she came to love
and be beloved by Lincoln, partly due to her unrelenting efforts
to nurture the boy and support his dreams. Though not literate
herself, she had brought with her a few books. One was *Robinson
Crusoe,* the stirring story of a castaway who survived because of his
courage and self-reliance. Another was *The Arabian Tales,* a collection of folk tales mostly told in prose but some in riddles and verse,
which Lincoln read, reread, and used to entertain his friends.[9]
(Dennis Hanks told a story, which may be more legend than fact,
of Thomas berating Abe for the book, which he thought was noth-

ing but "a pack of lies." "Mighty fine lies," Abe answered, "mighty fine lies.") Yet another book was *Webster's Speller,* one of many such guides Abe had on hand in school and at home. It is unclear how much they helped him. He made spelling mistakes all his life, such as writing "appologies," "opertunity," "immancipation," "Anapolis," "apparant," "inaugeral" . . . the list goes on. Some scholars believe these mistakes show that Lincoln was dyslexic, others, that he sometimes wrote in haste, but, at least as important, they showed that Lincoln cared less about how to write them correctly than about how the words sounded when spoken aloud.

Though these books might not have helped Lincoln to fully master written English, they pulled him out of the depression caused by his mother's death. They "reignited Abraham's love of learning,"[10] his stepmother said. Books were scarce on the frontier, so he read carefully rather than extensively. He memorized a good deal of what he read: "When he came across a passage that Struck him," his stepmother recalled, "he would write it down on boards if he had no paper & keep it there till he did get paper— then he would re-write it—look at it [and] repeat it" to himself until "it was fixed in his mind to suit him he . . . never lost that fact or his understanding of it."[11] Sarah stressed, too, that Abe was not just interested in reading but in writing, and in his writing, clarity of expression was all important. She said that Abe "seemed pestered to give Expression to his ideas and got mad almost at [any] one who couldn't Explain plainly what he wanted to convey." Lincoln had the same recollection that as a boy he was never satisfied "until I had put it in language plain enough, as I thought, for any boy I knew to comprehend." (His interest in writing and speaking plainly intensified with age.)

"Abe was getting hungry for book[s], reading everything he got his hands on," Dennis Hanks recalled, as did other friends and schoolmates of Lincoln.[12] "I never seen Abe after he was twelve 'at he didn't have a book some'ers 'round," he explained. Abe's stepmother agreed: "He read all day the books he could lay his hands on. He read diligently—studied in the day time—went to bed early—got up Early and then read."[13] "He'd put a book inside his shirt an' fill his pants pockets with corn dodgers," Dennis Hanks remembered,

an' go off to plow or hoe. When noon come he'd set down under a tree, an' read an' eat. In the house at night, he'd tilt a cheer by the chimbly, an' set on his backbone an' read. I've seen a feller come in an' look at him, Abe not knowin' anybody was round, an' sneak out again like a cat, an' say, "Well, I'll be darned." It didn't seem natural, nohow, to see a feller read like that.[14]

Abe read so much that he allegedly once told a friend that "he had read through every book he had ever heard of in that county, for a circuit of 50 miles."[15]

First cousin John Hanks remembered that when Lincoln returned home after work, he "would go to the cup-board, snatch a piece of corn bread, take down a book, sit down in a chair, cock his legs up as high as his head, and read."[16] Abe's contemporaries, David Herbert Donald wrote, "attributed prodigies of reading to him." Lincoln supposedly told his cousin Dennis Hanks, "The things I want to know are in books," adding, according to biographer Carl Sandburg, "My best friend is the man who'll git me a book I ain't read."[17] Books were, for him, "the main thing" in life.

Looking back in 1859, as he was mounting his presidential campaign, Lincoln wrote a letter to Jesse Fell, a close friend of David Davis, who was in charge of his campaign and a fellow lawyer. Fell had written Lincoln to inquire about whether the two might be related, and Lincoln wrote back that, while he doubted they were, "when I came of age I did not know much. Still somehow I could read, write, and cipher to the Rule of Three."[18] In a campaign biography, which he ghostwrote in 1860, Lincoln declared, "All told, the aggregate of his schooling did not amount to one year."[19] If Lincoln was underselling his formal education, it was not by much.

Abe's stepmother, Sarah, was careful not to blame Lincoln's father, Thomas, for ending Abe's formal education, recalling, "As a usual thing, Mr. Lincoln never made Abe quit reading to do anything if he could avoid it. He would do it himself first."[20] That was not how Dennis Hanks remembered it. He said Thomas thought his son wasted too much time on books, "having sometimes to slash him for neglecting his work by reading."[21] That is how Lincoln remembered it, too.

When neighbors complained to Thomas about Abe's propensity to tell jokes and stories rather than to do the fieldwork they were paying him for, additional beatings might follow. Though Thomas was over average height, he was burly, and Dennis Hanks remembered that Lincoln's "father would sometimes knock him a rod."[22] It was unsurprising that when asked what he wanted to be when he grew up, Lincoln said, "NOT a carpenter or a farmer like my father." It was little surprise that he yearned for "emancipation" from his father.

II

Several of the books in Lincoln's home were not unusual for a frontier working family of the 1820s. Indeed, because of the lessons, examples, virtues, and religious fidelity that they instilled, they were among the most popular of the time, especially among young families.

The first, of course, was the King James Bible, which Abe's mother, Nancy, had introduced him to. It was said that by the age of ten, Lincoln recited Bible verses and sermons for his friends and family. His stepsister, Matilda Johnson, said, "When father & mother woud go to church, . . . Abe would take down the Bible, read a verse—give out a hymn—and we would sing. . . . he would preach & we would do the Crying—sometimes he would join in the Chorus of Tears."[23] She further recalled, "One day my bro John Johnston caught a land terrapin—brought it to the place where Abe was preaching—threw it against the tree—crushed the shell and it Suffered much—quivered all over—Abe preached against

Cruelty to animals, Contending that an ant's life was to it, as sweet as ours to us. . . ."[24]

While his siblings and friends were impressed when he quoted or satirized biblical passages or verses, it is unclear how much of the Bible Lincoln actually read. He may have simply been well versed in the sections he liked best. Dennis Hanks said that Abe "didn't read the Bible half as much as [is] said." Hanks thought the Bible "puzzled" Lincoln, "especially the miracles. He often asked me in the timber or sittin' around the fireplace nights, to explain scripture."[25] Lincoln's stepmother agreed: "Abe read the bible some, though not as much as [is] said: he sought more congenial books—suitable for his age."[26] He certainly didn't seem to engage with the scriptures in total solemnity. Once she asked him to entertain guests by reading aloud from the Bible. Unhappy to do this, the boy began reading through passages as quickly as he could; the more she admonished him, the faster he read. Finally, as Michael Burlingame notes, "In exasperation, she grabbed a broom and chased him out of the cabin, much to his relief. Another time, he read aloud from the Book of Isaiah, playfully interpolating passages from Shakespeare."[27]

Such incidents are early signs of the skepticism Lincoln had about the Bible as fact, yet he never doubted the powers of its language and stories. "In regard to this Great Book," he reputedly wrote in response to several former slaves who had given him a copy of the Bible in 1864, "I have but to say, it is the best gift God has given to man. All the good the Savior gave to the world was communicated through this book. But for it we could not know right from wrong. All things most desirable for man's welfare, here and hereafter, are to be found portrayed in it."[28] Though he often questioned the scriptures as fact and ridiculed organized religion, he recognized its words and lessons as profound and enduring, not just for him but also for the townspeople and country folk who became his public.

While the Bible was apparently the first book Abe had tried to read, the second, *Aesop's Fables,* was equally if not more influential. The collection of 725 morality tales was written in Greek by a former slave in the sixth century BCE. As one of Abe's friends recalled, "He kept the Bible and *Aesop's* always within reach, and read them over and over again."[29] Abe read the book so often, he memorized much of it. (Decades later, he would make powerful use of Aesop's fable about the Lion and the Four Bulls: "A kingdom divided against itself cannot stand.")

A childhood friend, James Ewing, recalled,

> *I doubt if he ever told a story just because it was a story. If he told an anecdote it was to illustrate or make clear some point he wanted to impress. He had a marvelous aptitude for that—to illustrate the idea he wanted to convey. He was a wonderful observer, and he had the rare ability to remember what he had seen and heard and read, so as to apply such information to the point of anything that struck him as ludicrous. . . . He applied this wonderful gift of observation and appreciation of humor to a situation or to something which somebody had said.*[30]

Judge Owen Reeves, a friend from Lincoln's law-practicing days, agreed: "I heard Lincoln tell hundreds of anecdotes and stories, but never one that was not told to illustrate or give point to some subject or question that had been the theme of conversation, or that was not suggested by an anecdote or story told by someone else."[31]

In addition to the Bible and Aesop, Lincoln loved reading about the lives of the American Founders and their struggles to break free from British tyranny and to establish the United States. He had little taste for novels. He reputedly tried to read *Ivanhoe* but never finished. There are records of James Fenimore Cooper's novels in Lincoln's possession when he was a boy, but even as an adult he owned books by none of the great novelists of his time, such as Charles Dickens and Nathaniel Hawthorne. Instead, it was the history of politics that captured his imagination at an early age. He discovered "during my earliest days of being able to read Parson Weems' popular biography of George Washington, *A History of the Life and Death, Virtues and Exploits of General George Washington.*"

Weems's biography, first published in 1800, was more sermon

than actual history. It aimed to popularize the view of Washington as leading the colonists to victory in the Revolution because his military tactics were distinctly "American" in their ingenuity, courage, and independence, and his fight was for "American" ideals. By 1809, the year of Lincoln's birth, the book was in its fortieth edition, and Weems had added more anecdotes and myths to further promote the impression that Washington was wiser and, in a way that could only be American, more virtuous. (For instance, Weems created the purely fictional tale of young George chopping down a cherry tree and then declining to lie about what he had done.) The book effectively deified Washington as "the hero," the "Demigod," and "the 'Jupiter Conservator,' the friend and benefactor of men."[32] Weems went so far as to not only depict Washington as smarter, braver, and more ingenious than anyone else at the time, but also as more successful with the ladies.

The book perpetrated myths about the martial potency of superior American rectitude. For example, Weems declared that Washington at the 1753 Battle of Monongahela was immune to "[s]howers of bullets" from Native Americans and had "that TRUE HEROIC VALOUR which combats malignant passions—conquers unreasonable self—rejects the hell of hatred, and invites the love into our own bosoms, and into those of our brethren with whom we may have had a falling out."[33] It is unlikely that Lincoln, at fifteen, would have realized the book was filled with "effectual truths" (Niccolo Machiavelli's terms for myths serving political purposes) that Weems hoped would define American values. At fourteen or fifteen, Abe discovered that a neighbor had another, well-regarded biography of Washington written by David Ramsay, a physician and one of America's first self-proclaimed historians. Yet Abe never tired of recalling how he had walked two miles to a neighbor's house to first borrow Weems's book and spent months working off the cost of the book after it had been damaged in the rain.

Weems had been a friend of Thomas Jefferson, and his story of Washington constructed an inspiring vision of a new nation founded on the values of charity, modesty, and bravery. In contrast, Ramsay was one of the first Americans who aspired to do genuine historiography, researching primary documents, correspondence, diaries, and other authentic texts from the founding era. However, Ramsay was not without his own political agenda. He had fought in the Revolutionary War, and he hoped to validate republicanism's elevation of public service as a means to inculcate appropriate virtues, such as duty, mercy, loyalty, justice, and self-lessness, in the governed as well as in the governors—as well as to reinforce national unity. Ramsay believed that Washington's story, more than any other, could inspire the values indispensable for a good life and a good society. Whereas Weems depicted Washington as a near-deity, Ramsay emphasized that he was a man, albeit a remarkable one with extraordinary qualities, including humility and courage. In addition, Ramsay depicted the new nation as bent on reform and determination to improve on English governance.

It was Weems's book that stuck with Lincoln. When he traveled to Washington for his first inaugural, Lincoln told the New Jersey Senate he could not forget Weems's account of Washington's struggles at Trenton—"the crossing of the river; the contest with the Hessians; the great hardships . . . I recollect thinking then, boy even though I was, that there must have been something more than common that those men struggled for."[34] Lincoln knew by then, if not well before, that Weems's book was as much fiction as real history, but it never bothered him. Decades before his political ascent, Lincoln understood that the myth could count even more than the facts because what people believed, what drove them to action, was what mattered most.

As late in his life as 1865, he told his friend and law partner William Herndon that the biographies of great men "are all alike. You might as well print up these biographies with blank titles and fill in the name of any subject that you please."[35] They "are written as false and misleading," he told Herndon, "never once hinting at [the subject's] failures and blunders." Those "failures and blunders" were as important to Lincoln as their successes, because he learned from them both. But instead, the biographies then available "can fill up eloquently and grandly at pleasure, thus commemorating a lie, an injury to the dying and to the name of the dead." Lincoln suggested that booksellers should have "blank biographies on their shelves always already for sale—so that, when a man dies, and his heirs—children and friends wish to perpetuate the memory of the dead, they can purchase one already written, but with blanks, which they can fill up eloquently and grandly at pleasure, thus commemorating [the] names of the dead."[36] Lincoln lived in an age without professional biographies and thus could shape his "tale" as he liked. If left unattended, his story would be up for grabs. Properly tended and with stretches of truth but not outright lies, it could be enduring.

Another book that Lincoln likely read was Lindley Murray's *Reader*, which was popular in homes and schools throughout Lincoln's youth. A strong supporter of the war for independence, Murray became a grammarian and editor of many books that were in large demand as manuals for instructing young children on reading, grammar, and writing. Issued at the end of the eighteenth century, Murray's *Reader* was full of classical moral axioms from the ancient Greeks and Romans as well as from British magazines like *The Spectator* and other Enlightenment moral sources. Such readings enriched the appreciation of young Lincoln for writing to convey moral instruction and for constructing allegories, fables, and stories.[37]

The Bible, Aesop, Murray, chronicles of the lives of public figures—Lincoln considered them dynamic, elastic tools to influence opinions, not accounts of facts and not to be interpreted as such. Another influential book that shaped Lincoln's understanding of the founding of the republic was William Grimshaw's *History of the United States*, first published in 1821. Grimshaw considered himself an objective historian and wrote many history textbooks that were widely used in Lincoln's day. This book, Grimshaw's most popular, was intended to inspire young people to adopt the values (as Grimshaw understood them to be) of the men who wrote the Constitution and won independence. It began with the discovery of the United States and ended with Florida's acquisition in 1819. It finished with an account of Northern states establishing laws for emancipation and prohibiting slavery. Grimshaw exhorted his readers, "Let us not only declare by words, but demonstrate by our actions, 'That all men are created equal, that they are endowed, by their Creator, with the same unalienable rights; that, amongst these, are life, liberty, and the pursuit of happiness.'"[38]

These stirring words were, of course, the self-evident truths set forth in the second paragraph of the Declaration of Independence. Lincoln memorized the entire document, and of all that he read in his life, its words might well have been the most influential. Indeed, he devoted his adult life to reminding the American people that these words were the promises that the American Constitution was established to ensure. In an 1856 speech, he described it as the "sheet anchor of our principles."[39] Two years earlier in Peoria, he had instructed his audience, "Let us readopt the Declaration of Independence, and with it the practices and policy which harmonizes with it."[40]

But not everything Lincoln read was as civic-minded. Humor delighted him; he mimicked the styles of the different speakers he heard and the writers he loved, and he adored telling funny stories and jokes. There was good reason to believe his gift of storytelling came naturally or at least ran in the family. Dennis Hanks recalled that Lincoln's father, Thomas, "could beat his son telling a story—cracking a joke."[41] Watching his father bring laughter to a room was a powerful example of how even someone whose reputation did not much include amusement and diversion could win the crowd in the right circumstance with skilled deployment of narrative.

Lincoln's cheekiness is evident in two poems he wrote when he was in his teens if not younger, penned in his arithmetic book. The first announced, "Abraham Lincoln / His hand and pen / He will be good / But God knows when," and the other, "Abraham Lincoln is my name / And with my pen I wrote the same / I wrote in both hast and speed / And left it here for fools to read."[42] While these rhymes may not suggest a sophisticated fascination with poetry, in fact Lincoln was genuinely entranced by the art. As a boy, he read Thomas Gray's "Elegy in a Country Churchyard," a poem he quoted often. As president, he once told campaign biographer John Locke Scripps, "It is a great piece of folly to attempt to make anything out of my early life. It can all be condensed into a single sentence, and that sentence you will find in Gray's Elegy: 'The short and simple annals of the poor.'"[43]

Besides the poems he encountered in the readers he had at school, Lincoln would have come across poetry in William Scott's *Lessons in Elocution,* another of the books that his stepmother had brought into their home. Around 1825, Lincoln began to read the Scott collection closely, which included Shakespeare, a writer who, along with Robert Burns, Lincoln quoted frequently throughout his life. (Later, as president, he wrote, "I think nothing equals *Macbeth.* It is wonderful.")

Lincoln's facility with rhyme and wit was not always used virtuously. Around the age of nineteen, he wrote "Chronicles of Reuben," based on incidents dating back to 1826, when Lincoln's sister married Aaron Grigsby, whose family were neighbors of the Lincolns. When Sarah died in childbirth, Lincoln blamed Aaron and the Grigsbys for waiting too long to call a doctor to save her. His bitterness increased when he was not invited to the joint wedding celebration of Aaron's brothers Reuben and Charles, who were married on the same day. Lincoln purportedly arranged, through a friend, for Reuben and Charles to be brought to the wrong bedrooms, where each other's wife awaited after the wedding party. Lincoln wrote a description of the incident in "Chronicles of Reuben" as payback. Patterned after biblical scripture, a prose narrative was followed by a poem about Billy Grigsby, another of Aaron's brothers, who, it says, "married Natty," who was another man. Lincoln's neighbor Joseph Richardson claimed that "Chronicles of Reuben" was "remembered here in Indiana in scraps better than the Bible."[44] The episode was among Lincoln's earliest exploitations of the power of public ridicule and the sway it conferred on those who wield it. What he called "the power to hurt," a phrase he borrowed from a Shakespeare sonnet, would remain useful, though not the "chief weapon in his rhetorical arsenal."

A final book that fed Lincoln's love of language and allegory was John Bunyan's *The Pilgrim's Progress*, first published in 1678 and quickly translated into over a hundred languages. By the time Lincoln's father gave him a copy of the book (borrowed from a neighbor), it was second only to the Bible in its popularity in the United States. The youngster may have first discovered it when Benjamin Franklin referred to it in his autobiography, another book Lincoln loved. When Abe first discovered Bunyan's tale, his "eyes sparkled, and all that day he could not eat, and that night he could not sleep."[45] Sometimes called (erroneously) the first lengthy prose narrative in English, *The Pilgrim's Progress* tells the story of an everyman known as Christian who flees the City of Destruction (this world) and is instructed to follow the narrow path to the Celestial City, "That Which Is to Become" as the full title puts it. The quest is filled with set pieces that became some of Lincoln's favorite fables, and the tale as a whole was inspiring to a young man who dreamed of beating the odds to become a hero.

III

The histories and biographies Abe read sparked an interest in politics, as did the nearly daily conversations in his house and in the neighborhood. Thomas was a proud Democrat, as were most of the people in Abe's family and the neighboring homes. They all felt betrayed by the federal government, which was tightening their credit and backing the big businesses gouging them. As is the wont of many children, Lincoln adopted his father's party affiliation.

Nineteenth-century America was not ripe with the sort of "nonprofit" organizations that now constellate our daily lives. A political career held the greatest promise for a young man who saw public service, not family or acquiring money, as his principal calling. To feed his interest in what he saw as this much more virtuous pursuit, Lincoln discussed politics every chance he had with friends and neighbors. In 1828, he went to work for Colonel

William Jones, a local postmaster only six years older than Lincoln. Jones hired Lincoln as a farmhand and as a land clerk in his tiny general store, where Lincoln sometimes slept, even though his family's cabin was only about a mile and a half away. Jones was a graduate of the first public university in Indiana, Vincennes University, and his store became a popular gathering place for young men to debate political issues. Henry Clay Whitney, an Illinois friend of Lincoln's, recalled Abe telling him, "The sessions were held in Jones' store, where the auditors and disputants sat on the counter, on inverted nail kegs, or lolled upon barrels or bags, while the wordy contest raged."[46] He further said Lincoln explained that "the questions selected for discussion were not concrete. At one time there would be a debate upon the relative forces of wind and water; at another, upon the comparative wrongs of the Indians and the negro; the relative merits of the ant and the bee; also of water and fire."[47] Dennis Hanks, who sometimes joined in the sessions, recalled that "Lincoln would frequently make political speeches to the boys; he was always calm, logical, and clear. His jokes and stories were so odd, original, and witty all the people in town would gather around him. He would keep them all till midnight. Abe was a good talker, a good reasoner, and a kind of newsboy."[48]

Similarly, Nathaniel Grigsby, who had made his peace with Lincoln by this time and had become, like Lincoln, a regular participant in the Jones's store debate, recollected that "we attended [the debates]—heard questions discussed—talked everything over and over and in fact wore it out."[49] Grigsby acknowledged, "We learned much in this way," and, as for Lincoln, Grigsby added (perhaps in a rose-colored recollection), "His mind and the ambition of the man soared above us. He naturally assumed the leadership of the boys. He read and thoroughly read his books whilst we played."[50] Grigsby found that "Lincoln was figurative in his speeches, talks, and conversations. He argued much from analogy and explained things hard for us to understand by stories, maxims, tales, and figures. He would almost always point [out] his lesson or idea by some story that was plain and near, as that we might instantly see the force and bearing of what he said."[51] Grigsby agreed that Jones was "Lincoln's guide and teacher in politics."[52]

In the 1828 presidential election, the Democrats' champion was John Quincy Adams's foe, Andrew Jackson, who had lost to Adams four years before. Lincoln and Dennis Hanks "went to political and other speeches" and heard "all sides and opinions." Hanks remembered that "Lincoln was originally a Democrat after the order of Jackson—so was his father—so we all were."[53] David Turnham and Elizabeth Crawford, who both knew Lincoln in Indiana, recalled Lincoln and his father as "Jackson men" when they left for Illinois in 1830.[54] Abe had even been heard singing, "Let auld acquaintance be forgot / And never brought to mind. / May Jackson be our president /And Adams left behind."[55] This was the song of a Jackson man.

However, something happened between 1828 and the next presidential election. Shortly after the former election, Henry Clay took the lead in assembling a new political party. He called it the Whig Party, a name he borrowed from the British political party that favored legislative rather than executive dominance, emphasizing the importance of Parliament over the king as the best and most effective representatives of the people. Dennis Hanks remembered that "Colonel Jones made [Abraham] a Whig."[56] Jones backed John Quincy Adams in 1828, and though Clay was not a candidate in that election, several contemporaries in Indiana remembered Lincoln's memorizing Clay's speeches, reciting them for his friends, and praising Clay, as the newspapers did, as "Harry of the West." Jones had a library, and he subscribed to several newspapers as well as having at least temporary access to all the others that came through his office as postmaster. In those days, newspapers were the lifeblood of American democracy. They were the principal sources of news and notable speeches, particularly on the national stage. They also generally adhered to their owners' politics. Jones introduced Lincoln to the *New York Telescope,* the *Washington Intelligencer,* the *Western Register,* and the *Louisville Gazette,* all of which were Whig papers. Thus informed and influenced, as Lincoln prepared to leave his father's home for good, his choice on which of the two men to follow was already taking shape.

IV

For most of Abraham Lincoln's life, Andrew Jackson and Henry Clay were the two titans fighting over the future of America. They both were born before the American Revolution ended, and both men were tall. Jackson was six one, thin and sinewy. Clay was six three, and those who came to hear his oratory often remarked how he slowly unfurled to his full height as he rose to make his point. Lincoln grew to be the tallest of the three. His stature—in both the literal and figurative sense—would one day come to dwarf both men.

Jackson was ten years older than Clay and had fought in the War of Independence. It was a well-known fact in Lincoln's era that Jackson and his brother Robert had been captured by the British in 1781, and while in captivity, both had famously refused to shine the boots of a British officer, who slashed Jackson's face, giving him a scar that he wore proudly for the rest of his life.

Clay and Jackson both came from modest backgrounds, though Jackson's family was poorer. Jackson's father had died before he was born, and Clay's father had died when he was four. Jackson's mother moved in with her sister's family, while Clay's mother remarried, so Clay grew up in a crowded home with siblings and stepsiblings. Jackson attended a one-room schoolhouse for a year before entering a local academy for a year or two, while Clay was put on a fast track to become educated in Richmond. Both men took great pains to emphasize—indeed, exaggerate—their humble origins,[57] which did not slow either of them down. They both studied law. Jackson did it almost entirely on his own: he briefly apprenticed for a lawyer in Salisbury, North Carolina, and then, with the help of several other local attorneys, was admitted to the

North Carolina bar in 1787. Clay secured a coveted clerkship with the Virginia Court of Chancery, and his neat handwriting caught the eye of one judge, the esteemed George Wythe, America's first law teacher, who had mentored both Thomas Jefferson and the great chief justice John Marshall. Because Wythe had a crippled hand, he could not write for himself, so he hired Clay, then sixteen, as his secretary for the next four years. Wythe's influence on Clay was profound. "To no man," Clay said in 1851, "was I more indebted by his instructions, his advice, and his example."[58] Clay said that he learned his lifelong opposition to slavery from Wythe.

For a short while, Jackson practiced law but quickly moved on to become a prosecutor, land speculator, and most important, protégé of one of Tennessee's most colorful and controversial leaders, William Blount. Blount had been one of North Carolina's delegates to the Constitutional Convention and a signatory to the Constitution. With Blount's support, Jackson became Tennessee's first attorney general and a delegate to Tennessee's constitutional convention in 1796. When Tennessee became a state the next year, Jackson was elected as its first and only representative in the U.S. House of Representatives. A year later, Jackson was elected to the Senate, arriving there shortly after the House had impeached his mentor, Blount, and the Senate then expelled him for having conspired with the British to take control of Spanish Florida in exchange for money that would have helped him to pay off the enormous debts he had incurred as a land speculator.[59] Jackson left the Senate less than a year later to return to Tennessee, where he was elected to the Tennessee Supreme Court, on which he served until 1804.

Clay was experiencing his own meteoric rise to power. Already established as a lawyer (who sometimes corresponded with Andrew Jackson on legal matters), he was elected to the Kentucky General Assembly at the age of twenty-six. In 1804, Aaron Burr, Jefferson's vice president, killed Alexander Hamilton in a duel and then headed west to make a fortune. He was in Kentucky when he was charged with treason. (Burr had apparently been in discussions with William Blount over some "land scheme.") Burr turned to Clay and, with co-counsel John Allen, convinced him that the real plot was a Republican vendetta against him. Clay got the charges dismissed when a key witness failed to appear at trial. As Clay, already renowned for his eloquence, later explained,

> Such was our conviction of the innocence of the accused, that, when he sent a considerable fee, we resolved to decline accepting it, and accordingly returned it. We said to each other, Col. Burr has been an eminent member of the profession, has been attorney general of the State of New York, is prosecuted without cause, in a distant State, and we ought not to regard him in the light of an ordinary culprit.[60]

After his successful defense of Burr, Clay was elected in 1806 to complete the U.S. Senate term of the incumbent, who had resigned in anger after losing reelection. Clay, then twenty-nine, was already the speaker of the state house of representatives but was four months shy of the minimum age of thirty required in the Constitution for service in the Senate. Nonetheless, the Senate seated him with no objection. As Clay was preparing to go to Washington, Burr contacted him to let him know he was about to be indicted a second time for treason.[61] Clay was concerned that he might have a conflict of interest due to his new Senate duties, but Burr assured him in writing that he was innocent of the charges that he had promoted the dissolution of the United States in any way.[62] When Clay appeared in court on Burr's behalf, the prosecution again conceded that an important witness had failed to appear. Clay angrily denounced the prosecution's shenanigans, and two days later the jury acquitted Burr.[63]

Eager to get to Washington to begin his life as a senator, Clay left almost immediately after procuring Burr's acquittal, but before he arrived in the nation's capital, President Jefferson got word to him that he was in possession of a coded message from Burr in which his treacherous schemes were evident. Jefferson issued a proclamation warning the nation of a military conspiracy and urging the capture of the traitors, including Burr. Once Clay arrived in Washington, he tried to reassure everyone that he had believed Burr was innocent, but at the White House, President Jefferson showed Clay some of the decoded documents demonstrating Burr's guilt. "It seems," Clay wrote his father-in-law, "that we have been much mistaken about Burr."[64] Clay refused to appear on Burr's behalf at his treason trial conducted later before the chief justice of the United States, John Marshall. Clay told people that Burr had "deceived" him and that he would not give Burr "an opportunity for deceiving him [another] time."[65] Clay's association with Burr would follow him for the rest of his life, a warning to other lawyers, including Lincoln, on the risks of both representing and becoming too closely associated with unpopular clients.

The legal careers of Jackson and Clay each consistently took a backseat to their political ambitions. In 1812, Clay supported the United States' declaration of war against Britain after it attempted to stop American traders from supplying Russian and American hemp to France for salt-resistant cordage for Napoleon's navy. The British forced captured American sailors into servitude and encouraged Native American attacks on American settlers on the frontier. Jackson enthusiastically joined the cause and became a hero as the prevailing general when his ragtag army of roughly five thousand American men defeated the larger, better financed and armed forces of the British in the Battle of New Orleans in 1815. In that case, he'd marched his band of volunteers to Washington from Tennessee but, once he arrived, he was ordered to disband his men. He refused. Instead, he funded their march back five hundred miles to Tennessee and eventually to New Orleans. Thinking he was as tough as the hardest wood they knew, they dubbed him Old Hickory. The name stuck.

Jackson had a hair-trigger temper. He fought two duels; in one of them, the man he killed had struck him first near the heart. The bullet was so close to his heart that surgeons decided they could not remove it without killing Jackson. He caned various enemies and made numerous challenges that were not accepted. He had two shootouts with the governor of Tennessee and another with his then aide-de-camp Thomas Hart Benton, whose bullet struck him in his left arm. Nevertheless, Benton later became a friend and political ally, and Jackson later would sometimes make a show of being angry to intimidate people he wished to be rid of.

In 1817, President James Monroe directed Jackson to lead military forces to rebuff Native American attacks in Spanish Florida. Jackson succeeded, but not before ordering the brutal killing of Native Americans in his charge as well as two British citizens who had traded with them. When it became clear that Spain could not defend or control the Florida territory, President Monroe, through his secretary of state, John Quincy Adams, negotiated the transfer of sovereignty over it from Spain to the United States. Monroe then placed Jackson in control of the territory as military governor.[66]

With the 1824 presidential election fast approaching, Jackson and Clay both tried to position themselves to succeed President Monroe, who was finishing his second term. Jackson got off the mark faster. A young man named John Eaton, a friend of Jackson's family in Tennessee, published a biography of Jackson in 1817, the first campaign biography in American politics. Young Abraham Lincoln would likely have read it. Eaton's book and Jackson's campaign portrayed the candidate as a war hero, defender not only of his country but the common man, unbeholden to the banks and businesses, and a champion of states' rights.[67]

For his part, Clay had become the youngest speaker of the U.S. House of Representatives in its history, having been elected to the position when he first entered the House in 1811. Up until 1820, Clay had been an agitator rather than a peacemaker in Congress. His introduction of the Missouri Compromise in 1820 changed all that. A slave owner, he had founded the American Colonization Society in 1817 to provide a solution for the slavery problem by buying slaves their freedom and then funding their travel back to Africa. Clay was the society's first president, Andrew Jackson its first vice president. Alarmed that sectional differences over slavery were threatening to rip the country apart, Clay saw an opportunity he could not resist: his chance to exercise leadership on the question of slavery. On March 3, 1820, the House approved Clay's proposed compromise—allowing Maine to be admitted into the Union as a free state and Missouri as a slave state. The proposal further barred the expansion of slavery north of the 36°30′ parallel, excluding Missouri. As the first successful attempt to broker a compromise between the proslavery and antislavery members of Congress, Clay's effort, including detailed accounts of his four-hour oration, delivered over two days, was widely reported by newspapers throughout the country.

Clay split each part of the compromise into separate bills, which were supported by different coalitions in the House. Clay earned the nickname the Great Compromiser for his kind of tactic, and he would use the same technique thirty years later in his crowning achievement as a legislator—the Compromise of 1850, yet another attempt to find a solution to the slavery problem that, in the end, merely postponed a final reckoning.

Over the course of two legislative days (and over 40 pages in the *Annals of Congress*), Clay delivered a speech that foreshadowed the eloquence to come. He set forth his belief in tariffs to protect fledgling American industry, in federal funds for internal improvements, and in a national bank that would give credit to help people improve the quality of their lives. "The object of the bill under consideration," he declared, "is to create the home market, and to lay the foundations of a genuine American policy."[68] In characteristic fashion, he proceeded to systematically address ten arguments made against the tariff, and concluded, "But there is a remedy, and that remedy consists in modifying our foreign policy, and in adopting a genuine AMERICAN SYSTEM. We must naturalize the arts in our country, and we must naturalize them by the only means which the wisdom of nations has yet discovered to be effectual—by adequate protection against the otherwise overwhelming influence of foreigners."[69] As described by Lincoln's biographer David Herbert Donald, "Clay's American System sought to link the manufacturing of the Northeast with the grain production of the West and the cotton and tobacco crops of the South, so that the nation's economy would become one vast interdependent web. When economic interests worked together, so would political interests, and sectional rivalries would be forgotten in a powerful American nationalism."[70]

From an early age, Clay and Jackson had each regarded himself as the true heir to the political vision of Thomas Jefferson; at the turn of the nineteenth century, they each shared Jefferson's skepticism of a strong federal government and commitment to expanding access to voting for ordinary American citizens. In time, Clay, unlike Jackson, became increasingly concerned that relying too much on the uninformed opinions of the vast electorate was in conflict with the interests of American businesses whose success was crucial for ensuring the nation's prosperity. Clay thus developed confidence in a potent national government that backed ambitious public works, such as a vast transportation infrastructure that would have included canals, roads, harbors, and navigation improvements, as well as funds for schools. Clay's vision, which became known as the American System, was in sharp conflict with Jackson's certitude that the hard push for economic development would give rise to rapacious businesses and banks that would gouge ordinary Americans. The mismatch in the two men's worldviews would, unsurprisingly, lead to a clash between them. While one crucial element of the American System, the Tariff of 1824, passed narrowly in the House, it was clear that without the support of the chief executive, progress would be severely limited.

In the 1824 election, the problem that both Jackson and Clay faced, however, was that they were just two of the four major candidates running for president. At that time, the predominant national party was the Democratic-Republican Party, which had won six consecutive elections. The only other major party, the Federalist Party, which had been the party of Washington and John Adams, had dissolved shortly after Adams lost the 1800 presidential election to his own vice president, Thomas Jefferson. After the Federalist Party melted away, the country entered what was commonly called the Era of Good Feelings, a deceiving nomenclature, as the 1824 election showed. The fact that there was one dominant political party did not mean every Democratic-Republican shared the same outlooks or allegiances. Unlike today's political parties, those in the antebellum era did not simply coalesce around an agreed-upon platform of policies. Instead, they were formed around regional interests and beliefs about whether the Constitution envisioned a strong, effective federal government or a significant realm of state authority over which the federal government had no say.

In 1824, four different men vied for the Democratic-Republican Party's mantle: William Crawford, who had served as Monroe's secretary of war and Adams's Treasury secretary but then suffered a debilitating stroke. His disability was an open secret likely to hinder support from outside his home state of Georgia. John Quincy Adams, the son of President Adams, had served in a series of prestigious public offices; he was the sitting secretary of state, the same office that James Madison had occupied when he was elected president the first time in 1808, and that James Monroe had occupied when he ran in 1816 to succeed Madison. Jackson quit after serving brief stints in both the Senate and the House. Clay, on his way to becoming the longest-serving speaker of the House of Representatives, was widely revered (and castigated) as a visionary and seasoned statesman. A fifth candidate, John Calhoun, a South Carolina senator and the most outspoken defender of slavery on the national stage, competed only for the vice presidency and therefore ensured that he would serve as the vice president for whomever was elected president.

None of the four major candidates seemed to have a good chance at winning the presidency, as none could muster anything more than moderate regional backing. Adams received support in New England and split the Mid-Atlantic region with Jackson but did not win the popular vote. Jackson and Crawford split the popular vote in the South, while Jackson and Clay split it in the Western regions. Jackson claimed a plurality of the national popular vote, but no one won a majority in the Electoral College.

Under the Twelfth Amendment, which had been ratified to prevent any confusion over how electoral votes should be counted for president and vice president, as had occurred in the election of 1800, the House of Representatives decided the outcomes of presidential elections unresolved in the Electoral College. Only the top three candidates in the popular election were eligible to be considered in the House. Because Clay had finished fourth in the popular vote, he was no longer eligible to be a candidate, but as speaker of the House, he was in charge of handling the vote to break the electoral stalemate. He threw his support to Adams—unsurprising given that Adams was the only other candidate who had supported the infrastructure program that Clay favored. Due to Clay's support and influence in the House, Adams became president, and he named Clay secretary of state.

Bitterly disappointed, Jackson threw one of the most famous fits of anger in American history. From the moment the House made Adams president to the 1828 rematch between Jackson and Adams, Jackson denounced wherever he went the arrangement between Adams and Clay as a "corrupt bargain."[71]

When the 1828 election came around, Jackson's grudge against both Adams and Clay was as strong as ever. This time, it was only Jackson and Adams who faced off, since as Adams's secretary of state, Clay was obviously in the president's camp. Nor were there any minor candidates to siphon votes from either of the major candidates. Besides recirculating Eaton's biography, Jackson initiated a style of campaigning that candidates continue to use today, including personal appearances, soliciting backing from newspapers, and reminding voters of Jefferson's endorsement in his earlier run for president.[72] Jackson worked with Martin Van Buren and others to build a new party apparatus called the Democratic Party to replace the deteriorating and defunct Democratic-Republican regime.

For his part, Adams was largely tone-deaf when it came to political acumen. Throughout his administration, he had not bothered to use his appointments to solidify support within his administration or Congress, nor had he bothered to curry favor with congressional leaders to get his legislative initiatives enacted. He had little to show for his four years in office except perhaps for his stubbornness and determination not to suffer fools gladly. And this time he faced not only an energized Jackson but a brand-new party.

In the months prior to the election, Jackson, with the help of a small band of close advisers that included the new vice president, Martin Van Buren, had established the Democratic Party, emphasizing its commitment to following the will of a majority of voting Americans. Clay replaced the name of the Democratic-Republican Party, which was all but defunct anyway, adopting the other half of the old party's moniker: he was now the candidate of the National Republican Party.`

Jackson's new party destroyed the competition. Jackson won the 1828 election with nearly 56 percent of the popular vote and over twice as many Electoral College votes as Adams had won in 1824. During the 1828 campaign, Calhoun, then the vice president to Adams, offered to serve as Jackson's vice president if he won. Jackson agreed. Many of the people who voted for him were Southern Democrats who also supported Calhoun, and both he and Calhoun opposed federal overreach into domains that, in their opinion, were properly within the jurisdiction of the states.

Everyone expected the 1832 presidential election to be the culmination of the feud between the Jacksonites and Clay men. It was not. Their ideological dispute was not put to rest until after their deaths, Lincoln's reelection, and the end of the Civil War.

Neither Jackson nor Clay anticipated that Vice President John Calhoun would attempt to thwart Jackson's reelection. Jackson and Calhoun had been increasingly at odds throughout most of Jackson's first term, as Calhoun and his allies in the Cabinet belligerently tried to push Jackson toward a more extreme defense of slavery and adoption of the doctrine of nullification (the entitlement of states to disregard any federal laws they considered to encroach upon their autonomy).

In February 1831, with the presidential election more than a year away, Calhoun hoped to sabotage Jackson's reelection by leaking a copy of a letter that President Monroe had written to Jackson back in 1818. The letter seemed to indicate that Monroe had not given Jackson clear authority to take the aggressive measures that he did to help the United States wrest Florida from Spanish control. Clay and other critics of Jackson—who continued to claim that he'd had Monroe's support—seized on the letter as further evidence of Jackson's tyrannical disposition.

Jackson restructured his Cabinet to rid himself of Calhoun's allies as well as the members whose wives had snubbed the wife of Jackson's friend, biographer, and war secretary, John Eaton—a purge seemingly driven more by personal slight than policy. Jackson asked Secretary of State Martin Van Buren, who had suggested the reorganization, to resign so that Jackson could appoint him as minister to Great Britain. To avoid a clash in the Senate, Jackson gave Van Buren a recess appointment as minister in August 1831. When in 1832 the Senate split over the nomination of Van Buren as ambassador, Calhoun, in his capacity as vice president, cast the tie-breaking vote to defeat it. In doing so, he made Van Buren a martyr for Jackson's cause.

As the turmoil in the Jackson administration subsided, Henry Clay tried to outmaneuver Jackson on a different front. Clay decided to turn the national bank—the centerpiece of his American System—into *the* major issue in the upcoming presidential election. Previously, on May 27, 1830, Jackson had vetoed the Maysville Road project.[73] The road would have connected Lexington to Maysville, Kentucky, a sixty-six-mile stretch that would have extended the Cumberland Road, the nation's first interstate highway, built between 1811 and 1837 and designed to extend from Cumberland, Maryland, to the Northwest Territory. Jackson argued that the Maysville project was unconstitutional because it was purely intrastate and therefore a matter to be addressed solely by Kentucky authorities.[74] Clay decided, along with Nicholas Biddle, the director of the national bank, to bring up the rechartering of the bank five years sooner than it needed to be, in order to force Jackson to again show the extent to which he opposed the concept.

Clay underestimated Jackson. Jackson killed the national bank in perhaps the most famous veto message in American history. In spite of the fact that the United States Supreme Court had previously upheld the national bank's constitutionality as a lawful exercise of Congress's authority under the Necessary and Proper Clause (because the law made collection and redistribution of federal money much easier), Jackson, in a message drafted by his then–attorney general Roger Taney, argued that this same clause vested him with independent authority to determine whether a law was "necessary" or not.[75] Jackson concluded that the national bank charter was not. Lambasting the bank as "corrupt," Jackson warned that "if this monopoly were regularly renewed every fifteen or twenty years on terms proposed by themselves, they might seldom in peace put forth their strength to influence elections or control the affairs of the nation." As for the Supreme Court's decision unanimously upholding the bank's constitutionality, Jackson declared that its decision "ought not to control the coordinate authorities of this Government." In a declaration that would guide every subsequent president, including Lincoln, Jackson wrote, "The Congress, the Executive, and the Court must each for itself be guided by its own opinion of the Constitution." The president, he explained, is "independent of both" the Court and Congress, particularly since the president is the only *national* leader elected by the voters. In the upcoming campaign, Jackson men would argue that the bank was corrupt and indeed was responsible already for one national economic downturn, the Crisis of 1819, when banks failed because they overextended their credit and then foreclosed mortgages, forcing people out of their homes and off their farms.

Besides baiting Jackson with the national bank, Clay baited him with barbs, theatrics, and satire every chance he had. He rebuffed Jackson's charge of "bargain and corruption" by producing a letter from then-senator James Buchanan, who denied Jackson's claim that Buchanan had told him about the deal struck between Adams and Clay at Jackson's expense. While Jackson's rhetoric was plain, direct, and often profane, delivered much like a sharp jab to the jaw of an opponent, Clay characteristically opted for more rhythmic, dramatic rhetoric, often sprinkled with alliteration or palilalia (repeating the same sentence or word) and rhetorical questions instead of declarative statements, such as when he implored the House to approve the Tariff of 1824 (a protective tariff designed to protect American businesses from cheaper British commodities):

> The object of the bill under consideration is to create this home market, and to lay the foundations of a genuine American policy. . . . Are we doomed to behold our industry languish and decay more

*and more? But there is a remedy, and that remedy consists in . . .
adopting a genuine AMERICAN SYSTEM. We must naturalize the
arts in our country, and we must naturalize them by the only means
which the wisdom of nations has yet discovered to be effectual—
by adequate protection against the overwhelming influence of foreigners.*

Jackson rejected flowery language and complex metaphors in
further contrast with Clay, who often indulged in convoluted,
intricate similes and allegories to make his point, as he once did
when, in the Kentucky state legislature, he likened the current capital, Frankfurt, to "an inverted hat" and "penitentiary" so difficult
to navigate that he wondered, "Who that got in, could get out?"

Clay also undertook several efforts during the campaign to bolster his own image and to cast Jackson in a bad light. Believing that
Jackson's veto of the national bank placed him on the wrong side
of federal power, Clay and Biddle authorized the printing and distribution of thirty thousand copies of the text of the veto. Around
the same time, Clay authorized George Prentice to write a campaign biography as a response to the one Eaton had written several
years before about Jackson. Prentice hailed Clay as a man who had
worked his way up from nothing, whose oratory was "the voice
of salvation in the country" and whose story exemplified American independence and virtue.[76] Prentice made no money from the
biography, in spite of the fact that twenty thousand copies were
sold. To help Prentice financially, Clay procured him a job as the
publisher of the *Louisville Journal* in 1830.

Dennis Hanks later suggested that Lincoln had read Prentice's
biography in his teens, but that was impossible, for it was not published and widely distributed until 1831, when Lincoln turned
twenty-two.[77] But Lincoln did not need to read the book when it
first appeared to know the details of Clay's life, as he avidly read
the *Louisville Journal*, which reliably celebrated Clay and published
his speeches and many of the same pro-Clay, anti-Jackson stories
that appeared in the Clay campaign biography. Through his daily
reading and political discussions and debates with his friends and

neighbors, Lincoln tightened his embrace of Clay's American System to such an extent that his fidelity to Clay's political vision became obvious to everyone he met.

V

Largely by accident, twenty-two-year-old Lincoln arrived in New Salem, Illinois, in late July 1831. It was a tiny town, with only about a hundred residents. He was on his second trip there when the flatboat he was steering got stranded near a dam. Denton Offutt, who had hired Lincoln to pilot the boat, lent him an auger, which Lincoln used to drill a hole in the boat that allowed the water to drain out, freeing the boat to sail past the dam. Impressed with Lincoln's ingenuity, Offutt offered him a job in his store. Over the next six years, Lincoln would find employment as a shopkeeper (Offutt's store failed within a year, and he turned out to be a con man rather than a legitimate businessman), a shop owner (the shop went belly up, saddling Lincoln with debt that took years to pay off), a soldier (who saw no combat), a hardworking land surveyor tasked with determining the boundaries of several neighboring towns (in the words of one nearby resident, "Mr. Lincoln had the monopoly of finding the lines, and when any dispute arose among the Settlers Mr. Lincolns Compass and chain always settled the matter satisfactorily"), a rail splitter, and a postmaster.[78] None of these jobs captured his imagination. Lincoln disliked physical labor, even when he was good at it. (Lincoln loved to tell friends, "[My] father taught [me] to work but never learned [me] to love it.") Instead, politics captured his interest and shaped his ambitions.

The work did not impress Lincoln much but the people did: He found the residents of New Salem friendly, welcoming a young stranger who had no ties there. He settled into town around the time newspapers were reporting John Calhoun's leaked letter from President Monroe. Lincoln's arrival also coincided with an off-year election for Congress, set for August 1, 1831. It was the first election in which Lincoln cast a vote. Voting at that time was done not in secret but by open declaration. Lincoln walked up to the station set up in the middle of town and announced that he was voting for James Turney, the pro-Clay, National Republican candidate. It was a brave thing to do, given that the community was largely Democratic, but Clay's vision of federal government had enthralled Lincoln.

The man responsible for tabulating the votes was Mentor Graham, the local schoolteacher. Legend has it that Graham asked Lincoln to help him when he discovered Lincoln could read and write.[79] More likely, Lincoln hung around the table for most of the day, chatting with anyone who listened. By the time the day was over, Lincoln's candidate had lost, but Abe had met most of the townsfolk.[80]

Becoming known in the community quickly produced a problem for Lincoln. Gangly at six four but not fierce, he seemed ripe pickings for a local gang, led by a bully, Jack Armstrong. Armstrong was shorter than Lincoln but known for his muscle and toughness. While reports conflict on what exactly happened between the two, they entered into a wrestling match. Much of the town is said to have come out to watch the two men wrestle to a draw (though legend has Lincoln defeating him). Up until that point, Armstrong, as far as anyone knew, had never lost a match. Lincoln's strength and confidence impressed the gang and particularly his opponent, who afterward became a lifelong friend.

John Todd Stuart, a prominent Whig lawyer from nearby Springfield, said the wrestling match "was a turning point in Lincoln's life."[81] It is unclear whether Stuart actually witnessed the event (he met Lincoln later when he was a land surveyor and again when they served together in the state militia), nor is it clear exactly what Stuart meant. In light of Lincoln's later political acumen, however, the match obviously served as an apt metaphor for Lincoln's canniness in waiting for people to undo themselves by overplaying their hands, as well as his gift for bringing enemies into alliance.

In New Salem, Lincoln boarded first with one of the town elders, then with Mentor Graham for several months. Almost certainly, Graham later exaggerated his importance to Lincoln's education. He had only rudimentary knowledge of math and was barely literate himself, yet he was the best the town had for a schoolmaster. In all likelihood, Lincoln sharpened his math under Graham's roof and studied other books Graham and others gave him. Graham claimed that, besides teaching Lincoln arithmetic, he persuaded him that a thorough knowledge of grammar was indispensable to an aspiring politician.[82] Graham supposedly told Lincoln, "If you ever expect to go before the public in any capacity, I think it is the best thing you can do."[83] Even if this were the counsel he gave Lincoln, it was unnecessary advice, given Lincoln's studies before this point, but probably on his own volition, Lincoln spent time reviewing Samuel Kirkham's *English Grammar,* a leading textbook of the era, as well as any newspapers he could get his hands on. Indeed, Graham is said to have observed that Lincoln's "text book was the Louisville Journal."[84] This has the ring of truth, for the *Louisville Journal* had become, under the leadership of Clay's man George Prentice, the best-selling and most popular newspaper in the west. It deepened and strengthened Lincoln's allegiance to Clay. John Rowland Herndon, a friend from New Salem, recalled Lincoln's idolization of Clay much as Dennis Hanks remembered. Recounting Lincoln's early life, Herndon later wrote, "Henry Clay was his favorite of all the great men of the nation. He all but worshipped his name."[85]

Jack Kelso, a book lover himself, was another man with whom Lincoln briefly boarded in New Salem. Kelso urged Lincoln to read Shakespeare and Robert Burns, both of whom Abe had encountered earlier when his stepmother had brought William Scott's anthology home. Another New Salem resident said Kelso and Abe "were always together—always talking and arguing."[86] The two would sit for hours on the bank of the Sangamon River and "quote Shakespear."[87]

Lincoln and Burns had similar backgrounds. They both grew up in working-class families, and both had lost their mothers in their formative years. Lincoln adored Burns, sometimes claiming him as his favorite poet. The motifs, patterns, and rhyme schemes of Burns's poetry were ingrained in Lincoln's mind, memorized by him as a youth in Indiana, and recited by him as an adult in New Salem and beyond. Much of Burns's poetry deals with themes of poverty, enlightenment, independence, honesty, and the use of reason. At the 1865 annual banquet of the Washington, D.C., Robert Burns club, Lincoln was asked to make a toast to the poet. He replied, "I cannot. . . . I can say nothing worthy of his generous heart and transcending genius. Thinking of what he has said, I cannot say anything which seems worth saying."[88]

David Herbert Donald suggests that "during his New Salem years [Lincoln] probably read more than at any time in his life."[89] Retreating to the woods, walking through the village with a book tucked under his arm, or sitting by the fire, Lincoln "read a great deal, improving every opportunity, by day and by night."[90] Another friend from New Salem remembered that Lincoln "used to sit up late of nights reading & would recommence in the morning when he got up but 'knew nothing of his reading a novel.'" Instead, "History & poetry & the newspapers constituted the most of his reading."[91] A fellow shopkeeper, William Greene, believed "he never saw anyone who could learn as fast as Lincoln."[92] His abundant reading led more than a few people to complain that Lincoln was lazy, particularly because he was often discovered reading a book rather than working.

Another significant influence on Lincoln's development in New Salem was a man named Bowling Green, whom Lincoln had voted for as the town's justice of the peace in 1831. Green, who'd easily won that election, was much older than Lincoln, rotund, jocular, and a fierce Jacksonian Democrat. Later, the two met as members of a local debating society. At its meetings, residents fenced over political topics, not just of local interest but also state and national. Young men of all kinds joined such gatherings, which were common around the country. (When working with George Wythe, Clay had participated in a Richmond debating society.) Like Mentor Graham, Green encouraged Lincoln to read and loaned him books, including some about the law. Green also let Lincoln come to his court to observe him at work. Initially just for fun, he let his visitor argue some make-believe cases, and rather quickly Green saw Lincoln's potential and encouraged him to study law. Later, in 1839, Lincoln did not support Green for a seat in the General Assembly. When Green died three years later, Lincoln returned to New Salem from Springfield for the funeral, one of the few times anyone saw him cry. Those who knew Lincoln and Green remembered their close relationship. One of them recalled, "Mr. L Loved Mr. Green as he did his father"[93] and said "that he owed more to Mr. Green for his advancement than any other Man."[94]

Less than a year after arriving in New Salem, though not yet gainfully employed, Lincoln felt sufficiently comfortable in the community to follow Jackson's example in both declaring his candidacy for public office and heading into battle. Clay had been too young to fight in the Revolutionary War and then too old and well established in civilian life to fight in any other war, but it was common for (white) American men between eighteen and forty-five years of age to volunteer in the state militias. Lincoln followed other townspeople in signing up for the Black Hawk War, a campaign authorized by President Jackson to rebuff the efforts of Sauk warrior Black Hawks to bar Americans from settling on traditional tribal lands. The general Jackson appointed to lead federal troops was his friend Zachary Taylor, with whom he had fought against the Seminoles in Florida. In his first monthlong tour of duty, Lincoln was elected captain by the popular vote of his troops. He signed on for two additional short tours. In his last one, he served as a private in the Independent Spy Corps that unsuccessfully tried to track down Chief Black Hawk in southern Wisconsin. It had been at the urging of Zachary Taylor that Lincoln reenlisted as a private, so as to have experience being both a leader and a follower in the war, experiences expected to burnish his credentials for office.

Lincoln took pride in his soldiering (though he enjoyed making fun of his tours by saying the only combat he had was with the mosquitoes) and particularly in his selection by his men as their first captain. In 1859, he told supporters, "I was elected a Captain of volunteers—a success which gave me more pleasure than any I have had since."[95] A year later, after he became the Republican Party's candidate for president, he reiterated that "he has not since had any success in life which gave him so much satisfaction."[96] As a soldier, a Washington or Jackson he was not, but he was "elated"[97] to have had the experience.

Lincoln made several lasting friendships during the Black Hawk War, two of which were significant influences on the rest of his life. The first was with John Todd Stuart. Stuart was only fifteen months older than Lincoln and was a lawyer well known throughout the state for his sharp and crafty courtroom tactics. After graduating from Centre College in Kentucky in 1826, Stuart traveled west to look for a place to settle and to practice law. On October 25, 1828, he moved to Springfield, Illinois, then a small, friendly town of around three hundred people. He opened a solo law practice, which continued to thrive as he added and changed partners over the next several decades.

When Lincoln first met him, Stuart already was a prominent Whig. He was tall, thin, "debonair," and "handsome."[98] He was slick, satirical, but charming—qualities that would be useful in greasing the legislative rails but also in leaving an oily stain on his reputation. Though Lincoln never developed a taste for alcohol or carousing (one legend suggests Stuart introduced Lincoln to prostitutes), both Clay and Stuart were renowned for their abilities to drink, swear, and gamble, and for their reputed sexual affairs.[99]

Though Stuart later outranked Lincoln in the Black Hawk War (he became a major), Stuart initially served under him when Lincoln was captain of a company that included Jack Armstrong and the rest of his New Salem gang. (Stuart had then reenlisted as a private, following, like Lincoln, the advice of Zachary Taylor to young soldiers that they broaden their military experience to include time serving both as a leader and as someone led.) Stuart, whose recollections of Lincoln are widely considered to be among the most reliable, recalled that, while Lincoln was the captain of his men, he "had no military qualities whatever except that he was a good clever fellow and kept the esteem and respect of his men."[100]

Stuart laughed, in recollection, that Lincoln and the others (himself included) tried but failed "to find out where the Indians were," even though "the Indians were all about us, constantly watching our movements."[101]

Another man Lincoln befriended during the Black Hawk War was another prominent Whig lawyer in the state, Orville Browning. Nearly three years older than Lincoln, Browning was Stuart's opposite: taller and stouter, more ponderous in his manner than Stuart, who was quick-witted and sharp-tongued. Stuart was charming, while most of Browning's contemporaries regarded Browning as pompous. Stuart was keen on having a good time wherever he went, while Browning was a homebody whose wife quickly befriended Lincoln as well. Though neither Stuart nor Browning lacked self-confidence, neither much enjoyed the thrust and parry of political campaigns. Browning in particular preferred not to dirty his hands if he could help it. The two men differed in other ways: Browning kept a diary, in which he recorded thoughts, conversations, and events throughout his life, and he liked to write letters, often offering advice or thoughts about contemporary issues. Stuart kept no such records, and his letters rarely engaged in extended political discourse. Yet Lincoln looked to both men for honest, unvarnished counsel. They rarely disappointed.

CHAPTER TWO

FINDING THE PATH TO CONGRESS
(1834-1844)

Within seven months of his arrival in New Salem, Abraham Lincoln announced his first campaign for office. Over the next seven years, he would become the most powerful Whig in the Illinois State House of Representatives.

Lincoln and many other young Whigs supported both the Missouri Compromise and the American System, which he backed until the day he died. Perhaps more than anything else, however, Lincoln was driven by another of Clay's ideas. Andrew Jackson championed the "common man," whom he believed he exemplified—the poor, illiterate, disenfranchised, hardworking men, who were born with nothing but worked with their hands and were the backbone of America, men born in poverty but could rise to the nation's highest elected office. In contrast, Henry Clay had coined the concept of "the self-made man," in 1832, as the centerpiece of his economic vision.[1] Clay's idea was that any man in America, no matter how modest his beginnings, could become economically productive, particularly if the government supported the development of business and internal improvements, such as roads, bridges, and tunnels, that linked the disparate parts of America into one land. It was an ideal that inspired young Whigs like Lincoln to seek to improve themselves through hard work, self-discipline, and social respectability.

In the 1840s, Lincoln declared himself a "self-made man," though he had been expressing the same idea in emphasizing self-reliance in campaign speeches and public pronouncements beginning in the 1830s. He had examples of self-made men all around him, men such as George Washington, Benjamin Franklin, Clay, and even Jackson. Although Lincoln did not win a seat in Washington as quickly as Clay or Jackson, he was not yet forty when he established his own law firm, became a leader in the Illinois Whig Party, and mounted a successful campaign for the U.S. House of Representatives. His chance to make his mark loomed closer.

I

In 1832, both Bowling Green and John Todd Stuart encouraged Lincoln to run for an open seat in the Illinois legislature. Following their advice and his own burning ambition, Lincoln declared his candidacy for the Illinois House of Representatives on March 9, 1832.[2] Clay had begun his political career in the Kentucky General Assembly, and Lincoln looked to start his own in a similar way. Lincoln, however, was three years younger than Clay had been when he first ran for office.

Lincoln's announcement of his first political campaign might not surprise anyone looking back at his life, but it likely surprised more than a few of the townsfolk. He had just turned twenty-three, but to the many who engaged with him, he talked about politics incessantly. Orville Browning recalled that "even in his early days [Lincoln] had a strong conviction that he was born for better things than then seemed likely or even possible." Browning added, "Lincoln's ambition was to fit himself properly for what he considered some important predestined labor or work."[34]

Lincoln's speech to announce his candidacy for the Illinois House was not his best, but it showed he was learning from Clay. According to Clay's 1831 campaign biography, his first campaign speech was brief and to the point: "He told his fellow-freemen that he was, indeed, young and inexperienced, and had neither announced himself as a candidate, nor solicited their votes; but that, as his friends had thought proper to bring forward his name, he was anxious not to be defeated."[5] Clay "gave an explanation of his political views, and closed with an ingenuous appeal to the feelings of the people."[6] From reading that speech, as well as Clay's other speeches or reports on them, Lincoln learned their characteristic flow. Usually, Clay began with some kind of self-deprecation, such as reminding his listeners of his "humble" origins or lack of the superior intellect that others had. He often marked the occasion or context of the speech, where it was happening and why, and praised his adversaries before launching into systematic evisceration— often filled with ridicule and satire—of his opponents' arguments. (Clay did not care about what happened to his speeches once he was done with them. He rarely drafted a speech beforehand; instead, he would write down a few ideas on fragments of paper, which he usually tossed away afterward.)

Clay could build a case as well as any lawyer, but it was his keen sense of humor and colorful images and delivery that helped to set his rhetoric apart. His humor was designed to endear himself and his message to his audience, but he often used humor to mock and belittle his opponents (his followers, for example, called Jackson "jackass" and "King Andrew"), even while professing the greatest respect for them. Lincoln's first speech emulated Clay's early no-nonsense approach and his focus on substance, but it was short on humor. In time, Lincoln would learn to meld the two in winning fashion. In this first speech, Lincoln launched quickly into his purpose, just as Clay would have done. In the short opening paragraph, he declared, "It becomes my duty to make known to you— the people whom I propose to represent—my sentiments with regard to local affairs."[7] His efforts to connect with the crowd were weak because they were too formulaic, and Lincoln would further study Clay's almost invariably successful methods for bonding with his audiences. Yet Lincoln's conclusion was pure Clay in its self-deprecation and expression of humility:

Every man is said to have his peculiar ambition. Whether it be true or not, I can say for one I have no other so great as that of being truly esteemed of my fellow men, by rendering myself worthy of their esteem. How far I shall succeed in gratifying this ambition, is yet to be developed. I am young and unknown to many of you. I was born and have ever remained in the most humble walks of life. I have no wealthy or popular relations to recommend me. My case is thrown exclusively upon the independent voters of this county, and if elected they will have conferred a favor on me, for which I shall be unremitting in my labors to compensate. But if the good people in their wisdom shall see fit to keep me in the back ground, I have been too familiar with disappointments to be very much chagrined.[8]

In time, he would learn to begin, as Clay (and classical orators) nearly always did, with what became his trademark expression of humility.

Lincoln voiced support for "the public utility of internal improvements," the core of Henry Clay's American System.[9] Elaborating, he spoke of the need for "roads and canals," a need for "more easy means of communication than we now possess" in Sangamon County, the need to investigate "the expediency of constructing a railroad from some eligible point on the Illinois river," meant to improve the means for navigating the Sangamon River, and a need to outlaw "exorbitant rates of interests" on loans.[10] All of these proposals came from Clay.

So, too, did Lincoln's statement that education is "the most important subject which we as a people can be engaged in."[11] Funding education was to Clay a crucial element of the American System. In declaring that he could "see the time when education, and by its means, morality, sobriety, enterprise, and industry, shall become much more general than at present," Lincoln was delivering a sharp rebuke to his father.[12] Indeed, he was not only demonstrating his support for a program a still-loyal Democrat like Thomas opposed but, in seeking public office, he was publicly seeking a life his father never had, a life like Clay's.

Lincoln enjoyed the rough-and-tumble of the political campaign. At a stop at a sale in the tiny town of Pappsville, Lincoln was about to take the stage when he noticed a friend of his was being heckled. As Lincoln spoke, a fight broke out. Lincoln quickly stepped off the stage, stepped into the middle of the fray, grabbed the main culprit by the neck and trousers, and threw him to the ground. The fighting suddenly stopped. Lincoln calmly walked back up to the stage and recommenced his speech—"Fellow citizens, I am humble Abraham Lincoln."[13]

Whether Lincoln's volunteering for the state militia in the midst of his campaign was simply a means of relatively stable employment after a stretch of professional insecurity or a tactical move to burnish his credentials as a candidate, not long after announcing he headed off to battle. When he returned from the Black Hawk War, the election was only a couple of weeks away. On August 2, 1832, he and the other candidates for the New Salem seat in the Illinois House assembled to give their final speeches before the election on August 6. Drawing on a *Sangamo Journal* story about President Jackson's veto of the national bank's rechartering, along with advice from John Todd Stuart, Lincoln spoke for thirty minutes in defense of the national bank (and how it fit into Clay's broader vision), denouncing Jackson's veto of its rechartering.

It was for naught. Though Lincoln had promoted his candidacy throughout the county, he lost the election, finishing eighth out of thirteen candidates. He took solace in the fact that, although he lost the county, he had won 277 votes out of the 300 cast in New Salem, which was predominately Democratic. Stuart won his election for the Illinois House. Shortly thereafter, Stuart was elected as the floor leader for the Whigs in the House, so while Lincoln had failed in his first campaign for office, he now had a close friend and political compatriot in a position to help him in future campaigns.

Besides doing well as a Whig in a sea of Jackson men, Lincoln had, without realizing it, impressed some important Whigs. Stephen Logan, already one of the state's best trial lawyers, had been in the audience more than once. Logan recalled his first sighting of the candidate, at a political rally at the courthouse in Springfield:

I saw Lincoln before he went up into the stand to make his speech. He was a very tall and gawky and rough looking fellow then . . . But

after he began speaking I became very much interested in him. He
made a very sensible speech. [Up] to that time I think he had been
doing odd jobs of surveying, and one thing and another. But one
thing we very soon learned was that he was immensely popular.[14]

The year was far from over. Lincoln returned to the hustings, an activity he had loved ever since he was a little boy delivering funny speeches and stories to his friends. This time, he was on a bigger stage: with the presidential election only a few months away, he visited nearby counties to campaign for Henry Clay, but here too the outcome he had hoped for did not come to pass: Illinois went for Jackson, as did the nation. Calhoun left the vice presidency for the Senate. To replace him, Jackson brought Martin Van Buren back from Great Britain to serve as his new vice president and political heir.

Jackson's triumph over Clay was not the victory for which states' rights enthusiasts had hoped. After the election, the South Carolina legislature passed a law that declared unconstitutional the 1828 and 1832 tariffs that Clay had helped pass to protect American manufacturers from their European competitors. South Carolina declared its entitlement, by virtue of being one of the states that had founded the Union, to disregard, or nullify, any federal enactment it deemed illegitimate. Jackson disagreed. On December 8, 1832, he issued a proclamation declaring that South Carolina did not have the power to nullify a federal law and that such power was "incompatible with the existence of the Union, contradicted expressly by the letter of the Constitution, unauthorized by its spirit, inconsistent with every principle on which It was founded, and destructive of the great object for which it was formed."[15] Jackson had taken a page from Clay (and Clay's rival Daniel Webster) in acknowledging that the Union took precedence over the states and that therefore the states could not undermine it. In agreement, the Illinois governor, a Democrat, issued a proclamation declaring nullification "treasonable."[16]

While Jackson and much of the nation were focused on his rebuttal to South Carolina's "Ordinance of Nullification," Lincoln and Stuart were discussing ways for Lincoln to follow Stuart into the state legislature. Looking back at Lincoln's growing reputation at that time, Stuart said, "Everyone who became acquainted with him in the campaign of 1832 learned to rely on him with the most implicit confidence."[17] In his earnest efforts to help his neighbors by doing all sorts of odd jobs and favors for them and to entertain locals with his stories, Lincoln endeared himself to the voters in his district. He had not abandoned his political aspirations after his defeat, and in this he looked again to Clay for inspiration. Clay, after all, had lost the presidential election in 1824, sat out the 1828 election because he was secretary of state, returned to the Senate in late 1831 to raise his stature and create a perch from which he could try to block Jackson's initiatives, and became the first sitting senator to secure a major party's nomination for the presidency, although Jackson won the 1832 presidential election by a large margin, winning 218 of the 286 electoral votes cast. Those losses had been hard on Clay—how could they not be? After all, elections are measured in a precise and quantifiable manner, and when he'd craved the broadest support, Clay had consistently been liked less than his opponents. Even when he had been in high office, his policy dreams had been muffled, if not sometimes suffocated. And yet such rejection did not deter him. Such resilience was another reason Lincoln revered him.

Inspiration was also more immediately at hand. Stuart believed Lincoln's political career, much like his own, was just beginning, and he encouraged Lincoln to try again.[18] Lincoln had learned from the loss and realized that he had to start earlier and campaign harder.

Lincoln's awed recognition of Clay's fortitude would help him again and again and again. Looking back at the path that he had taken to Congress and later the presidency, he recalled that the 1832 election was the only time "he had ever been beaten by the people."[19] In truth, he would lose eight different elections, but to be fair, six were lost before the general election, and his election to the presidency was not, strictly speaking, a winning popular election but one where the key votes came from electors, not directly from the public. Had he at any time surrendered to the trend, we would likely not know his name today.

II

In 1833, Lincoln had no job and no source of income. John Todd Stuart, Bowling Green, and Orville Browning each stepped up their encouragement for him to study law. Stuart was impressed that Lincoln, then twenty-four, had already acquired a reputation for "candor and honesty," as well as for his ability "in speech-making."[20] Lincoln might not yet be comfortable onstage speaking to large audiences as a candidate, but in smaller settings, where he was speaking to friends or townspeople, he had an ease about him that made people like and trust him. Stuart, as well as Lincoln, thought this boded well not just for a legal career, but also a future in politics. Lawyers needed to write well, study and learn the language of law, and just as important, be able to avoid or settle litigation, protect clients' assets, get along with everyone as much as possible, and be persuasive and compelling in making arguments before judges or juries. The more he refined these skills, the better Lincoln would become as both a lawyer and a politician.

Stuart's law partner in those days, Henry Drummer, recalled that

Lincoln used to come to our office—Stuart's and mine—in Springfield from New Salem and borrow law-books. Sometimes he walked but generally rode. He was the most uncouth looking man I ever saw. He seemed to have little to say; seemed to feel timid, with a tinge of sadness visible in the countenance, but when he did talk all this disappeared for the time and he demonstrated that he was both strong and acute. He surprised us more and more at every visit.[21]

Even without the encouragement, Lincoln must have been considering it as an option, since so many successful politicians had begun as lawyers. Of the ten men who had been elected to the presidency as of 1836, seven had studied law—Thomas Jefferson, James Madison, James Monroe, John and John Quincy Adams, Andrew Jackson, and Martin Van Buren. Clay was not a president but he, too, was a lawyer. For all these men, the law itself was not the end. It was a stepping-stone to public office. It was a source of income and well-placed contacts who were instrumental for political advancement. Besides his military service and the controversies that he sparked wherever he went, Jackson had also been a legislator and a state judge before becoming president. Clay's national prominence had begun with his law practice, but more important, that work led to his election to Congress, informed and sharpened his arguments and oratory, and facilitated his leadership in the House and Senate. Even Clay's stint as John Quincy Adams's secretary of state was made possible because of the notoriety he had attained through his oratory and other endeavors. As a "self-made" man eager to make his mark on the world, Lincoln knew his path had to be his own, and with each career advancement, he was pushing the negative example of his father further behind him and was coming closer to becoming what his father had never been nor could tolerate—a man like Henry Clay, an accomplished professional and public figure.

Moreover, the law would be meaningful work, likely the best he could expect to find, given the constraints of community and his own education, and it was not the manual labor he detested. Such a station would situate him nicely within the community, and he could make enough money to support a family and his ambition of succeeding in politics. As Stuart well knew, such income could also help Abe pay off the debts he had been trying to settle for years.

The problem was that studying law required both time and money. Lincoln doubted he had enough of either. The time it would likely require to become qualified to practice was at least a couple of years, if not more, especially because, as the junior lawyer, he would have to do all the scut work Stuart didn't want to do himself. Though Lincoln yearned to better himself, he worried that he still did not have the money needed to secure a legal education.

Aware that the son of another villager had earned a law degree from a college in Louisville, Lincoln lamented that he did not have "a better education" that might have allowed him to avoid such tuition.[22] Later, he wrote, seemingly with regret, that he "was never in a college or Academy as a student; and never inside of a college or academic building till since he had a law license."[23]

It being clear to him that full-time pursuit of a law degree was impossible, besides already planning another run for office in 1834, Lincoln needed to find a paying job. From Stuart and Green, he learned that the position of postmaster might be open. The current occupant, Samuel Hill—a friend of Lincoln's—was widely suspected of having committed fraud. Initially, Lincoln demurred when his companions suggested he take the position. He didn't want to become a candidate unless he could be sure he wouldn't be pushing a friend out of the office. Once it was clear that Hill would be ousted because he had neglected his duties, Lincoln allowed his name to be put forward.

Ironically, the person making the appointment was Andrew Jackson. Lincoln did not worry that his vote for Clay would cost him the job. His Democratic friends had vouched for him with the administration, knowing the appointment was going to be made based on their recommendations. It didn't hurt that there was no competition for the position. Over the next three years, Lincoln handled his responsibilities meticulously, as his friends expected of a man they regarded as trustworthy, diligent, and friendly.

III

Less than a year after his loss in his first race for the state legislature and only a month after becoming the local postmaster on May 7, 1833, Lincoln announced his second run for the Illinois House. Sessions were short, usually three to four months long, and paid less than $200 per session. He had turned twenty-four just two months before, but two paying jobs, as postmaster and land surveyor, supported his studying law in his spare time. If he won, he would be beginning his political career and have a third paying job, but more important, he would be on track to make a difference in the world.

This time, Stuart did not leave matters to chance. He took an even greater strategic role in helping Lincoln's candidacy. Both men understood that it had not been effective for Lincoln to just go around the county giving speeches. Instead, he set out on what a friend described as "his hand-shaking campaign, traveling all around the county to talk face-to-face with the voters."[24] Another friend said Lincoln was determined to curry favor with "all persons, with the rich or poor, in the stately mansion or log cabin."[25] Yet another contemporary recalled Lincoln had "great faith in the strong sense of Country People and he gave them credit for greater intelligence than most men do."[26] Lincoln figured, as he had seen in Stuart's campaigns, that personal connections were more important than party. He and Stuart shared Clay's vision of resilient, hardworking people who made their way through the world under their own steam. While he cozied up to the Whigs who had power and money, Lincoln did not forget where the actual votes came from. Bowling Green, a staunch Democrat, encouraged Lincoln to make the second run even though Lincoln was a Whig. At the same time, Browning decided to make a second run for office.

Lincoln's 1834 campaign, however, quickly ran into an unexpected problem. He learned that the Democrats were plotting to defeat Stuart by splitting the Whig vote. Lincoln told Stuart that Bowling Green and other Democrats had told him that they "would drop two of their men and take [Lincoln] up and vote for him for the purpose of beating [Stuart]."[27] Stuart devised a plan almost immediately. He thanked Lincoln for disclosing the plot to him and advised Lincoln that he "had [such] great confidence in [his own] strength" that "Lincoln [should] go and tell [the Democrats who approached him that] he would take their votes."[28] "I and my friends," Stuart recalled, "knowing their tactics, then concentrated our fight against one of their other men . . . and in this

⌈way⌉ elected Lincoln and myself."[29] Stuart suggested Lincoln's part
was to keep his mouth closed on what he and Stuart were planning
to do, and Lincoln did. Stephen Logan, later a law partner of Lincoln's, said Lincoln never pretended to be anything but the Whig
he was. He believed that "he made no concession of principle—
whatever. He was as stiff as a man could be in his Whig doctrine."[30]
Yet Lincoln gave no speeches for this election, allowing himself
to ride the current of support from his friends on both sides of
the aisle. This time, he won—his first victory in elected politics.
He learned the lesson that sometimes silence works better than
saying too much.

IV

Years later, after Lincoln's death, William Herndon wrote, "His
ambition was a little engine that knew no rest."[31] In the next few
years, following his first election to the Illinois House, as Lincoln
began to rise in Whig leadership in the state legislature, his ambition became more apparent to all around him, as it would simultaneously power his pursuit of a law license and successful
completion of his term as postmaster.

As Lincoln's star kept rising and eventually reached the presidency, Stuart was fond of saying, to anyone who would listen,
that he (Stuart) would be remembered "as the man who advised
Mr. Lincoln to study law and lent him his law books."[32] But while
Stuart took credit for Lincoln's formal licensing as a lawyer, there
were others, such as Bowling Green, who had helped by allowing
Lincoln to observe his work as justice of the peace. Just how much
truth there was in Stuart's later claims to have helped Lincoln's
education was eventually almost beside the point. Stuart may not
have been the only Lincoln friend who crafted his story in a manner most advantageous to his own reputation, just as Lincoln did.
John Scripps, who wrote the first published biography of Lincoln,

in 1860, said that Lincoln had studied law with "nobody."[33] Lincoln often said so himself to emphasize that he had taught himself law by reading classic legal texts, closely observing lawyers argue their cases in various local courts, and discussing the law with Stuart, Green, and other local attorneys and judges. In 1855, Lincoln responded to a young man asking him about apprenticeship by saying that "I did not read with anyone." He told his correspondent, "Get the books, and read and study them till, you understand them in their principal features, and that is the main thing." Lincoln emphasized "your own resolution to succeed, is more important than any other one thing."[34]

That said, Stuart had indisputably been instrumental in guiding Lincoln in taking several important early steps in law and politics. Stuart found in Lincoln someone whose ambition, intellectual curiosity, and natural intelligence fit well with his need for a partner to both oversee his legal affairs and take up the mantle of his leadership in the Illinois House when he left to make a run for the U.S. House of Representatives. In turn, Lincoln demonstrated his own "resolution to succeed," which drove him to learn not only the law but also the business of practicing law, which he was finally able to do on September 9, 1836, when the justices of the Illinois Supreme Court agreed to admit him to the bar. Lincoln moved to Springfield to begin his practice with Stuart. He was twenty-seven.

In Springfield, the two men spent nearly all their time together. They roomed together, even slept in the same bed for a while, and Stuart introduced Lincoln to the other Whig leaders who lived in Springfield. Several had also served in the Black Hawk War, and one even described Lincoln as "a very decent looking fellow," appropriately dressed in loose trousers, which was then the Whig uniform.[35]

Stuart was grooming Lincoln to take over his responsibilities as the leader of the Whig minority in the Illinois house, so he could devote all of his attention to his own run for Congress. Where Stuart went, Lincoln followed. As Stuart's political partner and protégé, Lincoln became quickly known around Illinois "as one of the most Devoted Clay whigs in all the State."[36]

Stuart did much more than introduce Lincoln to politicians—he introduced him to politics as it was actually practiced. He schooled Lincoln in the art of logrolling—trading favors for votes—so men of differing views or with different agendas could both get what they wanted. Thomas Ford, a Democrat who was governor of Illinois when Lincoln first entered the statehouse, confirmed that the Sangamon delegation was unusually large and included "some dexterous jugglers and managers in politics" and had a "decided preponderance in the log-rolling system in those days."[37] Lincoln once described the plight of the legislator in having to find solutions to a variety of disputes:

> One man is offended because a road passes over his land, and another is offended because it does not pass over his; one is dissatisfied because the bridge for which he is taxed crosses the river on a different road from that which leads from his house to town; another cannot bear that the county should go in debt for the same roads and bridges; while not a few struggle hard to have roads located over their lands, and then stoutly refuse to let them be opened until they are first paid the damages. Even between the different wards and streets of towns and cities we find this same wrangling and difficulty.[38]

Yet here Lincoln stood out, as Stuart recalled, for "refus[ing] to be sold. *He never had a price.*"[39]

When Lincoln made proposals of his own or on Stuart's behalf, they almost always followed standard Whig Party policies and therefore aligned with those that Clay championed on the national stage. Indeed, the first bill Lincoln introduced proposed public financing of the Illinois and Michigan Canal. Initially, Lincoln had wanted the canal to be built from the proceeds of sales of government lands (a popular Whig strategy), but he switched strategies when it became clear that his fellow Whigs actually favored having the state pay for the canal. He then worked with his fellow Whigs, as well as Gurdon Hubbard, a former legislator who had pushed hard for the canal in 1832 and failed but who helped Lincoln behind the scenes. After passing the bill, Lincoln dedicated its passage "to the untiring zeal of Mr. Stuart," whose "high minded and honorable way" had secured "the accomplishment of this great task."[40] Hubbard was not happy Lincoln failed to thank him, but Lincoln needed Stuart's help more.

For all the talk of his devotion to Clay, Lincoln did not always follow his lead. Sometimes he followed Jackson instead. In 1834, Andrew Jackson's threat to veto the rechartering of the Second Bank of the United States reopened a debate over establishing state banks in Illinois. Though Democrats believed that Whigs would oppose state banks, as Clay had, Lincoln saw the virtue of a state bank that "could allow the surplus capital of the rich to be invested and available to the industrious poor person so he might get ahead."[41] Lincoln joined a coalition of Whigs and Democrats to send instructions to their two senators back in Washington to remove states' funds from the national bank and place them in state-chartered banks.

Two years later, Lincoln again followed Jackson's lead. This time, he was running for his second term in the Illinois house. Because of reapportionment, the state legislature had increased the number of Sangamon County's representatives to seven. Unsure of how many people were running for the state legislature in 1836, the *Sangamo Journal* invited all the candidates running to "show their hands." Lincoln quickly wrote back, "Agreed. Here's mine!"[42] He then outlined the principles he was running on. Among the first, he said, "I go for all sharing the privileges of government, who assist in bearing its burthens. Consequently I go for admitting all whites to the right of suffrage, who pay taxes or bear arms [serving in the state militia], (by no means excluding females)." In the prior session of the legislature, 1834–1835, Lincoln had supported a successful move to ensure that the right to vote be extended to all white males of the age of twenty-one years and not just to those who owned real estate.[43] Such proposals echoed Jacksonian principles for empowering ordinary Americans with access to voting (even though the state constitution already had extended the vote as Lincoln and others were now lobbying for).

In another important early legislative success, Lincoln joined with a group of eight other legislators known as the Long Nine, a name given to them because they were all at least six feet tall and were Whigs who shared similar political principles. The Long Nine were instrumental in supporting various internal improvements throughout the state in exchange for moving the capital from Vandalia to Springfield, a growing center of business in the state. Such investments in infrastructure could be complex because they involved benefits like roads and canals that were specific to particular areas. Lincoln was disposed to strongly favor them, as such improvements were a central tenet of the Whig philosophy espoused by Henry Clay. Because Lincoln's delegation from Sangamon County was the largest in the state assembly, it had significant leverage when its members voted as a bloc. On January 24, Browning, with both Stuart's and Lincoln's input, took the lead on Springfield's behalf by introducing the bill in the Senate for the legislature to move the capital to a new permanent location. After a few weeks of debate and delay, the Senate approved the bill 24–13.

In February 1837, Lincoln introduced the proposal in the House. The leader of the opposition was a newly elected Democratic member, Stephen Douglas, an ardent supporter of Andrew Jackson. Douglas preferred that the capital stay where it was or move to Jacksonville. With Stuart's help behind the scenes to leverage the sizable Sangamon County coalition, Lincoln and his fellow members of the Long Nine traded votes to ensure that, after four rounds of voting, the legislature formally approved relocating the state capital just before the end of the legislative session. The Long Nine, Douglas said, "used every exertion and made every sacrifice to secure the passage of the bill."[44] As one of the Long Nine, Robert L. Wilson, later recalled,

> In these dark hours, when our Bill to all appearances was beyond recussitation [sic], and all our opponents were jubilant over our defeat, and when friends could see no hope, Mr. Lincoln never for one moment despaired; but, collecting his Colleagues to his room for

consultation, his practical common Sense, his thorough knowledge
of human nature, then made him an overmatch for his compeers and
for any man that I have ever known.[45]

If Wilson's recollections lapsed too much into romanticizing Lincoln's role, Orville Browning, never one quickly to give praise to others, credited the Long Nine, including Lincoln, in less mellifluous prose for "their judicious management, their ability, their gentlemanly deportment, their unassuming manners, their constant and uniting labor."[46]

Throughout, Stuart closely watched as Lincoln observed floor leaders wheeling and dealing. The record indicates further that there was no occasion on which Stuart and Lincoln were both absent from the legislature. He voted independently of Stuart on twenty-six out of one hundred and twenty-six votes, but overall, Lincoln hewed closely to the Whig party line.

Under Illinois law, the new capital would not become official until two years later, when the building projects were to be completed. Originally called Calhoun to honor the then senator, and later vice president John Calhoun, the town of roughly 1,300 changed its name to Springfield in 1832 when its namesake fell out of favor because of his increasingly fiery defense of states' rights and slavery. By 1836, the town became a Whig stronghold within a state that leaned Democratic. Unsurprisingly, after the state capital opened there officially in 1839, Springfield secured its place as the center of law and politics in Illinois.

Just as important for Lincoln, Springfield was Stuart's home and was where, in the same year as the vote to relocate the capital from Vandalia, the two friends began their law partnership. Besides the bond that Lincoln and Stuart had formed during the Black Hawk War, they were both from Kentucky. Stuart was a graduate of Centre College and already a fixture in Springfield as its best-known lawyer. He was popular with local judges and juries and instrumental in introducing Lincoln to a wider circle of people who became lifelong friends, supporters, and political allies. The firm—Stuart and Lincoln—had a modest but steady practice

involving many trivial legal contests, such as neighborhood quar-
rels, livestock disputes, crop damage, replevin of large animals,
and some divorces.[47] Because Stuart's focus was increasingly on
his congressional contest against Stephen Douglas for a seat in the
U.S. House of Representatives, Lincoln learned on his own how to
interview clients, identify and apply the relevant law, draft legal
documents, and collect fees. Lincoln grew comfortable with being
his own boss.

With the contest between Stuart and Douglas looming, Lincoln
pushed the practice of law aside to help. This provided him with a
front-row seat to assess the matchup. Douglas was known as the
Little Giant, because he was a man of small stature but gigantic,
ruthless ambitions. Stuart often asked Lincoln to substitute for
him in debating Douglas throughout the district and to write letters on his behalf to the local newspaper, the *Sangamo Journal,* and
to Whig leaders in the district who were critical of Stuart's foe.
His deputation was so entrenched that at one point Lincoln joked,
"We have adopted it as part of our policy here, never to speak of
Douglas at all. Isn't that the best way to deal with such a small
matter?"[48]

In the middle of the campaign, Stuart and Lincoln found themselves on opposite sides of a murder case prosecuted by Douglas.
The trial and the campaign confirmed their polar opposites in
personality. Stuart liked to ridicule his opponents through deftly
worded taunts (as well as through anonymous letters he authorized Lincoln and other supporters to publish that attacked Douglas's political views and ethics—or lack thereof). Douglas used
words the way other people used fists, to bluntly humiliate and
pound his opponents into dust. Sometimes it went beyond words:
the debates between Stuart and Douglas occasionally ended in
physical confrontations, the last of which included Stuart's grab-
bing Douglas in a headlock that Douglas broke by savagely biting
Stuart's thumb. Stuart proudly displayed the scar for the rest of his
life. Lincoln, unsurprisingly, was determined to avoid such confrontations should he ever stand against Douglas.

Douglas had his nickname, and Stuart did, too. He had become so well known for his unctuous politicking, indolence, and underwhelming campaigning (thus the need for him to rely so much on his law partner) that his fellow legislators took to calling him Jerry Sly, a moniker that followed him for the rest of his life. Democrats, unimpressed by the energy of his campaigning, gave Stuart a different name, Sleepy Johnny, and enjoyed comparing him to Rip Van Winkle awakening from his long slumber. "The last we saw of him, he was rubbing his eyes open at the corner of the street, with his arms raised above his head, giving a most portentous yawn," the Democratic paper said of him during his campaign against Douglas. "After all of his boasting that he was sure of six thousand majority," it said, "without stirring from home, we marvel much, that he should have had energy to arouse himself from his lethargy, and sufficient condescension to visit the people at all."[49]

Everyone understood that the immediate subject of their contest was the economy (though slavery was intertwined with that as it was with virtually everything else). Just a year before, the nation had entered into its first great depression. A likely cause could be traced to one of Andrew Jackson's last acts as president, the Specie Circular, which required that all payments for public land be made in the form of specie, or hard currency (coins, in this case, of gold and silver). The problem was that specie was hard to get, and on May 10, 1837, New York banks, unable to meet continuing demands for it, refused to convert paper money into hard currency in the form of gold or silver. Soon thereafter, the convergence of many factors (including restrictive lending policies in Great Britain and a sharp decline in cotton prices and mortgage payments), produced panic nationwide. More than a third of the nation's banks failed, credit became practically impossible to obtain to start new businesses, cotton prices fell dramatically, and the trade balance with England deteriorated. At the same time, a drought hit Illinois and caused many crops to fail.

Naturally, Democrats blamed Whigs for the economic collapse, and Whigs blamed Democrats. Van Buren, Jackson's self-selected successor, believed that overreach by the federal government was the principal cause of the depression, but another was the Whigs' contempt for republican values, particularly a lack of commitment to civic virtue and to the self-discipline required for saving money and avoiding debt. Arguing that the American people were too self-indulgent, Van Buren proposed to replace the national bank with a system of independent treasuries that would create an institutional choke point between the federal treasury on one side and the states and the private sector on the other.

In the Senate, Clay denounced Van Buren as out of touch with most American workers and his financial plan as ineffective, since it failed to invest in programs and projects that could help the people economically succeed to the point where they would be able to contribute to the solvency of the system as viable economic actors. Following Clay's lead, Stuart and Lincoln themselves argued that the solution to the nation's economic problems was the national bank, which they believed was instrumental to stabilizing the nation's credit and improving the handling of the federal government's financial affairs. Clay and his fellow Whigs in Congress, as well as Stuart and Lincoln on the campaign trail, also derided Jackson's—and now Van Buren's—consolidation of presidential power as usurping the legislative authority vested in Congress by the Constitution.

As is often the case in politics, crisis became an opportunity for finger pointing. Just as Van Buren had blamed Whig ideology, the Whigs capitalized on the economic collapse and Van Buren's inability to sustain Jackson's appeal as the champion of the common man. Lincoln published a series of anti-Douglas letters in Springfield's *Sangamo Journal* as Douglas and Stuart traveled around the district debating each other.

Lincoln later recalled the give and take of the debates, describing from the vantage point of 1854 the thrust and the parry as "a neatly varnished sophism . . . readily penetrated" by Stuart,[50] but recognizing in Douglas's response that "a great, rough non-sequitur was sometimes twice as dangerous as a well-polished fallacy."[51] When the congressional election ended in August 1838, almost everyone expected a Douglas victory, but it did not materialize. Instead, the final tally gave Stuart a narrow victory by thirty-six votes. Douglas complained that Stuart had won because of vote fraud, but he did not press the claim, because the fraud was not likely confined to just one side of the ledger. Stuart and Lincoln also won their murder trial. It was not a good summer for Douglas.

In the first two of his four successive terms in the state legislature, Lincoln voted consistently in line with Clay's positions on infrastructure expansion. Many of the improvements were to be state funded, while Lincoln, as a state legislator, was in the position of voting to instruct the state's senators on the issues that came before them. The votes, particularly those mandating state funding, came back to haunt Lincoln when the 1837 depression hit. With the tightening of credit and extensive foreclosures, the state faced insolvency, and Lincoln had to choose between finding a way to keep the state budget from hemorrhaging and sticking with Clay. He stuck with Clay.

V

Two of Lincoln's speeches from this period show the significant influence of Clay as his mentor. The first was "The Perpetuation of Our Political Institutions," given in 1837 to the Young Men's Lyceum of Springfield, an organization Stuart had been instrumental in bringing to the town. The timing of the speech coincided with several significant developments—the economic fallout; Stuart's upcoming 1838 campaign for the House; Lincoln's repeated attempts in the state legislature to curb slavery, including filing an official protest declaring "that the institution of slavery is founded on both injustice and bad policy" and a failed attempt to amend a pro-Southern resolution cautioning Congress against abolishing slavery in the nation's capital (his amendment would have affirmed the voters' right to abolish slavery);[52] Lincoln's concurrent campaign for reelection to the state legislature; the lynch-mob murder of Elijah Lovejoy, an abolitionist minister and newspaper publisher in Alton, Illinois; and the horrific burning of an African American in St. Louis. While it is customary to consider the speech as a verbose attack on Stephen Douglas, it responded to a theme that was common to all these developments: threats to the rule of law from unruly mobs stoked by populists or, as Lincoln suggested, a tyrant like Stephen Douglas or especially Andrew Jackson.[53] At the same time, the speech was an attempt at something grander than the kinds of newspaper letters and political attacks Lincoln had unleashed against his political opponents.

Early in the speech, Lincoln referred to an "ill-omen" developing within the nation by which, he said, "I mean the increasing disregard for law which pervades the country; the growing disposition to substitute the wild and furious passions, in lieu of the sober judgment of Courts; and the worse than savage mobs, for the executive ministers of justice."[54] (In his Lyceum Address, Lincoln used the terms *mob* or *mobs* eight times.)[55] "This disposition," he argued, "is awfully fearful in any community; and that it now exists in ours, though grating to our feelings to admit, it would be a violation of truth, and an insult to our intelligence, to deny. Accounts of outrages committed by mobs, form the every-day news of the times. They have pervaded the country, from New England to Louisiana."[56] Though "tedious" to recount the "horrors" of what these mobs did, he did recount that in Mississippi "white men, supposed to be leagued with the negroes" and with "strangers" "were seen literally dangling from the boughs of trees upon every road side."[57] Lincoln did not expressly mention Lovejoy's murder; he said without saying. Jackson's base and those who attended his rallies were poor and rowdy, and to his detractors Jackson was King Mob because of the undisciplined and unruly mobs he controlled, yet here too his nickname went unsaid, the allusions clear. No one, at the time or later, mistook that Lincoln was saying Jackson's supporters were responsible for the violence against Lovejoy and other abolitionists.

Lincoln mentioned the rise of tyrants nearly as often as the mobs. The "new reapers" who fomented this violence were "men of ambition and talents" who would "spring up amongst us. And, when they do, they will as naturally seek the gratification of their ruling passion, as others have so before them."[58] Overly ambitious men would not be satisfied with merely "a seat in Congress" or "a gubernatorial or a presidential chair."[59] "What! Think you these places would satisfy an Alexander, a Caesar, or a Napoleon? Never! Towering genius disdains a beaten path. It seeks regions hitherto unexplored. –It sees *no distinction* in adding story to story, upon the monuments of fame, erected to the memory of others."[60] In all likelihood, Lincoln's reference to "towering genius" was a swipe at Stephen Douglas, who was often ridiculed for his short stature but lofty aspirations. Lincoln went further to suggest that the tyrannical disposition, which he obviously thought both Jackson and Douglas had, "*denies* that it is glory enough to serve under any chief. It *scorns* to tread in the footsteps of *any* predecessor, however illustrious. It thirsts and burns for distinction; and, if possible, it will have it, whether at the expense of emancipating slaves, or enslaving freemen."[61]

If "some man possessed of the loftiest genius, coupled with ambition sufficient to push it to its utmost stretch, will at some time, spring up among us," Lincoln warned, "it will require the people to be united with each other, attached to the government and laws, and generally intelligent, to successfully frustrate his designs."[62]

Of course, Lincoln himself was intensely ambitious and well understood the lure of power. The difference was that he understood it through Clay's prism, that its exercise was valid only if used in line with the republican principles of the Founding Fathers. He was also expressing the hope that the people who were inclined to support aspiring despots had the intelligence and the courage to bring them down. They should prefer someone who came from nothing but made his own way by his own labor, someone like himself, a self-made man, just like Clay.

The nation's salvation depended on the public's appreciating that "passion has helped us," for it brought about the American Revolution, but it "can do so no more."[63] Instead, he said "Reason, cold, calculating, unimpassioned reason, must furnish all the materials for our future support and defence.[64] Let those materials

be moulded into *general intelligence, sound morality,* and in particular, *a reverence for the constitution and laws,"* which could lead the
American people to a point at which they could look back and
know that "we improved to the last; that we remained free to the
last; that we revered his name to the last; that, during his long
sleep, we permitted no hostile foot to pass over or desecrate his
resting place; shall be that which to learn that the last trump shall
awaken our WASHINGTON."[65] This distinction between passion
and reason was important not just to Lincoln but to Madison and
the Founders—and likely Lincoln, too—who took it from classical authors Cicero and Seneca. Aristotle was a major influence on
the latter two, but while he counseled plainness and directness
in writing Lincoln was, at that time, seemingly in the thrall of
the ornamentation common to the classical orators Aristotle criticized. Perhaps Lincoln read no Aristotle. More likely, he had not
yet settled on the realization that the more plainly he spoke the
more the public understood, or took to heart, his message.

Lincoln's Lyceum speech was one of hope for a better future. He
was denouncing overly ambitious men but expecting the nation's
salvation to depend on finding a new Washington, fearing tyrants
but remaining faithful to a single leader who could lead the nation
out of its current mess. Lincoln, as Clay had done, was characterizing the project of overcoming slavery, demagoguery, and lawless-
ness as a return to the values that were the foundation of America
and its Constitution. It was a distinctly idealistic objective, which
Whigs clearly shared. (The *Sangamo Journal* reprinted the speech
so it could be read more widely.) Jackson and his acolytes, Van
Buren and Douglas, were promising to decentralize power, give
more of it back to the states, and expect less from the nation's capital, but the Whig solution for maintaining the Union was respect
for the rule of law and the values enshrined in the grand language
of the Constitution.

In retrospect, it is clear that Lincoln was struggling in his Lyceum Address, as he had struggled in his maiden campaign speech,
to find his voice. Occasionally he repeated a short phrase and a
word—a popular rhetorical device at the time, as reflected in the

speeches of Clay and Webster. (It is a rhetorical technique that also appears throughout the Shakespeare corpus, which Lincoln had studied and enjoyed reciting to himself and others.) But that technique was overwhelmed by a greater ambition, his urgent need to make a mark on the world. His remarks were cluttered with the kind of ornamentation common to classical orations and references to Napoléon (twice), the Bible, and the Founders as "a fortress of strength" and "a forest of giant oaks" swept away by "the all-restless hurricane," which "left only, here and there, a lonely trunk, despoiled of its verdure, shorn of its foliage; unshading and unshaded, to murmur in a few gentle breezes, and to combat with its mutilated limbs, a few more ruder storms, then to sink, and be no more."[66] This was the insecurity of a twenty-nine-year-old trying to embellish his message (in this specific case, before a crowd assembled under "academic" pretenses), and the ornaments dangled poorly. He was not unaware that the forced grandiosity and formality poorly suited him. He was searching for language that could straightforwardly, memorably, and movingly capture the Whig project and his own role in advancing it.

Prentice's biography of Clay mentioned his eloquence on nearly every page, and Clay's friends and foes alike often remarked upon that quality. Clay was not stirring, fluent, and memorable in the same ways as his contemporary and rival, Daniel Webster of New England, was in the turning of a phrase or striking declaration. Instead, Clay's reputation was rooted in his humor and skill as a performer—the way he held his body, the way he used his voice and hands, the pauses in his speech, and the timing of his insertion of a jab here or a cutting remark there. It is likely, given the way in which Lincoln viewed Clay, that if the young man had been able to witness his "mentor" (again, Lincoln's term), he might have shed the ponderous attempts to show off in favor of something more honest and open. By the time he finally saw Clay in action, he was already transforming his earlier extrapolations of Clay's style into something all his own.

Lincoln's other signature speech delivered in these years was his December 26, 1839, address on behalf of the Whig candidate for president, William Henry Harrison, who was challenging the incumbent, Martin Van Buren. While Clay was not the candidate, the expectation among Whigs was that Harrison would follow the party line and therefore allow Congress, where Clay reigned, to take the lead in domestic affairs. The arguments for Harrison were, in other words, the same as they would have been if Clay had been the candidate, including discouraging the use of the presidential veto and deferring to the Cabinet and Congress. Perhaps Harrison's most important innovation was in the campaign he ran, including using women to spread his message and the first campaign slogan in a presidential race, "Tippecanoe and Tyler, too," a reference to his prowess as a fighter against Native Americans and to his running mate, a stalwart Democrat, who brought balance to the ticket. Clay's "American System" was no match as a rallying cry.

Michael Burlingame suggests that Lincoln's speech "became the Illinois Whig Party's textbook for 1840."[67] Unlike his Lyceum Address in front of a more detached crowd, the intent here was to mobilize his fellow Whig partisans. Lincoln was more comfortable in these settings than academic ones, perhaps because he was self-conscious about his lack of formal education, or perhaps he could be less self-conscious and more relaxed and candid with his fellow partisans. In any event, this speech demonstrated that Lincoln was evolving toward the rhetorical patterns that had made Clay a legend. First, he did not save his humility for the end as he had done in his first campaign speech but instead began the speech with self-deprecation (a Clay trademark) by noting that the crowd was smaller than it had been for previous speakers. "The few who have attended," Lincoln said, "have done so, more to spare me of mortification, than in the hope of being interested in anything I may be able to say. This circumstance casts a damp upon my spirits, which I am sure I shall be unable to overcome during the evening."[68]

Lincoln then launched into a searing attack on President Van Buren's subtreasury plan (creating independent subtreasuries for the collection and distribution of federal money), which, he said, would cause "distress, ruin, bankruptcy and beggary" by removing money from circulation.[69] After suggesting that the poor would be hit the hardest by Van Buren's economic policies, he defended Clay's favorite project, the national bank, which Lincoln claimed handled money more responsibly than run-of-the mill government officials did. He proceeded to lambaste Jackson, Van Buren, and Douglas as spendthrifts, deftly turning the criticism leveled against proponents of internal improvements—that they were fiscally reckless—against them. His attacks were unsubtle: he declared Douglas "stupid" and "deserving of the world's contempt" and his arguments "supremely ridiculous," while Van Buren was captive to the "evil spirit that reigns" in Washington.[70] The biblical allusions of the Lyceum speech were now decidedly on the underside and the prophetic. He described the nation's capital as hell, with "demons" running amuck and "fiendishly taunting all those who dare resist [their] destroying course"[71] before casting himself as heroic and his cause blessed by the Almighty:

> If ever I feel the soul within me elevate and expand to those dimensions not wholly unworthy of its almighty Architect, it is when I contemplate the cause of my country deserted by all the world beside, and I standing up boldly and alone, and hurling defiance at her victorious oppressors. Here, without contemplating consequences, before high heaven and in the face of the world, I swear eternal fidelity to the just cause, as I deem it, of the land of my life, my liberty, and my love.[72]

Lincoln's speech for Harrison was a paean not only to the candidate but his new profession. It sounded the theme that faith in law could bring order to the chaos that was the alternative, a theme that Lincoln reiterated throughout his career, and one that Stuart, Clay, and Webster all repeated as well. At the same time, Lincoln's rhetoric was evolving stylistically, his phrasing was plainer, more succinct, and therefore more memorable. His fellow Whigs praised the eloquence, passion, and reasoning of this speech.

This Lincoln was neither shy nor retiring. He was choosing a different path; as he spent more time in the legislature, he was becoming more comfortable with becoming sharply partisan, following Clay's example by destroying the competition before it could destroy him. Lincoln was testing the limits of such fierce attacks. He lived in a community—and a state—dominated by Democrats. First in New Salem and then in Springfield, most of Lincoln's friends and neighbors, even those he worked for and learned from, such as the lawyer Bowling Green and surveyor John Calhoun, were Democrats. Incredibly, Lincoln managed to remain cordial to his friends on the other side. His adept balancing of his friendships with the increasing need for partisanship was evident in December 1839 in a series of debates held in a church in Springfield between leading Democrats on the one side and leading Whigs on the other. As one Lincoln biographer relates, "In Springfield in the winter following, when the legislature was in session, a new form of campaigning sprang up. It was called the Three day debate."[73] This newly fangled debate was informal and lengthy, and yet observed some simple rules—respect for your opponent and no spurious attacks. Among the Whigs who regularly participated in the debates were several of Lincoln's closest friends in the bar, including Edward Baker, Orville Browning, and Stephen Logan. Among the regular, prominent Democratic participants were Stephen Douglas and Ebenezer Peck, who would later become a prominent Chicago lawyer and Republican leader. "The debate was so good natured, informal, and helpful that a request was presented that the format be repeated and similar debates were held in nearby towns in the spring of 1840."[74] Lincoln did more than merely hone his debating skills in these contests. The friendships and trust forged among the regular participants endured in spite of sharp political differences in the ensuing years. Nevertheless, though Lincoln engaged with Democrats, he was actually becoming less tolerant of the spirit of compromise. The more time he spent with devout Whigs, the more Whiggish he became. That would become increasingly evident in the stretch ahead.

VI

In April 1841, John Todd Stuart amicably ended his law partnership with Lincoln. It was a business decision, which was intended to benefit them both. Stuart realized that, given Lincoln's passion for politics, his junior partner might not always be able to cover the firm's work when Stuart was in Washington. He needed someone younger, for whom the law, not politics, was the driving passion. Stuart took on a young, Yale-educated Whig, Benjamin Edwards, who was from Springfield. Stuart suggested Lincoln partner with another Springfield lawyer, the aforementioned Stephen Logan.

Lincoln and Logan already knew each other well. Logan, then a state court judge, heard the first case Lincoln tried with Stuart. He had watched Lincoln's campaign speeches closely. Logan was well known as one of the state's premier legal minds and well respected in the Illinois bar, a reputation upheld despite the fact that he dressed sloppily, never wore a necktie, and dribbled tobacco juice while he spoke in court.[75] He was widely feared, with a reputation for being difficult and demanding. (Logan was once described as "snappy, irritable [and] fighting like a game fowl.")[76]

Lincoln became friendly with Logan while they both rode the circuit together, which involved traveling with a judge and another lawyer or two to handle disputes in the counties comprising a particular region of the state, in this case the First Judicial Circuit of Illinois. Lincoln and Logan were among the lawyers who rode the ten-county First Judicial Circuit until 1839, when Sangamon County was included in the newly created Eighth Judicial Circuit, which covered nine counties. Besides Logan, Lincoln met several other lawyers and judges in those days who remained close friends and political allies over the years, including David Davis, a friend from the state legislature, and Leonard Swett, a fellow lawyer in Springfield. Browning sometimes rode the circuit, as did another young lawyer, Abraham Jonas, the first Jew admitted to the Illinois bar. Jonas had begun his career as an apprentice to Browning.

Logan, as the historian Michael Burlingame has noted, was "a better lawyer" but "a worse politician than Stuart."[77] Certainly Stuart was not alone in thinking that Logan was the best attorney in Springfield. In 1843, the *Sangamo Journal* declared that Logan "is regarded as perhaps the best lawyer in the State."[78] The article noted that, while Logan's "voice is not pleasant," he had "a most happy faculty of elucidating, and simplifying the most obstinate questions."[79] Such illumination was appealing not only to judges and jurors but also to Lincoln.

Logan accepted Lincoln as a partner mostly because he had been impressed with his speaking abilities—a skill, Logan said, "exceedingly useful to me in getting the goodwill of the juries."[80] Yet Logan also quickly realized Lincoln's limitations as a lawyer. Logan could see that although Lincoln had read some legal texts, Lincoln had little meaningful experience in the actual practice of law. Logan learned that Stuart might have taught Lincoln how to sway a jury or negotiate a contract but would not have taught him other important legal skills—how to argue before judges, draft legal documents, and file pleadings. Logan knew Lincoln was "never a reader of law; he always depended more on the management of his case."[81] Indeed, Logan was not surprised to find that, even after a few years of practicing law, "Lincoln's knowledge of the law was very small."[82] However, he knew Lincoln "would work hard and learn all there was in a case he had in hand." In the end, Lincoln became, in Logan's estimation, "a pretty good lawyer though his general knowledge of law was never very formidable."[83]

Under Logan's stern tutelage from 1841 to 1843, Lincoln "tr[ied] to know more and studied how to prepare his cases."[84] Logan instructed Lincoln on how to use the legal materials he had read to make his arguments stronger. While Stuart was skillful at cajoling and charming juries, Logan instead taught Lincoln how to better organize his arguments and to integrate the facts and the law of a case in order to persuade judges and juries. Logan required Lincoln to observe him in court, paying special attention to his arguments before juries. After watching, Lincoln stated "that it was his greatest ambition to become as good a lawyer as Logan,"[85] who was "the best [trial] lawyer he ever saw."[86] Lincoln never praised Stuart in such a manner. He admired how Logan could "make a nice distinction in law, or upon the facts, more palatable to the common understanding, than any lawyer he ever knew."[87] The lessons learned from Logan, including how to frame or adapt arguments to different audiences, would turn out to be important not only in Lincoln's development as a lawyer, but also as a politician.

When the Illinois Supreme Court and federal courts moved to Springfield in 1839, thanks to his partnership with Logan, Lincoln was an ideal position to expand his own practice.[88] As Michael Burlingame notes,

> *Of the 411 Supreme Court cases that Lincoln appeared in during his twenty-four-year legal career, a substantial number were tried during his brief partnership with Logan. In response to the hard times following the Panic of 1837, Congress enacted a short-lived bankruptcy law in 1841 to relieve debtors, many of whom enlisted the services of Logan and Lincoln. They handled seventy-seven such cases, more than any other firm in Springfield and the fourth largest of any firm in the state.*[89]

Lincoln continued to ride the circuit in order to earn additional money, and as he became more familiar with the law's complexities, he, like other lawyers at the time, had to deal with the jarring reality that judges varied in quality. Some judges knew and cared about the law, while many others did not. Outcomes often did not depend on the facts but on a judge's politics. In the nineteenth century, judges rose to the bench because of their political connections, rarely because of their acumen or distinction as lawyers. Their courts were often poorly run and their decisions erratic. Lawyers were representing different parties constantly, so an opponent one day could be an ally the next or vice versa. The legal system was much like the rest of the Old West at that time, where order did not strictly follow the law.

Lincoln was equipped to fare well in such a world. A common description of him as a lawyer in those years was that he was plainspoken, and this quality, along with his natural and well-honed penchant for storytelling, made him more effective with juries than judges. It also made him more effective on the hustings. Logan pushed Lincoln harder to not just repeat the points made by others but to make his arguments in his own words. Lincoln's 1840 textbook critique of Van Buren's fiscal policies demonstrated that he was improving in the clarity and coherence of his political attacks, putting them into plainer language and simpler metaphors that he could more effectively hammer home.

Even as Lincoln earnestly took Logan's counsel to heart, they both understood the extent to which partisan forces could shape outcomes. Perhaps no incident better illustrates this than Lincoln's unsuccessful effort as the Whig minority floor leader in the Illinois House to stop a Democratic initiative to remake the state's highest court in order to favor Democratic objectives. Stephen Douglas, four years younger than Lincoln, led this effort.

In 1839, the Illinois Supreme Court's decision in *Field v. People of the State of Illinois, ex rel. John McClernand*, provoked Democrats in the state legislature to introduce a court-packing plan.[90] Thomas Carlin, a Democrat, was elected governor of Illinois in 1839. Upon assuming office, Carlin nominated John McClernand as secretary of state, even though the office was occupied and had no set term. The current occupant, Alexander Field, was a Whig who had no intention of leaving. The Whigs and a small number of Democrats blocked the nomination on the ground that the governor could not remove Field without the legislature's approval. (A similar argument would be made later by Radical Republicans in response to President Andrew Johnson's dismissal of his secretary of war Edwin Stanton.) Carlin waited for the legislative recess, then named McClernand as the acting secretary of state. Field still refused to leave office and filed a lawsuit to block the governor. The trial judge, a Democrat, ruled against Field, who appealed the judgment to the state supreme court. The state supreme court at that time had four justices: three Whigs and a Democrat. One Whig recused himself, the other two voted to reverse the trial judge, and the Democrat voted to affirm the trial judge's decision.[91]

That is how things stood until 1840 when Democrats increased their control of the Illinois legislature. The Illinois state constitution was unclear on whether foreigners without U.S. citizenship were entitled to vote, but by 1840 more than nine thousand had joined the ranks of the Whigs, while barely a thousand supported the Democrats. The Democrats filed a lawsuit barring the noncitizens in the state in voting in statewide elections and then followed the lead of Stephen Douglas, whom the legislature had appointed secretary of state and who urged the legislature to expand the size of the state supreme court from four to nine. As a result of the 1840 elections, the newly constituted state legislature, with a Democratic majority, would have the power to appoint five new justices. In 1841, it did that, bringing the total number of Democratic appointees on the state supreme court to six against three remaining Whigs. Lincoln and thirty-four other Whig representatives in the state legislature denounced the scheme "as a party measure for party purposes," which manifested "supreme contempt for the popular will."[92] Lincoln's hand in crafting the attack is evident in the characteristic wordplay of the response, the repetition of *party* to underscore the clever use of *supreme* to ridicule what the party did. The Whigs further argued that the Democrats' "party measure" undermined "the independence of the Judiciary, the surest shield of public welfare and private right" and set a "precedent for still more flagrant violations of right and justice."[93] (Here again was Clay-like repetition to underscore the magnitude of what was lost—not just a particular right of the people but a safeguard of all rights.)[94] In April, an anonymous letter, likely written by Lincoln (it was similar in style and content to his public remarks), suggested that the bill probably passed the legislature because one member had delivered his vote in exchange for his appointment as clerk of the state supreme court.[95] The letter went further to ridicule Douglas, who was one of the five new additions to the state supreme court (thereby becoming the youngest person ever appointed to the court), dubbing him a hypocrite for publicly opposing life offices but then accepting one when he was appointed.[96]

With the newly reconstituted supreme court to back him, the

Democratic governor again fired Field. Field appealed the decision, as he had done before, but this time the court was literally stacked against him. He lost the appeal and his job.

It was not the last time that Lincoln witnessed the Democrats use their majority on the state supreme court to ratify the party's power and agenda. In 1843, the Illinois legislature enacted a law that allowed white men who were residents but not citizens of the state to vote in state elections. The Illinois Supreme Court, including Douglas, upheld the policy, this time in a move that seemed to undercut Douglas's and the Democrats' earlier concerns about noncitizens voting. And this time, Lincoln objected, believing that the law was designed to make it easier for Democrats from other states to reside just long enough in Illinois to tip elections in Democrats' favor. He argued that only citizens should be allowed to vote, but simply because he felt the Democrats were trying to manipulate elections. Broadening the entry into citizenship was consistent with Clay's American System, an attempt to maximize the contributions people could make to the productivity of the United States. Lincoln believed that this expansion of the vote was designed for partisan purposes, not democracy and not the economy.

In later years, Lincoln rarely mentioned the threats to judicial independence posed by the partisan court stacking, but it is possible he never felt the need to do so. He agreed with one of the foundations of Jacksonian democracy, the spoils system—the practice of giving plum appointments in return for political favors and campaign donations. As New York Democratic governor William Learned Marcy baldly explained the idea in the Senate in 1832: "They see nothing wrong in the rule that to the victor belong the spoils of the enemy."[97] It was not a new idea to give allies and friends the offices that would have gone to the opposition had it won the election, and Marcy lived by the code through three terms as governor until he finally lost reelection to an ambitious young Whig, William Seward, in 1838. Seward practiced the same philosophy as Jackson and Marcy, distributing rewards to patrons and friends when he came into power, just as Lincoln himself would try to do when the time came.

VII

The 1841–1842 session was Lincoln's fourth and last full term as a representative in the Illinois House. He had a few good reasons for leaving the legislature. The first was that his chances for reelection were becoming increasingly slim. The success of the Democratic Party in Illinois, built on Jackson's popularity and appeal as the champion of the working man, along with some popular policies, such as ending the national bank, likely accounted for the Whigs' and Lincoln's dwindling margins of victory. Aside from riding a wave of support resulting from its policy of allowing resident aliens to vote in state elections, Lincoln and his fellow Whig legislators were running out of ways to help the state avoid insolvency from paying for all the internal improvements that they had gotten the state legislature previously to approve. Lincoln's support in each election was less than in the previous one, and there was no reason to think the trend would reverse.

In leaving the state legislature, Lincoln was following the lead of his former partner and mentor John Todd Stuart, who had abandoned his seat in the Illinois House to run his congressional campaigns. Lincoln wrapped up his work in the state legislature in time to make a run for the Whig nomination for Illinois's Seventh District in the U.S. House of Representatives, but he entered too late and lost his party's nomination to a friend and distant cousin of Mary Todd's, John J. Hardin.

Hardin was a popular, handsome newspaper editor, a rival to Lincoln for leadership of the Whig Party in the district. Lincoln was also indebted to him politically and personally. As Lincoln was completing his service in the statehouse, the Democrats and Whigs were yet again embroiled in heated argument over the fate of the national bank. James Shields, a prominent Democrat who was the state auditor, publicly weighed in against the national bank shortly after the Democrats had swept the statewide elections in 1842. In response, Lincoln followed a tack he had before, publishing anonymous letters (in this case attacking Shields) in the

Sangamo Journal. As Sidney Blumenthal relates, "One of the telltale characteristics of Lincoln's writings and speeches throughout his career was his appropriation of the rhetoric of his opponents to turn against them."[98] Lincoln had learned the technique from both Clay and Stuart, but as he soon learned, he had yet to master it. It went one step too far in the way he demeaned Shields.

Shields was greatly offended by the numerous attacks made against him in the anonymous letters. The writer insinuated that Shields supported Mormonism, a new Christian denomination whose ten thousand members had settled in Illinois after being expelled by mobs and militias from their villages in New York and Missouri because of their biblical revisionism, fervent abolitionism, amity with Native Americans, prosperity, and insularity. The letters called Shields a "fool," a "liar," and a bumbling lover bound to marry one of the lovely ladies making fun of him.[99] Shields wanted an apology or a retraction and pressed the publisher of the *Sangamo Journal* to disclose the author of the offending letters. The editor relented.[100] It was Lincoln.

Shields challenged him to a duel. As the man challenged, Lincoln had the choice of weapons. He chose broadswords. He figured Shields was more adept at pistols than he was with swords and that his long reach would give him an advantage over his much shorter challenger. Lincoln even hoped Shields might withdraw from the fight once he realized the disadvantage. But Lincoln didn't know that Shields had trained as a fencer and was quite comfortable—indeed, confident—in a swordfight.

Hundreds of people reportedly gathered to watch the two men square off on the field of battle. There is more than one story about what happened next. The most popular is that, moments before the duel commenced, Hardin and other "would-be pacificators" rode up on their horses and pleaded with both men to put their dispute before a disinterested panel.[101] In another account, Hardin reached both men before weapons were drawn and persuaded them to drop the fight, and both quickly agreed. Whatever the truth, Lincoln subsequently looked back upon the affair only with embarrassment, as a cautionary tale about how far to go in taunting an opponent anonymously in the press. Whenever anyone, even Mary Todd, later referred to the matter, Lincoln refused to discuss it and quickly changed the subject.

Much to Lincoln's chagrin, Hardin won the general election for the congressional seat in 1842. Afterward, Lincoln persuaded other Whig Party leaders in the state to adopt "rotation in office," which Andrew Jackson and the Democrats had introduced at the federal level in 1828. In every election or appointment cycle, a political party would rotate new men as candidates for each office. Jackson defended the practice as implementing the will of the electorate, and Lincoln adapted it to ensure broader participation and inclusion of party loyalists in elections.

When Hardin secured his party's nomination for the House seat he occupied, he had appeased his fellow Whigs by agreeing to take the seat on the condition that he would rotate out of it in 1844 to allow another Whig to compete for the nomination that year. Unfortunately for Lincoln, Edward Baker, a friend but an equally ambitious Whig, took the nomination for the House seat next, won the general election in 1844, and served in the congressional seat until he resigned in 1846 to enlist as an officer in the Mexican War. The resignation became necessary to avoid a conflict with the Constitution's incompatibility clause, which bars an "officer of the United States" from serving in Congress.[102] Baker was a colonel in the local militia and therefore was technically an army officer, an "officer of the United States." When Baker resigned his seat, it became open for Lincoln to compete to fill.

VIII

Lincoln's politicking in the late 1830s and early 1840s did not interfere with his courting eligible women in Springfield. He was popular in Springfield social circles, as a young "self-made man" on the rise, and on November 4, 1842, he married Mary Todd. She was ten years younger than Lincoln and was soon expecting their first child, Robert Todd Lincoln, who was born on August 1, 1843.

Neither Stuart nor Browning expressed love, or much if any like, for Mary Todd. Both opposed the match, and neither ever changed his mind. We cannot know what, if anything else, either man said to Lincoln when they must have spoken in person about Mary Todd. We do not even know how often they did. From correspondence and observations of Stuart, Browning, and Joshua Speed, whom Lincoln had befriended when they both lived in Springfield, we can gather that all three men knew that Lincoln's relationship with her was often tempestuous and unstable, even before they married. After once breaking off his engagement with her, Lincoln wrote to Stuart, "I am now the most miserable man living," adding, "If what I feel were equally distributed to the whole human family, there would not be one cheerful face on the earth. Whether I shall ever be better I cannot tell; I awfully forbode I shall not. To remain as I am is impossible; I must die or be better, it appears to me."[103] Though Lincoln was sad other times, Speed felt obliged "to remove razors from his room [and] take away all the knives and other such dangerous things."[104]

Speed warned Lincoln that he would die unless he pulled himself together. Lincoln responded that he was unafraid to die but for his regret "that he had done nothing to make any human being remember that he had lived" and that he had not yet done anything to make the world a better place.[105] More than a year later, Lincoln still could not decide whether he should marry Mary Todd. "Before I resolve to do the one thing or the other," Lincoln confided to Speed, "I must regain my confidence in my own ability to keep my resolves when they are made."[106] Lincoln took his own advice to heart; he was characteristically slow and deliberate in making decisions, but once he made them, he stuck by them. As Mary Todd years later remembered, he was "a terribly firm man when he set his foot down—none of us—no man no woman Could rule him after he had made up his mind."[107] It was a rigidity bound to make things worse at home, since unlike Abraham, Mary had an aristocratic background and usually got her way. But his stubbornness would ultimately provide crucial ballast on a much greater stage.

Though Lincoln went on to do what he thought best, eventually marrying Mary Todd, Stuart and Browning each continued to talk of her as untrustworthy, histrionic, and difficult, and they both described Lincoln as unhappy on the day of his marriage. (Lincoln reputedly said, in response to a question about what he was about to do, that he was headed "to hell."[108]) Stuart and Browning believed that Lincoln's difficult, turbulent relationship with his wife might have explained why he spent so much time away from home riding the circuit and working on campaigns. Stuart and Browning also believed Mary Todd was more ambitious for Lincoln than even Lincoln was, though it seems reasonable to assume that Lincoln welcomed a wife who shared his ambitions or had even grander ones for him than he might have imagined for himself. In their opinion, however, she was the aggressor in the relationship, and possibly had seduced Lincoln on the eve of their marriage as a way to make it impossible for him to escape. (Robert Todd Lincoln, their first child, was born slightly less than nine months after the date of their wedding.)[109]

For Lincoln, there was another likely incentive: Henry Clay was a friend of the Todd family, a fact that undoubtedly would have pleased Lincoln. Mary Todd's father, Robert Todd, had been a law student of Henry Clay's at Transylvania University in Lexington, Kentucky, and later a business partner and political ally of Clay's. Mary Todd had supposedly met and socialized with Clay when she was younger and was so loyal to Clay that she became known as "a violent little Whig." For all of her faults, Mary Todd was fiercely in Lincoln's corner, calling their marriage "our Lincoln party." If Lincoln's confidence lagged, and it did, she was there to refuel their joint ambitions.

On January 16, 1844, Charles Dresser, the minister who had performed their wedding ceremony, sold Lincoln and Mary Todd a one-and-a-half-story, five-room cottage at Eighth and Jackson Streets. Jackson was one of five Springfield streets named for former presidents. Every day Lincoln took his short walk from his home to his office at Sixth and Adams, the State Capitol building at Fifth and Adams, or both, he almost certainly would have thought of the nation's early presidents and their ongoing presence in his and the nation's life.

While Lincoln was settling into married life in 1844, he ended his two-year law partnership with Stephen Logan and began a new one with a younger partner, William Herndon. Lincoln's impending family responsibilities and his interest in running for Congress explain why he chose to partner with a younger lawyer—he needed someone else to run the office while he campaigned for the House. As Lincoln learned from working with both Stuart and Logan, a younger partner could be given more of the firm's work, freeing the senior partner to drum up more business or plan a political run. According to Henry Clay Whitney, who rode the circuit with Lincoln, Lincoln had explained to Herndon that he had taken him "in partnership on the supposition that he was not much of an advocate, but that he would prove to be a systematic office lawyer; but it transpired, contrary to his supposition, that Herndon was an excellent lawyer in the courts and as poor as himself in the office."[110]

IX

On June 8, 1845, Andrew Jackson died. On his deathbed, he forgave his political enemies, including Clay. His final words were to his family, "Be good children, all of you, and strive to be ready when the change comes."[111] (Legend has it that Jackson had a parrot, which erupted in profanity the moment Jackson died.) At the time of his death, Democrats hailed his legacy as one of America's strongest presidents, an American hero who opposed corruption and an overreaching federal government and was the relentless champion of democracy and the common man.

The Whigs were not inclined to forgive Jackson for anything.

Daniel Webster, who had aligned with Jackson in opposing secession, was the only prominent Whig leader to eulogize Jackson. In a speech at the New-York Historical Society, he called Jackson a man "of dauntless courage, vigor, and perseverance" who sometimes had shown "wisdom and energy."[112] Clay said nothing. Yet again following Clay's lead, Lincoln was silent.

Like Clay, Lincoln considered Jackson a tyrant, whose twelve vetoes were more than those of any other president up until that time. His dismissals of his Cabinet and of Whigs throughout the government to serve his own political whims and his appointments of friends and allies to replace Whigs whom he had dismissed had seriously damaged the American Constitution. Clay and Lincoln believed that Jackson should have been held accountable, even in death, for his bad acts, including deliberately using inflammatory rhetoric to stoke vicious and frenzied actions by his supporters. They likely would have agreed with the assessment of Jackson made by the visiting French diplomat Alexis de Tocqueville, who in his monumental two-volume book *Democracy in America* declared, "General Jackson is the majority's slave; he yields to its intentions, desires, and half-revealed instincts, or rather he anticipates and forestalls them."[113] On his travels throughout the United States, Tocqueville had closely observed Jacksonian democracy in action, famously deriding it as nothing but the tyranny of the majority.[114] Lincoln and his fellow Whigs agreed.

Another likely reason that Clay and Lincoln said nothing positive in reacting to the news of Jackson's death is that, although the former president was gone, the worst parts of his legacy were still alive, including the man now in the White House. Jackson's protégé James K. Polk, newly into his first term, gave every indication that he was prepared to be the same kind of president in terms of policies, vicious partisanship, and constitutional outlook. "Young Hickory," as those who supported Polk called him, fully commanded the Whigs' attention. Besides being something for which they each might have once wished, Jackson's death was almost an afterthought when Polk took the oath of office in 1845, such was Polk's ideological fidelity.

Clay and Lincoln remained, however, shocked that it was not Clay in the Oval Office. In 1844, Clay had run against Polk as the Whig candidate for president; it was his third run as a major-party candidate for the nation's highest office. Lincoln had been selected as an elector for Clay in the Electoral College, and the entire Whig Party believed that Clay's chances had never looked better.

True, in 1840, the innocuous and bland William Henry Harrison beat out Clay to become the first Whig president, but he died a month into office, elevating to the presidency his vice president, John Tyler, who had left the Democratic Party to be Harrison's running mate. Whigs didn't trust him, and Democrats hated him for helping the Whig Party win the presidency.

The Whigs' distrust of Tyler turned out to be for good reason. Even though as a senator he had harshly criticized Jackson for his tyrannical behavior, as president he acted a lot like his former target—vetoing infrastructure investments and tariffs, and defending states' rights at every turn. Clay led the movement to oust Tyler from the Whig Party.

Tyler saved his most outrageous act for the end. As a lame duck (both parties hated him), he was seriously considering taking the initiative on perhaps the biggest issue confronting the country. Mexico continued to be perceived as a threat to the Union, since it claimed Texas on the basis of the original Spanish explorations by Alonso Álvarez de Piñeda, Álvar Núñez Cabeza de Vaca, Francisco Vásquez de Coronado, and Luis de Moscoso de Alvarado, the area rightly passing to Mexico upon independence in 1821. The Mexican government maintained it would go to war if the United States tried to take it. Americans who had become leaders in Texas asked for its incorporation into the Union, and Democrats like Jackson and Tyler looked favorably upon the request because its admission would add an additional slave state. From the sidelines, Jackson, still the most powerful voice in the Democratic Party, encouraged Democrats to take it. "Texas must be ours," he proclaimed. "Our safety requires it."[115]

Clay did not yet know who the Democrats might nominate to oppose him in the early months of 1844, but he was aware that

Martin Van Buren, who had lost the presidency to Harrison in 1840, was hoping to become the Democratic nominee again. Clay wanted to take a stand on the Texas issue before Van Buren could, and on April 17, 1844, with the election nearly six months away, he made the fateful decision, as his biographer Robert Remini wrote, to write "the letter on Texas that ultimately destroyed his presidential bid."[116] Clay began the letter by reviewing the history of the region. Originally, the territory of Texas was part of the land that was included in the Louisiana Purchase in 1803. But according to Clay, that area had been given back to Mexico in the Adams-Onis Treaty, negotiated in 1819, several years before he became John Quincy Adams's secretary of state. Hence, in Clay's judgment, it was absurd, "if not dishonorable," he wrote, to attempt to resume title to Texas as if it had never ceased being a part of the United States.[117] Claiming title to it would be an act of war, he predicted: "Of that consequence there cannot be a doubt. Annexation and war with Mexico are identical."[118]

Clay understood that Jacksonians were not averse to going to war, but "I regard all wars as great calamities," he wrote, "to be avoided, if possible, and honorable peace as the wisest and truest policy of this country."[119] While some Democrats believed that defeating Mexico in a war would not be difficult, Clay cautioned that Great Britain or France might come to its aid because each was "jealous of our increasing greatness, and disposed to check our growth" and eager to find a way to "cripple us."[120] Even if for some reason Mexico agreed to allow the United States to annex Texas, Clay argued that its annexation would be opposed by a "considerable and respectable portion of the Confederacy."[121] (*Confederacy* was a word that the framers' generation had used to mean the union, or compact, of states comprising the United States.) Hence, the annexation of Texas would produce discord and turmoil within the nation and possibly endanger the Union. He expected the likely Democratic nominee, Martin Van Buren, to also oppose annexation, and added, "We shall therefore occupy common ground."[122]

Clay's letter appeared in the *National Intelligencer* on April 27, 1844, the same day that the *Washington Globe,* the Democratic

newspaper begun with Andrew Jackson's blessing and led by
Francis Blair, published a letter from Van Buren to Congressman
William Hammett of Missouri expressing opposition to the annexation.[123] No one knew how Clay could have anticipated Van
Buren's position, but the simultaneous publications and nearly
identical content of the two letters quickly became problematic
for Van Buren. The similarities produced widespread concern that
Clay had once again struck a "corrupt bargain," this time to attain
the presidency. Jackson was one of the first to predict that the letter and an editorial Clay wrote in its defense would destroy Clay's
bid, noting that they made Clay "a dead political Duck."[124] But he
also said that, upon reading Van Buren's letter that "it was impossible to elect him."[125] Consequently, Jackson gave his support for the
Democratic nomination to his protégé James Polk, who had been
the governor of Tennessee after having served four years in Clay's
previous position as the speaker of the House of Representatives.
Clay had easily secured the Whig Party's nomination in April 1844
before his letter appeared, while Polk became the surprise winner
on the ninth ballot for the Democratic nomination.

In an attempt to mollify the increasing number of voters in favor of annexation, Clay tried to help himself by writing a series of
letters to clarify his position. The more he wrote, the worse it got,
and as he kept trying to stand on both sides of the Texas question
the more his prevarications played into the hands of Democrats
who used it to show that Clay's ambition was out of control. As
the campaign neared its conclusion, Clay wrote to the publisher of
the *Washington National Intelligencer* on October 1, 1844. His letter
reaffirmed his opposition to the annexation of Texas and complained that he was being misunderstood around the country and
said that he would no long speak publicly on national issues until
after the election.

That left a month in which Clay fell (largely) mute while the
Democrats relentlessly hammered him for his persistent shifts,
supposed hypocrisies, and overabundant ambition for the presidency. Whigs tried to counter by using many of the tactics they
had used effectively in 1840 to get Harrison elected, including put-

ting on numerous events to entertain prospective voters and plastering Clay's name on nearly anything that could be given away. They thought it clever to keep asking "Who is James K. Polk?" as a constant refrain to underscore his mediocrity and anonymity.[126] But their candidate was effectively nowhere to be seen.

In the end, it was a close election: Neither Polk nor Clay received a majority of the popular vote, though Polk had a slightly higher plurality, but only 105 electoral votes went to Clay as opposed to 170 for Polk. Clay won eleven states, but Polk won fifteen, including Indiana and Illinois, the two states where Lincoln had campaigned for Clay. Yet the election was even closer than it appeared. Clay lost New York by only 5,100 votes, and a win there would have made the difference between Clay as president and Clay as a three-time loser.

Lincoln had genuinely expected Clay to win this time and doubted the result, at least for a while. Less than a year after the loss, Lincoln wrote an acquaintance, "If the Whig abolitionists of New York had voted with us last fall, Mr. Clay would now be president, Whig principles in the ascendant, and Texas not annexed; whereas by the division, all that either had at stake in the contest, was lost."[127] Of course, it could also be said that if Clay had not kept shifting his position on Texas during the election, he might have been president. He also might have won if he had he given any thought on how to handle the third-party candidacy of James Birney of Michigan, who ran as the Liberty Party candidate, committed to abolition. He could not have kept him off the ballot, but he could have devised a strategy that might have prevented or minimized the damage it caused by siphoning votes away from Clay in key Northern states.

Lincoln never forgot the lessons of the 1844 presidential campaign: it was the closest Clay ever would come to winning the presidency, and he lost the office he most coveted through a series of mistakes and oversights, as well as through the inability of many Whigs to see beyond the moment. If, for example, Whig abolitionists had backed Clay, a lifelong opponent of slavery who owned slaves, they would have had a president much friendlier to the cause. If Clay had allowed his career and particularly his party to speak for him rather than publish letters where he said more than he had to and kept exposing himself to more trouble, he might have been president. If Clay had disposed of the Liberty Party or the appeal it had to abolitionist voters, he might have been president. If Clay had not compromised himself by shifting positions on the most important issue of the day, he might have won. Ironically, the man famous for his oratory was brought low by his own carelessness with words.

Later, as a candidate for the House and still later for the presidency, Lincoln would be a model of self-control, relentlessly staying on message and otherwise saying nothing, while party leaders, the party faithful, and Lincoln's surrogates reassured constituents that he was their best bet. Clay's mistakes would be among the most lasting lessons Lincoln learned from his mentor.

The Constitution gave the president-elect a six-month transition period between the election in November and inauguration in March. (The Twenty-Second Amendment, adopted in 1933, shortened the time between election and presidential inauguration from March to January 20.) True to character, President Tyler decided not to be idle during those six months. Working with his secretary of war, John Calhoun, the lame duck president developed a plan to get both houses of Congress to approve a treaty that Calhoun had negotiated with Texas for its annexation. Once they did, all that would be left to be done would be for Congress to admit Texas as a new state and for the next president to sign the bill authorizing Texas's admission into the Union. The ensuing battles, which carried over to the newly installed Polk administration, would shape Lincoln's presidency and years in Congress.

CHAPTER THREE

CLAY MAN IN THE HOUSE
(1844-1850)

Henry Clay's loss to Polk in the 1844 presidential contest was devasting. It meant much more than the end of his quest for the presidency and much more than a political defeat for the Whig Party. It was a constitutional defeat for the nation.

Jacksonian Democrats viewed the president as the central constitutional actor; he gave the orders, and it was Congress's duty to follow them. Whigs like Clay and Lincoln saw this as a distortion of the proper constitutional order, a seizure of power at the expense of Congress. They viewed the legislative branch as the locus for the great debates on the Constitution and the future of American democracy; in Congress, the great issues were to be deliberated on and resolved. Once resolved through policy, it became the president's duty to implement the policy formulated by Congress. Democrats, on the other hand, saw the president, and particularly Jackson, as the embodiment of democracy and the most legitimate servant of the popular will. Whigs countered that presidents should defer to their Cabinets and that Jackson's dismissal of his was a flagrant abuse of power. Democrats believed that presidents should *direct* their Cabinets and that Jackson's blanket dismissal was his prerogative.

Henry Clay, "the Great Compromiser," embodied the great spirit of democracy at work in Congress and, in the perspective of the Whigs, the country. With Clay at its helm, Congress could set the terms of national debate (not just Whigs, in theory) could shape the priorities of the nation, and thus rise to the challenges the framers had left this generation to handle. Yet with the only Whig elected president dying ridiculously early in his administration, the members of the Whig Party had yet to see their vision ever put fully into practice.

From the time that Lincoln first cast a ballot in a presidential election, in 1832, to his vote for Clay in 1844, only once had he backed the winning candidate. At thirty-five, he had experienced only a single month of a Whig presidency. There had been a long string of Democrats—Jackson followed by Van Buren, Tyler (whom no one mistook for a Whig), and now Polk—all destroying Clay's and Lincoln's shared vision of constitutional ideals. Theirs was a gloomy view: Jackson's eight years as King Mob, who tried to dominate Congress, had been followed by four years of Van Buren bumbling through the nation's first depression in a failed effort to extend that legacy. Harrison never had a chance to stifle the tidal wave of Democratic control, while Tyler had realized the Whigs' worst fears, turning out to be a staunch Jacksonian. Tyler and Clay had once been friends, but they were foes by the time Tyler left office.

If Henry Clay were president, he might have been able to restore the balance of power that his Democratic predecessors had upended, but of course he was not. Now Lincoln watched from afar as yet another Democrat, Polk, prepared to lead the nation for another four years in exactly the wrong constitutional direction.

Lincoln's interest in running for Congress was thus motivated by concerns more important than merely satisfying his personal ambitions. True, he was ambitious, and election to Congress was another step on a path to achieve national fame, but this would have been equally true for all of Lincoln's contemporaries who were no less ambitious and aspired to the same offices. Lincoln ran for the House because he desperately needed to be a part of the solution, which included the larger debate on the great constitutional issues of the day, to assert the Whig conceptions of presidential and congressional power and national unity, and to push back against—if not thwart—Jacksonian domination of the presidency and, thereby, of the entire federal government, including Congress. He did not run merely for fame or fortune or even for rewards to his district. He was heeding Clay's call. He ran to secure Clay's American System.

But, on March 4, 1845, it was James K. Polk who took the oath of office as the eleventh president of the United States. His rise to power marked a corresponding decline in the influence of the Whig Party. Polk became president with solid Democratic majorities in both the House and the Senate, yet his single four-year term turned out to more momentous than he ever dreamed. It coincided with the most serious military conflict that Americans had been involved in since the War of 1812. As a candidate and member of Congress, Lincoln attentively studied how a president actually handled a war and its aftermath, coming to believe, like Clay and so many other Whigs, that Polk had abused his power in starting the war, misleading Congress in the process.

In 1847, Lincoln joined what was called a "president-making Congress," because it coincided with the 1848 presidential election and many of its members were eyeing the presidency for themselves or their party. From his desk near the back of the chamber, and in the hallways and rooms of the Capitol, Lincoln watched as the most prominent politicians of the day vied to become (or to control the selection of) the next president, so consumed with their own ambitions they missed what was happening on the floor of the House in front of them. Many Democrats and Republicans who fought in Congress from 1847 to 1849 were still battling a decade later—some on the battlefield—over the fates of the Union and slavery. Jefferson Davis was in the Senate, as Clay returned for a platform to fight for at least one last time for the Whig nomination for president and the preservation of the Union. Initially, Lincoln aligned with Clay, but he soon felt compelled to choose between backing Clay or the same vision championed by a different man, Zachary Taylor, a Whig who had a much better chance of being elected president in 1850. He chose Taylor.

I

James K. Polk justly earned his nickname, "Young Hickory." Though he was a half foot shorter than Jackson and thicker around the middle than his sinewy mentor, no one was more loyal to Jackson than Polk. Jackson's temperament had been forged in war, while Polk's had been forged in more refined, less violent venues, the halls of state legislatures, the House of Representatives, and the Tennessee governor's mansion. Though Polk lacked Jackson's fiery demeanor, he was no less a partisan, and—like Jackson—never forgot a slight. Jackson exploded in anger; Polk quietly seethed and plotted revenge.

Polk was born in North Carolina in 1795, grew up in central Tennessee, and attended college at the University of North Carolina. After graduation, he went back to Tennessee, where family connections and his background as a college debater earned him a law apprenticeship with a powerful Tennessee Democrat and attorney, Felix Grundy. With Grundy's help, he became clerk to the Tennessee Senate in 1819. Four years later, he was elected to the Tennessee House. Once there, he increasingly sided with Andrew Jackson, who was beginning in earnest his campaign for the presidency, instead of Grundy, in skirmishes over what instructions the Tennessee state legislature should have been giving its senators in Washington on such issues as the national bank and land reform. Polk married into a well-connected political family, for whom Jackson was a close friend and was affectionately known as Uncle Andrew. In 1823, Polk successfully led an effort to break a deadlock in the Tennessee legislature over Jackson's appointment as a senator; Polk's intervention helped to tip the balance in favor of Jackson, who was then able to add the title, if not the experience, of U.S. senator to his military accomplishments. In 1825, at twenty-nine, Polk was elected to the U.S. House of Representatives. His first speech, proposing to replace the Electoral College with the popular vote, fell largely on deaf ears. He had greater success in advising Jackson in the run-up to the 1828 presidential election, after which President Jackson never missed a chance to assist Polk's rise in the party and in Congress.

In Jackson's second term, he was so grateful for Polk's loyalty and his ability to calmly but firmly advance his policies in Congress that in 1833 Jackson placed Polk in charge of the congressional battle over the national bank's future. Jackson arranged for Polk to be appointed chair of the House Ways and Means Committee, which meant that any issue relating to the national bank would have to go through him. Polk promptly directed investigations into corrupt practices of the bank's president, Nicholas Biddle, and wrote a minority report approving of Jackson's transferring federal funds from the national bank to state banks as a way to thwart Biddle's control. Polk supported Jackson's opposition to the bank and to Clay's beloved internal improvements, while Jackson reciprocated by rallying support for Polk to win election as speaker of the House in 1835. (The two men must have been delighted to have Polk now in the position that Clay once held in the House.) As speaker, Polk was instrumental in supporting Jackson's and Van Buren's fiscal policies and in 1836 implementing the controversial "gag rule." The Constitution guarantees to citizens the right to "petition the government for a redress of grievances," and in the first few decades of the nineteenth century, Congress was receiving thousands of petitions, most of them sponsored by the Anti-Slavery Society. In response to these petitions, the House adopted the gag rule—a series of resolutions barring the House from hearing or taking any action on any of these petitions. The rule stayed in effect until former president John Quincy Adams, who, after leaving office, had been a member of the House in 1830, eventually assembled a coalition of Northern and Southern Whigs and Northern Democrats to get it repealed in December 1844.

After seven terms in the House, Polk faced a turning point. He was concerned how he could fulfill his own presidential ambitions, and staying in the House felt like a handicap, as no speaker had ever been elected president. Instead, he ran for governor of Tennessee. As it turned out, Polk served only a single term before being defeated in the Whig wave of 1841 in response to the nation's first great depression, which had begun on Van Buren's watch and torpedoed his presidency.

Neither Jackson nor Polk forgot the importance of Jackson's support and mentoring in Polk's career. Particularly as a presidential candidate, Polk frequently consulted Jackson. After his victory in the 1844 election, Polk exchanged letters with Jackson on appointments and the new administration's priorities. In January 1845, Polk traveled to Jackson's home to speak personally with the former president before heading to Washington for his inauguration. Jackson was pleased that Polk intended not only to have a geographical diversity of Cabinet officers but also to secure a promise from anyone serving in his Cabinet to forgo any presidential ambitions while serving Polk as president. Aware that he might be viewed as merely Jackson's tool, Polk made clear to the party, congressional leaders, and the public that he intended not just to keep anyone else or Congress from hijacking his administration (as Calhoun had tried to do with Jackson), but also to prevent Jackson from doing the same. More than once, Polk declared that he alone would be responsible for everything his administration did.

As a candidate, Polk shrewdly made another vow—that he intended to serve only a single term. Harrison had been the first president to pledge serving only a single term but died long before he could complete even those four years. Polk and Harrison had nothing else in common, but such an affirmation made sense for them both: Harrison could appease voters concerned about a possible Whig tyrant in the White House, Polk's lame duck status immediately distinguished him from Jackson, who had run in three successive elections, and from Clay, who in 1844 was making his third run for the presidency. Thus, Polk reassured undecided voters that as president he would pursue what he believed to be the right policies and not just the expedient ones that would have been good for him or, at least in theory, his chosen successor.

Polk distinguished himself further from Jackson—and all previous presidents—by announcing only four ambitious goals as president. In both the campaign and in his Inaugural Address, Polk declared his plans (1) to lower tariffs (and thereby avoid funding or supporting the internal improvements favored by Whigs); (2) to complete the acquisition of the Oregon territory; (3) to acquire from Mexico the Western territories of California, New Mexico, and Texas; and (4) to reestablish the independent subtreasury system that Van Buren had championed.[1] Polk's concerns about the Oregon territory and those that he wanted to secure from Mexico reflected his overriding determination to realize the nation's "manifest destiny" (a term that publisher John Louis Sullivan had coined in an unsigned editorial in *The United States Magazine and Democratic Review* in the summer of 1845) to control as much of the continent between Mexico and Canada as possible. Yet acquiring total control of the Oregon territory (comprising the present-day states of Oregon, Washington, and Idaho, along with parts of Montana, Wyoming, and British Columbia) would not be easy, since the United States and Great Britain had signed a treaty in 1818 providing them both with "joint occupation," a settlement that lasted until 1846. Meanwhile, Mexico laid claim to the lands of California and New Mexico, as well as Texas, in spite of a declaration of independence by the Republic of Texas in 1836. Polk prioritized the quest to control the Oregon territory, whose northern boundary was the latitude line 54°40′, memorialized the slogan "Fifty-Four Forty or Fight."

Acquiring Texas was complicated. It was unclear what if any diplomatic mechanism the United States could use to annex Texas legitimately. A treaty seemed unlikely because of Mexican intransigence—perhaps through some other agreement, since a portion of Texas had proclaimed its independence from Mexico, and therefore Mexico's assent to its acquisition was unnecessary. As a candidate, Polk expressly approved Tyler's annexation of Texas and had even helped behind the scenes during the transition to secure the passage of the annexation resolution. Tyler teed up the issue nicely for Polk by publishing his plan on the annexation of Texas on the evening before Polk's inauguration, though resistance persisted; Democrats, such as John Calhoun and Jefferson Davis, made clear they supported the acquisition because Texas would likely be admitted into the Union as a slave state, tipping the balance of power in Congress in favor of the proslavery forces. Whigs would never consent to any such thing.

In his Inaugural Address (the second longest after Harrison's), President Polk characteristically left no doubt where he and the country stood. He declared that two states "have taken their positions as members . . . within the last week,"[2] referring to Florida and Texas. He emphasized that he was as committed as Jackson had been to maintaining the Union. "Every lover of his country," he proclaimed, "must shudder at the thought of the possibility of its dissolution, and will be ready to adopt the patriotic sentiment, 'Our Federal Union—it must be preserved.'"[3] These last words had been taken verbatim from Andrew Jackson's toast given at the annual Democratic Jefferson Dinner on April 13, 1830, as a retort to John Calhoun's nullification efforts. In sealing the divide between the two men, Calhoun had responded with his own toast, "The Union, next to our liberty, most dear. May we always remember that it can only be preserved by distributing equally the benefits and burdens of the Union."[4]

Still, after using Jackson's phrase, Polk followed with a robust defense of slavery. "It is a source of deep regret," he declared, "that in some sections of our country misguided persons have occasionally indulged in schemes and agitations whose object is the destruction of domestic institutions existing in other sections— institutions which existed at the adoption of the Constitution and were recognized and protected by it."[5] Polk then returned to the Texas question, repeating points made by Tyler and Calhoun: "Texas was once a part of our country—was unwisely ceded away to a foreign power—is now independent, and possesses an undoubted right . . . to merge her sovereignty as a separate and independent state in ours."[6] Without mentioning Tyler, Polk offered to "congratulate my country that by an act of the late Congress of the United States the assent of this Government has been given to the reunion."[7] He expressed equally steadfast support and recognition of "the right of the United States to that portion of our territory which lies beyond the Rocky Mountains."[8] Knowing that Britain laid claim to the Oregon territory and that its ambassador was in the crowd, Polk repeated the pledge of the Democratic platform: "Our title to the country of the Oregon is 'clear and unquestionable,' and already are our people preparing to perfect that title by occupying it with their wives and children."[9] Polk reminded constituents that to his administration "belongs the duty of protecting them adequately wherever they may be upon our soil."[10] Echoing Jackson's declaration less than a decade before, he concluded that "in his official action he should not be the President of a part only, but of the whole people of the United States."[11]

Polk's messages, then and while at the White House, were clear (though not always trustworthy, in Lincoln's judgment), and so was his determination. He intended to work hard, and he did: He personally reviewed every unit of the government, including every department, and carefully monitored all his appointees to ensure they were sticking to his policies. His approach to Cabinet selection helped to ensure allegiance within his administration to him and his priorities, though the exception was Polk's appointment of James Buchanan as his secretary of state. Jackson opposed the appointment, because he thought Buchanan had been in league with Adams's "corrupt bargain," which had denied him the presidency in 1824. Polk did not trust Buchanan, even though Buchanan had served as Jackson's minister to Russia. Polk, however, respected Buchanan's experience in foreign affairs (which also included serving as chairman of the powerful Senate Committee on Foreign Affairs). Even more important, Polk wanted to lower the tariff and needed Buchanan's support in Pennsylvania, a state that favored protective duties for manufacturing and mining operations. Though Buchanan initially favored Polk's appointing him to the Supreme Court to fill a vacancy Tyler had failed to fill before the end of his term, Buchanan eventually relented and agreed to become Polk's secretary of state. It came as no surprise that the most difficult relationship Polk had with any Cabinet officer was with Buchanan, whose allegiance always appeared to be in service to his own rather than the president's ambitions.

For much of his first year in office, Polk tried to avoid being drawn into war on two fronts: one on the western coast against the British over the Oregon territory and another in Mexico over California, New Mexico, and Texas. Polk had declared in his Inaugural Address that the United States' claim to the entire Oregon territory was "clear" and "indisputable," and in his end-of-the-year message to Congress in 1845, he requested a joint resolution approved in both chambers to notify the British of the termination of the joint occupancy agreement. Not surprisingly, these bold declarations angered the British. Knowing that such a resolution would bring the two nations closer to war and that Britain's naval power was far superior to that of the United States, Congress debated the issue for months. Eventually, in April 1845, Congress settled on a relatively mild resolution calling upon the parties to settle the matter amicably. Even so, Polk welcomed the result, which, he believed, strengthened his negotiating position. With most of its ships on the eastern seaboard, Britain did not relish the hardships of moving the bulk of its force to the opposite coast nor the likely war that might ensue if it did. Fourteen months later, on June 18, 1846, the Senate ratified a treaty that Buchanan had negotiated with the British to transfer the Oregon territory into American hands and establish its northern boundary at the 49th parallel, which had become Polk's fallback once it was clear the British would never agree to the 54th. (Slogans were designed to win campaigns, not bind a president once in office.)

In the meantime, Polk was discovering that avoiding war with Mexico was much trickier than it had been with the British. Though Polk worked tirelessly on each matter, the job of annexing Texas presented far more problems. Securing congressional approval of the acquisition of Texas—something he desperately wanted—was one thing, but finishing the job turned into a bloody drama.

By the end of 1845, Texans had voted overwhelmingly in favor of annexation, which Congress had approved as well. While Texas planned a constitutional convention on the matter, Mexican authorities refused to budge. Polk then arranged for John Slidell, who spoke Spanish and was a loyal Democratic supporter of Polk in the

House, to travel to Mexico in December 1845 to offer $25 million in exchange for Texas. Mexico had been notoriously unstable since the country had secured its independence from Spain in the 1820s. While the government initially refused to receive Slidell, his arrival coincided with a successful revolution by nationalist forces. The new government, led by General Mariano Paredes y Arrillaga, was even less receptive to Slidell's overtures to purchase Texan independence and more determined to prevent Texas from becoming part of the United States.

Next, Polk hit upon the idea of trying to turn the instability of Mexico's government to his advantage. He arranged safe passage for Mexico's deposed president, General Antonio López de Santa Anna, to travel from exile in Cuba into Mexico, where Santa Anna promised that, in exchange for $30 million for himself, he would arrange for Mexico to sell all contested property to the United States at a reasonable price. Nevertheless, once successfully returned to Mexico in August 1846, Santa Anna reneged on his promise to Polk and instead declared his intention to lead Mexican forces to defend against U.S. aggression.

In the meantime, Polk's success in convincing Texas to take steps to support American annexation brought war closer. Polk was aggressive not just in his rhetoric, but also in his use of power, particularly his use of the military. After Polk accepted Texas's claim of the Rio Grande as its boundary with Mexico, Mexican leaders threatened to attack the Texas frontier. Having promised to protect Texas as soon as they had accepted annexation, Polk dispatched a naval squadron along the Gulf Coast and moved several thousand troops from the Louisiana border to the northern edge of the disputed boundary zone, granting permission to the commanding general to move south if he thought necessary. General Zachary Taylor did just that.

In choosing Taylor to lead the United States' defense against Mexican aggression, Polk bypassed several more-senior officers, including Winfield Scott, the commanding general of the U.S. Army. His reasoning was with a less exotic battlefield in mind:

Polk distrusted Scott because he was an unapologetic Whig. Taylor, commander of the army's Western Division, appealed to Polk because he had a long-standing reputation as apolitical. Perhaps more important, Jackson had recommended Taylor to Polk. Jackson had grown close to Taylor and developed great respect for his leadership when he had served as a commander under Jackson in fighting the Seminole tribe in Florida.

Zachary Taylor had been a soldier all his adult life. Born in 1784 to relatively wealthy landowners in Virginia, he grew up in Kentucky and joined the army in 1808, a year before Lincoln was born. For the next several decades, Taylor rose steadily in the ranks. He commanded troops as a captain in the War of 1812, serving under Andrew Jackson. He was a colonel in the Black Hawk War in 1832 and a brigadier general in the Second Seminole War in 1837. In 1840, he was assigned to a post in Louisiana, where he settled on a large estate, with more than one hundred slaves, in Baton Rouge. His Southern heritage and support for slavery made him popular in the South—and with Polk. All along the way, Taylor's men revered him for many distinctive qualities—keeping his head under fire, always being willing to listen to and trust his officers and troops in the heat of combat, his ingenuity and undefeated record in major battles, and his honest and unassuming nature. His soldiers dubbed him Old Rough and Ready based on his gruff language and preference to dress plainly during battle rather than in uniform.

As Taylor and his army moved deeply into the disputed territory between the Rio Grande and Nueces River, he heard rumors of possible attacks coming from both north and south of where his troops were placed. Taylor decided to investigate. He sent one group of soldiers farther south to determine whether a threat was coming from that direction. The group reported back that there was none. Taylor then sent other troops north under the command of Captain Seth Thornton, who encountered Mexican troops who had crossed the Rio Grande. The Mexicans attacked his men. They retreated quickly and sent word of the attack to Taylor. In turn,

Taylor relayed the news to Polk, who informed Congress of an attack on American forces, one that he considered unprovoked.

As Polk prepared to ask Congress for a declaration of war, Taylor and his army of roughly four thousand men found themselves under assault. As they had done in attacking Thornton's units, Mexican troops again crossed the Rio Grande, this time to directly challenge Taylor's forces. The American army decisively defeated Santa Anna's forces on two successive days in the battles of Palo Alto (May 8) and Resaca de la Palma (May 9). On May 13, 1846, Congress agreed, at Polk's request, to declare war against Mexico. Polk's Cabinet wanted him to order American forces to take all of Mexico, but Polk rejected their advice. He defined the American objective as securing Texas.

For the victories in Palo Alto and Resaca de la Palma, Taylor received a brevet promotion to major general and a formal commendation from Congress. Back in the states, he became a popular hero and was promoted to the full rank of major general. The national press began comparing Taylor to both George Washington and Andrew Jackson, whose status as military heroes became stepping stones to the presidency, but he quickly rejected the comparison, saying, "Such an idea never entered my head, nor is it likely to enter the head of any sane person."

Taylor did not rest on his newly won laurels. He led his troops south across the Rio Grande and advanced toward the city of Monterrey. They captured the city on September 22–23, and Taylor, on his own initiative, then granted the Mexican army an eight-week armistice.

This grant of respite infuriated Polk, who was already disturbed by Taylor's growing popularity. He pressed Taylor to push his advantage harder, but Taylor angered Polk further by writing a letter, which found its way into the press, criticizing both the president and his secretary of war, William Marcy, for their handling of the conflict. In response, Polk ordered Taylor to confine his actions to those necessary for defensive purposes and transferred Taylor's best troops to the army led by Winfield Scott, relenting on his earlier decision to bar any Whig commanders for the sake of electoral victory.

In February 1847, Taylor learned that Santa Anna was mobilizing an army to attack his diminished forces. Taylor marched his troops south into a narrow pass that made an attack by Mexican forces difficult. As Taylor was ordering his men forward, the Mexican army attacked. Santa Anna had intercepted an American letter acknowledging Taylor's depleted forces and tried to press the advantage against the smaller American contingent. In the ensuing Battle of Buena Vista during February 22–23, 1847, Taylor's troops won a significant victory over a Mexican army that outnumbered the Americans nearly four to one.

The headlines praising Taylor for his unexpected victory at Buena Vista were the last straw for Polk. He was angrier at Taylor than ever before. He considered that his order to Taylor to assume only a defensive position meant that Taylor should no longer take the initiative in the fight against Mexico. Polk relieved him of his command. Taylor spent the next few months in Mexico waiting for leave to return home, which came in November. His victory in the Battle of Buena Vista was his last battle.

Scott oversaw several more victories against Mexican forces, which brought the Mexican government to the bargaining table and the war to its end. The Mexicans agreed to settle the dispute through the Treaty of Guadalupe Hidalgo, signed on February 2, 1848. It granted to the United States more than 500,000 square miles of new territory, including land that now makes up part or all of eight Western states—California, Texas, Arizona, Utah, Nevada, Colorado, New Mexico, and Wyoming. With that one stroke, Polk expanded the United States nearly one-third in size. In return for the lands Mexico ceded to the United States, the United States paid Mexico $15 million and agreed to assume $3.25 million in debts that Mexico owed to American citizens.

Taylor returned to the United States a hero, but many did not come back. Nearly 14,000 Americans lost their lives, the largest number of casualties that the United States suffered in any military conflict until the Civil War.

II

As Polk was struggling to avoid war with the British on the west coast and to end the war with Mexico, Abraham Lincoln's attention was elsewhere. He was intent on going to Congress.

Ron Keller suggests, in his study of Lincoln's years in the Illinois House, that "something awakened in Lincoln in 1839 and 1840. His stature as a Harrison presidential elector, his visibility and attention as a statewide spokesman for Whig policy, and his leadership position in the state legislature instilled in him a certain consciousness."[12] Whether the interest in higher office came earlier, later, or at that time, Lincoln undoubtedly had it. Many years later, Herndon agreed that by 1840, Lincoln "had begun to dream of destiny."[13] Lincoln's runs for the state legislature, beginning in 1832, suggest that his "little engine" of ambition (as Herndon called it) had been changing hard for some time.

Oratory was instrumental to Lincoln's success, as it had been for Clay's. Since Lincoln had been a boy, he had worked tirelessly on his delivery. He was not just parroting Clay and Webster, another of America's greatest orators, but identifying what worked best for each of them and adapting their techniques and language to fit his needs. Beyond the debates he'd had with Stephen Douglas in Stuart's run for Congress and later during the 1840 presidential campaign as a Whig elector, he constantly honed his speaking style when campaigning across the state. After delivering what he considered to be a subpar performance in one debate with Douglas, he urged the organizers to give him a second chance. Witnessing the next debate between the two men, Lincoln's friend Joseph Gillespie recalled, "I never heard and never expect to hear such a triumphant vindication as he then gave Whig measures or policy. He never after to my knowledge fell below himself."[14]

At another Whig rally in 1840, an observer noted that Lincoln "discussed the questions of the time in a logical way, but much time was devoted to telling stories to illustrate some phase of his argument, though more often the telling of these stories was resorted to for the purpose of rendering his opponents ridiculous."[15] At yet another campaign event, a reporter observed that Lincoln was "highly argumentative and logical, enlivened by numerous anecdotes, [and was] received with unbounded applause."[16] Even some Democrats recognized the power of his delivery, as one did in observing that Lincoln "always replies [jokingly] and in good humor . . . and he is therefore hard to foil."[17] Robert Wilson, who studied Lincoln as he studied others and would have a bright political future himself, said that Lincoln "seemed to be a born politician. We followed his lead; but he followed nobody's lead. It almost may be said that he did our thinking for us."[18]

While there is no doubt Lincoln was blessed with considerable natural talents, he made it all look easy because he had studied classical speeches and Clay's oratory for so long and, having diligently practiced his techniques, had begun mastering the art of adapting them to suit his purposes. By 1846, Lincoln was convinced that his turn for national office had finally come: his cousin-in-law John Hardin had served a term in Congress from 1843 to 1845, after which his competitor Edward Baker had served most of a two-year term in the House. If the Whig Party followed the Jacksonian practice of rotation, which Lincoln repeatedly urged, 1846 was his year.

Lincoln left nothing to chance. He had learned the hard lesson more than once that politics was do or die and early entry into a race was essential for victory. The first thing Lincoln had to do was to secure the Whig nomination, which he did not expect to be difficult as long as Hardin deferred to the practice of rotation. But while Lincoln liked the idea of rotating after a single term, Hardin did not, and he told everyone, including Lincoln, that he objected to following it in this election cycle. He argued that the party should back the most deserving man, who he thought was himself.

Hardin had several advantages over Lincoln. He had gained notoriety as the commander of five hundred state militia troops who had restored law and order in the Utah territory after the murder of Mormon prophet Joseph Smith. (In 1844, Orville Browning successfully defended five men charged with Smith's murder.) And of course, Hardin had previously defeated Lincoln and Logan for the Whig nomination for the U.S. Senate in 1845.

Lincoln proceeded to use some of the same tactics and wiliness that he had learned from Stuart to outmaneuver Hardin. Stuart had been ruthless in his victory over Stephen Douglas in 1838, and Lincoln now followed suit. He wrote to several newspapers to push Hardin's candidacy for Illinois governor to get him out of the way for the congressional seat. When Hardin read the papers, he announced his refusal of the invitation and blamed Lincoln for the ploy. Once Lincoln's friend, Hardin was now his rival. Hardin proposed a direct primary of Whig voters in the district with the candidates restricted to electioneering only in their home counties. Lincoln objected and instead stood by the convention system, "the old system," as he wrote Hardin.[19] Under this system, Whigs gathered in each district to nominate their preferred candidates. He had early experience with this system and knew how to lobby the delegates in each district, so he could ensure that, by the time a general nominating convention was held, most districts would be supporting his nomination. He knew most of the likely attendees in each district anyway and felt confident he could persuade them to adopt the practice of rotation, which would work to his benefit. At the same time, Lincoln did what he and Stuart had done before, rallying the support of friendly newspaper editors in the district as well as reminding Whig elected officials, who would serve as convention delegates, "Turn about is fair play,"[20] by which he meant that rotation, giving someone else a chance to win the office, was the right thing to do. Hardin angrily wrote Lincoln that he had never consented to a deal for rotation and defended his proposal for a direct primary. Lincoln, rather disingenuously, responded that he had not been trying to force Hardin out of the race and that Hardin's accusations against him were an "utter injustice."[21]

As they argued back and forth, Lincoln continued to work the conventions throughout the district. As the voting in each convention began to be tallied, it was clear that Lincoln had the requisite support sewn up, and Hardin grudgingly withdrew his name in February 1846. In the Whig convention on May 1, each of the district delegations pledged its vote to the nominee whom a majority of its delegates had supported. With Hardin presiding, the delegates from every district cast their votes for Lincoln. The convention then adopted a platform for supporting a strong tariff to fund internal improvements. It did not mention either the Oregon territory negotiation with Britain or the mounting tensions over Texas.

As Lincoln was wrapping up his party's nomination for the House, Democrats were still scrambling for a candidate. Eventually, they settled on Peter Cartwright, a Methodist preacher who was an ardent supporter of Jackson and his policies.

Thirteen days after the Whig convention nominated Abraham Lincoln as its candidate for the House, Congress declared war on Mexico. While questions about the cost and purpose of the war were debated throughout much of the country, they were of less concern in Illinois. A number of Lincoln's friends and associates enlisted, including both Hardin and Baker as colonels. Rather than focusing on Lincoln's opposition to the war, Cartwright and his fellow Democrats leveled against Lincoln the same kinds of charges that Polk and the Democrats had successfully directed against Clay in the 1844 presidential election: They charged him with immorality and even infidelity, a man of no religion and no principle. Lincoln countered by traveling throughout the district, meeting voters, telling stories whenever possible, and pushing the need for a tariff to fund domestic improvements. He barely mentioned the war.

Shortly before the election, Lincoln responded to Cartwright's constant charges of immorality by adopting a strategy he had employed before. In Whig newspapers, he published a handbill in which he denied that he had ever been critical of religion and declared that he had great respect for it.[22] In fact, Lincoln's friends knew he had questioned religion generally and Christianity in particular, but during the election they supported the stance he was taking.

Lincoln and Cartwright crisscrossed the district, and at one revival meeting that he was leading, Cartwright spotted Lincoln in the crowd. (Lincoln enjoyed attending opponents' rallies, both to rattle them and to study them.) From the lectern, Cartwright pointed his finger and shouted at Lincoln, "If you are not going to repent and go to heaven, Mr. Lincoln, where are you going?" Without missing a beat, Lincoln responded, "I am going to Congress, Brother Cartwright."[23] As was true for Clay, Lincoln's quick wit was invariably his best asset.

In response to Cartwright's persistent charge that he was a man of no religion and no principles, Lincoln published a handbill in league with abolitionists—who were, in Cartwright's opinion, responsible for the unrest and turmoil throughout the nation. Instead of publicly engaging with the charge, Lincoln agreed to be interviewed by two abolitionists. They were pleased with his answers and his record defending people harboring fugitive slaves. They spread the word that he was sympathetic to their cause.

On August 3, 1846, Lincoln won the congressional race by the largest margin that a Whig had ever captured in the district, 6,340 votes to Cartwright's 4,829, with a smaller number going to a third-party candidate. Lincoln was finally headed to Washington.

III

Only thirty-nine when he won election to the House, Lincoln was already well known as Old Abe among his friends and neighbors. For the past ten years, as he walked the path, nearly daily, from his home to his law office or the capitol, he appeared to them as a dotty old man. With his victory in hand, Old Abe was spotted yet again in the streets, mulling over the issues likely to come before him. He had to make the most of his short term, since he expected to be rotated out in two years.

While Stuart had been especially helpful to Lincoln in securing election in the past, he was planning to run for a seat in the Illinois Senate, and—opening a breach with Lincoln that would intermittently separate them from each other for the next several years—Stuart had no time to help Lincoln ready himself for Congress. Nor did he have to. Through his eight years in the Illinois House, Lincoln had already learned what he needed to know to succeed in the U.S. House, already understood from firsthand experience and Stuart's tutelage the importance of logrolling and working within the party system for advancement. From his law practice with Logan and his successes in campaigning for the state and national legislatures, Lincoln had developed increasing confidence in being able to tutor himself on the issues that he knew had to be mastered. He also had plenty of time to prepare himself, as there was a long hiatus between the date of his election to the House on August 3, 1846, and his swearing in more than a year later on December 6, 1847.

The Mexican War was fought and won before Lincoln stepped into the House, but the peace had yet to be brokered, and many of the dead had yet to be buried. The losses were felt keenly in Lincoln's world. Among the local heroes brought back to be buried in Illinois was John Hardin, who had died leading a counterattack at the Battle of Buena Vista. Many of Lincoln's fellow Whigs in Illinois had thought Hardin, not Lincoln, had the greater promise to become a leader on the national stage. Instead, it was Lincoln who would serve from 1847 to 1849 as Illinois's only elected Whig in the nation's capital. Of Illinois's representatives in the House, all seven were Democrats except for Lincoln, and the two senators, Stephen Douglas and Sidney Breese, were Democrats.

As the sole Illinois Whig in the Thirtieth Congress, Lincoln was alone, but he had the self-assurance he needed to do the job. In July 1847, he made his first trip to Chicago as congressman-elect to speak at the River and Harbor Convention, where about 2,500 delegates, mostly Whigs, had gathered to protest Polk's 1846 veto of an appropriations bill for rivers and harbors, most in the Great Lakes region. This would be the largest crowd Lincoln had ever addressed. They assembled under a large pavilion, about a hundred feet square, near the center of downtown Chicago. Many of the Whig Party's most prominent leaders gathered there, including Edward Bates, a St. Louis lawyer, who was chosen as the convention president. Others in attendance were Horace Greeley, the editor of the *New York Tribune,* a staunch critic of Jackson and an ally of Henry Clay, and two other notable New Yorkers—the political boss Thurlow Weed and his protégé, William Seward, New York's governor from 1839 to 1842. Lincoln was undaunted, though newspapers and even some fellow Whigs made fun of his appearance. His brief speech made a small but positive impression on Greeley, who reported that "in the afternoon, Hon. Abraham Lincoln, a tall specimen of an Illinoisan, just elected to Congress from the only Whig district in the State, was called out, and spoke briefly and happily in reply to" David Dudley Field, a prominent Democrat who had braved the convention to speak against internal improvements.[24]

While Lincoln's brief speech confirmed his credentials as a Clay Whig, his mentor was not there. He was secluded at his estate in Ashland, Kentucky. The old, defeated warrior, just a few months before, had received the shattering news of the death of his son, Henry Clay Jr., in the Mexican War.

IV

One question that has intrigued students of Lincoln is whether he and Henry Clay ever met in person. Historians generally think the answer is no or, if so, only briefly and in passing. There is reason to believe they might have met once, though there is better reason to think the claim that they did is suspect. In the spring prior to his move to Washington, Lincoln received a copy of a book, *The Life and Speeches of Henry Clay,* with an inscription written in Clay's hand, reading "To Abraham Lincoln: With constant regard to friendship H. Clay Ashland 11 May, 1847."[25] There is no correspondence, nor any other record, indicating whether the two men had met before Clay sent him the volume; every indication is that they had not.

Lincoln was, however, in Clay's presence at least once. In November 1847, before Lincoln and his family traveled to Washington, they stopped in Lexington, Kentucky, to visit Mary Todd's family. While there, Lincoln attended a speech given by Henry Clay. Lincoln's father-in-law, Robert Smith Todd, a former student and longtime friend of Clay's, brought both Lincoln and Mary Todd to Clay's speech. With Lincoln in the crowd, Clay sat on the stage next to Todd. As vice chairman of the event, Todd introduced Clay, who then gave a long-awaited speech heralding his return to the national stage. The subject was Polk's war.

The speech was delivered nearly nine months after Clay's son had died during the Battle of Buena Vista. Young Clay's death must have been on the old man's mind as he set forth his case against the Mexican War, in what became known as the Market Street Speech, because of its location. (Clay's bitterness over the death of his son was surely compounded by Polk's insensitivity in not sending any condolence to the grief-stricken family.)

Clay's speech was remarkably long, even for him, taking more than two and a half hours to deliver.* At the outset of the speech, Clay characteristically played down the occasion by saying that it was nothing more than his civic duty as a private citizen to voice his concerns about the war. Everyone, including Lincoln, would have thought that Clay was considering, if not already committed to, another White House run. Indeed, shortly after receiving the news of his son's death, Clay had written to a friend, "Up to the Battle of Buena Vista, I had reason to believe that there existed a fixed determination with the mass of the Whig party, throughout

---

* At seventy, Clay must have appreciated the fact that, as he stood in the cold, his speech was longer than the ninety-minute Inaugural Address given by the first Whig president, William Henry Harrison, in the icy rain that led to the pneumonia that, along with typhoid fever, killed him.

the U.S., to bring me forward again. I believe that the greater portion of that mass still cling to that wish, and that the movements we have seen, in behalf of [General] Taylor, are to a considerable extent superficial and limited."[26] Clay was further stung by the betrayal of his former political ally and fellow Kentucky senator John Crittenden, who was already backing Zachary Taylor in the upcoming presidential election. Lincoln did not need to know Clay's private thoughts to know the direction of the old man's ambitions.

Lincoln must have listened closely to Clay's speech. Having studied Clay's oratory all his life, he would have quickly recognized that its structure was classic Clay, beginning with a comment on the occasion itself, a "dark and gloomy" day, for it was overcast and raining when he spoke. Clay likened the day to "the condition of our country, in regard to the unnatural war with Mexico."[27] This war, he said, was not like other wars the country had fought, even the Revolutionary War: "This is no war of defence, but one of unnecessary and offensive aggression. It is Mexico that is defending her fire-sides."[28] Clay made no direct reference to the death of his son, but rather raised a startling rhetorical question, asking, "Who have more occasions to mourn the loss of sons, husbands, brothers, fathers, than Whig parents, Whig wives and Whig brothers, in this deadly and unprofitable strife?"[29] He reminded his audience that the crucial defect of the war was that it did not derive proper authorization from Congress, as it should have in the Whig conception of separation of powers, but rather from the president. "Who," he asked, "in the free government is to decide upon the objects of a War, at its commencement, or at any time during its existence? Does the power belong to the Nation, to the collective wisdom of the Nation in Congress assembled, or is it solely vested in a single functionary of the government?"[30]

Again characteristically, Clay proceeded systematically to lay out the problems with the war and its supposed rationale. First, he complained that Congress had not set forth the objective of the war as it should have. Second, he rejected the conquest of Mexico as a legitimate basis for waging war. Third, he worried that there had not yet been a settlement of the war. Clay expected the United

States to be paid money for its expenses, though it was not until early the next year that the United States would receive a lot of territory and money from Mexico in exchange for ending the war. Fourth, Clay denounced "any desire, on our part, to acquire any foreign territory whatever, for the purpose of introducing slavery into it."[31] Fifth, Clay reminded his audience of the "unmixed benevolence" of turning to the project of "gradual emancipation," which he, as a charter member of the American Colonization Society, had long advocated as a solution to the problem of slavery.[32] These resolutions included a commitment: "That we do, positively and emphatically, disclaim and disavow any wish or desire, on our part, to acquire any foreign territory whatever, for the purpose of propagating slavery, or of introducing slaves from the United States, into such foreign territory."[33] Under this plan, the federal government would assist with purchasing the freedom of African American slaves and arranging for them to relocate "back to their homelands" in Africa. Clay concluded with eight resolutions designed to emphasize the right and duty of Congress to investigate the origins of the Mexican War—"to determine upon the motives, causes, and objects of any war, when it commences, or at any time during the progress of its existence."[34] This commitment was a restatement of a proposal made on August 6, 1846, by Representative David Wilmot, a Pennsylvania Democratic member of the House, to ban slavery in any of the territory acquired from Mexico in the Mexican-American War. Wilmot introduced his proposal as a rider to a $2 million appropriations bill intended for the final negotiations to settle the war. The House passed the bill on August 6, 1847, but it failed in the Senate. The proposal became known as the Wilmot Proviso. The congressman repeatedly tried to attach it as a rider to each new appropriations bill for the settlement of the war, but each time it failed in the Senate.

Lincoln never offered an opinion, public or private, on the Market Street Speech. There are, however, reports that after the speech, he spoke briefly with Clay, who, one person alleges, invited him to dinner at his Ashland estate near Lexington. Many

years later, Usher Linder, a Democrat who had sometimes practiced law with Lincoln on the circuit, recalled details of the dinner. He said that Lincoln had told him that

> though Mr. Clay was most polished in his manners, and very hos-
> pitable, he betrayed a consciousness of superiority that none could
> mistake. [Lincoln] felt that Mr. Clay did not regard him, or any other
> person in his presence, as, in any sense, on an equality with him. In
> short, he thought that Mr. Clay was overbearing and domineering,
> and that, while he was apparently kind, it was in that magnificent
> and patronizing way which made a sensitive man uncomfortable.[35]

Sidney Blumenthal notes that Alexander McClure, a prominent
Republican supporter of Lincoln's, suggested in his biography of
Lincoln that at the dinner, "Clay was courteous, but cold . . . Lincoln
was disenchanted; his ideal was shattered."[36]

Clay was well known for being arrogant and full of himself (his
opponents always chastised him for being elitist), but he was also
renowned for being a charming host, though he might not have
been that evening, given the recent death of his son, his preoc-
cupation with the upcoming presidential election, or both. Moreover, while Linder was a friend of Lincoln's before he entered
Congress, he later became a Democratic ally of Stephen Douglas,
and his characterization of Clay aligns perfectly with the Democratic critique directed successfully against Clay in the 1844 presidential campaign. Also, Linder's recollections portray Lincoln as
a simpleton who was incapable of developing nuanced appraisals
of the men he was dealing with. Lincoln knew that between him
and Clay the one of them who had a political future in 1847 was
likely the incoming congressman from Illinois and not the former
speaker. (Linder's son fought for the Confederacy, which suggests
a further affinity on Linder's part to construct a negative image of
Lincoln or the man whom he idolized, Clay.)

Lincoln revered Clay's oratory and career, but he never thought
of Clay as perfect. At that point, in 1847, he considered Clay a

three-time loser in seeking the presidency and past his prime. Lincoln was clear-eyed in his appraisal of men, not someone to be put off by the charm or demeanor of a host or ally. Clay's speech probably confirmed that Clay's best days were behind him; it was not Clay's finest by a long shot. The humility seemed disingenuous, as when he proclaimed near its beginning, "I have come here with no purpose to attempt a fine speech, or any ambitious oratorical display. I have brought with me no rhetorical bouquets to throw into this assemblage." It was too late in Clay's career to lower the crowd's expectations on the one thing everyone knew as Clay's greatest strength, his oratory, and his long technical critique of how Polk, not the Mexicans, had started the war likely surprised no one in attendance. Lincoln, who had closely studied Clay's rhetoric, would have recognized the flaws in Clay's performance.

Lincoln was always closely watching people and learning from them. Sometimes, he learned the most by observing the failures of others. Clay's strengths had molded him, but Lincoln's genius in selecting his mentors was in his capacity to distinguish the deficiencies of the admirable and the admirable traits of the deficient. Lincoln could learn from all of them. He despised much of Jackson's despotic conduct, but he still found several elements in Jackson's strong leadership worth emulating. Clarity was one of them; consistency, another. It was not in Lincoln's nature to follow someone mindlessly or blindly but rather to learn what he could later put to his own purposes.

Less than a month after Clay's Market Street Speech, Lincoln was sworn in to Congress. He soon would use many of the same arguments Clay had made in Lexington to denounce Polk and the Mexican War on the floor of the House. One of Clay's tragic flaws, Lincoln knew, was he had trouble seeing how others saw him or how they could sometimes see through his artifice, a fault that was on display early in the Market Street Speech when he referred to the support that he naïvely thought he still had within the Whig Party. Lincoln knew Clay had no such support, because a rising star in the party already had it.

V

When Lincoln arrived in Washington, in the winter of 1847, it was his first visit to the nation's capital. Five years earlier, Charles Dickens had declared Washington a "City of Magnificent Intentions," with "spacious avenues that begin in nothing, and lead nowhere; streets, mile-long, that only want houses, roads, and inhabitants; public buildings that need but a public to complete."[37] Another British writer, Alexander MacKay, said that "at best, Washington is but a small town, a fourth-rate community."[38]

On entering the Capitol itself, Lincoln saw "an immense lantern, towering over the dome of the rotunda," six feet in diameter inside an eight-foot mast.[39] That impressive rotunda was lined with paintings telling the story of America—from John Vanderlyn's *The Landing of Columbus* (just installed in January 1847) to John Trumbull's magnificent depictions of the Second Continental Congress's reception of the Declaration of Independence, Lord Cornwallis's surrender at Yorktown, and General Washington's resigning his commission. When Lincoln and other House members walked into the Hall of Representatives, they passed through a door with a portrait above of the folk hero Daniel Boone fighting a tomahawk-wielding Native American. Elsewhere within the Capitol was the Supreme Court library and the nation's largest collection of books, the Library of Congress, which would give Lincoln opportunities to advance his knowledge. Marble busts of American leaders— George Washington, Thomas Jefferson, John Marshall, Andrew Jackson, and even Martin Van Buren, among others—stood on pedestals throughout the room. Lincoln could not have helped but notice the absence of any important Whig among them.

Lincoln arrived three days late for the Thirtieth Congress, which had convened on December 6, 1847. It had taken seven days for Lincoln, Mary Todd, and their two sons to trek from Springfield to Lexington to Washington as a family, a break with the traditional practice that members of Congress left their families back home. Once in the capital, Lincoln was eager to get to work.

Initially, the family stayed at the city's most storied hotel, Brown's Indian Queen Hotel, though it lacked whatever grandeur it had once had. It had been the site of the inaugurations of James Madison, James Monroe, and John Tyler. For nearly thirty-five years, it had been the place where the members of the Supreme Court boarded, including Chief Justice John Marshall, during each term of the Court.

Finding Brown's too expensive, Lincoln moved his family to a more affordable boarding house managed by Mrs. Ann Sprigg, the widow of a clerk of the House of Representatives, and near the home of Duff Green. In 1827, Jackson had asked Green to start a partisan newspaper in Washington, the *United States Telegraph*, defending his administration, and Green became a member of President Jackson's band of unofficial advisers, which his opponents called the "kitchen Cabinet." Green often ate at Mrs. Sprigg's.

Within a few weeks after taking his seat, Lincoln sent his family back to Springfield. He told them they "hindered me some in doing business."[40] Mary Todd was not unhappy, because she felt bored and alone in Washington. Later, Lincoln wrote to her that he missed her and wished she and the children were with him, but they never returned while he was in Congress.[41]

Lincoln was the youngest of the members of Congress in his boarding house. He befriended them all, regardless of party. They debated the issues before the House, often finding common ground on many of them. Late into the evenings, Lincoln entertained them with his storytelling and countless humorous asides. He forged especially close relationships with three of the men, each of whom would have a major impact on his term in Congress and career afterward.

One was Joshua Giddings, an Ohio Whig, who would push Lincoln to oppose slavery altogether. Orlando Ficklin, a Democrat from Ohio, recalled that "Lincoln was thrown in a [boarding house] with Joshua R. Giddings," described as a "tall man, of stout proportions, with a stoop in his shoulders, the face marked, and the hair gray." Ficklin suggested that it was in this company that Lincoln's "views crystallized, and when he came out from such association he was fixed in his views on emancipation."[42] In his two-year term, Lincoln consistently sided with Giddings against the extension of slavery. Less than three weeks after Lincoln was sworn in to the House, he cast one of his first votes in support of Giddings's motion to refer an antislavery petition to the Judiciary Committee. In July and August, Lincoln voted with Giddings on thirteen of fourteen roll call votes on the question of allowing slavery in the territories. In the one deviation, Lincoln supported the suspension of House rules to permit consideration of a joint resolution declaring it expedient to establish civil government in New Mexico, California, and Oregon. Giddings opposed the resolution because he worried that New Mexico might be forced to unite with Texas, which was strongly proslavery. The two men disagreed further on the need for extremist tactics in fighting against the slave power, Giddings disposed to support any measure in opposition no matter how radical but Lincoln inclined to compromise and to oppose any resistance to the rule of law, no matter how wrong the law.

Another friend at the boarding house was Pennsylvania's David Wilmot, who was as virulently antislavery as Giddings. Wilmot's reputation as a fierce abolitionist grew each time he tried to attach his proviso as a rider to any appropriations pertaining to the territories acquired in the Mexican War.

Lincoln proclaimed that he had voted for the Wilmot Proviso "at least" forty times in his single term in the House, but in fact he had voted for it or its equivalent only about five times.[43] The exaggeration came in handy for Lincoln when he was battling for the Republican nomination for president against fiercer opponents of slavery, including Seward of New York and Salmon Chase of Ohio. In 1847, however, Lincoln's moderate position on slavery— opposing its extension but not its continuation—was enough to lump him together with Giddings and Wilmot so that their boarding house became known as Abolition House.

Another new friend who lived nearby and hung out with Lincoln and his housemates was "a little slim, pale-faced, consumptive man," Alexander Stephens, a congressman from Georgia and a fellow Whig. Stephens and Lincoln formed a lasting bond, rooted in friendship, mutual antipathy for the Mexican War, hatred of Polk, and a shared interest in forging a compromise on slavery that would keep the country unified. Stephens was one of the few people Lincoln trusted as a confidant, even from the beginning. Stephens said that "he was as intimate with Lincoln as well as with any man except perhaps" Robert Toombs, one of Georgia's two senators.[44] Toombs and Stephens remained Whigs as long as they could before the national divisions over slavery pushed them both to side with secession. Lincoln respected Stephens, his favorite Southern Whig, and his respect was reciprocated. On February 2, 1848, Lincoln wrote Herndon that Stephens, with a voice that reminded him of Logan, "has just concluded the very best speech, of an hour's length I ever heard. My old, withered, dry eyes, are full of tears yet."[45] In it, Stephens shredded the basis for the Mexican War. He declared, "The principle of waging war against a neighboring people to compel them to sell their country, is not only dishonorable, but disgraceful and infamous."[46] Lincoln and Stephens joined in looking for a promising Whig to take back the White House in 1848, as well as for a way out of the bloody civil war they both wanted to avoid.

In terms of his pre-Washington experience, Lincoln was not alone in the House. Two hundred other representatives were also new in town. Two-thirds of the House members had served in their state legislatures. Shortly after being sworn in, he wrote Herndon,

> As to speech-making, by way of getting the hang of the House, I made a little speech two or three days ago, on a post office question of no general interest. I find speaking here and elsewhere about the same thing. I was about as badly scared, and no worse, as I am when I speak in court. I expect to make [another speech] within a week or two, in which I hope to succeed well enough, to wish you to see it.[47]

The focus of Lincoln's first speech was on a matter he well understood as a former postmaster. It involved a dispute over a postal contract for the Great Southern Mail, which carried mail by rail through Virginia to Washington but charged outrageously high rates for its service. Once its contract expired, the U.S. Post Office opted for a less expensive carrier, but Congressman John Botts of Virginia, the chair of the House Committee on Post Offices and Roads (of which Lincoln, too, was a member), introduced a bill to force the postmaster general to renew the contract with the Great Southern Mail. Lincoln's humor and grasp of the facts were immediately apparent. He demonstrated how under the contract the postmaster

> took the most expensive mail coach route in the nation. He took the prices allowed for coach transportation on different portions of that route and averaged them, and then built his construction of the law upon that average. It came to $190 per mile. He added 25% of that rate and offered the result to this railroad company. The gentleman from Virginia says this was wrong [for the Postmaster General to do]: I say it was right.

It was not a huge sum, but it was the principle of gouging the public that animated Lincoln, who insisted the carrier was entitled to "just compensation" but not to a rate that exceeded the one for traveling across the state of New York on a steamboat or the one for traveling between Cincinnati and Louisville. His humor was aptly employed. When confronted with the fact that he was out of order to reference the committee proceedings on the issue, he laughed and said he never could stay in order for long. Even the *Congressional Globe* recorded the fact that there was a "laugh" when he referred to "the lawyers in this House (I suppose there are some)."[48] In the evening he joked with his housemates about it. One guest later wrote, "I recall with vivid pleasure the scene of merriment at the dinner after his first speech in the House of Representatives, occasioned by the descriptions by himself and other of the Congressional mess, of the uproar in the House during its delivery."[49]

Lincoln supported every internal-improvement measure proposed, including not just the first bill he spoke about on the House floor but also the resolutions upholding all the measures that had been proposed at the Chicago River and Harbor Convention to improve navigation on the nation's lakes and rivers.

Lincoln noticed two significant differences from his prior legislative experience. In the Illinois state legislature, leadership mattered; people usually followed what their party leader told them to do. In the House, it mattered less. Lincoln was a loyal party man, which is what the House leadership cared about most, but he was, after all, expected to serve only a single term, and none of the leaders had any leverage over how he voted. Generally, he voted the party line, but not always. When Lincoln was not speaking or working behind the scenes on drafting legislation and crafting coalitions, he was studying his colleagues. (He was present for 97 percent of House votes, compared with the House average of 74 percent.)

In the vote for speaker, Lincoln supported Robert Winthrop while Giddings stayed out of the selection process so he could avoid being held responsible for its result. Winthrop was not a Clay Whig. He was a protégé of Daniel Webster, one of Clay's fiercest competitors for the heart and soul of the Whig Party. Winthrop had no patience or interest in Lincoln's penchant for hanging around the House post office to entertain people with his stories. The new congressman lacked a college education like Winthrop, who had attended Harvard, or Webster, who had studied at Dartmouth. In Illinois, Lincoln had been one of the more literate people, but in Washington he was not.

More significant, Winthrop had voted for the Mexican War, which placed him and Lincoln at sharp odds. Lincoln was faithful to the Whig conception of executive power, which included the president deferring to Congress and to the Cabinet. Though he opposed the war, Lincoln made clear his support for the troops in the field.

As Lincoln knew, the Whig Party was divided into regional camps. Northern Whigs tended to favor more internal improvements, because they would benefit their states more. Southern Whigs were more concerned with stifling executive tyranny but often disagreed about opposing slavery. In Congress, these differences were readily apparent, and they weakened the Whigs, then destroyed the party less than a decade after Lincoln left the House.

No one knows precisely why Lincoln became such a determined, persistent, and vocal critic of the Mexican War. Back home, Democrats relentlessly attacked him for his stand, which one might have thought would have endeared him to ardent Whigs, but it didn't, because in his district so many had fought in the war or had family or friends who did, and to them it seemed a just cause. Lincoln may have been haunted by the deaths of Henry Clay Jr., John Hardin, Daniel Webster's son Edward, or the more than thirty-five thousand other soldiers and civilians on both sides who died of combat, battlefield diseases, and collateral damage. At any rate, something led him to question the lawfulness of the war. He might have thought it was simply good Whig politics or perhaps a way to emulate, if not ingratiate himself with, Clay, as he would likely have seen in Polk's push for war the same kind of abuse of power he and Clay had seen in Jackson, particularly in his slaughter of American natives and efforts to kill the national bank. As Lincoln fashioned a censure of Polk for his misleading the nation into war, his model was Clay's 1834 censure of Jackson for illegally transferring federal deposits from the national bank in an effort to destroy it. The Senate's later expungement of the resolution in January 1837 did not erase its passage from the memories of Clay and his supporters.

Lincoln's first order of business, upon his arrival in the House, was to follow through on Clay's critique of the war. Barely a month after hearing his idol denounce the Mexican War for more than two hours, Lincoln did the same on the House floor. According to Polk, Mexican forces on American land had provoked the United States into war by firing and spilling American blood first. On December 22, 1847, with Mary Todd in the balcony, Lincoln introduced eight resolutions to demand from Polk the exact "spot" of "soil" where "the blood of our citizens was so shed."[50] Lincoln's spot resolutions were legalistic in their fixation on the precise location Mexico started the war, as if Lincoln figured that all he had to do to defeat Polk was call attention to the weakest spot in his argument. Though the failure of these resolutions was often used to taunt Lincoln, on January 3, 1848, only six weeks after Clay's speech in Lexington, the House approved a resolution not unlike Lincoln's declaring that the Mexican War had been begun "unnecessarily and unconstitutionally."[51] Lincoln voted in favor of the resolution, which barely passed 85–81. When House Democrats tried to expunge the House's censure resolution of Polk, as the Democrat-controlled Senate had expunged Clay's censure of Jackson a decade before, the House rebuffed their efforts 105–95. Polk thus became the only American president to be, in effect, censured twice by the House.

Before that second vote, however, on January 12, 1848, nearly two months to the day since Lincoln had listened to Clay in Lexington, Lincoln took to the floor of the House to deliver a major speech denouncing Polk's war.[52] His main purpose was to respond to Polk's end-of-the-year message in December 1847 defending his order for American forces to take the initiative. In his Market Speech, Clay had described Polk's "order for the removal of the army," which placed it in harm's way, as "improvident and unconstitutional."[53] Lincoln made the same point in simpler, more direct language.

In Lexington, Clay had focused his attack on Polk's displacing Congress from its "right and duty" (a phrase he repeated more than once for emphasis) to determine the objects of the war, a position that aligned perfectly with the Whig conception of separation of powers.[54] But Clay framed the attack within broader discourses on both separation of powers and the history of war. Lincoln left out a disquisition on war and the nature of government and instead focused on the abuse of presidential authority, particularly Polk's duplicity and incompetence. Sometimes, Lincoln spoke like the lawyer that he was, repeatedly crafting his arguments as if they were being made in a court of law, referring to the need for "evidence" to support Polk's shifting justifications for the war.[55] At other times, Lincoln sounded like the partisan he also was, even as he dismissed that he or others criticizing the war were engaged in "mere party wantonness."[56] Clay cultivated his oratory in the halls of Congress, where he cast his rhetoric to fit the occasion and the audience. Lincoln cultivated his oratory not just in the courtrooms of Illinois but also in diligently refining arguments down to their basics for an audience of farmers and laborers. Clay rarely distilled his argument down to a single memorable sentence. His message came through the overall flow of his speech, but the same could not be said of Lincoln. Lincoln made his points directly, unvarnished, and crystal clear: "I propose to try to show," he declared on the floor of the House, that "the whole of this—issue and evidence—is, from beginning to end, the sheerest deception."[57] Indeed, in the paragraph in which this line appears, Lincoln three times refers to Polk's "deception."[58] The rhetorical trick of repeating the word or idea that the speaker wishes his audience to take away did not originate with Clay, but he was among those whose mastery of the technique Lincoln followed.

Lincoln, following Clay's lead, focused on Polk's inconsistent, disingenuous statements about the objectives of the war and particularly for indemnifying the conflict. "How like the half insane mumbling of a fever-dream, is the whole war part of his late message!" Lincoln exclaimed rhetorically.[59] While Clay's Mexican War speech lacked his usual caustic asides or analogies, Lincoln inserted his relentlessly: "The President is resolved under all circumstances to have full territorial indemnity for the expenses of the war; but he forgets to tell us how we are to get the excess after those expenses shall have surpassed the value of the whole of the Mexican territory."[60] Lincoln continued, "So, again, he insists that the separate national existence of Mexico, shall be maintained; but he does not tell us how this can be done after we shall have taken all her territory. Lest the questions I here suggest be considered speculative merely, let me be indulged a moment in trying to show they are not."[61]

But here Lincoln had gone too far, undoubtedly emboldened with the verbal thrashings he had been using in campaigns over the preceding decade. The harsh attack on character was a tendency Lincoln indulged but had to break or his prospects for future office might suffer. In trying to outdo Clay in the chamber where he served for eleven years, almost all of which were as speaker, Lincoln managed to do the opposite—lose crucial support back home. He was already thinking, early in his term, of retaining the office "if nobody wishes to be elected," but he squandered his chance by insulting his Democratic friends who backed Polk and by denouncing a war many of his own constituents supported.[62] Nevertheless, to his credit, any loss in popularity back home did not deter him from speaking out.

Lincoln also stuck with the oratory and humor that got him to the House. Just as Clay often did, Lincoln routinely incorporated stories into his speech to illustrate his points, stories that he must have tested on his friends in the boarding house and the House post office. (Mrs. Spriggs's home offered him a similar opportunity in Washington as he had back in Springfield where he roomed with friends and allies.) Lincoln remembered to work the stories into his speech:

> I know a man, not very unlike myself, who exercises jurisdiction over a piece of land between the Wabash and the Mississippi; and yet so far is this from being all there is between those rivers, that it is just one hundred and fifty-two feet wide, and no part of it much within a hundred miles of either. He has a neighbor between him and the Mississippi—that is, just across the street, in that direction—whom, I am sure—he could neither persuade nor force to give up his habitation; but which, nevertheless, he could certainly annex, if it were to be done, by merely standing on his own side of the street and claiming it, or even sitting down and writing a deed for it.[63]

He inserted another analogy to underscore the president's duplicity: "I have sometimes seen a good lawyer, struggling for the client's neck, in a desperate case, employing every artifice to work round, befog, and cover up, with many words, some point arising in the case, which he dared not admit, and yet could not deny."[64] Lincoln concluded by asserting that "after all this, this same President gives us a long message, without showing us that, as to the end, he himself has even an imaginary conception. As I have said before, he knows not where he is. He is a bewildered, confounded, and miserably perplexed man." In these partisan broadsides, Lincoln was delivering in plain, simple, direct language the essence of the problem before the nation, a president who lied, and was vindictive and out of his depth.

In some ways, Lincoln was repeating the mistakes of his Lyceum Address, trying too hard to impress his audience through volume and parlance, and thus having trouble finding the right tone. In a paragraph that later came to haunt him, Lincoln told the House:

> *Any people anywhere, being inclined and having the power, have the right to rise up, and shake off the existing government, and form a new one that suits them better. . . . Any portion of such people that can, may revolutionize, and make their own, of so much territory as they inhabit. More than this, a majority of any portion of such people may revolutionize, putting down a minority, intermingled with, or near about them, who may oppose their movement. Such minority, was precisely the case, of the tories of our own revolution. It is a quality of revolutions not to go by the old lines, or old laws; but to break up both, and make new ones.*[65]

As Michael Burlingame explains, "Lincoln may have been trying to curry favor with Southern Whigs resentful of Northern congressmen, like John Quincy Adams, who had denied the legitimacy of the Texas revolution of 1836."[66] Among those closely listening was his friend Alexander Stephens, who was already helping Zachary Taylor, a fellow Southerner and the hero of the Mexican War, win the Whig Party's nomination for president.

Despite Lincoln's oratorical growing pains, Stephens, thinking back to that speech and others Lincoln delivered in the House, said, "Lincoln always attracted and riveted attention of the House when he spoke," because "his manner of speech as well as thought was original."[67] It might have been Lincoln's effort to make his speech plainer, less adorned with the high-sounding rhetorical flourishes of Henry Clay, that captured listeners' attention. In Stephens's judgment, Lincoln "had no model."[68] In fact, Lincoln had his models; he just did not follow them robotically or thoughtlessly.

Certainly on the policy front, Lincoln followed Clay in supporting "compensated emancipation," the American Colonization Society's goal of purchasing the freedom of African American slaves and transporting them back to the countries of their origins.[69] This plan was intended to compensate the Southern slave owners but did not include any provision to compensate the people enslaved for their labor or suffering. Nor did it provide an option for the enslaved in America to become citizens or to obtain any of the property they worked to build. Yet Lincoln did follow Clay, as well as Giddings and Wilmot, in opposing the extension of slavery. Historian Kenneth Winkle notes that, in "his first year in Congress, Lincoln's voting record on slavery adhered closely to the statement of principles that he had enunciated a decade earlier," as a member of the Illinois house and on the hustings helping Stuart, other Whigs, and of course his own elections. "He supported every anti-slavery measure that came before the House, most of which called for the abolition of slavery or the slave trade in the District of Columbia."[70] Though Giddings thought Lincoln somewhat "timid" in opposing slavery in his early days in the House, Lincoln joined in voting to remove from the agenda and table all the pro-slavery measures.[71] He was a consistent supporter of Whigs' efforts to stop slavery from spreading westward.

However, in the second year of his term, Lincoln broke with his friends Giddings and Wilmot, who were plotting to "blow the Taylor party sky-high."[72] The two men brooked no compromise on slavery, whereas Lincoln, by temperament, training, and emulation of Clay, favored compromise on this most difficult, divisive issue. In the judgment of Giddings and Wilmot, slavery was immoral and therefore had to be stopped. Lincoln knew it was wrong, but as a pragmatist, like Clay, he had not concluded that law and morality must be one and the same in this case. He saw the law more as a policy to be incrementally walked back. It might be less than perfect but could be (re)shaped, step-by-step, into something better with the support and consensus of the voting public.

Giddings was disappointed that Lincoln had voted with a majority of the House to table a motion to abolish the slave trade altogether in the District of Columbia. Lincoln went against Giddings again shortly thereafter, but this time he had Wilmot on his side, in opposing an initiative introduced by Giddings to require a vote to decide the fate of slavery in the District of Columbia. Again, Lincoln was on the winning side. His preference when torn between following Clay and extremists, like his housemates, was to follow Clay, who regarded the "extension" of slavery as the most immediate problem facing the country. Indeed, after returning to Congress in 1848, "Lincoln never again voted to support discussion of abolition in the District." He did support the Wilmot Proviso, though not as much or as often as he claimed.[73]

Another inspiration for Lincoln was not far from his seat in the back of the House—John Quincy Adams, the former president and longtime ally of Clay who was now in his seventeenth year in Congress. Adams had made a mess of his presidency by not caring to rotate out of his administration disloyal Democrats who backed Jackson or his initiatives. (He did not take the time and did not have the temperament or interest to do so.) Adams was a staunch critic of slavery and repeatedly condemned the Mexican War as being waged by slaveholders so they could extend slavery into the territories. Adams agreed with Lincoln's vigorous attacks on Polk and the war.

On February 21, 1848, in the midst of a debate on honoring army officers who had served in the Mexican War, Adams collapsed at his seat from a massive cerebral hemorrhage. Because Lincoln sat in the farthest row in the back, he was near where Adams fell, and he likely was among the first House members who rushed to the former president's aid. Adams died two days later. Speaker Winthrop acknowledged the political alliance between Lincoln and Adams by naming Lincoln as a member of the House committee responsible for making the arrangements for the funeral.

Three days later, on February 24, 1848, Congress held a joint meeting of the House and the Senate to honor Adams. Lincoln, Stephens, and Howell Cobb, who would soon become the speaker, were among the House members present, while Jefferson Davis, Hannibal Hamlin, and Andrew Johnson were among the senators attending. Stephen Douglas was there, too, having entered the Senate as Illinois's newest senator less than three months after Lincoln had been sworn into the House. Lincoln, John Calhoun, Thomas Hart Benton, and Chief Justice Roger Taney were among the pallbearers. Clay was unusually silent, though he and Adams had long been political allies and Clay had served with distinction as his secretary of state, not to mention being the man whose support gave the presidency to Adams in 1824. Adams was one of the few people Lincoln met who had known the Founders personally. Adams had believed that the principles of the Declaration of Independence were the foundation of the Constitution and that slavery violated those principles, and, among those carrying him to the Congressional Cemetery to be laid to rest, only Lincoln was in the position to pick up the mantle Adams would no longer carry.

VI

From the moment Lincoln entered the House of Representatives, he was immersed in the 1848 presidential election, as was nearly every other member of Congress. Joshua Giddings described it as "a President-making Congress," because the upcoming contest overshadowed everything else being done in the House.[74] Lincoln had not backed a winner in a presidential election since William Henry Harrison in 1840; he was desperate to do so now, and it was clear to him who could win this time around (Zachary Taylor)— and who could not (Clay). Even the formerly staunch Jacksonian Duff Green backed Taylor. Green told Lincoln that Taylor would have the support of not only the Whigs but also Democrats who, like himself, had preferred Calhoun over Jackson because of Calhoun's much stronger support for the rights of states over the federal government on questions relating to slavery. Lincoln was also approached by John J. Crittenden, one of Kentucky's two Whig senators and a longtime friend of Clay's. Crittenden had frayed their relationship when he backed Taylor early for the 1848 presidential election. Crittenden had served as a member of both the House and the Senate and as attorney general for presidents Harrison and Tyler. He was governor-elect of Kentucky in 1848 but was promoting Taylor this time, to Clay's never-ending condemnation. Even though Clay came to Washington in 1848 to solicit support for yet another run, Lincoln, too, had already cast his lot with Taylor.

In fact, Lincoln was a Taylor man before he arrived in Washington. On August 30, 1847, Whig leaders, who were attending a state constitutional convention, gathered at the home of Ninian W. Edwards (married to Stuart's younger partner) to discuss the upcoming presidential campaign. Lincoln explained to the group that the purpose of the meeting was to choose "some other man than Henry Clay as the standard bearer of the Whig party." Lincoln suggested Taylor was the man and urged "the necessity of immediate action," because, "if the Whigs did not take Taylor for their candidate" for president, then "the Democrats would," because Taylor had appeal as a war hero. If this sounded as if victory was the main thing that mattered to Lincoln, it was, because that is what he had learned by watching Clay repeatedly fail in his bids for the White House. Lincoln reportedly told the group that "the Whig party had fought long enough for principle, and should change its motto to success."[75]

After the Mexican War, Taylor was often compared to Washington and Jackson, which boded well for his chances, but he also differed from them in important ways. He bore little physical resemblance to either of them; both were tall, while Taylor was only five eight, with a thick, powerfully built frame, long arms, short bowed legs, and an angular face. He was not known for his eloquence, but his men adored him for his candor and courage under fire. From the time he had become a captain and commander of Fort Knox in Kentucky, he developed a knack for outmaneuvering the enemy. In the War of 1812, Taylor led the defense of Fort Johnson and the first land victory of the war and thereby earned

Jackson's respect and gratitude. More than thirty years later, he continued to lead his troops successfully against the odds. Washington had not only been the commanding general of the army but the presiding officer at the U.S. Constitutional Convention. Jackson had at least been a member of both the House and Senate, a judge, a member of his state's constitutional convention, and an experienced politician, not to mention a three-time nominee for president. Taylor had had no career other than the military (except as an occasional land speculator and owner of a large plantation).

Taylor was among the first to acknowledge that he had never been involved in politics, indeed had never bothered with it before. The absence of any record, particularly any that indicated affinity for the Whig Party, turned off party leaders early in the campaign. Caleb B. Smith of Indiana said that he expected Taylor would have no success among Northern Whigs; Senator Willie Mangum of North Carolina agreed.[76]

One of the bonds between Lincoln and Stephens was their shared confidence that Taylor would win the presidency in 1848. Stephens liked the fact that Taylor was a slaveholder and was therefore expected to be sympathetic to the interests of slaveholders, while Lincoln found Taylor's apolitical history appealing: he could not be attacked, as Clay had been, for shifting positions to suit the current needs, but he could instead be sold as being above politics. Together, Stephens and Lincoln formed the first Congressional Taylor Club, which was dominated by Southerners. They called themselves the Young Indians, corresponded with Whigs around the country, and gave speeches on behalf of Taylor, both on the floor of the House and wherever else they were needed and could go.[77] The Young Indians agreed that the fact that Taylor had not been a lifelong politician distinguished him from Clay as well as Polk or whomever else might become the nominee of the Democratic Party. Lincoln and Stephens worried that Polk might reconsider serving for only a single term, but they believed that Taylor's record made him a more compelling figure than either Clay or Polk.

The Young Indians shared plenty of advice with Taylor, perhaps too much—Taylor hated being lectured to and treated as though he knew nothing of the world. Lincoln suggested that Taylor should announce his intention to endorse a national bank if Congress were to pass a bill establishing one, recommend a higher protective tariff to fund internal improvements, pledge not to use his veto power, and seek to acquire no territory from Mexico "so far South, as to enlarge and aggravate the distracting question of slavery."[78] In April 1848, Taylor made the decision to publicly identify himself as a Whig, then went further to denounce wars of conquest (even agreeing that the Mexican War had been unconstitutional) and proclaim his willingness to sign Whig economic measures into law if they were enacted by Congress.

At the Whig national convention held in Philadelphia on June 7, 1848, Lincoln attended as a delegate for Taylor. It was the quickest nominating convention in history. Conventions usually lasted a few days, but this one met for only a single day, choosing Taylor as its candidate for president on the fourth ballot. Clay finished a distant second. The Whigs endorsed no platform after the delegates recognized that Taylor could be hurt only if he allowed himself to be pinned down on the issues. The delegates chose a longtime faithful Whig, New York congressman Millard Fillmore, as their vice presidential candidate.

All of this was in dramatic contrast to the Democratic convention a month before. Polk's strategy not to include presidential hopefuls in his administration worked relatively well for maintaining unity and support within that circle, but it did not help the Democrats. Polk's determination to serve only a single term left the incumbent out of the 1848 presidential election and no obvious successor. In the end, the Democratic Party settled on Lewis Cass, who had served as a brigadier general in the War of 1812, governor of the Michigan Territory, Jackson's secretary of war, and Polk's floor leader for three years in the Senate. The nomination split the party. Cass was a strong proponent of popular sovereignty—the notion that each state's voters should decide for themselves whether their states would be slave or free, but Democrats who opposed slavery opted to back Martin Van Buren, who had emerged as the presidential candidate for the new Free Soil Party, formed as an alternative to the Whigs and the Democrats after the latter refused to endorse the Wilmot Proviso at their convention. Recoiling from their party's endorsement of popular sovereignty, the radical faction known as Barnburners (who opposed extending slavery) joined with antislavery Whigs and members of the Liberty Party, which had supported abolition, to form the new party. Its major principle was steadfast opposition to the extension of slavery into the Western territories.

Shortly after the Whig Party nominated Taylor as its presidential candidate in 1848, Lincoln took to the House floor to support the nomination and destroy Cass. He defended Taylor's promise to use the veto power sparingly, a pledge that had become part of Whig orthodoxy to distinguish their candidates from Jackson, who had used it to thwart the national bank. Lincoln praised Taylor's willingness to defer to Congress, because it aligned with "the principle of allowing the people to do as they please with their own business."[79] He expressed the hope that Taylor would oppose "the extension of slavery" into the territories and sign any bill with the Wilmot Proviso.[80]

Throughout the speech, Lincoln exhibited a growing mastery of Clay's tools of gesticulation and ridicule, yet he also demonstrated an improved alliance of barb and instruction. His humor and storytelling set him apart from the other, more somber members of Congress, and he relished the attention it brought him. Lincoln's "sparkling and spontaneous and unpremeditated wit" entertained Daniel Webster, when they met occasionally for breakfast on Saturdays, as well as other "solid men of Boston" in Congress.[81] When Maine's senator Hannibal Hamlin came to the House chamber and asked the newspaper man Ben Perley Poore who the speaker was that was entertaining the House and the galleries, Poore said that it was Abe Lincoln, known as "the champion story-teller of the Capitol."[82]

For much of the speech, Lincoln stuck again with what had brought him to the House. He ridiculed the Democrats and their candidate mercilessly. Responding to the claim made the day before by a Georgia congressman that the Whigs had "deserted all our principles, and taken shelter under General Taylor's military coat-tail," Lincoln accused the Democrats of having used "the ample military coat tail" of Andrew Jackson:

> *Like a horde of hungry ticks you have stuck to the tail of the Hermitage lion to the end of his life; and you are still sticking to it, and drawing a loathsome sustenance from it, after he is dead. A fellow once advertised that he had made a discovery by which he could make a new man out of an old one, and have enough of the stuff left to make a little yellow dog. Just such a discovery has General Jackson's popularity been to you. You have not only twice made President of him out of it, but you have had enough stuff to make Presidents of several comparatively small men since; and it is your chief reliance now to make still another.*[83]

Lincoln satirized nearly everything about Cass—his military record, which Lincoln suggested was comparable to his own experience in the Black Hawk War dodging mosquitoes; his waffling on the Wilmot Proviso; his financial records when he was governor of the Michigan Territory from 1813 to 1831; and especially his corpulence.[84] Lincoln ridiculed the Democratic candidate's "wonderful eating capacities," which enabled him to consume "ten rations a day in Michigan, ten rations a day here in Washington, and near five dollars worth a day on the road between the two places!"[85] He warned his colleagues never to stand between Cass and food.[86]

After Congress adjourned on August 14, 1848, Lincoln stayed in Washington to help the Whig Executive Committee of Congress organize the national campaign. It was Lincoln's first chance to see inside the operations of a national campaign, overseen by Clay's onetime friend Senator Crittenden of Kentucky.

Under the direction of Crittenden, Lincoln corresponded with party leaders and distributed copies of his speeches and those by other Whigs in defense of Taylor. He instructed young Whigs back home to take the initiative, to get involved and push for Taylor: "You must not wait to be brought forward by the older men," he wrote during the campaign, "For instance, do you suppose that I should ever have got into notice if I had waited to be hunted up and pushed forward by the older men[?]"[87] It was a telling instruction from Lincoln. He had learned from his own political experience not to be overly dependent on others but rather to depend primarily on himself to advance his own interests and those of his party. Here was his foundational self-made man proposition through a civic filter. At the same time, Lincoln persistently counseled the candidate and his surrogates to stay on message, avoiding Clay's folly in opining on nearly everything (often in contradictory ways), instead emphasizing that Taylor leaned in the direction of the Whigs and their basic principles but was no zealot.[88]

Over the next two months, Lincoln followed his own advice to the young Whigs. He went out among the voters and stumped for Taylor. Besides writing letters and (usually anonymous) opinion pieces and helping to organize support from Washington, Lincoln rallied supporters in Maryland, New England, and Illinois. In Massachusetts, he shared the stage with one of New York's senators, William Seward. As someone who had vied with Fillmore to control the Whig Party—and its spoils—back home, Seward could not have been happy to see his rival on the national ticket at his own expense. Nonetheless, he gave a rousing speech in opposition to slavery. But Seward was not impressed with Lincoln's performance, criticizing to a correspondent Lincoln's "rambling storytelling speech, putting the audience in good humor, but avoiding any extended discussion of the slavery question."[89] Lincoln was more complimentary in return. "I have been thinking about what you said in your speech," Lincoln told Seward after their joint appearance. "I reckon you are right. We have got to deal with this slavery question, and got to give much more attention to it hereafter than we have been doing."[90]

Throughout the campaign, Taylor's doubtful allegiance to Whig principles troubled party leaders, especially when he accepted a nomination for president from a group of dissident South Carolina Democrats, which had formed their own mini-convention in protest over the party's nomination of Cass. Thurlow Weed, the powerful Whig boss of New York and Seward's mentor, threatened to call a mass meeting of New York Whigs to denounce their party's candidate. Fillmore wrote Taylor directly in mid-August to point out the dangers of accepting support from the opposition party. On September 4, Taylor replied to Fillmore in a letter much like one that he had written previously to his brother-in-law Captain John S. Allison of Louisville, Kentucky, who had shared the letter with the public. The first missive, dated April 22, 1848, was prompted by the need for Taylor to push back an effort by Clay to claim the party's nomination. In it, Taylor acknowledged that he was not sufficiently familiar with many public issues to pass judgment on them. He said, "I reiterate [that] I am a Whig but not an ultra Whig. If elected, I would not be the mere president of a party—I would endeavor to act independent of party domination, & should feel bound to administer the Government untrammeled by party schemes."[91] He then promised, in good Whig fashion, to limit his vetoes to "cases of clear violation of the Constitution," since "the personal opinion of the individual who may happen to occupy the executive chair ought not to control the action of Congress upon questions of Domestic policy."[92] This statement meant he would follow Congress on questions of the tariff, currency, and internal improvements. The letter achieved its purpose of reassuring Whig voters of Taylor's commitment to their party's basic principles.

In the second letter that Allison shared with the public, Taylor complained that people had not properly understood what he had been trying to say in the first. He pointed out that all who had served with him in the Mexican War knew that he was a Whig in principle. Moreover, even while a commanding general in Mexico, he had been nominated for president by informal, popular assemblies of Whigs, Democrats, and Native Americans but had declined the endorsements in order to avoid appearing to be partisan. Taylor continued to insist that he was not a partisan candidate but would be the president of all the people, promising not to impose indiscriminate, politically motivated personnel changes nor to coerce Congress with vetoes of constitutional legislation. He wrote separately to reassure Crittenden that this would be the last letter he intended to write during the campaign. The correspondence held the party in line, at least through election day.[93]

The split within the Democratic Party nearly guaranteed Taylor a victory in the general election, but his final margin of victory was thin, with 1,360,000 popular votes to 1,220,000 for Cass and 291,000 for Van Buren. Taylor's margin over Cass in the Electoral College was more decisive, 163–127. Van Buren did not carry a single state, but his 120,000 votes at home in New York provided Taylor his victory margin there. Taylor carried all of New England except Maine and New Hampshire, plus the three Middle Atlantic states and the four border ones. In the South, he carried four (Kentucky, Louisiana, Florida, and Georgia) of the seven. He won four states that Clay had lost—Georgia, Louisiana, New York, and Pennsylvania, but in the Midwest he lost Michigan, Indiana, Missouri, and Illinois. He carried no Western states, and the Democrats controlled a majority of seats in both the House and the Senate. For the fifth straight time in five presidential elections, Abraham Lincoln had yet to deliver Illinois to his preferred candidate.

VII

The end of Lincoln's single term in the House coincided with Zachary Taylor's inauguration. In the five months between election day in November 1848 and Taylor's inauguration on March 5, 1849, Lincoln reaffirmed his strong attachment to Clay's American System in his support for internal improvements and opposition to slavery. But whatever hope he had to retain his place in Congress was firmly dashed when Herndon wrote to inform him that Lincoln's friend and former partner, Stephen Logan, wanted the seat.[94] Lincoln stood down. Democrats tightly controlled Illinois's Senate seats, so Lincoln looked elsewhere. He told several friends, perhaps in an effort to save face and appear still to be politically relevant, that he had declined an offer to serve as the head of the Land Office in the newly created Department of the Interior, even though the position paid the handsome salary of $3,000 annually.[95]

For one of the few times in his political career, Lincoln was indecisive, unsure of what to do. He wrote Mary Todd that "having nothing but business—no variety—[Washington] has grown exceedingly tasteless to me." Worrying about his sons, he asked her, "Don't let the blessed fellows forget father."[96] He felt her beckoning him home—"How much, I wish instead of writing, we were together this evening"—but Lincoln stayed in Washington.[97]

With Taylor's inauguration speedily approaching, Lincoln realized that there was still a vacancy in the position of leading the Land Office, which oversaw the administration of federally owned lands throughout the nation and the territories.[98] Whoever ran the Land Office would have a significant say over the extent to which the United States allowed or barred slavery in federal territories.

Throughout the first several months of 1849, Lincoln, in his capacity as a member of Congress, forwarded names for the administration to consider for the job. On March 11, he and Edward Baker, with whom he had long competed for leadership of the Sangamon County Whigs, visited the office of Secretary of the Interior Thomas Ewing, an Ohio Whig who had once been close with Henry Clay. One object of the meeting was to secure the commissionership of the Land Office for an Illinois candidate, but neither Lincoln nor Baker made a recommendation for either of the two leading candidates—Cyrus Edwards, a lawyer who once ran for governor of Illinois, or J.L.D. Morrison, a Mexican War veteran and Democratic member of the Illinois house—to be chosen. Several of Lincoln's friends urged him to break the deadlock by competing for the office himself, but he declined, telling them that he refused to compete for the appointment unless Taylor denied it to Edwards, a friend who was also the brother of Ninian Edwards, the husband of Mary Todd's sister.[99] The stalemate broke when another candidate, Justin Butterfield of Chicago, said he wanted the office. Lincoln knew Butterfield as an accomplished lawyer and fellow Whig but an active partisan for Clay, not Taylor. Butterfield's lack of efforts on Taylor's behalf prompted Lincoln to write Ewing that "[Butterfield] is my personal friend, and is qualified to do the duties of the office but of the quite one hundred Illinoisians equally well qualified, I do not know one with less claims to it."[100] Lincoln thought it absurd that Taylor would give the position to someone who had neither backed the Taylor campaign nor worked for it, as Lincoln had done, and he was offended when Ewing said Butterfield was his choice for Land Office commissioner.[101]

At this point, Lincoln decided to compete for the office himself; perhaps he had been planning to all along. He sent appeals to his friends in Illinois and surrounding states, asking them for support and urging them to contact Taylor personally.[102] Taylor and Ewing made clear that they wanted the office to go to a Clay man, and Clay backed Butterfield. Lincoln was displeased, particularly with the administration's seemingly perverse interest in rewarding not those who had served the campaign, but rather those who had served past party leaders. Ewing instead offered Lincoln the prestigious governorship of the Oregon Territory. Lincoln declined.

In fact, he wanted the position. John Todd Stuart had encouraged Lincoln to pursue the opportunity, since it likely meant that once Oregon became a state, Lincoln would be assured, as so many other territorial governors had been, of returning to Congress as one of the new state's two senators. Cass, for example, had once briefly been the governor of the Michigan Territory before his congressional career.

The opposition to the move came from Mary Todd. The Oregon Territory was a dangerous place, which she didn't want to visit, much less move to. Indeed, John Gaines, the man whom Taylor appointed to the spot Lincoln had turned down, lost two daughters to sickness as they were traveling to Oregon from Kentucky.

Lincoln, however, never expressed regret about moving back to Springfield. As the only Whig in Illinois's national delegation, he had a special role to play back home. Indeed, he had explained in letters asking friends to back him for the Land Office job that such concern had held him back before deciding to apply late for the position.

But there may have been another reason for Lincoln's turning down the assignment. Lincoln had watched how Jackson had used patronage to unify his party and administration and how Adams's mismanagement of it had doomed his presidency. Now it seemed that Taylor was abandoning the Jackson strategy in a hopeless attempt to placate his opponents. Lincoln confessed to his friend Joshua M. Lucas, a clerk in the Land Office, however, that he was personally hurt because the Taylor administration flattened his wishes "in the dust merely to gratify" Clay and his followers.[103] Lincoln had been a Clay man all his adult life, but he didn't follow Clay mindlessly, as his support of Taylor made clear. Instead, he learned from Jackson's success and Clay's failures. Jackson and Van Buren had invented the spoils system as a way to reward their allies and supporters. Jackson had promised that system as a candidate and used it to win a second term in the White House. Lincoln had told friends that Butterfield's appointment was "an egregious political blunder," because of the negative repercussions he believed it would have on loyal Whigs who expected patronage in exchange for their support.[104] As former Whig Committee chairman Dr. Anson G. Henry had asked rhetorically, "Who ever heard of Butterfield as a Whig, until the fight was over?"[105]

In public and in communications with the Taylor administration, Lincoln was careful to keep his complaints to himself (or close friends). He wanted to be a good party man, and so he put on a brave front and threw his support behind the appointment of Butterfield. He knew that was what his fellow Whigs expected, and what he often pleaded with them to do, and that the needs of the floundering Whig Party in Illinois were of greater urgency than infighting over how Taylor handed out appointments. Even when he learned that Secretary Ewing had likely removed two letters of recommendation from his file in an effort to weaken his candidacy, Lincoln stood by the appointment of Butterfield and the administration. As he told David Davis, then an Illinois state judge, "I hope my good friends everywhere will approve the appointment of Mr. B. in so far as they can, and be silent when they cannot."[106] A fractured party had kept Cass out of the White House in 1848, done the same to Clay and the Whig Party in 1844, while a unified party had kept Jackson in charge of it. If Taylor was not careful, Lincoln worried, the Whigs would lose the power that he and his fellow ardent supporters had won for them in 1848.

Of course, it was sadly ironic that Taylor wanted a Clay man to get the position of leading the Land Office, given that Lincoln had turned away from his mentor for the campaign. Lincoln had expected that his service to Taylor would have counted as the most important thing in his appointment, but in spite of his protestations that he was a genuine, long-standing Clay man, Lincoln was not considered enough of one to appease either the wounded Clay or the victorious Taylor.

Besides staying for Taylor's inauguration, Lincoln attended one of the major inaugural balls on the evening of March 4, 1849. He spoke briefly with Taylor's former son-in-law Jefferson Davis, now remarried. We know little about whom else he talked to, but among the other guests were Robert E. Lee, who had served on General Winfield Scott's staff in Mexico, and an unhappy President Polk. After initially resisting meeting President-Elect Taylor, Polk welcomed him to the White House and held a dinner in his honor. The former president left town the day after the inauguration for his first vacation in years.

Unfortunately, it did not end well. The work of the presidency had taken a toll on him, and within three months of leaving office, Polk died, his body exhausted after four hard years in office and vulnerable to the cholera that killed him. Less than a decade before, sixty-eight-year-old William Henry Harrison, at the time the oldest person ever elected president, died barely a month into office. Over the span of a few weeks, he had been weakened by pneumonia and ultimately succumbed to typhoid fever. Polk, who had been the youngest person elected president, was now the youngest former president to have died. He was fifty-three. Clay greeted the news of Polk's death in the same way he had greeted news of Jackson's—he said nothing. Following suit, Lincoln remained silent.

Taylor's inauguration was the first that Lincoln ever attended. He had no idea if he would ever attend another or whether he would ever return to Washington. The knowledge that there was another Whig in the White House was of little solace to him. A presidential victory that depended on the thinnest of margins of victory in a state—New York—that Whigs should have had firmly in their column was cause for concern, not celebration. While a Whig presidency should have brightened his future, clouds had swept in. Essentially a political neophyte, Taylor was an enigma to Whigs and Democrats alike. Taylor could barely stay on script during the general election. The prospect that Seward and Weed would write the only script he might follow in the future did nothing to relieve Lincoln's concerns or those of most Whig loyalists.

The fragile coalition that brought Taylor into office was further cause for concern. As historian Kenneth Winkle noted, "Lincoln clearly recognized the strange Whig brew [electing Taylor] that included free soil Democrats, writing 'all the odds and ends are with us—Barnburners [an antislavery faction in New York], Native Americans, Tyler men, disappointed office seeking locofocos [a radical faction opposed to any financial policies they deemed antidemocratic], and the Lord knows what.'"[107] Keeping that coalition together or forging a more solid one to ensure Taylor's reelection was likely impossible. It could not be done without a compelling vision of the future that no one in the party had yet put together. Lincoln was not sure how to do that. Before he could chart that larger course, he had to do something else, something he had not done since he walked into New Salem—find a path that could take him from obscurity to political relevance.

CHAPTER FOUR

LEARNING FROM FAILURE
(1849-1856)

If a man can learn from his failures, he can go far. Lincoln did.

On returning home, Lincoln had much to be proud of. From humble origins, he had risen in eighteen years to significant heights—serving four terms in the Illinois House, becoming a Whig leader in Illinois in all but name, and serving a term in the U.S. House when the nation elected its second Whig president. In that span, Lincoln met two presidents (Polk and Taylor), traded stories with Daniel Webster, debated the Democrats' rising star Stephen Douglas more than once, and stood on the floor of the House to make speeches that made that grand old institution shake with laughter. He had been friendly with the last giant of the founding era, John Quincy Adams, and had stood shoulder to shoulder with such Democratic luminaries as John Calhoun, Thomas Hart Benton, Chief Justice Taney, and Jefferson Davis. He had a volume of Henry Clay's speeches inscribed by the man himself, and he had even seen Clay in person. He'd made a reputation in Washington and Illinois as a storyteller extraordinaire, and now he had more than a handful of impressive encounters to recount.

Nevertheless, Lincoln's failures were mounting. His effort to secure a second term in the House had come to naught, and he was returning to a community in which he had managed, through his strident partisanship, to alienate many Democratic friends and old Whig allies. He had made some useful friends in Washington, but none among the party's leadership. To add insult to injury, back to practicing law just before he left the capital, he argued his only case before the Supreme Court, a dispute in which Lincoln represented the defendant, who argued that Illinois law barred an action against him brought by a nonresident of the state. In an opinion by Chief Justice Taney, the Court ruled 6–1 against Lincoln's client. The only vote Lincoln got came from John McLean, a Jackson appointee who had many Whiggish sympathies. A few years later, Lincoln would try in vain to get the Whig Party to nominate McLean as its presidential candidate. Lincoln had many fences to mend in Springfield, besides trying to revive a law practice that Herndon had managed to keep afloat in his absence. Herndon recalled that, upon his return to Springfield, Lincoln was despondent that he had not done more: "How hard—Oh how hard it is to die and leave one's Country no better than if one had never lived for it!"[1]

From nearly half a continent away, Lincoln watched the nation's leaders fumble in their efforts to find solutions to the mounting problems of nativism—a movement among the "native born" (those born in the United States) against new immigrants (as well as actual natives)—and slavery that were dividing the nation. Yet Lincoln never fully disappeared from the national stage in the 1850s. When he made his bid to take center stage in 1859 and 1860, he was not the same man who had returned to Springfield ten years before. He was still Abe Lincoln, the father, husband, storyteller, and loyal Whig, but this reconstructed Lincoln was more moderate, more deliberate, more contemplative, more disciplined, less biting in his wit, and more eloquent. Like Clay and Jackson, he never quit, but he went further than either of them to work on himself and find his voice and a vision for the nation's future that picked up where Clay's had ended.

I

It is unknown exactly when the people closest to Lincoln began noticing he was cold and ruthlessly ambitious. There is a point in every great history of Lincoln when this side—perhaps his core—is noted, but ambition is always retrospectively obvious when the subject is a president. Lincoln's intense craving for higher office may have become most apparent in the 1850s, but it is likely to have been there all the time, and his brief stint in Washington likely made him more eager than ever to be at the center of the political action.

Among Lincoln's friends and mentors, there was surprising consensus on what they perceived the adult Lincoln to be. Herndon said of Lincoln in the 1850s, "Mr. Lincoln never had a confidant, and therefore never unbosomed to others. He never spoke of his trials to me or, so far as I knew, to any of his friends."[2] After Lincoln's death, Herndon wrote, "Even after my long and intimate acquaintance with Mr. Lincoln I never fully knew and understood him. [He] was the most reticent and mostly secretive man that ever existed; he never opened his own soul to one man . . . even those who [were] with him through long years of hard study and under constantly varying circumstances can hardly say they knew him through and through."[3] In short, Herndon said, "He never touched the history or quality of his own nature in the presence of his friends."[4] Leonard Swett, a fellow lawyer in Springfield, agreed that "beneath a smooth surface of candor and an apparent declaration of all his thoughts and feelings," Lincoln was a private man, who "handled and moved men remotely as we do pieces on a chessboard."[5] He added that Lincoln was a "remorseless trimmer with men; they were his tools and when they were used up he threw them aside as old iron and took up new tools."[6]

John Todd Stuart, who had known Lincoln all his adult life, agreed as well. He said, "L[incoln] did forget his friends. There was no part of his nature which drew him to acts of gratitude to his friends." He observed as well that there was in Lincoln "his want of passion—Emotion" that accounted for Lincoln's "peculiar constitution—this dormancy—this vegetable constitution."[7] The word "peculiar" appears in the assessments of many associates and family members. David Davis, who traveled the law circuit with Lincoln and later led his presidential campaign, said,

"Lincoln was a peculiar man. [He] never asked my advice on any question. . . . [Lincoln] had no Strong emotional feelings for any person—Mankind or thing. He never thanked me for any thing I did."[8] Lincoln's sister-in-law Elizabeth Edwards said, "I knew Mr L well—he was a cold man—had no affection—was not Social—was abstracted—thoughtful."[9] Herndon's experiences, too, confirmed that "Lincoln was undemonstrative" and "somewhat cold and yet exacting."[10]

Many other allies, from as far back as New Salem and later in Washington, spoke of Lincoln's circumspection. Gustave Koerner, the leader of the large German American population in southern Illinois, supported Lincoln all his adult life but thought he was not "really capable of what might be called warm-hearted friendship."[11] Pennsylvania journalist Alexander McClure wrote, "Mr. Lincoln gave his confidence to no living man without reservation. He trusted many, but he trusted only within the carefully-studied limitation of their usefulness, and when he trusted, he confided, as a rule, only to the extent necessary to make that trust available."[12] After years of working with Lincoln, McClure concluded, "Neither by word nor expression could anyone form the remotest idea of his purpose, and when he did act in many cases he surprised both friends and foes."[13] John Bunn, who knew Lincoln in his Springfield years, recalled that Lincoln "had his personal ambitions, but he never told any man his deeper plans and few, if any, knew his inner thoughts. What was private and personal to himself he never confided to any man on earth. When men have told of conversations with Lincoln in which they represent him as giving out either political or family affairs of a very sacred and secret character, their tales may be set down as false."[14] A fellow chess player observed that

> While playing chess [he] seems to be continually thinking of something else. Those who have played with him say he plays as if it were a mechanical pastime to occupy his hands while his mind is busy with some subject. He plays what chess-players call a "safe game."

*Rarely attacking, he is content to let his opponent attack while he
concentrates all his energies in the defense—awaiting the opportunity of dashing in at a weak point or the expenditure of his adversary's strength.*[15]

There may have been no man who was closer to Lincoln than
Orville Browning. Like Stuart, he had known Lincoln all his adult
life. Contrary to Herndon's assessment, Lincoln often confided
with Browning, indeed, perhaps more with him than anyone else,
with the likely exception of Mary Todd. On looking back over his
long relationship with Lincoln, Browning said, "Our friendship
was close, warm, and I believe sincere. I know mine for him was,
and I never had reason to distrust his for me. Our relations to my
knowledge were never interrupted for a moment."[16] Yet Browning
also saw Lincoln's ambition and cunning.

True, many of these critical appraisals of Lincoln seem like sour
grapes, the negative reactions of people who may not have gotten
all they had wanted from Lincoln, or perhaps attempting, after Lincoln's death, to bring him down a peg or two. Yet these comments
remain credible because they came from people who thought of
themselves as Lincoln's allies and friends. They might have felt betrayed because they were not as important to Lincoln as they had
hoped or because Lincoln had not confided in them as much as
they would have liked as he paved his path to the presidency. They
wished to be remembered for their own impact on history.

Many of Lincoln's closest associates, perhaps all, missed the fact
that Lincoln was fundamentally a pragmatist. Principles, history,
ideas, people—he thought of them all as tools. Lincoln's hero, Clay,
had been the same, earning his most common nickname, the Great
Compromiser. Yet it would be wrong to say that the only thing
Lincoln refused to compromise was his own career prospects. His
moral imperative, particularly when it came to opposing the extension of slavery, remained his compass. The surprise for many
was how fiercely he pushed others out of the way on his relentless
quest for power.

II

Lincoln was ever mindful of his failures and limitations. In 1850, he began a lecture to young law students with the candid acknowledgment (reminiscent of Clay's trademark self-deprecation) that "I am not an accomplished lawyer. I find quite as much material for a lecture in those points wherein I failed, as in those wherein I have been moderately successful."[17] He thought, too, of how Stephen Douglas, younger than he, had surpassed him in Illinois and national politics. In 1852, he acknowledged, "Douglas has got to be a great man, and [be]strode the earth. Time was when I was in his way some; but he has outgrown me and [be]strides the world; and such small men as I am, can hardly be considered worthy of his notice; and I may have to dodge and get between his legs." Four years later, he was still thinking of Douglas, remembering that

> twenty-two years ago Judge Douglas and I first became acquainted. We were both young then; he a trifle younger than I. Even then, we were both ambitious; I, perhaps, quite as much so as he. With me, the race of ambition has been a failure—a flat failure; with him it has been one of splendid success. His name fills the nation; and is not unknown, even, in foreign lands. I affect no contempt for the high eminence he has reached.[18]

Lincoln wanted such eminence for himself. He put in the hard work on behalf of a cause bigger than himself for the good of the country and acclaim seen through the prism of Lincoln's faith in the self-made man. He was not going to quit because he had yet to achieve such eminence. Jackson did not despair in 1824 that he had lost his chance to be president. To the contrary, he redoubled his efforts and became the standard of success for other presidents to measure themselves. Though Clay had lost the presidency three times, he never quit, either. As Lincoln was leaving Washington, Clay was returning to Washington to try, in spite of his frail health, for at least one more chance to lead the Senate and make another run for the presidency.

Though Lincoln was back in Springfield, he had not lost his ambition. As Browning noted, Lincoln was "always a most ambitious man."[19] In Browning's judgment, Lincoln hoped "to fit himself properly for what he considered some important predestined labor or work . . . , that he was destined for something nobler than he was for the time engaged in."[20]

In the last paragraph of the first speech Lincoln gave announcing his arrival on the political stage, he had declared, "Every man is said to have his peculiar ambition. Whether it be true or not, I can say for one that I have no other so great as that of being truly esteemed of my fellow men, by rendering myself worthy of their esteem. How far I shall succeed in gratifying this ambition, is yet to be developed." For the rest of his life, he reminded audiences of his "humble" origins ("I was born," he said, "and have ever remained in the most humble walks of life."[21]) while acknowledging his "ambition" to serve. In his 1838 speech to the Young Men's Lyceum in Springfield, Lincoln denounced the hazards of political ambition, which enticed men to become tyrants. Years later, as president, he understood the importance of braving those hazards for the greater good. As he counseled General McClellan, "If we never try, we shall never succeed."[22]

The most common theme of Lincoln's counsel to others in the 1850s was that success required relentless determination on behalf of a noble cause. These are what drove men like Jackson, Clay, and Lincoln to make an enduring mark on the world. Failure was not going to stop Lincoln's "little engine" of ambition. Not trying was, for Lincoln, the greatest failure of all.

A yearning for something better, some grander objective, was deeply engrained within him. In 1850, Lincoln further suggested to young law students, "The leading rule for the lawyer, as for every man of every other calling, is diligence. Leave nothing for tomorrow which can be done today."[23] In 1855, Lincoln counseled Isham Reavis, who had written for advice on becoming a lawyer: "If you are resolutely determined to make a lawyer of yourself, the thing is more than half-done already."[24] Nearly three years later, he wrote another aspiring lawyer, William Grigsby, "If you wish to be a lawyer, attach no consequence to the place you are in, or the person you are with; get books, sit down anywhere, and go to reading for yourself. That will make a lawyer of you quicker than any other way."[25] In 1858, James Thornton wrote to Lincoln seeking his assistance in training John Widmer, who was aspiring to practice law even though he was older than the usual young man starting the study of the law. Lincoln declined the chance "to be a suitable instructor for a law student." His advice for Widmer was that "he reads the books for himself without an instructor. That is precisely the way that I came into the law. Let Mr. Widmer read Blackstone's Commentaries, Chitty's Pleadings—Greenleaf's Evidence, Story's Equity, and Story's Equity Pleadings, get a license, and go to the practice and keep reading."[26] Lincoln told Reavis, "I did not read with anyone."[27] (So much for John Todd Stuart, Bowling Green, and Stephen Logan.)

Lincoln's reading always extended beyond the law. In the 1850s, he reread the poetry and other books he loved best. John Hay, one of Lincoln's secretaries when he was president, recalled that Lincoln "read Shakespeare more than all the other writers together."[28] As president, Lincoln told the actor James Hackett, "Some of the plays I have never read; while others I have gone over perhaps as frequently as any unprofessional reader. Among the latter are Lear, Richard Third, Henry Eighth, Hamlet, and especially Macbeth."[29] Lincoln read *Macbeth* aloud when the mood suited him, and also the somber, sad poetry of Robert Burns, *Pilgrim's Progress,* and William Knox's *Mortality* (sometimes called by its most famous line, "Oh Why Should the Spirit of Mortal be Proud?").

Pilgrim's Progress was a perrennial Lincoln favorite. Its hero, Christian, finds salvation only after surmounting the challenges of corruption run wild in the world and the countless temptations that routinely bring men down—most of all, the sin of pride. Lincoln also loved *Hamlet* and the *Henry* plays, but one reason he might have preferred *Macbeth* was that its actual hero does not die in the end. Macbeth is an antihero whose ambition leads to his downfall. In the end, it is "the good Macduff," the quiet man who is respected when he speaks and is devoted to his country above all else, who prevails. Lincoln also loved Shakespeare's *Julius Caesar,* another tragedy in which an overambitious main character dies. Brutus says, "As he was ambitious, I slew him."[30] There were lessons in these literary masterpieces, and Lincoln relished not only their language, but also their precepts.

In the 1850s, Lincoln scoured newspapers daily, perhaps more voraciously than ever before. He was rarely home, spending most of his time in his law office, reading, writing briefs, arguing cases in court, and traveling the circuit. Sometimes, Herndon thought, he was too wrapped up in his own thoughts and business. Herndon had to hear everything Lincoln read, too: "Lincoln never read any other way but aloud," he complained. "This habit used to annoy me beyond the point of endurance."[31] Herndon recalled that, when he arrived at work at seven every morning, he found Lincoln reading newspapers. "Lincoln's favorite position when unraveling some knotty law point was to stretch both of his legs at full length upon a chair in front of him," wrote Herndon, and "in this position, with books on the table nearby and in his lap, he worked up his case."[32] Herndon was also annoyed when Lincoln's "brats," Tad and Robert, came to visit the office. They tore books, newspapers, and legal materials apart, and even peed on the floor.[33]

Others noted Lincoln's absorption in the written word. "He would pick up a book and run rapidly over the pages, pausing here and there," remembered a clerk, "at the end of an hour—never, as I remember, more than two or three hours—he would close the book, stretch himself out on the office lounge, and with hands under his head, and eyes shut, he would digest the mental food he had just taken."[34]

One of the few books Lincoln read in its entirety in the 1850s was Euclid's classic work on geometry, an odd choice for most people, but Lincoln believed it would sharpen his mind, making his thinking more logical, rigorous, and organized as he figured out the next steps he had to take in order to return to the nation's capital. Given that the former land surveyor thought of others as objects to be moved and replaced, the study of angles and congruencies might have offered an interesting, if literally tangential, amusement.

III

William Herndon was few people's idea of a good lawyer. In Lincoln's earlier associations, he had been the junior partner to older and more experienced men—John Todd Stuart and Stephen Logan. Stuart was gone for much of the time, serving in Congress while Lincoln ran both the law office and Stuart's campaigns. With Logan, Lincoln again ran the law office, but the partnership foundered in part because both of them wanted to run for Congress. Herndon was different. He was as surprised as anyone when Lincoln invited him to be his partner. He came from a long line of opiniated Jacksonian Democrats; worse, he was an angry drunk and notoriously untactful. He alienated Mary Todd for life when in 1837 he first met her at a ball where he asked her to dance but told her that she "seemed to glide through the waltz with the ease of a serpent." She never forgave him for the comparison.

Lincoln chose Herndon because he was "a laborious, studious young man . . . far better informed on almost all subjects than I have ever been."[35] He loved books as much as Lincoln did and had one of the best private libraries in Springfield, filled with law books; the works of English historians; translated writings of great Western philosophers, such as Kant; works on political economy by John Stuart Mill and Henry Carey, who had built the economic underpinnings of Clay's American System; and great literature, including the works of Shakespeare and everything that the essayist Ralph Waldo Emerson and the preacher Henry Ward Beecher, among others, had published. It is unclear how much of the library Lincoln read, though, later as president, he was evidently familiar with Carey's economics and Beecher's sermons. Lincoln and Herndon were opposites who in theory complemented each other: Lincoln believed in "cold, calculating, unimpassioned reason," while Herndon was intuitive, disposed, in Lincoln's estimation, to "see the gizzard of things."[36] Herndon claimed that he could predict the future in his bones, leading Lincoln to joke regularly upon seeing him first thing in the morning, "Billy—how is your bones philosophy this morning?"[37] Herndon knew how to keep the books and, perhaps most important, was eager to serve Lincoln's every need.

The First Judicial Circuit, which Lincoln had crisscrossed numerous times as a young lawyer, had been merged with the Eighth Circuit, which consisted of fifteen counties. Lincoln was the only lawyer who rode the entire territory. His old friend David Davis, who had become a judge in 1848, oversaw the circuit and did almost all the judging in it. The other lawyers who frequently rode with Lincoln—Leonard Swett, Nathan Judd, Ward Hill Lamon, and sometimes Stephen Logan and Orville Browning—were lifelong friends and fellow Whigs. (John Stuart Todd sometimes rode the circuit as well, though his political leanings were shifting, and he was increasingly aligning more with the Democrats, including his onetime rival Stephen Douglas.)

Lincoln enjoyed the travel and the camaraderie, and his companions came as close as any group to being Lincoln's kitchen Cabinet. Nevertheless, every evening Lincoln kept up an active correspondence. His work took him away from Springfield and his family for nearly six months of the year: three months each spring and each fall. Looking back at that time many years later, Davis said, "Lincoln was as happy, as happy as he could be."[38]

Lincoln's law practice was varied. Some cases were much bigger than others, some clients had more money than others, some cases were more interesting than others, and Lincoln immersed himself more deeply into some cases than others. He represented masters seeking the return of their slaves and slaves who had escaped their bondage. By the late 1850s, Lincoln started avoiding fugitive-slave cases, though in 1856 or early 1857 he agreed to help a woman whose son faced enslavement in New Orleans by raising money to secure the young man's freedom. During the 1850s,

Lincoln and his partner "appeared in at least 133 cases concerning railroads—sometimes representing the roads, and sometimes opposing them. The most famous of these cases involved the Illinois Central Railroad; Lincoln & Herndon, as attorneys for the railroad, received what was then the enormous fee of $5000 for their services. It, like all other fees, was divided equally between the partners."[39] They also represented small banks, debt collectors, spouses in divorce cases, and once Orville Browning, who had tripped on a Springfield sidewalk and broken his leg. (Lincoln took Browning's case all the way to the Illinois Supreme Court, which ruled for Browning because the city had a duty to keep its streets safe and well maintained.)

Almost 10 percent of Lincoln's cases were in the federal courts, including the one Supreme Court case Lincoln argued, which he lost. Lincoln's practice also included murder trials. In one, he sharply questioned a witness's certainty in identifying the defendant on the night of the murder, successfully casting doubt by presenting an almanac that showed that there was very little moonlight on the evening in question. He won another murder trial, shortly before he secured his presidential nomination, by convincing a jury that his client had killed another man in self-defense.

Lincoln's practice with Herndon also included advising clients on all sorts of matters not involving litigation, including writing deeds, registering land, paying taxes, and drafting contracts and wills. In all, over the course of a legal career extending from 1837 until shortly before he left for his inauguration in 1861, Lincoln handled more than five thousand cases. Despite their many hours together, Lincoln's preoccupations about politics, law, and literature confounded Herndon. Herndon was an inveterate optimist, who believed "in the universal progress of all things, especially of man's up going."[40] Lincoln agreed, but he was far more contemplative. Though he often broke his silence with stories and laughter, these were not enough to put Herndon at ease. He found Lincoln "incommunicative—silent, reticent, secretive—having profound policies—and well laid—deeply studied plans."[41]

Everyone, including Lincoln, agreed that he did his best work with juries. Otherwise, the lawyers who practiced with Lincoln regarded him as a good lawyer but not a great one. For example, his friend Henry Clay Whitney, who rode the circuit with him, said that Lincoln "was not more than ordinarily successful for a first-class lawyer."[42] One court observer compared Lincoln to Norman Purple, who served on the Illinois Supreme Court from 1845 to 1848, by saying that Purple,

> *in intricate questions, is too much for [Lincoln]. But when Purple makes a point, which cannot be logically overturned, Lincoln avoids it by a good-natured turn, though outside the issue. Lincoln's chief characteristics are candor, good nature, and shrewdness. He possesses a noble heart, an elevated mind, and the true elements of politeness.*[43]

Another friend said that Lincoln "did not stand at the head of the bar, except as a jury lawyer."[44] Herndon, too, did not consider Lincoln to be a first-rate lawyer, believing he was "very deficient" in some ways because he "never thoroughly read any elementary law book" and "knew nothing of the laws of evidence—of pleading or of practice. [He] was purely and entirely a case lawyer—nothing more."[45]

Later, when Lincoln was considering the old Pennsylvania pol Simon Cameron for his Cabinet, he said, "I suppose we could say of General Cameron, without offence, that he is not 'Democrat enough to hurt him.' I remember people used to say, without disturbing my self-respect, that I was not lawyer enough to hurt me."[46] Likening himself to swine scavenging for acorns in a forest, Lincoln described himself as "only a mast-fed lawyer."[47] According to one of the residents Lincoln visited when he rode the circuit, Lincoln was "aware of his inferiority as a lawyer" and was always ready to acknowledge it "with a smile or a good-natured remark."[48]

Herndon saw weakness in Lincoln's inability to be more deeply philosophical in his reasoning. In describing how to confound Lincoln, Herndon said, 'If you wished to be Cut off at the knee, just go at Lincoln with abstractions—glittering generalities—

indefiniteness—mistiness of idea or expression." In response to abstract thinking or arguments, Lincoln would "become vexed and sometimes foolishly so."[49] Yet, Lincoln chided Herndon, "Billy don't shoot too high—shoot low down, and the common people will understand you."[50] Rather than a defect, crafting his rhetoric to be understood by "the common people"—reminiscent of both Jackson's oratory, direct and unvarnished, and Clay's, which employed humorous and enlightening analogies—was one of Lincoln's defining strengths.

Lincoln's limitations as a lawyer were, however, dramatically exposed in two of his most important cases. In the first, *Todd Heirs v. Wickliffe,* Lincoln represented his father-in-law, Robert Todd, in an 1848 lawsuit. Todd claimed that Robert Wickliffe had illegally taken the property of Todd's cousin, Polly, after she had become Wickliffe's second wife. Polly's father had left her his vast estate, and her first husband and her son had died before Polly married Wickliffe. When she died, Todd claimed that he was the rightful heir to her estate under her father's will, which stipulated that if she had no living heirs, the estate would be split among the descendants of her father's brothers, which included Todd. The will, however, had disappeared. Lincoln represented the Todds, who maintained that Wickliffe had coerced Polly into marrying him and giving her estate to him in exchange for the right to purchase the release of her family's two slaves. (As a married woman, she had no right to purchase or sell property herself.) It appeared to be an open secret that one of the two slaves was in fact Polly's son, who would have been an heir to her estate but for the fact that as an African American slave he was not recognized as a person under Kentucky law, let alone having any of the rights or privileges of a propertied white man, including the right of inheritance. Lincoln tried mightily to prove the existence of the will, but in spite of testimony from John Todd Stuart's father that he had seen the document, the Kentucky courts, including the Kentucky Supreme Court, ruled in favor of Wickliffe as the heir to Polly's fortune. Lincoln knew that the case broke Robert Todd's heart, and did not speak of it for years.

In the other case, Lincoln had the rare opportunity to measure his talents against one of the nation's most famous and highly respected lawyers, Kenyon College–educated Edwin Stanton of Cincinnati. In 1855, the Great Reaper Trial, as it was known, pitted a nearly penniless inventor from Rockford, Illinois, John Manny, and his partners against the wealthy industrialist Cyrus McCormick, Chicago's largest employer. Both Manny and McCormick manufactured agricultural reapers, and both held several patents, although the foundational patent on the McCormick reaper had expired, and therefore the original invention had entered the public domain and was free for anyone to use. McCormick sued Manny in federal court in Chicago, seeking to put him and his partners out of business. The Manny team decided it needed a lawyer who knew the local judge and the local law, and they hired Lincoln as their counsel. However, the case was eventually moved for the judge's convenience to Cincinnati, and Manny's team, who no longer had as much need for Lincoln's limited utility, brought in Edwin Stanton, later Buchanan's attorney general, as their main counsel. No one told Lincoln. Because it was a high-profile dispute and Manny's team had promised Lincoln one of the largest remunerations he was ever promised as a lawyer, $5,000, he spent considerable time on the matter, and was excited to travel to Cincinnati to make the closing arguments in the case. When Lincoln arrived, Stanton took one look and asked an associate, "Where did that long-armed baboon come from?"[51] Lincoln was told his services were no longer needed. He remained to watch Stanton argue the case. So impressed with Stanton's command of the law and the facts was he that Lincoln told his co-counsel, Ralph Emerson, "I am going home. I am going home to study law."[52] Emerson pleaded, "Mr. Lincoln, you stand at the head of the bar in Illinois now! What are you talking about?"[53] Lincoln responded, "I do occupy a good position there, and I think I can get along with the way things are done there now. But these college-trained men, who have devoted their whole lives to study, are coming West, don't you see? And they study their cases as we never do. They have got as far as Cincinnati now. They will soon be in Illinois."[54]

After a pause, he added, "I am as good as any of them, and when they get out to Illinois I will be ready for them."[55]

Lincoln's varied clientele reflected his pragmatism to earn his living however he could. Whereas Stephen Douglas was perfectly aligned with the interests of big business, Lincoln's practice showed him how the law worked on the ground, its effect on everyday citizens, not just elite businesses. All of this deepened his ability to identify with the common man, because he did legal work for common men nearly every day. It also sensitized him to how status often drove the application of the law, as he saw firsthand that women and slaves were not afforded the Constitution's promise that all would be treated equally.

Lincoln's law practice kept him busy—so busy that he claimed he did not have time to travel a hundred miles to attend the funeral of his father, who died on January 17, 1851. Lincoln's stepbrother had urged him to see Thomas before he died, as well as to come to the funeral, but Lincoln demurred. "Say to him that if we could meet now, it is doubtful whether it would not be more painful than pleasant." In his life, Lincoln never had a kind or positive word to say about his father and, in the years after his death, spoke of Thomas only in ways that elevated his own image of himself as a "self-made man" and denigrated his father as squandering his own chances to make something of himself.

The skills Lincoln refined as a lawyer were as important for his political career as they were for his legal practice. In 1850, he advised young law students, "Discourage litigation. Persuade your neighbors to compromise whenever you can. Point out to them how the nominal winner is often the real loser—in fees, expenses, and waste of time. As a peacemaker the lawyer has the superior opportunity of being a good man. There will still be business enough." He stressed, "Never stir up litigation. A worse man can scarcely be found than one who does this. Who can be more nearly a fiend than he who habitually overhauls the register of deeds in search of defects in titles, whereon to stir up strife, and put money in his pocket? A moral tone ought to be infused into the profession which should drive such men out of it."[56] In 1842, Lincoln had told the Springfield Washington Temperance Society, "When the conduct of men is designed to be influenced, persuasion, kind, unassuming persuasion, should ever be adopted. It is an old and true maxim that a 'drop of honey catches more flies than a gallon of gall.' So with men. If you would win a man to your cause, first convince him that you are his sincere friend."[57]

Such advice was not as obvious as it might seem. Litigation was almost universally understood as the path for a lawyer to achieve prominence in his community or state, and thus it was quite tempting to treat the other side as the enemy. A focus on mutual accommodation clashed with the combative nature of the court system and couldn't yield the thrill of winning and crushing your opponent. Lincoln told young law students about the importance of both persuasion and preparation, two skills crucial to success in law and politics. In 1850, he explained, "Extemporaneous speaking should be practiced and cultivated. It is the lawyer's avenue to the public. However able and faithful he may be in other respects, people are slow to bring him business if he cannot make a speech."[58]

One observer said, "He never considered anything he had written to be finished until published, or if a speech, until it was delivered."[59] Lincoln also "habitually studied the opposite side of every disputed question, of every law case, of every political issue, more exhaustively, if possible, than his own side. He said that the result had been, that in all his long practice at the bar he had never once been surprised in court by the strength of his adversary's case—often finding it much weaker than he had feared."[60]

The advice Lincoln gave in 1850 applied to everything he said and did in politics and law, both before and especially later—to be a "peacemaker," to speak with "a moral tone," to find "compromise" whenever possible, to listen to what others had to say, to learn from his failures, to study opponents' arguments for their strengths and weaknesses, to persuade people with "a compliment," and to make friends of your enemies. This was the creed of Clay.

IV

In 1850, Henry Clay looked at the first year of Zachary Taylor's presidency and concluded, "I have never seen such an administration. There is very little co-operation or concord between the two ends of the avenue. There is not, I believe, a prominent Whig in either house that has any confidential intercourse with the Executive."[61] Lincoln agreed. There was no prominent Whig leader in the Cabinet, either. Taylor left filling vacancies within the administration to his Cabinet heads, but they felt little loyalty to Taylor or his policies. The ensuing disorder was compounded by Taylor's refusal to abide by one of the central tenets of the Whig Party—presidential deference to the will of both Congress and the Cabinet. In one of his few acts as president before he died, William Henry Harrison had rejected that practice as well, arguing that he, not they, was elected to office and their job was to advise him, not the other way around. Taylor followed suit, and, like Harrison, found himself at odds with his own Cabinet.

Taylor's next move infuriated the Whig faithful further. As a presidential candidate, Taylor had gone to great lengths to assure loyal Whig voters that he shared their principles of governance. In the second of the letters that he had written to his brother-in-law John Allison, he had reemphasized that "I am not prepared to force Congress, by coercion of the veto, to pass laws to suit me or to pass none."[62] Though Taylor had made no reference to those principles in his first public statement following his election, many Whigs were reassured by his declaration in his Inaugural Address that "it is for the wisdom of Congress itself, in which all legislative powers are vested by the Constitution, to regulate [various] matters of domestic policy."[63] He went further, at that time, to say, "I shall look with confidence to the enlightened patriotism of that body to adopt such measures of conciliation as may harmonize conflicting interests and tend to perpetuate the Union which should be the paramount object of our hopes and affections."[64]

However, in Taylor's first—and, as it turned out, only—Annual Message to Congress (delivered at the end of 1849), the fears of his Whig constituents were fully realized: rather than wait for any lead or signal from Congress, he laid out the bold proposal for Congress to admit California and New Mexico separately as new states into the Union and leave each of them to decide how they would handle the issue of slavery. While Taylor recognized that Congress had complete discretion to condition the admission of a new state on any basis it chose, he made clear that his proposal should be the first and only order of business in Congress. Everyone knew that if Taylor's plan were followed, it would tip the balance of power in Congress in favor of antislavery forces, because both California and New Mexico were disposed to endorse antislavery constitutions and the Senate was at that time evenly split between slave and free states. Many Whigs liked the idea of weakening the slaveholders' power in Congress, but they liked even less that Taylor was demanding that they do as he directed. Taylor stuck by his proposal because he believed it would avoid, rather than provoke, a nasty fight in Congress over extending slavery into the territories. He thought the plan had the further advantage of respecting popular sovereignty, because it would have allowed each territory to choose for itself in its constitution whether to allow or prohibit slavery. Most members of Congress, including Stephen Douglas (who considered himself the principal champion of popular sovereignty), objected to Taylor's plan because they either disagreed with it substantively or objected to his making demands of Congress rather than following its will. House leaders refused to take any action on his plan, while Senate leaders refused to act on hundreds of his nominations to positions requiring confirmation. Taylor then set a record for making the most recess appointments by any president till then.

New Mexico responded to the president's plan immediately by applying for statehood under an antislavery constitution. Texas authorities had other ideas. To expand the domain of slavery, they threatened to acquire, by force if necessary, all the New Mexico land east of the Rio Grande, including Santa Fe. They declared that it belonged to them and threatened civil war if the United States tried to stop them.

The prospect of war didn't deter Taylor; indeed, it strengthened his resolve. In February 1850, he met with Southern leaders in Congress and warned them that anyone "taken in rebellion against the Union, he would hang . . . with less reluctance than he had hanged deserters and spies in Mexico."[65] He sent federal troops to Santa Fe and directed the colonel in charge to prepare his men to rebuff any invasion of New Mexico. These soldiers kept Texas forces at bay. Taylor made clear that if Texas made any aggressive move to capture any portion of New Mexico, he would lead federal troops in response.

Taylor tried repeatedly but unsuccessfully to persuade Democrats and Southern Whigs that his plan was the best possible compromise because it gave "the North substance of the Wilmot Proviso but without forcing the South to swallow it as a formally enacted principle."[66] The fact that Taylor's chief defender in the Senate was Andrew Jackson's old Democratic ally, Thomas Benton of Missouri, revealed how much Taylor's leadership violated Whig principles. Not only that, but Southern Whigs, led by Henry Clay, responded with their own proposal that included a fugitive slave law. Whereas Clay favored a compromise that helped the slave power, Taylor did not, objecting that it would have drawn the federal government into supporting slavery and would have ripped the Union apart. Southern Whigs, including Clay, were outraged by Taylor's threatened veto of the compromise because they believed, in accordance with Whig orthodoxy, that a president should veto only measures that are clearly unconstitutional, and since the Constitution at that time recognized slavery (for example, in calculating the populations of congressional districts), their legislation did not exceed that threshold. Accordingly, on May 21, 1850, Clay formally broke with Taylor, arguing that Taylor, "entertaining that constitutional deference to the wisdom of Congress which he had professed, and abstaining from any interference with its free deliberations, ought, without any dissatisfaction, to permit us to consider what is best for our common country."[67]

Taylor again warned Southern Democrats that they would be worse off if they failed to support his proposal. He argued that their opposition ran the risk of motivating Congress to approve the Wilmot Proviso to ban slavery in any of the territory acquired from Mexico in the Mexican-American War, which, he made clear, he believed was constitutional. In response, several Southern Democrats, including Mississippi senator Jefferson Davis, declared their opposition to Taylor's plan because it enabled antislavery forces to become a majority in the Senate—after already controlling the House—and thus provided a back door through which to enact the Wilmot Proviso. When other Southern Democrats, including John Calhoun in his last statement on the floor of the Senate, threatened secession rather than accept Taylor's plan, the president issued his own threat—to use military force to stop any secession movement. He was ready to stop Texas aggression *and* Southern secession.

By the spring, with war threatening on two fronts, Taylor's problems with his Cabinet worsened. On one front, Taylor had managed, in his lame efforts to secure geographical balance, to exclude anyone from the Northeast. Thus the only contingent that actually supported his plan, the Northern Whigs, were absent from his team. At the same time, a scandal of unprecedented proportions threatened to rip Taylor's administration apart. His attorney general, Reverdy Johnson, had authorized Treasury Secretary William Meredith to pay the full amount of the interest on a claim that the Galphin family had made against the U.S. government for wrongfully seizing control of their family estate in Georgia in 1773. When it became known that just before Taylor took office Congress had enacted a law directing that the interest should be five times the size of the principal and that half of the principal and half of the interest were owed to Taylor's war secretary, George Crawford, for his legal services on behalf of the Galphin family, a public outcry arose. The matter festered for months, while the House considered censuring not only Cabinet members Johnson, Crawford, and Meredith but perhaps also Taylor. Under intense pressure from Congress to get his administration in order, Taylor considered firing his entire Cabinet to remove any appearance of corruption within his administration. Not satisfied with that response from Taylor, some House leaders considered initiating an impeachment inquiry against Taylor for allowing such corruption to fester in his administration.

With a stalemate in Congress over the admission of the two new states and a threat of impeachment hanging over him, Taylor reluctantly made plans to reorganize his administration. It included dismissing his entire Cabinet. Taylor knew Jackson and Tyler had each removed their Cabinets entirely, so neither he nor his able attorney general, Reverdy Johnson, had any doubts that he had the power to do the same thing. As Taylor prepared to go public with his plans, he died unexpectedly from either a stomach virus or cholera on July 9, 1850.

Back in Springfield, Abraham Lincoln was as stunned as most Whigs to read about the turmoil in Taylor's Cabinet and the impasse over admission of California and New Mexico. Now Taylor's death left nearly all of them speechless—all but Lincoln. He was one of the few prominent Whigs who took the opportunity to eulogize the late president. He put aside his disappointment in not securing an appointment with the Taylor administration, the debacle of Taylor's Cabinet appointments, and the anger over Taylor's break with Clay. Long overshadowed by the tribute that he would offer his idol Henry Clay two years later, this eulogy reveals Lincoln's significant affinities for his subjects—a fondness sculpted by a great deal of selective recall. In its fourth sentence, Lincoln notes that Taylor's "youth was passed among the pioneers of Kentucky, whither his parents emigrated soon after his birth; and where his taste for military life, probably inherited, was greatly stimulated."[68] Lincoln said nothing of the time Taylor spent in any other state, particularly Louisiana, where he had owned a plantation with slaves.[69]

Zachary Taylor came into prominence and the presidency because of his military career, and Lincoln's eulogy was devoted almost entirely to that period of his life. Yet the fact that Polk picked Taylor as his initial commanding general in Mexico went without comment.[70] Nor did Lincoln mention that Polk had removed Tay-

lor from command in Mexico.[71] Instead, Lincoln recalled the fallen
heroes in that "last battle" of Taylor's, including Henry Clay's son,
as well as John Hardin, his onetime friend and rival. "Passing in
review, General Taylor's military history, some striking peculiari-
ties will appear."[72] For Lincoln, the first was this:

> No one of the six battles which he fought, excepting perhaps, that of
> Monterey, presented a field, which would have been selected by an
> ambitious captain upon which to gather laurels. So far as fame was
> concerned, the prospect—the promise in advance, was, "you may
> lose, but you can not win." Yet Taylor, in his blunt business-like
> view of things, seems never to have thought of this.[73]

Lincoln found most significant the fact that "it did not happen to
Gen. Taylor once in his life, to fight a battle on equal terms, or
on terms advantageous to himself—and yet he was never beaten,
and never retreated. In *all*, the odds was greatly against him; in
each, defeat seemed inevitable; and yet *in all*, he triumphed."[74] Lincoln did not have to mention the race for the presidency, since that
turned out just as every other battle in Taylor's life did, with Taylor prevailing in the end. "Wherever he has led," Lincoln noted,
"while the battle still raged, the issue was painfully doubtful; yet
in *each* and *all*, when the din had ceased, and the smoke had blown
away, our country's flag was still seen, fluttering in the breeze."[75]

   Though Lincoln had never seen, much less participated in, an
actual battle, he recognized the greatness in Taylor as a military
commander. Lincoln declared, "General Taylor's battles were not
distinguished for brilliant military maneuvers; but in all, he seems
rather to have conquered by the exercise of a sober and steady
judgment, coupled with a dogged incapacity to understand that
defeat was possible." Here was Lincoln delivering his greatest ac-
clamation for Taylor and most aspirational for himself and the
country. "His rarest military trait," Lincoln said of Taylor, "was
a combination of negatives—absence of excitement and absence
of fear. He could not be flurried, and he could not be scared."[76] At
the precise time in Lincoln's life when he had reason to be scared

that he might never win another campaign or achieve the fame he desperately desired, Taylor's capacity to be unafraid of failure was inspiring.

> *It is perhaps enough to say—and it is far from the* least *of his honors that we can* truly *say—that of the many who served with him through the long course of forty years, all testify to the uniform kindness, and his constant care for, and hearty sympathy with, their every want and every suffering; while none can be found to declare, that he was ever a tyrant anywhere, in anything.*[77]

The "tyrant" was a not so subtle reference to the difference between Taylor and the Democratic presidents who preceded him. He was not, in other words, disposed to be a Jackson, a Tyler, or a Polk. Lincoln did not yet know all of the men who would look back with gratitude at Taylor's "uniform kindness" as a leader and mentor, but on the night of Taylor's inauguration, he had likely met two of them—Robert E. Lee and Jefferson Davis.

While Lincoln was serving in the Black Hawk War, Taylor's daughter, Sarah, had met and fallen in love with Davis, who was a lieutenant under Taylor's captaincy. Taylor opposed the marriage because he thought the life of a military wife would be too hard for her. Davis then resigned from the service, and the two were married with Taylor's blessing in 1835. For their honeymoon, Davis brought her back to his family's plantation in Mississippi, where she contracted malaria and died just a few months later. Taylor vowed never to forgive Davis, but after nearly a decade of not speaking to each other Davis returned to the army, the two men reconciled, and Davis served with distinction under Taylor during the Mexican War. Later, shortly after Taylor's election, Taylor told Davis to follow his personal and constitutional convictions without fear of losing Taylor's respect. In turn, Davis was one of the three senators who planned Taylor's inauguration, though he vigorously opposed his policies. The two kept in touch, though they never discussed politics. Davis was at Taylor's bedside when he died, and he persuaded House leaders to put aside the movement to censure Taylor after his death. In his eulogy, Davis defended Taylor's proposals on California and New Mexico as the only way to preserve the opportunity for Congress to peaceably settle the boundaries of Texas and New Mexico.

Lincoln could only stand in awe of a man who could earn the allegiance of a fierce proslavery senator like Davis *and* a fierce abolitionist like William Seward, who was widely believed to be Taylor's closest confidant. Taylor was the model of a man who could separate politics from the personal in order to maintain bridges across the chasm of political differences defining his time.

Having lauded Taylor for his military prowess and lack of any pretensions or arrogance, Lincoln moved next to the "point of time" when "Taylor began to be named for the next Presidency."[78] He noted, "The incidents of his administration up to the time of his death, are too familiar and too fresh to require any direct repetition."[79] Thus Lincoln was able to gloss over most of the chaos of Taylor's fifteen months in office. After all, Lincoln said, "The Presidency, even to the most experienced politicians, is no bed of roses; and General Taylor like others, found thorns within it."[80] In apparent acknowledgment of the hostile House he faced at the time of his death, Lincoln observed, "No human being can fill that station and escape censure. Still I hope and believe, when General Taylor's official conduct shall come to be viewed in the calm light of history, he will be found to have deserved as little as any who have succeeded him."[81]

Of Taylor's death, Lincoln could not help but wonder "what will be its effect, politically, upon the country."[82] Lincoln knew, as did the nation, that Taylor's death elevated to the presidency an old-line Whig—his vice president, Millard Fillmore. "I will not pretend to believe," Lincoln expressed hopefully, "that all the wisdom, or all the patriotism of the country, died with General Taylor."[83] Yet as a close student of the news printed in the Whig papers that he religiously read, Lincoln expected Fillmore to be hard pressed by Clay, Douglas, and others to bend too far in favor of the slave power to spare the country from a civil war. "I fear," Lincoln then said, "the one great question of the day, is not now so likely to be partially acquiesced in by the different sections of the Union, as it would have been, could General Taylor have been spared to us."[84]

This was as far as Lincoln would go to opine on the choices facing the nation, but he knew whom he hoped would come up with the answer. "Yet, under all the circumstances, trusting to our Maker, and through his wisdom and beneficence, to the great body of our people"—meaning, in Lincoln's parlance, the Congress— "we will not despair, nor despond."[85]

Before Lincoln closed with a quotation from the Gospels, a hymn from Isaac Watts (the "Godfather of English hymnody"), and several stanzas from one of Lincoln's favorite poets, William Knox,[86] he reminded the audience of its "duty." He repeated the word three times,[87] emphasizing that he expected "the American people" and their leaders to undertake it now that Taylor was dead.[88] In closing, Lincoln said,

> The death of the late President may not be without its use, in reminding us, that we, too, must die. Death, abstractly considered, is the same with the high as with the low; but practically, we are not so much aroused to the contemplation of our own mortal natures, by the fall of many undistinguished, as that of one great, and well known, name. By the latter, we are forced to muse, and ponder, sadly.[89]

The prospect of "duty and death" bracketed not just Taylor's life but Lincoln's own.

V

Millard Fillmore was not much better known in 1850 than he is today. Taylor had barely given a second thought to the selection of his vice president, either at the convention or once they were in office. Their differences defined their relationship, or lack of one. At six feet, Fillmore towered over the short, weather-beaten ex-soldier. As a young man, Fillmore was regarded as strikingly handsome with brown wavy hair. Now his hair was white, and he had gained nearly fifty pounds. Prior to his selection as a compromise choice to be Taylor's running mate, Fillmore had chaired the powerful House Ways and Means Committee. He tried but failed to win the House speakership in 1841 and New York's governorship in 1844. He arrived at the convention as New York's comptroller, a particularly unimpressive credential, but his nomination appealed to the delegates because he came from a rival wing of the party to that of Seward and Weed, who were already suspected of having positioned themselves to wield enormous influence within the administration if Taylor won. Thus Fillmore's addition to the ticket was not only supposed to broaden Taylor's support within the party but check the oversized influence of Seward and Weed.

The plan failed. Taylor ignored Fillmore. Taylor didn't listen to him; indeed, he barely let Fillmore say anything in his presence.

Fillmore knew Lincoln but was unimpressed. As a modestly successful lawyer from the West, Lincoln offered little of interest to the new president as a one-term Whig congressman who told entertaining stories. Thurlow Weed, ever the self-interested matchmaker, arranged the first meeting between the two men in Albany, New York, on September 24, 1848. Weed had been instrumental to Taylor's nomination, and he was the force behind William Seward's rise in American politics. As someone who appeared aligned with Seward, Lincoln knew he could expect little if anything from Fillmore. Whatever was discussed at their meeting, it led nowhere. Lincoln never bothered to apply for any government post while Fillmore was president. The two men did not meet again in person until after Fillmore left office.

Even before Taylor died in July 1850, Fillmore had told Clay that he intended, if he got the chance, to cast the tie-breaking vote to approve a compromise bill. Taylor had been angry over the news, for Fillmore seemed disposed to accept *any* compromise with the slave power, not just the one Clay forged. On the day that Fillmore became president, he told Daniel Webster, then a key proponent for compromise in the Senate, that he was withdrawing Taylor's plan and was willing to accept any reasonable agreement approved by Congress. On Fillmore's second day as president, the entire Cabinet resigned in protest over his pledge to Webster. Fillmore promptly accepted the resignations and underscored them by appointing Webster as secretary of state. Webster accepted the job, correctly realizing that his selection was a confirmation of Fillmore's intention to approve any legislative compromise. Others in Congress read it the same way.

Meanwhile, Clay was trying to craft an agreement through a series of separate bills, each appeasing a different constituency in Congress. Nearly six months before Taylor died, he began introducing these bills in a set of speeches that would eventually be remembered as among his greatest. Clay ended the first of them with a clever piece of theater. He said a man had come to his hotel earlier that day and given him "a precious relic."[90] Clay paused, and then produced it: "It is a fragment of the coffin of Washington."[91] He rhetorically asked if this relic was a bad omen for the Union and answered, "No, sir, no."[92] It was instead a warning for Congress "to beware, to pause, to reflect before they lend themselves to any purposes which shall destroy the Union which was cemented by his exertions and example."[93] The story never happened, but it served Clay's purposes.

Two months later, Clay delivered one of his longest and most memorable speeches, on February 5, 1850, to plead for a compromise to save the Union. Warning that disunion was an imminent threat, he begged his fellow senators, from both the North and the South, to "pause" before they took the final steps to destroy the Union through bloody civil war. He proposed a plan that had three goals: to settle once and for all every question arising from the problem of slavery; to craft a compromise that required neither side to sacrifice its core principles; and to ask the opposing sides to make concessions, "not of principle," but "of feeling, of opinion."[94] He called for mutual forbearance from both parties, saying that flexibility would not be easy for either side, but asking each to overcome their distrust of the other or face the destruction of all they held dear.

The next day, Clay returned to the floor of the Senate to continue pressing his case. He again predicted that secession—as threatened by John Calhoun and Jefferson Davis—meant civil war. Staring at Fillmore, who was sitting as the presiding officer of the Senate, Clay declared,

> I am directly opposed to any purpose of secession or separation. I am for staying within the Union, and for defying any portion of this confederacy to expel or drive me out of the Union. I am for staying within the Union, and fighting for my rights, if necessary, with the sword, within the bounds and under the safeguard of the Union. . . . Here I am within it, and here I mean to stand and die.[95]

He then concluded his two-day plea:

> I implore gentlemen, I adjure them, whether from the South or the North, by all that they hold dear in this world—by all their love of liberty—by all their veneration for their ancestors—by all their regard for posterity—by all their gratitude to Him who has bestowed on them such unnumbered and countless blessings—by all the duties which they owe to mankind—and all the duties they owe to themselves, to pause, solemnly to pause at the edge of the precipice, before the fearful and dangerous leap is taken into the yawning abyss below. . . .[96]

Clay took one final pause, again looking directly at Fillmore. "I implore, as the best blessing which Heaven can bestow upon me, upon earth, that if the direful event of the dissolution of the Union is to happen, I shall not survive to behold the sad and heart-rending spectacle."[97]

On March 7, 1850, Daniel Webster joined the debate with one of his finest speeches—some think the greatest he ever gave—declaring at the outset that "I speak today for the preservation of the Union."[98] After pleading for compromise between the contending sides, he concluded,

We have a great, popular, constitutional government, guarded by
law and by judicature, and defended by the affections of the whole
people. No monarchial throne presses these States together, no iron
chain of military power encircles them; they live and stand under
a government popular in its form, representative in its character,
founded upon principles of equality, and so constructed, we hope,
as to last forever.[99]

These words foreshadowed an even more troubling declaration
he would make later, after Taylor's death, which would effectively
acknowledge the legitimacy of secession.

Less than a week later, Seward denounced the slave power on
the floor of the Senate in the harshest terms possible. He acknowledged that the Constitution provided for the recapture of fugitive
slaves. "But," he proclaimed, "there is a higher law than the Constitution. God's law made slavery immoral and illegitimate, and
God's law commanded Christians to disobey laws, which they
deemed unjust. Slavery subverted democracy."[100] Seward scolded
his colleagues, "I confess that the most alarming evidence of our
degeneracy which has yet been given is found in the fact that we
even debate such a question."[101] He dismissed threats of secession as
mere bombast. Inflaming the public did not bother him; he dared
Congress to do the only sensible, principled, moral thing to do—
end slavery. "Whatever choice you have made for yourselves," he
suggested, "let us have no partial freedom; let us all be free."[102]

At the end of March 1850, South Carolina's John Calhoun died of
tuberculosis. Earlier that month, he had foreseen the kinds of arguments Seward and others would make and had asked a colleague,
James Mason of Virginia, to read what became Calhoun's last statement on the floor of the Senate. Characteristically, it was a zealous
defense of both slavery and secession. The prospect of war did not
bother Calhoun. "What is it that has endangered the Union?"[103] Calhoun had Mason ask rhetorically. The answer, for Calhoun, was the
persistent meddling from outsiders with the "equilibrium between"
the North and the South on the institution of slavery at the time of
the founding.[104] The movement to admit California into the Union

threatened to create "disequilibrium" in the Union. "If you admit her under all the difficulties that oppose her admission," Mason read, "you compel us to infer that you intend to exclude us from the whole of the acquired territories."[105] Calhoun argued that slavery was sanctioned in the original Constitution and therefore was indissoluble. The "agitation" over slavery portended disunion, and Calhoun wrote that the only protection against such disunion was to guarantee slavery in the acquired territories.[106] He went further to call for a constitutional amendment to restore to the South "in substance the power she possessed of protecting herself before the equilibrium between the sections was destroyed by the action of this government."[107] He ended by suggesting that it was entirely up to the North, not the South, to decide if and how the "great questions" could be settled.[108] He emphasized that it should all be done on the South's terms: "If you who represent the stronger portion" of the Union "cannot agree to settle" these questions "on the broad principle of justice and duty, say so; and let the states we represent agree to separate and part in peace."[109] But, he warned at the end, "If you are willing we should part in peace, tell us so, and we shall know what to do when you reduce the question to submission or resistance."[110]

When Calhoun died, several senators praised his legacy, while many others denounced it. Henry Clay rose to honor Calhoun as a long-serving Senate colleague. Calhoun and Clay had battled against each other more than once for the presidency, but they had been allies early in their careers in agreeing on the need for internal improvements to connect the states more tightly together in one Union. "He has gone!" Clay cried, and lamented further,

> No more shall we witness from yonder seat the flashes of that keen and penetrating eye of his, darting through this chamber. No more shall we be thrilled by that torrent of clear, concise, compact logic, poured out from his lips, which, if it did not always carry conviction to our judgment, commanded our great admiration.[111]

Clay cleverly used words beginning with hard *c* to subliminally reinforce a connection between adjective and subject. Webster praised Calhoun's "undoubted genius,"[112] while Democratic senator Thomas Hart Benton warned his colleagues, "He is not dead, sir—he is not dead. There may be no vitality in his body but there is in his doctrines. Calhoun died with treason in his heart and on his lips. Whilst I am discharging my duty here, his disciples are disseminating his poison all over my state."[113]

But, in the weeks immediately after Taylor's death, compromise was elusive, even though Fillmore was now in the White House. Fillmore was dull but a true Clay Whig, who was disposed to follow the lead of Congress. Clay's health continued to deteriorate, and on July 22, he rose for the last time to speak in his beloved Senate. Though physically drained, Clay held the floor for more than four hours. He did not begin with his usual dramatic flourish but instead peppered his comments with harsh asides directed at opposing senators. His tone was patronizing and angry because Southern senators had made no concessions to save the Union but instead offered fiery defenses of slavery and their right to maintain it and extend it into U.S. territories. Defending the omnibus bill that he had offered as the only option that could bridge the gulf between the contending sides, Clay said, "There is neither incongruity in the freight nor in the passengers on board of our 'Omnibus.' . . . We have no Africans or Abolitionists in *our* 'Omnibus,' no disunionists or Free Soilers, no Jew or Gentile. Our passengers consist of Democrats and Whigs," who had abandoned their customary antagonisms to confront the nation's crises like patriots.[114]

At one point, Clay condescendingly asked, "What is a compromise? It is a work of mutual concession—an agreement in which there are reciprocal stipulations."[115] Uncharacteristically, but desperate for compromise, Clay defended slave-owners as benign masters, not monsters, and depicted abolitionists as the genuinely bad actors "who would go into the temples of the holy God and drag from their sacred posts the ministers who are preaching his gospel for the comfort of mankind,"[116] and who, "if their power was equal to their malignity, would seize the sun of this great system of ours, drag it from the position in which it keeps in order the whole planetary bodies of the universe, and replunge the world in chaos and confusion to carry out their single idea."[117] Yet he also accused the extremists in the South of inviting war by arbitrarily interpreting the Constitution to suit their own interests. He pleaded: "All we want is the Constitution."[118] These extremists, too, had lost the ability to listen to the other side, he said, and their desire to carry slaves westward to the south of the Missouri Compromise line exposed the fundamental and hopeless contradiction at the heart of their position: "You cannot do it without an assumption of power upon the part of Congress to act upon the institution of slavery; and if they have the power in one way, they have the power to act upon it in the other way."[119]

Not even Clay knew that he was making the point that years later Lincoln would push Douglas to acknowledge in their great debates. In this debate, Clay asked (as Lincoln would almost a decade later), "What more can the South ask?" The compromise Clay was struggling to forge would have killed the Wilmot Proviso, strengthened the Fugitive Slave Law, and barred abolition in the District of Columbia. His omnibus bill remained, in his judgment, everyone else's best option; it "was the vehicle of the people, of the mass of the people."[120] Without the bill, Clay warned, there would be war not only in Texas and New Mexico but throughout the country, for "the end of war is never seen in the beginning of war."[121] Americans wanted no war, he declared. "The nation . . . pants for repose, and entreats you to give it peace and tranquility."[122] Reaching for one final metaphor, he declared, "I believe from the bottom of my soul, that the measure is the re-union of this Union. I believe it is the dove of peace, which, taking its aerial flight from the dome of the Capitol, carries the glad tidings of assured peace and restored harmony to all the remotest extremities of this distracted land."[123] He collapsed in his seat as the galleries erupted in tears and applause.

Nevertheless, the speech failed to garner a majority for any of the bills that Clay had introduced to save the Union. Dejected and sick, Clay left Washington for a desperately needed break.

At that point, Stephen Douglas, the Senate's youngest member, jumped into the center of the national stage. Clay had tried for half a year to move the Senate to compromise, but Douglas, chair of the powerful Senate Committee on Territories, needed only a few days to broker the deal that the Senate approved. It was not his speech that saved the compromise, though he had spoken for two days in March to assert his solution to the dilemma facing the country—popular sovereignty, letting a majority of white men in each territory and state decide whether to allow or bar slavery. Rather, it was his indefatigable work behind the scenes, tirelessly lobbying as the younger Clay had done, stitching together the different Senate coalitions that eventually supported each of the five bills that formed the historic Compromise of 1850. Clay, having briefly returned to Congress, worked by Douglas's side to help House leaders overcome opposition from Northern Whigs to approve the Compromise of 1850. With this success, the thirty-seven-year-old Douglas had established himself as a force to be reckoned with.

There were two essential steps left to put the Compromise of 1850 into effect. First, Fillmore signed each of the five bills into law—admitting California as a new state, abolishing the slave trade in the District of Columbia, organizing New Mexico and Utah as new territories, resolving the border dispute between Texas and New Mexico, and approving the Fugitive Slave Act of 1850, which required that all escaped slaves, upon capture, be returned to their masters with the cooperation or assistance of officials and citizens of the free states. Together, Fillmore declared, these bills provided "a final settlement of the dangerous and exciting subjects which they embraced."[124]

Now law, the compromise required a second element, no less important, to go into effect: enforcement. Fillmore, as well as Secretary of State Webster, promised to vigorously enforce each provision, including the Fugitive Slave Act. When nine Northern states enacted personal-liberty laws later in 1850 and in 1851, forbidding state officials from enforcing the Fugitive Slave Law or using their jails in fugitive slave cases, both Fillmore and Webster defended the law and denounced any such resistance as treacherous and no different from nullification and secession, both of which were unconstitutional. Fillmore even promised to send federal forces to Boston to put down riots against the law. In May 1851, the president and his Cabinet took the first train from New York City to the shores of Lake Erie to celebrate completion of the Erie Railroad. Along the way, Webster made side trips to defend the Fugitive Slave Law in Syracuse, Buffalo, and Albany. In Syracuse, he denounced the refusal to abide by federal law as treason.[125] In Buffalo, he used the metaphor of a "house divided" to stress the need to "preserve the Union of the States, not by coercion—not by military power—not by angry controversies . . . but by the silken chords of mutual, fraternal, patriotic affection."[126] He reminded his audience that the Constitution, through its supremacy clause, made federal law supreme over state resistance, and he promised "to exert any power I had to keep that country together."[127]

In June 1851, Webster went a step too far in his argument. In Virginia, he likened the Constitution to a compact, in which no party had the freedom to ignore a provision that the parties had originally consented to. "A bargain cannot be broken on one side, and still bind the other side," he said.[128] "I am as ready to fight and to fall for the constitutional rights of Virginia as I am for those of Massachusetts. . . . I would no more see a feather plucked unjustly from the honor of Virginia than I would see one so plucked from the honor of Massachusetts."[129] Naturally, slaveholders took his language to mean that Webster supported secession, inferring from the Constitution's barring any federal interference with slavery before 1808 as tacit recognition of its legality. Webster spent the remainder of his life trying to explain that he never intended that interpretation and that his wording, though infelicitous, was nonetheless consistent with his support of the Union.

The secession question was intertwined with the issue of the extent to which the Constitution permitted the federal government to regulate slavery or required protecting it. Back in Illinois, Lincoln followed the great debate in newspapers and correspondence. Lots of people, including Whigs like Lincoln, read Webster's comments as accepting the legitimacy of secession and confirming what Lincoln and many other Whigs had suspected for some time—that there could be no real compromise possible between those who demanded abolition and those who demanded the protection and expansion of slavery. As early as 1850, John Todd Stuart had told Lincoln that he predicted that soon all men would have to choose between abolitionism and the Democratic Party.[130] Stuart leaned in favor of joining the latter, while Lincoln agreed the choice had to be made but added, "in an Emphatic tone" that, "when that time comes my mind is made up. The Slavery question can't be compromised."[131]

VI

Henry Clay never returned to the Senate. Too ill to go home to Lexington, he died of tuberculosis in Washington on June 29, 1852. Rutgers College president Theodore Frelinghuysen, a former senator and Clay's running mate in 1844, delivered his eulogy in Washington. In Congress, members of each chamber rose to pay homage to Clay, and there was an outpouring of eulogies around the country.

Eight days after Clay's death, Abraham Lincoln delivered his. Eulogies are notoriously unreliable, for they tend to accentuate, if not overstate, the positive. (Clay's for Calhoun was a perfect example.) This was no doubt as true of the eulogies given by former colleagues and rivals of Clay as it was of Lincoln's. His eulogy was noteworthy because it was his highest-profile address since leaving Congress, and he had arranged, with editors he'd known in Illinois and met in Washington, for the eulogy to be printed in newspapers all over the country.

Today, people read Lincoln's eulogy of Clay for how it illuminates Lincoln's vision of himself. At the time he delivered and distributed it, his tribute reminded Whigs that he was still alive and well and one of them. Speaking at a podium at the front of the Hall of Representatives in the Illinois state capitol, he could not have asked for a more dramatic setting, and if anyone had previously missed Lincoln's persistent declarations of fealty to Clay, they could not miss them now.

Lincoln began, as Clay and classical funeral orations character- istically did, with an acknowledgment of the circumstances. Just a few days before, the nation had celebrated Independence Day, and Lincoln recognized that Clay's life and the life of the United States had nearly been identical; Clay was born one year after indepen- dence from Britain. As Lincoln remarked, "The infant nation, and the infant child began the race of life together. For three quarters of a century they have travelled hand in hand. They have been companions ever."[132] He reminded his audience of the crucible that the nation had been fused in then, perhaps not unlike those that forged Clay's character in public life. "The nation has passed its perils, and is free, prosperous, and powerful. The child has reached his manhood, his middle age, his old age, and is dead. In all that concerned the nation the man ever sympathized; and now the na- tion mourns for the man."[133]

Lincoln knew that the Whig Party, which Clay had built and Lincoln had long supported, was splintering under the stress of the Compromise of 1850. It had always been Clay's greatest aspi- ration to place the country's needs above his own and those of his party. Lincoln cleverly elaborated on that theme by quoting from "one of the public Journals, opposed to him politically."[134] What followed (the longest quote from another source in the eulogy) underscored Clay's patriotism.

> Ah, it is at times like these, that the petty distinctions of mere party disappear. We see only the great, the grand, the noble features of the departed statesman. . . . Henry Clay belonged to his country—to the world; mere party cannot claim men like him. His career has been national, his fame has filled the earth, his memory will endure to the last syllable of recorded time.[135]

Lincoln quoted the journal's description of the distinctive attri- butes of Clay's patriotism, noting that his "character and fame are national property."[136] This portion echoed the epitaph that Clay had written for his tombstone: "He knew no North, no South, no East, no West, but only the Union, which held them all in its sacred circle, so now his countrymen will know no grief, that is not as widespread as the bounds of the confederacy."[137]

Lincoln, drawing his words from the same source, recalled Clay's remarkable career of public service, trying to bring peace whenever and wherever he could, through deeds and words: "'His eloquence has not been surpassed. In the effective power to move the heart of man, Clay was without an equal.'" In the fights that threatened to rip the Union apart, he "'has quelled our civil commotions, by a power and influence, which belonged to no other statesman of his age and times.'"[138]

Lincoln then began to sketch in his own words the course of Clay's life. He remarked that "Mr. Clay's lack of a more perfect early education, however it may be regretted generally, teaches at least one profitable lesson: it teaches that in this country, one can scarcely be so poor, but that, if he will, he can acquire sufficient education to get through the world respectably."[139]

He proceeded to review Clay's career, beginning with his legal studies, his law practice, election to the Kentucky state legislature, election to the U.S. Senate, reelection to the Kentucky House of Representatives, selection as the speaker of the Kentucky House, service again for the remainder of an open term in the Senate, election to the U.S. House of Representatives, selection as speaker there, commissioner for negotiating an end to the war with Britain in 1819, reelection to the House, reselection as speaker, selection as secretary of state, his return to the practice of law, and reelection to the U.S. Senate more than once. Through that remarkable public career, Lincoln suggested, "there never has been a moment since 1824 till after 1848 when a very large portion of the American people did not cling to him with an enthusiastic hope and purpose of still elevating him to the Presidency. With other men, to be defeated, was to be forgotten; but to him, defeat was but a trifling incident."[140]

Lincoln found Clay as averse to quitting as Taylor (and of course

himself). "Even those of both political parties, who have been preferred to him for the highest office, have run far briefer courses
than he, and left him, still shining, high in the heavens of the political world. Jackson, Van Buren, Harrison, Polk, and Taylor, all rose
after, and set long before him."[141] (Lincoln did not bother including
Tyler or Fillmore, neither of whom had been elected; they were
accidents.) Clay, unlike any other man in Lincoln's estimation,
"was surpassingly eloquent; but many eloquent men fail utterly;
and they are not, as a class, generally successful. His judgment
was excellent; but many men of good judgment, live and die un-
noticed. His will was indomitable; but this quality often secures to
its owner nothing better than a character for useless obstinacy."[142]
Taken together, these qualities "are rarely combined in a single individual; and this is probably the reason why such men as Henry
Clay are so rare in the world."[143] Emulating Clay meant refining
several attributes, not just one; a great and inspiring leader like
him led not only through example, but through excellence in judgment, eloquence, and determination. Clay had shown Lincoln—
and now Lincoln was trying to show his fellow Whigs—that party
was only part of a man, that party came second, after allegiance to
the nation and its perpetuity. The Whig Party could fracture, but
the nation had to endure.

Clay's eloquence required further comment. Webster's excellence as an orator was based on his beautiful declarations, while
Calhoun's oratory was based on the remorseless logic of his arguments. But Clay's rhetoric, Lincoln suggested,

> *did not consist, as many fine specimens of eloquence do, of types
> and figures—of antithesis, and elegant arrangement of words and
> sentences; but rather of that deeply earnest and impassioned tone,
> and manner, which can proceed only from great sincerity and a
> thorough conviction, in the speaker of the justice and importance
> of his cause.*

In fact, as many had concluded, Clay's eloquence was a matter of
theatrics. Lincoln nonetheless suggested that Clay stood out as an

orator because "no one was so habitually careful to avoid all sectional ground. Whatever he did, he did for the whole country."[144] Even if this was not entirely true of Clay, Lincoln understood that a great orator seeks to cast his rhetoric on a higher plane for a higher purpose than mere partisan interest. In the case of Clay, Lincoln said that higher purpose was "a deep devotion," like Clay's, "to the cause of human liberty—a strong sympathy with the oppressed everywhere, and an ardent wish for their elevation."[145]

Where there was division, Clay relentlessly looked for unity. Lincoln surveyed Clay's uncanny knack at working out deals to avert disaster and achieve compromise. Lincoln brought up the decades-earlier controversy about the admission of Missouri into the Union as a slave state, which would have thrown off the equilibrium between proslavery and antislavery forces in Congress. He recalled Thomas Jefferson's remembrance that "'this momentous question, like a fire bell in the night, awakened, and filled me with terror. I considered it at once as the knell of the Union.'"[146] Just starting his storied career in Congress at that point, Clay forged the Missouri Compromise, which allowed for the admission of Maine (a free state) with Missouri (a slave state) at the same time. The deal also provided that, except for Missouri, slavery was to be excluded from the Louisiana Purchase lands above an imaginary line drawn by Congress at 36°30' north latitude.

Lincoln saved the most difficult subject—slavery—for the last several paragraphs of his eulogy. He knew that Clay had arranged for his own slaves to be gradually released after his death and that he was committed to gradual emancipation all of his life. Indeed, Lincoln said, Clay "did not perceive, that on a question of human right, the negroes were to be excepted from the human race."[147] Lincoln suggested that even though Clay owned slaves, he "did not perceive, as I think no wise man has perceived, how it could be at once eradicated, without producing a greater evil, even to the cause of human liberty itself."[148] Lincoln said Clay had the virtue of not being at either extreme in the slavery debate, while it was the extremists at both ends who raised the specter of disunion.[149] No one missed the obvious fact that, in 1852, Lincoln still thought of Clay's vision—an indissoluble Union and gradual emancipation—as his own.

Without naming his target, Lincoln took direct aim at the pro-slavery theologian Alexander Campbell, who had sneered at the "declaration that 'all men are created free and equal'" and dismissed it as not being in his Bible.[150] Lincoln identified that position with Calhoun and others who had contempt for "republican America. The like was not heard in the fresher days of the Republic,"[151] and he contrasted such hateful rhetoric "with the language of that truly national man, whose life and death we now commemorate and lament."[152] He quoted from a speech given in 1827 by Clay to the American Colonization Society, of which Clay had long been president. In it, Clay responded to the critics of the society who defended slavery and opposed their efforts to return enslaved African Americans to their native lands. Lincoln quoted Clay: "If they would repress all tendencies towards liberty, and ultimate emancipation, they must do more than put down the benevolent efforts of this society. They must go back to the era of our liberty and independence, and muzzle the cannon which thunders its annual joyous return. They must renew the slave trade with all its train of atrocities."[153] Worse, he said,

> they must blow out the moral lights around us, and extinguish that greatest torch of all which America presents to a benighted world— pointing the way to their rights, their liberties, and their happiness. And when they have achieved all those purposes their work will be yet incomplete. They must penetrate the human soul, and eradicate the light of reason, and the love of liberty. Then, and not till then, when universal darkness and despair prevail, can you perpetuate slavery, and repress all sympathy, and all humane, and benevolent efforts among free men, in behalf of the unhappy portion of our race doomed to bondage.[154]

Lincoln could not match Clay's eloquence in his final two paragraphs, so he did not try. Instead, he returned to Clay's hope for "a glorious consummation" when slavery could be abolished.[155] "And

if, to such a consummation, the efforts of Mr. Clay shall have con-
tributed, it will be what he most ardently wished, and none of his
labors will have been more valuable to his country and his kind."
Lincoln ended by reminding his audience—thanks to the pending
distribution of his speech, an audience well beyond that in front of
him—that the nation still was "prosperous and powerful" in part
because of Henry Clay.[156] "Such a man," he declared, "the times
have demanded, and such, in the providence of God was given us.
But he is gone. Let us strive to deserve, as far as mortals may, the
continued care of Divine Providence, trusting that, in future national emergencies, He will not fail to provide us the instruments
of safety and security."[157]

Lincoln concluded with repeated references to the divine, putting both Clay and himself on the side of the angels. Lincoln never
mentioned the enormous impact Clay had on him, at least not in so
many words, nor did he quote from Clay's more recent speeches,
particularly his last. Instead, he quoted most extensively Clay's earlier speeches, no less great than his later ones, to show the longevity and consistency of Clay's thoughts. Those, of course, had been
the speeches that Lincoln had studied and recited for years.

On October 24, 1852, four months after Clay passed away, Daniel Webster died. The "great triumvirate," Calhoun, Clay, and
Webster—the three men who had dominated national discourse
for decades—were no more. Though Lincoln admired (and often
modeled) Webster's oratory, he made no public eulogy for Webster
as he had done for Taylor and Clay. Lincoln knew Webster and
respected him, but he never felt the ideological kinship to him that
he had felt for Clay. Besides the fact that Webster was from Massachusetts and the other two from his home state, Webster had
been more equivocal over secession than either Taylor or Clay had
been. Clay and Taylor had always opposed secession, and while
Clay had supported a fugitive slave law, he did so in the spirit of
compromise, without the zeal with which Webster had defended
it. Webster's ambiguous rhetoric over secession and his over-the-
top support for the Fugitive Slave Act of 1850 confirmed Lincoln's
reluctance to honor him. Silence would be Lincoln's farewell.

VII

As the presidential election of 1852 approached, Lincoln was aware that the Whig Party was nearly defunct. Regional differences had weakened it, but the Compromise of 1850 finished the job once and for all. It divided the Whigs into proslavery and antislavery camps, with fault lines so deep they made Fillmore the first sitting president to fail to receive his party's support for another term. Instead, the party eventually agreed on its fifty-third ballot to nominate the old general Winfield Scott for president. Scott's long-standing opposition to slavery discouraged support anywhere in the South, while the Whig Party's discrepant decisions not to embrace the Compromise of 1850 and not to denounce slavery in its platform cost it support throughout the North. That left Scott only one option—to lose, which he did, to a lackluster former senator, Franklin Pierce of New Hampshire. Pierce likened himself to Jackson and Polk, though unlike either of them, he had a drinking problem so severe that his wife had forced him to leave the Senate to deal with it. It was the last time the Whig Party nominated a presidential candidate.

Besides his drinking problem, Pierce had other distinctions, none good. He had never sponsored any bill in his two terms in the House or his single term in the Senate. He had been a major in the Mexican War but was discharged early after he was injured when his horse fell on him. When informed her husband had won the Democratic Party nomination for president, Pierce's wife fainted. She was convinced that if he won the general election, the pressure of the presidency would lead him to start drinking again. Though Pierce called himself Young Hickory of the Granite Hills to invoke the notion that he was Andrew Jackson reincarnated, his most notable attribute was that he was a "doughface," a proslavery Northerner.

As Pierce took the oath of office, Lincoln was still practicing law in Springfield and traveling the circuit, but he watched with consternation as Stephen Douglas, Jefferson Davis, and now President

Pierce, yet another pretender to Jackson's legacy, pushed a scheme that brought the nation closer to civil war.

It began with a constitutional conundrum that Pierce largely made for himself and had to confront. Strict constructionists, like Pierce, who claimed to read powers-granting provisions of the Constitution narrowly, had argued that Congress lacked the power to restrict slavery in the territories, but as a candidate Pierce had promised to uphold the Compromise of 1850, which, in reauthorizing the original Missouri Compromise, had barred slavery from the federal territories it covered. Kansas and Nebraska had been included in the land that the French sold the United States in the Louisiana Purchase, and by 1854, farmers, ranchers, and prospectors were moving out west to seek their fortunes. The surge intensified the pressure for organizing territorial governments in Kansas and Nebraska to the extent that it became impossible for Pierce and Congress to ignore it. Abolitionists wished for the two areas to be free, while their opponents wanted to extend slavery into both, but the Missouri Compromise stood in their way, and although Pierce urged vigorous enforcement of the Fugitive Slave Law, he did not want to revisit the Missouri Compromise. He left the crafting of a solution to Congress, meaning the Senate Committee on Territories.

The committee's chair, Stephen Douglas, devised a plan to promote western expansion and facilitate construction of a transcontinental railroad. (As a lawyer for the rail lobby, he received sizable payment for his support.) Douglas figured the vast plains west of the Missouri could be broken into two new federal territories, Kansas and Nebraska. The most controversial part of the original plan was his proposal to organize the new territories in accordance with the principle of popular sovereignty, which he believed had the greatest promise of averting civil war.

In pushing his plan, Douglas initially tried to avoid saying anything about repealing the Missouri Compromise. He hoped to report out of his committee a bill giving settlers in Kansas and Nebraska the right to draft their own state constitutions at the time of statehood. Southern Whigs and Douglas's fellow Democrats told him, however, that they would not support the proposed bill unless it allowed the local residents of both Kansas and Nebraska to vote on the slavery question during the territorial years. This required repealing the Missouri Compromise, which had forbidden the extension of slavery west of Missouri.

Until then, neither Pierce nor his Cabinet knew about the negotiations over Douglas's bill. Douglas pushed a reluctant Pierce to meet with his Cabinet on Saturday, January 21, 1854, to discuss whether the Missouri Compromise should be repealed. With the backing of a majority of his Cabinet, Pierce agreed to bring the question to the Supreme Court, which he expected, given that a majority of its members had been appointed by Democratic presidents (including himself), to declare the Missouri Compromise unconstitutional on the ground that stripping slave-owners of their property (slaves) when they traveled through the territory violated the Fifth Amendment's ban on seizing property without due process of law. This way, Pierce could let the Court take the heat for dismantling the Missouri Compromise.

Later on January 21, Douglas consented to a repeal of the Missouri Compromise to maintain the support of Southern Whigs and Democrats for his planned Kansas-Nebraska bill. Because he had to present the amended bill on Monday or have the bill take a backseat to other pressing legislative business, he realized that he had to meet Pierce the next day, Sunday, January 22, 1854. But Pierce had vowed, after becoming president, not to do any business on Sunday. It was a decision prompted by tragedy. When Pierce, his wife, and their son, Bennie, were traveling to Washington for the inauguration, their train car derailed, and, as it fell off an embankment, Pierce and his wife watched with horror as their only son was crushed to death in front of them. Pierce's wife swore never to forgive him, and Pierce promised to attend church services and spend his time contemplating spiritual matters and doing penance on Sundays. To get a meeting then, Douglas had to find some way—or some person—to persuade the president to make an exception to his rule.

Early on Sunday morning, Douglas went to the one man he believed could persuade Pierce to do business on a Sunday, Pierce's

secretary of war, Jefferson Davis. Davis had become Pierce's closest and most trusted confidant. Douglas brought a group of Southern Democrats with him to Davis's residence, where they met with
Davis and stressed the need to move as quickly as possible on the
Kansas-Nebraska Act. They persuaded Davis to intercede with
Pierce and arrange a meeting with him for later that same day.
Democrats had the numbers in both the House and the Senate to
get the bill passed, but they needed Pierce's support to ensure that
it would become law. If Pierce were inclined to keep his promise to
abide by the Missouri Compromise, Douglas and Davis knew they
did not have the numbers to muster the two thirds necessary to
override his veto. Moreover, Democrats held only a bare majority
in the House, and they needed every vote to get the bill through.
Davis and the group then went to the White House, and Davis met
privately with Pierce to urge that he meet with them. After some
hesitation, Pierce agreed.

Douglas and Davis took charge of the meeting. They explained
to Pierce that the bill was consistent with the constitutionally protected rights of states and slaveholders and the ideal of popular
sovereignty. Pierce agreed to support their bill. Determined that
Pierce not change his mind, Douglas took the extraordinary step
of asking him to write out in his own hand the portion of the bill
repealing the Missouri Compromise. Pierce agreed.

The wheels in motion, Pierce did everything he could to get
the law approved in both chambers of Congress. The Democratic
majority was so large in the Senate that passage was a virtual certainty, but Pierce made support for the bill a test of loyalty in the
House, vowing to withhold patronage and other favors from any
Democrat who opposed it. When the dust settled, a coalition of
more than half the Northern Democrats and most Southerners
approved the bill 35–13 in the Senate and 113–100 in the House.
Eight days later, on May 30, 1854, Pierce signed it into law.

The Kansas-Nebraska Act instantaneously transformed the
constitutional landscape. By repealing the Missouri Compromise,
Pierce and Congress were rejecting one of Henry Clay's greatest
achievements, which had been predicated on the principle that had

been a fixture of American law since Monroe's administration—
that Congress had the authority to bar slavery in the territories.
The new law was based on the entirely different constitutional
principle of popular sovereignty. This new principle was largely
untested; it had worked in Nebraska, where the people overwhelmingly voted to oppose slavery, but Pierce, Douglas, Davis, and the
majority of each chamber in Congress were betting that popular
sovereignty would solve the fight over slavery in Kansas.

The bet was a bust of monumental proportions. Southern Democrats and the people of Nebraska were content to leave the slavery question to the people of Kansas, but the Democratic Party
suffered huge losses in the midterm elections. In the congressional elections, Democrats lost in every antislavery state except California and New Hampshire. Losing more than fifty seats in the
House, the Democrats went from having a solid majority in the
first two years of Pierce's administration to a minority in what
would be its last two years. While Democrats actually increased
their number of seats in the Senate, the gains were illusory: The
coalition that had brought Pierce to the White House was shattered, and Democrats who opposed slavery would flee the party.

Lincoln did not know all the machinations that had brought
the new law into being, but even from as far away as Illinois, he
knew the Kansas-Nebraska Act was a disaster for everything that
he, Henry Clay, and the Whigs had worked to attain over decades.
Just two years before, Lincoln's eulogy for Clay had placed the
Missouri Compromise that he had forged in 1820 at the center of
his legacy. Now, two years later, the Whig Party and the Missouri
Compromise lay in ruins. It was unclear what or who would take
their place. A new antislavery party, the Republican Party, had
been established in Ripon, Wisconsin, on March 20, 1854, but it
was too soon to tell whether it would have any more success than
the Free Soil and Know-Nothing parties.

Still thinking of himself as a Whig, Lincoln would not let the
destruction of one of Clay's greatest legacies go unaddressed. On
October 3, 1854, Douglas appeared in the House of Representatives hall at the Illinois state capitol to defend the Kansas-Nebraska

Act. Lincoln waited in the wings as Douglas spoke and shouted at Douglas once he finished that he wished to offer a rejoinder the next day on the same stage. Douglas reluctantly agreed.

Lincoln arrived well-prepared, reading his well-researched speech, dissecting Douglas's arguments one by one. Douglas sometimes interjected, but the interjections did not disrupt Lincoln's flow, establishing a pattern they would repeat during their 1858 contest for the Senate. On this occasion, as he would twelve days later in Peoria, Illinois, when Douglas repeated the same speech, Lincoln meticulously explained his objections to popular sovereignty and the Kansas-Nebraska Act, defended the Missouri Compromise, and explained his moral objections to slavery. As he did so, Lincoln did not mince words; the very opening of his speech reflected his awareness of the damage just done to a rightfully valued centerpiece of Clay's legacy.

He began, "The repeal of the Missouri Compromise, and the propriety of its restoration, constitute the subject of what I am about to say."[158] He laid out the foundations of the Missouri Compromise, which he said had been based on the Northwest Ordinance, a law enacted by the Continental Congress that prohibited slavery in the new territories that were eligible for admission into the Union as new states. Lincoln explained how the Northwest Ordinance had been fashioned in 1787 by Thomas Jefferson, "the author of the Declaration of Independence, and otherwise a chief actor in the revolution; then a delegate in Congress; afterwards twice President; who was, is, and perhaps will continue to be, the most distinguished politician in our history."[159] Seventeen when Jefferson died in 1826, Lincoln was drawn to Jefferson's political philosophy before becoming a fan of Clay. Indeed, Lincoln rarely thought of one without the other, Jefferson having crafted the vision that Clay and then Lincoln followed. "All honor to Jefferson," Lincoln declared later in 1859:

> a man who, in the concrete pressure of a struggle for national independence by a single people, had the coolness, forecast and capacity to introduce into a merely revolutionary document, an abstract

*truth, applicable to all men and all times, and so to embalm it there,
that today, and in all coming days, it shall be a rebuke and a stum-
bling block to the very harbingers of reappearing tyranny and oppression.*[160]

Back in Peoria in 1854, Lincoln did not need to tell his audience that Jefferson was the same man whose political vision
Henry Clay had always argued he was attempting to champion.
Lincoln said it was "with Jefferson" that "the policy of prohibiting
slavery in new territory originated"[161] and that the Missouri Compromise followed the example of the Northwest Ordinance, which
the Continental Congress adopted on July 13, 1787. "The Missouri
Compromise," he reminded his audience, and Douglas, "had been
in practical operation for about a quarter of a century, and had received the sanction and approbation of men of all parties in every
section of the Union."[162] Lincoln then recounted the history of that
landmark agreement, which had been fashioned by "Henry Clay,
as its prominent champion," and argued that popular sovereignty
could not supersede the Northwest Ordinance and the Missouri
Compromise.[163] Lincoln conveniently left out of his history that
the fugitive slave clause in the original Constitution had first appeared in the last article of the ordinance.

If the North and South were locked into positions on slavery
that they could not easily or naturally abandon, Lincoln wondered,
what national solution to the problem was feasible? He confessed
that he did not know the answer, even while maintaining that
owning slaves was immoral: "When southern people tell us that
they are no more responsible for the origin of slavery, than we; I
acknowledge the fact. When it is said that the institution exists;
and that it is very difficult to get rid of it, in any satisfactory way, I
can understand and appreciate the saying. I surely will not blame
them for not doing what I should not know how to do myself. If all
earthly power were given me," he confessed, "I should not know
what to do, as to the existing institution."[164] Lincoln acknowledged
further the problems with some proposed solutions: "My first impulse would be to free all the slaves, and send them to Liberia,—to

their own native land. But a moment's reflection would convince me, that whatever of high hope (as I think there is) there may be in this, in the long run, its sudden execution is impossible."[165] He explained that "if [the slaves] were all landed there in a day, they would all perish in the next ten days; and there are not surplus shipping and surplus money enough in the world to carry them in many times ten days."[166] Unable to go beyond what he thought the people of the state could accept, he asked, "What next?"

> Free them and make them politically and socially, our equals? My own feelings will not admit of this; and if mine would, we well know that those of the great mass of white people will not. Whether this feeling accords with justice and sound judgment is not the sole question, if indeed, it is any part of it. A universal feeling whether well or ill-founded, cannot be safely disregarded. We cannot, then, make them equals. It does seem to me that systems of gradual emancipation might be adopted; but for their tardiness in this, I will not undertake to judge our brethren of the south.[167]

Nevertheless, try as he might not to be critical of Southerners as a people (he had many close friends from the South, including Alexander Stephens), Lincoln had gently suggested that the likeliest solution to the problem was the one that Clay had championed—gradual emancipation—for which Lincoln had praised Clay at the end of his eulogy[168] but which conflicted with the desires of Southern leaders to protect slavery, not to do away with it.

Lincoln went further to dismiss the disingenuous arguments Douglas had made that conditions in Kansas and Nebraska were somehow not hospitable to the entry of slavery; it was a poison that Lincoln knew would threaten to consume these territories if given the chance, and Douglas had given them that chance. Lincoln acknowledged, "The doctrine of self-government is right—absolutely and eternally right—but," he added,

> it has no just application, as here attempted. Or perhaps I should rather say that whether it has such just application depends upon

whether a negro is not or is a man. If he is not a man, why in that
case, he who is a man may, as a matter of self-government, do just
as he pleases with him. But if he is a man, is it not to that extent, a
total of destruction of self-government, to say that he too shall not
govern himself? When the white man governs himself that is self-
government, but when he governs himself, and also governs another
man, that is more than self-government—that is despotism.[169]

Lincoln attacked the morality of slavery itself, just as Henry Clay had done and he had lauded Clay for doing. He argued that the slaves were people, not animals, and consequently possessed certain natural rights. "If the negro is a man, why then my ancient faith teaches me that 'all men are created equal'; and that there can be no moral right in connection with one man's making a slave of another."[170] He continued, "No man is good enough to govern another man, without that other's consent. I say this is the leading principle—the sheet anchor of American republicanism."[171]

It followed, then, that the extension of slavery into the territories and, prospectively, "to every other part of the wide world, where men can be found inclined to take it," was equally wrong.[172] So was Douglas's "declared indifference, but as I must think, covert real zeal for the spread of slavery."[173] Lincoln had arrived at the core of his political and constitutional belief, derived from the Declaration of Independence, that "all men are created equal."[174] He explained that the Founders understood that slavery was wrong.[175] For practical reasons they could not outlaw it altogether at the time of the founding, but instead they "hedged and hemmed it in to the narrowest limits of necessity."[176] Indeed, they did not allow the word *slavery* in the Constitution but permitted only indirect references to it, "just as an afflicted man hides away a wen or cancer, which he dares not cut out at once, lest he bleed to death; with the promise, nevertheless, that the cutting may begin at the end of a given time."[177]

Lincoln added, "The Missouri Compromise ought to be restored. For the sake of the Union, it ought to be restored."[178] Whether it be the Whigs or whatever party took their place, Lincoln was urging his fellow citizens "to elect a House of Representatives which will vote its restoration. If by any means, we omit to do this, what follows?"[179] Lincoln predicted it would not be just the possibility of slavery being approved in Kansas, but something worse—the loss of "the SPIRIT of COMPROMISE; for who after this will ever trust in a national compromise? The spirit of mutual concession—that spirit which first gave us the constitution, and which has thrice saved the Union" (by the Northwest Ordinance, the Missouri Compromise, and the Compromise of 1850), the spirit of Henry Clay, would be lost again.[180] Lincoln denounced slavery again for its immorality:

> *Slavery is founded in the selfishness of man's nature—opposition to it, is his Love of justice. These principles are in eternal antagonism; and when brought into collision so fiercely, as slavery extension brings them, shocks, and throes, and convulsions must ceaselessly follow. Repeal the Missouri compromise—repeal all compromises— repeal the declaration of independence—repeal all past history, you still can not repeal human nature. It still will be the abundance of man's heart, that slavery extension is wrong; and out of the abundance of his heart, his mouth will continue to speak.[181]*

Lincoln devoted much of the rest of the speech to exposing the fallacies of Douglas's claims that the principle of popular sovereignty developed from or was perfectly consistent with the principles of the earlier compromises over slavery. (Nearly every one of Clay's great speeches had a list of reasons for opposing any alternative to Clay's proposals.) Douglas claimed that the Compromise of 1850 established the precedent for the new law, but Lincoln argued that "the North consented to" allowing Utah and New Mexico to decide whether to come into the Union "with or without slavery as they shall then see fit,"

> *not because they considered it right in itself; but because they were compensated. . . . They, at the same time, got California into the Union as a free State. This was far the best part of all they had*

struggled for by the Wilmot Proviso. They also got the area of slavery somewhat narrowed in the settlement of the boundary of Texas. Also, they got the slave trade abolished in the District of Columbia. For all these desirable objects the North could afford to yield something, and they did yield to the South the Utah and New Mexico provision.[182]

The Kansas-Nebraska Act, in contrast, gave the North nothing in exchange for its support.

A major problem Lincoln had with the principle of popular sovereignty was his fervent belief that some things were not negotiable and should not be subject to majority rule or popular decision, including the morality of slavery, which a growing number of Americans believed was wrong. Another problem was that popular sovereignty "enables the first FEW, to deprive the succeeding MANY, of a free exercise of the right of self-government."[183] It was not just bad that Douglas and the Democrats were arguing that popular majority could decide who was human and who was property, but that once those popular majorities had their way, they could keep others subjugated.

Next, Lincoln considered "whether the repeal, with its avowed principle, is intrinsically right."[184] He likened the fight over the extension of slavery to a fight in which the South considered any compromise a defeat. Reverting to a useful metaphor, Lincoln explained, "It is as if two starving men had divided their only loaf; the one had hastily swallowed his half, and then grabbed the other half just as he was putting it to his mouth."[185] He further illustrated the injustice and absurdity of the South's position by pointing to how the South wanted both to allow jurisdictions to permit slavery but to bar jurisdictions from keeping slaveholders out. In this way, the South could never lose and the North could never win.[186]

At the end, Lincoln contested Douglas's claim that Clay and Webster were not in agreement with Lincoln's side of the argument:

They were great men; and men of great deeds. But where have I assailed them? For what is it, that their life-long enemy, shall now make profit, by assuming to defend them against me, their life-long friend? I go against the repeal of the Missouri compromise; did they ever go for it? They went for compromise of 1850; did I ever go against them? They were greatly devoted to the Union; to the small measure of my ability, was I ever less so? Clay and Webster were dead before this question arose; by what authority shall our Senator say they would espouse his side of it, if alive? Mr. Clay was the leading spirit in making the Missouri compromise; is it very credible that if now alive, he would take the lead in the breaking of it? The truth is that some support from whigs is now a necessity with [Douglas], and for thus it is, that the names of Clay and Webster are now invoked. His old friends have deserted him in such numbers as to leave too few to live by.[187]

Never before had Lincoln reached such rhetorical heights, and never before had he focused so clearly and forcefully on the immorality of slavery. He would return to these themes throughout the remainder of the decade, time and again defending and aligning himself with Clay's legacy of compromise and long-held belief that the framers did not design the Constitution to protect slavery but rather to allow the federal government to regulate, even abolish, slavery.

Douglas stayed for the entire Peoria speech and quickly charged onto the stage the moment it ended. In front of hundreds of people, Douglas challenged Lincoln to a debate. Lincoln surprised him by immediately asking Douglas to debate him in Peoria on October 16. After some hesitation, Douglas agreed to speak the same day as Lincoln but not at the same time; he spoke in the afternoon, before the evening, when Lincoln was scheduled to speak. In Peoria, Lincoln gave the same speech he had given in Springfield, but this time he wrote it out in full for publication over a week's issues in the *Illinois State Journal*.[188] After Peoria, he delivered the same speech in Urbana. The speeches reminded the voters of Illinois that Lincoln was a political force to be reckoned with. They belied any notion that he had retired from politics. Indeed, he had never left it.

CHAPTER FIVE

BECOMING PRESIDENT
(1856-1860)

There was no dramatic moment when Lincoln suddenly became the man who would be the mythic, beloved president he became. It is tempting to think there must have been some epiphany, such as when in the 1850s Lincoln awoke with a start to tell his friend Judge Dickey, "This nation cannot exist half slave and half free." But Lincoln was not a man of fits and starts. He was invariably cautious, deliberate in his actions, probing issues from every angle until he was content that he fully grasped them. The 1850s were no exception.

For much of the nation, the biggest and most sudden political surprise of the 1850s was Abraham Lincoln. The split between North and South had been coming for decades. Pierce's actions led to violence between abolitionist and proslavery forces in Kansas. This conflict known as "Bleeding Kansas" was an intermittent five-year guerrilla war begun when proslavery militias sacked the abolitionist town of Lawrence, burning down the Free State Hotel and destroying the presses of the two newspapers. (One of the raiders suffered the only known death, killed by a piece of the collapsing hotel.) The violent strife was a precursor of the bloodletting that would spread outside the state's borders and culminate in the Civil War.

Meanwhile, in the 1850s, a new generation of leaders fought their elders to take control of each of the major parties. Stephen Douglas's star continued to rise, but it was older Democrats, like James Buchanan of Pennsylvania (a man Clay dismissed as one of the "subordinates of Democracy" unworthy of his respect), Clay's former friend John Crittenden of Kentucky, and New Jersey's William Dayton, who vied for the Democratic Party's soul. As the Whig Party collapsed, William Seward, Salmon Chase, Edward Bates, and Orville Browning were among the prominent former Whigs intent on establishing a new party devoted to the abolition of slavery. Both of the major parties traced their origins to Thomas Jefferson, Democrats taking their name from half of Jefferson's Democratic-Republican Party, while part of the Whig diaspora claimed the other half.

Few gave a second thought to Lincoln. He was in Illinois, practicing law, while the man who outmaneuvered Douglas to win the presidency in 1856, James Buchanan, was hastening civil war. Buchanan had snatched the Democratic Party's nomination from Pierce, but like the man he'd defeated, he backed slave-owners' interests over those of abolitionists, placed federal power on the side of slavery, and blamed abolitionists for the nation's troubles. The spirit of compromise had died with Henry Clay.

In Washington, all eyes were on Douglas. Democratic leaders saw him as a lock on the 1860 presidential nomination as long as he won his reelection campaign in 1858. They saw no reason for him to be worried about reelection, but Douglas did. He was acutely aware—and well informed—that Lincoln was still a popular Whig leader in their home state and Lincoln was planning a run against him.

In declaring himself a "flat failure," Lincoln had lowered any expectations that he would ever be returning to the national stage.[1] Yet those closest to Lincoln—David Davis, Orville Browning, William Herndon, Leonard Swett, to name a few—knew that lowering expectations served Lincoln's political purposes. His self-deprecation reinforced his "humble" image as a "self-made man." This was, of course, fully in line with Clay's presentation of himself, a braid of himself and a lot of tactics. Drawing on one of Aesop's most famous fables, Lincoln was the tortoise in this race; the others, the hares, bounding ahead to the audition. Few saw Lincoln coming. Douglas did. They both knew that, in a nation that was tearing itself apart, only the man in the middle could mend it.

I

In the aftermath of the violent fallout from the Kansas-Nebraska Act and Pierce's strong backing of slavery interests in Kansas, Lincoln faced two immediate challenges. The first was to decide what, if any, role he expected to play in helping to establish a new political party to replace the Whigs. Lincoln's second challenge was figuring out his future in politics. Surmounting both challenges was crucial for Lincoln's newest ambition—a run for the Senate.

In making his decision, Lincoln was following the examples set much earlier by both Jackson and Clay. With the Whig Party dead, Lincoln had to find a new political home. "The man who is of neither party is not—cannot be, of any consequence," Lincoln said of Clay in his eulogy.[2] Jackson, too, showed that the path to higher office could be traveled only with a unified party behind a candidate with a clear constitutional vision that voters could rally around.

As Leonard Swett recalled, Lincoln "believed from the first, I think, that the agitation of Slavery would produce its overthrow, and he acted upon the result as though it was present from the beginning. His tactics were, to get himself in the right place and remain there still, until events would find him in that place."[3] John W. Bunn, a fellow Whig partisan and a merchant who funded Lincoln's campaigns, found that "Lincoln was a practical politician, but he was not altogether like many other practical politicians. He had his personal ambitions, but he never told any man his deeper plans and few, if any, knew his inner thoughts."[4] In the absence of such disclosures, Lincoln can best be judged on the basis of what he did; his actions revealed his plans. He recognized that he needed the support of a party system, so it was to the party that he knew best that he next turned: the remains of the Whig Party apparatus and the party faithful in Illinois.

Perhaps by design, Lincoln managed not to be in Springfield in October 1854, when the newly formed Republican Party held its convention there. Its platform urged an end to slavery in all federal territories and a repeal of the Fugitive Slave Act of 1850. The Republicans, including Browning, designated Lincoln as a member of the state central committee. Ever sensitive to keeping the different factions within the party as happy as possible, Lincoln pointedly neither accepted nor declined membership on the committee. He likely shaded the truth when he told a friend on the committee, "I have been perplexed some to understand why my name was placed on that committee. I was not consulted on the subject; nor was I apprized [sic] of the appointment, until I discovered it by accident two or three weeks afterwards."[5] Thinking about his run for the Senate the next year, Lincoln explained, "I supposed my opposition to the principle of slavery is as strong as that of any member of the Republican party; but I had also supposed that the extent to which I feel authorized to carry that opposition, practically, was not at all satisfactory to that party."[6] Lincoln still saw himself as closely aligned with the policies of Clay, as he made clear in his Peoria speech, where he had declared that he was not opposed to the elimination of slavery in all the territories but that he still accepted the Fugitive Slave Act (as long as it was narrowly enforced), even expressing sympathy for Southerners. While Democrats who opposed slavery were turning their backs on Pierce, Buchanan, and Douglas, they were still nominally Democrats, who were nonetheless interested in leaving the party for a more sympathetic base. Lincoln was trying to walk a narrow path that could win old-time Whigs, former Whigs who called themselves Republicans, and Democrats, as well as those who were aligned with the anti-immigration, anti-Catholic Know Nothing Party. He retook a seat in the Illinois House to remind the legislators that he was one of them but perhaps alienated some of his constituents when he resigned it after only twenty-two days to focus on joining Douglas in the Senate.

The new Illinois state legislature, assembled on January 1, 1855, would select the state's next senator. Lincoln told Herndon that he thought he had twenty-five members committed to his candidacy for the Senate. His difficulty was that he needed twenty-five more to get the majority needed from the assembly to secure the open seat, but he was unsure where he could find the requisite support. It would be difficult, given that almost half the general assembly were Democrats who likely supported the incumbent congressman James Shields, an old rival. Lincoln enlisted support from Judge David Davis and Stephen Logan, who had been elected to the Illinois House of Representatives; they estimated on the day of the election in the legislature that Lincoln was only three votes away from the majority he needed to win. What they did not count on was the support enjoyed by Lyman Trumbull, who had bolted the Democratic Party in opposition to the Kansas-Nebraska Act and was currently a member of the U.S. House of Representatives.

On the initial ballot, Lincoln had forty-five votes to Shields's forty-one; Trumbull had five votes, and one member went for Governor Joel Mattison. After six ballots, nothing changed, but on the seventh, the Democrats who supported Shields shifted their support to Mattison, and by the ninth ballot, Lincoln's support had dwindled to fifteen, Trumbull's had increased to thirty-five, and Mattison's forty-seven was only three away from the majority he needed for election to the Senate. Lincoln, still loyal to his Whig roots, made the snap judgment to ask his supporters to vote for Trumbull on the tenth ballot. They did, and Trumbull handily won the seat. Lincoln later confessed, perhaps with false modesty, that "a less good humored man than I, perhaps would not have consented to it."[7] Privately, Lincoln was angry and so, too, was David Davis, who made known that he distrusted Trumbull as "a Democrat all his life—dyed in the wool—as ultra as he could be."[8] Mary Todd was so outraged that she refused to speak to Trumbull's wife, who once had been a friend of hers.

Lincoln learned from this defeat, as he had learned from every one of the ups and downs in his career. If Clay, Jackson, and Taylor had anything in common besides their strong fidelity to the Union, it was the fact that they were always planning their next move. Lincoln had said as much in his eulogies for Taylor and Clay, but he had once again moved too slowly in corralling support. He had begun hustling for the Senate seat a year earlier, but it was not early enough. Taking this lesson to heart, Lincoln did what he had not done before: he began moving quickly and decisively to position himself well in advance to run for the state's other Senate seat in 1858. On the night after Trumbull won his seat, the Anti-Nebraska Democrats (those opposed to allowing slavery by popular sovereignty), who were gratified that Lincoln had made an appearance that same evening to show his support for his former foe, pledged in return to support him in the next Senate race. Lincoln was also able to get two other Anti-Nebraska Democrats, Norman Judd and John Palmer, both from Chicago, to pledge their future support. Each had previously bankrolled successful Whig candidates.

Thus backed, Lincoln focused on challenging Douglas's reelection bid in 1858. Lincoln had more than two years to plan, but he understood that victory required him to choose his party as soon as possible. Jackson first ran for the presidency as a Democratic-Republican in 1824, while Clay first ran as the candidate for the National Republicans in 1832. In 1833, a year after he had run against Jackson, Clay founded the Whig Party as a foil to Jackson and a base for future runs for the presidency. With the Whig Party in ruins, the Republican Party was the logical place for Lincoln to go, but in 1855 and 1856 it had not yet become the home for all the opponents of slavery.

Lincoln pondered his options, as revealed in a letter to his friend and fellow former Whig Joshua Speed of Kentucky. Speed told Lincoln of his strong opposition to slavery and his view that the Union ought to be dissolved if Kansas declared itself proslavery, and his hope—against all the evidence to the contrary—that Kansas might still find a way to vote itself a free state. Lincoln struck a pragmatic tone in response. He wrote that he thought of the Kansas-Nebraska Act "not as a law, but as violence from the beginning,"[9] a brazen effort to force the spread of slavery. He disagreed with Speed's view that the men enforcing the law were the problem; Lincoln believed that "the way it is being executed is quite as good as any of its antecedents. It is being executed in the precise way which was intended from the first; else why does no Nebraska man express astonishment or condemnation?"[10] Not yet ready to fully commit to the newly minted Republican Party, Lincoln answered Speed, who'd asked where he stood. "I think I am a whig; but others say there are no whigs, and that I am an abolitionist."[11] He reminded Speed, "When I was in Washington I voted for the Wilmot Proviso as good as forty times, and I never heard of anyone attempting to unwhig me for that. I now do more than oppose the extension of slavery."[12] In fact, Lincoln voted for the Wilmot Provision far fewer times, but his support for it was consistent. Next, he ruled out the Know Nothing Party. Indeed, he had hosted Fillmore when he came through Springfield in June 1854. Fillmore was then considering his third-party run for the Know Nothing Party's presidential nomination, but Lincoln had no interest in joining a party that did not stand on the same principles on which the Whig Party had been founded. As Lincoln explained to his friend Speed in 1855,

> I am not a Know-Nothing. That is certain. How could I be? How can any one who abhors the oppression of negroes, be in favor of degrading classes of white people? Our progress in degeneracy appears to me to be pretty rapid. As a nation, we begin by declaring that "all men are created equal." We now practically read it "all men are created equal, except negroes, and foreigners, and Catholics." When it comes to this I should prefer emigrating to some country where they make no pretense of loving liberty—to Russia, for instance, where despotism can be taken pure, and without the base alloy of hypocrisy.[13]

On February 22, 1856, Lincoln and Herndon secured invitations as two nonjournalists to attend the conference of Anti-Nebraska newspaper editors who were planning for the upcoming presidential election later that year. With Lincoln's input, the conference drafted a declaration that called for restoring the Missouri Compromise, upholding the constitutionality of the 1850 Fugitive

Slave Law, and promising noninterference with slavery in the states where it currently existed. It is little wonder that Lincoln would fit in so comfortably with the group; they were endorsing Clay's compromises of 1850. At the same time, the group endorsed Free Soil doctrine, recommending that freedom be guaranteed in federal districts and territories but not abolished by force in slave states. Free Soilers urged religious toleration and opposed restrictive changes in immigration laws. Though the conference avoided calling itself Republican, it was in all but name, and it planned at a statewide convention in Bloomington, Illinois, on May 29, 1856, to formalize the establishment of the Republican Party in Illinois. On the night of the final banquet for the program, the editors toasted Lincoln "as the warm and consistent friend of Illinois, and our next candidate for the U.S. Senate."[14]

Also, in May 1856, Herndon, who was a member of the Anti-Nebraska state committee at the editors' gathering, had published a call for a meeting of Sangamon County residents opposed to the Kansas-Nebraska Act so they could select delegates for the Bloomington convention. With Lincoln's knowledge, Herndon signed both Lincoln's name and his own as potential delegates. Reputedly, when the signatures became public, an irate John Todd Stuart stormed into Lincoln's law offices to ask whether Lincoln had actually signed the form. Lincoln was not around, but his partner Herndon, who was, said he had done it without consulting Lincoln, to which Stuart replied, "Then you have ruined him."[15] Stuart did not believe that a Republican, especially a radical one (and he thought all Republicans were radicals), could win a statewide election in Illinois.

Herndon and Lincoln thought otherwise, but the former had anticipated that Lincoln might have wanted to try to have it both ways—to be able to take advantage of aligning with the party growing in popularity but at the same time appear not to be taking a leading role or sanctioning its major activities. When he telegraphed Lincoln to alert him that he had signed him up as a delegate to the Republican convention, and that conservative Whigs like Stuart were not happy with his new radicalism (in 1856, Stuart campaigned for Fillmore in the hopes of thwarting "Black Republicans," those within the party who supported giving the vote to African Americans, which, he said, would "array the North against the South"), Lincoln responded briefly, "All right; go ahead. Will meet you—radicals and all."[16]

On May 29, 1856, Lincoln was one of around 270 delegates to attend the Bloomington convention, which formally recognized the formation of the Republican Party in Illinois. Whigs of all stripes were there, as were Anti-Nebraska Democrats and abolitionists. Lincoln's friend John Palmer, a former Democrat, was the presiding officer. Orville Browning was there, too, representing conservative Whigs. Browning's plan was to move the old Whig Party in the state, as he told Lyman Trumbull, "under the control of moderate men, and conservative influences, and if we do so the future destiny of the State is in our own hands—victory will inevitably crown our exertions. On the contrary if rash and ultra counsels prevail all is lost."[17] Recognizing upon his arrival that there had been few advance plans, Browning took the initiative to undertake one of the most important functions there: drafting a platform for the new party. This was a delicate task, the kind that required tact and coordination and finding consensus and common ground. It was easy to think Browning might not be well suited for the task, since his self-assurance and tendency to take himself too seriously might have rubbed more than a few people the wrong way. But this was a job for a good lawyer and astute politician, and Browning was both in spite of whatever other limitations he had. He understood this was his moment, and he shined. He called fifteen to twenty delegates to his room to get their input and find out what they could each accept and not accept. Then he prepared two platform resolutions that everyone, including Lincoln, could endorse: that the Constitution and the nation's institutions guaranteed that "we will proscribe no one, by legislation or otherwise, on account of religious opinions, or in consequence of place of birth,"[18] and that Congress

possesses the full power to prohibit slavery in the territories; and that while we maintain all constitutional rights of the south, we

also hold that justice, humanity, the principles of freedom as ex-
pressed in our Declaration of Independence, and our national con-
stitution, and the purity and perpetuity of our government, require
that the power should be exerted to prevent the extension of slavery
into territories heretofore free.[19]

On the afternoon of May 29, Browning made two speeches to
the convention, each emphasizing the values of compromise and
moderation. He was there to speak to the "old Clay-Whigs" and
he did. He told the delegates that they could choose no better role
model than Henry Clay. As Browning's biographer Maurice Baxter
describes the speech,

He read extracts from the speeches of Henry Clay from his first en-
trance upon public life to the close of his career, all of which proved
him to have been steadfastly and uniformly opposed to the spread of
slavery into free territory, and that he had still been upon the na-
tional stage when his great measures of pacification—the Missouri
Compromise—was ruthlessly violated, his voice and vote would
have been the same in 1854, that they were in 1820.

Baxter surmised that "Browning's emphasis of old-line Whig tra-
ditions [was] cogently and eloquently expressed and made a strong
impression on his colleagues."[20]

It was no accident that Browning looked to Clay as the model
for the new party. He knew—nearly everybody did—that Clay was
Lincoln's idol and the idol, in all likelihood, of most of the people
in attendance. Browning had met Clay in 1844, and as he noted
in his diary, "I was never more charmed with a man. So plain, so
unaffectedly kind, so dignified, so unaustatious [*sic*], so simple in
his manners and conversation, that he is irresistibly fascinating."[21]
This meeting apparently arose shortly before the 1844 presidential election, at which time Clay might well have been at his most
charming—and in need of support among old-line Whigs like
Browning, John Todd Stuart, and Lincoln. Ostensibly, as Browning recalled, no introduction was required—Clay knew who he

was and where he was from. As Baxter describes the meeting, "This evidence of personal interest reinforced Browning's belief that Clay had all the personal qualities with which he had been credited. He appeared younger than Browning had expected, as he was in good health, and above all he was not the least bit egotistical. Undoubtedly he ought to be President of the United States." Browning's perception of Clay is more consistent with what we know about Clay than Linder's description of Lincoln's meeting, when Clay was described as obnoxious and pretentious. Browning's view is also consistent with what we know about *him;* much more than Lincoln, he would have been receptive to such flattery and attention.

Browning was determined to revive the spirit of Clay for the convention. If "compromise" and "moderation" were the hallmarks of a Republican, as Browning told the delegates, Lincoln fit that bill. Indeed, Browning often assumed Lincoln was a conservative like himself, prone to upholding internal improvements funded by tariffs and restrictions on slavery but not to undertake more radical notions, such as complete abolition. Lincoln did nothing to dissuade Browning of that impression.

Though Lincoln had little role in organizing the convention, the delegates called him as the last major speaker. Everyone knew Lincoln had his eye on the Senate. He gave a short preview of the positions he expected to take in the upcoming election. As Judge John Scott, who witnessed the speech, recalled, Lincoln had "an expression on his face of intense emotion seldom if ever seen upon any before. It was the emotion of a great soul. Even in statu[re] he appeared great. A sudden stillness settled over the body of thoughtful men as Lincoln commenced to speak. Everyone wanted to hear what he had to say." Scott concluded that this was "the speech of his life in the estimation of many who heard it. [It] was a triumph that comes to few speakers. It was an effect that could be produced by the truest eloquence."[22] Such effusive praise was possible not only because of Lincoln's passionate delivery (while he was sometimes reserved in personal conversation, once onstage he came alive, particularly in front of a friendly audience) but also the content—

content that Lincoln, unlike that of his other speeches, did *not* want to reach a wide audience. This became the "lost speech," which Lincoln purposely had not written down but suppressed, because he and party leaders thought it too radical for publication and nationwide distribution. Lincoln had learned from Clay's mistakes of often speaking too much and inconsistently to appease a broad constituency, and he and his supporters were careful not to allow anything but moderation to define Lincoln.

What made the speech so touchy was that Lincoln identified slavery as the principal cause for the nation's problems, defended the idea of a union that opposed the extension of slavery, and closed by reiterating the declaration that Daniel Webster famously had given in 1830 in response to the South Carolina nullification movement, which maintained states had the authority to nullify or refuse to follow federal laws they did not like: "Liberty and Union, now and forever, one and inseparable."[23] This was the Webster whom Lincoln preferred to remember. In beginning his life as a Republican, Lincoln reminded those listening, and those later reading reports about his remarks in friendly newspapers, that while he might have been newly repackaged as a Republican, he remained, contrary to what Douglas said in 1854, faithful to the commitment of Clay and Webster to the Union.

II

While Lincoln was maneuvering for the Illinois Republican nomination for the Senate, Pierce's plans were falling apart. Kansas was in turmoil. Even before the Kansas-Nebraska Act had gone into effect, proslavery residents of Missouri had flooded into Kansas. They congregated mostly in the southern part of the territory, where, on March 30, 1855, they voted to select a proslavery legislature. That legislature quickly passed a statute outlawing antislavery activities. Meanwhile, antislavery residents had congregated in the northern portion of Kansas, where they formed a government of their own, which they called the Topeka Constitution. With the state divided into two halves and both sides beginning to engage in physical confrontations and violence, Pierce publicly sided with the proslavery faction and announced he would send federal forces into the territory if necessary to enforce their claims. He blamed Kansas's problems on the "inflammatory agitation" of outsiders, code for abolitionists.[24] He also declared that the Constitution protected the rights of slaveholders and agreed to station federal troops at Forts Leavenworth and Riley to serve as needed by the territorial governor.

Several violent outbreaks further grabbed national attention, one in the U.S. Senate and others in Kansas. On May 19–20, 1856, Massachusetts senator Charles Sumner delivered what he believed was the most important speech of his career. He was right, but for the wrong reason. He not only spoke of a widespread conspiracy involving the Pierce administration to force slavery on the prospective new state of Kansas, but he directed insults—including vulgar sexual imagery—against Douglas and particularly South Carolina senator Andrew Pickens Butler. Two days later, Congressman Preston Brooks, a cousin of Senator Brooks, marched into the Senate and confronted Sumner at his desk. Without warning, he repeatedly beat Sumner with his walking stick. Sumner was stunned and blinded by the first few blows, after which he wrenched his desk from its mooring and fell onto the Senate floor. Several senators watched him bleeding profusely and unconscious on the floor but did nothing, having been told to back off by Brooks's gun-waving accomplices, two other Southern representatives. By the time some senators stopped the beating, Sumner was nearly dead. Although reelected in 1857, the damage from the beating kept him out of the Senate until 1859. As for Brooks, an investigating committee recommended his expulsion from the House, but instead he was merely censured. Brooks resigned from the House but was reelected.

Back in Kansas, on May 21, 1856, proslavery forces ransacked the city of Lawrence, which had been settled largely by antislavery forces from Massachusetts. In response, a rabid abolitionist named John Brown led a closely knit band of followers who killed five pro-slavery settlers north of the Pottawatomie Creek during the night of May 24 and early morning of May 25. From there, Brown led his antislavery forces at two other battles in Kansas. They wreaked havoc wherever they could, culminating in an attack on a federal armory in Harper's Ferry, Virginia. Within thirty-six hours, a Marine force led by Robert E. Lee captured Brown. Brown was tried and convicted of treason and hanged on the spot. Democrats used Brown as the symbol of the radical abolitionism they said was destroying the country.

III

Bleeding Kansas was the central issue in the 1856 presidential election. On the Democratic side, Stephen Douglas made his first serious effort to grab the presidential nomination, as did three other prominent contenders, including Pierce, Lewis Cass of Michigan, and Polk's secretary of state, James Buchanan. At the Democratic nominating convention held in Cincinnati, Ohio, in 1856, Pierce struggled to keep his candidacy alive for as long as he could, but he withdrew his name from contention as his support fell precipitously in the late rounds of balloting. In being denied renomination, Pierce became the first elected president not to be renominated for a second term in office. Douglas's candidacy lasted until the seventeenth ballot, but he ultimately withdrew to avoid a contest with Buchanan that would have amplified internal tensions within the party and hurt its chances to keep the White House in the fall. Douglas's decision was made easier because, at forty-three, he expected to be a viable candidate in the next election. Buchanan, who had been out of the country during Pierce's administration, serving as the ambassador to Great Britain, had the advantage of not having been involved with any of the administration's decision making on the Kansas-Nebraska Act or the civil war in Kansas.

The Republicans held their first presidential nominating convention in Philadelphia, Pennsylvania. Republicans, in the judgment of Lincoln and Orville Browning, were looking for a candidate who was not conservative or at least not perceived as one, and both men supported John McLean, who had served Monroe and John Quincy Adams as the nation's first postmaster general and whom Andrew Jackson had appointed to the Supreme Court in 1829. (The American Party, which was merely the nativist Know Nothing Party under a different name, nominated Millard Fillmore, who still had great appeal to conservatives in both of the major parties.) Lincoln told Trumbull that McLean's nomination "would save every whig, except such as have already gone over hook and line" to the Democrats. He explained, "I am in, and shall go for any one nominated unless he be 'platformed' expressly, or impliedly, on some ground which I may think wrong."[25] Lincoln had never failed to back his party's presidential nominee, but for the first time, he kept the door open to withholding support from the nominee. The most important thing was to do nothing that could hurt his chances for the upcoming Senate election.

The nomination did not go to McLean, who was actively campaigning in spite of the fact that he was still sitting as a justice on the Supreme Court. Nor did the nomination go to either Seward or the newly elected governor of Ohio, Salmon Chase; as strident abolitionists, neither man held any appeal with the conservative wing of the Whig or Republican Parties. Instead, the nomination went to John Frémont, a former soldier and explorer. In 1848, he had been court-martialed for mutiny and insubordination in a dispute over who was the rightful governor of California, but Polk, in his last full year as president, commuted his sentence and reinstated him to his rank as major in the army so he could resign without disgrace. Frémont had led five expeditions into California, earning him the nickname the Pathfinder. He had also been U.S. senator from California and governor of the Arizona Territory. His appeal derived in part from his having been a Democrat until he resigned from the party in protest over the Kansas-Nebraska Act. A dashing figure himself, he had by his side his wife, Jessie, renowned as the beautiful, fiery daughter of Jacksonian Democrat Thomas Hart Benton. Drawing on the appeal of Frémont's views and persona, the party adopted as its slogan for the presidential campaign, "Free speech, free press, free soil, free men, Fremont and victory!"

Having chosen Frémont to lead the ticket, the delegates looked for a solid Whig as vice president. The early frontrunner was former New Jersey senator William Dayton, but when Illinois delegates complained that the party was overlooking the middle of the country, they arranged for John Allison of Pennsylvania to put Lincoln's name into consideration for vice president because he was "the prince of fellows, and an Old-Line Whig."[26] Illinois delegate William Archer seconded the nomination, saying that he had known Lincoln for thirty years and had always found him "as pure a patriot as ever lived."[27] Lincoln's friend John Palmer added his endorsement, saying, "We [in Illinois] can lick Buchanan any way, but I think we can do it a little easier if we have Lincoln on the ticket with John C. Fremont."[28]

By the time Lincoln was nominated for vice president, more than half the delegates were already committed to other candidates. In fact, Lincoln was not even present; he was riding the circuit. Dayton eventually won the contest with 253 votes, but the 110 votes cast for Lincoln placed him second in a crowded field of fifteen candidates. When David Davis told him the news that he had been nominated for vice president, Lincoln joked, "I reckon that ain't me; there's another great man in Massachusetts named Lincoln, and I reckon it's him."[29] Yet Lincoln felt the sting once again of beginning a campaign too late as he had done in prior congressional races, especially since this time, without any effort, he had finished second within his party for its nomination for the second-highest office in the land. With effort, he could improve vastly on that.

As the 1856 general election approached, Pierce tried to manage Bleeding Kansas in a way that would not hurt his party's nominee, Buchanan. This meant he continued to put all the force of the national government behind the proslavery forces and enforced their selection of a proslavery constitution for Kansas. In response, Republicans focused on only a single issue: that the Democrats and their attachment to slavery were fanning the flames of war. Democrats argued that Pierce was enforcing the principle of popular sovereignty only when it favored slavery and Buchanan would simply continue to maintain the same policy. When the dust settled, the arguments of both parties did not matter: The election split almost entirely between free states and slave states. Buchanan won seventeen states, including all but one of the slave states, while Frémont won only nine states and received no electoral votes in ten of the fourteen slave states. Buchanan won the electoral vote, 174–114.

In Illinois, Republicans did better than expected, winning four congressional districts, more than the Whig Party ever had in a single election. Lincoln and Browning were vindicated in their belief that a former Democrat had a better chance than a former Whig at winning an election against a Democrat when their candidate for the House, William Bissell, won.

Lincoln had grown experienced in counting votes for logrolling in the state legislature and later in nominating conventions and general elections. In 1856, the Democrats had won the presidential election again in Illinois, but the margin between Buchanan and Frémont was less than 4 percent in the popular vote. By Lincoln's estimate, the results in Illinois and the nation were equally promising for the Republicans: Buchanan had won in both because the opposition vote was split between Frémont and Fillmore. Fillmore received over 15 percent of the popular vote in Illinois, nearly four times the margin separating the two major candidates there.

After the election, in December 1856, Lincoln told a gathering of Chicago Republicans that the path to the White House was clear—if all the factions that opposed Buchanan could "let [their] past differences, as nothing be" and could agree that the proposition that "all men were created equal" was "the 'central idea' in our political public opinion," then the Republicans could win the next presidential election.[30] This might have sounded to many as obvious but hard to do, but for Lincoln, the essential thing that

winning required was maintaining party unity, a theme he had
been stressing since Clay lost in 1844. For that strategy to work,
Lincoln believed only a moderate could draw enough support to
win in 1860. But the political landscape could change a lot over the
next four years. And it did.

IV

As a lawyer, Lincoln well understood that Supreme Court appointments were among the most important legacies of a president.
Once on the Supreme Court, justices—appointed for life—decided
the great constitutional questions of the day and would continue
to do so for decades thereafter. They were beyond any political retaliation, and if one shared the same constitutional outlook as the
president who nominated him, he would entrench that view into
the fabric of American constitutional law for many years after that
president left office. The Constitution empowers every president
to nominate justices—and everyone in the Senate with the authority to confirm or reject them based on their merit or outlook.

When Lincoln was a child and until his early twenties, the man
who presided over the Court was John Marshall, who had studied
law with the same man whom Clay had studied with. Marshall
came to the Court appointed by the last Federalist president, John
Adams, and remained as its leader long after Adams's Federalist
Party died. In his nearly thirty-five years as chief justice, Marshall
wrote many of the Court's landmark opinions upholding the Federalist vision of a strong national government. Among these, as
Lincoln and Douglas knew, was the opinion in 1819 upholding the
national bank, a centerpiece of Clay's American System. Marshall
died in 1835 renowned as the Great Chief Justice for having maintained comity on the Court and having raised its stature among
the grand institutions of the federal government. To replace Marshall, Jackson appointed his closest political adviser and attorney

general, Roger Taney. Taney had written Jackson's famous veto against Clay's national bank in 1832.

Now, in a case that began in a small courtroom in St. Louis, the Court had to make the biggest decision it had ever confronted, and the nation knew and watched. At issue in the case, called *Dred Scott v. Sandford,* was the constitutionality of Clay's other singular achievement, the Missouri Compromise. Rumors swirled through Washington that, in preparing for his inauguration, Buchanan had written the two Pennsylvania justices on the Court—John Catron and Robert Grier—asking them to side with the slave-owners challenging the agreement Clay had forged in Congress. Taney assured Buchanan, both before his election and on the day of his inauguration, that he had the case well in hand and that Buchanan and the Democratic Party should expect a welcome outcome. The decision in *Dred Scott* came down on March 6, 1856, just two days after Buchanan was sworn in. The timing confirmed the widespread suspicion that the Court's majority was conspiring to take the heat off Buchanan by solving the slavery debate on its own.

At stake in the case was the fate of Dred Scott, a Missouri slave, who had been taken by his owner, army surgeon Dr. John Emerson, first to Rock Island, Illinois, whose constitution prohibited slavery and which, as Lincoln repeated in more than a few speeches, was covered by the Northwest Ordinance, which prohibited slavery. From Illinois, Emerson took Dred Scott to Fort Snelling, in the Minnesota Territory, where slavery had been barred by the Missouri Compromise. After returning to Missouri, Emerson died. Scott found a lawyer—former Jacksonian Democrat Montgomery Blair, the son of Francis Blair, who had helped Jackson found the Democratic Party and had run the newspaper that became Jackson's mouthpiece while he was in office. (The entire Blair family left the Democratic Party in protest over the Kansas-Nebraska Act.) Before the Supreme Court, Montgomery Blair argued that Scott had been made free as a result of having traveled through Illinois and the Missouri Territory, both of which had stripped him of whatever slave status he had and regarded him as a free man.

Though each of the nine justices wrote an opinion, Taney wrote the official opinion for the Court, joined by six others. First, the Court ruled that Scott was not entitled to sue, because only citizens of the United States could bring a lawsuit in federal court and as a slave (and descendant of Africans), he was not eligible to be a citizen of the United States. Taney declared that the framers never intended for Negroes to be citizens of the United States, that, at the time of the founding, enslaved Negroes were considered "so far inferior, that they had no rights which the white man was bound to respect," and the framers had not included them in either the Declaration of Independence or the Constitution.[31]

Second, the chief justice ruled that traveling through the Missouri Territory did not transform Scott into a free man because the Missouri Compromise was unconstitutional. On behalf of the majority, Taney explained that slaveholders had a constitutional right to own slaves under the Fifth Amendment of the Constitution, which protected their property from being deprived without "due process of law."[32] Taney said the law did just that, and so he undid another centerpiece of Clay's legacy.

If there was anything novel in what the Court did, it was the decision holding that the Constitution affirmatively protected slavery. But, in terms of reasoning, there was little new in the opinion insofar as Lincoln was concerned. He knew the arguments against the Missouri Compromise well; he had studied that law and followed the congressional debates on its constitutionality for more than two decades. He had also discussed it at length in his eulogy for Clay and again in his speeches in Springfield and Peoria in 1854. Well versed in the arguments that slave-owners and Democratic senators like John Calhoun and Jefferson Davis had been making for years about the rights of slave-owners to own slaves, Lincoln saw no reason to rehash these arguments now. Much as he had once joked to a friendly audience, with respect to the Kansas-Nebraska Act, he said that he "could not help feeling foolish in answering arguments which were no arguments at all."[33]

On the merits, Lincoln agreed with the two dissenting opinions. One was written by Justice John McLean, whom Lincoln and Browning had supported more than once for president. They knew McLean supported the Missouri Compromise, so it was no surprise that he dissented. Jackson never said much about the Missouri Compromise after it had been signed into law by President Monroe in 1820, but everyone presumed his silence meant he supported it. As a Jackson appointee, McLean was a living reminder of the complexity of Jackson's legacy.

The other dissent came from Benjamin Curtis, a Fillmore appointee. Lincoln agreed with his argument that slavery was wrong and violated the promise of the Declaration of Independence. Curtis had taken the dramatic step, after issuing his dissent, of resigning in protest of the decision, but he merely enabled Buchanan to replace him with Nathan Clifford, a virulently proslavery zealot whom the Senate approved 26–23, the closest Senate confirmation vote up until that time.

The *Dred Scott* decision placed the Court at the center of national attention. Lincoln, too, closely followed the case. As a lawyer, he was disposed to be respectful of the Court, even when he knew it was wrong. Indeed, he had previously defended the Court as the ultimate arbiter of disputes over slavery. During the 1856 presidential campaign between Buchanan and Frémont, Lincoln said, "The Supreme Court of the United States is the tribunal to decide such questions," and he pledged for the Republicans that "we will submit to its decisions; and if [the Democrats] do also, there will be an end to the matter."[34]

Lincoln did not publicly discuss the decision until Douglas gave him the opportunity. In June 1857, Douglas returned to Illinois to defend the "honest and conscientious" justices in the majority and to condemn any criticism of the Court as a "deadly blow at our whole republican system of government."[35] He heartily agreed with Taney's claim that African Americans belonged to "an inferior race, who, in all ages, and in every part of the globe . . . had shown themselves incapable of self-government."[36] It was not uncommon for people who defended slavery to claim it kept white women safe from African American men or for the people who owned slaves to rape and terrorize them, and Douglas, true to form, warned his constituents that Republicans favored the "amalgamation, between superior and inferior races."[37]

On June 26, 1857, Lincoln delivered a speech at the State House in Springfield, responding to Douglas's defense of the *Dred Scott* decision. He began, as expected, by declaring that the decision was "erroneous. We know the court that made it, has often over-ruled its own decisions, and we shall do what we can to have it over-rule this. We offer no resistance to it."[38] As he had in the past, Lincoln was expressing the classical respect for a Supreme Court decision that might have been wrongly decided; there was a right way to undo it and a wrong way. The wrong way was to disobey it. Clay would never have sanctioned flouting the rule of law, and Lincoln would not go that far either. Instead, Lincoln was reminding his audience that the Constitution provides legitimate ways to over-rule or undo erroneous decisions, including persuading the justices that they erred. Another, but not one mentioned in Lincoln's speech, was appointing justices who would move the Court in a different direction.

Lincoln knew that Douglas, once the beneficiary of the Illinois Democrats stacking the state supreme court to rule as they would like, was being hypocritical in his praise for the Court. Rather than being guilty of any such hypocrisy himself, Lincoln instructed his audience on how weak the opinion was and built the case for its overruling. In January 1856, Lincoln acknowledged that, if the Court had decided that Dred Scott was a slave, he "presumed, no one would controvert its correctness."[39] Now, in June, he went further: "If this important decision had been made by the unanimous concurrence of the judges, and without any apparent partisan bias, and in accordance with legal public expectation, and with the steady practice of the departments throughout our history," it would be "factious, nay, even revolutionary," not to accept it.[40] But that was not this case. When a majority of the justices—overruling numerous precedents and ignoring Justice Curtis's historical evidence showing many states had recognized African Americans as citizens—extended their ruling to an entire race, it was wrong. Thus, the decision lacked the attributes of a decision commanding respect from the other branches and the American people. Without saying so, Lincoln was questioning *this* Court's legitimacy.

Lincoln did not end his speech there. He was especially troubled by Chief Justice Taney's claim that neither the Declaration of Independence nor the Constitution was ever intended to include African Americans. Lincoln declared that Taney was doing "obvious violence to the plain unmistakable language of the Declaration," which used to be thought by Americans to include everyone.[41] In this, Lincoln agreed with Clay that the Declaration's great pronouncements were intertwined with the Constitution. But the Declaration, on Taney's reading, "is assailed, and sneered at, and construed, and hawked at, and torn, till, if its framers could rise from their graves, they could not at all recognize it."[42] Not only was Lincoln making it clearer that the decision lacked legitimacy, but he never again spoke of the Taney Court with any respect; indeed, he largely ignored it.

The legitimacy of the Court's decision, Lincoln argued, was eroded further by Taney's appearing to be in league with Douglas and the Democratic Party. For many Radical Republicans, this was the heart of the matter, though it took Lincoln time to get there. He charged that Taney and Douglas had allied to oppress Negroes and conspired to perpetuate and extend slavery. Lincoln argued that Douglas tried to make the oppression tolerable by suggesting Republicans wanted to have sex with black women. Lincoln rejected "that counterfeit logic which concludes that, because I do not want a black woman for a slave I must necessarily want her for a wife."[43] The authors of the Declaration of Independence never intended "to say all were equal in color, size, intellect, moral developments, or social capacity," but they "did consider all men created equal—equal in 'certain inalienable rights, among which are life, liberty, and the pursuit of happiness.'"[44]

In the end, this was not a radical speech, though many would charge that it was. Instead, it was Lincoln's return to the familiar ground that he shared with Clay (and the old-line Whigs he was wooing) that the Constitution must be understood in light of the promises of the Declaration of Independence.

V

Lincoln was determined in 1858 not to repeat the mistake he'd made too many times before, as when he'd last delayed entering the Senate race. Now he wasted no time. Two years before the Senate election, he hustled to consolidate support within the Republican nominating convention in Illinois. In August 1857, he began encouraging his fellow Republicans to do something "now," in letter after letter, to take control of the Illinois state Senate. It was imperative Republicans capture the Senate because it would be selecting whoever took Douglas's seat.[45]

Clay had never been subtle in his campaigns. Regardless of whether he was running for the House, Senate, or presidency, he let the world know his views loudly, if not clearly. It was not always a winning strategy for Clay; he peaked too soon in every presidential race he ran. Particularly with respect to securing his party's nomination for the Senate, it would be better, Lincoln thought, to consolidate support within his party while the opposition was preoccupied with smoothing the path for its nominee—this time, Douglas, of course—to win the general election. "Let all be so quiet that the adve[r]sary [Douglas] shall not be notified," Lincoln advised his supporters.[46]

It came as some surprise to Lincoln, however, that continuing violent instability in Kansas transformed the upcoming midterm election. Lincoln characterized what unfolded next as "the most exquisite farce ever enacted."[47] President Buchanan wanted to rush the admission of Kansas through Congress, but Free Soil voters in Kansas decided not to participate in the selection of a state Constitution because they thought it was rigged against them. As a result, only about 2,200 voters out of the registered 9,000 showed up at the state convention held in Lecompton, Kansas, which was charged with drafting the new constitution for Kansas. Unsurprisingly, the delegates drafted a proslavery constitution, which guaranteed not only that roughly two hundred slaves already in Kansas would remain in bondage but that their offspring would also be slaves. The new constitution further provided that it could not be amended for seven years. Then the delegates rejected Buchanan's advice that they should ratify the constitution they had just drafted. Instead, they arranged for a referendum on the question of whether more slaves could be introduced into the state.

Eager for the Kansas crisis to be settled, Buchanan preemptively kept pushing for the convention to ratify a new constitution. When the delegates produced the proslavery constitution he had been pushing, their Lecompton Constitution, Buchanan sent it to Washington for Congress's approval.

To nearly every Democrat's surprise, Douglas split with Buchanan and decided to oppose the Lecompton Constitution. Because it was increasingly likely he would face Lincoln in the Senate race set for the next year, he understood that Lincoln would press him mercilessly on the defects of the drafting of the document—particularly on the failure to follow the prescribed path for admission, which required a statewide referendum on the new constitution. Douglas was already losing public support over the Kansas-Nebraska Act, Bleeding Kansas, and the *Dred Scott* decision, each becoming an albatross he had to shake in order to keep his seat and begin his presidential campaign.

The Lecompton Constitution violated Douglas's principle of popular sovereignty because the inhabitants of the Kansas Territory had not actually exercised their right to choose their own form of government. It was a constitution foisted on the people of the state rather than one they had approved themselves. Douglas knew that—and publicly acknowledged as much. He vowed to lead the Senate fight against the Lecompton Constitution. He denounced it as "a flagrant violation of popular rights in Kansas" and a violation as well of "the fundamental principles of liberty upon which our institutions rest."[48] He stressed that he took issue not with the proslavery constitution itself but rather with the process by which it was adopted.

The announcement permanently estranged Douglas from Buchanan, who began to systematically remove all Illinois patronage from Douglas's control. However, much to Lincoln's chagrin,

Douglas's opposition to the Lecompton Constitution pleased Republicans nationwide, and Senator Trumbull went further to suggest that, despite the fact that Douglas was a Democrat, the Illinois Republicans nominate him as their candidate for the Senate. When Republicans from outside the state appeared to rally around the idea, Lincoln and his supporters pushed back hard. In the meantime, Democratic newspapers in Illinois proclaimed Lincoln unelectable and touted the possible candidacy of "Long John" Wentworth, a former congressman and editor of the *Chicago Democrat* newspaper, for the Senate.

Lincoln responded to Wentworth's threat as he had done to Hardin's efforts years before to derail his candidacy for the House. There were eighty-seven seats up in the state legislature, and Lincoln had to find a way to secure at least a majority of the Republicans. With Lincoln's approval, his supporters went to county conventions to secure support for his nomination to the U.S. Senate. With that in hand, they arranged, for only the second time in American history, to have a statewide convention held for the purpose of nominating a candidate for the Senate. Lincoln had not been moved to do so by any principle but rather, as he said, "more for the object of closing down upon this everlasting croaking about Wentworth, than anything else."[49]

When Republicans convened at the statehouse for their convention on June 16, 1858, the outcome was preordained. First, because Lincoln and other Republicans approved the platform Browning had drafted for the prior convention establishing the party, they asked him to draft an almost identical platform for this one. After nominating candidates for state treasurer and superintendent of education, they turned to the business of deciding on a nominee to challenge Douglas. The Lincoln forces quickly played their hand: state legislator Norman Judd and his Chicago delegation unfurled a banner declaring, COOK COUNTY IS FOR ABRAHAM LINCOLN. As the delegates applauded, a member from the Peoria delegation moved for the convention to adopt the motto, "Illinois Is for Abraham Lincoln." The resulting momentum crushed Wentworth's chances, and the convention moved unanimously to nominate Lincoln as "the first and only choice of the Republicans of Illinois for the United States Senate, as the successor of Stephen A. Douglas."[50]

Lincoln secured the nomination at five o'clock on June 16, 1858. The convention scheduled Lincoln's acceptance speech for three hours later. He had been working on his draft for some time, and now he read the draft to Swett, Herndon, Palmer, and other campaign advisers. They all agreed on one thing—the speech would end Lincoln's political career. Herndon thought it was powerful but was delivering the wrong message at the wrong time. Years later, Leonard Swett blamed the speech for Lincoln's defeat: "Nothing could have been more unfortunate or inappropriate; it was saying first the wrong thing, yet he saw it as an abstract truth, but standing by the speech ultimately would find him in the right place."[51] Herndon agreed that the speech proved, in spite of what he thought, to be helpful to Lincoln in the long run: "Through logic inductively seen, Lincoln as a statesman, political philosopher, announced an eternal truth—not only as broad as America, but covers the world."

Lincoln rejected the advice to modify his message or give a more palatable one. He told them, "The proposition [set forth in it] is indisputably true . . . and I will deliver it as written. I want to use some universally known figure, expressed in simple language as universally known, that it may strike home to the minds of men in order to rouse them to the peril of the times."[52] Lincoln had found the self-assurance to speak in his voice. He could not merely compromise for the sake of compromise. Even Clay had understood that there had to be inviolable values—the Union was one, and opposing the extension of slavery was another. Clay did not write his speeches to be read but to be delivered. Lincoln was writing his speech to be both delivered and read later in newspapers. His message had to be simple and direct so anyone reading in a newspaper or reciting it aloud could experience the power of his words. So confident was Lincoln, he invited reporters to cover the speech and spread its message wherever their papers were sold. It soon became known as the House Divided Speech.

Lincoln arrived at the Illinois State Capitol at eight o'clock on

June 16, 1858. It was a familiar venue where he had spoken dozens of times. He began with an homage to Daniel Webster. No one delivered more powerful opening lines than Webster, and Lincoln knew every word of Webster's famous reply to Senator Robert Hayne of South Carolina, the defender of nullification. Rising from his desk in the old Senate chamber, Webster had opened his rebuttal with an elegant metaphor:

> *When the mariner has been tossed for many days in thick weather, and on an unknown sea, he naturally avails himself of the first pause in the storm, the earliest glance of the sun, to take his latitude, and ascertain how far the elements have driven him from his true course. Let us imitate this prudence, and before we float farther, refer to the point from which we departed, that we may at least be able to conjecture where we are now.*[53]

Lincoln adapted that same opening for his midwestern audience, in an early but significant sign of his mastery of simplifying flowery declarations and complex arguments. He wasn't speaking to other senators but to the people of his state. He wasn't aiming to reach the educated elite but the farmers and laborers he'd spoken to for more than two decades in countless campaign rallies. Fancy images were inauthentic. If his message reached them, it could move them to move their representatives and party leaders in Lincoln's favor.

Lincoln's opening thus did not mince words: "If we could first know where we are, and whither we are tending, we could better judge what to do, and how to do it."[54] From there, he reminded his listeners that "we are now in the fifth year, since a policy was initiated, with the avowed object, and confident promise, of putting an end to slavery agitation."[55] However, under the Pierce and Buchanan policy of appeasing the slave power,

> *that agitation has not only, not ceased, but has consistently augmented. In my opinion, it will not cease, until a crisis shall have been reached, and passed. "A house divided against itself cannot*

stand." I believe this government cannot endure, permanently half
slave and half free. I do not expect the Union to be dissolved—I do
not expect the house to fall—but I do expect it will cease to be di-
vided. It will become all one thing, or all the other.[56]

It was common for Lincoln to try out his speeches and arguments
on different audiences and keep tinkering with his message until
the last moment. The metaphor of a house divided was not new, either for Lincoln or his audience. For those who had read Aesop's fables, it was familiar. For those who read the Bible, it was even more
so. As David Herbert Donald suggests in his biography of Lincoln,
the phrase "was familiar to virtually everybody in a Bible-reading,
churchgoing state like Illinois; it appeared in three of the Gospels."[57]
Lincoln had used the image as early as 1843 in urging party solidar-
ity among the Whigs, and abolitionist Wendell Phillips had used it
in his speeches condemning slavery in the 1840s and later. The idea
behind the metaphor as he now used it, that slavery and freedom
were incompatible, had been a standard part of the abolitionists' argument for decades. Webster, too, had used the phrase. As Donald
determined, "As early as 1855, after his first defeat for the Senate,
[Lincoln] raised the question with a Kentucky correspondent, 'Can
we, as a nation, continue together permanently—forever—half
slave and half free?'" During the Frémont campaign the next year,
he used the same phrase again, and in December 1857 Lincoln used
it in another speech he was drafting. The persistent use was clever
rhetorically, for it immediately made his audience feel smart and
identify Lincoln as one of them.

Having shared that powerful message at the outset, Lincoln then
launched into a broadside against the Democrats. Just as he had
done in his response to Douglas on *Dred Scott,* Lincoln returned
to his theme that the Democrats likely were conspiring as their
next step in protecting slavery to get a Supreme Court opinion that
extended the logic of the decision to bar any state from keeping
someone entering with his slaves. "But when we see," he told his
fellow Republicans, "a lot of framed timbers, different portions of
which we know have been gotten out at different times and places

and by different workmen—Stephen, Franklin, Roger, and James, for instance—and when we see these timbers joined together, and see they exactly make the frame of a house or mill, all these tenons and mortices exactly fitting, and . . . not a piece too many or too few," it was impossible not to think the four men had worked from the same plan.[58] With that, Lincoln accused Douglas, Pierce, Taney, and Buchanan of working in concert to strengthen and extend the slave power. He predicted, "Put that and that together, and we have another nice little niche, which we may, ere long, see filled with another Supreme Court decision, declaring that the Constitution of the United States does not permit a state to exclude slavery from its limits. [Such] a decision is all that slavery now lacks of being alike lawful in all the States."[59]

VI

By 1858, Lincoln and Stephen Douglas had been debating each other for more than two decades. They had done it before large crowds, as in Peoria, and smaller ones, as when Lincoln substituted for Stuart in 1838. They had debated in taverns and at conventions. Lincoln had watched Douglas debate John Todd Stuart throughout the district in 1838, and he followed Orville Browning's debates with Douglas in their 1843 contest for the House. In 1838, Stuart had bested Douglas, barely, while five years later Browning had lost, albeit by a respectable margin. In debates, Stuart matched Douglas's combative style with his own brand of sarcasm and taunting, while Browning, with considerably thinner skin than either Stuart or Lincoln, had reached a different agreement beforehand with Douglas "not to violate with each other the courtesies and proprieties of life; and not to permit any ardor or excitement of debate to betray us into coarse and unmanly personalities; [and] the compact was well and faithfully kept on both sides . . . Not one unkind word or discourteous act passed between us."[60]

Lincoln and Douglas had met so often on the battlefield of politics that they had developed a long-standing respect for each other. When Lincoln received the Republican nomination for senator, Douglas told a newspaper man, "I shall have my hands full. He is as honest as he is shrewd."[61] Despite the stinging rebukes Lincoln directed at Douglas in the aftermath of the Kansas-Nebraska Act and Bleeding Kansas, he openly acknowledged Douglas outshone him:

> Senator Douglas is of world-wide renown. All of the anxious politicians of his party, or who have been of his party for years past, have been looking upon him as certainly, at no distant day, to be the President of the United States. They have seen, in his jolly, fruitful face, post offices, land offices, marshalships and Cabinet appointments, chargeships and foreign missions, bursting and sprouting forth in wonderful exuberance, ready to be laid hold of by their greedy hands. On the contrary, nobody has ever expected me to be President. In my poor, lean, lank face nobody has ever seen that cabbages were sprouting out.[62]

Elevating Douglas while exuding humility about himself was perhaps tongue-in-cheek, but it was a persistent note Lincoln sounded, inspired by Clay's frequent use of the same device. Herndon agreed with Lincoln's assessment of Douglas:

> I always found Douglas at the bar to be a broad, fair, and liberal-minded man. Although not a thorough student of the law his large fund of good commonsense kept him in the front rank. He was equally generous and courteous, and he never stooped to gain a case. I know that Lincoln entertained the same view of him. It was only in politics that Douglas demonstrated any want of inflexibility and rectitude, and then only did Lincoln manifest a lack of faith in his morals.[63]

Lincoln considered Douglas "a very strong logician, that he had very little humor or imagination; but where he had right on his side very few could make a stronger argument; that he was an exceedingly good judge of human nature, knew the people of the state thoroughly and just how to appeal to their prejudices and was a very powerful opponent, both on and off the stump."⁶⁴ Joseph Gillespie recalled that Lincoln "always admitted that Douglass [*sic*] was a wonderfully great political leader and with a good cause to advocate he thought he would be invincible."⁶⁵

Yet Lincoln hesitated to meet Douglas in debate because he feared that Douglas would stack the crowds with his supporters. With encouragement from Browning, Herndon, Judd, Swett, and others in his kitchen Cabinet, Lincoln warmed to the idea when he realized that standing on the stage with Douglas would improve his own stature and that reporters from Republican-friendly papers would report favorably on his performance. Lincoln saw his advantage. He extended an invitation in writing to Douglas, who initially resisted because Lincoln would tower over him and the debate might raise Lincoln's profile, but the prospect of statewide and national coverage persuaded him to accept.

Even after Douglas accepted the invitation and the two men scheduled debates in seven sites around the state, Lincoln knew he had little chance of winning. With Douglas's help, Democrats in the state legislature had reapportioned seats to enable greater representation from the Southern, more Democratic districts. Douglas thus entered the contest in a strongly favorable position; he just had to keep the Democratic districts in line, while Lincoln had the nearly impossible task of not just carrying the legislators from his districts but somehow snatching a few from Democratic ones.

Lincoln meticulously examined the results of the past election district by district, which indicated where he should focus his energies. He gave only four speeches in the northern portion of Illinois, four in the south, and otherwise spent his time rallying voters in the central part of the state, where he hoped to generate significant support. He thought his best chance to keep the old Whigs in his camp was to remind them that he was a Henry Clay Whig, as past election patterns made it clear Clay played well in many of the districts he would be traveling through.

The problem was that Douglas was claiming the mantle of Henry Clay for himself. When Lincoln discovered that John Crittenden, the powerful Kentucky senator who had been Clay's protégé, had followed Clay in joining the National Republican Party in 1828, and was openly encouraging people to support Douglas, Lincoln wrote him. Lincoln had worked closely in the 1848 presidential campaign with Crittenden, whom he thought was as devout a party man as himself and Clay. Crittenden responded that in fact the report was true and that he felt a Douglas reelection was "necessary as a rebuke to the Administration, and a vindication of the great cause of popular rights and public justice."[66] Crittenden was no friend to him or to Clay's memory.

Because of the reapportionment in the state, Douglas had everything to lose in the election. On the one hand, if defeated, Lincoln would be in no worse position than he was now—out of office trying to find his way back in, but likely to have enhanced his reputation and gained national stature by going toe-to-toe with such a daunting and powerful opponent. Douglas was fully aware he could not let his guard down; no one else had more experience debating Douglas than Lincoln.[67] Lincoln never let his guard down, either. Whatever his respect for Douglas may have been, he distrusted his opponent, whom Lincoln regarded as dangerous, unprincipled, and underhanded.

In the seven debates, the two men covered many familiar subjects, ranging from popular sovereignty to slavery to Douglas's charge that Lincoln and his party were dangerous radicals. They were fighting for the middle, and that meant they were fighting over who had claim to Clay's legacy. Douglas claimed that he was Clay's true heir, since Clay and Douglas had both declared themselves devotees of Jefferson, and Douglas had worked with Clay to fashion the Compromise of 1850. Lincoln scoffed at the idea, and the outsize influence of Clay, particularly in Lincoln's thinking, was central in every debate. At the first debate, in Ottawa, Illinois, Lincoln quoted Clay more than forty times, more than he cited any founder or prior president. Together, the two combatants quoted and mentioned Clay nearly a hundred times. These numbers do not include the many times Lincoln was talking about an idea or concept he learned from Clay. Six years after Lincoln delivered his eulogy of Clay, championing his legacy was crucial to the future of his campaign and the country. Proclaimed Lincoln,

> *Henry Clay, my beau ideal of a statesman, the man for whom I fought all my humble life. [Clay] once said of a class of men who would repress all tendencies to liberty and ultimate emancipation, that they must, if they would do this, go back to the era of our Independence, and muzzle the cannon which thunders its annual joyous return; they must blow out the moral lights around us; they must penetrate the human soul, and eradicate there the love of liberty; and then, and not till then, could they perpetuate slavery in this country!*[68]

Clay had been one of the first party standard bearers to insist that following the Constitution meant keeping faith with the Declaration of Independence, a position he regarded as the conservative one, derived from Thomas Jefferson, designed to keep the Union intact, and to preserve the ideals of both texts. (George Prentice rammed this point home in his 1830 campaign biography of Clay.) In repeatedly re-sounding that same theme, Lincoln hoped to cast Douglas and other Democratic leaders such as Taney and Buchanan as the radicals bent on destroying the promises and ideals of the nation's founding documents. Rather than intend to protect slavery, the framers had drafted a Constitution that gave the federal government the power to regulate it, Lincoln argued. It was Lincoln and the Republican Party that took the conservative position to follow the original scheme of the Constitution and preserve its original ideals and promises; it was Douglas and his cohorts who were willing to destroy the country's founding commitments for the sake of appeasing slave-owners.

In the second debate, held in Freeport, Illinois, Lincoln said,

> *Yet as a member of Congress, I should not with my present views, be in favor of endeavoring to abolish slavery in the District of Co-*

lumbia, unless it would be upon these conditions: First, that the abolition should be gradual. Washington, Jefferson, and Clay had all endorsed the principle of gradual emancipation. Second, that it should be on a vote of the majority of qualified voters in the District; and third, that compensation should be made to unwilling owners.[69]

These conditions were familiar to many people in the crowd who remembered Clay, who had pushed for them in 1850.

Douglas answered that "Clay was dead, and although the sod was not yet green on his grave, this man [Lincoln] undertook to bring into disrepute those great Compromise measures of 1850, with which Clay and Webster were identified."[70] Indeed, Douglas pointed out,

up to 1854 the old Whig party and the Democratic party had stood on a common platform so far as this slavery question was concerned. You Whigs and we Democrats differed about the bank, the tariff, distribution, the specie circular, and the sub-treasury, but we agreed on this slavery question and the true mode of preserving the peace and harmony of the Union. The Compromise measures of 1850 were introduced by Clay, were defended by Webster, and supported by Cass, and were approved by Fillmore, and sanctioned by the National men of both parties.[71]

No one had to remind listeners that Lincoln was back in Springfield practicing law when this was happening. Everyone knew it was Douglas, not Lincoln, who had made the compromise a reality. "Thus," Douglas said,

they constituted a common plank upon which Whigs and Democrats stood. In 1852 the Whig party, in its last National Convention at Baltimore, indorsed and approved these measures of Clay, and so did the National Convention of the Democratic party, held that same year. Thus, the old line Whigs and the old line Democrats stood pledged to the great principle of self-government, which

guarantees to the people of each Territory the right to decide the
slavery question for themselves.

Now Douglas twisted the knife:

In 1854, after the death of Clay and Webster, Mr. Lincoln, on the
part of the Whigs, undertook to Abolitionize the Whig party, by
dissolving it, transferring the members into the Abolition camp
and making their train under Giddings, Fred Douglass [who had
escaped slavery to become one of the nation's most eloquent and out-
spoken advocates for eradicating slavery], Lovejoy, Chase, [Hiram]
Farnsworth [of Kansas], and other Abolition leaders.[72]

This back-and-forth did not impede Lincoln's strategy in the second debate. Following the advice of *Chicago Tribune* managing editor Joseph Medill, Lincoln pressed Douglas to answer four loaded questions: First, would Douglas favor the admission of Kansas before it had the requisite number of inhabitants as specified in the law controlling the admission of new states? Second, could the people of a territory, such as Kansas, "exclude slavery from its limits prior to the formation of a state Constitution?"[73] Third, would Douglas follow a Supreme Court decision declaring that the states could not exclude slavery from their limits? Fourth, did Douglas favor acquisition of new territory "in disregard of how such acquisition may affect the nation on the slavery question?"[74] Douglas dodged most of Lincoln's questions, but it was the second one that Lincoln and his cohort thought was key, and Douglas took the bait. He answered that the passage of "unfriendly legislation" could keep slavery out of any federal territory because "slavery cannot exist a day or an hour anywhere, unless it is supported by local police regulations."[75] He added that "the people of a Territory had the lawful power to exclude slavery, prior to the formation of a [State] Constitution."[76] This was a full embrace of his principle of popular sovereignty, and Lincoln and the Republicans would use it relentlessly as a wedge to divide Douglas from Buchanan and other proslavery Democrats.

Lincoln also jumped on Douglas's defense of *Dred Scott*. His first rejoinder came in response to Douglas's assertion that he was prepared not to comply with the *Dred Scott* decision. Lincoln pressed Douglas to take a position on the issue that *Dred Scott* contradicted Douglas's principle of popular sovereignty. After ruling in *Dred Scott* that the federal government could not bar slavery from the territories, the next likely step for the Court would be to forbid states from outlawing slavery, presenting Douglas with the dilemma of choosing either the decision *or* his principle of popular sovereignty. Something had to give. Forcing Douglas into a corner, Lincoln argued that "there is nothing that can divert or turn him away from this decision. It is nothing to him that Jefferson did not so believe. I have said that I have often heard him approve of Jackson's course in disregarding the decision of the Supreme Court pronouncing a national bank constitutional. He says, I did not hear him say so . . . though it still seems to me that I heard him say it twenty times."[77]

Lincoln cast Douglas as lacking any principled fidelity to the Supreme Court or the law. He reminded those in attendance that Douglas had endorsed the Democratic Party's 1856 platform that opposed Jackson's stance on the national bank. Because the Supreme Court had upheld the constitutionality of the national bank, Jackson's opposition placed him at odds with the Court. But then Lincoln called attention to the 1840s when Douglas took the lead in Illinois to overturn the "decision of the Supreme Court of Illinois, because [Democrats] had decided that a Governor could not remove a Secretary of State."[78] Lincoln pointedly added, "I know that Judge Douglas will not deny that he was then in favor of overslaughing that decision by adding five new Judges, so as to vote down the four old ones. Not only so, but it ended in the Judge's sitting down on that very bench as one of the five new Judges to break down the four old ones."[79]

Lincoln returned to this theme in the fourth debate, at Charleston, Illinois. He pressed his audience not to forget "the fact that [Douglas] was one of the most active instruments at one time in breaking down the Supreme Court of the state of Illinois, because it had made a decision distasteful to him—a struggle ending in the remarkable circumstance of his sitting down as one of the new Judges who were to overslaugh that decision—getting his title of Judge in that very way."[80] Douglas ignored the taunt. Instead, he reiterated his position that Supreme Court decisions are final, and his duty was to follow them, regardless of whether he agreed with them or not.

In the fifth debate, held in Galesburg, Lincoln reminded his audience of the ideals Clay had fought for all of his life, saying, "I can express all my views on the slavery question by quotations from Henry Clay." He proceeded to do that, once again recalling

> that Mr. Clay, when he was once answering an objection to the Colonization Society, that it had a tendency to the ultimate emancipation of the slaves, said that "those who would repress all tendencies to liberty and ultimate emancipation must do more than put down the benevolent efforts of the Colonization Society—they must go back to the era of our liberty and independence and, so far as in him lies, muzzling the cannon that thunders its annual joyous return—they must blot out the moral lights around us—they must penetrate the human soul, and eradicate the light of reason and the love of liberty.[81]

Lincoln was repeating portions of his eulogy but to good effect with the old Clay Whigs in his audience.

In the sixth debate, held in Browning's hometown of Quincy (though Browning's law practice had taken him out of town), Lincoln again pressed the point that he, not Douglas, was the true heir of Clay. "I wished to show," he said, "but I will pass it upon this occasion, that in the sentiment I have occasionally advanced upon the Declaration of Independence, I am entirely borne out by the sentiments of our old Whig leader, Henry Clay, and I have the book to show it from; but because I have already occupied more time than I needed to do on the topic, I pass over it."[82] Lincoln held the book of Clay speeches in his hand, his personal bible. In responding to Douglas's opposition to thinking of the Declaration of Independence as setting forth the promises and ideals that the Constitution was designed to implement, Lincoln said,

> *The Judge has taken great exception to my adopting the heretical statement in the Declaration of Independence, that "all men are created equal," . . . [but] I have only uttered the sentiments that Henry Clay used to hold. Allow me to occupy your time a moment with what he said. Mr. Clay was at one time called upon in Indiana, and in a way that I suppose was very insulting, to liberate his slaves, and he made a written reply to that application, and one portion of it in these words,*

which Lincoln then quoted,

> *What is the foundation of this appeal to me in Indiana, to liberate the slaves under my care in Kentucky? It is a general declaration in the act announcing to the world the independence of the thirteen American colonies, that "men are created equal." Now, as an abstract principle, there is no doubt of the truth of that declaration, and it is desirable in the original construction of society, and, in organized societies, to keep it in view as a great fundamental principle.*[83]

Douglas responded with a lengthy protest that he had been consistently attached to both Clay and his principle of popular sovereignty. In 1850, he "was supported by Clay, Webster, Cass, and the great men of that day" when he included within the Compromise of 1850 provisions that allowed for the entry of California into the Union as a free state but created the new territories of Kansas and Nebraska, which would decide for themselves whether to be slave or free. Thus, he said, he held on to the same principles "in 1854, and in 1856, when Mr. Buchanan was elected President." His audience no doubt thinking of the eventual break between Buchanan and Douglas over the Lecompton Constitution, Douglas continued, "It goes on to prove and succeeds in proving, from my speeches in Congress on Clay's Compromise measures, that I held the same doctrines at that time that I do now, and then proves that by the Kansas and Nebraska bill I advanced the same doctrine that I now advance."[84]

Held before a largely Republican audience, the final debate featured yet another struggle for each man to show he was more faithful to Clay's legacy. Douglas, Lincoln said, now "brings forward part of a speech from Henry Clay—the part of the speech of Henry Clay which I used to bring forward to prove" that Clay's Whigs were not radical abolitionists. The audience laughed. Lincoln paused, and then said, "I am somewhat acquainted with Old Line Whigs. I was with the old line Whigs from the origin to the end of that party; I became pretty well acquainted with them, and I know they always had some sense, whatever else you could ascribe to them." Again, the audience howled in laughter. Lincoln then read an even broader excerpt from Clay's speech on the Declaration and slavery than he had at previous debates. "'That declaration, whatever may be the extent of its import, was made by the delegations of the thirteen states,'" Lincoln quoted, continuing,

> In most of them slavery existed, and had long existed, and was established by law. It was introduced and forced upon the colonies by the paramount law of England. Do you believe, that in making that declaration the States concurred in it intended that it should be tortured into a virtual emancipation of all the slaves within their respective limits? Would Virginia and other Southern states have ever united in a declaration which was to be interpreted into an abolition of slavery among them?[85]

Lincoln's objective was to cast Clay—and thereby himself—as no radical on abolition. Rebutting any claim that either he or Clay was radical, he quoted Clay further:

> I desire no concealment of my opinions in regard to the institution of slavery. I look upon it as a great evil, and deeply lament that we have derived it from the parental Government, and from our ancestors. I wish every slave in the United States was in the country

of his ancestors. But here they are; the question is, how can they be best dealt with? If a state of nature existed, and we were about to lay the foundations of our society, no man would be more strongly opposed than I should be, to incorporating the institution of slavery among its elements.[86]

It could be no surprise to Lincoln that later Frederick Douglass, who had escaped slavery to become a renowned abolitionist, proclaimed Lincoln completely complicit with the slave-owners. Friends like Giddings and Wilmot also looked away. But their extremism would not win Lincoln this election—or the next he had his eyes on. Only moderation could.

Yet, as Lincoln saw it, moderation did not mean complicity. He reminded the audience of "the real issue in this controversy," the conflict "on the part of one class that looks upon the institution of slavery as a wrong, and of another class that does not."[87] He concluded,

That is the issue that will continue in this country when these poor tongues of Judge Douglas and myself shall be silent. It is the eternal struggle between these two principles—right and wrong— throughout the world. They are the two principles that have stood face to face from the beginning of time; and will ever continue to struggle. The one is the common right of humanity and the other the divine right of kings.[88]

The unmistakable inference was that the framers' legacy ran straight through Jefferson to Clay and to Lincoln himself. And in referencing the distinction between "the common right" of people and that of kings, Lincoln was returning to a notion that Clay had made so many times that it earned him another nickname, the Great Commoner. Lincoln began with Clay and ended with him.

Douglas held nothing back. He attacked Lincoln's assertion that he was an old-line Whig. "He was not," Douglas flatly declared. "Bear in mind that there are a great many old Clay Whigs down in this region. It is more agreeable, therefore, for him to talk about

the old Clay party than it is for him to talk Abolitionism."[89] Douglas pointed out that Lincoln said nothing about being an old-line
Whig when he was campaigning in Democratic districts. Douglas
gleefully asked listeners if they had read a speech from General
James Singleton, who was widely known as a friend of Clay's. The
audience said, "Yes, yes," and cheered. "You know," Douglas said,
"that General Singleton was for twenty-five years the confidential friend of Henry Clay in Illinois, and he testified that in 1847,
when the Constitutional Convention of this State was in session,
the Whig members were invited to a Whig caucus at the house of
Mr. Lincoln's brother-in-law, where Mr. Lincoln proposed to throw
Henry Clay overboard and take up General Taylor in his place, giving as his reason, that if the Whigs did not take up General Taylor
the Democrats would." The crowd cheered more loudly, as Douglas
warmed to his point:

> *Singleton testifies that Lincoln, in that speech, urged, as another
> reason for throwing Henry Clay overboard, that the Whigs had
> fought long enough on principle and ought to begin to fight for suc-
> cess. Singleton also testifies that Lincoln's speech did not have the
> effect of cutting Clay's throat, and that he (Singleton) and others
> withdrew from the caucus in indignation. He further states that
> when they got to Philadelphia to attend the National Convention
> of the Whig Party, that Lincoln was there, the bitter and deadly
> enemy of Clay, and that he tried to keep him (Singleton) out of the
> Convention because he insisted on voting for Clay, and Lincoln was
> determined to have Taylor.*

The crowd again laughed and applauded. "Singleton says that
Lincoln rejoiced with very great joy when he found the mangled
remains of the murdered Whig statesman lying cold before him.
Now, Mr. Lincoln tells you that he is an old line Clay Whig!" The
cheers and laughter got louder. "General Singleton testifies to the
facts I have narrated, in a public speech which has been printed and
circulated broadcast over the State for weeks, yet not a lisp have

we heard from Mr. Lincoln on the subject, except that he is an old line Whig."[90]

Douglas went in for the kill:

> What part of Henry Clay's policy did Lincoln ever advocate? He was in Congress in 1848–9, and when the Wilmot Proviso warfare disturbed the peace and harmony of the country, until it shook the foundation of the Republic from its center to its circumference. It was that agitation that brought Clay forth from his retirement at Ashland again to occupy his seat in the Senate of the United States, to see if he could not, by his great wisdom and experience, and the renown of his name, do something to restore peace and quiet to a disturbed country. Who got up that sectional strife that Clay had to be called upon to quell? I have heard Lincoln boast that he voted forty-two times for the Wilmot proviso, and that he would have voted as many times more if he could.

The crowd laughed. "Lincoln is the man, in connection with Seward, Chase, Giddings, and other Abolitionists, who got up that strife that I helped Henry Clay put down." The crowd erupted in tremendous applause. Douglas could have ended there, but he did not.

> Henry Clay came back to the Senate in 1849, and saw that he must do something to restore peace to his country. The Union Whigs and the Union Democrats welcomed him the moment he arrived, as the man for the occasion. We believed that he, of all men on earth, had been preserved by Divine Providence to guide us out of our difficulties, and we Democrats rallied under Clay then, as you Whigs in nullification time rallied under the banner of old Jackson, forgetting party when the country was in danger, in order that we might have a country first, and parties afterwards.

The record indicates the crowd proclaiming, "Three cheers for Douglas."[91]

The outcome of the election was never in doubt. The Republican newspapers all applauded Lincoln for having done more than hold his own with the Little Giant. Even so, on a cold election day in November 1858, Republicans won the popular vote, but because of the apportionment scheme they did not take control of the state senate. Because of that scheme, as David Donald explains, "Republicans, who received about 50 percent of the popular vote, won only 47 percent of the seats in the house, while the Democrats with 48 percent of the popular vote gained 53 percent of the seats."[92] In the election within the state legislature on the choice of Illinois's next senator, held on January 5, 1859, Douglas was the clear winner. He had 54 votes to Lincoln's 46. Douglas was headed back to the Senate, and in an ideal position to launch a serious presidential bid just two years away in 1860.

After the debates, Lincoln reviewed transcripts to provide edited versions that sympathetic newspapers ran. He excised the three times he had used the N-word for the parts of the country that tended toward abolition and made other edits to ensure readers in more conservative regions got the versions that fit their politics.

Beyond Illinois, Lincoln boosted his stature, while the harder Douglas kept harping on his close alliance with Clay, the more it weakened his support among Democrats. Democrats had never been a particularly harmonious party, but those who had not yet fled to the Republican Party or others were largely strong advocates for the maintenance and extension of slavery. John Todd Stuart, for example, had become a Democrat by the time Lincoln was in Congress. In 1856, he had supported Buchanan for president. Nevertheless, as the 1860 presidential election neared, he was growing increasingly frustrated with the party's pandering to the slave-owners. Lincoln was not going to get his vote (which went to third-party candidate John Crittenden), but neither would Douglas. It was clear to both Douglas and Lincoln that as 1858 turned to 1859 and 1860, Democrats like Stuart were wasting their votes or flocking to the Republican Party. Either way, the Republican nominee would be the beneficiary of the exodus.

Abraham Lincoln relied heavily on books and his mentors for guidance or support, as shown here in a lighter moment in a photograph taken in the midst of the Civil War (1863). *Alexander Gardner*

Lincoln was ten at the time of this first-known portrait of Henry Clay, then a member of the House of Representatives. *Transylvania University*

MAJ. JOHN T. STUART.

John Todd Stuart in his prime as a Whig leader and successful lawyer. *Northern Illinois University*

Cartoon from 1832 election lampooning Andrew Jackson for acting like a king trampling the Constitution (1832).
Library of Congress, LC-DIG-ppmsca-15771

Jackson subduing Clay in the 1832 election and sewing his mouth shut (1834).
Library of Congress, LC-DIG-ds-00856

PROCLAMATION

BY ANDREW JACKSON, PRESIDENT OF THE UNITED STATES. Dec. 10, 1832.

WHEREAS a convention assembled in the State of South Carolina, have passed an ordinance, by which they declare, " That the several acts and parts of acts of the Congress of the United States, purporting to be laws for the imposing of duties and imposts on the importation of foreign commodities, and now having actual operation and effect within the United States, and more especially," two acts for the same purposes, passed on the 19th of May, 1828, and on the 14th of July, 1832, " are unauthorized by the Constitution of the United States, and violate the true meaning and intent thereof, and are null and void, and no law," nor binding on the citizens of that State, or its officers; and by the said ordinance, it is further declared to be unlawful for any of the constituted authorities of the State, or of the United States, to enforce the payment of the duties imposed by the said acts, within the same State, and that it is the duty of the legislature to pass such laws as may be necessary to give full effect to the said ordinance :

1828, ch. 55.
Vol. iv. p. 270.
1832, ch. 227.
Vol. iv. p. 583.

And whereas, by the said ordinance, it is further ordained, that, in no case of law or equity decided in the courts of said State, wherein shall be drawn in question the validity of the said ordinance, or of the acts of the legislature that may be passed to give it effect, or of the said laws of the United States, no appeal shall be allowed to the Supreme Court of the United States, nor shall any copy of the record be permitted or allowed for that purpose, and that any person attempting to take such appeal shall be punished as for contempt of court :

And, finally, the said ordinance declares that the people of South Carolina will maintain the said ordinance at every hazard ; and that they will consider the passage of any act, by Congress, abolishing or closing the ports of the said State, or otherwise obstructing the free ingress or egress of vessels to and from the said ports, or any other act of the Federal Government to coerce the State, shut up her ports, destroy or harass her commerce, or to enforce the said acts otherwise than through the civil tribunals of the country, as inconsistent with the longer continuance of South Carolina in the Union ; and that the people of the said State will thenceforth hold themselves absolved from all further obligation to maintain or preserve their political connection with the people of the other States, and will forthwith proceed to organize a separate government, and do all other acts and things which sovereign and independent States may of right do :

And whereas the said ordinance prescribes to the people of South Carolina a course of conduct in direct violation of their duty as citizens of the United

Jackson's 1832 proclamation against South Carolina's threatened nullification and secession. *Law Library of Congress*

Portrait of Jackson
by Edward Dalton
Marchant (1840).
Courtesy of the Union
League Legacy Foundation

SCENE AT THE CAPITOL.

Cartoon depiction of Jackson thrashing his would-be assassin (1835).
Library of Congress, from "Shooting at the President!: The Remarkable Trial of Richard
Lawrence, for an Attempt to Assassinate the President of the United States"

Portrait of William Henry Harrison by Thomas Wilcocks Sully (1840).
Courtesy of the Union League Legacy Foundation

Portrait of Clay by
John Neagle (1843).
*Courtesy of the Union League
Legacy Foundation*

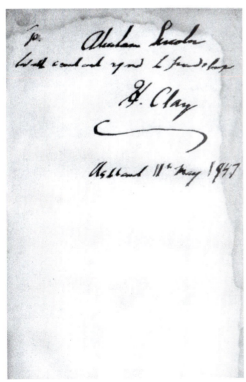

Clay's inscription on a set of
his speeches given to Lincoln.
*Courtesy of Ashland, the Henry
Clay Estate, Lexington, Kentucky*

Famous depiction of Clay enthralling the Senate with his defense of the Compromise of 1850, with Millard Fillmore presiding as president of the Senate (1855). *Library of Congress, LC-DIG-pga-05850*

Portrait of Zachary Taylor by Robert Street (1850).
Courtesy of the Union League Legacy Foundation

Portrait of Fillmore by unknown artist (1850).
Courtesy of the Union League Legacy Foundation

POLITICAL INTELLIGENCE.

The Illinois papers come to us filled with reports of the Douglas and Lincoln debates. The audiences are of the largest kind—ten, fifteen, and even twenty thousand people gathering together to witness the sparring, if they cannot hear it, between the Judge and his antagonist. The last great debate came off on Wednesday at Quincy, and so vast was the collection of people that the town was overflowing, the hotels crowded, and many private houses completely filled with strangers. Senator Douglas seems to have made a regular triumphal march from Augusta to Quincy. At Camp Point, on the route—a small town of about one thousand inhabitants—he was met by a cavalcade of military, bands of music, and citizens gathered from that and the adjacent towns. In front of the station-house a splendid bonfire was flaming, and hundreds of torches were carried in the streets. Every house in the town was illuminated, presenting altogether one of the finest spectacles witnessed during the campaign. He arrived at Quincy in the evening, and was received there by a large torchlight procession. On either side of the immense procession by which Senator Douglas was escorted to his hotel—the Quincy House—stood in line hundreds of men holding up to view appropriate and gorgeous transparencies. The evening reception was large and brilliant. Upon the day of the speech the people came into Quincy displaying hickory poles and flags, until the town looked like a forest of hickories. There were present at the meeting upwards of fifteen thousand people, consisting of men of all parties, Democrats and Republicans. But the enthusiasm is by no means all on one side—the Lincoln men are equally as hopeful and determined as their opponents, and make just as vigorous exertions.

A report on the Lincoln-Douglas debates printed on October 23, 1858, in the *National Intelligencer*, the Republican-leaning and leading newspaper published in the nation's capital. *National Intelligencer*

Painted to show the widespread support of national leaders for the Compromise of 1850, this group portrait did not originally include Lincoln. On the eve of the Civil War, it was redone to insert Lincoln at the center in place of John Calhoun but kept Clay, who had died nearly a decade before, seated directly to Lincoln's right to reflect his significant influence on the new president (circa 1861).
U.S. Senate Collection

The two known photographs of Orville Browning, one taken in the late 1850s or early 1860s *(left)*, and the other when he was a senator from Illinois *(right)*. *Courtesy of the Lincoln Museum, Fort Wayne, Indiana. Reference Number: 2578* (left). *Library of Congress, LC-DIG-cwpbh-01588* (right).

Browning's audacious September 17, 1861, letter to President Lincoln. *Library of Congress, Manuscript Division, Abraham Lincoln Papers*

A drawing of Lincoln showing his draft of the Emancipation Proclamation to his Cabinet, with the official portrait of Jackson in the background (1864).
Library of Congress, LC-DIG-pga-02502

Portrait of Lincoln by Edward Dalton Marchant (1863).
Courtesy of the Union League Legacy Foundation

Drawing of Taylor
(circa 1848).
Library of Congress,
LC-USZ62-71730

Photograph of Ulysses
Grant (June 1864).
Library of Congress,
LC-USZ62-1770

Lincoln's October 24, 1864, meeting with the abolitionist Sojourner Truth, who had waited for hours to meet the president and recalled of Lincoln, "I never was treated by anyone with more kindness and cordiality than were shown to me by that great and good man." *Library of Congress, LC-USZ62-16225*

Lincoln's second inaugural, with John Wilkes Booth among those looking down upon Lincoln from a White House portico over Lincoln's left shoulder. *Library of Congress, LC-USA7-16837*

VII

After the debates, Lincoln told a friend that they "gave me a hearing on the great and durable question of the age, which I could have had in no other way; and though I now sink out of view, and shall be forgotten, I believe I have made some marks which will tell for the cause of the civil liberty long after I am gone."[93] He told another friend, "The fight must go on. The cause of civil liberty must not be surrendered at the end of one, or even, one hundred defeats."[94] If the great question of the day was the future of slavery, and if Douglas was the likely Democratic nominee for president, Lincoln knew—and anyone reading newspapers reporting the debates around the country knew—that everyone could see there was only one Republican in the country who had stood on the same stage as Douglas for seven straight debates and given as good as he got. It was unimportant that Lincoln was not in the Senate to debate Douglas. He had already more than held his own with Douglas in public, while the other contenders for the 1860 presidential election had not done as well when they had the chance in Congress.

Notwithstanding the themes he had sounded in his debates with Douglas, the Lincoln of 1858 and 1859 was not a starry-eyed follower of Clay, nor was he unmindful of the genuine challenges facing his party and the country. He had learned from not only Clay's successes but his failures. As far back as 1852, Lincoln suggested that the "signal failure of Henry Clay, and other good and great men, in 1849, to effect anything in favor of gradual emancipation in Kentucky, together with a thousand other signs, extinguishes that hope utterly." He added, "Not a single state" had abolished slavery since the founding era. "That spirit," he said, "which desired the peaceful extinction of slavery has itself become extinct. [The] Autocrat of all the Russias will resign his crown, and proclaim his subjects free republicans sooner than will our American masters voluntarily give up their slaves." As for the ultimate fate of slavery, Lincoln said, "The problem is too mighty for me." He said "peaceful, gradual emancipation" was no longer a viable option in the United States.[95]

Lincoln the pragmatist would not say this out loud in his debates with Douglas or in public. If he had, it would have ended all hope of his appealing to anyone who did not want to embrace slavery as the most outspoken Southerners did. The Kansas-Nebraska Act, Bleeding Kansas, and the splintering of the Democratic Party presented the newly established Republican Party with an opening it could exploit—and Lincoln planned to do so.

Rather than back away from pushing for the extension of slavery for the sake of another last-minute compromise to save the Union, Jefferson Davis did the opposite. On July 6, 1859, he proclaimed, "There is not probably an intelligent mind among our own citizens who doubts either the moral or legal right of the institution of African slavery."[96]

A little more than two months later, Lincoln removed doubt about where he stood on the great issue of the day, fully casting aside whatever despair he had in 1852. In a speech on September 16, 1859, to a largely pro-Chase audience in Columbus, Ohio, Lincoln again relied on Clay as his guide, reusing one of his favorite quotes of Clay telling "an audience that if they would repress all tendencies to liberty and ultimate emancipation, they must go back to the era of our independence and muzzle the cannon which thundered its annual joyous return on the Fourth of July; they must blow out the moral lights around us." This imagery plainly appealed to Lincoln, who now adapted it to the task at hand of calling

> attention to the fact that in a preeminent degree these popular sovereigns are at this work; blowing out the moral lights around us; teaching that the negro is no longer a man, but a brute; that the Declaration has nothing to do with him; that he ranks with the crocodile and the reptile; that man, with only body and soul, is a matter of dollars and cents. I suggest to this portion of Ohio Republicans, or Democrats [that] there is now going on among you a steady process of debauching public opinion on this subject.[97]

Here Lincoln might well have been intentionally doing something far more pragmatic—he may have been saying different things to different people, tailoring his message to appeal to his audience (as Douglas had charged in the debates). Lincoln could have been testing the waters, feeling out who else might share this bleak opinion. He sounded conservative to Browning and Stuart but not to Giddings and Wilmot. He was creating a big enough tent of supporters to include not just the old-line Whigs but the growing masses who opposed slavery. Lincoln was not without principle. He drew a line at secession; Clay always had opposed it, just as Jackson did in 1832. He found the bridge that connected the two.

With Jackson and Clay as his inspirations, Lincoln set his sights squarely on the presidency. More nationally prominent, better-known national figures than Lincoln, such as Seward, Chase, and former Missouri attorney general Edward Bates, were maneuvering to secure the Republican Party's nomination for president in 1860. With the date of the mid-May convention fast approaching, the leading contenders gave barely if any thought to the rangy Westerner who had never impressed them in person. Among party leaders, he was widely regarded as a second-tier candidate at best.

However, losing the Senate race to Stephen Douglas made Abraham Lincoln a national figure. Lincoln knew how Jackson, Taylor, and Clay had each made of their failures and built their presidential campaigns by rallying the support not just of the public but also party leaders and the press. He began following—if not perfecting—that path immediately after his debates with Douglas, mailing copies of his speeches and debate transcripts and highlights around the country to friendly newspapers, old friends, and political contacts he was keen to nurture.

The object of Lincoln's efforts was to insert himself into the presidential race, already well underway, beginning with his speech at the Cooper Union Institute in New York on February 27, 1860. Lincoln had stumped for Republican candidates throughout the North, and he was hopeful, after Republican successes in state and local races in key midterm elections, that his candidacy for the presidency held greater promise. The Cooper Union appearance (part of a series of lectures that winter) gave Lincoln the opportunity to move from a second-level candidate to the front ranks. The speech held the prospect of enabling him to secure the support of the Republican elite in New York City, as well as to audiences all over the Northeast and New England through the favorable newspaper coverage that Lincoln was carefully cultivating at the same time.

In the biggest race of his life, Lincoln projected his moderate self. He understood that he was in William Seward's home state but that the sponsors of the debate came from the anti-Seward wing of the party and therefore favored Salmon Chase. Lincoln could expect few sympathetic supporters in the audience, but having lived for more than two decades in a state and county dominated by Democrats, Lincoln was used to being around people who didn't support him. He stayed out of the fight between Seward and Chase. Instead, he would let them knock each other off in their quest for the nomination. Meanwhile, he would seek to reach the broad middle of the Republican Party as well as Anti-Nebraska Democrats.

A further challenge was not to be overshadowed by the two men who were scheduled to appear before him in the series of lectures scheduled at the Cooper Union Institute that winter. The first was Frank Blair of Missouri, Jackson's longtime friend and a Democratic Party founder who left the party over its embrace of slavery. The former editor of the Democratic Party's favored newspaper, Blair focused his remarks on attacking slavery. The second speaker was Cassius Clay of Kentucky, a cousin of Henry Clay's and an ardent foe of slavery. He attacked slavery just as relentlessly as Blair. Lincoln was on the card because the organizers felt that all three of the speakers—Blair, Clay, and Lincoln—would help Chase by weakening enthusiasm for Seward.

Lincoln had no problem denouncing slavery—he had done that before. But in the Cooper Union address, he felt the need to do something he had not yet done with the diligence it required: deep research on the founding. Following Logan's example as well as his own advice and experiences, he hit the books. Because the material he found was so copious, he made sure he had the written research before him when he spoke. It was only the second time he used a manuscript when giving a major speech.

Lincoln pushed his appearance back to two weeks after his birthday, February 27. He made it later than the organizers had wanted so there would be less time between the event and the Republican national convention scheduled for early that summer. He spent hours in the Illinois State House's library, just across the street from his law office, and, between court appearances, he pored over the history of the Constitution, the Northwest Ordinance, and Jonathan Elliot's multivolume set of the debates on the Constitution in the various state ratifying conventions.

The speech was unique in two ways: its tone and its substance.

Its first section reflected Lincoln's long hours in the law library. Responding to Douglas's claim that the nation's Founders had endorsed popular sovereignty, Lincoln conceded that "our fathers, when they framed the Government under which we live, understood this question just as well, and even better, than we do now."[98] He then demonstrated something he had not done in debates with Douglas but which was enormously effective in this rarified setting: He examined the actions of the signers of the Constitution to establish that "our fathers," about whom Douglas spoke so reverentially, actually supported the power of Congress to regulate slavery in the territories. Systematically going through votes on such measures as the Northwest Ordinance (signed no less than by George Washington, he emphasized), Clay's Missouri Compromise, and the acts that Congress took to organize the Mississippi and Louisiana territories, Lincoln showed that of the thirty-nine men who signed the Constitution, twenty-three had had opportunities to vote on federal authority over slavery in the territories; of them, twenty-one voted to ban slavery from the territories. Turning to the remaining sixteen Founders of the Constitution who never had the chance to participate in the later votes, Lincoln argued that fifteen of them had opposed slavery and left "significant hints" that they would have voted to restrict it from the territories if given the chance to do so.[99] He figured the framers lined up thirty-six to three in favor of the power of Congress to regulate slavery in the territories. He mentioned Douglas by name only five times but pronounced the names of the signers of the Constitution thirty-nine times, George Washington's name eight times, and Thomas Jefferson's name twice.

In the next part of his speech, Lincoln appealed to the South, not unlike the way his mentor Henry Clay had tried to many times on the floor of the Senate. Lincoln hoped to convey to Southerners that *he* was no threat to them. It was Southerners who insisted on straying from the legacy of the framers. Placating the South with half measures like popular sovereignty would abandon the intentions of the framers, Lincoln argued, and it would fail, he predicted, because nothing short of federal activism on behalf of slavery would satisfy Southern demands. If there was a breach coming, it would be the South's fault, not the North's.

Concluding, Lincoln proclaimed the immorality of slavery. He could not entirely ignore the crowd in front of him. So, he felt comfortable saying, "If slavery is right," then "all words, acts, laws, and constitutions against it are themselves wrong, and should be silenced, and swept away. All they ask, we could readily grant, if we thought slavery right."[100] But slavery was not right. Only a platform like the Republican one, based on the idea that slavery was wrong, was morally and politically right. Lincoln concluded by imploring his fellow Republicans not to delude themselves into "groping for some middle ground between the right and wrong," which did not exist, but instead to "have faith that right makes might, and in that faith . . . dare to do our duty as we understand it."[101]

This conclusion sounded less like Clay and more like Lincoln's fiery House Divided Speech, but Lincoln deftly broadened his appeal. He did not emphasize compromise, though his tone was respectful when discussing Southerners, several of whom were long-standing friends of his. Nor did he openly stress Clay's name as he'd done in his debates with Douglas. Douglas was barely in the speech; he didn't need to be. This, after all, was not Illinois, or Clay country. These Republicans were more radical than the

Democrats he lived with. These were Seward and Chase folks, not Lincoln men. Convincing this crowd of his reverence for Clay was unnecessary and unproductive. Clay was there, to be sure, albeit in spirit and Lincoln's arguments, as well as the presence of Clay's cousin. (Abe even chose to stay at the Astor, the same place where Webster and Clay had each spoken.) It was enough that Lincoln knew he was following in the footsteps of his mentors. Lincoln cast the speech in such a way that scholars, to this day, do not agree on whether it was conservative or moderate. The difficulty of pinning it down as one or the other proves that it achieved Lincoln's aim of appealing to both. Yet, in the end, the substance and style were much closer to Clay than to Seward, Chase, or Owen Lovejoy, son of Elijah Lovejoy and a popular abolitionist preacher.

As Harold Holzer explains in his study of the Cooper Union speech and its consequences, Lincoln's delivery was more refined and sophisticated than ever before. Lincoln did not use the same tropes that he had used to reach the voters in Illinois, instead aligning himself and his arguments unmistakably with the framers. As Holzer notes, "Having identified thirty-nine framers whose slavery votes cry out for analysis, [Lincoln] w[ould] repeat the number thirty-nine for emphasis twenty separate times in a parallel burst of repetition for effect."[102] As Holzer notes further, on federal authority to regulate slavery, Lincoln repeated fifteen times the sentence, "Our fathers, when they framed the government under which we live, understood this question just as well, and even better than we do now."[103] Lincoln used the phrase *our fathers* at least five times in the speech and the word *fathers* nine times.[104] Further driving home this message was the technique, used by Clay and Webster, of "alternatively parallel and contradictory double phrasing—the device of antiphony—to neatly set up his audience for his arguments."[105] Antiphony is an old form of singing in which voices alternate, like the recitations in church when the leader of the congregation reads a line followed by the congregation reading another, back and forth, until the end. And so, Lincoln used the phrase *you say* repeatedly to introduce some of the arguments of the Southerners threatening secession but then following each time with his blunt denials, which were sure to resonate with the crowd and those later reading the speech.

When Lincoln finished, the audience erupted in thunderous applause. *New York Times* editor Henry Raymond declared Lincoln a national leader of "preeminent ability" and New York's second choice for the Republican nomination after Seward.[106] Mason Brayman, a Democrat from Springfield, who knew Lincoln from his early days as a lawyer for the Illinois Central Railroad, had called on Lincoln before the speech and agreed to stand in the back of the hall and signal if he could not hear Lincoln's voice. No signal came. Brayman reported that Lincoln's voice, like his rhetoric, filled the room.

The next evening, Lincoln visited the offices of the *New York Tribune* to correct proofs of the speech that would appear in the newspaper the next day. In the weeks that followed, several other newspapers throughout the Northeast and back home reprinted his address. The New York press was the most productive and powerful in the nation, and it churned out favorable news about his Cooper Union address that was read widely throughout the region. Lincoln and his friends distributed copies of the speech as far and wide as they could, while Lincoln began a speaking tour in the Midwest and Northeast. The local *Illinois Journal* produced pamphlets of the speech that it sold in bulk to Republican clubs throughout the country.

VIII

Nothing in Lincoln's life led him to doubt that he had the credentials to become president. Many presidents had been *less* qualified. In his lifetime, he had seen Zachary Taylor, a man with no political experience whatsoever, win his party's nomination and the presidency. Winfield Scott had been the Whig candidate in 1852. He had seen men with considerable political experience—Henry Clay and Lewis Cass—fail repeatedly to win the presidency and losing each time to a candidate with a less impressive record of service to the country. He had seen a man with no meaningful political experience—Franklin Pierce—win the presidency and then stumble so badly in office that he couldn't even win his party's nomination for reelection. He had seen a man with perhaps the most extensive résumé of any politician yet running for the presidency, James Buchanan, fail so miserably as president that he became the second to have no chance even to secure his party's nomination. He remembered his friend and colleague John Quincy Adams, with a résumé as good as Buchanan's, finish his one term without a single legislative accomplishment. Neither of the leaders often considered greatest, Washington and Jackson, had had any executive experience in political office before becoming president. Lincoln did not have their military records, but neither did his likely Democratic opponent. Like Clay, Lincoln had no executive experience, but Douglas could barely claim more, a lackluster stint of less than three months as Illinois's secretary of state.

There were three things a successful candidate needed, and Lincoln had them all. The first was a compelling vision of the Constitution and the future of the country. Vision had trumped experience in nearly every presidential election so far, and when experience seemed to matter (as with John Quincy Adams and Buchanan), it mattered less than the winning candidates' politics, which were simply vision wrapped differently. Lincoln asserted his vision strongly in his debates with Douglas and even more clearly at Cooper Union.

Next in importance was campaign organization. Lincoln's was a patchwork of the networks he had used to disseminate his Cooper Union speech; his many contacts with Republican-friendly newspapers throughout the country; the candidates he'd helped in Illinois, the Midwest, and Northeast; and his close-knit band of supporters in Illinois. While Lincoln was speaking at Cooper Union, his people were already on the ground in Chicago preparing for the upcoming Republican convention. Powerful local newspapers were already singing his praises. The state's largest newspaper, the *Chicago Tribune,* run by his friend Joseph Medill, continued to quote Lincoln's speeches freely and praise his every move, as it had done during his Senate campaign.

The final factor was party unity. Lincoln had been preaching its importance for decades. He had pushed for it in 1844 with Clay, 1848 with Taylor, 1852 with Scott, and 1856 with Frémont. Taylor was the only one of those who had won, but like Harrison, he had died at the beginning of his term. There was no Whig legacy to run on. Taylor's short-lived victory of 1848 was the only high point before the Democratic decline of Pierce and Buchanan. In this election, the winner was likely to be whichever party could hold itself together in addressing the monumental issue of slavery. In his run against Douglas for the Senate, Lincoln had actually gotten more votes than Douglas because the Republican Party organization within the state had performed better than its Democratic counterpart. Districts could swing states, and Republicans held the advantage in organization in almost every region of the country except the South.

The Democratic Party was splintering well before 1860. Douglas's split with Buchanan in 1857 foreshadowed the difficulty that their party faced in the run-up to the 1860 election. If Douglas got the Democratic nomination as expected, it was hardly a sure thing that his party would follow him. It had been hemorrhaging since 1854. But if Lincoln secured the Republican nomination, it was likely that the party would follow.

This brought Lincoln back to his ground game in Chicago. It was no accident the convention would be held in Lincoln's backyard. When the Republican National Committee met in New York on February 22, 1860, Lincoln's friend Norman Judd came as the representative of Illinois. When Seward's and Chase's representatives could not agree on a city, with Seward's pushing for New York City and Chase's pushing for an Ohio city, Judd astutely proposed a neutral city, which he reminded everyone was in a state—Illinois—that Republicans needed to win but had no serious candidate of its own in the mix for the nomination. The committee agreed on Chicago. Judd never mentioned Lincoln but had given him a boost with the convention being held on his turf and not that of any other serious contender.

Earlier in February, newspapers in Springfield and Chicago had endorsed him for president. In New York, Richard McCormick, a member of the Young Men's Republican Union, had generated admirable publicity for Lincoln. Lincoln had won his first election with the wily veteran John Todd Stuart calling the shots. This time, Stuart was not at Lincoln's side; instead, Lincoln turned to the sharpest political veteran available, the Yale-educated judge and former legislator David Davis. With the help of Norman Judd, Joseph Medill, John Palmer, Ward Hill Lamon, William Herndon, and Leonard Swett, Davis had been maneuvering months before the Republican convention to produce a Lincoln victory. For Davis and his team, the Republican convention became just another local convention to orchestrate. Lincoln gave Davis broad instructions on what to do, while Davis handled the day-to-day logistics.

Lincoln was the favorite of the Illinois delegation but of no others, which meant if matters were decided on the first ballot, he would lose. Lincoln and Davis figured the best strategy, just had been the case writ small with Lincoln's Cooper Union speech, was to stay alive as a candidate for as long as possible in the hope that the front-runners—Seward, Chase, and perhaps Bates—knocked each other off. Seward and Chase were well known, but they were well known not as moderates or Clay men but as radical abolitionists. Bates was a weak moderate but had the support of Horace Greeley, the founder and editor of the influential *New York Tribune*. The strategy was for Lincoln's surrogates to get each of the major candidates' camps to consider Lincoln as their second choice. James Polk had entered the 1844 convention in a similar position and come out the winner, albeit after he was proposed a compromise candidate on the eighth ballot.

Despite the odds, Bates mattered because Missouri was a swing state. He had served in the state house and for one term in the U.S. House of Representatives, and he had been a well-regarded Missouri attorney general and a well-known, popular figure in national Whig circles. Bates—and Jackson's friends, the Blair family—had significant sway over whichever way the delegation leaned at the convention.

To get the Missouri delegation to make him their second choice, Lincoln turned to Orville Browning. Browning had come to Illinois by way of Missouri, and he was a longtime friend of Bates, a founder of the Republican Party in Illinois and an old-line Whig who revered Clay. Before the Illinois Republican convention gathered, the two met, and Browning told Lincoln he was going to push for Bates but would back Lincoln as a second choice if Bates faltered. That sounded fine with Lincoln, who was already, unbeknownst to Browning, working through his surrogates to solidify support among Illinois Republican delegates before the convention began. Lincoln's plan worked—Bates's nomination was dead on arrival, and Browning backed Lincoln in Bloomington.

Lincoln asked John Palmer to help Davis in Chicago. Palmer had become a leader in the state's Republican Party after leaving the Democratic Party in disgust over its virulent proslavery orientation. Palmer instructed the Illinois state delegates to use all "honorable" means to secure Lincoln's nomination for president. When the time came to select at-large delegates to the convention from Illinois, Lincoln urged Browning's inclusion. He trusted Browning to support the state convention's decision once Bates collapsed as a candidate. Lincoln had expected Bates to fold early.

Davis, Medill, and Palmer began pushing the Illinois delegates to support Lincoln well before the convention started. With Davis in charge and everyone aware of Lincoln's strategy, Lincoln went to Springfield to await developments. It was no secret that Seward was the front-runner. *Harper's Weekly* published on its cover a large engraving of the eleven leading candidates for the Republican Party's nomination, with Seward as the largest in the center and the others in smaller portraits around the edges. Seward and his mentor Thurlow Weed had been lobbying delegations at least as hard and as long as Lincoln's team. Months before the convention, Seward went from one Republican stronghold to another trying to drum up support for his candidacy.

However, his case did not match the story that Lincoln's friends were spreading throughout the convention to persuade state delegations to accept Lincoln as a second choice to whomever was their first choice. It was the story of a self-made man, who was born in a log cabin, had no schooling, worked as a farmhand, split rails for a living, and taught himself law. An honest man and a man of the people he was. The nicknames "Old Abe" and "Honest Abe" made the rounds of the convention. Lincoln could not be sure that slogans would win him the presidency, but his aides left nothing to chance, coming up with the memorable image of Lincoln as "The Railsplitter." He had bested Douglas in their debates. It didn't matter that these assertions were not all literally true. Lincoln had learned that what people believed was more important than what actually happened. Myths and stories moved people, and if some exaggeration was part of the game—and it was on all sides—at least it could be applied for noble purposes.

Davis assigned Browning and Swett to lobby state delegations to make Lincoln their second choice, an effective strategy ensuring that once a front-runner stumbled, Lincoln would get his delegates—and Lincoln figured rightly that Seward and Chase each would stumble. On the day before the convention began, Browning wrote in his diary, "By request I went in company with Judge Davis [to] meet and confer with the Maine delegation; and at their solicitation made them a speech. Also called upon the delegation from New Hampshire. At night we received a message from the Massachusetts delegation, and called upon them at their rooms."[107] Davis and John Palmer, who later would be elected governor of Illinois, called on the New Jersey delegation. Each delegation agreed to Lincoln as a second choice. On that same day, Davis and Jesse Dubois wired Lincoln, "We are quiet but moving between heaven & earth. Nothing will beat us but old fogy politicians. The heart of the delegates are with us."[108] When Davis was not lobbying state delegations, he met with several old political friends, whom he enlisted to support Lincoln. He persuaded his friends in the Indiana delegation to go with Lincoln, though when they broke his way a rumor began that Davis had promised Caleb Smith, a prominent Indiana lawyer, a Cabinet post. Davis denied it. Lincoln denied it. In 1861, Lincoln selected Smith as his interior secretary.

On the second day of the convention, Browning and Gustave Koerner, a fellow member of the Illinois bar and Lincoln's connection with the growing German American population, were following Francis Blair of Missouri, who visited the Indiana, New Jersey, and Pennsylvania delegations to urge their support for his friend Bates. Browning and Koerner urged each of the delegations to support Lincoln as a second choice. The two met again with the three delegations later in the evening and secured their agreement to favor Lincoln as a second choice if the time came. Although Lincoln telegraphed Davis, "I authorize no bargains and will be bound by none," and attempted, without success, to have a note delivered to Davis declaring, "Make no contracts that bind me," Davis ignored the instructions. Instead, he engaged in negotiations with Simon Cameron of Pennsylvania, one of the weaker contenders for the nomination, that left the impression, at least to Cameron and his supporters, of a deal to eventually deliver his delegation, if needed, to Lincoln in exchange for a Cabinet post.[109]

On May 18, 1860, the balloting began. In keeping with the custom at the time, Judd put Lincoln's name into nomination in a single sentence: "I desire, on behalf of the delegation from Illinois, to put in nomination, as a candidate for President of the United States, Abraham Lincoln, of Illinois."[110]

On the first ballot, Seward led with 173½ votes, while Lincoln followed with 102. Cameron had 50½, Chase had 49, and Bates had 48. It was a surprise to find Seward lacked a majority, but an even bigger surprise that Lincoln was in second place on the first ballot. Lincoln had not only won the unanimous support of the Illinois and Indiana delegations but also unexpected support 7–1 from New Hampshire, where he had spent much of March. Seward received the remaining vote in New Hampshire, while Connecticut gave two of its votes to Lincoln and none to Seward.

On the second ballot, Lincoln showed considerably more momentum than anyone else. Seward gained 11 votes but was well short of the 233 required for the nomination. Lincoln gained 79 because of defections from Rhode Island, New Hampshire, Pennsylvania, and Chase's home state of Ohio, where Lincoln had campaigned hard in the months preceding the convention. After the second ballot, Lincoln was only 3½ votes behind Seward.

The end was clearly in sight. Seward had nowhere to go but down, and the other state delegations were breaking for Lincoln. On the third ballot, more defections from Rhode Island, Pennsylvania, and Maryland went for Lincoln, and he was suddenly only 1½ votes shy of the majority needed for nomination. Ohio, which had elected Chase as both a senator and governor, struck the final blow—it moved four votes over to Lincoln. With that, it was over. Seward's supporters could add as well as anyone else and asked for the nomination to become unanimous. Many Seward delegates cried in shock, while Lincoln's supporters, at Davis's direction, carried out a life-size portrait of the winner onto the stage. The convention chose Hannibal Hamlin, a senator from Maine and a founder of the Republican Party, as Lincoln's running mate.

Nathan Knapp, chairman of the Scott County, Illinois, Republican Party, had been assigned the job of keeping Lincoln informed back in Springfield. At the end of each of the first two ballots, he telegraphed the news to Lincoln. The future president was lounging in a chair at the Springfield telegraph office when Knapp's next telegram came: "We did it. Glory be to God." Lincoln accepted congratulations from all around the crowded room and from the wall of people outside. He was calm. "I knew this would come when I saw the second ballot," he told well-wishers.[111] On breaking free from the crowd, he said, "Well gentlemen there is a little woman at our house who is probably more interested in this dispatch than I am."[112] Koerner, Swett, and Judd were among those telegraphing Lincoln not to come to Chicago. Davis, too, instructed Lincoln, "Don't come here for God's sake. Write no letters and make no promises till you see me."[113] Lincoln did not come, and he made no promises.

Two weeks before the Republican convention, the Democrats had deadlocked and adjourned without naming a nominee. They were more sharply divided than ever before. Eventually, they split into two camps, one favoring the party's eventual nominee, Stephen Douglas, and the other calling itself the Southern Democratic Party, which backed Buchanan's vice president John Breckinridge of Kentucky for president. (The incumbent president, Buchanan, urged Democrats to back his vice president.) Once again, when it came to Lincoln, Douglas was magnanimous. Although James Russell Lowell joined other Eastern newspaper and magazine editors bemoaning the fact that Seward had not won, Douglas told a group of Republicans in Washington that they had made no mistake: "Gentlemen, you have nominated a very able and a very honest man."[114] John Bell, a former House speaker and senator, ran as the candidate for the Constitutional Unionist Party, whose agenda was to appeal to disenchanted Whigs who wished to take no stance on slavery. The Republican Party and several publishers produced hastily assembled campaign biographies of Lincoln, some including the text of the Cooper Union Institute address.

With several Southern states threatening to secede if Lincoln won and the Democratic Party severely split, Lincoln's assignment was simple to understand but difficult to execute: If the Republican Party remained unified as Lincoln urged and he helped them by keeping his own counsel to Taylor to stay silent during the general election (as Lincoln wished Clay had in 1844), the election was his. It was one of the more difficult challenges he ever faced. Surrogates, like Browning, did the talking, as they fanned out to lobby or reassure old Whigs or former Democrats to go with Lincoln. The five months between Lincoln's nomination in May and election day in early November 1860 were the longest stretch in his political life when he made no public appearances and gave no speeches. Lincoln's surrogates, including Browning, who lobbied old Whigs, made his case around the country. But Lincoln stayed silent. In spite of his repeated protestations of being "bored-bored badly" (as Herndon related after Lincoln's death), he followed the advice he had given to Taylor—and party leaders were giving to him—to keep his mouth shut. He did, and the reward was the presidency.

"HE WAS ENTIRELY IGNORANT NOT ONLY OF THE DUTIES, BUT OF THE MANNER OF DOING BUSINESS" (1860-1861)

In the 1860 election, Abraham Lincoln's lack of executive and congressional experience was both an asset and a liability. On the one hand, it helped him win the presidency. Nearly half a continent away from the disputes ripping Congress apart and several hundred miles from the violence in Kansas, he never had to directly confront, much less vote on, any of the policy proposals bandied back and forth in either place. His lack of a record made him a small target, overshadowed by first-tier candidates Seward and Chase, who took most of the hits. People knew Lincoln by his words, not his actions. In some parts of the country, they had read about him. In the Northeast, many people knew him from his campaigning. (He gave 175 speeches in the run-up to the presidential election.) Newspapers combed his statements for clues, but most journalists concluded he was weak and out of his depth. Indeed, the editors of many newspapers had grave doubts that Lincoln was good at anything else than telling stories and jokes.

Lincoln might have been gratified to know that he didn't have to do much to exceed most people's expectations of him. Unlike Jackson, he came into office with no mandate. Though he had won the election with 180 electoral votes, he had won less than 40 percent of the popular votes, a lower proportion than Jackson had won in 1824. (Lincoln's three opponents together won 123 electoral votes.) With the fragmentation of the Democratic Party, Douglas finished far worse than expected, and Lincoln's closest competitor turned out to be Buchanan's vice president John Crittenden, who had run as the nominee of the Southern Democrat party. Douglas won only one state (Missouri) and received the fewest number of electoral votes of the three candidates running against Lincoln.

Lincoln was the first president to be elected from outside the South since William Henry Harrison's victory two decades before, the first president to come from the West since Jackson (since Tennessee at that time was considered Western frontier), and one-half of the first successful national ticket that did not have a Southerner on it. He was, and still is, the only president who ever argued a case before the U.S. Supreme Court. He was the first man elected president without carrying a single Southern state.

None of these distinctions boded well for Lincoln. Lack of support and lack of experience were hardly a winning combination. He certainly wasn't overconfident. He admitted to his friend Robert Wilson of Pennsylvania "that, when he first commenced doing [his] duties, he was entirely ignorant not only of the duties, but of the manner of doing business" of the presidency.[1]

Where would Lincoln find guidance? He was flooded with advice from everywhere. He listened to much of it but heeded little of it. He declined the proffered services of those hoping to dominate him. He didn't want to make Taylor's mistake of letting Seward and Weed dominate him. Responding to the charge that he would just be Andrew Jackson's puppet, Polk had said, "I intend to be myself the President of the United States." He also said, "I prefer to supervise the whole operations of government myself rather than entrust the public business to subordinates, and this makes my duties great."[2] Lincoln had been there to see how he had done it. Polk was determined to be at the center of his administration. He seldom took others into his confidence and rarely sought the advice of even his closest friends. That same description fit perfectly Lincoln's management style. This was apparent in how Lincoln had secured his party's nomination, by establishing personal bonds with those working for him, emphasizing their loyalty to him and not some greater cause, often tasking more than one person to do a job, while all the while keeping his own counsel. Polk's model was Jackson. Yet Lincoln knew that the intense four years of Polk's presidency had killed him.

Where else, besides the five presidents he had met—Polk, Taylor, and Buchanan while they were president, and Van Buren and Fillmore after each had left office—and the few Illinois men he brought to Washington, would Lincoln turn for guidance and counsel? Where he always did. He looked first to his own experiences and then to the men who had been president before him. Washington was a patrician, both as a general and as president. He rarely consorted with the common man. Adams and his son were both arrogant Harvard graduates. Their arrogance doomed each of them to one term, as it alienated friends and enemies alike. Jefferson, Monroe, and Tyler had all studied at the College of William & Mary; Madison was a graduate of Princeton and had studied abroad; Polk graduated from the University of North Carolina; Pierce, from Bowdoin College in Maine; and Buchanan from Dickinson College, Roger Taney's alma mater. Lincoln lacked the military experience of Washington, Jackson, Harrison, and Taylor, but of these only Andrew Jackson had professed to champion the common man and opposed secession steadfastly. There was much to learn from all of these, both in what to emulate and what not to do. Yet the one figure Lincoln shared most with was Jackson, a self-made man who earned his nickname Old Hickory because he was tough as hardwood and who opposed secession. Lincoln was prepared to align himself with those qualities and that kind of leader.

Once he became president-elect, Lincoln still saw himself as a champion of the common man. In Springfield, he left his office door open so friends and neighbors could drop in to speak with him. He patiently mingled with crowds in the street, not just in Springfield but wherever he went. This was Lincoln's milieu. His philosophy of governance came from Jackson. Jackson had believed his election made him a leader for all Americans, rich and poor, black and white, male and female, free or enslaved. Lincoln believed that to the extent he had authority, it came from the Constitution and "We the People" who had ratified it. Lincoln enjoyed the large crowds as he had in his home state. The people ruled, and he basked in their delegated power. He could joke and tell stories, and he could learn.

Yet he quickly felt the weight of the great responsibilities of his office. In the nearly four months between Lincoln's election as president on November 6, 1860, and his inauguration on March 4, 1861, seven Southern states, beginning with South Carolina on December 20, declared their secession from the Union, and reports spread of South Carolina's intention to capture the two federal forts overlooking Charleston Bay in South Carolina, Sumter and Moultrie, as well as nearby forts in Florida. Browning recalled that, in July 1861, Lincoln "told me that the very first thing placed in his hands," as he first entered his presidential office, was an urgent report from Major Robert Anderson, the ranking officer in charge of Fort Sumter who was worried they were on the verge of being attacked by South Carolina to remove any federal presence. The soldiers of the two Charleston Bay forts had gathered in Fort Sumter for safety and Major Anderson feared "the impossibility of defending or relieving Sumter" if it were attacked.[3] Anderson's report included a message from Winfield Scott, the Union Army's commanding general, warning the president that there was "no alternative but a surrender."[4] Lincoln told Browning that "of all the trials I have since I came here, none begin to compare with those I had between the inauguration and the fall of Fort Sumter. They were so great that could I have anticipated them, I would not have believed it possible to survive them."[5] With so much chaos around him and considerable intrigue among those angling for appointments, Lincoln looked to the past for guidance on how to be president.

I

Once Lincoln won the presidency, one of his first thoughts was of Andrew Jackson. According to Gideon Welles, a Democrat-turned-

Republican newspaper editor who supported him throughout the 1860 campaign, Lincoln told him that on election night he had dreamed of "what his predecessors had done" when faced with crises—especially Jackson.[67]

Anyone paying attention to the political clashes preceding the Civil War, as Lincoln did, knew that threats of nullification, secession, and invasion of federal territory were not new. In the Hartford Convention of 1814, members of the soon-to-be defunct Federalist Party endorsed resolutions urging secession of the Northern states, among other things, in response to the continuing war with England and the domination of the federal government by a string of presidents from Virginia—four of the first five American presidents were Virginians. Like most everyone else, Lincoln knew that, as soon as the Fugitive Slave Law had been signed into law, the leaders of Northern states urged its nullification. He knew of William Lloyd Garrison's call in 1844 (much repeated later) for Northern states' separation from the Southern states that supported slavery. And he knew that Zachary Taylor and Millard Fillmore had confronted violent threats of treachery against the federal government, Taylor sending federal troops to the border when the Texas Republic threatened to invade New Mexico, and Fillmore sending even more troops to dissuade Texas forces from entering federal territory. Lincoln was well versed in the arguments made by Fillmore and Webster on the constitutional obligations of Northern states to comply with the supremacy of federal law. South Carolina's threats to storm two federal forts invited similarly strong responses. Jackson and Taylor had each denounced those threatening disunion as rebels and traitors. Lincoln followed their lead.

As Lincoln sought to determine what steps to take in order to quiet the brewing insurrection, he struggled to have the time to think and to be alone. He was mobbed throughout the day with people seeking jobs and favors and by journalists. The Democratic-leaning *New York Herald* sent Henry Villard to cover Lincoln during the 1860 presidential election and ensuing transition. Villard had covered him in the 1858 Senate campaign. A proud, upstanding German American, Villard was as patrician as he looked. After

Lincoln lost to Douglas in the 1858 race for the Senate, Villard said he grew to like Lincoln but did not respect him, both because of his "inborn weakness" as a candidate and his penchant for telling off-color stories and jokes.[8]

In later years, Villard changed his opinion. He especially delighted recounting the story of how he met Lincoln

> *accidentally about nine o'clock on a hot, sultry night, at a flag railroad station about twenty miles west of Springfield, on my return from a great meeting at Petersburg in Menard County. [Lincoln] had been driven to the station in a buggy and left there alone. I was already there. The train that we intended to take to Springfield was about due. After vainly waiting for half an hour for its arrival, a thunderstorm compelled us to take refuge in an empty freight car standing on a side track, there being no buildings of any sort at the station. We squatted down on the floor of the car and fell to talking on all sorts of subjects. It was then and there he told me that, when he was clerking in a country store, his highest ambition was to be a member of the state legislature.*

Lincoln paused, then confessed, "Since then, of course, I have grown some, but my friends got me into this BUSINESS [the Senate race]. I did not consider myself qualified for the United States Senate, and it took me a long time to persuade myself that I was." With a laugh, he told Villard, "Now to be sure I am convinced that I am good enough for it; but, in spite of it all, I am saying to myself every day, 'It is too big a thing for you; you will never get it. Mary insists, however, that I am going to be Senator and President of the United States, too." Villard wrote that, at this point, Lincoln "followed with a roar of laughter, with his arms around his knees, and shaking all over with mirth at his wife's ambition. 'Just think,' he exclaimed, 'of such a sucker as me President!'" Lincoln's aside—to Villard more of a hedge than an admission of Lincoln's plans— confirmed the journalist's suspicion that Lincoln might be using the Senate race as a springboard for a run for the presidency.

In mid-November 1860, Villard's opinion of Lincoln was still fixed, though it would eventually yield. "I doubt Mr. Lincoln's capacity for the task of bringing light and peace out of the chaos that will surround him,"[9] he reported. He conceded Lincoln was "a man of good heart and good intention" but concluded that "he is not firm. The times demand a Jackson."[10] Many people worried that Lincoln was too frivolous, and Villard agreed, finding Lincoln's "phrases are not ceremoniously set, but pervaded with a humorousness and, at times, with a grotesque joviality that will always please. I think it would be hard to find one who tells better jokes, enjoys them better, and laughs oftener than Abraham Lincoln." Such demeanor might amuse a crowd, but Villard joined the many people who had yet to see any Jackson in him.

Even so, Lincoln won, forcing further assessment. On December 21, 1860, Villard wrote,

> *Mr. Lincoln is known to be an old Henry Clay Whig. He calls the immortal Kentuckian his "beau ideal of a statesman." That his position in reference to the secession issue . . . is the identical one occupied by Mr. Clay in 1850, with regard to the then threatened nullification by South Carolina of the Compromise Measures of that year, will be seen by the following quotations from a letter and speech written and delivered by his prototype during the same period.*[11]

Villard excerpted at length from Clay's final Senate speech, urging his colleagues to endorse the Compromise of 1850. Villard explained at the end of his report, "I have quoted these two passages, for the special reason that Mr. Lincoln has used them within my own hearing, in explanation of his position, to visitors."[12]

Yet as Villard focused on how Clay might influence the priorities of the new president, two days later, on December 23, 1860, he elaborated on his doubts about Lincoln's capacity to rise to the demands of the presidency, suggesting that,

> *although unaccustomed to shape both resolution and execution according to the dictates of [Lincoln's] own clear judgment—to*

measure and pass upon the merits of things with the aid of his own moral and intellectual standard—the efficacy of this guide, demonstrated by his success in life, never produced conceit enough to induce him to overlook altogether the ideas, motives, arguments, counsels and remedies of others. On the contrary, a coincidence of his own views with those of the master spirits of his and previous ages is always greeted by him with great satisfaction and consciousness of increased strength. No one can be more anxious to fortify his position by precedents. No one rejoices more in the knowledge of reflecting the sentiments of the statesman and patriots that illuminate the pages of the history of his country.[13]

If there were any doubt who these "master spirits" were, Villard told the world as he recalled Lincoln's steadfast opposition to secession. Lincoln, he said, would not "content himself with supporting his position by democratic authorities" but persistently quoted Clay at length.[14]

Lincoln often wrote to supporters who shared his reverence for Clay. Daniel Ullman, a New York Whig, sent Lincoln a bronze token he had fashioned for the "first citizen of the school of Henry Clay" to be elected president. He praised Lincoln as "a true disciple of our illustrious friend."[15] Lincoln wrote back "to express the extreme gratification I feel in possessing so beautiful a memento of him, whom, during my whole political life, I have loved and revered as a teacher and leader."[16] When the *Richmond Dispatch* got wind of the medal and Lincoln's letter praising Clay, its editors wrote, "His teacher! His leader. Henry Clay the teacher of Mr. Lincoln. What lesson of Henry Clay had he learned? Where does he follow his leader's footsteps?"[17]

As Lincoln prepared for the presidency, he thought, too, of Zachary Taylor's brief presidency marked by a standoff between him and Congress over the administration's priorities. At the invitation of former president Millard Fillmore, he stopped in Buffalo, New York, for two nights. On the first night, Fillmore made sure people saw them together so that it could be reported around the country that they were united. A Buffalo newspaper got the message: "Mr.

Lincoln's ground, most firmly taken, is that he is to be president of the American people and not of the Republican Party." (Unfortunately, Fillmore abandoned that stance and his friend Henry Clay's fierce opposition to secession, when a year later he proclaimed Lincoln, deep into the project of saving the Union, a "tyrant [who] makes my blood boil." In 1864, he voted for McClellan rather than Lincoln, whom he charged with leading the country to "national bankruptcy and military despotism.")[18]

Less than three weeks before his inauguration, Lincoln reassured an audience in Pittsburgh of "the political education" he had received from Taylor that "strongly inclines me against a very free use of" the veto and any other means of usurping congressional authority.[19] "As a rule," Lincoln explained, "I think it better that congress should originate, as well as perfect its measures, without external bias."[20] He was reassuring his Pittsburgh audience and former Whigs elsewhere that, above all else, he remained, in spirit, a faithful Whig, It was the same message he had repeatedly urged Taylor to make during the 1848 presidential campaign.

In his next speech on February 22, 1861, in Philadelphia, Lincoln delivered a passionate expression of his vision of the Constitution. Standing in front of Independence Hall, the president-elect declared,

> I have never had a feeling politically that did not spring from the sentiments embodied in the Declaration of Independence. I have often pondered over the dangers which were incurred by the men who assembled here, and framed and adopted that Declaration of Independence. I have pondered over the toils that were endured by the officers and soldiers of the army who achieved that Independence. I have often inquired of myself, what great principle or idea it was that kept this Confederacy so long together.

In substance, these words were another reminder that the Declaration of Independence had, from an early age, made a lasting impression on Lincoln, whose study of the founding and attachment to Clay cemented his belief that it was a founding document that had enduring significance for America. In terms of style, Lincoln's repetitions and use of the first person to gain momentum was straight out of Clay's handbook, but now he was going beyond his mentor in his "shooting low," as he had once encouraged Herndon—speaking without complexity or much detail but using plain and sometimes poetic terms that the crowds listening or reading could understand and remember. (Grappling with the rigor of Euclid's axioms and theorems, Lincoln might have seen their relevance for speaking to the public, to do so plainly, directly, in a straight line, so to speak.)

For Lincoln, the principle at stake, he explained further in Philadelphia, was "liberty, not alone to the people of this country, but, I hope, to the world, for all future time." He asked rhetorically, "Can this country be saved on that basis," and answered his own question, "If it can, I will consider myself one of the happiest men in the world, if I can help to save it. If it cannot be saved without giving up that principle [of the Declaration of Independence], it will be truly awful. But if this country cannot be saved without giving up that principle, I was about to say I would rather be assassinated on this spot than surrender it." He underscored his point at the end: "I have said nothing but what I am willing to live by and, if it be the pleasure of Almighty God, die by."[21]

The closer Lincoln got to the White House, the more he thought about the troubles ahead. Many of those observing and listening to him were surprised that, even at this juncture, he stubbornly insisted the Southern states were merely bluffing. During Frémont's campaign in 1856, Lincoln had proclaimed, "All this talk about dissolution of the Union is humbug, nothing but folly. We do not want to dissolve the Union; you shall not." In 1860, he wrote to a correspondent that he had received "many reassurances [from] the South that in no probable event will there be any formidable effort to break up the Union. The people of the South have too much of good sense, and good temper, to attempt the ruin of the government."[22] Certainly he did not expect a war.

Upon arriving in Washington, Lincoln met with delegates from the Peace Convention, 131 politicians from fourteen free states and seven Southern states. They had come to the capitol in early February to forge a compromise to avoid the war and had agreed on a plan that included a proposal to amend the Constitution to prevent the extension of slavery in all new federal territories. When the delegation met with Lincoln to share their plan, his face lit up when he was introduced to a Democratic member of the House named James Clay—Henry Clay's son. Lincoln told him, "Your name is all the endorsement you require. From my boyhood the name of Henry Clay has been an inspiration to me."

While he was dismissing the likelihood of a real battle between secessionist forces and the United States, he was probably still mulling over his Cabinet selections, although he told close aides that he had been pondering the issue since the evening he was elected president. Many years later, he told his private secretaries, John Hay and John Nicolay, "When I finally bade my friends good-night . . . I had substantially created the framework of my Cabinet as it now exists."[23] Nicolay, the former newspaperman, and Hay, the Brown-educated legal apprentice, said in retrospect that in assembling a Cabinet, Lincoln sought "a council of distinctive and diverse, yet able, influential, and representative men, who should be a harmonious group of constitutional advisers and executive lieutenants—not a board of regents holding the great seal in commission and intriguing for the succession."[24] Whig orthodoxy, or what was left of it, held that the Cabinet should be a check on a president, not just an esteemed team of advisers. Whigs expected presidents to follow their Cabinet, but every Whig president before Lincoln—Harrison, Taylor, and Fillmore—had rebelled against that idea once they were in office.

The men Lincoln assembled bore only a faint resemblance to the romantic Whig ideal of a Cabinet, as described by Nicolay and Hay. Its members were neither harmonious in spirit nor particularly loyal to Lincoln, and Lincoln had no intention of deferring to their judgments. He picked prominent men who had considerable experience and reflected some geographical balance, but they were men who had been important in the formation and development of the Republican Party. Three—Seward, Bates, and Chase—had been his rivals for the presidency, and at least one other, Simon Cameron, was anything but a team player. Nonetheless, Lincoln brought them together to facilitate party unity, which he had insisted to Taylor and Crittenden should be the preeminent concern in assembling a Cabinet.

Though Lincoln had told his Springfield friends he expected to return some day, he could not be sure whether that would ever happen. Before leaving Springfield on February 11, 1861, Lincoln stopped to spend a few days with his stepmother. During the break, he visited his father's grave for the first and last time. He never told anyone about what he thought, did, or said there, yet it is not hard to imagine he might have looked down at his father with a mixture of satisfaction and sadness. He could only be proud of how far he had come but must have been sad to remember that his father had never seen greatness in him. It had to be enough that he had proven his father wrong.

II

As plans were being formulated to respond to the impending crises, Lincoln still had to complete his Inaugural Address. In preparation, he consulted only four documents. Besides Jackson's proclamation on South Carolina, he studied Henry Clay's final speech in the Senate, Daniel Webster's 1830 reply to South Carolina senator Robert Hayne's assertion of the state's entitlement to nullify federal laws, and the Constitution, which he quoted or referenced over twenty times in his address. Lincoln considered Webster's speech the single "greatest oration" in American oratory (indeed, Jackson was among those who openly delighted in Webster's speech at the time, which he had "expected" to demolish Hayne's arguments), and Clay's speech eloquently set forth the case against secession as Jackson had done in his proclamation.

For his Inaugural Address, Lincoln solicited direct advice from only a few people. Seward was one. The evening before he received the dispiriting news from Anderson about Charleston, Lincoln had met with Seward. The two men differed in many ways. Lincoln towered over Seward; Seward was mercurial, temperamental, hyperactive, disheveled, but occasionally courtly, while Lincoln was steady, plodding, and unpretentious. Short, with a bulbous nose, Seward was always in motion. They had encountered each other for the first time in 1848 when they were both campaigning for Taylor. From then through the first few months of Lincoln's presidency, Seward did not hide his disdain for Lincoln. He freely shared his harsh criticisms of Lincoln with allies and friends, such as Horace Greeley. Indifferent to decorum, he was even condescending and patronizing in Lincoln's presence.

For much of the transition, Lincoln had let Seward act as a spokesperson for the administration. But as late as the night before the inauguration, Seward was still playing hard to get. If he could not have the throne, he desperately wanted to be the power behind it and was pressing Lincoln to give him the power he craved, including picking the rest of the Cabinet.

Though irritated by the persistent push of Seward and Weed to micromanage his administration, Lincoln needed Seward in his Cabinet. Seward had significant support among abolitionists in the Northeast and would bring lots of votes with him if he joined the Cabinet. As Adams had done with Henry Clay, Lincoln offered his rival the top position in the Cabinet, the post of secretary of state, widely regarded as a stepping-stone to the presidency itself.

Orville Browning was his other sounding board. After Browning had helped to build a bridge between Lincoln and the Bates delegates, he remained a confidant throughout the election and transition. Thomas Hicks, the portrait painter for whom Lincoln sat after winning the Republican nomination for president, wrote that "the one man, in those days, who was always with" Lincoln, "with whom he advised, in whom he confided, with whom he talked over the Constitution of the United States in its relations to slavery, the condition of the South, and the mutterings of slave-owners, whose views accorded with his own, whom he held by the hand as a brother, was Orville H. Browning of Quincy."[25] Hicks continued, "When he and Browning met together, they discussed with thoughtful consideration many events which might occur, among which were the threatening of an unnecessary civil war, the cruelties of which, fortunately, could not be foreseen, in those peaceful days, by his friends and neighbors in the quiet town of Springfield."[26]

As the inauguration approached, Lincoln begged Browning to join the train trip he and his entourage planned to take from Springfield to Washington. Never one to like trains, much less crowded ones, Browning agreed to go as far as Indianapolis. Once there, Lincoln asked Browning back to his room in the Bates House (named for the prominent banker Hervey Bates, no relation to Edward), then the grandest hotel in downtown Indianapolis. Once there, Lincoln retrieved his traveling bag, from which he extracted a draft of his Inaugural Address. He asked if Browning "would not read it over, and frankly tell him my opinion of it."[27] After a quick review, Browning told Lincoln that it seemed "able, well considered, and appropriate."[28] Browning added, "It is, in my judgment, a very admirable document."[29] Lincoln was pleased, but asked Browning to take a closer look but "under promise" that he would speak only to Lincoln about it.[30] Browning agreed "to take it back with me, and read it over more at my leisure."[31] He told Lincoln, "If I see anything in it that I think ought to be changed, I will write to you from home."[32]

As Seward and Browning read over the draft, they unsurprisingly saw the influence of Clay and particularly Jackson. It made eminent sense. Lincoln had, after all, within a few days of the election reviewed Jackson's 1833 proclamation denouncing South Carolina's threat of nullification and secession. As Harold Holzer notes in his study of Lincoln's transition, "It came as no surprise that another visitor to Springfield found Lincoln on November 14th 'reading up anew' on the history of Andrew Jackson's response to the 1832 Nullification Crisis. While he made no effort to conceal 'the uneasiness which the contemplated treason gives him,' Lincoln assured his guest that, like Jackson, he would not 'yield an inch.'"

Nearly three decades before Lincoln's inauguration, Jackson had issued the first presidential statement to reject secession categorically, and it helped to widen the split between him and his vice president, John Calhoun, once and for all. Jackson's proclamation was an important step in a long, intense series of moves in which Calhoun would challenge his commitment to states' rights. They had clashed over appointments as well as when they gave dueling toasts at the end of April 1830. And so it was no surprise when two weeks after Jackson issued his proclamation, Calhoun resigned to protest the president's failure to embrace nullification doctrine, which held that states were entitled, especially when they were a political minority, to block proposals from the more powerful federal government that violated their rights. Some historians believe Jackson's proclamation was largely motivated by his confidence that the issue driving nullification—a rise in tariffs—would not hurt his own plantation and that the threat would in any event eventually vanish. Nevertheless, the proclamation's substance went further to systematically dismantle Calhoun's doctrine. As a result, it had become popular among many Republican leaders, especially those who once had been Jacksonian Democrats.

As his last act as vice president, Calhoun thwarted Jackson's plan to move Van Buren from secretary of state to ambassador to Great Britain, allowing Jackson to keep Van Buren on his team but to have fulfilled his announced aim of reorganizing the Cabinet. When the nomination came to the Senate, Calhoun cast the tie-breaking vote to defeat the nomination. Elated at the outcome, Calhoun said, "It will kill him dead, sir, kill him dead. He will never kick, sir, never kick."[33]

The celebration came too soon. Because Jackson had given Van Buren a recess appointment as the ambassador to Great Britain, Van Buren was in England when he learned the news of his rejection by the Senate. British royalty assured Van Buren that the defeat would make him a martyr. Jackson agreed, "The people will properly resent the insult offered to the Executive, and the injury intended to our foreign relations, in your rejection, by placing you in the chair of the very man whose casting vote rejected you." Jackson was right: After removing Calhoun as vice president, Jackson placed Van Buren on his ticket in the 1832 election.

These events were well known to Lincoln and his growing inner circle, and they were undoubtedly on Seward's mind as he advised Lincoln in the closing days of the transition. Though he had not yet committed to serving as secretary of state, Seward could not keep his hands off the draft. He meticulously scoured it and made over fifty suggestions, almost all of which reflected his concern that the new president not antagonize the South or provoke civil war.[34] Seward's object was conciliation. His suggestions helped Lincoln to soften the tone of the address.

On February 9, shortly before Lincoln and his entourage left for Indianapolis, Browning and Lincoln met privately in Lincoln's room at the Chenery House, a hotel in Springfield, his own house already having been rented out. According to Browning, they "discussed the state of the Country expressing our opinions fully and freely."[35] According to Browning's notes, the president-elect

> agreed entirely with me in believing that no good results would follow the [Peace] convention now in session in Washington, but evil rather, as increased excitement would follow when it broke up without having accomplished anything. He agreed with me no concession by free States short of a surrender of everything worth preserving, and contending for would satisfy the South, and that [Kentucky Senator John Crittenden's] proposed amendment to the Constitution [barring the extension of slavery to any new federal territories] in the form proposed ought not to be made, and he agreed with me that far less evil & bloodshed would result from an effort to maintain the Union and the Constitution, than from disruption and the formation of two confederacies.[36]

That Lincoln took the time to sit with Browning and subsequently adopted his suggestions suggest that he was not merely echoing Browning's ideas to please him.

Back in Quincy, Browning wrote to Lincoln on February 17, 1861, with his suggestions on the draft.[37] Browning agreed with

Lincoln on the arguments against secession and proposed that Lincoln's first move—to send supplies or arms to Sumter—would likely induce South Carolina to attack the fort, "and then the government will stand justified before the entire country, in repelling that invasion, and retaking the forts."[38] He advised, "Without an aggressive act by the federal government, the South would appear to be in an unjustified position."[39] Lincoln agreed and therefore dropped a clause from the draft in which he had threatened to lead the federal government into reclaiming seized federal property in the South. (Lincoln's private secretaries later described Browning's suggestion as "the most vital change in the document.")[40] With Browning's help, Lincoln had his commitments set forth more clearly and also had a plan for addressing the immediate threats to disunion stirring in South Carolina almost a month before his inauguration—one nearly identical to the plan that he eventually followed.

Lincoln's Inaugural Address was only the second presidential declaration formally opposing secession. On this occasion, as had been the case for Jackson, the most direct threat to the Union came from South Carolina. The inaugural crowd was substantial, so consumed were the citizens in and around the District of Columbia with worry about the country's future. As Lincoln had prepared to take his oath of office, dozens of newspapers in the North, Midwest, and West urged Lincoln to emulate Jackson. Lincoln made it clear he had gotten the message. His speech was a unique blend of Jackson, Clay, and Webster, with vivid imagery and language inspired by decades of reading Shakespeare and other poetry. Lincoln had adapted Jackson's arguments and even some of its wording to the current crisis facing the nation. Jackson had declared that the "most important" of the Constitution's "objects" was "'to form a more perfect Union.'"[41] Lincoln said, "One of the declared objects for ordaining and establishing the Constitution, was 'to form a more perfect Union.'"[42]

Pierce and Buchanan had blamed the impending hostilities on Northern abolitionists. Like Clay and Jackson before him, Lincoln blamed them on Southern secessionists, agreeing with Jackson that nullification and secession undermined the all-important object of maintaining the Union. In Jackson's words, "I consider, then, the power to annul a law of the United States, assumed by one State, incompatible with the existence of the Union, contradicted expressly by the letter of the Constitution, unauthorized by its spirit, inconsistent with every principle on which it was founded, and destructive of the great object for which it was formed."[43] Lincoln echoed him: "But if destruction of the Union, by one, or by a part only, of the States, be lawfully possible, the Union is less perfect than before the Constitution, having lost the vital element of perpetuity."[44]

Because the objective of the Constitution was to "form a more perfect Union," Jackson had suggested that nullification was treason.[45] "To say that any State may at pleasure secede from the Union, is to say that the United States is not a nation."[46] Lincoln went further, arguing, as Webster had in his widely known debate with Hayne, that secession, like nullification, promised "anarchy."[47] Like Jackson and Webster, Lincoln refuted the right of a political minority to refuse to abide by a majority's will, the right that Calhoun and Hayne had insisted made secession legitimate. Lincoln said, "If a minority . . . will secede rather than acquiesce, they make a precedent which, in turn, will divide and ruin them; for a minority of their own will secede from them, whenever a majority refuses to be controlled by such minority."[48] Lincoln then asked rhetorically, "Why may not any portion of a new confederacy, a year or two hence, arbitrarily secede again. [All] who cherish disunion sentiments, are now being educated to the exact temper of doing this."[49] *Disunion* was a word that was common to Jackson, Clay, and Webster; they had used it in each of their famous declarations opposing nullification. Lincoln's use of the same term linked his message to theirs.

Jackson had gone further to refute the idea that the sovereignty of states was absolute. With some irony, he listed many of the same constitutional provisions that John Marshall had used in some of his most famous cases to support the basic idea that the Constitution itself and any federal laws consistent with it were the supreme law of the land. One plain inference from this foundational concept was that states were not preserved certain rights under the Constitution, but rather that states could not impede lawful federal action.

Next, Lincoln again echoed Jackson in emphasizing that the Constitution derives its authority not from the states but from the people of the United States, who were the country's principal sovereign. Jackson had declared,

> The people of the United States formed the Constitution, acting through the State legislatures, in making the compact, to meet and discuss its provisions, and acting in separate conventions when they ratified those provisions; but the terms used in its construction show it to be a government in which the people of all the States collectively are represented. We are ONE PEOPLE in the choice of the President and the Vice President.[50]

Describing the Constitution as a "contract" that could not be nullified by a single party, Lincoln agreed with Jackson that "the Chief Magistrate derives all his authority from the people."[51]

Jackson had emphasized that his "oath" and presidential duties required him to stand firmly against nullification or any other effort that threatened to undermine the Union and the Constitution, declaring that it "is the intent of [the Constitution] to PROCLAIM [that] the duty imposed on me by the Constitution 'to take care that the laws be faithfully executed,' shall be performed to the extent of the powers already vested in me by law, or of such others as the wisdom of Congress shall devise and Entrust to me for that purpose."[52] Near the end of his proclamation, Jackson reiterated the point but with greater clarity and bluntness:

> I rely with equal confidence on your undivided support in my determination to execute the laws—to preserve the Union by all constitutional means—to arrest, if possible, by moderate but firm measures, the necessity of recourse to force, and if it be the will of Heaven that the recurrence of the primeval curse on man for the shedding of a

brother's blood should fall upon our land, that it not be called down
by any offensive act on the part of the United States.[53]

With greater brevity, now President Lincoln declared in a steady
voice, "I therefore consider that, in the view of the Constitution
and the laws, the Union is unbroken; and, to the extent of my ability, I shall take care, as the Constitution itself expressly enjoins
upon me, that the laws of the Union be faithfully executed in all
the States."[54] Drawing on Jackson's widely accepted view of the
president as an agent of the people, Lincoln continued, "Doing this
I deem to be only a simple duty on my part; and I shall perform it,
so far as practicable, unless my rightful masters, the American people, shall withhold the requisite means, or, in some authoritative
manner, direct the contrary."[55]

Jackson had ended his proclamation by declaring that whether
conflict would result from the nullification effort was up to South
Carolina. "There is yet time," he pleaded, "to show that the descendants" of the great leaders from South Carolina that had joined the
American Revolution, "will not abandon that Union, to support
which so many of them fought and bled and died. I adjure you . . .
to retrace your steps."[56]

In a similar vein, Lincoln concluded thus: "Fellow-citizens! the
momentous case is before you. On your undivided support of your
Government depends the decision of the great question it involves,
whether your sacred Union will be preserved, and the blessing it
secures to us as one people shall be perpetuated."[57] Lincoln again
reminded Americans that whether there would be bloodshed was
up to those pressing for secession. "In your hands, my dissatisfied
fellow countrymen," he told the people of the South, "and not in
mine, is the momentous issue of civil war. The government will
not assail you. You can have no conflict, without being yourselves
the aggressors."[58]

In his final Senate oration Clay had appealed "to all the South"
and for the country "to elevate ourselves to the dignity of pure and
disinterested patriots."[59] Clay called upon both sides to forget the
"bitter words, bitter thoughts [and] unpleasant feelings" in their

debate over the great compromise.[60] "Let us go," he admonished them, "to the altar of our country and swear, as the oath was taken of old, that we will stand by her; we will support her; that we will uphold her Constitution; that we will preserve her Union." Now Lincoln twice invoked past patriots as reminders of the sacrifices made on behalf of the Union by asking, "To those, however, who really love the Union may I not speak?"[61]

Having immersed himself in the records of the founding, Lincoln mentioned "precedent" three times in asking all the key players in the national drama to think hard about the consequences of their actions. Lincoln asked a series of questions, urging Clay-like caution: "Why should there not be a patient confidence in the ultimate justice of the people? Is there any better, or equal hope, in the world? In our present differences, is either party without faith of being in the right?"[62]

In his character, acting abruptly was less an option for Lincoln than for either Jackson (who used his mercurial temper to intimidate people) or Clay (whose eagerness to please sank his chances in the 1844 presidential election). Lincoln said, "If it were admitted that you who are dissatisfied, hold the right side in the dispute, there still is no good single reason for precipitate action."[63] Instead, he urged, in an appeal reminiscent of Clay: "Intelligence, patriotism, Christianity, and a firm reliance on Him, who has not yet forsaken this favored land, are still competent to adjust, in the best way, all our present difficulty."[64] Unlike Clay, Lincoln did not use the word *compromise* at the beginning of his great speech, but he held open its prospects by devoting eight of the first nine paragraphs of his address to reiterating his determination not to force the end of the institution of slavery. No one could miss his point—indeed, that *was* his point. Lincoln would not make Polk's mistake of lying about who fired the shot that started a war. Instead, Lincoln let the world know that it was entirely up to the Southern states to decide whether or not there would be a conflict.

Clay famously declared that the compromise he proposed "is the re-union of this Union. I believe it is the dove of peace, which, taking its serial flight from the dome of the Capitol, carries the glad tidings of assured peace and restored harmony to all the remotest extremities of this distracted land."[65] Lincoln's more famous conclusion dropped Clay's well-worn use of the dove of peace as well as the word *harmony*, though harmony was the theme of his address. Browning and Seward had worked on his message, but they could not match Lincoln in the precision and elegance of his language. He concluded:

> *We are not enemies, but friends. We must not be enemies. Though passion may have strained, it must not break our bonds of affection. The mystic chords of memory, stretching from every battle-field, and patriot grave, to every living heart and hearthstone, all over this broad land, will yet swell the chorus of the Union, when again touched, as surely they will be, by the better angels of our nature.*[66]

III

Of course, an Inaugural Address is just one thing every president has to do. A more fundamental task was to figure out how to act as the president. The only chances Lincoln had to observe a president in action were when he watched Polk for two years and the day of Taylor's inauguration, when he briefly saw both him and Taylor together. Lincoln had the sophistication and intelligence to find some things he could learn from Polk, particularly his hard work ethic, announcing his goals clearly, sparingly sharing the strategies on how to achieve those goals, and monitoring his subordinates to ensure they were doing what he wanted done. Having condemned Polk for his deceptions and watched from afar as the corruption of Taylor's Cabinet nearly brought down his administration, Lincoln was determined to keep faith with the American people by being a model of integrity. The country had had to endure the divisive, mean-spirited rhetoric of all the Democratic presidents in Lincoln's lifetime—Jackson, Polk, Pierce, Buchanan. Lincoln's job was to bring people together, not turn them against each other. That called for a tone of moderation.

Perhaps most important, Lincoln understood that, as president, he had to be a model for all Americans, not just for the present generation, but also for those who followed. When he was repeatedly being admonished to be like Jackson, he took that to mean that he had to find a way to be himself but limit his propensity for compromise, not relenting on his determination to keep the Union intact, rising to the occasion, defending the Constitution, and being responsible and attentive to all Americans, not just his friends and fellow Republicans. Lincoln had to find a way to be Lincoln but act like Jackson.

Finding his way took time and effort. For many observers, the election had made Lincoln humbler and more serious. The time was over for his biting wit, ridicule, off-color jokes, saucy stories, and partisan sniping. "Even before his inauguration," it was reported, "Abe is becoming more grave. He doesn't construct as many jokes as he did." His first crucial speeches—his Inaugural Address and mid-March message opening Congress's special session on the war—"were marked by an earnest appeal to patriotism and a sober explanation of why the rebellion had to be put down; only some quaint expressions and an occasional touch of irony hinted at the author's underlying sense of humor."[67] Lincoln stuck with the same strategies that had gotten him this far, particularly at the convention—listening carefully and patiently to all who sought his counsel or were giving theirs to him, treating everyone with respect, being honest rather than crafty as Polk and Jackson had often been, gathering as much data as he could before making decisions, taking no single person as a confidant but instead tasking several of them with the same mission to ensure the job got done, and standing firm, when he had to, like Jackson and Clay in defending the Union.

If Lincoln failed, his mistakes would become lessons to avoid for his successors, assuming there would be any. If he succeeded, his achievements would be sung for the ages. Either way, history was the judge—a theme he repeatedly emphasized throughout his life. His friend Joshua Speed recalled Lincoln's letter, written in the depths of his despair in 1841, "that he had done nothing to make any human being remember that he had lived—and that to connect his name with the events transpiring in his day & generation and so impress himself upon them as to link his name with something that would redound to the interest of his fellow man what he desired to live for."[68] In 1858, Lincoln had observed,

> In the first place, let us see what influence [Douglas] is exerting on public opinion. In this and like communities, public sentiment is everything. With public sentiment, nothing can fail; without it nothing can succeed. Consequently, he who molds public sentiment goes deeper than he who enacts statutes or pronounces decisions. He makes statutes and decisions possible or impossible to be executed.[69]

In his narrative, he was the man at the center of the storm, the man whose charge was to lead the country out of harm's way or die trying. Long after he was gone, people would be reading his story, as he had read and reread Washington's. In his story, Lincoln would be what Clay had helped him to be—the ultimate manifestation of the self-made man, who had risen from "humble" beginnings to the highest office in the land. In this story, Lincoln was the hero, even if it killed him. The most important thing was how he would be remembered, Lincoln often said. Later, in the thick of the war in 1862, he told Congress, "In times like the present men should utter nothing for which they would not willingly be responsible through time and in eternity."[70]

Much of Lincoln's days were structured, which irritated him when he felt too constrained but pleased him when it gave him something worthwhile to do. With the Inaugural Address done, the next business on his agenda was to meet with his Cabinet. (In those days, the White House was open to the public and, like the presidents preceding him, he had to thread his way through people lobbying for appointments whenever he ventured into public areas.) After many twists and turns, he had settled on the seven men who were on his list all along, reshuffling them until he finally found the combination that suited his desires and their ambitions. Finalizing the Cabinet was Lincoln's strongest demonstration of, as then New York senator William Marcy memorably put it in 1828, "the Jackson 'doctrine' that to the victors belong the spoils[,] universally the creed of all politicians."

A long-standing tradition in American politics held that the "most prominent" remaining member of the victorious party should be offered the State Department. In 1824, Adams offered that post to Henry Clay, just as Jefferson had appointed his preferred successor, James Madison, in 1808 and Madison in turn had appointed his onetime rival James Monroe to it in 1816. Everyone, including Lincoln, knew that Seward was the logical choice, and though bitter after the election, Seward grudgingly agreed to exchange letters with Lincoln. Knowing Hannibal Hamlin, his vice president, had been good friends with Seward in the Senate, Lincoln asked him to hand Seward a letter. It read, "With your permission, I shall, at the proper time, nominate you to the Senate, for confirmation, as Secretary of State for the United States."[71] Lincoln later told Weed, men "[like] a compliment."[72] But, Weed told Lincoln he thought while Seward had earned the post, he would decline it, as Seward did in response to the letter given to him by Hamlin. Weed reminded Lincoln that when William Henry Harrison had been elected president in 1840, Harrison had traveled to Clay's home for advice. Lincoln might have known the truth, which was quite different—Harrison tried mightily to avoid Clay during the transition, because he did not want to be cornered into giving Clay power over his administration. When Harrison traveled to Kentucky to speak with friendly crowds and supporters, Clay was already lying in wait for him. The conversation did not go well. Clay stormed out when Harrison insisted that he, not Clay, was the president. Weed urged Lincoln to do the same, and travel to Seward's home in Auburn, New York. Gideon Welles said, "Mr. Lincoln declined to imitate Harrison."[73]

Instead, Lincoln kept up his correspondence with Seward, and slowly, letter by letter, coaxed him into a political friendship they both began to enjoy. Seward warmed up when Lincoln sent a personal note asking him for his help in selecting the rest of the Cabinet. Unhelpfully, he replied with an ultimatum for taking the job—denying Chase, Blair, and Welles anything in the Cabinet. Having once been Democrats, Seward distrusted them all. Lincoln ignored the demand, telling Nicolay he could not "afford to let Seward to take the first trick."[74] After the inaugural ceremonies on March 4, 1860, Lincoln again approached Seward, again with a compliment, that he needed him as secretary of state. The personal touch made the difference, and Seward agreed. Seward could not sit on the sidelines with the fate of the country at stake, and Lincoln did not want him there beyond his control.

Lincoln planned to fill his Cabinet partly with men from states he needed to win again. Welles, from New England, fit the bill. Having once been a close ally of Martin Van Buren, he became a Republican, opposed to slavery and a supporter of Lincoln throughout the presidential election. Welles came with considerable credibility as a reasonable man with good judgment and experience, as a former newspaper reporter, Connecticut comptroller, and Jacksonian Democrat, who left the party to protest the Kansas-Nebraska Act. Lincoln called him Father Neptune because he was big, burly, and had the longest beard Lincoln had ever seen and thus reminded him of the King of the Sea. His experience as chief of the Bureau of Provisions and Clothing for the Navy under Polk made him a logical choice for Navy Secretary. Lincoln loved having Father Neptune nearby; Welles enjoyed Lincoln's stories more than anyone else in the Cabinet.

Coming from Ohio with a long record of opposing slavery, he seemed an inevitable choice, but Lincoln did not want him: a large man with an oversize ego, Chase had his followers, but most people found him overbearing. Yet Lincoln found Chase irresistible because Chase no more wanted Seward in the Cabinet than Seward wanted him. This made for a perfect match, in Lincoln's judgment, as they would either balance or cancel each other out. Chase had just won a seat in the Senate and was reluctant to give it up. Again, it was a matter of coaxing. It became harder because Simon Cameron was telling newspapers Lincoln had offered him the post of secretary of the Treasury. Eventually, Lincoln convinced Chase to take the Treasury by presenting it as a chance to keep an eye on Seward.

Slight, bespectacled, and with a beard that rivaled Welles's, Edward Bates appealed to Lincoln for many reasons. He had been urged upon Lincoln by many friends in the Republican Party, especially Browning and Horace Greeley, who served briefly in the House of Representatives and was the powerful publisher of the *New York Tribune* (the most widely circulated newspaper in the United States at that time). Bates had been a Whig, like Lincoln, and had been offered but turned down Fillmore's invitation to be his secretary of war. Since Lincoln wanted a Republican leader from the West in his Cabinet, Missouri's Bates was perfect. Lincoln needed Missouri again in his corner. Bates had been Missouri attorney general and had a distinguished career in the Missouri bar; he seemed a perfect fit as U.S. attorney general.

There were three posts left to fill. One was the Department of the Interior. With few clamoring for it, Lincoln thought of Caleb Smith. The two men had served together in the House, and Smith came from Indiana, another state he wanted to keep in his column. Lincoln, however, seemed to be one of the few men who thought Smith was up to the job—indeed, he hesitated because he agreed with Villard that Smith was not just dumb, but "worse than mediocrity."[75] He could not, however, ignore the debt he felt to him, because Indiana had come to his aid at the convention when he most needed it and Smith had been the man to make that happen. He needed a Midwesterner in the Cabinet. Smith was the man.

Nor could Lincoln ignore Simon Cameron's persistence in demanding a Cabinet post. Cameron felt entitled to the Treasury and insisted on reminding Lincoln at every opportunity that the post should be his, forcing Lincoln to tell him, "Since seeing you things have developed that make it impossible for me to take you into the Cabinet."[76] Ongoing rumors of his corruption made Lincoln less enthusiastic, but the constant pressure he felt from Pennsylvania because Cameron had campaigned around the state showing everyone Lincoln's December 31 letter offering him a post (while concealing the January 3 letter rescinding it) shamed him into offering Cameron a Cabinet position. Mindful that he should avoid writing letters that could came back to haunt him, Lincoln grudgingly appointed the Pennsylvanian secretary of war.

This left the position of postmaster general, perhaps the least prestigious Cabinet office, yet it was essential for ensuring communications with the army if war came. Lincoln chose Montgomery Blair, a member of the influential Blair family of Missouri. There was much to be said for Blair's inclusion. He had been well respected as a lawyer, including as a counsel for Dred Scott. He had served in multiple offices before, including as a judge of the Court of Common Pleas, U.S. solicitor in the Court of Claims, and U.S. district attorney. Blair knew his way around government. The problem was that no one but Lincoln liked Montgomery Blair. A graduate of West Point, Blair wanted to be war secretary, but he was a notorious malcontent and contrarian, who, as the journalist Noah Brooks once wrote, "was a relentless mischief-maker, . . . and he was apparently never so happy as when he was in hot water or making it hot for others."[77] He quickly alienated nearly everyone he worked with in the new administration.

It was no accident that more than half of Lincoln's Cabinet had been Jacksonian Democrats. Though neither he nor the Blairs respected Franklin Pierce (much less agreed with him on anything), his was the only Cabinet yet in American history to have stayed intact for a president's entire term. Its endurance lent the weak-willed Pierce a semblance of stability. Montgomery Blair's brother, Francis Preston Blair Jr., recalled that Buchanan had "put in his cabinet his enemies—men who felt his nomination a blow to their ambitions."[78] He remembered that Buchanan also "named former Democratic presidential candidate Lewis Cass secretary of state, and Fillmore's [Cabinet] had such luminaries as Daniel Webster, Thomas Corwin, and John Crittenden"[79]—the result hardly harmonious. Lincoln took a page from the experiences of both presidents. He had met Fillmore even before he became president in 1850 and met Buchanan at his own 1861 inauguration, making a total of five presidents he had met before he took the oath of office.[80] He assembled a Cabinet of political leaders from throughout the nation and the political spectrum—with the obvious exception of extremist Democrats like Jefferson Davis, who had been sworn into the presidency of the Confederacy two weeks before Lincoln's inauguration.

The composition of Lincoln's Cabinet should have left no doubt about his determination to be inclusive and his confidence in managing a group that included men who looked down on him. (When told that Chase thought he was bigger than Lincoln, Lincoln said, "Well, do you know of any other men who think they are bigger than I am? Because I want to put them all in my cabinet.") It also showed his intent to follow Jackson and his strong stance against secession. To remove any lingering doubt, one of Lincoln's first decisions upon entering the presidential office was to choose a portrait to hang over the mantelpiece. He asked for the official portrait of Andrew Jackson to be placed there. For the rest of Lincoln's presidency, Jackson looked down upon the long table where the Cabinet met every Tuesday and Friday. The choice of Jackson was obvious; it underscored Lincoln's recognition of Jackson's impassioned and effective commitment to preserving the Union. Lincoln used the portrait to great effect throughout his presidency.

Lincoln hung no portraits of Henry Clay. He didn't need any. He walked, talked, and thought about Clay nearly all the time; he had told Clay's son as much, and nearly anyone else who listened, as he quoted Clay at length. If Lincoln needed any further reminder of his mentor, he had his collected speeches nearby on his desk, alongside his beloved Shakespeare, available whenever he needed comfort or inspiration from either.

As people looked to the White House, they saw something never seen before—a president who was committed to following Jackson and Clay. The question they all had was, could he pull it off?

IV

Two days after Lincoln's inauguration, the Senate confirmed all seven members of his Cabinet, as well as John Nicolay's nomination as Lincoln's principal secretary. As had been done in prior administrations, the Cabinet met at a twenty-five-by-forty-foot hard walnut table in the president's office. Lincoln's third secretary, William Stoddard, described what visitors saw when they entered his office: "Folios of maps leaned against the walls or hid behind the sofas. Volumes of military history and literature came and went from various libraries and had their days of lying around the room or on the President's table."[81] The Jackson portrait hung just to Lincoln's left whenever he sat at the head of the Cabinet table. On the same wall hung a photograph of John Bright, a staunch supporter of the United States in Parliament since he entered in 1843 and a steadfast opponent of slavery. Lincoln said of Bright, "I believe he is the only British statesman, who has been unfaltering in his confidence in our success." (His admiration for Bright was evident when, after his assassination, his wallet was found to contain a clipping of a letter Bright sent to Horace Greeley about Lincoln's reelection, praising it for showing that "Republican institutions, with an instructed and patriotic people, can bear a nation safely and steadily through the most desperate perils.")

Lincoln called the Cabinet into his office for their first meeting on March 6. He did not mention Fort Sumter. On March 9, he brought the Cabinet back together to share new information he had just received on the growing threat in South Carolina. After a vigorous discussion, every Cabinet member except Montgomery Blair urged Lincoln to abandon the fort.[82]

Blair was unhappy, a circumstance that came as no surprise to anyone who knew him. Lincoln had long counted among his friends Blair's father, Francis Sr., the stern head of a political dynasty. In spite of their long attachment to the Democratic party, the Blairs had fled the party in opposition to slavery and the Kansas-

Nebraska Act. Buchanan hired him to serve as the government's top lawyer for the Court of Claims but then fired him because of his strident opposition to the administration's proslavery positions. Though he opposed slavery, Blair was not sufficiently extreme in his support of the rights of African Americans to appease Radical Republicans, nor was he as conservative as many other Republicans who were less absolutist in opposing slavery and more eager to avoid war. As a result, Blair ended up as postmaster general, not in the center of power but near it. Noah Brooks reported, "Blair, though a good Postmaster General, was the meanest man in the whole government."[83]

True to form, Blair did not hesitate to take advantage of Lincoln's trust. After the March 9 Cabinet meeting in which he found himself virtually alone on the question of how to handle Sumter, he threatened to resign. As Navy Secretary Welles related, Blair was "determined not to continue in the Cabinet if no attempt were made to relieve Fort Sumter."[84] Welles was as close to a natural ally to Blair as there could be. Like Blair, he had been a Jacksonian Democrat who had left the party to oppose slavery and had enthusiastically supported Lincoln during the 1860 election. Both men distrusted Seward: In the 1860 convention, Welles, who'd led the Connecticut delegation, had been unsettled by Seward's continuing attempts to undermine Lincoln throughout the convention, even after Lincoln had secured the nomination. Blair agreed with Welles that Seward was more interested in promoting his own career than in helping Lincoln succeed or protecting the Union. Now, with the Cabinet favoring appeasement and opposing any effort to salvage the fort, Blair told Welles he was disgusted and was considering resigning.

Blair's father, Francis, from whom Montgomery undoubtedly got his stubbornness, was apoplectic. On March 11, Francis Blair barged past Lincoln's secretaries to speak with Lincoln. He was determined that Lincoln not accept Montgomery's resignation but to resolutely stand by Fort Sumter. Lincoln knew Francis well enough to understand why he was agitated, and he anticipated what was coming. Everyone knew that the old man revered Jackson; Francis had been not only one of his earliest supporters but also a close adviser to President Jackson. After leaving the Democratic Party in 1854, Francis became one of the first prominent figures to join the founding of the Republican Party. In the 1860 Republican convention, the elder Blair initially supported his fellow Missourian Bates but quickly threw his support behind Lincoln when it became clear Bates had no chance. He was also, like Clay and Jackson, a strong advocate of gradual emancipation as a solution to the problem of slavery.

Now, with Jackson peering down from his portrait on Lincoln and Blair, and with Gideon Welles taking notes, the elder Blair "entered his protest against the non-action, which he denounced as the offspring of intrigue."[85] Francis told the president bluntly that doing nothing to save Fort Sumter was "virtually a surrender of the union."[86] Not mincing words, he told Lincoln, "It would be treason to surrender Sumter, sir. . . . If you abandon Sumter, you will be impeached!"[87] Welles observed that Blair's "earnestness and indignation aroused and electrified the President; and when, in his zeal, Blair warned that the abandonment of Sumter would be justly considered by the people, by the world, by history, as treason to the country, he touched a chord that responded to his invocation."[88] As Welles recalled, "The President decided from that moment that an attempt should be made to convey supplies to Major Anderson, and that he would reinforce Sumter."[89] It was the same basic plan that Browning had proposed to Lincoln in February before the president-elect left Illinois for his inauguration in Washington.

A letter Francis wrote to his son Montgomery the next day, March 12, confirms that Jackson was on his mind when he urged Lincoln to be firm. Though expressing misgivings about having spoken so bluntly to the president, Francis mentioned Jackson twice. First, he told Montgomery that the current situation demanded an "exposition" from the president "even more than the one faced by Genl Jackson" in 1832.[90] He also stressed that "there never was an occasion when an eloquent appeal by the President to the people like that of Genl Jackson in the crisis of 1832, could be of more use."

Lincoln agreed. On March 13, he met with Montgomery and his brother-in-law Gustavus Fox, a former Navy lieutenant. Fox had tried but failed to persuade President Buchanan to let him lead a reconnaissance expedition to learn firsthand the best ways to resupply Fort Sumter. Fox asked for permission to do this now. Lincoln said he would consider it.

Two days later, on March 15, the president requested each Cabinet officer to respond in writing to the question of whether it was "wise to attempt" to deliver provisions to Fort Sumter.[91] Blair was the only Cabinet member to unequivocally answer yes.[92] Seward and Chase were not alone in worrying Lincoln had "no conception of his situation."[93]

Lincoln was determined not to authorize any "hasty action" but instead wanted, as Welles noted, "time for the Administration to get in working order and its policy to be understood."[94] Lincoln talked further with Fox, then sent him to Charleston to get a look firsthand at Fort Sumter and the environs, get a sense of its vulnerability to attack, and assess the most feasible way to resupply it. Because Seward had insisted that the people of South Carolina might not want war, Lincoln asked Stephen Hurlbut, an old friend and fellow lawyer from Illinois, and Ward Hill Lamon, a former law partner who opposed abolition, to travel to Charleston and assess public opinion as best they could.[95] Lincoln understood the utility of spies, as George Washington had relied on Native Americans for intelligence during the French and Indian War and put together an underground spy network to track the British forces during the Revolutionary War. Winfield Scott had used a band of Mexican outlaws and army engineers to collect intelligence on the enemy's movements during the Mexican War, and Lincoln himself had served for twenty days with a special company of spies and scouts doing reconnaissance during the Black Hawk War.

Just as Lincoln's agents left to gather the intelligence he needed to firm up his plans, Lincoln heard again from Francis Blair. The senior Blair could not refrain from pushing Lincoln again. On March 18, he wrote the president to reiterate his demands that Lincoln should stand firmly by Fort Sumter, and he warned Lincoln that Seward was a "thoroughly dangerous counselor."[96]

Lincoln barely had time to digest Blair's opinions. He received more compelling news from both Fox and Hurlbut. Fox reported the logistics required to reinforce Sumter, while Hurlbut shared the disquieting news that the South Carolinians whom he had seen and spoken with "had no attachment to the Union."[97]

As the time for action grew short, Lincoln had to work with what he had. After a state dinner on March 28, he asked his Cabinet to remain so that he could update them on the Sumter situation. He shared a letter he had received from General Scott suggesting that it would be necessary to abandon Fort Sumter as well as Fort Pickens on the Florida coast in order to "soothe and give confidence to the eight remaining slave-holding States, and render their cordial adherence to the Union perpetual."[98] Montgomery Blair "smelled a rat." He erupted, denouncing Scott for "playing politician" rather than acting as a general.[99] Everyone present understood Blair's reference was really directed at the secretary of state.[100] By the end of the meeting, Seward alone stood by the general, while the rest of the Cabinet, much to Blair's surprise, agreed with him.[101]

On March 30, Lincoln reassembled the Cabinet to hear his plan to shore up the fort. With his mind made up, Lincoln marched back and forth at the front of the room as if he were addressing a jury, musing aloud about the arguments for and against his authorizing Fox to lead a clandestine expedition to bring supplies to the fort. The presentation worked. Over the objections of Seward and Interior Secretary Caleb Smith, who'd chaired Indiana's delegation at the Republican convention, the Cabinet approved the plan, though some members were concerned about whether Lincoln was up to the job ahead.

V

The execution of the plan was less than perfect. Lincoln bypassed Simon Cameron, the aging war secretary, who had missed most of the important meetings of the Cabinet and seemed clueless when it came to talking substance or strategy. Instead, Lincoln turned to Seward, Welles, and Blair, all men of action. The problem was that Seward and Welles disliked each other, while no one much liked talking with Blair. Lincoln still did not trust Seward, but he insisted on being involved, and Lincoln saw no problem with allowing him a say.

Seward was convinced Sumter was a lost cause. At the March 30 Cabinet meeting, he had proposed as an alternative to Fox's expedition that a separate one, led by Captain Montgomery Meigs, the army engineer in charge of the construction of the Capitol, should be sent to reinforce Fort Pickens as a base in Florida. Seward followed up the meeting with a memorandum pressing Lincoln to abandon the Fox expedition in favor of the Pickens expedition.[102]

Lincoln stuck by his commitment to Fox's resupply effort. On his own volition, Seward reached out to the official commissioners that the Confederate States of America (the name adopted by the eleven secessionist states) had sent to Washington, in order to determine if any peaceful settlement was possible. Nothing came of the scheme, except for distancing Seward even further from the Navy venture Welles was organizing.

On April 1, Lincoln directed Fox to lead an expedition down to Charleston to deliver supplies to Fort Sumter.[103] Unfortunately, Lincoln did not realize that he had already ordered the ship he was intending to be sent with Fox down to Florida—the *Powhatan*, the most powerful in the Navy. Shortly after the March 30 Cabinet meeting, Seward had brought a large stack of papers for Lincoln to sign, but Lincoln was busy and signed them without carefully looking. One authorized the *Powhatan* down to Florida.

When Welles discovered on April 1 that Seward had tricked

Lincoln, he hauled Seward to the White House. It was midnight, but the president had not yet gone to bed. Welles demanded Seward explain himself to Lincoln. Seward reminded the president of his plan to send reinforcements to Fort Pickens and of the fact that the president had signed the paper authorizing them. Welles recalled Lincoln "took upon himself the whole blame, said it was carelessness, heedlessness on his part" and that "he ought to have been more careful and attentive."[104] Then, with both Seward and Welles beside him, Lincoln directed that the *Powhatan* be redirected so that it could support Fox's expedition.[105]

Seward kept hedging his bets. He telegraphed Lieutenant David Porter, who was in command of the *Powhatan*, to change course from sailing to help Fort Pickens in Florida and instead go to Charleston to support Fox and Fort Sumter. But he signed the telegraph "Seward," and Porter declined to follow the direction, since an order from the secretary of state could not countermand the original order, which had been signed by the president. As a result, the *Powhatan* remained on its path to Fort Pickens. This left Fox's Sumter expedition so seriously weakened that Fox's wife, Virginia, the daughter of Levi Woodbury, a former Navy secretary and Supreme Court justice, later described Seward's actions as "cruel treachery."[106]

On April 6, with Fox hoping in vain that his expedition could have been kept secret, President Lincoln agreed to follow Browning and Seward's counsel to send a messenger from the State Department to inform the governor of South Carolina of Lincoln's intention to send supplies, but not arms and ammunition, to Fort Sumter. Lincoln had no illusions about the response. His spies had informed him about the local population's and government's hostility toward the Union. Still fresh in his mind was Browning's prediction, made a month before his inauguration, that South Carolina would attack any effort to resupply the fort. Other advisers agreed that such an attack was highly likely under the circumstances.

On April 12, the governor of South Carolina did just as Lincoln and Browning had predicted and Seward feared. With Fox and his two hundred reinforcements watching helplessly, South Carolina began bombarding Fort Sumter with heated cannonballs and Fort Moultrie with shots from forty-three guns and mortars. Thirty-six hours later, Major Anderson and his garrison at Sumter surrendered.

The bombardment of Fort Sumter signified the beginning of the Civil War, though many people hoped reconciliation was still possible. Over the next few weeks, newspapers and politicians from both parties, including Seward, pressed Lincoln not to engage Southern forces. But Lincoln knew the die had been cast. When a delegation from Maryland, including the governor and the mayor of Baltimore, urged him to avoid war, he pointed to the portrait of Jackson and praised the former president as an exemplar of the firmness he now needed to emulate. Lincoln vowed to emulate Jackson's "manliness" in doing whatever was necessary to protect the Union and the fort.[107]

On April 22, Lincoln responded to the Maryland delegation's letter, pointing out that they were urging him to "ask for peace on any terms, and yet you have no condemnation for those who are making war on us."[108] He told them, "You would have me break my oath and surrender the Government without a blow. There's no Washington in that—no Jackson in that—no manhood nor honor in that."[109] Now that war was imminent, Jackson was Lincoln's north star, not Clay.

South Carolina's attack rallied both sides to their respective causes. On April 15, Lincoln had implemented a draft for seventy-five thousand recruits and called for a special session of Congress to convene on July 4 to discuss the impending war.[110] While support for Lincoln and the Union was never stronger throughout the North, each of the four Southern states that had not yet seceded promptly did so—North Carolina, Virginia, Arkansas, and Tennessee. They joined the Confederacy, which moved its capital to Richmond. Meanwhile, Lincoln communicated with governors throughout the North and met with state delegations, as well as on April 14, for several hours, with Stephen Douglas, who was eager to help the president put down the rebellion of the secessionist states.

Lincoln read Douglas the proclamation calling for conscripting seventy-five thousand troops. Douglas said, "Mr. President, I cordially concur in every word of that document, except instead of a call for seventy-five thousand men I would make it two hundred thousand. You do not know the dishonest purposes of those men [the rebels] as I do."[111] He then showed Lincoln "the strategic points which" needed to be "strengthened for the coming contest."[112]

VI

Before the special session of Congress began on July 4, 1861, Lincoln made one of the most controversial decisions of his presidency. On April 27, he suspended the writ of habeas corpus to give military authorities the power to jail anyone without due process whom they deemed to be a traitor or who was stirring rebellion between Washington and Philadelphia. The decision provoked a confrontation between Lincoln and Jackson's old friend, Roger Taney, the chief justice who had sworn Lincoln into office. Ironically, here too, Lincoln's model was Jackson.

Maryland was the problem. It was split over whether to secede or to help the Union, particularly whether to assist, even indirectly, with fortifying the capital. At the April 15 meeting with the delegation from Baltimore, President Lincoln did more than refuse to agree to their pleas that he prohibit his troops from moving anywhere through Maryland.[113] With Welles taking notes, Lincoln explained further, "Our men are not moles, and can't dig under the earth; they are not birds, and can't fly through the air. . . . Go home and tell your people that if they will not attack us, we will not attack them; but if they do attack us, we will return it, and that severely."[114]

The situation quickly worsened, as Marylanders destroyed railroad bridges and cut telegraph lines linking Maryland to the

North. On April 25, reinforcements finally arrived in Washington, but Maryland remained a hotbed of Confederate sympathizers, who impeded federal troops and encouraged interference with their operations.

Unrestricted by habeas corpus, federal authorities, led by General George Cadwallader, promptly imprisoned a man named John Merryman for recruiting, leading, and training a drill company in Maryland in service of the Confederacy. Lincoln had yet to ask Congress to declare war but instead to treat those joining or helping the rebellion as traitors, the same way Taylor and Fillmore had regarded forces rebelling against federal authority.

Imprisoned in Fort Henry in Baltimore, Merryman directly appealed to Chief Justice Roger Taney for a writ of habeas corpus. (Because Taney also sat as a U.S. Circuit Judge in Maryland, the petition came directly to him.) The writ would have required Merryman's jailer to come before Taney to explain Merryman's confinement. But when the chief justice ordered the writ to be delivered to General Cadwallader for an answer, the general ignored it. Incensed by Cadwallader's insolence, Taney granted the writ and issued an opinion explaining his reasoning and denouncing Lincoln's actions as unconstitutional. Taney stated that only Congress had the power to suspend the writ and rejected Lincoln's argument that he had the authority to suspend habeas corpus while Congress was in recess. Taney did not directly order Lincoln or anyone in his administration to release Merryman, making it easier for Lincoln to ignore Taney's opinion. Besides having questioned the legitimacy of Taney's *Dred Scott* decision, he was now open to the charge of flouting the Constitution and Taney's authority.

Yet the precedent for suspending habeas corpus and ignoring Taney's decision came from Andrew Jackson. After victories over the British in Mobile, Alabama, Major General Jackson directed his forces to head off the British army before it could take New Orleans. Arriving before the British, Jackson declared martial law on December 14, 1814. Over the next few months, he imposed a curfew, censored a newspaper, came close to executing two deserters, arrested a hundred soldiers for mutiny and desertion, banished Frenchmen newly naturalized as American citizens, ignored the Louisiana governor's order to stand down, and jailed a congressman, a federal judge, and the federal district attorney. When he was confronted with a writ of habeas corpus, he ignored it. Even though the Supreme Court had declared not long before that only Congress had the authority to suspend the writ of habeas corpus, Jackson decided, on his own initiative, to suspend it.

When news finally reached Jackson that a peace treaty had been signed to end the hostilities, he relented and grudgingly lifted his order of martial law. After he was released, John Dick, the federal district attorney, demanded that Jackson be brought before the federal judge, Dominick Hall, to explain himself. When Jackson appeared before Judge Hall, he refused to answer any of his questions. Judge Hall found Jackson in contempt of court and fined him $1,000.

It was not in Jackson's nature to forget any slight—in this case, the fine. Nor did he forget how Henry Clay had opposed any efforts in Congress to erase the fine. Jackson never gave up trying to have the fine wiped from his record. After leaving the presidency, he got Polk's help to expunge the fine once and for all. Shortly before he died, he asked Congress to reimburse the fine. On January 8, 1844, Congress, on the twenty-ninth anniversary of the Battle of New Orleans, agreed.

There was no mention of Jackson's declaration of martial law in George Prentice's 1831 biography of Clay, for Clay was out of the country at the time as a member of the presidential commission charged with negotiating the Treaty of Ghent, which ended the War of 1812. But Lincoln must have known about Congress's indemnifying the fine, since he was eyeing a run for the House as early as 1844 and discussion of the reimbursement was one of the most heated debates of the time.

There was also the 1832 Supreme Court opinion, written by Chief Justice John Marshall, that invalidated a Georgia criminal statute that prohibited non–Native Americans from being present on Native American lands without a license from the state of

Georgia. Marshall explained that the statute was unconstitutional because only the federal government had the authority to interact with Native American tribes. Jackson is said to have responded, "John Marshall has made the decision, now let him enforce it."[115] Although the comment is probably apocryphal, both Jackson and the State of Georgia ignored the decision.

Marshall had not addressed any order directly to either Jackson or the State of Georgia, even though the State of Georgia was holding three missionaries in its prison for having violated the state statute. Rather than abide by the decision, Georgia authorities pressed the federal government to remove the Cherokees from the land that the state was trying to regulate on its own terms. If Jackson did not enforce the Court's decision, none of the missionaries would be released. Though the missionaries were released a year later, Jackson did nothing in the interim. Having lived among diehard Democrats for decades, Lincoln likely often heard praise for Jackson's defiance of the Court's order in 1832. Lincoln was even more familiar with Jackson's veto of the national bank, which rejected any obligation on the president's part to follow the Supreme Court's earlier decision upholding the national bank's constitutionality.

By the time Lincoln was president, there was no question he knew of Jackson's actions and approved of them. In an address to Congress on July 4, 1861, and in his first Annual Message delivered at the end of the year, he asked the same rhetorical question, "Are all the laws, but one, to go unexecuted, and the government itself to go to pieces lest that one be violated?"[116] He answered the question the same way in both, by citing the only precedent he thought was relevant—the 1844 refund demonstrating Congress's approval of Jackson's declaration of martial law, including his suspension of habeas corpus.

Throughout the remainder of 1861, Congress debated but never reached consensus on the conditions required for the suspension of habeas corpus, and Lincoln felt compelled further to suspend the writ on the Florida coast and in the area between New York and Philadelphia. In the meantime, Henry May, who had been elected as a Democratic representative to the House from Maryland in the 1850s, returned to the House in 1861 as a member of the newly formed Unionist Party, which was made up of former Whigs who wanted to stop the movement toward secession over the issue of slavery.

After the special session of Congress that Lincoln had called, May was taken into custody, without charges or recourse to habeas, on suspicion of treason. His crime was strongly objecting to Lincoln's war policies. Eventually, May was released, and in December 1861, he returned to his seat in the House, where he sponsored a bill that would have made it impossible for someone charged with a federal offense to be incarcerated pending trial and conviction. In 1863, the Senate approved a law incorporating May's bill, which became known as the 1863 Habeas Corpus Suspension Act. President Lincoln signed the act into law, which effectively ratified his suspension of habeas corpus. As Jackson had done, Lincoln had found a way for Congress to vindicate his actions.

VII

Stephen Douglas had done everything he could to fight secession and the prospect of war. While Lincoln remained silent in public during the 1861 presidential campaign, Douglas did not. Knowing he would lose the general election, Douglas did what he could do to preserve the Union, campaigning vigorously, particularly in the South, denouncing secession, and urging reconciliation and compromise everywhere he went. (He was spending the night in Mobile, Alabama, when he received the news that he had won only a single state in the election, while other Democratic candidates had won fourteen, and Lincoln had won with eighteen states and a majority in the Electoral College.) Shortly after the election, Douglas met with Lincoln and pledged his support to his success and preservation of the Union. On Inauguration Day, he dramatically stepped forward to hold Lincoln's hat and cane while Lincoln read his Inaugural Address. With Lincoln's knowledge and gratitude, Douglas traveled widely throughout the spring and early summer of 1861 to rally support among conservatives for maintaining the Union, but on the Southern portion of his trip he contracted typhoid fever, and on June 3, 1861, Douglas died in the arms of his wife. Upon hearing the news, Lincoln told those around him that he and Douglas "are about the best friends in the world" and then, in a rare display of emotion, burst into tears.[117] The two men had known each other—and battled each other—for three decades. The only public statement from the administration came from Simon Cameron, who had known Douglas for years as a colleague in the Senate. In a rare moment of leadership of the War Department, Cameron issued a circular to be distributed throughout the Union Army announcing "the death of a great statesman, [a] man who had nobly discarded party for his country."[118] The Little Giant was forty-eight years old.

The vacancy in the Senate opened up an opportunity that Lincoln quickly seized. With the Illinois legislature in recess, the state's Republican governor, Richard Yates, in consultation with Lincoln, made the appointment of Orville Browning on June 12, 1861, to take Douglas's seat. In early July, Lincoln's spirits visibly lifted when Browning arrived. Senator Trumbull presented Browning's credentials to the Senate, and in his first act as a senator, Browning joined his new colleagues in eulogizing Douglas. He emphasized the fact that Douglas had "placed patriotism above partisanship."[119]

In characteristic fashion, Lincoln unburdened himself to Browning. Browning did not hesitate to speak his mind, though the two did not always agree. Well before Browning arrived in person in July, he had written to the president, on March 26, saying, "You should not permit your time to be consumed, and your energies exhausted by personal applications for office."[120] Lincoln demurred: "I must see them."[121] Yet Lincoln was happy to see Browning, upon his return to Washington, sharing the news that "The plan succeeded." According to Browning, Lincoln explained,

"They attacked Sumter—it fell and thus, did more service than it otherwise could."[122]

Upon Fox's return from South Carolina, Lincoln appointed him an assistant secretary of the navy and consoled him, "You and I both anticipated that the cause of the country would be advanced by making the attempt to provision Fort Sumter, even if it should fail; and it is no small consolation now to feel that our anticipation is justified by the result."[123] During the special war session, Lincoln told Congress the same thing: "No choice was left but to call out the war power of the Government; and so to resist force, employed for its destruction, by force, for its preservation."[124] Perhaps Lincoln was trying to make the best of a bad situation, to control the narrative as best he could, or to show that he had not followed Polk's haste in making war. In any event, both the fact that South had fired the first shot and the perception of the Union's needing to use force to defend itself against the South's aggression broadened Lincoln's public support in the critical early days of the Civil War.

Nicolay recalled that Browning and Lincoln conversed daily in the White House and frequently rode and dined together, sometimes in the company of Lincoln's wife and family. Browning was more than a friend—he knew what Lincoln needed to become his best. Over the next year and a half until the end of Browning's Senate term in January 1863, no senator visited the White House more than he did. He became Lincoln's "eyes and ears" and leading spokesman in the Senate.[125] It was good to have someone to confide in, especially as Lincoln could not have imagined the challenges that were about to beset his family and the growing conflicts that he had to address with the Cabinet, Supreme Court, Republican Party, and Congress as well as on the bloody battlefields in the South.

CHAPTER SEVEN

COMMANDER IN CHIEF
(1861-1864)

Of the nation's first sixteen presidents, Abraham Lincoln had nearly the least military experience. Only two—John Quincy Adams and Martin Van Buren—had less: each had none. Lincoln's brief experience fighting mosquitoes in the Black Hawk War could hardly match the experience and know-how of the legendary generals Washington, Jackson, and Taylor. Aside from the presidents who, in office had directed or prior to taking office, ordered the slaughter of Native Americans—Washington, Jackson, Harrison, and Taylor, among them—James Madison was the only president before Lincoln to serve as commander in chief in a war on American soil, but his fleeing the White House in 1814 to escape the invading British soldiers, who torched the place in retaliation for the U.S. invasion of Ontario two years before, hardly set the example of the kind of commander Lincoln needed to be.

Lincoln had the complicated example of Polk's management—or as he saw it, the mismanagement—of the Mexican War. He had criticized Polk for his lies in starting the war, his poor relations with Congress during the war, and especially his refusal to share information about the war with Congress, based on "the important principle, always heretofore held sacred by my predecessors," to decline congressional requests for internal executive branch documents.[1] Polk cited Washington as a precedent for his defiance, but John Quincy Adams, with Lincoln sitting nearby, had thundered on the House floor, "Although the very memory of Washington, by everybody in this country, at this time (and by none more than myself), is reverenced next to worship—the President was wrong in that particular instance, and went too far to deny the power of the House; and as to his reasons, I never thought they were sufficient in that case." The case involved the Jay Treaty (forged in 1795 to avert further war with England, settle debts, and provide a framework for peaceful relations), and Adams noted that friend and foe alike in Congress opposed Washington's refusal to release documents.[2]

In Congress, Lincoln never seriously objected to Polk's vigorous use of his war powers once the conflict had begun, even though Polk had battled with his two Whig generals, Taylor and Scott. Moreover, Democrats were at the top of Lincoln's list when appointing generals and other officers at the start of the war, among them John McClernand, who had served with Lincoln in Congress and been an outspoken critic of the Wilmot Proviso; Benjamin Butler, who had supported Jefferson Davis for the Democratic nomination for president in 1860; and John Dix, who had served as Buchanan's secretary of the Treasury. As president, "Lincoln never accused Democratic generals of sabotaging the war effort. Polk rarely mentioned Taylor and Scott without making such an accusation."[3] Though some Democrats were disasters in the field, Lincoln's appointments pleased their constituents.

The war was not the only thing Lincoln confronted in these tense years, just the most important. Not only were he and his army fighting to save the Union, but he and Congress were also refashioning America, just as Clay had envisioned it. In signing legislation authorizing two companies to build a transcontinental railroad to link East to West, Lincoln was effectuating a central component of Clay's American System, reassuring former Whigs and others in the mainstream that Clay's dream was still alive. He reshaped the Supreme Court, as only George Washington and Andrew Jackson had done before him. In repeatedly clarifying the objectives of the war, Lincoln did the opposite of what Polk had done in obscuring the reasons for the Mexican War. Lincoln revised the objectives several times, finally settling on the abolition of slavery as instrumental to maintaining the Union.

I

Shortly after the war began, Lincoln told his personal secretary John Hay that "the central idea pervading this struggle is the necessity that is upon us, of proving that popular government is not an absurdity. We must settle this question now, whether in a free government the minority have the right to break up the government whenever they choose."[4] Secession "is the essence of anarchy," he echoed Jackson and Clay on another occasion, "for if one state may secede, then others could claim the same entitlement until no government and no nation were left."[5] In his war message to Congress in July 1861 as well as his Annual Message delivered at the end of that same year, Lincoln made clear that the objective of the war was no less than to preserve the way of life of the United States. "This is essentially a people's contest. On the side of the Union it is a struggle for maintaining in the world that form and substance of government whose leading object is to elevate the condition of men to lift artificial weights from all shoulders; to clear the paths of laudable pursuit for all; to afford all an unfettered start, and a fair chance in the race of life."[6]

The way of life that Lincoln declared the Union Army was protecting was nearly identical to the political and economic vision underlying his primary mentor's conception of the American ideal. Lincoln said, "This just and generous, and prosperous system . . . opens the way for all—gives hope to all, and energy, and progress, and improvement of condition to all."[7] The freedom to become a "self-made man" was now the fundamental ideal that Lincoln believed the war was being fought to protect.[8] The foundation for linking the purpose of the war to eradicating bondage was thus set.

However, Lincoln encountered two immediate problems as commander in chief. First was his lack of any experience and knowledge of war compared with the fanatical leader on the other side. Jefferson Davis was a graduate of West Point and had distinguished himself as a colonel during the Mexican War. Only self-education could fill the absence of anything in his background to match that. Throughout the transition and his presidency, Lincoln tackled the study of military strategy as thoroughly as he had taught himself grammar, the law, geometry, and land surveying. "I am never easy now," he once explained, "when I am handling a thought, till I have bounded it north and bounded it south and bounded it east and bounded it west."[9] Herndon had witnessed firsthand that Lincoln "not only went to the root of a question, but dug up the root, and separated and analyzed every fiber of it."[10] John Hay, too, saw how Lincoln "gave himself, night and day, to the study of the military situation." Carl von Clausewitz, a Prussian general and military theorist, published posthumously his masterwork, *On War,* in 1832, the same year Lincoln was entering the state legislature. The leading treatise in the field, it defined war as "the continuation of politics by other means." Lincoln agreed. He was witness to the breakdown in American politics that brought the contending sides to war. He "pored over the reports from the various departments and districts of the field of war. He held long conferences with esteemed generals and admirals, and astonished them by the extent of his special knowledge and the keen intelligence of his questions."[11]

Lincoln familiarized himself, too, with the relevant precedents as to the basic chain of command set forth in the Constitution; commanding generals reported to the secretary of war, who in turn reported to the president. Ironically, this principle of civilian control of the military had been cemented most clearly and recently by Jefferson Davis when he served as Pierce's secretary of war after he had had trouble getting the army's commanding general, Winfield Scott, to report to him rather than directly to the president. Pierce came down on the side of Davis and the idea of civilian control of the military, ordering Scott to report to Davis, who in turn would report to him. Unhappily, Scott obliged.

As war secretary, Jefferson Davis had overseen the improvement of American weaponry and the professionalization of the army. Because of him, the Union Army was in better shape in 1861 than it had ever been before, with more advanced training and arms, as well as new standards for promotion that were designed to ensure that the best officers moved up the ranks. Lincoln was also familiar with presidents reprimanding or removing incompetent or disloyal generals, as Polk claimed to have done with Taylor, who had defied his orders, and as Monroe had done with Jackson during the Seminole War for going well beyond his instructions.

As Lincoln explained to Nicolay, Simon Cameron had been of no help in the struggle threatening the Union. He was, Lincoln said, "utterly ignorant and regardless of the course of things . . . Selfish and openly discourteous to the President. Obnoxious to the Country. Incapable either of organizing details or conceiving and advising general plans."[12] Because Cameron was inept and often absent and unreachable, Lincoln requested Scott on April 1, 1861, to "make short, comprehensive daily reports to me of what occurs in his Department, including movements by himself, and under his orders, and the receipt of intelligence."[13] Getting the facts was essential. Lincoln would visit battlefields at least a dozen times over the course of his presidency, lifting men's spirits when he could and assessing the progress of war, as well as the men he had charged to end it. Lincoln respected the opinions of experts, the experienced military personnel in the field. Scott was fussy, as everyone said, but Lincoln did not care about his demeanor, and trusted his opinion more than that of anyone else around him. Nothing was more important than winning the war, so Lincoln put aside his ego to find and listen to the generals who could end it as soon as possible.

Scott's place in the firmament did not last long, however. It soon became clear that advanced age, poor health, lack of energy, and the inability to mediate disputes about scenarios and plans among his generals made it impossible for Scott to run the army competently, and he submitted his resignation letter to Lincoln. At first, Lincoln did not accept it. He was unsure who would be a suitable replacement. Scott urged Lincoln to name his chief of staff, Henry Halleck, as his replacement. Nicknamed Old Brains because of his high forehead and supposedly high intellect, Halleck had returned to the army only a couple of months before, after having served for many years as an expert on mining law in California. The other possible choice was General George McClellan, who was pushing hard for the job. After the Union forces led by Brigadier General Irvin McDowell, who reported to Scott, had been routed in the First Battle of Bull Run, in Virginia on July 21, 1861, Lincoln summoned McClellan from western Virginia, where his forces had been more successful than Scott's men had been. On July 26, the Department of the Shenandoah (under Major General Nathanial Banks) was merged with McClellan's Division of the Potomac (which included the Department of Northeast Virginia, led by McDowell, and the Department of Washington, under Brigadier General Joseph Mansfield). The merged forces became known as the Army of the Potomac, the Union's principal army in the eastern theater of the war. However, McClellan almost immediately began clashing with Scott and Halleck over nearly everything, from tactics and strategy to how much autonomy he had. Aware that Scott was ailing, McClellan met with senators to lobby for Scott's position and to force the old general out. On October 18, 1861, Lincoln and his Cabinet accepted Scott's resignation, which he had resubmitted, and on November 1, Lincoln named McClellan as his replacement. On turning the entire command over to McClellan, the president confided to him his concerns that "this vast increase of responsibility . . . will entail a vast labor upon you."[14] McClellan assured him, "I can do it all."[15]

II

At thirty-four, dashing George McClellan was the youngest man ever placed in command of the U.S. armed forces. Though his appointment raised great hopes for a swift Union victory, it took only four months for him to be a greater disappointment than Scott. He had graduated second in his class at West Point but was egotistical and duplicitous. As Lincoln's biographer Michael Burlingame suggests, "Compounding his paranoia was a streak of narcissism, predisposing him to envy, arrogance, grandiosity, vanity, and hypersensitivity to criticism."[16] An unabashed Democrat, McClellan was a protégé of Jefferson Davis, but Lincoln figured this might be an asset for McClellan in anticipating the enemy's moves. Lincoln was wrong. Except for Davis, McClellan had a record of holding every other one of his superiors in contempt, especially Lincoln. McClellan told his wife that Lincoln was his "inferior" and "nothing more than a well-meaning baboon."[17] When in Lincoln's presence, he could barely look at him, much less speak with him.

McClellan's downfall was more than a story of his own undoing. It was the story of Lincoln's mastering one of the most important but most underestimated powers of the presidency—the power to remove, without any other branch's approval, badly performing, disloyal, or incompetent executive branch officials. In this case, Lincoln had plenty.

III

Of the six presidents preceding Andrew Jackson, only one—Thomas Jefferson—had been confident that the Constitution gave the president the authority to remove Cabinet and other officers whenever he saw fit, regardless of whether the Senate had confirmed them. Proclaiming himself the heir to Jefferson, Jackson had taken this practice to a new level. He called it the principle of rotation, which merely meant that a president could dismiss officials and replace them with his friends and allies as he pleased.

Whig presidential candidates campaigned hard against the principle of rotation. They thought it was just a cover for the spoils system and that it encouraged corruption and undercut the professionalism and expertise that they hoped would distinguish the federal workforce. Nevertheless, when Whig presidents got into office, they followed Jackson's model anyway. Subsequently, Tyler and Fillmore, the latter perhaps the most faithful Whig president of all, removed their Cabinets when they didn't obey. Zachary Taylor removed nearly two-thirds of his predecessors' political appointees during his first year in office and was on the verge of replacing his entire Cabinet before he died.

Once he became president, Lincoln did Taylor one better and orchestrated the largest turnover in office of any president up until that time. He was convinced that his power to remove men not performing as he liked was his most important weapon for winning the war. He not only rotated Buchanan's entire Cabinet out of office, but also removed 1,457 men from the 1,639 offices to which he was entitled to make nominations. Like Jackson, who had introduced the spoils system to American politics, Lincoln understood, as David Herbert Donald observed, that "patronage is one sure way of binding local political bosses to the person and principles of the President, and for this reason [Lincoln] used and approved the spoils system . . . Lincoln's entire administration was characterized by astute handling of patronage."[18] Lincoln proudly declared that his administration "distributed to [its] party friends as nearly all the civil patronage as any administration ever did."[19] Yet, removal was easier for Democrats who believed that the Cabinet was supposed to serve the president rather than for the Clay Whigs, who believed that the president served the Cabinet. Many of the former Whigs in Lincoln's Cabinet—like Seward and Chase—would have preferred that Lincoln do no tinkering with the Cabinet but instead defer to them on whether or not to dismiss any of their underlings.

When it was Lincoln's turn to decide whether or not he had the unilateral authority to remove a Cabinet officer, he deftly tried to follow the Jackson model but paid lip service to the Whig orthodoxy of seeking congressional acquiescence, if not approval. Whigs, particularly Henry Clay, had made this a central tenet of their party in response to what they perceived as Jackson's arrogation of congressional authority through his overabundant use of the veto and determination to impose his will on the entire executive branch, beginning with the Cabinet. If presidents allowed their Cabinets to give them direction on domestic issues, their function would become more confined to carrying out the will of either the Cabinet or Congress. Either way, the president would be contained.

Harrison was president only for thirty-one days, but from the first day of his administration, Harrison bristled at suggestions from his Cabinet, including Secretary of State Daniel Webster, that decisions should be made by majority rule, with each Cabinet member having a single vote and with the president having only a tie-breaking vote if the Cabinet was deadlocked. Harrison opposed the idea, and he kept his temper in check until the day when Webster told him that the Cabinet had rejected his preferred candidate John Chambers for the position of governor of Iowa and instead appointed someone else more to their liking. After a few seconds of awkward silence, Harrison motioned for a piece of paper on which he wrote a few words. He asked Webster to read the message to the Cabinet; the message was succinct: "William Henry Harrison, President of the United States." Harrison then rose to his feet and angrily told the Cabinet, "—And William Henry Harrison, President of the United States, tells you, gentlemen, that . . . John Chambers shall be Governor of Iowa."[20]

In early January 1862, Lincoln did not formally seek his Cabinet's approval to dismiss Cameron. In this case, several Cabinet members—Seward, Welles, and Chase—had been encouraging Lincoln for some time to remove him, making this a decision unlikely to backfire. Besides concerns about inadequacy, corruption, and mismanagement of the department's finances (the War Department was commonly described as "the lunatic asylum," with generals running things as they saw fit), Lincoln ordered Cameron to withdraw a declaration that he had made in December 1861 announcing, in an effort to ingratiate himself with Radical Republicans, the emancipation of all rebel-owned slaves. Congressional leaders were aghast that Cameron had done this on his own volition, so when the time came to replace him, Lincoln had the tacit approval of Congress. Lincoln cushioned the dismissal with an offer to appoint Cameron U.S. minister to Russia.

Lincoln's choice of Edwin Stanton to replace Cameron surprised nearly everyone. Besides insulting Lincoln when they first met in 1855, Stanton had been a loyal Democrat who had served as Buchanan's attorney general and a confidant of McClellan. (McClellan's reference to Lincoln as a "baboon"—or "gorilla," as sometimes quoted—likely was borrowed from Stanton.) However, Lincoln needed competence, decisiveness, intelligence, and energy at the helm of the War Department, not a friend. The appointment would be a helpful bridge to Democrats who wanted a voice in the administration, and Stanton, renowned for his integrity and relentless commitment to excellence and organization, would bring to the War Department sorely needed administrative leadership. Indeed, Lincoln knew Stanton had tried during the last months of the Buchanan administration to do what his boss refused to do: find a way to help the federal forts under siege in South Carolina and Florida. That effort, though unsuccessful, strengthened his suitability for the appointment.

Stanton enjoyed considerable support with influential leaders in both parties. Both Seward and Chase favored his appointment. Even Cameron is said to have approved of, or at least taken credit for, Stanton's replacing him. The Senate confirmed Stanton on January 15, 1862, eight days after Cameron's removal.

A few months later, Cameron's nomination as minister to Russia was delayed while a congressional committee considered censuring him for financial mismanagement in the War Department. Once Lincoln assured committee members that Cameron was not responsible for the irregular procurement practices of his department thus far in the war, the committee relented, and Cameron was confirmed.

While Cameron lasted less than a year in his new position, Stanton did all that Lincoln wanted and more. Over the next few years, he brought order, high standards, and efficiency to the War Department. He continued in the position until after Lincoln's death, when Andrew Johnson fired him. His dismissal became a basis for Johnson's impeachment, because Congress, after the assassination, had, based on its distrust of Johnson, modified its

Tenure in Office law to require Senate approval as a condition for removal of a Cabinet officer.

Rotation in office was not the only Jacksonian principle that Lincoln followed as president. He met with his Cabinet only when he felt the need. Even then, he used it primarily as a sounding board. He never expected the Cabinet to be harmonious or loyal, but he did expect it to be helpful. Thus he was following the predominant models of presidential-Cabinet relations. (The composition of Pierce's Cabinet had remained steady, although it sometimes tried to impose its will on the president.) Jackson and Polk, too, had used their Cabinets primarily as sounding boards and to rally support for and help in implementing the president's policies.

One dramatic episode illustrating this help arose in December 1861, when two Confederate envoys, James Mason and John Slidell (the latter of whom Polk had previously used to try to settle the tensions with Mexico before the conflict) were seized from the British passenger ship *Trent* by Union officers. Fears of a large-scale conflict with Britain had already prevented the sale of bonds in Britain to finance the war against the Confederacy. Lincoln consulted his Cabinet, as well as Browning, who wrote a memorandum for Lincoln on the points of international law involved in the removal of Mason and Slidell from the *Trent*. Browning initially favored using force if Britain "were determined to force a war upon us," but he and Lincoln eventually "agreed that the question was easily susceptible of a peaceful solution if England was at all disposed to act justly."[21] Browning urged a policy of conciliation—in order to ensure that the Union was fighting just one war at a time—which Lincoln followed in settling the matter by letting the two envoys go.

Lincoln handled the Cabinet much as he had dealt with temperamental foes in legislatures, courtrooms, and conventions. Perhaps illustrating the truth of John Todd Stuart's insight that Lincoln's wrestling match with Sullivan was a turning point in his life, Lincoln surprised opponents by using their own arguments and actions against them. He followed the same pattern in deciding to remove McClellan as the general in chief.

In his four-month tenure, McClellan strained Lincoln's patience to the breaking point. Less than two weeks after he was appointed supreme commander, Lincoln, Seward, and Hay visited his home on the evening of November 13, 1861, to check on the progress of the war. Told McClellan was out, the trio waited for his return. After an hour, McClellan returned through a different entrance. Notified by a servant that Lincoln and two others were waiting for him, McClellan made no comment and went straight to his room. After another half hour of waiting, Lincoln inquired again about McClellan's availability and was told that he had already retired for the evening. Hay was outraged and begged Lincoln to do something about such insolence, but Lincoln responded that it was "better at this time not to be making points of etiquette and personal dignity."[22] Lincoln could put up with anything if it helped to end the war as soon as possible. Although McClellan raised morale, the troops under his command lay fallow and inactive. Lincoln understood that victory meant destruction of the enemy and that such destruction required engagement and bloodshed, which McClellan appeared to want to avoid as much as possible.

McClellan's constant requests for more men and resources with no apparent plan for using them was testing everyone's patience. On January 10, 1862, Lincoln called a meeting of top generals and directed them to formulate a plan of attack. Claiming illness, McClellan refused to attend. Lincoln told the council, "If General McClellan did not want to use the army, [I] would like to borrow it."[23] When word reached McClellan that Lincoln was moving ahead with plans without him, he came to Washington and met with Lincoln and the other generals on January 12. Reluctantly, McClelland revealed—for the first time to Lincoln—his plan of attack. It entailed transporting the Army of the Potomac by ships to Urbanna, Virginia, on the Rappahannock River, for the purpose of outflanking the Confederate forces near Washington. From there, McClellan explained, the army would proceed to capture Richmond. Even when pressed, McClellan refused to give any further details to either Lincoln or his newly appointed secretary of war, Edwin Stanton.

Though pleased that McClellan had a plan, both Lincoln and Stanton were dubious. On January 27, 1862, Lincoln issued General Order No. 1, which specified that the "Land and Naval forces" should move "against the insurgent forces" on or before February 22.[24] Four days later, Lincoln issued a supplemental order directing the Army of the Potomac to move against the railroad supplying the Southern forces gathering at Manassas, Virginia, south of the national capital, on or before the same date. McClellan replied with a twenty-two-page letter objecting in detail to the president's plan and defending his own. By early March, McClellan and his superiors were at a standstill, and no engagement had yet occurred. Frustrated, Lincoln told a congressman, "If General Washington, or Napoleon, or General Jackson were in command on the Potomac they would be obliged to move or resign the position."[25]

Lincoln was hardly the only one whose patience had run out. Often during his visits to the White House, Browning scanned the battle maps lying around while Lincoln reviewed news from the front. They frequently talked about Lincoln's difficulty in finding the right general to lead Union forces and particularly about McClellan. Browning was present when Lincoln was visited by the members of the Joint Congressional Committee on the Conduct of the War, formed in December 1861, to tell Lincoln to push McClellan harder or fire him. After tepidly defending McClellan, Lincoln agreed. McClellan's friend and ally Edwin Stanton agreed, too.

McClellan had been overjoyed by the news of Stanton's appointment, which he had hailed as "a most unexpected piece of good fortune."[26] After just a few days in office, however, Stanton had a remarkable turnaround after seeing for himself how McClellan had mismanaged the situation. At the time, he wrote the president, "As soon as I can get the machinery of office going, the rats cleared out, and the rat holes stopped, we shall move. This army has got to fight or run away."[27] He added, "[The] champagne and oysters on the Potomac must be stopped."[28] On March 11, Lincoln met with his Cabinet, which agreed to end McClellan's short tenure as the commanding general of the Union forces. Lincoln told McClellan his dismissal was necessary so as to devote attention to leading the Army of the Potomac to rebuff Lee's invasion of Maryland.

Despite McClellan's mixed results, Lincoln named McClellan on September 12, 1862, to command "the fortifications of Washington, and all the troops for the defense of the capital."[29] A majority of the Cabinet sharply disagreed, declaring "our deliberate opinion that, at this time, it is not safe to entrust Major General McClellan the command of any of the armies of the United States."[30] Lincoln, backed by Browning and Stanton, felt that no one could do the job better than McClellan. The final clash between McClellan's forces and Lee's occurred at the Battle of Antietam in September 1862. The fight was a draw, after which Lee withdrew to the South.

Shortly after visiting an openly disrespectful McClellan on the battlefield in October, Lincoln had had enough. He ordered Stanton to remove McClellan from command entirely. Upon receiving the message of dismissal from Stanton, McClellan swore to his wife that Lincoln and Stanton "have made a grave mistake."[31] When the old political warrior from Missouri Frank Blair visited Lincoln on November 6, 1862, to protest McClellan's removal from general command of the Union's forces, Lincoln told Blair, "I said I would remove him if he let Lee's army get away from him, and I must do so. He has got the 'slows,' Mr. Blair."[32] Just as Taylor had used his dismissal as a platform to run for president, McClellan did the same. Once he was out from under Lincoln's command, he began assembling his own run for the presidency. He would be Taylor to Lincoln's Polk.

Lincoln struggled to maintain a good working relationship with members of Congress, many of whom fervently believed that they, not Lincoln, knew best how to win the war. The conflicts came to a head near the end of 1861.

In two long caucus meetings on December 16 and 17, Republican senators voted to press for a reorganization of Lincoln's Cabinet to secure "unity of purpose and action."[33] They were outraged over the outcome of the Battle of Fredericksburg, fought December 11–15, 1861, during which the Army of the Potomac incurred casualties three times as heavy as those incurred by Lee's forces. The senators blamed Seward for the debacle. This conveniently dovetailed with Treasury Secretary Chase's eagerness to get rid of Seward, who, Chase kept telling them, wielded too much influence over Lincoln, just as had he done over Taylor.

When he was a member of the House, Lincoln had urged transparency and candor upon the White House. But he would not allow himself, any more than Polk ever did, to be subservient to the House or Senate or a small band of senators, even if from his own party. On the evening of December 16, a messenger, accompanied by Senator Preston King of New York, brought to him a curt letter of resignation from Seward. Seward's son, Frederick, included his resignation as an assistant secretary of state. Lincoln rushed to Seward's house. Seward and his son were packing for their return to New York. Seward was adamant that Lincoln accept his letter of resignation in order to ease relations within the Cabinet and between the president and Congress. Lincoln strongly disagreed and urged Seward to stop packing. He asked Seward to keep the matter confidential until he had a chance to address the charges of discord in the Cabinet. Seward reluctantly agreed.

On the afternoon before the evening of his scheduled visit with the Senate delegation on December 18, Lincoln met with Browning, who reported what had happened during the Senate caucus meetings on the two days prior. Browning explained that he had defended Lincoln and Seward during the meetings, and that had Lincoln "caved in" and accepted Seward's resignation, he would have risked losing control over his administration.[34] Browning encouraged Lincoln to consider reorganizing his Cabinet. Lincoln listened, but regardless of the merits, the brewing insurrection within the members of his own party in the Senate infuriated him. He nearly shouted at Browning, "What do these men want?"[35] Lincoln then answered his own question. "They wish to get rid of me, and I am sometimes half disposed to gratify them. . . . Since I heard last night of the proceedings of the caucus I have been more distressed than by any event of my life."[36] Lincoln confided, "We are now on the brink of destruction. It appears to me that the Almighty is against us, and I can hardly see a ray of hope."[37]

When the senators arrived at the White House that evening, Lincoln had devised a plan. He kept the news of Seward's letter to himself. Once inside with the caucus, Lincoln listened patiently to the senators' complaints "attributing to Mr. S[eward] a luke-warmness in the conduct of the war, and seeming to consider him the real cause of our failures."[38] Benjamin Wade of Ohio blamed Lincoln for entrusting the conduct of the war to "men who had no sympathy with it or with the cause."[39] Wade further blamed Republican defeats in the recent midterm elections on the fact that the president had placed the direction of military affairs "in the hands of bitter and malignant Democrats."[40] William Fessenden of Maine added "that the Cabinet were not consulted as a council—in fact, that many important measures were decided upon not only without consultation, but without the knowledge of its members."[41] Fessenden denounced Seward for undue influence over the war's management and McClellan for being "pro-slavery," "sympath[izing] strongly with the Southern feeling," and unfairly blaming the administration for its failing to support the army.[42]

At this point, Lincoln interrupted. After years of experience in trying cases and in debating formidable foes, Lincoln had no intention of letting the opposition pull off a filibuster with no rebuttal; he had never allowed that to happen in court, and he would not allow it here. Producing a large stack of papers, he slowly read for more than half an hour the letters that he had written to McClellan to demonstrate his long-standing commitment to helping him and the war effort. His recitation caught the senators by surprise; they had no ready response. Without committing himself, the president invited the senators back the next evening, and they agreed.

The first thing the next morning, Lincoln assembled his Cabinet with the exception of Seward. He informed them of Seward's resignation and the visit from the group of nine senators representing the Republican caucus. "While they believed in the President's honesty," he told them, "they seemed to think that when he had in him any good purposes Mr. S[eward] contrived to suck them out of him unperceived."[43] Lincoln asked the Cabinet to return that evening "to have a free talk."[44]

When everyone arrived at seven thirty that evening, both the senators and Cabinet members, sans Seward, were surprised; neither group had had any idea that the other was coming. With senators and Cabinet members sitting uncomfortably across the table from each other, Lincoln delivered a long statement, commenting "with some mild severity" on the resolutions presented the evening before by the senators and explaining that whenever possible he consulted the Cabinet about important decisions but that he alone made the decisions, especially on matters of military strategy and command.[45] He said that members of his Cabinet sometimes disagreed but all supported a policy once it was decided, that he was "not aware of any divisions or want of unity," and that Seward was a valuable member of his administration.[46] Then Lincoln turned to the Cabinet members and asked them "whether there had been any want of unity or of sufficient consultation."[47]

Lincoln did not have to say what everyone present already understood, that he had just put Chase on the spot. Lincoln was aware that Chase had been the one who told the senators that Seward was the main source of all the problems in the Cabinet. Lincoln understood that if Chase now agreed with Lincoln, he would lose face with the senators who were present, but if Chase disagreed openly with the president, he would then lose the president's confidence. With all eyes on him, Chase took a moment to compose himself. He began defensively by saying "that he should not have come here had he known that he was to be arraigned before a committee of the Senate."[48] With everyone waiting for more, Chase grudgingly agreed "that questions of importance had generally been considered by the Cabinet, though perhaps not so fully as might have been desired" and that there was no want of unity in the Cabinet.[49] The meeting went on until one in the morning, but no one left thinking there would be any change in the Cabinet.

Much embarrassed, Chase came to the White House the next day to submit his resignation. "I brought it with me," said Chase.[50]

"Let me have it," Lincoln said, as he took the letter from Chase.[51] "This . . . cuts the Gordian knot," he said. "I can dispose of this subject now."[52] When Stanton, who had asked to be present, offered his own resignation letter to allay fears that senators were most concerned about the lack of progress in the war, Lincoln swiftly declined and ordered him to "go to your Department. I don't want yours."[53] He placed both in his desk.

In recounting the developments later to Senator Ira Harris of New York, Lincoln shared an anecdote about the time when he was a boy and had learned how to carry pumpkins while riding horseback. "I can ride on now. I've got a pumpkin in each end of my bag," he told the senator.[54] Lincoln had letters of resignation from his two most ambitious and meddling Cabinet secretaries—Seward and Chase—in hand, but he had no present intent to cash in on either. Later, he told Browning that he felt that he had shown that "he was master" of his administration and the senators "should not" have attempted to seize control of his Cabinet.[55]

A year after the Cabinet crisis of December 1862, Lincoln told John Hay, "I do not see how it could have been done better. I am sure it was right. If I had yielded to that storm and dismissed Seward the thing would all have slumped over one way and we should have been left with a scanty handful of supporters. When Chase sent in his resignation I saw that the game was in our hands and I put it through."[56] In the intervening months, before Lincoln made that comment to Hay, he and his Cabinet had addressed an even larger challenge in 1863—deciding whether the objective of the war was to maintain the Union, abolish slavery, or both.

IV

Settling on the objective of the war proved nearly as vexing for Lincoln as figuring out how to end the conflict. In his Inaugural Address, Lincoln had declared that the preservation of the Union was the war's objective—that and nothing more. He later explained that preserving the Union meant guaranteeing all Americans the freedom to become self-made men, on whatever terms they wished.

Lincoln appreciated both the legal and political problems with that objective. Lincoln was unsure whether the North could remain unified if the Union's fate turned on abolishing slavery. As a constitutional matter, it was unclear after the *Dred Scott* decision how far the federal government could go in regulating slavery. Was it barred from doing anything? Could it abolish slavery in the states, or at least the District of Columbia? Could it take intermediate steps, such as abolishing the slave trade in the District of Columbia or perhaps arrange funding to buy slaves and send them to Africa? Or was it required to allow slavery to spread without interference?

As Lincoln's presidency progressed, Radical Republicans and the members of the congressional Joint Committee on the Conduct of the War argued that the preservation of the Union and the abolition of slavery were intertwined. This raised other important but more technical questions: Did the president have inherent authority over the spread of slavery or status of those deemed slaves in the South, and to what extent did the president require congressional support or authorization to take any action, even as commander in chief, to interfere with slavery? As president, Lincoln worked through these questions step by step, and oftentimes with congressional approval or acquiescence, on a path that led him to issue one of the most famous executive actions in American history. In retrospect, the journey was not surprising. Through it all, Lincoln kept faith with the ideals he'd professed for decades as a Clay Whig.

The first step did not require Lincoln's participation at all. In fact, it was taken in the opposite direction of emancipation. In March 1861, Lincoln had insisted in his First Inaugural Address that the war's objective was to maintain the Union and pledged not to interfere with the institution wherever it existed. The next day Confederates routed Union forces in the First Battle of Bull Run.

On July 25, 1861, all but a handful of senators voted in favor of a resolution sponsored by Representative John Crittenden of Kentucky (he entered the House after running unsuccessfully for the presidency in 1860) and Senator Andrew Johnson of Tennessee, the only senator from a seceded state who remained loyal to the United States and remained in Congress after the Civil War began. The resolution affirmed that the war was being fought not for the purpose of "overthrowing or interfering with the rights or established institutions of those States," but only "to defend and maintain the supremacy of the Constitution and to preserve the Union."[57]

The second step toward emancipation occurred a few weeks later. On August 6, Congress passed the First Confiscation Act. The law authorized the federal government to seize the property of all those participating directly in rebellion. The law purposefully avoided specifying the permanent future status of confiscated slaves, but few expected that they would be returned to slavery. Without comment, Lincoln signed the bill into law. Privately, he told Browning, who had voted for it, that "the government neither should, nor would send back to bondage such as came to our armies."[58] Lincoln's position might have reflected the fact that as a former Whig he was prepared to approve laws that were not clearly unconstitutional and that this law, in any event, was consistent with the long-standing international law of war, which authorized the seizure of any property, including slave property. Also, the confiscation of slaves was a move long considered and contemplated among Whigs like Clay, for years, though they had envisioned the possibility of compensating slaveholders. With war, such financial niceties were no longer on the table.

The next step toward emancipation occurred shortly after Lincoln had signed the First Confiscation Act. Lincoln had assigned John Frémont, as head of the Department of the West, to clear Missouri of secessionists. On August 30, 1861, Frémont issued a proclamation in which he implemented martial law in the state of Missouri and ordered the confiscation of property, including slaves, of those who were resisting the Union. The reverberations of his proclamation were felt all the way back to Washington, where Lincoln countermanded Frémont's declaration on September 11,

1861. Lincoln had many problems with Frémont's proclamation. One was his conviction that the action breached the chain of command; Lincoln believed any such order should have become from the president, not a general, assuming that the president even had the power to do so. Lincoln ordered Frémont to revise his proclamation to conform to the Confiscation Act.

As soon as he learned of the trouble, Browning wrote to Lincoln on September 17, expressing his hope that the rumors he was hearing that the Cabinet had disapproved the proclamation were unfounded. He went further, in brash terms, to say that all truly loyal citizens favored the Frémont proclamation and that rescinding it would be demoralizing to many pro-Union Americans. He pointed out that there was no statute providing for the court-martial of enemies caught behind one's own lines, but everyone, he argued, agreed that such treatment of spies and traitors was perfectly legal. Why should those who rejected the Constitution expect to enjoy its protection? He said Frémont should receive unqualified backing from the administration, and he hoped the rumors that Lincoln planned to replace him were untrue. He went further to say, "I do think measures are sometimes shaped too much with a view to satisfy men of doubtful loyalty, instead of true friends of the Country. There has been too much tenderness towards traitors and rebels. We must strike them terrible blows, and strike them hard and quick, or the government will go hopelessly to pieces."[59]

Browning's tone stunned Lincoln. In a strongly worded letter on September 22, Lincoln outlined his national strategy. It was one of the longest letters he ever wrote in defense of his policies. Lincoln responded that Frémont's proclamation was "purely political and not within the range of military law or necessity."[60] He said that a general might, on grounds of military necessity, seize property from the enemy, including slaves, but "when the need is past, it is not for him to fix their permanent future condition."[61] He added, "You speak of it as being the only means of saving the government. . . . Can it be pretended that it is any longer the government of the U.S.—any government of Constitution and laws,— wherein a General, or a President, may make permanent rules of property by proclamation?"[62] It was Congress's job to make such rules, Lincoln reminded his fellow former Whig. He recalled that Congress had actually passed the Confiscation Law in the previous session before Frémont issued his proclamation and that Browning had voted for the law. To be consistent, Browning should give up his "restlessness for new positions, and back me manfully on the grounds upon which you and other kind friends gave me the election, and have approved in my public documents."[63] In other words, sometimes Lincoln just needed Browning to be a friend, not a mentor—or worse, a nag.

Another reason for the order countermanding Frémont's proclamations, Lincoln wrote, was the need to keep the border states in the Union. He told Browning there was a great deal of credible evidence that Kentucky would have gone over to the Confederacy if Frémont's proclamation had stood. "I think to lose Kentucky is nearly the same as to lose the whole game. Kentucky gone, we can not hold Missouri, nor, as I think, Maryland. These all against us, and the job on our hands is too large for us. We would as well consent to separation at once, including the surrender of this capitol [sic]."[64] Yet Lincoln emphasized that he did not issue his order to Frémont simply because of his concerns about Kentucky, for Kentucky had not yet taken a position. Lincoln assured Browning that he had no thought of removing Frémont because of what he had done and that he hoped it would not be necessary to remove him for any other reason.

Self-awareness was not one of Browning's strengths. He did not relent. On September 30, he responded with a long statement about the nature of war and its relationship to Frémont's proclamations. He believed that the law of nations furnished sufficient authority to use every advantage to weaken the enemy. He cited authority showing that a government waging war against a public enemy had a preexisting right to use the enemy's property to secure the ends of war. The people of Kentucky, Browning said, would not have opposed confiscation of horses to aid the federal troops; therefore, they should not oppose the confiscation of slaves when the exigencies of war and the preservation of the Union re-

COMMANDER IN CHIEF (1861–1864)

quired it. He disagreed with Lincoln that the matter fell within the sphere of Congress's authority and therefore he felt no need to reconcile his arguments now with those he had had when Congress enacted the Confiscation Act. He did not take this situation as a pretext for breaking with the administration, and he assured Lincoln that he would continue to support him without wavering from his zeal for the welfare of the Union. He told Lincoln that he tried to share his suggestions in a friendly and helpful spirit.

Lincoln did not respond, instead leaving the matter where he thought it then belonged: Congress. Later, in October, when a congressional committee investigated Frémont's activities, Browning was invited to attend, but he declined. When he learned the committee had uncovered substantial corruption in Frémont's headquarters, he made clear his belief that it was not Frémont's fault but the fault of scoundrels whom Frémont naïvely allowed to be a part of his operations.

The next moves toward emancipation came largely from Congress. Each were critical steps in reshaping the objective of the war. In December 1861, as more escaped slaves joined Union forces as they pushed more deeply into the South, a solid Republican vote in the House passed a new resolution overriding the Crittenden-Johnson one that had disavowed any antislavery purpose for the war. In his end-of-the-year message to Congress, Lincoln felt the need to address a question that kept coming back to him—and indeed had been a theme in Browning's letters. This was the question of what initiative, if any, the president should take in helping to abolish slavery. The Whig in Lincoln called for deferring entirely to Congress, but as president, he appreciated that doing so was not a well-thought-out constitutional strategy.

Instead, Lincoln said, "In considering the policy to be adopted for suppressing the insurrection, I have been anxious that the inevitable conflict for this purpose shall not degenerate into a violent and remorseless revolutionary struggle."[65] This statement might have simply reflected Lincoln's judgment on the awfulness of war. It could also have been a subtle reference to a hope that the conflict would entail only a minimal transformation of the South and no serious damage to the Southern economy or quality of life. He also rejected the idea that "the Presidency conferred upon me an unrestricted right to act officially upon this judgment and feeling"[66] that slavery was immoral. However, as Civil War historian James McPherson notes, "Another passage in the same message—little noticed at the time—pushed that minimum beyond what it had been when he signed the Confiscation Act the previous August. He referred to the contrabands affected by that law as 'thus liberated'— that is, free people who could not be returned to slavery."[67] Lincoln had effectively adapted Clay's and Jackson's long-standing project of gradual emancipation to the circumstances at hand.

In March 1862, Lincoln and Congress took two further steps closer to demanding an end to slavery altogether. Though no one had any idea how much longer the war might last, Lincoln had begun to see that Clay's dream of gradual emancipation was within the federal government's grasp. On March 6, Lincoln sent a special message to Congress recommending passage of a joint resolution offering financial assistance to any state "which may adopt gradual abolishment of slavery."[68] Congress passed the resolution, with Republicans unanimously in favor and almost all Democrats opposed. Then, a week later and a mere two days after Lincoln sacked McClellan, on March 13, 1862, Congress gave the Union Army the right to take any and all personal property from rebellious persons, including slaves. It declared that captured slaves would not be returned to their owners, and it prohibited army officers, under pain of court-martial, from returning escaped slaves to their masters. Lincoln signed the bill.

On April 16, 1862, Lincoln signed yet another law, this time a bill that abolished slavery in the District of Columbia. He undoubtedly recalled, as did many of those who had supported him, that an identical law had been part of Clay's Compromise of 1850. The act was also a repudiation of the Supreme Court's decision in *Dred Scott,* which had ruled the Missouri Compromise unconstitutional because it violated the right of slave-owners to their so-called property.

By May, the Union appeared to have the momentum in the war.

It had scored a series of victories everywhere but Virginia. Building on that momentum, Congress, on June 19, 1862, prohibited slavery in all current and future territories, though not in any current or future states. Again, without comment, Lincoln quickly signed the bill. Once Congress had decided to ignore the *Dred Scott* ruling, Lincoln obviously had no problem in going along.

On July 16, 1862, Congress took another significant move toward emancipation when it passed the Confiscation Act of 1862, which made court proceedings available to slaves freed from owners who had engaged in the rebellion. The next day, Lincoln took the extraordinary step of preparing a veto message, the first of his presidency. He objected to the forfeiture of property beyond the life of the person convicted of treason. He suggested that this was tantamount to "corruption of blood" or visiting the sins of the father on his descendants.[69] Lincoln found fault in the law's use of in rem proceedings in the courts, which would authorize forfeiture based on status as a slaveholder but not necessarily on evidence of treason. He thought that the defendants in such proceedings required more time to prepare their defenses than the act allowed. In response to the prospect of a veto, Senate leaders hurriedly drafted a resolution to bar the punishment exacted by the law to "work a forfeiture of the real estate of the offender beyond his natural life."[70] They did not have time to fix any other parts of the act. Lincoln signed the law. His sensibilities as a Whig kept him from interfering too much in the legislative process and to enforce the policy enacted by Congress even if it was not perfect and he did not completely agree.

In late July, he convened his Cabinet to announce his intention to issue an emancipation proclamation. He likely based his reasoning on a pamphlet published by a lawyer he had appointed to the Interior Department, which argued that the laws of war "give the President full belligerent rights" as commander in chief to seize all enemy property, including slaves, being used to wage war against the United States.[71] Blair opposed going all the way, while Seward counseled Lincoln to postpone doing it "until you can give it to the country supported by military success."[72] Seward's advice "struck me with very great force," Lincoln admitted later.[73] He pocketed the proclamation until a more propitious moment.

Lincoln continued to correspond over the next few months with Browning, who was campaigning for reelection in Illinois. Earlier that year, in February, Browning had spent hours with the Lincolns after the death of his eleven-year-old son, William. Mary Todd was inconsolable, while Browning kept a close watch on Lincoln, who he knew was prone to bouts of melancholy. Come fall, he vigilantly did what he could to keep Lincoln's spirits up as the war dragged on. Browning reassured Lincoln, "The skies do not appear to be very dark to me. . . . Be of good cheer—hold yourself up to the work with a pure mind, and an eye singled to your country's good, and all things will yet be well."[74] Though Lincoln heard otherwise from other sources, Browning further assured him that "the hearts of the people were never warmer for you than they are today."[75]

In the summer of 1862, the pressure on Lincoln to take a strong stand against slavery grew. Both Browning and Lincoln's old friend Joshua Speed of Kentucky counseled him further on the propriety of issuing an emancipation proclamation. Speed had previously advised Lincoln on how to keep the border state of Kentucky in line, having urged him the previous year to overrule Frémont's proclamation. But now Lincoln wrote to Speed, "I believe that in this measure . . . my fondest hopes will be realized."[76]

On September 22, 1862, a few days after the Battle of Antietam and one year to the day after Lincoln had written to Browning that he doubted a general or president had the authority to issue an emancipation proclamation, Lincoln did just that, declaring slaves in rebellious states "forever free" as of January 1, 1863, unless those states returned to the Union within one hundred days.[77]

When he heard the news, Browning, perfecting his contrariness, was skeptical. Upon his return to Washington in November, he promptly had an extended conversation with Lincoln. In the intervening year, the two men exchanged positions, Browning now explaining his objections and Lincoln defending what he had once criticized Frémont for doing. Browning believed that Lincoln's planned proclamation had had a disastrous effect on the Republicans during the 1862 midterm elections and that withdrawing or altering it would bolster the war effort. When Lincoln showed no interest in changing his mind, Browning met with Seward, who had been persuaded by Lincoln that the proclamation made sense because of the signal it sent other countries. Seward told Browning that the proclamation showed other countries the true purpose of the war—the United States' commitment to abolishing slavery. Browning could do nothing to stop the proclamation from going into effect, as planned, on January 1, 1863.

One month before the deadline, on December 1, 1862, Lincoln delivered his Annual Message to Congress. He devoted most of it to the recommendation of a constitutional amendment to offer federal compensation to states that abolished slavery before 1900. The proposal confused many of his fellow Republicans in Congress, but Lincoln's proclamation was aimed only at freeing slaves in the ten states in rebellion. Yet, in his message, just a month before the effective date of the Emancipation Proclamation, he continued to refine his articulation of the war's purpose and admonished Congress, "The dogmas of the quiet past, are inadequate to the stormy present."[78]

Lincoln warned all that the issue was not only about legality but legacy. "In times like the present men should utter nothing for which they would not willingly be responsible through time and in eternity," admonishing his audience that

> we can not escape history. We of this Congress and this Administration will be remembered in spite of ourselves. No personal significance or insignificance can spare one or another of us. The fiery trial through which we will pass will light us down in honor or dishonor to the latest generation. We say we are for the Union. The world will not forget that we say this. We know how to save the Union. The world knows we do know how to save it. We, even we here, hold the power and bear the responsibility. In giving freedom to the slave

we assure freedom to the free—honorable alike in what we give and
what we preserve. We shall nobly save or meanly lose the last best
hope of earth. Other means may succeed; this could not fail. The
way is plain, peaceful, generous, just—a way which if followed
the world will forever applaud and God must forever bless.[79]

The repetitive contrasts—freedom and bondage, remembering and forgetting, the individual and the nation, achieving eternal praise or admonishment—were more than echoes of Shakespeare and Burns. Lincoln was doing more than using the rhetorical techniques of the great poets. He was adapting them to the present circumstances by using the plain language of Americans to profound effect. His speech was all the more powerful because he was sharing with the world his own mindset, the urgency of contemplating each day, each step, in light of how history would judge him and judge them all.

Slightly more than two weeks later, on December 17, 1862, General Ulysses Grant, then leading Union forces in the vast Department of the Tennessee, surprised nearly everyone when he issued General Orders No. 11. When Grant caught wind that peddlers were cheating army officers under his command (and Jewish cotton traders might have been doing the same to his father), he ordered that "Jews, as a class, violating every regulation of trade established by the Treasury Department, and also Department [of War] orders, are hereby expelled from the Department."[80] Grant directed all Post Commanders within twenty-four hours of receipt of his order "to see that all of this class of people are furnished with passes and required to leave, and anyone after such notification, will be arrested and held in confinement until an opportunity occurs of sending them out as prisoners unless furnished with permits from these headquarters."[81]

While anti-Semitism was common in that era, a government command based on it was not. Henry Halleck, who was then the general in chief for all the Union forces, had concerns about Grant even before the order. A West Point graduate and veteran of the Mexican War, Grant had resigned from the army in 1854 because he saw little future in it and "the vice of intemperance had not a little to do with my decision to resign."[82] Elihu Washburne, an Illinois congressman and long-standing friend of Lincoln's, had found a place for Grant to return to the army shortly after the outbreak of the Civil War, first as a captain, then as a colonel, and eventually as a brigadier general on the western front. In less than two years, Grant had steadily risen in rank because the armies he led delivered much more often than not, including as recently as October 3–4, 1862, when troops from his Army of the Mississippi had repelled Confederate forces in Corinth, Mississippi.

The timing of Grant's General Orders No. 11 was awkward, both for him and the Jews who were its subject. It overshadowed Union victories just a little more than a week before at Chickasaw Bayou and Vicksburg, the last Southern stronghold on the Mississippi. At the same time, the Treasury Department desperately wanted to restore business within occupied areas to win back the inhabitants' loyalty, while the War Department worried that Southern profits from the cotton trade could be used to help the Confederate troops. The inflammatory directive barred Jews from residing or doing any business within the area under Grant's command. Grant's adjutant, Colonel John Rawlins, urged Grant not to promulgate the order, but Grant responded, "They can countermand this from Washington if they like."[83]

In fact, General Orders No. 11 was the most sweeping anti-Semitic governmental action ever undertaken in American history, but less than seventy-two hours after Grant issued it, implementation was delayed because several thousand troops led by Major General Earl Van Dorn attacked his forces at Holly Springs and General Nathan Bedford Forrest's cavalry attacked Grant's troops from the opposite direction, destroying fifty miles of railroad and telegraph lines. Communications to and from Grant's headquarters were disrupted for weeks because of the attacks, and therefore news of the order spread slowly, if at all. Once Grant and his officers got around to enforcing the order, in Paducah, Kentucky, on December 17, 1862, Jewish merchants and residents who were being expelled sent a telegram to Washington protesting their expulsion. In all likelihood, Lincoln never saw Grant's telegram until after January 1, if he ever saw it.

On the afternoon of New Year's Day, Lincoln attended a reception at which he shook hands for more than three hours. Afterward, he invited several members of his Cabinet back to his office to watch him sign the ground-breaking Emancipation Proclamation. When he first tried, his hand trembled so badly that he set the pen down, because "all who examine the document hereafter will say, 'He hesitated.'"[84] Lincoln was determined not to have that happen. To those gathered, he said, "I never in my life felt more certain that I was doing right than I do in signing this paper."[85] He added, "If my name ever goes into history, it will be for this act, and my whole soul is in it."[86] He tried again, but his hand still trembled. "The South had fair warning," he said as he looked at those assembled, "that if they did not return to their duty, I should strike at this pillar of their strength. The promise must now be kept, and I should never recall one word."[87] He then signed the proclamation without a tremor. "That will do," he said.[88] With the stroke of the pen, Lincoln had taken the biggest step ever taken to advance what he believed Jefferson and Clay had set in motion.

In spite of the historical magnitude of what Lincoln had done, the press was much more focused on reporting the effects of Grant's General Orders No. 11. Telegrams flooded the White House in protest, and two days after Lincoln signed the Emancipation Proclamation, he supposedly had a meeting on January 3, 1863, with Cesar Kaskel, a merchant from Paducah, Kentucky, and Congressman John Gurley of Ohio. In their presence, he is said to have immediately agreed to reverse Grant's order. On that day, Halleck telegraphed Grant that "[a] paper purporting to be General Orders No. 11, issued by you December 17, has been presented here. By its terms, it expels all Jews from your department. If such an order has been issued, it be immediately revoked."[89] On January 7, Grant's headquarters answered, "By direction of the General-in-Chief of the Army at Washington, the General Order from the Head Quarters expelling Jews from this Department is hereby revoked."[90] As

he had done with Frémont and Cameron, Lincoln vigilantly enforced the chain of command. When outraged Jewish leaders met
with Lincoln that same day, he reassured them that "to condemn
a class is, to say the least, to wrong the good with the bad. I do not
like to hear a class or nationality condemned on account of a few
sinners."[91] On January 21, Halleck transmitted to Grant Lincoln's
reaction to his order: "The President has no objection to your expelling traders & Jew peddlars, which I suppose was the object of
your order, but as it in terms Proscribed an entire religious class,
some of whom are fighting in our ranks, the President deemed it
necessary to revoke it."[92]

Luckily for Grant, Republican Elihu Washburne, his longtime
supporter, was still a member of the House and still on Grant's side.
"Your order touching the Jews has kicked up quite a dust among
the Israelites," he wrote Grant.[93] "They came here in crowds and
gave an entirely false construction to the order."[94] Extolling Grant
as "one of our best generals," Washburne successfully got a House
censure motion tabled by a narrow margin of 56–53.[95] In the Senate, Republicans, sticking by the president and war effort, defeated
a similar motion to censure Grant 30–7.

While Lincoln did not share, at least in writing, any further
thoughts about the connection between the proclamation and his
reversal of Grant's order, some things were clear. President Lincoln
never spoke of it again after the censure effort died down, while
Grant did only to express deep remorse over his order. We also know
that in the frantic period from mid-December 1862 through the first
few days of January 1863, Lincoln was deeply immersed in thinking
about the legal and political ramifications of the ideal of equality
as expressed in the Declaration of Independence and protected by
the Constitution. We know, as well, that the actions Lincoln took
during the winter of 1862 reconfigured the objective of the war.
It was now the army's task, with about seven thousand Jews and
a growing number of African Americans in its ranks (eventually
reaching more than 170,000), to finish the job of suppressing the
rebellion against the ideal of equality that Clay had long extolled

and Lincoln had been enforcing, culminating in the Emancipation Proclamation and revocation of Grant's anti-Semitic order.

V

It was hard for Lincoln to escape the gloom of war, but he tried. Having entered the presidency with acute awareness of the stakes involved in the days, weeks, and months ahead, Lincoln took time, when he could pull himself away, to do things that had always helped to keep him steady and buoyant.

Two episodes illustrate how he used humor to lighten spirits. As his Cabinet members walked into his office for the meeting at which he first sprang the news of his planned emancipation proclamation, Stanton recalled that they found the President

> *reading a book of some kind, which seemed to amuse him. It was a little book. He finally turned to us and said: "Gentlemen, did you ever read anything by Artemus Ward? Let me read you a chapter that is very funny." Not a member of the cabinet smiled; as for myself, I was angry, and looked to see what the President meant. It seemed like buffoonery to me. He, however, concluded to read a chapter from Artemus Ward, which he did with great deliberation, and, having finished, laughed heartily, without a member of the Cabinet joining in the laughter. "Well," he said, "let's have another chapter," and he read another chapter, to our great astonishment. I was considering whether I should rise and leave the meeting abruptly, when he threw his book, heaved a sigh, and said: "Gentlemen, why don't you laugh? With the fearful strain that is upon me, day and night, if I did not laugh I should die, and you need this medicine as much as I do."[96]*

Later that year, on July 24, Martin Van Buren passed away. Lincoln had said nothing when Jackson and Polk had died, but this was different. Lincoln was now president and thus felt obliged to express gratitude for another president's distinguished career, especially one who, like Van Buren, had defended the Union to his last breath. More important, Lincoln had met Van Buren—indeed, he was the first president Lincoln met. A little more than a year after leaving office, Van Buren planned to visit Democrats in Springfield on June 14, 1842, but he got stuck in Rochester, Illinois. When "the leading Democrats" in Springfield learned he was delayed there, they "hurried out to meet the distinguished visitor."[97] As Herndon wrote years later about the meeting, "Knowing accommodations at Rochester were not intended for, or suited to the entertainment of an ex-President, they took with them refreshments in quantity and variety, to make up for all the deficiencies." They also brought with them their state representative, Abraham Lincoln, "whose wit was as ready as his store of anecdotes was exhaustless." Lincoln, then in his last term in the Illinois legislature, dressed in his haphazard way, towered over the short Van Buren, dressed as dandily as ever. Throughout the evening, Lincoln had "a constant supply (of stories), one following another in rapid succession, each more irresistible than its predecessor." Van Buren had his own anecdotes about politics in New York when he was a young man and his early interactions with the likes of Alexander Hamilton and Aaron Burr. The laughter lasted late into the evening, when Van Buren excused himself "because his sides were sore from laughing."[98] Later, he said he had never "spent so agreeable a night in my life."[99]

Lincoln did not forget that evening either. Nor did he forget that, in 1860, Van Buren had cast aside party loyalty and voted for him as president, in the hope that the election could avert civil war. It did not, but Van Buren kept looking for ways to help keep the Union intact until he fell ill for the last time. "[Out] of tribute to their friendship and respect to a former president," particularly one who stood by the Union, Lincoln "ordered a special military salute in Van Buren's honor."[100] As Lincoln had told young lawyers many years before in Springfield, a foe can become a friend. Indeed, they could come from the unlikeliest places, even among those closest to Jackson, Henry Clay's greatest foe.

VI

Democratic appointees had dominated the Supreme Court for nearly all of Lincoln's professional life. Jackson had made five appointments to the nine-member Court, while the three Whig presidents—Harrison, Taylor, and Fillmore—together had managed to make only one. Adding insult to injury, Fillmore's appointee, Benjamin Curtis, left the Court to protest *Dred Scott* and had his seat filled by Buchanan.

No president other than George Washington had as many vacancies to fill in as short a time as Lincoln did: Washington appointed ten justices (including one recess appointment), all of whom were strongly committed to the Constitution and the Federalist vision of a strong national government. Jackson's five appointees transformed the Court to become more protective of private property and state autonomy from federal dominance, but their installments had stretched out over a much longer period than it took Lincoln. Within fifty-two days of taking the oath of office as president, Lincoln had three vacancies on the Court to fill, and he got a fourth in his second year in office. Lincoln had even come into office with a vacancy to fill; one seat had been open since Justice Peter Daniel died nearly one year to the day before Lincoln's inauguration, and the Senate had not allowed Buchanan to fill it. A second opened when Justice John McLean died during Lincoln's first month in office; and the third seat became vacant less than a month after McLean died, when Justice John Campbell, appointed by Pierce, left the Court to return to his home state of Alabama to join the Confederacy. With solid Republican majorities in both the House and Senate, Congress took the unprecedented step in 1862 of adding a seat to the Court to enable Lincoln to increase his influence.

Lincoln did not, however, move quickly to make the appointments. Facing several court actions against the war, Lincoln sought justices who strongly supported the war effort and therefore would be disposed to uphold the constitutionality of what he had done to save the Union. Lincoln followed Congress's preferences in filling two of the seats because of his long-standing belief as a Whig in deferring to that body and because he knew the Senate was likely to use the same criteria as he in filling the seats. For the other two, he did what Jackson had done: he used the appointments to reward political allies and fortify his base.

Initially, Lincoln focused on filling McClean's seat, perhaps because it was the easiest. At that time, Congress had divided the federal courts system into several circuits, and it configured the Supreme Court so that its seats were aligned with those circuits. Supreme Court seats therefore had geographical requirements, and because McClean came from Ohio, Lincoln looked to Ohio for his replacement. Lincoln was happy to do this, because he owed Ohio Republicans for their vital third-ballot support at the Republican national convention in 1860. The Ohio congressional delegation was already lobbying Lincoln hard to appoint Ohio lawyer Noah Swayne. Indeed, before he died, McClean had recommended that Lincoln appoint Swayne to replace him. Lincoln was impressed with Swayne as an abolitionist; he had freed his slaves and moved from Virginia to Ohio in protest over Virginia's strong support for slavery. Swayne's Virginia ancestry probably would have appeal to the border states, as would the fact that Jackson had appointed him to serve as the U.S. attorney in Ohio. Lincoln also liked the fact that Swayne had served in both chambers of the Ohio legislature. Three days after Lincoln nominated Swayne, the Senate confirmed him by the overwhelming vote of 38–1.

Lincoln turned his attention next to the seat that had been vacated due to Peter Daniel's death. Daniel came from Virginia, which Lincoln felt no need to appease, especially with a seat on the Supreme Court. Instead, he deferred almost completely to Congress. Indeed, the lobbying done by members of Congress and Western political leaders to fill the seat was unprecedented. Western governors, Iowa's attorney general, and the entire Iowa Supreme Court all favored Iowa's most distinguished lawyer, Samuel Freeman Miller. Moreover, 129 of 140 House members and all but 4 senators petitioned Lincoln to appoint Miller. No one from Iowa or born west of the Appalachian Mountains had ever been appointed before the Court, and Miller's credentials were impeccable—a native of Kentucky, he had studied medicine at Transylvania University in Lexington, which Lincoln knew well as the alma mater of his father-in law. Miller had been a physician for a dozen years before he earned his law degree. He'd left Kentucky for Iowa in protest over slavery and been one of the first and most prominent Republicans there, and he had done everything a lawyer could do in his community, state, and region in a distinguished practice. Within thirty minutes of Miller's nomination to the Court, the Senate unanimously confirmed him.

This left the fourth Supreme Court vacancy to fill. Lincoln felt no more need to appease the South in filling Campbell's vacated seat than he had in filling Daniel's. Free to do as he pleased, he very quickly narrowed his choice down to two people, his old friend and state judge David Davis and Orville Browning. Both had been precluded from consideration for the initial vacancy for circuit reasons and for the second after Lincoln decided it should go to someone from the far Western states. Both Davis and Browning wanted the seat desperately, and each lined up considerable support to strongly lobby Lincoln. On April 9, 1862, Browning had written sheepishly asking for the appointment, but his wife wrote soon thereafter to press his case harder. Lincoln understood Browning's confirmation might have been easier than Davis's because other senators tended to support one of their own when nominated to the Cabinet or other offices. But Lincoln did not relish losing Browning to the Court. He also appreciated that Davis's constitutional views were more in line with his own than Browning's were. Lincoln wanted to please them both, and at one point he considered appointing Davis to the Cabinet and Browning to the Court. It was a difficult choice, as reflected in the fact that Lincoln left the seat vacant for more than a year as he decided what to do.

After Lincoln's death, Herndon shared a note that he said Lincoln had written during the year when he considering whom to appoint to take Campbell's seat on the Court, confessing, "I do not know what I may do when the time comes, but there has never been a day when if I had to act I should not have appointed Browning."[101] So why did Lincoln not choose Browning for the Court? Herndon suggests that there was a crucial turn of events when Lincoln's friend Leonard Swett met with Lincoln in August 1861 to discuss the appointment. Swett had pushed Lincoln in writing to make the appointment, but he supposedly told Lincoln that "he could kill 'two birds with one stone' by appointing Davis."[102] Believing Lincoln owed him and other supporters in Illinois a favor for their support, Swett explained that he "would accept [the appointment of Davis] as one-half for me and one-half for the Judge; and that thereafter: If I or any of my friends ever troubled him, [Lincoln] could draw [Swett's] letter as" proof he made the appointment to pay Swett back for his support.[103] Swett said Lincoln nodded in agreement and responded, after some thought, "If you mean that among friends as [the letter given to him by Swett] reads I will take it and make the appointment."[104] Swett later said that the meeting was the decisive moment sealing Davis's appointment.

The fact that at least a year went by after the meeting before Lincoln actually made the appointment weakens Swett's claim. Browning enjoyed strong support in the Senate, and he had other prominent political leaders backing him, including Attorney General Edward Bates. In March 1862, Noah Swayne told Bates of a rumor going around that Caleb Smith, the secretary of the interior, might get the open seat, but Bates said he did not believe that it would interfere with Browning's being nominated to the Court. Later that same day, Bates wrote in his diary, "Nobody [I] think objects to Browning—He is a proper man."[105] Nevertheless, throughout this same time, Davis's friends were flooding Lincoln with letters hailing him, Lincoln's former partner Stephen Logan among them. John Todd Stuart wrote to Lincoln that the appointment would be especially "pleasing to the circle of your old personal friends."[106] As it was, the numbers of supporters pushing for Davis exceeded those pushing for Browning. In the longest letter Lincoln ever received on the question of which person to choose, a Bloomington, Illinois, Republican told Lincoln that the public favored Davis, but beyond that, fairness demanded that Lincoln appoint Davis because of the large political debt Lincoln owed him, and because the public would question Lincoln's generosity and benevolence if he did not appoint him. Davis had obviously done more to get Lincoln elected than Browning, and he had made it crystal clear that nothing but appointment to the Supreme Court would satisfy him. Lincoln also knew, as Stuart and Joseph Medill had stressed to him, that as someone who had been a highly respected judge, Davis had a temperament better suited for appointment to the Court. Lincoln did not need reminding that Browning was mercurial.

Davis was not perfect. He was a hard-core partisan who complained Lincoln did not listen to him enough. This time, however, Lincoln listened. On December 8, 1862, he nominated Davis to the Supreme Court. Later that day, the Senate confirmed him.

Losing the nomination was hard for Browning, and the month got more difficult for him. He had lost his Senate seat when Democrats, in control of the Illinois state legislature, chose to replace him with William Richardson, a former House member, longtime friend of Lincoln and Browning, and campaign manager for Douglas during the 1856 and 1860 Democratic presidential nominating conventions. As Browning looked back at the past few months, he confided in his diary, "The counsels of myself and those who sympathize with me are no longer heeded. I am despondent, and have but little hope left for the Republic."[107]

All three of Lincoln's first three appointments to the Court subsequently supported his administration when challenges to his actions or policies came before it. In one of the first and most important, known as the *Prize Cases* because the fate of prizes of war was at issue, the question was whether Lincoln as president had the authority to order the seizure of vessels bound to or from Confederate ports prior to July 13, 1861. On July 13, Congress declared a state of insurrection, but Lincoln had ordered the blockade before then, in April. On March 10, 1863, the Supreme Court, by the narrow margin of 5–4, announced its decision agreeing with Lincoln. It noted, "As a civil war is never publicly proclaimed [as a formal matter] against insurgents, its actual existence is a fact in our domestic history that the court is bound to notice and to know."[108] Given the circumstances, the Court reasoned that the president was "bound to meet [the war] in the shape it presented itself, without waiting for Congress to baptize it with a name."[109] The four dissenting justices included Chief Justice Taney. All three of Lincoln's appointees joined with two Democratic appointees to form the five-member majority. Their decision ultimately provided the legal foundation for Lincoln's use of emergency powers to combat the seven Southern states' rebellion against the United States. Historian Mark Neely described the case as "the most important Supreme Court decision of the Civil War."[110]

On the same day that the Court decided the *Prize Cases,* the Senate unanimously confirmed Lincoln's fourth nominee for the Court, Stephen Field of California. Browning was not a candidate for the fourth seat, since Congress had created it solely for a new circuit, which included the states of California, Oregon, and later Nevada. Field was an obvious choice, since he had served with distinction as both an associate justice and chief justice of the California Supreme Court, graduated first in his class at Williams College, and was a Democrat who had been a leader in keeping California within the Union. His brother, David Dudley Field, was a prominent advocate for legal reform and abolition. He had helped Lincoln win the presidential nomination in 1860 but became one of the president's sharpest critics among Radical Republicans, who thought he moved too slowly and timidly in eradicating slavery. The nomination did not silence Dudley Field, but it did reward his brother, who was a fierce defender of the Union and had strong political support, including from California's governor, Leland Stanford.

The fact that Field's appointment was the first in history in which a president had crossed major party lines was no accident. With his chances for reelection the next year already looking bleak, Lincoln had to attract some Democratic support to have any chance at all.

VII

The war dragged on. Both the North and South were desperately looking for ways to expand the numbers of their troops. With only limited success, the Confederacy kept trying to expand the requirements for service in the military but met with resistance, sometimes violent. Lincoln had signed a series of laws, including as recently as March 3, 1863, to register all males between twenty and forty-five, including aliens with the intention of becoming citizens, by April 1. Lincoln had once opposed Douglas's and the Democrats' efforts in Illinois to count aliens as residents, but of the 168,649 people the draft reached, nearly two-thirds were substitutes and around seventy-five thousand were actually pressed into service. These numbers barely told the story. Over the course of the war, nearly 2,000,000 troops fought for the Union as compared with 750,000 for the Southern Confederacy.

Each conflict seemed more momentous than the ones before. This was especially true when, on July 1–3, 1863, the Union Army of the Potomac and Robert E. Lee's Army of Northern Virginia clashed at Gettysburg in one of the most decisive battles of the war. Many believed that the fate of the Civil War was at stake.

Such a clash had been a long time coming, and each side had steadily amassed the forces to annihilate the other. Lee had rarely been beaten and never caught and had thus become the bane of the Union's existence. Having defeated Union forces at the Battle of Chancellorsville in Virginia, Lee turned his army northward toward Pennsylvania, only the second time after Antietam that Lee brought his forces north. He aimed to bring the war into the Union's own territory and once and for all break the will of stubborn Northern politicians who had been pressing for continued engagement with the South. Lee would bring with him nearly seventy-five thousand troops.

Aware that Lee's forces were heading North for what might be a decisive confrontation with the Union, just three days before one of the most famous battles of the Civil War, Lincoln decided to replace the Army of the Potomac's commanding general, Joseph Hooker, who had been reluctant to reengage with Lee's forces after his defeat at Chancellorsville. Lincoln's replacement was General George Meade. A career army officer, Meade had graduated from West Point and distinguished himself in the Second Seminole War and the Mexican War. Meade had been a temporary commander at the Battle of Antietam in September and was one of the few commanders who were not disgraced by the South's slaughter of Union troops at the Battle of Fredericksburg in mid-December 1861.

The two great armies clashed near the small Pennsylvania town of Gettysburg, where they fought for nearly three days in a battle won by the Union but that produced more than fifty thousand casualties, the largest number of any battle in the entire Civil War. So many men died on both sides that, in the aftermath, both Lee and Meade tried in vain to submit their resignations to their respective leaders, Lee because the fighting had ended so badly (and, to many, inexplicably) and Meade because of the catastrophic losses at Gettysburg compounded by his failure to prevent Lee and his army from escaping across the Potomac and retreating back into Virginia. When Meade told Lincoln that Lee and his men had fooled his entire Army of the Potomac by leaving campfires lit to give the appearance that the Confederate troops were still encamped and not escaping stealthily at night, Lincoln exploded in anger, "We had them within our grasp. We had to only stretch our hands and they were ours."[111] On July 14, he drafted a letter he never sent to Meade explaining his "deep distress" over Meade's failure to have his men pursue Lee's army and crush them once and for all.[112]

On the evening of November 18, 1863, more than four months after the battle, Lincoln visited Gettysburg for the first time. He arrived as one of a handful of speakers to dedicate the battlefield to honor fallen soldiers on both sides of the conflict. He came knowing that, in spite of some notable Union victories that year, no end to the war was in sight. Still, after so many bloody battles, the carnage at Gettysburg was unprecedented. His speech was planned as the shortest and billed in the program merely as "Dedicatory Remarks."[113] The featured speaker was Senator Edward Everett of

Massachusetts, who was scheduled to deliver the only "oration" for the program.[114] Everett was a close ally and protégé of Daniel Webster, and widely considered to be nearly Webster's equal at great oratory. In 1860, conservative ex-Whigs placed him on the ticket of the Constitutional Union Party as a vice presidential candidate alongside its presidential candidate John Bell, but in winning thirty-nine electoral votes they mostly ensured Lincoln's victory and Douglas's loss. Everett was among the many people Lincoln welcomed the chance to work with to save the Union, in spite of past political differences.

At Gettysburg, Everett spoke, as people often did at such ceremonies, for more than two hours. In contrast, Lincoln's "few remarks," as he called them, lasted for three minutes. There is no consensus on precisely every word Lincoln said. At least five different versions of the text written in Lincoln's hand can be found, and Lincoln might have deviated from the written text when he spoke. Even the four stenographers working the event did not report the same words.

The first notable thing about Lincoln's speech is what it was not. Classical funeral orations, like those he delivered for Clay and Taylor or like what Everett delivered at the Gettysburg ceremony on November 19, were of epic duration. In contrast, Lincoln spoke only 271 or 272 words (scholars agree on that range). Historical accuracy was one of the aims of those longer orations, as were marking the significance and achievements of the person(s) who had died. Everett did all of that and more, mentioning the names of many of the fallen, incanting them like the tolls of a bell. Lincoln, on the other hand, never mentioned slavery or the names of any soldiers who died there.

Lincoln's deliberate concision reflected the maturation of his rhetorical development. Clay's most famous speeches were lengthy, and the vast majority of Lincoln's, ranging from the Lyceum Address to each of his debates with Douglas, also had been protracted. Long, complex sentences were common in Clay's speeches and many of Lincoln's earlier ones. It is not because their thoughts were unclear or meandering. They were made this way, as exemplified in the last sentence of the Gettysburg Address, because they were to be spoken slowly and to be heard. The delivery was the important thing. However, as president, Lincoln had come to believe that the public had no patience for long speeches. More important, they would remember nothing from them.

From his decades of studying the orations of his predecessors, Lincoln had come to appreciate the adage, often attributed to the French intellectual Blaise Pascal, "If I had more time, I would have written a shorter letter." Just a few weeks before his address in Gettysburg, Lincoln had referred to the character Polonius in *Hamlet*, the same Polonius who had admonished his son, Laertes, "Brevity is the soul of wit" just before Laertes left for school.[115]

Brevity was a virtue, particularly for presidents—it could be easily grasped, recorded, and disseminated. In his eulogy for Clay, Lincoln had praised Clay's "impassioned tone," but for Lincoln that tone required plainness of language. Lincoln understood that both Clay and Webster each had achieved their unique kind of eloquence, but neither was plain nor concise in his rhetoric. A marriage of condensation and majesty was harder to achieve than prolixity but more memorable, and worthy of the grandeur of both the office and the message. As president, Lincoln no longer spoke as a partisan. As president, he was obliged to act in the best interests of all Americans and the Constitution. He expected people would be reading and reciting his remarks aloud, as he had done himself for other speeches for decades.

Everett came to Pennsylvania to do what was expected in commemorating battlefields; this was not his first time to do so. But honoring the battle was secondary to Lincoln. He had his eyes on the bigger picture. What had happened at Gettysburg was monumental, but it was only part of the larger Civil War, which remained unsettled.

Politics were not entirely absent from his remarks, but they were, Lincoln believed, of a greater strategic nature that went beyond party. Pennsylvania's governor, Andrew Curtin, was facing a difficult reelection campaign, and Lincoln's own reelection was less than a year away. From where (and when) he stood in Gettysburg, his prospects looked no better.

The questions he had not yet answered to the public's satisfaction were, What were these soldiers dying for? and, Why reelect the president?

The answer came in the opening line of his remarks, "Four score and seven years ago our fathers brought forth, on this continent, a new nation, conceived in Liberty, and dedicated to the principle that all men are created equal."[116] This was why they had gathered, this was why they had fought, and this was why the Union needed to win, an imperative that Lincoln affirmed in his next sentence: "Now we are engaged in a great civil war, testing whether that nation, or any nation so conceived and so dedicated, can long endure."[117] Later, many newspapers expressed outrage that Lincoln had redefined the purpose of the war to include the securement of not just liberty but equality, but for Lincoln this objective reflected the inextricable connection between the Constitution and the Declaration of Independence, a synthesis that Clay had spent his adult life advocating and Lincoln his life learning. Lincoln was not oblivious to the fact that the founding document of the nation was the Constitution, not the Declaration, but he understood the Constitution as built on the foundation of the Declaration. Lincoln had also come to understand, through his own painful struggles and those of the nation, that his link was not widely appreciated in the United States. With newspapers taking down every word he spoke, this was his moment to commemorate this "new birth of freedom" to the American people.[118]

Webster had built a concept of constitutional union that was precisely what Lincoln saw himself, his administration, and the Union Army defending. Lincoln still recited to himself lines from Webster's extraordinary response to Hayne's promotion of the dangerous doctrine of nullification, and his argument that Americans had come together as one people long before the Constitution was drafted and ratified. "It is," Webster had declared in his reply, as he stared at Calhoun as presiding officer of the Senate, "the people's Constitution, the people's government, made for the people, made by the people, and answerable to the people."[119] (Lincoln was fond of this phrasing. In the 1850s, he had filed a copy of an address from a speech by abolitionist Theodore Parker, in which Parker defined democracy as "Direct Self-government, over all the people, for all the people, by all the people.") Rejecting Hayne's notion that all federal power came from the states, Webster declared, "We are here to administer a Constitution emanating immediately from the people, and trusted by them to our administration. It is not the creature of State governments."[120]

Lincoln had adopted the same understanding of democracy, using similar words, in his First Inaugural Address. Tracking the same argument that Webster repeatedly had made, Lincoln explained, in his First Inaugural, that it was the people, not the states, who existed before the Constitution, and that the people, not the states, were the source of authority for the Constitution itself. If that were not the case, Lincoln explained, "The United States [would] be not a government proper, but an association of States in the nature of a contract merely."[121] He further declared, "Descending from these great principles, we find the proposition that, in legal contemplation, the Union is perpetual, confirmed by the history itself."[122] Lincoln then traced, as Clay and Webster each had done, the lineage of the Constitution back to its original source, the Declaration of Independence. He had no intention of delivering the same history lesson again in his brief Gettysburg remarks. Instead, he asked the thousands gathered there (and the thousands more who would read the remarks) to "resolve that the dead shall not have died in vain" and "that the nation shall, under God, have a new birth of freedom, and that government of the people, by the people, and for the people, shall not perish from the earth."[123] Here was Webster's great articulation of American democracy reformulated into simpler, more easily remembered words. Complexity of language remained, but to create rhythmic momentum, as, for example, Lincoln's contrasting "those who here gave their lives" with the "people," whose system of government "shall not perish from the earth." Lincoln juxtaposed references to the past and the present,

the dead and the living at Gettysburg, and the "new birth of freedom" against the Founding Fathers' bringing forth "the great task
remaining."[124] Lincoln concluded his remarks with an eighty-two-
word sentence in which devotion to a worthy cause is forcefully
driven home:

> *It is rather for us to be here dedicated to the great task remaining*
> *before us—that from these honored dead we take increased devotion*
> *to that cause for which they gave the last full measure of devotion—*
> *that we here highly resolve that these dead shall not have died*
> *in vain—that this nation, under God, shall have a new birth of*
> *freedom—and that government of the people, by the people, for the*
> *people, shall not perish from the earth.*

Edward Everett's most famous student, Ralph Waldo Emerson,
immediately recognized (as did Everett) the grandeur of Lincoln's
sparse remarks: "His brief speech at Gettysburg will not easily be
surpassed by words on any recorded occasion."[125] Emerson's assessment was in stark contrast to how he had judged Daniel Webster's
speech defending the Compromise of 1850 and its noxious fugitive slave provision for the sake of maintaining the Union. "Let
Mr. Webster for decency's sake," Emerson had declaimed, "shut
his lips once and forever on this word [liberty]. The word liberty in
the mouth of Mr. Webster sounds like the word love in the mouth
of a courtesan."[126] With less grandiosity than Webster, Lincoln declared, "The world will little note, nor long remember, what we
say here," though of course it did because of his eloquence that
day.[127] Lincoln stood where Clay and Webster had dreamed of
standing, and had the confidence, born from years of speaking to
ordinary citizens, to take what those two great orators had done
and do it better. Lincoln had taken the additional step, urged for
decades by Clay and Webster, to equate liberty and union. Here
was a renewed understanding of its foundation in the fulfillment
of the promises made in the Declaration of Independence. Lincoln
understood that Clay and Webster each had achieved their unique
kind of eloquence, but neither spoke to the common folks that

Lincoln had always aimed to reach. Jackson and Taylor were each brief in most of their orders and proclamations, but neither was eloquent. Lincoln aspired to be both concise and eloquent.

Lincoln was not the kind of man who left things to chance. He polished his sparse remarks as if they were fine diamonds. Rather than having written them on an envelope at the last moment, he was seen working on them, off and on, for days, and he had seen Everett's lengthy oration the evening before the day of the dedication. Days before the event, Lincoln asked aides to review schematics of the area. They did the evening before his remarks and reported to him the size of the space and how the speech would carry.

Up until the late 1850s, Lincoln mostly relied on his handwritten notes, not a manuscript, when making a speech, though he used manuscripts for special occasions—the House Divided and Cooper Union speeches as a candidate for office, and as president the First Inaugural and now the Gettysburg Address. He continued to rework his draft as late as the evening before. He did not work alone. According to the daughter of William Slade, an African American who had long been Lincoln's servant and traveled with him,

> the president locked himself in his [hotel] room with only Slade present. He then began carefully to weigh every thought and carve every word in the address which has become so famous. After writing a sentence or so he would pause, and read the piece to Slade. He would then say, "William, how does that sound?" Slade, who by this time was quite a critic, would express his opinion. This went on until all was completed and he then sent for his secretary and others to hear it. Having received the praise and criticism of his messenger Lincoln felt that even the most ordinary person would understand his speech.

This account resonates with Lincoln's own account of how he wrote. As he told a law student in the 1840s, "I write by ear. When I have got my thoughts on paper, I read it aloud, and if it sounds all right I just let it pass."

The next day after Lincoln delivered the speech, John Hay

confirmed in his diary how "the President, in a fine, free, way, with more grace than is his wont, said his half dozen words of consecration."[128] While virtually every contemporary report on the event focused primarily on Everett's address, Everett well understood the significance of what Lincoln had said in his three minutes. The day after the event, Everett wrote Lincoln a short note: "I should be glad, if I could flatter myself that I came as near the central idea of the occasion in two hours, as you did in two minutes."[129] Everett's praise was not mere puffery. He had helped to write some of Webster's finest speeches and had a distinguished career in his own right—member of the House of Representatives, governor of Massachusetts, and president and professor of Greek Literature at Harvard before becoming secretary of state and senator. No one had better claim to being the leading expert on American oratory than Edward Everett.

Lincoln's remarks at Gettysburg became known as an "address" because their level of eloquence was much higher and commanding than that of other presidential remarks. As compared with their all too frequent off-the-cuff, forgettable remarks, Lincoln's were aimed at something different. They reflect the continuing influence of the theater, Shakespeare, classical tropes, and the Bible on his rhetoric. The influence of each of these may be found at length in the voluminous literature on Lincoln's remarks. Clay was best remembered for his theatrics, the most distinctive part of his speeches. The words on the page often seem flat. But Clay was a man on fire when he spoke, and it was that spectacle that many of his colleagues in Congress considered to be what distinguished his oratory.

Lincoln did something no president had done—not Washington, not Jefferson, not Jackson. He had spoken in the rhythm of America's greatest poetry. Lincoln captured in the brevity of his rhythmic verse the solemnity of the recommitment of the nation and its people to the "unfinished work" of the war, which was to ensure that the Constitution guaranteed liberty and equality to all citizens.[130]

As Shakespeare's Henry IV rallied his "band of brothers" to his cause, Lincoln hoped to do the same for his. Certainly Lincoln's words could rally the people to fight for a worthy cause, but the fighting still had to be done. Less than a week after Lincoln's remarks at Gettysburg, the Union scored another significant victory. On November 23–25, the Union Army, under the command of Major General Ulysses Grant, broke the Confederate siege of Chattanooga, ultimately forcing the Confederate Army, under the command of General Braxton Bragg, to retreat back into Georgia. The defeat was so severe that Jefferson Davis recalled Bragg to Richmond. After a draw one week later at Mine Run Creek, Confederate forces under the command of General James Longstreet, a veteran of the Mexican War and Gettysburg, lay siege to Knoxville, where Union forces successfully rebuffed their attack. The rebels were losing more than they were winning, but the final defeats were yet to come as Lincoln let his army do the talking.

VIII

The war dominated the shifting political landscape. The longer it lasted, the more trouble Republicans were facing at the polls, particularly in the border states. The shifts were apparent even in Illinois, where Republicans had never dominated. One of the casualties was Orville Browning. In early 1863, he left the Senate, only to return to Washington as a partner in a small law firm. He would be of less use to Lincoln as an outsider who was becoming increasingly pessimistic about the progress of the war and Lincoln's chances for reelection. The two interacted often, but Browning never felt that Lincoln sought his counsel as much as he once had, perhaps because Lincoln likely figured that Browning could not offer the inside information he once had and was not supportive of his reelection or policies.

In another sign of the shifting political landscape, John Todd Stuart was returning to Washington, as the representative for the eighth district of Illinois, though this time no longer as a Constitutional Unionist but as a Democrat. He had won the seat in a close contest against Leonard Swett, his and the president's old friend. Lincoln backed Swett but did not campaign back home out of respect for Stuart. Stuart won.

The shifting political landscape, exemplified in the exodus of Browning and the return of Stuart, underscored several problems Lincoln was facing at the end of his last full year in office before the presidential election. The first was that Lincoln was more isolated and alone. In spite of their differences, Browning and Lincoln always could speak honestly with each other, though Lincoln had become increasingly less likely to hear Browning's advice. Lincoln and Stuart still enjoyed a cordial relationship, undoubtedly a testament to how each managed to keep the vicious politics of the time from becoming overly personal to either of them. Stuart knew as well as Lincoln did that Browning provided better and stronger emotional support than he could. Browning provided better insight into substantive matters, too. Stuart simply was not the deep, pondering thinker that Browning was, nor was he the strong party man that either Browning or Lincoln was. (Browning was a Whig for decades and a Republican until 1869; Stuart had switched parties more than once and was not returning as a member of either of the two major parties.) Though Stuart remained friendly with Lincoln and informed him of House activities, he was no longer someone on whom Lincoln could count for support.

Of even greater import, Lincoln could not escape the ever increasing toll of the protracted war. In 1863, the Union had scored major victories at Gettysburg and Vicksburg, but they were achieved at an awful cost in lives and did not seem to have brought the war any closer to an end. None of Lincoln's generals had succeeded yet in doing what he kept insisting be done—fulfill the objective that Scott had identified at the outset of the war: crushing the enemy to death. Lee had lost multitudes of men, including some of his ablest generals (Stonewall Jackson had died of pneumonia eight days after being shot at the Battle of Chancellorsville), but others were out there, including Lee, as elusive and dangerous as ever. The Union still had superior resources, including far greater manpower, but Lincoln stayed up late into the night, wracked with insomnia, worrying about how his army's advantages were being squandered.

Lincoln worried, too, about what he should do if the war ever ended. Even before the Union victories of 1863, he and his allies were trying to look ahead. In early January 1863, ground was broken in Sacramento to begin construction on the transcontinental railroad approved the prior year by Congress. More than a few, including Lincoln, thought that the Union could develop a policy on reunion sufficiently appealing to the war-weary Southerners that it might erode their continued will to fight. Lincoln, after all, had never been a prosecutor or long-serving judge. He was a dealmaker, having learned the art from Stuart and through decades of negotiations in courtrooms, conventions, and legislatures. On December 8, 1863, he delivered his last Annual Message to Congress before the beginning of the presidential election year, a message that was pervaded by a surprisingly strong sense of optimism for the future. He acknowledged that the Emancipation Proclamation, which began the year on a forceful note, had been "followed by dark and doubtful days."[131] Yet Lincoln's confidence had not been shaken. Now, near year's end, "the crisis which threatened to divide the friends of the Union is past."[132] He reassured Congress and the nation that the African American troops who were joining the army were "as good soldiers as any" and had helped convert many opponents to supporters of emancipation.[133] (There was precedent: African American soldiers served under Washington during the war of independence and served under Jackson at the Battle of New Orleans.) Referring to developments in Maryland and Missouri, which were both exempt from his Emancipation Proclamation, Lincoln said that neither of those states "three years ago would tolerate any restraint upon the extension of slavery into new territories," but they "only dispute now as to the best mode of removing it within their own limits."[134]

Lincoln had attached to his Annual Message a document entitled Proclamation of Amnesty and Reconstruction.[135] He explained

that, under his constitutional authority to grant pardons for of-
fences against the United States, he was offering "full pardon" and
restoration of property "except as to slaves" to former participants
in the rebellion who would swear an oath of allegiance to the
United States and to all laws and proclamations concerning emancipation.[136] When the number of voters taking the oath in any state
equaled 10 percent of the number of people who had voted in 1860,
this contingent could reestablish a state government to which Lincoln promised executive recognition.

While the Republican majority in the House had narrowed as
a result of the most recent midterm elections, the party had expanded its margin of control in the Senate. This was not all good
news for Lincoln. As 1863 gave way to 1864, Lincoln faced a Congress in disarray, with Republicans sharply split into radical and
conservative camps. Radicals wanted to be tougher on the rebels,
while conservatives urged Lincoln to be more solicitous of the
South than he had been thus far. He defended his amnesty plan as
a wartime measure to weaken Southern resistance. A "tangible nucleus" of loyal citizens was all that was needed to spark a state's re-
entry into the Union. In March 1864, he continued to insist that he
had made his proposal "to suppress the insurrection and to restore
the authority of the United States,"[137] but the plan went nowhere.

As the legislative session of 1863–1864 was drawing to a close,
there was little good news for Stuart to report to Lincoln. Congress seemed paralyzed until Radical Republicans mustered sufficient support in early July for both chambers in Congress to
approve an alternative to the scheme Lincoln was proposing.

The alternative had been put together by Representative Henry
Winter Davis and Senator Benjamin Wade. Davis called their proposal "the only practical measure of emancipation proposed in
this Congress."[138] It required, as a first step in the reorganization
and readmission of any Southern state into the Union, that it commit to completely abolishing slavery. The bill specified further
that 50 percent, rather than the 10 percent Lincoln suggested, of
the 1860 voters must participate in elections to reorganize their
respective state governments. In addition, the bill required that

the electors in any constitutional conventions in any of the states that attempted to secede take a different oath than the one Lincoln had proposed, which had merely entailed swearing future fealty to the Union. Instead, the Wade-Davis bill required them to take an "iron-clad" oath swearing that they had never voluntarily borne arms against the United States or aided the rebellion. Those taking the oath would be blaming the state's leaders for the rebellion.

The bill posed a dilemma for Lincoln. Its substance was not just more extreme than anything he had previously supported, but the bill also would have forced him into aligning himself and his administration with the Radical Republicans, a realignment that chafed Lincoln's moderate impulses and would almost certainly doom his chances for reelection, which depended on maintaining support from his base: Democrats who supported the Union and moderate Republicans.

A faithful Whig was supposed to follow the lead of Congress. With his December message, Lincoln was reversing that arrangement and, in doing so, violating again a central tenet of Clay's conception of the presidency. For much of 1863, Lincoln had avoided that dilemma. Congress had ultimately been inert, and he had already exploited a defect in the Whig orthodoxy, which never spoke to executive power in wartime. Lincoln was convinced that as commander in chief in the midst of the war he could take measures like the Emancipation Proclamation, aimed to press the war to its end. Come the middle of 1864, he had no choice but to confront the dilemma of whether to revert to the Whig philosophy of deferring to Congress or help his own reelection. It was not a hard choice. He chose the presidency and his political future over fealty to Congress or one faction within his party. He chose to let the Wade-Davis bill die through a pocket veto, a rarely used means of killing a bill passed at the end of a congressional session by simply not signing it. Any further pretense that Lincoln was a Clay Whig was discarded. Lincoln had surpassed his mentor in both word and deed.

CHAPTER EIGHT

FINAL ACT
(1864–1865)

In 1860, Abraham Lincoln was the youngest man ever elected president. When he won reelection four years later, he became the youngest president to be reelected. He remained the youngest person to have won two successive presidential terms until the only other president from Illinois, Barack Obama, was reelected nearly 150 years later.

Entering what most people expected would be his final full year as president in 1864, Lincoln was fifty-five. It was thirty-two years since Lincoln had cast his first vote in a presidential election when Henry Clay, then fifty-five, lost his first presidential race to Andrew Jackson. At fifty-five, Jackson had suffered a physical breakdown. With two bullets lodged in his body, he was completely exhausted from years of intensive military campaigning. He was coughing up blood, and his body shook uncontrollably. After several months of rest, he began to recover. Two years later, he mounted his first serious run for the presidency, in 1824. At fifty-five, Zachary Taylor was still in the army, working his way to becoming a colonel in the Black Hawk War in 1832.

Now Lincoln squarely faced not just the question whether he would be reelected or not, but how he would be remembered. If the Union won the war, he could be sure to be remembered as one of America's great presidents. If the Union did not, he would likely be its last.

Throughout the first seven months of 1864, Lincoln, like most others, expected his term to end with the election of McClellan as president. He was preparing to lose, and he would have been hard-pressed to identify his accomplishments—not the closure of the war, not the establishment of the Republican Party, and neither the Thirteenth Amendment nor the formal abolition of slavery. His rhetorical flourishes might have died with him, considered an oddly poetic interlude in the Union's final years.

Perhaps his legacy would have been nothing more than that of all the other one-term presidents in the nineteenth century. Americans like a winner. They rarely fete the loser. If Lincoln failed to be reelected, he would likely have been remembered, if at all, for mishandling the war, placing the once heavy-drinking Grant (whom Mary Todd called a "butcher"[1]) in charge of the Union Army, and handing over to his successor a federal government depleted of precious resources and lives squandered on behalf of an impossible dream. Slavery would have endured and almost certainly expanded within and beyond the United States, subjugating entire new classes of people to bondage, bigotry, and avarice. Lost would have been Lincoln and Clay's dream of sealing the connection between the Declaration of Independence and the Constitution. Had Lincoln failed to be reelected, his idolization of Clay and Taylor would have mattered little; to the extent he had any right to be viewed as their heir, he would have been a failed one, no more successful than either of them had been in averting civil war and saving the Union. Lincoln and the Republican Party would not be seen as the architects of "a new birth of freedom"[2] but as the precipitating causes of the greatest destruction ever brought against whatever was left of the Union.

In 1864, Orville Browning was no apologist for Lincoln. He shared with his friend Edward Cowan his opinion of Lincoln as president: "I faithfully tried to uphold him, and make him respectable; tho' I have never been able to persuade myself that he was big enough for his position. Still, I thought he might get through, as many a boy in college, without disgrace, and without knowledge; I fear he is a failure."[3] Browning was far from alone in that harsh assessment.

It is tempting to think of Lincoln's final acts as preordained. As the presidential election of 1864 approached, he increasingly spoke of fate, divine will, and his belief that events were controlling him, not the other way around. This kind of talk might simply have been his way of hedging his bets or adopting Clay's strategy of projecting humility. At the same time, Lincoln did not question success. He took it in stride, just as Clay, Jackson, and Taylor had each done. However, through the summer of 1864, nothing happening in the war indicated Lincoln would repeat Jackson's feat of reelection.

Still, even if greater forces shaped events, Lincoln never waited passively for events to break his way. As often as Lincoln spoke of forces beyond his control, he stubbornly thought of himself as a "self-made man" and wondered aloud how future generations would judge his presidency. No one knows what Lincoln thought late at night when he sat alone in his office with the portrait of Jackson hanging overhead, but it seems possible that he, as well as loyal friends like the Blairs, would have recalled the simple fact that Jackson had not gotten as far as he had without doing the hard work. Neither Jackson, nor Clay, nor Taylor left anything to chance.

For most of the preceding three years, Lincoln had put into place an organization that would help him win reelection, but like Jackson and Clay, he renamed his party to serve his purposes. But some things had not changed. Browning and Stuart both continued to visit, albeit not as frequently, and the Blairs, particularly Jackson's old friend Frank Blair, peppered Lincoln with advice. The Jackson portrait never moved.

A remarkable feat in Lincoln's final year in office was congressional approval of the joint resolution to submit the Thirteenth Amendment to the states for ratification. Though not formally required for the amendment process in the Constitution, Lincoln signed a copy of the Thirteenth Amendment to emphasize his approval and the responsibility he felt. If things turned around for Lincoln and the country during the last year of his presidency, it is because Lincoln, the people in his administration and in the army, and the voters who stood by them all helped to turn them around.

Lincoln's reelection brought him unprecedented relief and a new boost of confidence. For the first time, he consulted almost no one on his major speeches, including his last. Long attentive to the lessons his mentors had set for him, now he sometimes ignored them. Had he paid closer attention, he might have lived longer.

I

By March 1864, Lincoln had had enough. No matter how much experience his commanding generals had, they were not hastening the war's end. Even though it was an election year, he again had to make a change at the top. Voters needed to know that he was not passively waiting for fate to turn his way, in spite of his remark that his "policy was to have no policy."[4] This comment did not mean Lincoln was clueless but rather, like Taylor, determined to be flexible and not going to commit himself to say or do anything more than he needed to. He was not going to telegraph his strategy to the enemy, and he was determined not to share his final plans with the Cabinet or members of Congress until ready.

Henry Halleck had been the general in chief since 1862, but Lincoln—and many Republicans in Congress—had wanted Halleck out for some time. Lincoln told John Hay that after McClellan's failure Halleck had requested that he "be given full power and responsibility to run the Union army on that basis till [Major General John] Pope's defeat [at the second Battle of Bull Run August 28–30, 1862]; but ever since that event, [Halleck] had shrunk from the responsibility whenever it was possible."[5] In the interim, one name for Halleck's replacement repeatedly came to the attention of the President. It was certainly not Meade, whom Lincoln never forgave for not chasing Lee's army when it was in retreat. It was the commanding general of the Army of Mississippi, Ulysses Grant. Grant had graduated from West Point but left the army when it appeared he could rise no higher than second lieutenant after the Mexican War. He tried his hand at business but had no better luck there. When the war broke out, he persuaded his congressman, Eli Washburne, to find a place for him in the army, and Washburne did. Grant had been working his way up since then, steadily rising in the ranks because of a string of victories and his stubbornness to keep at it until the enemy relented or was crushed.

Before making any final decision, Lincoln reached out to Washburne for confirmation that Grant was up to the task. "All I know of Grant," Lincoln told Washburne, "I have got from you. I have never seen him. Who else besides you knows anything about Grant?"[6] In particular, Lincoln wanted a sense of his personal ambitions, particularly whether he was inclined to mount a presidential run.

Washburne told the President that he should talk to J. Russell Jones, the U.S. Marshal for Chicago, who was from Grant's hometown, Galena, Illinois, and corresponded regularly with Grant. Jones wrote Grant and asked him whether he had interest in running for the presidency, as many members of Congress were hoping. Grant answered directly, "I already have a pretty big job on my hands, and my only ambition is to see this rebellion suppressed. Nothing could induce me to think of being a presidential candidate, particularly so long as there is a possibility of having Mr. Lincoln re-elected."[7] When Jones visited Lincoln in February 1864, Lincoln asked him whether Grant wanted to be president. Jones showed him Grant's letter. "My son," Lincoln responded, "you will never know how gratifying that is to me."[8]

Assured that Grant's focus would be on winning the war and not his political fortunes, Lincoln lent his support to the bill circulating in Congress to revive the rank of lieutenant general. It was a rank that no American commander other than George Washington ever had. The measure passed the House (117–19) on February 1, 1864 and the Senate (31–6) on February 26, and Lincoln signed it into law on February 29, 1864. Halleck then wrote to Grant to inform him that Lincoln had signed his commission as lieutenant general and to "report in person to the War Department as soon as practicable."[9]

Grant immediately headed east from his headquarters in Nashville, and his arrival in Washington on March 8, 1864, was characteristic of the man, turning up with his son but with no fanfare, fancy uniform, or welcoming committee. Once he signed into his hotel, the word spread so quickly that by the time he returned to his room from dinner an invitation to the White House was waiting for him. Grant immediately ventured back out still wearing the rumpled clothes that he had traveled in. He was quickly rushed into the East Room, where a reception was taking place. The room fell silent when he arrived, as all eyes turned toward him. Lincoln feigned surprise, exclaiming, "Why, here is General Grant! Well, this is a great pleasure, I assure you."[10] Grant blushed as the esteemed guests, including Seward and the First Lady, greeted him warmly. He followed Seward's suggestion to stand on a sofa to acknowledge the applause. "For once at least," a newspaper reporter wrote, "the President of the United States was not the chief figure of the picture. The little, scared-looking man who stood on the crimson-covered sofa was the idol of the hour."[11]

After the ceremony, Lincoln took Grant upstairs for a private meeting. Lincoln explained, with some humility, that he had not been a soldier and had no special expertise in military affairs. He told Grant of his impatience with the procrastination of previous commanders and the pressure from Congress that had forced him into issuing direct orders to them, as he had done with both McClellan and Halleck. As Grant recalled, Lincoln said,

> He did not know but they were all wrong, and did know that some of them were. All he wanted or had ever wanted was someone who would take the responsibility and act, and call on him for all of the assistance needed, pledging himself to use all the power of the government in rendering such assistance. Assuring him that I would do the best I could with the means at hand, and avoid annoying him or the War Department, our first interview ended.[12]

The next day, Lincoln held a small ceremony for Grant to meet the Cabinet. He met Lincoln again two weeks later, after he had had a chance to visit his generals in the field and assess their readiness to do what Lincoln wanted—take the war to the enemy and never cease hounding them until the war was done.

Grant's ambition was to emulate Zachary Taylor. No one admired Taylor more than Ulysses Grant. Nearing the end of his life in 1883, he confided in his memoirs, "There was no man living who I admired and respected more highly" than Zachary Taylor.[13] During the Mexican War, Grant served directly under Taylor and ever since modeled himself on the future president. Jean Edward Smith, notable biographer of Grant's, wrote, "What few recognized was that Grant's attitude had been nurtured fifteen years earlier in Mexico watching the way Zachary Taylor operated."

Similarly, Grant's biographer Ron Chernow observed, "In describing Taylor, Grant provided a perfect description of his own economical writing style: 'Taylor was not a conversationalist, but on paper he could put his meaning so plainly that there could be no mistaking it. He knew how to express what he wanted to say in the fewest well-chosen words.'"[14] Like Taylor, Grant was not disposed to making fancy, high-sounding pronouncements; both men were direct, succinct, to the point. "He is a copious worker and fighter," Lincoln said, "but a very meager writer, or telegrapher."[15] In 1850, he had noted that "General Taylor's battles were not distinguished for military maneuvers; but in all, he seems rather to have conquered by a sober and steady judgment, coupled with a dogged incapacity to understand that defeat was possible. His rarest military trait, was a combination of negatives—absence of excitement and absence of fear. He could not be flurried, and he could not be scared."[16] If there was a difference between Grant and Taylor, it was that Grant, for many years, loved having a drink or several. That never seemed to bother Lincoln, who made his confidence well known, saying Grant "doesn't worry and bother me. He isn't shrieking for reinforcements all the time. He takes what troops we can safely give him . . . and does the best he can with what he has got."[17] In 1863, there arose a story, probably apocryphal, that when commanders asked Lincoln to remove Grant because of his excessive drinking, Lincoln said that "if anyone could find out what brand of whiskey Grant drank, he would send a barrel of it to all the other commanders."[18]

In addition to how he framed his orders, Grant admired Taylor's style, because he didn't trouble

> *the administration much with his demands, but was inclined to do the best he could with the means given him. If he had thought that he was sent to perform an impossibility, he would probably have informed the authorities and left them to determine what should be done. If the judgment was against him he would have gone on and done the best he could . . . without parading his grievance before the public.*[19]

Grant also admired Taylor's understanding of the role of the soldier; Taylor, he said, "considered the administration accountable for the war, and felt no responsibility resting on himself other than the faithful performance of his duties."[20] He even donned a linen duster and a battered civilian hat, like those that Taylor liked to wear.

George Meade, who had served with Grant under Taylor, observed that Grant "puts me in mind of old Taylor, and sometimes I fancy he models himself on old Zac."[21] Perhaps most important, Grant had witnessed firsthand how Taylor carried the fight to the enemy, precisely what Lincoln wanted and what he intended to do. Grant's sense of duty was evident in his unquestioning acceptance of Lincoln's policies on emancipation and the recruitment of Negro troops. Unlike McClellan and Buell, Grant dismissed whatever personal doubts he may have had and pitched in wholeheartedly. When Halleck instructed him to assist Lorenzo Thomas, the adjutant general, in enlisting freed slaves, Grant said frankly, "I never was an abolitionist, nor even what could be called antislavery. [However,] you may rely upon it I will give him all the aid in my power. I would do this whether arming the negro seemed to me a wise policy or not, because it is an order that I am bound to obey and I do not feel that in my position I have a right to question any policy of the government."[22]

Once in charge, Grant did what Taylor would have done, wasting no time or words. Grant wanted the entire Union Army to move in a coordinated fashion after the enemy, squeezing and chasing them relentlessly so that Lee and Joseph Johnston couldn't use their troops to help each other. He told Meade, "Lee's army will be your objective point."[23] Taking over Grant's former command was his friend William Sherman, whose foster father was the powerful Ohio politician Thomas Ewing, who had been a close friend of Clay's. Grant told Sherman "to move against [Joseph] Johnston's army in the south, to break it up and get into the interior of the enemy's country as far as you can, inflicting all the damage you can against their resources."[24] Sherman summarized the simple strategy: "He was to go for Lee and I was to go for Joe Johnston. That was the plan."[25]

The next year would be the busiest and bloodiest of the war. Grant and the Union Army would be tested as Lee's forces again brought the fight northward. Like other commanders before him, Grant faced organizational, structural, and personnel problems, including the fact that many units in the army were constrained to reporting directly to the secretary of war, not the commanding general. Lincoln assured Grant that, although he could not turn over control of these forces directly to Grant, "there is no one but myself that can interfere with your orders, and you can rest assured that I will not."[26]

Other than that, as Grant told his friend Jones, he knew he had "a pretty big job on my hands."[27]

II

From even before his first day in office, Lincoln worried about how the war might end, and if his army was victorious, how to mend the Union. Nearly a year before his second inaugural, on March 26, 1864, at a time when he expected to lose reelection, he met with a trio of Kentucky dignitaries—Albert Hodges, editor of the *Frankfort Commonwealth,* Archibald Dixon, a former senator from Kentucky, and Kentucky's governor, Thomas Bramlette. They came to protest the enlistment of former slaves as soldiers. Lincoln gave them "a little speech," explaining why he felt obligated to change from his inaugural promise not to interfere with slavery to his decision to issue an emancipation proclamation. Acknowledging they had been persuaded by his remarks, Hodges asked Lincoln for a copy of what he had said. Lincoln's remarks were extemporaneous, but he promised to send a letter with his thoughts written down.

Nine days later, Lincoln sent the letter. He explained the series of events that had "driven [him] to the alternative of either surrendering the Union, and with it, the Constitution," or of arming Southern slaves.[28] He said, "I hoped for greater gain than loss; but of this, I was not entirely confident."[29] At the end of his missive, he wrote, "I add a word which was not in the verbal communication." He emphasized, "In telling this tale I attempt no compliment to my own sagacity. I claim not to have controlled events, but confess plainly events have controlled me." He added further that "after the end of three years struggle the nation's condition is not what either party or any man devised or expected. God alone can claim it."[30] His addendum reflects Lincoln at his best, manifesting both humility and piety in "telling this tale" of everything he had done thus far as president. Casting himself in the tale as an instrument of God could do him no harm in Kentucky.

Lincoln's two overriding concerns throughout 1864—ending the war and winning reelection—informed his judgment and leadership in virtually everything he did. It was no accident that Lincoln was the first incumbent president to be renominated by his party for president since Martin Van Buren in 1840. He had used his patronage to bolster party support, and he now decided to reorganize his Cabinet to solidify his support within the Republican Party.

Salmon Chase had been a thorn in Lincoln's side since joining the Cabinet. In 1863–1864, he was angling for the Republican nom-

Chapter 2

General Principles of Insecticide Toxicology

2.1. EVALUATION OF TOXICITY

Toxic interactions of any chemical and any given biological system are dose related. At extremely high concentrations, most chemicals have toxic effects on biological systems. The toxicology of poisonous chemicals can be termed the "science of doses."

The toxicity of insecticides (or of any toxicant, for that matter) to a particular organism is usually expressed in terms of the LD_{50} (for "lethal dose"). This value represents the amount of poison per unit weight which will kill 50 % of the particular population of the animal species employed for the tests. The LD_{50} is commonly expressed as milligrams per kilogram (mg/kg) or occasionally as milligrams per body (e.g., mg/female fly). In some cases, the exact dose initially given to the insect cannot be determined but the concentration of the insecticide in the external media can, so that the LC_{50} is used. For example, the toxicity of insecticides to mosquito larvae or fish is commonly assessed by the concentration of the toxic compound in water that will kill half the animals exposed for a specified period of time.

The term LT_{50} is also frequently used; it represents the time required to kill 50 % of the population at a certain dose or concentration. This method of assessing toxicity requires relatively few individuals and therefore is often employed for field tests where the possibility of collecting a sufficient number of individuals is limited (e.g., the World Health Organization's standard test for German cockroach colonies). In limited instances, the rate of knockdown of insects becomes a more important criterion for assessing the efficiency

17

of insecticides than the rate of kill; in such cases, the median knockdown dose and knockdown time, KD_{50} and KT_{50}, are used. There are cases where killing or knockdown does not constitute the desired criterion. For instance, in tests for chemosterilants the idea is not to kill the test insects but to sterilize them without reducing their vigor. The chemical in question is measured according to its effects on fertility and fecundity. The ED_{50} and EC_{50} (effective dose and effective concentration) are used to describe results of such tests. The method of obtaining these values and its mathematical implications will be discussed later.

2.1.1. Toxicity Tests Against Insects and Other Invertebrates

There are several ways to administer insecticides to an animal. The most commonly employed method for insects is *topical application*, where the insecticide is dissolved in a relatively nontoxic and volatile solvent, such as acetone, and is then allowed to come in contact with a particular location on the body surface. Usually combinations of a constant amount of solvent with varied concentrations of the insecticide are used for this purpose in order to keep the area of contact, as well as the effect of solvent, constant. Although the results obtained with the topical application procedure can be a very reliable indication of the relative contact toxicity of any insecticide to a certain animal, the method is not sufficient by itself to indicate the actual amount entering the animal's body. When knowledge of the exact amount of insecticide inside the body is required, the *injection method* is usually employed. The insecticide is commonly dissolved in carrier material, such as propylene glycol or peanut oil, and injected into the body cavity, e.g., intraperitoneally. For insects, injection is usually made at the abdominal sterna or the intersegmental regions, avoiding the longitudinal center line so as not to injure the abdominal nerve cord. The needle is held in position for a while and then pulled away gradually in order to avoid bleeding due to internal pressure.

These standard methods cannot be used in certain cases where the insect's mode of life or its morphological arrangement conflicts with the testing method; for instance, dipterous larvae cannot withstand the skin injury caused by the injection method, and topical application cannot deliver a sufficient quantity of insecticide. A number of specially designed testing methods are available for many of these unorthodox cases (Busvine, 1971). For instance, the *dipping method* is used for dipterous larvae. The insects are simply picked up with a pair of forceps and dipped into the insecticide preparation, which is either a suspension in a solvent such as acetone or methyl ethyl ketone or an emulsion in an emulsifier such as Triton X-100. The

LC_{50} rather than the LD_{50} is generally used to express results with the dipping method, but of course the meaning of LC_{50} here is quite different from that in, say, the mosquito larvae test, since the insects are removed from the insecticidal preparation. Assessment of the proper range of reliability becomes important in the dipping test; for instance, the mortality rate often does not increase beyond a certain point because of the limitation of the insecticide's low solubility or limited amount that can be suspended (e.g., Labadan, 1965).

The *contact method* or *residual exposure method* is another way of exposing insects to an insecticide. The insecticide in a solvent is applied to the container or the panel surface where insects are to walk or rest. The solvent is evaporated by rotating the container or the panel, so that the insecticide is evenly spread over a known area. Oppenoorth (1959) found a linear relationship between the results obtained with the contact method and topical application; by his procedure, the contact LD_{50} values are approximately 20 times greater than the topical application LD_{50} values. There are a few distinct deviations from this rule. The results with DDVP, which has a high vapor pressure, are one example; they suggest that this insecticide penetrates the insect's body in vapor form and that even with the contact method a relatively large amount can be picked up without appreciable effects on the insect.

Based on the various types of practical control measures and the insects' mode of life, there are several ways of testing the effectiveness of insecticides toward various insect species. The *leaf-dipping method* for testing two-spotted spider mites, the *grain fumigation method* for stored-product pests, and the *feeding method* for various larvae are typical examples. In all cases, the results are relative and are compared with the known effects of conventional insecticides to assess the potential of a chemical as a commercial insecticide.

In order to assess the susceptibility of any insect population to a certain poison, probit units of percent mortalities are customarily plotted against a logarithmic scale of dosages. It has been empirically observed that in many biochemical and physiological processes equal increments in effect are produced only when the stimulus is increased logarithmically, i.e., by a constant "proportion" (rather than constant amount). Bliss (1935) was the first to propose that this method of plotting the logarithms of dosage vs. probits of percent mortality would yield a straight line. Further discussion is available in several specialized books (Swaroop and Uemura, 1956; Finney, 1949, 1952; Busvine, 1971).

A population of insects is expected to exhibit a wide range of individual variation in susceptibility to insecticides. It is therefore necessary to find a reliable means to express the overall effectiveness of insecticides on a given insect population. This is particularly important when one has to decide, for instance, whether there is an insecticide-resistant population in the field by comparing its susceptibility data with those of other insect populations normally found elsewhere. Like any other distribution of individual variation within a homogeneous population sample, the insects' response in terms of susceptibility to an insecticide is expected to form a type of "normal distribution" curve. According to the distribution curve, the number of individuals which respond (e.g., death by acute poisoning criteria) at the LD_{50} dose is expected to be maximal, and the number of individuals with high degrees of resistance or susceptibility is expected to be small.

In an ordinary toxicological test, however, it is far easier to determine the percentage of surviving (or killed) ones in the total insect population at a certain time or at a certain dose than to assess the number (or percent) of individuals belonging to a particular susceptibility range. When the survival or mortality percentages are plotted (i.e., the susceptibility data are plotted cumulatively) on a dosage–mortality graph, the response of a normal homogeneous population of insects to an insecticide should be expressed in the form of a symmetrical sigmoid curve. The following is a brief summary of the statistical methods employed by the World Health Organization (Swaroop, 1957) for malaria control work.

2.1.1a. Graphical Method for Estimating LD_{50}

The first step is to find the logarithm of the appropriate concentration value of the insecticide. Each recorded mortality percentage is then converted to the probit value. When these points are plotted, they should fall along a straight line (if the probit log sheet is used, untransformed data can also be plotted, but these values must be converted for subsequent statistical calculations).

The method for graphically obtaining LD_{50} values is different when there is natural mortality among the controls. In such cases, the data are analyzed by regarding the values of adjusted mortality (by Abbott's formula) as if they were actually observed mortality values and as if there were no natural mortality, e.g.,

$$\frac{x - y}{x} \times 100$$

where x is the percent survival in the untreated controls and y is the percent survival in the treated insects. This approximation is permissible when the control mortality is small (less than 20%) or is based on a large number of observations.

Once the points have been plotted, a probit regression line is drawn freehand. With actual data, a line is drawn by disregarding the 0 and 100 points and fitting the line to the points around LD_{50} as well as possible.

Litchfield and Wilcoxon's method (1949) is used in conjunction with this procedure to increase the accuracy of fitting a line to the points on the graph. This method further increases accuracy in estimating LD_{50} values.

2.1.1b. Goodness of Fit (χ^2 Test) and Confidence Limit

The goodness of fit of the probit regression line to the data may be assessed numerically (Wilcoxon and McCallan, 1937; Swaroop, 1957). The calculation is as follows.

1. Read from the probit log paper the difference between the observed data (original data) and the expected mortality percentage data (that on the line on the graph) at each dose level.
2. Compute the quantity d for each concentration level:

$$d = \frac{(\text{observed} - \text{expected percentage mortality})^2}{(\text{expected mortality}) \times (100 - \text{expected percentage mortality})}$$

3. Add the total d values and multiply the sum by the average number of insects at each concentration. This gives χ^2:

$$\chi^2 = (\text{total } d) \times (\text{average number insects per dose})$$

The value of χ^2 measures the goodness of fit of the probit regression line to the points. The smaller the value of χ^2, the better the fit. The computed value of χ^2 should be compared with the critical χ^2 table. To use this table, the degree of freedom must be found, which is the number of concentration levels minus 2. If the value of χ^2 exceeds that given in the table, the line is not a good fit to the observed data. A new line should then be fitted by eye to reduce the larger values of d and χ^2 again computed. This is repeated until χ^2 is below the value given in the table.

If the data are heterogeneous, sometimes a satisfactory line cannot be drawn to give a χ^2 value below that given in the table. In such cases, the line that gives the lowest value of χ^2 is used.

Following is a procedure for estimating the upper and lower confidence limits within which the LD_{50} is expected to lie with 95 % (or 99 %) probability.

1. Read from the probit regression line the values of LD_{16} and LD_{84}.
2. Using those figures, calculate S:

$$S = \frac{LD_{84}/LD_{50} + LD_{50}/LD_{16}}{2}$$

3. From the original data, find the total number of insects tested at the dose levels between LD_{16} and LD_{84}. This is N.

4. Using the values of S and N, find f, which is the factor by which LD_{50} should be multiplied or divided to obtain, respectively, the upper and lower confidence limits at 95% probability:

$$\log f_{95} = \frac{2.77}{N} \log S$$

For confidence limits at 99% probability, use the formula

$$\log f_{99} = \frac{3.641}{N} \log S$$

5. Find the upper and lower confidence limits:

$$\text{upper limit} = \text{original } LD_{50} \times f$$

$$\text{lower limit} = \text{original } LD_{50} \div f$$

In most laboratory tests, these graphical treatments of the toxicological data are adequate. In more elaborate cases, mathematical estimation of LD_{50} data is possible. The most commonly used method is that of Finney (1952).

2.1.1c. Symptomatological Observation

The first step in assessing the toxic effect of a poison is to observe the physical and behavioral responses of the poisoned animal. These responses are the basis for an important pharmacological classification of insecticides.

At median doses, nerve poisons characteristically induce the appearance of symptoms in four stages: (1) excitation, (2) convulsions, (3) paralysis, and (4) death. The typical narcotic fumigants cause only the three stages excitation, paralysis, and death, and irritant vapors lack the stage of paralysis (Brown, 1951).

Often the first symptom of poisoning appears after a certain lapse of time. This is called the latent period and is common among the stomach poisons. The excitation period is also preceded by a period when the insect appears restless. It is at this stage that the insect often exhibits typical "cleaning" movement; this consists of cleaning the antennae or other parts of the body by mouth. In larvae, the preexcitation period is not so clear. Klinger (1936) described the pyrethrin poisoning process in caterpillars as follows: "fast, restless locomotion with dorsum raised, the prolegs strongly prehensile, the head moving side to side, the mandibles snapping, and food being regurgitated." Mosquito larvae exposed to dieldrin show curling-up and telescopic movements.

The excitation period can usually be recognized by the insect's frantic running or flying movements, depending on its mode of life. Honeybees become highly agitated on contact with chlordane or pyrethrins and fly about frantically. At the end of the excitation period, honeybees show ataxia, with zigzag flight and locomotion. Cockroaches, at the late excitation period of nerve poisoning, usually exhibit "locomotive instability." These observations give valuable information on the nature of intoxication in the animal. Prolonged latent periods, for instance, generally indicate metabolic conversion of the compound *in vivo* to more toxic material. Comparison of the relative length of the poisoning stages indicates the nature of the poisoning, such as paralytic or stimulatory (excitant) action.

There are a number of techniques which have been utilized to aid symptomatological evaluation of insecticide action. The most commonly used technique is manometric and respirometric assay. Because of their small size, insects can be placed directly into a Warburg flask, for instance, to study the effects of the toxicant on their respiratory mechanisms (e.g., Harvey and Brown, 1951). Another convenient method is the assessment of effects on heartbeat. The heart of the American cockroach (*Periplaneta americana*) can be directly observed dorsally through the thoracic and abdominal terga, or it can be isolated for observation of direct action *in situ* (Krijgsman and Krijgsman, 1950). It is known that most cholinergic agents have direct action on the insect heart. Electrophysiological techniques are often utilized because of the relative ease of using insect material for this type of experiment (Narahashi, 1963). For instance, the abdominal nerve cord, which represents the central nervous system in *P. americana*, can be directly assayed by external silver–silver chloride electrodes. The afferent input can be studied by simply puffing air currents on the cerci, which are loaded with mechanoreceptor sensilla, and then transmission through synaptic processes can be studied by utilizing their well-defined ganglia.

2.1.2. Toxicological Evaluation in Higher Animals

The toxicity evaluation process in the high animals is different from that in lower animals since the number of available animals usually is limited. For instance, it is not difficult to use a few thousand *Drosophila* or *Daphnia* for a single toxicity test, but it is not economically or otherwise practical to use a few hundred mammals (even small ones such as mice or other rodents) for the evaluation of a single toxicity test. While the processes involving LD_{50} determination are identical once the mortality figures (or any other criteria used for evaluation) have been obtained, the emphasis in mammalian toxicology is on more qualitative aspects of poisoning rather than on quantitative ones. The limitation in number has necessitated several adjustments in order to ascertain the validity of toxicity determinations in higher animals. The second characteristic of toxicological tests in higher animals is that in most cases the overriding concern in such studies is to evaluate safety for man.

2.1.2a. Selection of Test Animals

In many animal experiments, the individual and colony variabilities become very problematical in securing reproducibility of the tests. Particularly important are genetic differences in susceptibility to disease and toxic compounds. It has been generally considered that 25% of the variability in animal experiments is due to genetic variations and 75% to environmental influence (Sabourdy, 1961; Hurni, 1970). This is based on the results of germ-free animal tests. Every expert thus recommends the use of homogeneous animal populations. According to Hurni (1970), there are basically two types of breeding systems which have provided scientific communities with relatively consistent animal materials: (1) strict inbreeding and (2) outbreeding. For inbreeding, the animal strain must be bred for at least 20 generations, strictly involving brother × sister or offspring × parent crossings. Inbred strains are altered by mutation or can go through genetic divergence. Thus, to attain stabilization, any offshoot stock of inbred strains must be propagated no longer than three to five generations. After that, natural changes in gene composition could occur. Introduction of pairs from a recognized "primary-type colony" would become necessary thereafter. (The International Committee on Laboratory Animals, ICLA, in cooperation with the World Health Organization, looks after this matter.)

For outbreeding, a number of inbred strains are genetically mixed. This step assures heterogeneity with respect to heterosis and other genetic expressions in the resulting F_2 hybrids. This method is particularly successful in yielding consistent strain characteristics whenever the number of individuals maintained for the parental stock inbred strains is large.

Stabilization and standardization of environmental factors are also important. The most serious factor influencing the outcome of toxicity tests is the state of health of the experimental animals; various diseases have been known to affect them. The most important development in this regard is the introduction of SPF (specific pathogen–free) animals for scientific use. SPF animals are free of the pathogenic microorganisms and parasites frequently found in their particular species (there are 31 and 21 such disease-causing organisms, respectively, listed for mice and rats alone; Hurni, 1970).

Provided that the animals are healthy and of acceptable genetic composition, other important factors influencing the toxicity of insecticides in higher animals are age, sex, nutrition, and other rearing conditions (Durham, 1969). It is well known that newborn animals are much more sensitive to toxic compounds than adult animals, even after adjustment for body weight differences. This is probably because newborn animals often completely lack certain detoxification processes for insecticidal chemicals, such as microsomal mixed-function oxidase systems; these are known to develop rapidly after birth. It is possible that other factors, such as the ratio of the size of the total nervous system to the body weight, the ratio of food intake to the body weight, the body surface ratio, and susceptibility to diseases, could also play important roles. In the case of old animals, this last criterion is an important consideration. The sex difference in susceptibility to toxicants and drugs in the rat has been well documented. In general, female rats are more susceptible on a per kilogram basis, particularly to organophosphate insecticides, but there are a number of pesticidal chemicals to which males are particularly susceptible. In any event, the LD_{50} figures must be accompanied by specification of the method of administration and sex to be complete. It is clear that the nutritional state of the animal has profound effects on its susceptibility to chemicals, particularly to lipophilic insecticidal chemicals. In brief, animals on nutritionally proper or excessive diets have the best resistance to pesticidal chemicals. This is at least partly due to protection of the target organ by fat and protein and to proper functioning of the liver enzyme systems in animals raised on high-protein diets. This point will be further discussed in subsequent chapters.

2.1.2b. Acute Toxicity Data

In assessing the safety of any poisonous chemical for higher animals, the first task is to determine the acute oral LD_{50} value, a simple expression of the degree of toxicity that can be understood by general scientists (DuBois and Geiling, 1959). Chemicals can be conveniently classified according to their LD_{50} values. They are (1) "extremely toxic" (LD_{50} on the order of 1 mg/kg or less), (2) "highly toxic" (1–50 mg/kg), (3) "moderately toxic" (50–500 mg/kg), (4) "slightly toxic" (0.5–5 g/kg), (5) "practically nontoxic" (5–15 g/kg), and (6) "relatively harmless" (more than 15 g/kg). Such categorization, though convenient, is only a rough standard.

It is well acknowledged that the route of administration of toxicants strongly influences the LD_{50} data, and hence a description of the method of application is usually included along with the LD_{50} values. The most commonly employed method of administering pesticidal compounds is by the oral route. This is because oral ingestion is expected to be the most likely route of entry for insecticidal compounds other than for people occupationally engaged in pesticide spraying or manufacturing processes. There are two major methods of oral administration. One is to add the pesticidal compound to the animal's diet and the other is to give the toxicant directly through a stomach tube or in capsules. The former method is less accurate for estimation of the actual oral dose taken by the animal, but is less strenuous to the animal. Another problem is the difficulty of preparing homogeneously pesticide-impregnated feed. Generally, a volatile solvent and vegetable oil are used in this process. The technique of prior fasting is often used to promote ingestion of all the food given, but fasting animals cannot be recommended as experimental subjects. With the direct method, the vehicle used to carry the pesticide becomes rather important. For stomach tube methods, vegetable oil and polyethylene glycol (and other glycols) are frequently used. In any event, the volume of orally administered material should not exceed 2–3% of the body weight, because the laxative effect of oil could spoil the experiment (Balzs, 1970). Administration of pesticides either in concentrated forms for liquids or in capsules for dry preparations (e.g., Tucker and Crabtree, 1970) has been done. Generally speaking, this method of application results in much slower gastrointestinal absorption. In the case of highly irritant or toxic compounds, particularly liquids, such a method of application is not recommended, because of possible damage to the stomach and emetic effects.

For rodents in particular, intraperitoneal injection is commonly employed in acute toxicity studies. In many cases, toxicities of pesticides are best manifested when this route of administration is used, e.g., for determining the relative toxicities of a related series of pesticide analogues. Often, however, the method may induce shock reactions in the animal, particularly when the pesticide is an irritant. Also, care must be taken to choose the right vehicle and to use the proper speed of injection in the same manner as for intravenous injection.

Intravenous injection is the most direct method of application. It is generally agreed that the route that carries the actual toxicant most rapidly to the bloodstream is the one that causes the greatest toxic action on the animal. However, stringent precautionary measures must be taken with regard to injection. The vehicles used are emulsions containing 15–20% of either vegetable oil or polyethylene glycol 300, or both, and an emulsifying agent and water, preferably isotonic saline. In any event, the volume for rapid intravenous injection should be limited to 0.1–0.5 ml for rodents, and the speed of injection must be slow.

Acute dermal toxicity and acute inhalation toxicity tests are the two most important methods employed in assessing the safety of pesticidal chemicals for workers. Dermal toxicity tests are conducted on compounds suspected to cause acute poisoning in man. In animal experiments, the most widely used method is that of Draize (1955) for rabbits. A number of organophosphates have been examined by this method. To study their effects, an albino rabbit weighing 2–3 kg is shaved around the abdomen and back, and the chemical in question is painted (dry powder is moistened with isotonic saline to prepare a paste) over the area, which is then covered with either a rubber sleeve or cotton gauze held in place with a wire screen. Gaines (1960) has measured the dermal toxicities of organophosphates and chlorinated hydrocarbon insecticides in rats using xylene as a solvent and applying the solution at 0.0016 ml/kg body weight to a shaven area 3 by 4.5 cm over the top of the shoulder. Hayes *et al.* (1964) painted the right hand and forearm of human volunteers with 2% dust, 2% emulsion, or 47.5% emulsifiable concentrate of parathion for 70–90 min at 80–103°F to study the dermal toxicity of this compound to man.

Acute inhalation toxicity tests can be conducted in two ways: by using either a static or a dynamic air flow system. In the former procedure, the animal is kept in a closed chamber and exposed to one or several spray or aerosol applications. The simplest chamber is an enclosed glass container with small openings for inlet and outlet. Gage (1970) illustrates a number of devices that are utilized in inhalation toxicity tests. In the dynamic air flow procedure, the animal is maintained in a chamber which is constantly supplied with an air current containing a constant amount or level of pesticidal compound. For nongaseous pesticides in the normal range of experimental temperatures, introduction of the sample in the form of mist or dust becomes a necessity. The methods for vaporization, atomization, etc., have been discussed by Lehman *et al.* (1964).

A key question in interpreting the results of inhalation toxicity tests is the distinction between inhalation toxicity and dermal toxicity. Efforts have been made in the past to assess the relative importance of these two routes of entry by first estimating the total entry by either gas chromatography or urinary product analyses and then calculating the amount of respiratory absorption by respiratory exposure (i.e., consideration of total inhalation volume, pesticide concentration, and absorption coefficients). The amount of dermal absorption is estimated as the difference between the total absorption and the inhalation absorption and therefore represents a minimal value (Durham and Wolfe, 1963, 1972). The respiratory volume for a man doing light work is on the order of 36 m³/day. By employing such a comparative absorption test, Hartwell *et al.* (1964) and Hayes *et al.* (1964) were able to conclude that the toxicity of parathion is greatest through the inhalation route, despite a long-standing belief that parathion is most dangerous through skin penetration in man. Hartwell *et al.* found the general ratio of parathion toxicity in man to be inhalation 10, oral 3, and dermal 1. The level of inhalation toxicity can be expressed either as mg/m^3 (preferred for the static tests) or as parts per million (ppm). In limited cases, the actual amount of pesticide taken in through inhalation can be estimated and exposure levels in terms of mg/kg/hr can be substituted for the other criteria for expressing inhalation toxicity. Recommended threshold limit values for various pesticides in the breathing air are summarized in Table 2-1.

TABLE 2-1. Recommended Threshold Limit Values[a] for Selected Pesticides in the Breathing Air of Working Environments

Pesticide	Parts per million (ppm)	Milligrams per cubic meter (mg/m^3)
Aldrin—skin	—	0.25
Arsenic and compounds (as As)	—	0.5
Calcium arsenate	—	1
Camphor	—	2
Carbaryl (Sevin®)	—	5
Carbon disulfide—skin	20	60
Carbon tetrachloride—skin	10	65
Chlordane—skin	—	0.5
Chlorinated camphene—skin	—	0.5
Chlorobenzene (monochlorobenzene)	75	350
Chloropicrin	0.1	0.7
Cyanide (as CN)—skin	—	5
2,4-D	—	10
DDT—skin	—	1
DDVP—skin	—	1
Demeton—skin	—	0.1
1,2-Dibromoethane (ethylene dibromide)—skin	25	190
1,2-Dichloroethylene	200	790
Dichloroethyl ether—skin	15	90
Dibrom® (naled)	—	3
Dieldrin—skin	—	0.25
Dinitro-o-cresol—skin	—	0.2
Endrin—skin	—	0.1
EPN—skin	—	0.5
Guthion®	—	0.2
Heptachlor—skin	—	0.5
Hydrogen cyanide—skin	10	11
Lead arsenate	—	0.15
Lindane—skin	—	0.5
Malathion—skin	—	15
Methoxychlor	—	15
Methyl bromide—skin	20	80
Nicotine—skin	—	0.5
Paraquat—skin	—	0.5

TABLE 2-1. Continued

Pesticide	Parts per million (ppm)	Milligrams per cubic meter (mg/m³)
Parathion—skin	—	0.1
Pentachlorophenol—skin	—	0.5
Phosdrin® (Mevinphos®)—skin	—	0.1
Phosphine	0.3	0.4
Pyrethrum	—	5
Ronnel[b]	—	15
Rotenone (commercial)	—	5
Sodium fluoroacetate (1080)—skin	—	0.05
2,4,5-T	—	10
TEPP—skin	—	0.05
Thiram®	—	5
Warfarin	—	0.1

American Conference of Governmental Industrial Hygienists (1966).

[a]Threshold limit values refer to airborne concentrations of substances and represent conditions under which it is believed that nearly all workers may be repeatedly exposed, day after day, without adverse effects.
[b]Tentative value.

2.1.2c. Chronic Toxicity and Other Nonacute Toxicity Tests

The importance of chronic toxicity studies of pesticidal compounds cannot be overemphasized. In pesticide research there are three different objectives and therefore three designs of study: (1) studies for true chronic toxicity effects on tissues and organ function, (2) studies for secondary effects such as carcinogenicity, teratogenicity, and mutagenicity, and (3) studies for "no-effect" levels. These three testing programs should be independently designed from the beginning, since the dose levels, as well as the criteria for evaluation, are very different. In the true chronic toxicity studies, the purpose of the test is to find any undesirable, harmful effects of a given substance (Benitz, 1970). The dose chosen is generally high, and signs of illness and gross morphological and physiological changes are expected and watched. In experiments designed to study secondary effects, the period of administration ranges from 3 to 18 months and doses are chosen so as not to damage the health of the animal. In usual pesticide experiments, the doses are chosen at levels a few times higher than those expected to be present in food for human consumption. In any case, a definite evaluation criterion is set (e.g., liver carcinoma for carcinogenicity, chromosomal aberrations for mutagenicity) from the beginning. The time span of the experiments is usually long, generally the total life span of the experimental animal. The experiments for "no-effect" levels include some chronic toxicity tests, but they are different in that the most sensitive signs of intoxication and effects are adopted as the judging criteria, and the most sensitive animal species is selected. The rationale of such a design is the assumption that human susceptibilities to pesticidal chemicals are equivalent to those found in the most sensitive animal species.

In addition to these three basic tests, there are a few nonacute tests which are pertinent to pesticide research. Usually these tests are not standardized, being designed specifically for a particular compound. Three examples of such tests are cited below.

For studying the effects of organophosphates on vision, Upholt *et al.* (1956) applied 2 drops of 0.1 % solution of TEPP in peanut oil to the eye and found maximal miosis, a decrease in light perception, and increased depth of focus. The test has a practical implication. For instance, pest control operators' pilots could experience difficulties in landing their planes after spraying organophosphates (a case of this has been reported by Quinby *et al.*, 1958).

A skin sensitization test was conducted on naled (Dibrom®) and four other pesticides by Edmundson and Davies (1967). They employed patch tests on human subjects who had a history of exposure to naled and the other pesticides and on controls who did not have any previous exposure. The patches were attached to the inner surface of the forearms of the volunteers for 24 hr. The results clearly indicated that only naled had the ability to induce "sensitization."

For testing the allergic purpura reaction, Nalbandian and Pearce (1965) employed the indirect basophil degranulation test for *p*-dichlorobenzene (Di-chloricide,® mothball) sensitized workers. The test has the advantage of not exposing the patients to the hazard of recurrence.

In true chronic toxicity studies, the criterion for toxicity can be death or any other harmful effects. The general relationship of acute and chronic toxicities in terms of killing effects is that cumulative effects are observed for generally persistent pesticides, while in degradable pesticides the toxic effects do not carry over. Tucker and Crabtree (1970) found that the daily doses of degradable insecticides required to induce chronic toxicity are only slightly lower than the oral LD_{50} value for acute poisoning. This generalization can be dangerous for unknown compounds, however, since some irreversible damage can be inflicted on various enzyme, tissue, and organ functions even by degradable pesticides, which increases the potential danger of chronic poisoning. The case in point is the irreversible reaction ("aging") of organophosphates with cholinesterase and certain other esterases. Irreversible interaction of DFP, TOCP, or certain other organophosphates and a particular esterase could cause demyelination, leading to the "delayed ataxia" syndrome in animals (see Chapter 4, on mode of action, and Chapter 11, on health hazards).

The most frequently employed administration method is oral feeding with the animal's diet. This method is least strenuous in the long run, and also can simulate the conditions under which man and wildlife are exposed to pesticides in the form of food residues. In limited cases, both capsules (e.g., Tucker and Crabtree, 1970) and the stomach tube (e.g., Innes *et al.*, 1969) have been used to ascertain the exact doses administered. With regard to the duration of the tests, the lifespan of the experimental animals is a prime consideration in the experimental design. The rate of cell turnover is generally faster in small animal species within the same animal group, reflecting their shorter life spans. Thus it is most convenient in lifetime chronic studies to choose small animal species, such as the mouse, for which 2–4 years is the equivalent of 60–80 years of human life. Benitz (1970) has made a convenient conversion table for human equivalency in five experimental animals (Table 2-2). The duration of the tests also depends on the kind of information being sought, and therefore there are no set rules. In most cases, life span studies for mice and 6–18 months of chronic studies for larger animals are adopted.

As for the criteria of toxic effects, both clinical observation (including biochemical testing) during the experimental period and postmortem (or end-of-experiment autopsy) examination of the animals are employed. While the criteria are different from compound to compound, the most frequently utilized clinical observations are mortality, body weight gain or loss, amount of food intake, behavior changes, functional tests such as liver enzyme tests (see Chapter 11) and serum cholinesterase tests, hematological examination, and analysis for pesticide residues in blood and urine (mostly for the metabolic products). For postmortem studies, the most frequent observations are measurement of body weight, organ weight, skeletal length, and size; morphological changes in vital organs, particularly the liver and kidneys; presence or absence of hemorrhage; and pesticide levels through residue analyses.

After considering all available data on each compound, "no-effect" levels are determined for human safety by regulatory governmental agencies, including world-wide regulatory bodies such as WHO (World Health Organization) and FAO (Food and Agriculture Organization). Both groups are being organized under the United Nations to establish guidelines. Although there are no set rules in deriving the "no-effect" level for pesticidal chemicals, past experience indicates that there are a few rules generally followed by the experts:

1. "No-effect" levels are assessed as "daily intake" values mostly through food ingestion.

TABLE 2-2. Time Relationships Between Drug Exposure and Life Span of Various Experimental Animals, with Time Equivalents in Man

Duration of study in months	Rat		Rabbit		Dog		Pig		Monkey	
	Percentage life span	Human equivalent in months	Percentage life span	Human equivalent in months	Percentage life span	Human equivalent in months	Percentage life span	Human equivalent in months	Percentage life span	Human equivalent in months
1	4.1	34	1.5	12	0.82	6.5	0.82	6.5	0.55	4.5
2	8.2	67	3.0	24	1.6	14	1.6	14	1.1	9
3	12	101	4.5	36	2.5	20	2.5	20	1.6	13
6	25	202	9.0	72	4.9	40	4.9	40	3.3	27
12	49	404	18	145	9.8	81	9.8	81	6.6	53
24	99	808	36	289	20	162	20	162	13	107

From Benitz (1970).

2. In most cases, the data for the most sensitive system in the most sensitive species are adopted as the minimum-effect values.
3. Man is regarded as equivalent to the most sensitive species.
4. Human data, if available, take precedence over any other animal data.
5. Such minimum-effect values are generally divided by 100 (the "safety factor") to arrive at acceptable daily intake values.

The criteria for "effects" vary from compound to compound. Generally recommended data (Fitzhugh, 1965) are

1. Acute LD_{50} and LC_{50} data on experimental animals.
2. Chronic toxicity data on more than two species, e.g., rat, 2 years; dog, 5 years.
3. Metabolism studies and determination of toxic forms.
4. Biochemical data including enzyme studies.
5. Reproduction studies (recommended for three generations).
6. Human studies, if possible (e.g., organophosphate and blood cholinesterase level studies).

This scheme requires further elaboration. Metabolic data should also include the nature of the residues commonly occurring in food commodities as well as the toxic impurities which might be present in the technical preparations of the pesticide. The enzyme studies can be used for the compounds for which the mode of action has been well defined. For all cholinesterase inhibitors, for instance, the task is relatively simple. For other classes of compounds, however, assessment of "no-effect" becomes difficult. Often criteria such as liver damage and clinical symptoms (which are the result of biochemical lesions) are adopted. In view of the likely probability that biochemical lesions occur at much lower pesticide levels than the ones that cause gross morphological and clinical effects (as is the case with organophosphates), it is clear that such criteria should be replaced by other more sensitive systems which reflect the direct action of the chemical at molecular levels. Moreover, even for those chemicals for which the mode of action has been elucidated, the possibility of other side-effects should not be overlooked. Thus it is apparent that the accuracy and validity of any "no-effect" evaluation are determined by the adequacy of the biochemical assay criteria.

A few examples quoted by Fitzhugh (1965) are useful in gaining some practical insight. For chlorbenside, lindane, and methoxychlor, the "no-effect" levels in the diet for the rat were determined to be 20, 25, and 200 ppm, the equivalent of 1.0, 1.25, and 10.0 mg/kg/day doses. Introduction of the safety factor (i.e., division of these doses by 100) gives an acceptable daily intake value of 0.01, 0.0125, and 0.1 mg/kg/day for these compounds, respectively. However, as Table 2-3 shows, in some instances the "safety

TABLE 2-3. Examples of Calculation of the Acceptable Daily Intake Values for Three Important Pesticides[a]

Animal	"No-effect" level[b] (ppm)	Equivalent daily intake (mg/kg)	"Acceptable daily intake" decided (mg/kg/day)	Remarks on why "safety factor" of 100 was not adopted
Parathion				
Rat	1	0.05		Consideration of human data
Dog	<1	<0.025	0.005	and cholinesterase data;
Man	0.05 mg/kg/day[c]	0.05		factor of 10 adopted
Diazinon				
Rat	5	0.25	No decision[d]	Lack of human cholinesterase
Dog	0.25	0.006		data; factor of 100 too impractical
Captan				
Rat	1000	50		Insufficient metabolism data,
Dog	4000	100	0.1	but generally safe use records
Pig	480	19		

From Fitzhugh (1965).

[a]In each case, some practical consideration was taken into account.
[b]Cholinesterase inhibition data for parathion and diazinon.
[c]Human experiment for 2 months.
[d]Later agreed to be 0.002 mg/kg/day.

factor" of 100 has not been adopted by FAO–WHO, for the various reasons listed in the remarks column.

The recent "acceptable daily intake" values for other important pesticides are cited by Duggan and Corneliussen (1972) and are listed in Table 11-15. In essence, the basis for determining acceptable daily intake values has been that such values should reflect the daily doses which during an entire lifetime appear to be without appreciable risk. "Without appreciable risk" is understood to imply "as a matter of practical certainty." Thus practical considerations, such as the data on occupationally exposed human subjects, epidemiological studies, and frequency of appearance as residues in food, are also important. The methods of approach for carcinogenicity, teratogenicity, mutagenicity, and other biochemical tests will be described in Chapter 11.

2.1.2d. *Observation of Toxic Symptoms*

In higher animals qualitative observations of toxicological symptoms become very important. Such observations often yield valuable information about the site of poisoning, mechanisms of action, defensive responses of the animal, delayed action, and sensitization, and even suggestions for possible treatment. Despite this importance, there is no set procedure that can be utilized exclusively for pesticidal compounds. In most cases, pesticide toxicologists have followed symptomatological observation procedures similar to those developed for drugs (e.g., Campbell and Richter, 1967),

TABLE 2-4. General Scheme for Observation of Symptoms of Insecticide Poisoning

System	Examination	Signs of toxicity
CNS and somatomotor	Body movements	Twitch, tremor, ataxia, convulsion, paralysis, fasciculation
	Muscular tone	Rigidity, flaccidity
	Behavior (experimental animal)	Restlessness, jumping, general motor activity, vocalization
	Reaction to stimuli (animal)	Response in the open field, swimming ability, reaction to falling and to sudden light and sound
	Clinical sign (man)	Headache, dreams and poor sleep, perspiration, nervousness, dizziness, test on reflex
Autonomic	Pupil size	Myosis, mydriasis
	Secretion	Salivation, lacrimation
Respiratory	Nostrils	Unusual discharge or movements, rhinorrhea
	Character of breathing	Bradypnea, dyspnea, yawning
	Clinical sign (man)	Constriction of chest, cough and wheezing
Ocular	Eyelids and eyeball	Ptosis, exophthalmos
	Clinical sign	Pain on accommodation, dimness, lacrimation, conjunctival injection
Gastrointestinal	Signs in experimental animals	Shape of abdomen (contraction, swelling due to flatulence), diarrhea, vomiting
	Clinical sign	Anorexia, nausea, vomiting, diarrhea
General side-effects	Acute poisoning	Temperature, skin texture and color, cardiovascular effects, cyanosis, jaundice
	Chronic poisoning	Food intake, body weight, tumor, disease, sleep time

which are generally useful in determining the class-of-action group such as sympathomimetic, sympatholytic, parasympathomimetic, parasympatho lytic, or central-acting (e.g., convulsant). The most important criteria are ptosis (drooping eyelids), salivation, lacrimation, effects on pupil size (myosis and mydriasis), piloerection, temperature changes in paw and rectum, and central effects such as convulsion.

The majority of insecticides are nerve poisons, and thus the CNS effects are most carefully observed. Observations of insecticide effects on the autonomic nervous system have also been helpful in elucidating the action mechanisms of these chemicals *in vivo*; the most celebrated case is organophosphate and carbamate insecticides, which show clear parasympathomimetic activities. A few insecticides are known to be respiratory posions. As far as the acute poisoning symptoms are concerned, most insecticidal chemicals elicit typical CNS stimulant effects. In some cases, gastrointestinal upset can be very severe (e.g., acute organophosphate poisoning), or respiratory effects, such as asthma or bronchopneumonia-like symptoms, may be predominant. Key symptomatological observations pertinent to insecticide poisoning are summarized in Table 2-4. Examples of poisoning descriptions for each group of insecticidal compounds are also presented in Chapter 11.

2.2. ANALYTICAL METHODS FOR INSECTICIDES

2.2.1. Chromatographic Analysis of Insecticide Residues

Analysis of insecticide residues poses an entirely different type of problem for toxicologists, for these residues are present in extremely small quantities in generally heterogeneous materials including biological materials. From the beginning, the challenge for residue chemists has been the lowering of the detection limits for residues (Zweig, 1968). Essentially the process of residue analysis consists of three major steps:

1. Extraction and "cleanup" to initially eliminate the bulk of unwanted material from the samples.
2. Separation of most of the insecticide derivative in question by chromatography.
3. Detection of the chemical in question at the highest possible sensitivity without interference by other substances.

There are a number of excellent books and publications available in this particular field, so just the bare essentials of the principles of residue analyses will be presented here. More detailed descriptions can be found in Gunther and Blinn (1955), Gunther (1962), Zweig (1963, 1964, 1968),

Zweig and Sherma (1972), Burke (1965), MacDougall (1971), Van Middelem (1971), Burchfield and Johnson (1965), Thornburg (1966), and Barry *et al.* (1963).

2.2.1a. Sampling, Extraction, and Cleanup

The importance of proper sampling has been stressed by many workers. In essence, unbiased random sampling techniques, those often adopted by agronomists, are satisfactory for this purpose provided that sufficient quantities of samples are taken and that proper checks are made on untreated areas or general samples, when necessary, to insure accurate relative assessment of the residue situation. Checks must be made as well to insure the validity of the particular residue analysis technique through fortification or "spiking" with known quantities of authentic reference compounds. The samples thus collected should be analyzed as soon as possible, but if necessary they can be wrapped in clean aluminum foil, unused wax paper or other pretested material and stored at $-20°$F. Water loss, degradation of unstable residues, and changes in extractabilities are the three major concerns. In extreme cases, parallel control experiments are necessary to correct for the changes that are caused by these processes during storage. After sampling, the prime concern of residue chemists must be the homogeneity of the mixing within one sample, which is normally limited to 1000 g at most.

For extraction of residues, there are two important factors to consider: (1) whether the solvent is pure and (2) whether the solvent is the proper one for maximum extraction efficiency. Some solvents may contain interfering and reactive impurities or pesticide residues. Other frequent sources of contamination are rubber, plastic linings (especially avoid polyethylene products), silicone grease, detergents on glassware, and mineral oils. Choice of the proper solvent is very important because, generally speaking, insecticides are soluble in both polar and nonpolar solvents and sparsely soluble in aqueous solutions. Thus relatively polar solvents are used for oily or fatty samples and nonpolar (or mixtures of nonpolar and polar) solvents for nonfatty, aqueous substances. Solvent mixtures are used for many biological materials with lipid–protein complexes to elute as much residue as possible. There are a few universally adaptable solvent systems. Thornburg (1963) describes a series of solvent extractions using hexane–isopropanol (or benzene–isopropanol), acetonitrile, and finally redistilled methylene chloride. Zweig and Sherma (1972) cite a universal extraction method employing 2 ml/g propylene carbonate which is unsatisfactory only for milk.

For fatty substances (cheese, peanuts, fatty tissue of animals), methanol or ethanol is recommended for the original extraction. For nonfatty substances and those having high to medium moisture content, acetonitrile is used along with some water. Addition of water to dry samples (for instance, dry soil samples) tends to increase the extractability of residues (Barry *et al.*, 1963). Benzene and hexane have been exclusively utilized for extraction of chlorinated hydrocarbon and organophosphate insecticides. In all cases, recovery tests must be conducted on the combination of samples and the solvents to assure the validity of the experiments.

There are two major procedures for general cleanup: the solvent–solvent or solvent–aqueous solution partitioning method and the column absorption method. (A third method, which utilizes distillation processes, is less frequently employed.) The most frequently used partitioning technique is the acetonitrile–hexane (or acetonitrile–petroleum ether) partitioning system, which allows residues to be partitioned into the acetonitrile phase, leaving fatty materials in the hexane phase. N,N-Dimethylformamide is often used in place of acetonitrile. Usually the hexane phase is reextracted several times to assure completeness. Evaporation of acetonitrile, or re-extraction into hexane or any other volatile solvent after the addition of water to the acetonitrile phase and subsequent evaporation, usually accomplishes the task of concentration. The most frequently used column material is Florisil®, since this material can handle a large quantity of material as compared to alumina, for example, while retaining some adsorption characteristics which are useful in cleanup processes for pesticide residues. Other materials occasionally used are silica gel, activated charcoal, activated alumina, and celite.

2.2.1b. Separation and Detection of Residues

By far the most popular method of residue analysis is the gas–liquid chromatographic technique (GLC), but thin-layer chromatographic techniques (TLC) are becoming increasingly popular, as well. Paper chromatographic (or thin-layer cellulose) techniques have been used for polar groups of pesticide residues. Both thin-layer and paper chromatographic methods can be used to supplement the GLC method, since they are suitable for mass production and for analyses of labile compounds in addition to being based on an inherently different principle of separation than the GLC method; this facilitates qualitative, positive recognition of a substance against reference standards. Although GLC systems offer both excellent resolution plus sensitivity and relative ease of operation plus reproducibility, they have been frequently abused (Robinson, 1970). The problems encountered most often are misidentification of the compound, lack of information on degradation of the residues [e.g., in most cases the peak(s) thought to be the result of endrin injection is not endrin at all but ketoendrin and/or

TABLE 2-5. Selectivity and Sensitivity of Various Detectors for Gas Chromatography

Detector[a] (synonym) and modification	Frequency of use	Principles	Selectivity	Selectivity index	Sensitivity (ng)
Electron capture (EC or ECD)	+++	All electron-capturing substances are ionized by electrons to be attracted to anode	Cl, other halogens, and other electron-capturing substances including S, O, N compounds	Medium to low	0.0001–1
Flame photometer	+	Use of filter on flame emission; 526 mμ for P and 394 mμ for S	P, S (less sensitive)	Specific	1
Thermionic (sodium thermionic or stacked flame)	++	Modified hydrogen flame detector with salt-tip electrode	P and halides (halides often the sources of interference)	Fairly specific	1–10
Electrolytic conductivity	+	Pyrolizer in H_2 over N to form NH_3, which changes conductivity	N (also analyzes Cl on oxidative mode)	Specific	10–100
Microcoulometer (MCGC)	++	Combustion: oxidation for halogens or reduction for phosphorus, and titration	S, P, and halogens (except F)	Fairly specific	100–1000

[a]Another type of detector showing promise is the microwave emission detector, used for phosphorus residues (Bache and Lisk, 1966).

endrin aldehyde formed by thermal decomposition], and misjudgment of the quantity of the residues owing to insufficient cleanup treatments or to ignorance of the nonlinearity of responses by some of the detectors.

The most celebrated case of misidentification was that of PCB (polychlorinated biphenyls, stable plasticizer mixtures) peaks for DDT–R (DDT and related compounds including DDT metabolites and impurities in technical products). The problem was originally spotted by Jensen (1966). Reynolds (1969) also examined this problem and concluded that PCB peaks, particularly the one which coincides with DDD, can be mistaken for DDT–R on GLC systems. Since PCBs tend to migrate nearer to the solvent front in various thin-layer chromatographic systems (or in critically performed Florisil® column methods), separation of PCBs from DDT–R is possible. The incident suggests that complete reliance on one or even two GLC systems for identification of residues is dangerous. Use of two or more GLC systems, coupled with at least one TLC analysis (preferably preceding the GLC analyses), is desirable.

With regard to the quantitative aspect, problems can arise from insufficiently purified samples. Impurities can shift the GLC positions, in addition to their influence on sensitivity. In any event, critical examination of the quantity–detector-response relationship for the same sample by spiking is necessary to ascertain the validity of the method. Even the quantities of the standard samples must be examined, since improper use of microsyringes can cause either excess (by the holdup volume in the needle portion) or insufficient (by leaking from the plunger space) injection.

The characteristics of the GLC detectors currently in use are summarized in Table 2-5. The choice of detectors depends on their selectivities, sensitivities to the type of residues for study, and the ease or reliability of their operation.

2.2.2. Other Confirmatory Techniques for Residue Analysis

As discussed above, chromatographic analyses often are not sufficient to positively identify insecticide residues. Moreover, chromatographic methods are basically for separation of organic chemicals, and as such can be used only for known chemicals for which authentic reference standards are available. In practice, however, insecticidal compounds are metabolized or converted into many unknown or ill-defined products. Three basic approaches have been employed to study such products: chemical, spectroscopic, and biological–biochemical assay methods.

2.2.2a. Chemical Reactions

Among chemical assay techniques, the most commonly used ones are colorimetric assays. Some colorimetric reactions can be highly sensitive (e.g., the diazo blue reaction against 1-naphthol for carbaryl residues) and specific. This subject has been quite adequately covered by Gunther (1962) and Beckman *et al.* (1963). In addition, the structural and chemical properties of residual compounds can be altered by numerous standard chemical reactions to facilitate elucidation of their molecular composition. Useful reactions for insecticide analyses include dehalogenation on liquid ammonia and sodium (Beckman *et al.*, 1958); dehydrochlorination on alcoholic alkali; esterification to increase both volatility and detectability (e.g., with pentafluoropropionate and trichloroacetate); oxidation of P$=$S to P$=$O for further cholinesterase assay (Voss and Geissbuehler, 1967), or OH to ketone, aldehyde, or acid (for functional group analyses); bromination or hydrogenation over PtO_2 with H_2 for unsaturation; and epoxidation of cyclodiene insecticides. In most cases, these reactions are well-known standard chemical techniques which happen to have found a use in pesticide analyses because of the particular molecular characteristics of the chemicals involved. The other chemical method applied to insecticide residue analysis is total element: mostly chlorine (Lisk, 1960) and phosphorus (Dunn *et al.*, 1963). The limits in sensitivity and the quality of information have lately discouraged use of this technique.

2.2.2b. Spectroscopic Methods

Spectroscopic analyses have been extensively utilized in metabolic studies. In residue analysis their usefulness is usually limited by their general insensitivity and requirement for purity, but a few methods have found some usefulness. Mass spectroscopy and GC–mass spectroscopic techniques have been getting a great deal of attention (Widmark, 1972). The detection limit of the ordinary GC–mass spectroscopic technique is generally on the order of 1 μg. However, Baughman and Meselson (1973) used a time-averaging device along with a high-resolution mass spectroscope and so extended its sensitivity for tetrachlorodibenzo-*p*-dioxin (TCDD) to less than nanogram quantities (or samples in ppt ranges have been measured). This sort of computer analysis of the GC–mass spectroscopic results is no doubt going to be the method of choice for future residue analysts, since it gives both quantitative and qualitative (i.e., positive identification) data. One drawback is the cost of such a setup.

Another sensitive spectroscopic analytical technique is the fluoroscopic method (MacDougall, 1962, 1964). Generally speaking, fluoroscopic assays are limited in the number of compounds applicable, for the compounds suitable for assay by this method must have the properties of absorbing electromagnetic energy and then releasing at least a part of it in the form of light (UV to visible). When the release is accomplished almost instantaneously the chemicals are called fluorescent, while the chemicals which require a much longer time period for release are called phosphorescent. The fluorescing types of compounds are steroids, polynuclear compounds, indol- and catechol-type chemicals, and polyaromatic compounds (e.g., naphthalene). Examples of the method are found in analytical processes developed for carbaryl metabolites, Co-ral®, and Guthion®. With the use of coupling agents, the method may become more widely applicable.

Similarly, Moye and Winefordner have tried a phosphorimetric technique on pesticide residues (Moye and Winefordner, 1965; Winefordner and Moye, 1965). They found it to be applicable at 5–10 mg/ml for carbaryl, Zectran®, Metacil®, Mesurol®, carbofuran, and UC 10854 (3-isopropyl phenyl *N*-methylcarbamate), nicotine, and nicotinine.

The major difficulty with both techniques is the frequent presence of interfering substances in biological materials that cause high background readings. This is a particular problem with plant materials, and the prior cleanup process must be carefully programmed.

Ultraviolet and infrared spectroscopy are somewhat less sensitive techniques than the fluorescence and phosphorescence methods. Also, a rather complete prepurification of the sample is required. Only a limited number of applications for insecticide residues are available (Blinn, 1964; Van Middelem, 1971). Similarly, there is only one example of NMR (nuclear magnetic resonance spectroscopy) analysis (Fukuto *et al.*, 1964) of residues, in which ^{31}P spectra were used to analyze fenthion in plants.

2.2.2c. Biological and Biochemical Assay

One of the most important considerations in analysis of residues is assessment of their potential biological damage. It is unfortunate that biological and biochemical assessment have not been really extensively applied to the field of residue analysis. Among biochemical assay methods, inhibition of cholinesterases by organophosphates has been most widely utilized for residue analysis (Archer, 1963). A recent outstanding development has been automated analysis of various residues against both serum and erythrocyte cholinesterases, with analytical accuracy and capability almost comparable to GLC analyses (Winter and Ferrari, 1964; Ott and Gunther, 1966; Voss and Geissbuehler, 1967; Voss and Sachsse, 1970). Indeed, such a method has been utilized in an extensive residue analysis of market food commodities (Renvall and Akerblom, 1971). The blood cholinesterases of some avian species, such as peacocks and quail, are extremely susceptible to insecticidal organophosphates, making the assays very sensitive. Unfortunately, there is no other biochemical assay system comparable to cholinesterase inhibition tests.

As for bioassay, probably the most complete reviews of the subject have been made by Dewey (1958) and by Sun (1963). In essence, bioassay animals are selected on the basis of high pesticide sensitivities and by the ease with which large numbers of them can be reared. Vinegar flies (including *Drosophila melanogaster*), houseflies, mosquito larvae, *Daphnia magna*, *Daphnia pulex*, and brine shrimp have been most extensively used for this purpose. Also, some fish species (guppies and goldfish) have been used for assay. Although insecticidal chemicals have widespread toxicity among many biological systems, true specific tests for any insecticidal chemical are difficult to develop, but some degree of specificity can be gained by combinations of chromatographic separation techniques. The agents causing the toxicity in the field are often known before the need for bioassay arises (i.e., a monitoring program after parathion spray in the field). In these cases, some degree of specificity can be expected: such methods are available for aldrin, dieldrin, and Phosdrin®. The basic mathematical approach for bioassay is identical to the one used for toxicity testings (LD_{50} determination), as explained before.

Perhaps the wisest use of bioassay for residue analysis, however, is as a supplement to chromatographic and other analytical methods. Even though bioassay methods cannot be used to identify specific insecticidal chemicals, insect bioassay systems are very specific for separating insecticidal chemicals from any other group of toxins and contaminants; after all, no other chemicals should be able to match the toxicity of insecticides to insects. Thus other interfering substances in instrumental analyses, such as PCBs, can easily be ruled out by this route. Such a collective specificity is rare and has indeed been useful in the final identification of residue problems.

2.3. REFERENCES

American Conference of Governmental Industrial Hygienists (1966). *Threshold Limit Values.* Secretary-Treasurer, 1014 Broadway, Cincinnati, Ohio 45202.

Archer, T. E. (1963). In *Analytical Methods for Pesticides, Plant Growth Regulators and Food Additives*, Vol. 1. G. Zweig, ed. Academic Press, New York, p. 373.

Bache, C. A., and D. J. Lisk (1966). *Residue Rev.* **12**:35. F. A. Gunther, ed. Springer-Verlag, New York.

Balzs, T. (1970). *Methods in Toxicology.* G. E. Paget, ed. Blackwell Scientific Publications, Oxford and Edinburgh, p. 49.

Barry, H. C., J. G. Hundley, and C. Y. Johnson (1963). *Pesticide Analytical Manual*, Vol. 1. Food and Drug Administration, U.S. Department of Health, Education and Welfare, Washington, D.C.

Baughman, R. W., and M. S. Meselson (1973). *Environmental Health Perspectives.* Issue No. 5, p. 27, National Institute of Environmental Health Sciences, Research Triangle, N.C.

Beckman, H. F., E. I. Ibert, B. B. Adams, and D. O. Skovlin (1958). *Agr. Food Chem.* **6**:104.

Beckman, H. F., R. B. Bruce, and D. MacDougall (1963). In *Analytical Methods for Pesticides, Plant Growth Regulators and Food Additives*, G. Zweig, ed., Academic Press, New York, Vol. 1, p. 131.

Benitz, K. F. (1970). In *Methods in Toxicology*. G. E. Paget, ed. Blackwell Scientific Publications, Oxford and Edinburgh, p. 49.

Blinn, R. C. (1964). *Residue Rev.* **5**:130. F. A. Gunther, ed. Springer-Verlag, New York.

Bliss, C. I. (1935). *Ann. Appl. Biol.* **22**:134.

Brown, A. W. A. (1951). *Insect Control by Chemicals*. John Wiley & Sons, London, p. 817.

Burchfield, H. P., and D. E. Johnson (1965). *Guide to the Analysis of Pesticide Residues*, Vol. 1. Government Printing Office, Washington, D.C.

Burke, J. A. (1965). *J. Assoc. Offic. Agr. Chemists* **48**:1037.

Busvine, J. R. (1971). *A Critical Review on the Techniques for Testing Insecticides*. Commonwealth Agricultural Bureaux, Slough, England, 345 pp.

Campbell, D. E. S., and W. Richter (1967). *Acta Pharmacol. Toxicol.* **25**:345.

Dewey, J. E. (1958). *J. Agr. Food Chem.* **6**:274.

Draize, J. H. (1955). *Food Drug Cosmetic Law J.*, p. 722.

DuBois, K. P., and E. M. K. Geiling (1959). *Textbook of Toxicology*. Oxford University Press, Oxford, 302 pp.

Duggan, R. E., and P. E. Corneliussen (1972). *Pesticides Monitoring J.* **5**:331.

Dunn, C. L., D. J. Lisk, H. F. Beckman, and C. E. Castro (1963). In *Analytical Methods for Pesticides, Plant Growth Regulators and Food Additives*, Vol. 1. G. Zweig, ed. Academic Press, New York.

Durham, W. F. (1969). In *Chemical Fallout*. M. W. Miller, G. G. Berg, and A. Rothstein, eds. Charles C. Thomas, Publisher, Springfield, Ill., Chap. 23, p. 433.

Durham, W. F., and H. R. Wolfe (1963). *Bull. World Health Org.* **29**:279.

Durham, W. F., and H. R. Wolfe (1972). *Bull. World Health Org.* **26**:75.

Edmundson, W. F., and J. E. Davies (1967). *Arch. Environ. Health* **15**:89.

Finney, D. J. (1949). *Ann. Appl. Biol.* **36**:187.

Finney, D. J. (1952). *Probit Analysis*, 2nd ed. Cambridge University Press, Cambridge, 318 pp.

Fitzhugh, O. G. (1965). In *Research in Pesticides*. C. O. Chichester, ed. Academic Press, New York, p. 59.

Fukuto, T. R., E. O. Hornig, and R. L. Metcalf (1964). *J. Agr. Food Chem.* **12**:169.

Gage, J. C. (1970). *Methods in Toxicology*. G. E. Paget, ed. Blackwell Scientific Publications, Oxford and Edinburgh, p. 258.

Gaines, T. B. (1960). *Toxicol. Appl. Pharmacol.* **2**:88.

Gunther, F. A. (1962). In *Advances in Pest Control Research*, Vol. 5. R. L. Metcalf, ed. Interscience Publishers, New York, p. 191.

Gunther, F. A., and R. C. Blinn (1955). *Analysis of Insecticides and Acaricides*. Interscience Publishers, New York, 696 pp.

Hartwell, W. V., G. R. Hayes, and A. J. Funckes (1964). *Arch. Environ. Health* **8**:820.

Harvey, G. T., and A. W. A. Brown (1951). *Can. J. Zool.* **29**:42.

Hayes, G. R., A. J. Funckes, and W. V. Hartwell (1964). *Arch. Environ. Health* **8**:829.

Henly, R. S., R. F. Kruppa, and W. R. Supina (1966). *J. Agr. Food Chem.* **14**:667.

Hurni, H. (1970). *Methods in Toxicology*. G. E. Paget, ed. Blackwell Scientific Publications, Oxford and Edinburgh, p. 11.

Innes, J. R. M., B. M. Ulland, M. G. Valerio, L. Petrucelli, L. Fishbein, E. R. Hart, A. J. Pallotta, R. R. Bates, H. L. Falk, J. J. Gart, M. Klein, I. Mitchell, and J. Peters (1969). *Natl. Cancer Inst. J.* **42**:1101.

Jensen, S. (1966). *New Scientist*, December 15 issue, p. 612.

Klinger, H. (1936). *Arb. Physiol. Angew. Entomol.* **3**:49, 115.

Krijgsman, B. J., and N. E. Krijgsman (1950). *Nature* **165**:936.

Labadan, R. M. (1965). Ph.D. thesis. Cornell University, Department of Entomology, Ithaca, N.Y.

Lehman, A. J., D. W. Fassett, H. W. Gerarde, H. E. Stokinger, and J. W. Zapp (1964). *Principles and Procedures for Evaluating the Toxicity of Household Substances.* Publication 1138, National Academy of Sciences, National Research Council, Washington, D.C.

Lisk, D. J. (1960). *J. Agr. Food Chem.* **8**:119.

Litchfield, J. T., and F. Wilcoxon (1949). *J. Pharmacol. Exptl. Therap.* **96**:99.

MacDougall, D. (1962). *Residue Rev.* **1**:24. F. A. Gunther, ed. Springer-Verlag, New York.

MacDougall, D. (1964). *Residue Rev.* **5**:119. F. A. Gunther, ed. Springer-Verlag, New York.

MacDougall, D. (1971). In *Pesticides in the Environment*, Vol. 1, Part II. R. White-Stevens, ed. Marcel Dekker, New York, p. 271.

Moye, H. A., and J. D. Winefordner (1965). *J. Agr. Food Chem.* **13**:516.

Nalbandian, R. M., and J. F. Pearce (1965). *J. Am. Med. Assoc.* **194**:238.

Narahashi, T. (1963). Properties of insect axons. In *Advances in Insect Physiology*, Vol. 1. J. W. L. Beament, J. E. Treherne, and V. B. Wigglesworth, eds. Academic Press, New York, p. 512.

Oppenoorth, F. J. (1959). *Entomol. Exptl. Appl.* **2**:216.

Ott, D. E., and F. A. Gunther (1966). *J. Assoc. Offic. Anal. Chemists* **49**:669.

Quinby, G. E., K. C. Walker, and W. F. Durham (1958). *J. Econ. Entomol.* **51**:831.

Renvall, S., and M. Akerblom (1971). *Residue Rev.* **34**:1. F. A. Gunther and J. D. Gunther, eds. Springer-Verlag, New York.

Reynolds, L. M. (1969). *Bull. Environ. Contam. Toxicol.* **4**:128.

Robinson, J. (1970). *Ann. Rev. Pharmacol.* **10**:353.

Sabourdy, M. A. (1961). Techniques with germ-free animals. Paper presented at the lecture tour organized by ICLA with the support of UNESCO and IAEA, September–October. (Indirectly cited from Hurni, 1970.)

Sun, Y.-P. (1963). In *Analytical Methods for Pesticides, Plant Growth Regulators and Food Additives.* G. Zweig, ed. Academic Press, New York, Vol. 1, p. 399.

Swaroop, S. (1957). *Statistical Methods for Malaria Eradication Programs.* World Health Organization, Geneva, Switzerland.

Swaroop, S., and K. Uemura (1956). *Probit Analysis.* World Health Organization, Geneva, Switzerland. (Mimeograph Malaria/178.)

Thornburg, W. W. (1963). In *Analytical Methods for Pesticides, Plant Growth Regulators and Food Additives*, Vol. 1. G. Zweig, ed. Academic Press, New York, p. 87.

Thornburg, W. W. (1966). *Residue Rev.* **14**:1. F. A. Gunther, ed. Springer-Verlag, New York.

Tucker, R. K., and D. G. Crabtree (1970). *Handbook of Toxicity of Pesticides to Wildlife.* Resources Publication No. 84, U.S., Denver Wildlife Research Center, Denver, Colo.

Upholt, W. M., G. E. Quinby, G. S. Batchelor, and J. P. Thompson (1956). *Am. Med. Assoc. Arch. Ophthalmol.* **56**:128.

Van Middelem, C. H. (1971). In *Pesticides in the Environment*, Vol. 1, Part II. R. White-Stevens, ed. Marcel Dekker, New York, p. 309.

Voss, G., and H. Geissbuehler (1967). *Med. Rijks Faculteit Landbouwetenschappen Gent* **32**:877.

Voss, G., and K. Sachsse (1970). *Toxicol. Appl. Pharmacol.* **16**:764.

Widmark, G. (1972). In *Environmental Quality and Safety*, Vol. 1. F. Coulston and F. Korte, eds. Georg Thieme Verlag Stuttgart, p. 78.

Wilcoxon, F., and S. E. A. McCallan (1937). Graphical method of probit analysis. *Contrib. Boyce Thompson Inst.* **10**:329.

Winefordner, J. D., and H. A. Moye (1965). *Anal. Chim. Acta* **32**:278.

Winter, G D., and A. Ferrari (1964). *Residue Rev.* **5**:139. F. A Gunther, ed. Springer-Verlag, New York.

Zweig, G., ed. (1963). *Analytical Methods for Pesticides, Plant Growth Regulators, and Food Additives*, Vol. 1. Academic Press, New York, 637 pp.

Zweig, G. (1964). *Chromatog. Rev.* **6**:110.

Zweig, G. (1968). *The Vanishing Zero, the Evolution of Pesticide Analyses.* Syracuse University Research Corporation, Syracuse, N.Y., 39 pp.

Zweig, G., and J. Sherma, eds. (1972). *Handbook of Chromatography*, Vol. II. Chemical Rubber Company Press, Cleveland, pp. 237–246.

Chapter 3

Classification of Insecticides

3.1. HISTORY AND GENERAL GROUPINGS OF INSECTICIDAL COMPOUNDS

The development of new insecticides in the past three decades has been so rapid that no single book can adequately cover all the insecticidal derivatives without the risk of being outdated as soon as it is published. It is not, therefore, the purpose of this chapter to describe all the details of insecticidal compounds. Rather, the intention is to classify these chemicals according to the chemical properties which make them insecticidal. Other descriptions are added when it is felt that such explanation would assist the reader in understanding the historic, economic, or chemical importance of the compounds.

Man's desire to control his environment has created many useful chemicals. Evolution of chemical insecticides (Table 3-1) essentially began with readily available materials such as arsenicals, petroleum oils, and botanical insecticides (e.g., nicotine, pyrethrin, rotenone). The first synthetic organic insecticides that appeared for public use were dinitro compounds and thiocyanates (Murphy and Peet, 1932; see also books cited below). Perhaps the most significant discovery leading to the proliferation of new synthetic insecticides was that of DDT. This unusual compound, 1,1,1-trichloro-2,2-bis(p-chlorophenyl)ethane, was first synthesized by Zeidler in 1874, but its insecticidal properties were first discovered in 1939 by Müller of Switzerland. The use of DDT revolutionized the control of insect pests. Other chlorinated hydrocarbon insecticides such as BHC, toxaphene, chlordane, aldrin, and dieldrin followed immediately thereafter. The second massive introduction of new insecticides was initiated by a German worker,

TABLE 3-1. History of Development and Use of Insecticides

Year	Chemicals and location	Remarks
900[a]	Arsenites in China	Era of natural products
1690	Tobacco used in Europe	
1787	Soaps used in Europe	
1800[a]	Pyrethroids in Caucasus	
1845	Phosphorus (inorganic) in Germany	
1848	Derris root powder in Malaya	
1854	CS$_2$ for fumigation (France)	Era of fumigants, inorganics,
1867	Paris green (US)	petroleum products
1868	Petroleum products (US)	
1874	DDT synthesized (Zeidler)	
1877	HCN as fumigating agent	
1880	Lime-sulfur (US)	
1883	Bordeau (France)	
1886	Pine resin for scales	
1892	Lead arsenate (US)	
1918	Chloropicrin (France)	
1932	Methyl bromide (France)	
1925	Dinitro compounds	Modern synthetic insecticides
1932	Thiocyanates	
1939	DDT insecticidal properties discovered (Müller)	
1941	2,4-D synthesized (US)	
1941	BHC (France)	
1942	BHC (United Kingdom)	
1944	Parathion (Germany, Schrader)	
1940–50	Aldrin, dieldrin, endrin, etc. (US)	
1945	Chlordane (Julius Hyman, US; Riemschneider, Germany)	
1947	Development of insecticidal carbamates, Geigy, Switzerland (Dimetan,® pyrolan, and Isolan®)	
1950	EPN (DuPont)	
1952	Malathion	
1953	Lidov: aldrin, dieldrin, Shell patent	
1958	Sevin® (carbaryl) (Union Carbide, US)	
1962	Rachel Carson's *Silent Spring* appears (US)	
1967	First hormone mimic (juvenile) insecticides (US)	Hormones and pheromones
1970	DDT trial begins (Sweden, US)	

[a]Years are approximations.

Gerhard Schrader, a pioneer in the chemistry and uses of organophosphorus insecticides. The number of organophosphorus compounds (OP compounds) used for insect control today is unmatched by any other group of insecticides, and undoubtedly many more will soon be on the market. The most widely used organophosphorus compounds include parathion, Systox®, malathion, EPN, diazinon, and DDVP (dichlorvos). The carbamates are the newest group of synthetic insecticides. These compounds (essentially the synthetic analogue of an important alkaloid poison, physostigmine or eserine) were first developed by the Geigy Company for insecticidal use in 1947 (see Gysin, 1952, 1954).

As already stated, it is not the purpose of this book to describe the use, characteristics, and commercial value of all existing insecticides. Readers who are interested in the properties and chemistry of each insecticidal compound should refer to any of the following publications:

Agricultural Research Service. *Suggested Guide for the Use of Insecticides to Control Insects Affecting Crops, Livestock, and Households.* Washington, D.C.: U.S. Department of Agriculture, 1965. (For use recommendations.)

A. W. A. Brown. *Insect Control by Chemicals.* New York: John Wiley and Sons, 1951. 817 pp. (For general information on principles.)

D. E. H. Frear. *Pesticide Index*, 3rd ed. State College, Pa.: 1965. 295 pp. College Science Publishers. (Compilation of data on chemicals.)

D. E. H. Frear, ed. *Pesticide Handbook—Entoma*, 24th ed. State College, Pa.: College Science Publishers, 1972, 279 pp. (Reference on manufacturers, formulations, and trade names.)

D. F. Heath. *Organophosphorus Poisons.* New York and London: Pergamon Press, 1961. (For organophosphates.)

R. E. Johnsen and W. M. Hanstbarger. *Handbook of Insecticides.* Fort Collins, Colo.: Colorado State University, 1966. (For use patterns.)

E. E. Kenaga (Entomological Society of America). "Commercial and Experimental Organic Insecticides." 1963 revision: *Bull. Entomol. Soc. Am. 9(2):* 67–103; 1969 revision: *Bull. Entomol. Soc. Am. 12(2):* 161–217. (For principles of nomenclature and accepted names.)

H. Martin, ed. *Insecticide and Fungicide Handbook*, 3rd ed. Oxford and Edinburgh: Blackwell Scientific Publications, 1969. 387 pp. (Crop oriented.)

R. L. Metcalf. *Organic Insecticides.* New York: Interscience Publishers, 1955. (For general and chemical information.)

R. D. O'Brien. *Toxic Phosphorus Esters.* New York: Academic Press, 1960. (Organophosphates.)

R. D. O'Brien. *Insecticides.* New York: Academic Press, 1967. 332 pp. (For general information on principles.)

G. Schrader. *Die Entwicklung über insektizider Phosphorsäure Ester.* Weinheim: Verlag Chemie, 1963. (Technical data on chemical properties of OP.)

TABLE 3-2. General Grouping of Insecticides

Class	Origin	Chemical group	Examples
Insecticides and acaricides	Synthetic—organic	Chlorinated hydrocarbons	DDT derivatives BHC Cyclodienes
		Organophosphorus compounds	Aliphatic Aryl
		Carbamates	Naphthyl Phenyl Heterocyclic Oximes
		Thiocyanates	
		Nitrophenols	
		Organofluorine compounds	Fluoroacetate
		Sulfonates, sulfides, sulfones	
		Fumigants	Methyl bromide
		Miscellaneous compounds	
	Synthetic—inorganic	Arsenicals Fluorides Mercurials	
	Natural products—organic	Botanicals	Nicotinoids Pyrethroids Rotenoids
		Microbials	Toxins Antibiotics

Activators or synergists	Mostly synthetic—organic	Synergists	Methylenedioxyphenyl compounds
Carrier or bulk material	Natural—organic Natural—inorganic	Petroleum products Dusts	
Attractants	Synthetic—organic	Food attractants Sex attractants	
Repellents	Synthetic—organic		
Growth regulators	Synthetic—organic	Hormone mimics	
Chemosterilants	Synthetic—organic	Alkylating agents	

W. T. Thomson. *Agricultural Chemicals*. Davis, Calif.: Simmons Publishing Co., 1964. (Compounds, use classification.)

W. T. Thomson. *Cropland Pesticide Application Guide for 1969*. Davis, Calif.: Thomson Publications, 1969. 567 pp. (Classification according to crops.)

Table 3-2 summarizes the general classification of insecticides. It is a modification of a generally accepted method (e.g., see, from above listing, Kenaga, 1963; Thomson, 1964; Brown, 1951; Metcalf, 1955). Since it groups insecticides according to chemical nature and origin (natural, synthetic, organic, and inorganic), it is rather comprehensive and useful. For all practical purposes, insecticides include activators and synergists, chemosterilants, and hormonal agents, which may not directly kill insects by themselves. However, for toxicological purposes, and in this book, hormonal agents, repellents, and attractants are not considered, since these three groups of compounds do not seem to cause toxicological problems at present.

In organizing this chapter, compounds have been selected according to their importance in toxicology and in environmental toxicology and not by their practical importance. They are discussed here because of their toxicological importance. Since most of the information needed for this chapter is either well established or available from commercial sources rather than from recent scientific work, most of it has been gathered from the publications listed above.

3.2. CHLORINATED HYDROCARBON INSECTICIDES

The chlorinated hydrocarbon compounds include such important insecticides as DDT, BHC, chlordane, and dieldrin. All compounds which belong to this group are characterized by (1) the presence of carbon, chlorine, hydrogen, and sometimes oxygen atoms, including a number of C—Cl bonds; (2) the presence of cyclic carbon chains (including benzene rings); (3) lack of any particular active intramolecular sites; (4) apolarity and lipophilicity; and (5) chemical unreactivity (i.e., they are stable in the environment).

These chemicals are often considered to belong to the group of organochlorine pesticides. However, their group characteristics make them very different from other organochlorine pesticides such as fumigants, chlorinated organophosphates (e.g., Dursban®), carbamates, and chlorinated aliphatic and aromatic acids. Therefore, *chlorinated hydrocarbon insecticides* is the preferred name.

Generally speaking, there are three major kinds of chlorinated hydrocarbon insecticides: DDT analogues, benzene hexachloride (BHC) isomers, and cyclodiene compounds.

3.2.1. DDT Analogues

DDT is one of the most important insecticides ever to appear on the market. Its chemical structure is

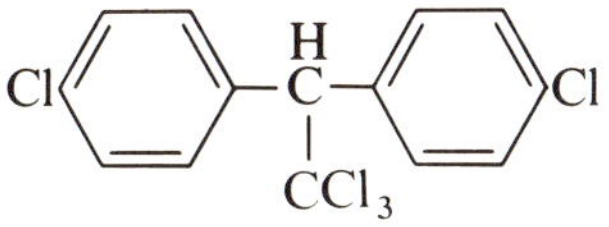

1,1,1-TRICHLORO-2,2-BIS(*p*-CHLOROPHENYL)ETHANE

It can also be called dichlorodiphenyltrichloroethane, which accounts for the name DDT.

Technical DDT is a white to cream-colored amorphous waxy powder, and pure DDT is a crystalline powder. The melting point of the pure compound is 109°C, and its setting point (crystallization on slow cooling) is between 103° and 105°C. Technical DDT has a melting point of 89°C and is considered a satisfactory product if its setting point is not below 88°C.

DDT is one of the most apolar compounds known to exist; hence it is soluble in most apolar organic solvents and is practically insoluble in water and cold ethanol. Its water solubility is less than 2 ppb (parts per billion). Since DDT is soluble in hot ethanol, it can be recrystallized by cooling ethanol in which DDT has been dissolved. The solubility of DDT in most solvents rises steeply with increase in temperature.

The average commercial product contains about 70% of the pure *p,p'*-DDT. Increasing the amount of chloral used in DDT preparation increases the purity of the product. The major contaminant of crude DDT is *o,p*-DDT, which is not as insecticidal as *p,p'*-DDT.

DDT is very stable, chemically as well as biochemically, except in the presence of alcoholic alkali which dehydrochlorinates it to form dichlorodiphenyldichloroethylene, known as DDE, which is nontoxic to insects. Traces of iron, aluminum, or chromium salts catalyze this reaction.

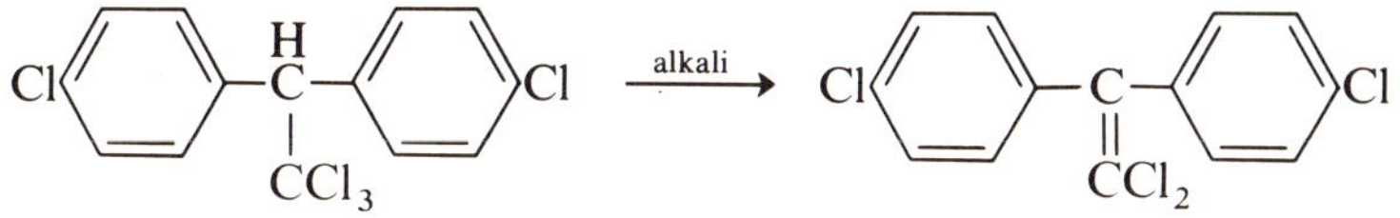

Solid-form DDT is not decomposed by sunlight or ultraviolet light, but residues of powdered DDT under field conditions with maximum surface exposure and ultraviolet irradiation are slowly decomposed into non-insecticidal components (Maqsud-Nasiv, 1953). Tropical temperatures and high relative humidity cause DDT dusts to lose their toxicity at faster rates, but hot, dry weather also increases loss, probably by raising its volatility. Pure DDT is stable to the action of heat and does not decompose below 195°C; however, technical DDT decomposes at about 100°C due to the presence of impurities. DDT has a low vapor pressure of 1.5×10^{-7} mm Hg at 20°C.

The toxicity of DDT to man is rather accurately known. A single dose of 10 mg/kg produces illness in some but not all people, and convulsions frequently occur at a dose of 16 mg/kg. Doses as high as 285 mg/kg have been taken without a fatal result. The oral LD_{50} of DDT for rats is 250 mg/kg.

TDE (DDD)

An important DDT analogue widely used to control various kinds of insect pests is TDE (DDD or Rhothane®).

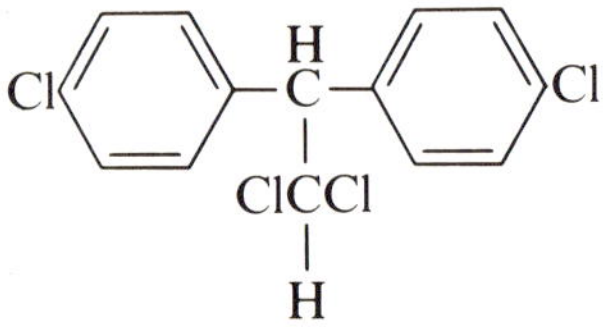

1,1-DICHLORO-2,2-BIS(*p*-CHLOROPHENYL)ETHANE

This insecticide is sometimes called dichlorodiphenyldichloroethane (hence DDD) or tetrachlorodiphenylethane (hence TDE).

Pure TDE is a white crystalline solid with a melting point of 109°C, and the technical product has a setting point of 86°C. It has a sweet odor. Its solubility is similar to that of DDT, but it is dehydrochlorinated in alkali at a slower rate than DDT.

In general, TDE is less effective than DDT for controlling insects, with some exceptions. It is superior to DDT for controlling black fly larvae, red-banded and fruit tree leafrollers, tomato and tobacco hornworms, and the Mexican bean beetle. It is 1/5 to 1/10 as toxic as DDT to mammals, with an oral LD_{50} in rats of 3400 mg/kg.

METHOXYCHLOR

Another important DDT analogue is methoxychlor (DMDT, or methoxy-DDT).

$$CH_3O - \underset{}{\bigcirc} - \overset{\overset{H}{|}}{\underset{\underset{CCl_3}{|}}{C}} - \bigcirc - OCH_3$$

1,1,1-TRICHLORO-2,2-BIS(*p*-METHOXYPHENYL)ETHANE

Methoxychlor is a white crystalline solid with a melting point of 89°C. Technical methoxychlor is a pale buff to gray flaky powder and has a setting point of 69°C. The technical product contains about 89% of the pure *p,p'*-isomer, with the remainder mostly the *o,p*-isomer. The water solubility is quoted as 0.1 ppm (Gunther *et al.*, 1968), and it is soluble in most organic solvents and subject to dehydrochlorination.

Methoxychlor is only 1/25 to 1/50 as toxic as DDT to mammals. It has an LD_{50} in rats of 6000 mg/kg, which makes it essentially nontoxic. It is not accumulated in fatty tissue or excreted in milk as DDT is. Therefore, methoxychlor is preferred to DDT for use on animals, in animal feed, and in barns. Since methoxychlor is more unstable than DDT, it has less residual effect. Compared to DDT, methoxychlor is more toxic to some insects and less toxic to others. It has a faster knockdown of houseflies than DDT.

DICOFOL (KELTHANE®)

Although mites and ticks (class Arachnida, order Acarina) are not insects, their control is within the realm of the entomologist; chemicals used for this purpose are called acaricides. Ovicides, which kill eggs, are also important since the life span of the adult mite or tick is short and a large proportion of the population is always in the egg stage.

When the use of DDT became widespread, mite and tick populations often increased since DDT killed their natural predators, the insects, while not harming them. Many acaricides were developed and tested during World War II as clothing and skin repellants and toxicants against the mite vector of scrub typhus.

There are several DDT analogues which are effective acaricides and nontoxic to insects. One of these, dicofol, is also a major metabolite of DDT from *Drosophila* and German cockroaches (*Blattella germanica*). Dicofol is also known by the commercial name Kelthane.®

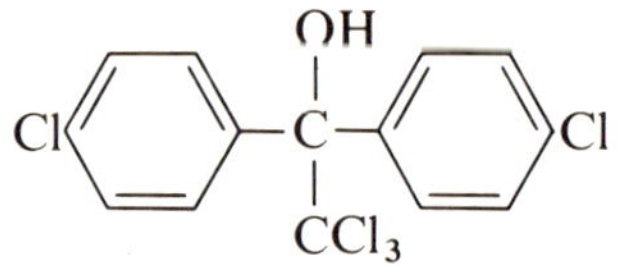

4,4′-DICHLORO-α-(TRICHLOROMETHYL)BENZHYDROL

Dicofol is a brown, viscous oil which is soluble in most organic solvents. It is slightly more soluble in water than DDT. It is moderately toxic, with an oral LD_{50} in rats of 575 mg/kg.

DMC (DCPC)

A DDT analogue effective against both eggs and active stages of many mite species is DMC, marketed as Dimite.[®]

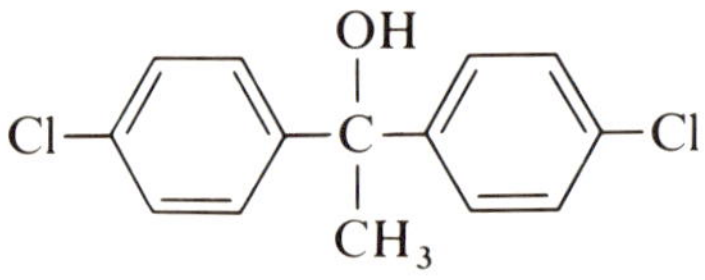

4,4′-DICHLORO-α-METHYLBENZHYDROL

DMC is a colorless, crystalline solid with a melting point of 70°C. It is decomposed by acid, is stable in alkaline solutions, and is soluble in organic solvents. It is not easily manufactured and is therefore expensive.

DMC, which has a moderate residual effect, kills slowly and produces a semiparalysis of the mite. DMC can also be used to inhibit the ability of DDT-resistant houseflies to dehydrochlorinate DDT. However, such flies are able to develop more enzyme to overcome the effect of DMC. Its oral LD_{50} in rats is 926–1391 mg/kg.

CHLOROBENZILATE[®]

Another DDT analogue which has been used as an acaricide is Chlorobenzilate.[®]

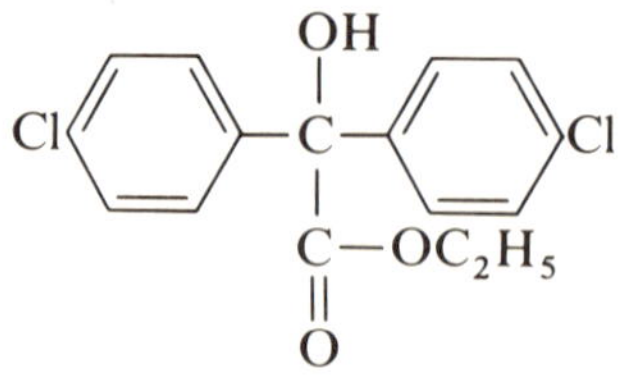

ETHYL 4,4′-DICHLOROBENZILATE

Chlorobenzilate® is a yellow crystalline solid with a melting point of 35–37°C. It is soluble in most organic solvents and slightly soluble in water. It is hydrolyzed by alkalis and strong acids. It is not generally toxic to insects. Residues of this material are known to be persistent in the field. Its mammalian toxicity (rat, oral LD_{50}) is 3100–4850 mg/kg.

3.2.2. Benzene Hexachloride

BHC, benzene hexachloride or 1,2,3,4,5,6-hexachlorocyclohexane, has the empirical formula $C_6H_6Cl_6$ and was first prepared in 1825 by Michael Faraday, who did not recognize its insecticidal properties. In 1912, Van der Linden discovered four isomers. In 1942, Dupire and Raucourt in France and Slade in England both discovered the insecticidal properties of BHC. The British group isolated the toxic γ-isomer and named it *lindane* in honor of Van der Linden.

Generally speaking, the isomers of BHC are relatively stable. They are stable to light, high temperature, hot water, and acid, although they are dechlorinated in alkali. Lindane is relatively soluble in water (10 ppm; Gunther *et al.*, 1968) for this group of compounds. It is approximately 100 times more volatile than DDT and therefore has a fumigant action. Its vapor pressure is 9.4×10^{-6} mm Hg at 20°C (Balsom, 1947).

BHC consists of a mixture of six chemically distinct isomers and heptachlorocyclohexanes and octachlorocyclohexanes. Crude BHC can be prepared by the chlorination of benzene in the presence of ultraviolet light (Slade, 1945). It is a grayish or brown amorphous solid with a characteristic musty odor. This odor gives an off-flavor to fruits and potatoes which somewhat limits its agricultural use. Crude BHC begins to melt at 65°C. The water solubility of crude BHC is in the range of 10–32 ppm.

The toxicity of BHC is proportional to the content of its toxic element, the γ-isomer. This isomer, by contact, stomach, or fumigant action, is about 50–10,000 times more effective than the other isomers (Back, 1951; Metcalf, 1947). It is toxic to mammals, with an acute oral LD_{50} in rats of 125 mg/kg (Sherman, 1948). Preparations that contain at least 99% of the γ-isomer are called lindane. Furthermore, the isomers of BHC have different actions. The γ- and α-isomers are stimulants of the central nervous system, with the principal symptom being convulsions. The β- and δ-isomers are depressants of the central nervous system. Like DDT, BHC is capable of rapid penetration of the insect cuticle; however, the biochemical reaction by which BHC is toxic differs from that of DDT.

There are eight theoretically possible isomers, and five of them, including a pair of optical isomers, have relatively strainless bonds and predominate in technical mixtures (Daasch, 1947). These isomers are present in technical BHC in approximately the following amounts:

α 65–70%
β 5–6%
γ 13%
δ 6%

The spatial configurations of the predominant isomers of BHC, indicating the type of bonds present on each, are

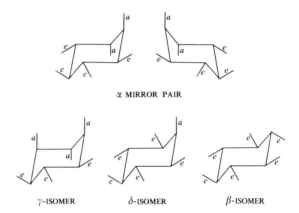

α MIRROR PAIR

γ-ISOMER δ-ISOMER β-ISOMER

There are almost no other chlorinated cyclohexane derivatives which can be used for insecticidal purposes; trichloro-, heptachloro-, octochloro-, and enneachlorobenzenes are practically nontoxic. Nor are hexamethyl- and hexaethylcyclohexanes good insecticides.

3.2.3. Cyclodiene Compounds

Cyclodiene compounds are the collective group of synthetic cyclic hydrocarbons consisting of such important insecticides as chlordane, heptachlor, aldrin, dieldrin, and endosulfan. A number of cyclodiene compounds are produced by the Diels–Alder reaction.

CHLORDANE (CHLORDAN)

Technical chlordane is a dark amber viscous liquid with a cedar-like odor. It contains approximately 60% chlordane, with the remainder being hexa-, hepta-, and nanochlor and other related dicyclopentadiene derivatives. Chlordane is insoluble in water and soluble in most organic solvents. It is susceptible to high temperature and alkaline treatments. It is compatible with most common insecticides.

Chlordane has two structural isomers, *cis-* and *trans*-octachloro-methano-tetrahydroindane (or β- and α-chlordane, respectively). The α-isomer (*trans*) is ten times less toxic than the β-isomer.

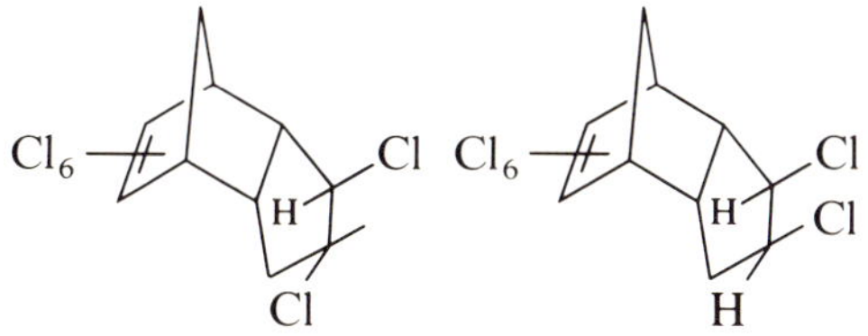

α *trans*-CHLORDANE   γ OR β *cis*-CHLORDANE

Chlordane is toxic to mammals, with an acute oral LD_{50} in rats of 225–590 mg/kg. It can be absorbed through the skin. *cis*-Chlordane is sometimes referred to as γ-chlordane by the industry.

HEPTACHLOR

The two isomers of chlordane can be separated by chromatographic adsorption on aluminum oxide (Brown, 1951), and it is then possible to obtain two further derivatives of chlordane—heptachlor and hexachlor. Heptachlor, a white crystalline solid, is four to five times more insecticidal than technical chlordane and is also more toxic than the β-isomer of chlordane (Brown, 1951). The acute oral LD_{50} of heptachlor in rats is 90 mg/kg.

Heptachlor is stable to heat up to 160°C, light, moisture, air, acids, bases, and oxidizing agents (Velsicol Corp., 1974). The melting point of the pure compound is 95°C, while the technical product has a melting range of 46–74°C.

In biological systems, heptachlor is converted to its epoxide and is stored in that form (Davidow and Radomski, 1953). Heptachlor epoxide is more toxic than heptachlor, so the epoxidation is a vital process to produce toxicity (Soloway, 1963). The epoxidation of heptachlor cannot be accomplished by simple chemical means, e.g., with peroxy acids, which are known to cause epoxidation of aldrin to dieldrin. Recently, a chromic acid oxidation method has been found to produce heptachlor epoxide from heptachlor in small quantities.

ALDRIN

Aldrin, a white crystalline solid with a melting point of 100–103°C, is a residual compound with a vapor pressure of 6×10^{-6} mm Hg at 25°C. Aldrin is almost insoluble in water (0.2 ppm; Gunther *et al.*, 1968) and soluble in most organic solvents. The technical product is a tan to brown solid with a small amount of supernatant liquid at 25°C. Its setting point is 49–60°C, and it contains approximately 78 % aldrin. Aldrin is stable to alkali and dilute acid. It is readily converted in plant and animal tissue and in the soil to its epoxide, dieldrin. Hence aldrin shows the same toxic effects as dieldrin, and both have very similar oral LD_{50}s in rats in the range of 55–60 mg/kg.

The spatial configuration of aldrin is

ALDRIN

Since it is *endo-exo*, it has the chair formation.

DIELDRIN (HEOD)

Dieldrin, which is the epoxy of aldrin, is one of the most persistent chemicals ever known. It has been used extensively since 1952, especially in situations where a long-lasting residual effect is advantageous.

Dieldrin has a melting point of 173°C, and its vapor pressure is 1.8×10^{-7} mm Hg at 25°C (less volatile than aldrin). The pure compound is an odorless, white crystalline solid, while technical dieldrin (usually about 76 % dieldrin) is a flaky tan solid with a setting point of 95°C. Dieldrin is as insoluble as aldrin in water (0.25 ppm; Gunther *et al.*, 1968) and less soluble in organic solvents. Only such procedures as treatments with strong acid or long exposure to intense ultraviolet light are known to decompose dieldrin. It is less apolar than aldrin.

Dieldrin can be absorbed through the skin and has an acute oral LD_{50} in rats of 60 mg/kg. It acts as a stimulant of the central nervous system.

The spatial configuration of dieldrin is

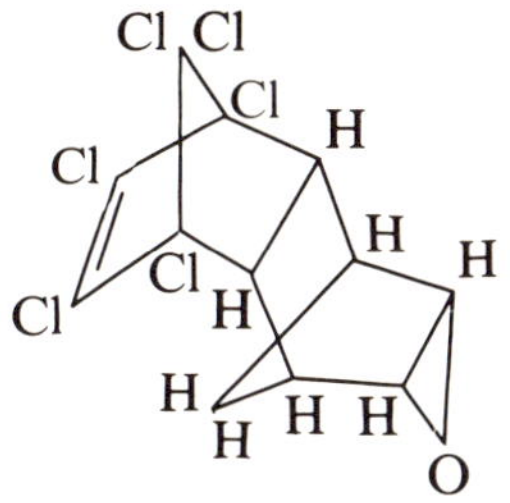

DIELDRIN

The toxic activities of aldrin and dieldrin resemble those of β-chlordane, heptachlor, and lindane. Three-dimensional studies of these compounds indicate that certain of the chlorine atoms and methylene groups occupy similar spatial configurations. Also, houseflies that show resistance to the aldrin, dieldrin, chlordane, heptachlor group also are resistant to lindane. Thus it appears that these compounds share a common mode of action (Busvine, 1964).

ENDRIN AND ISODRIN

The *endo-endo* isomer of aldrin is isodrin, which, like aldrin, is converted to its epoxide endrin, which in turn is the *endo-endo* isomer of dieldrin. Isodrin and endrin are less stable than their *endo-endo* isomers (Metcalf, 1955), and their toxic effects are similar to those of aldrin and dieldrin. Only endrin has been developed commercially (Martin, 1964).

Technical endrin is a light tan powder, while the pure compound is a white crystalline solid with a melting point above 200°C. Chemically it is very similar to dieldrin, but endrin can be easily degraded by heat and light. Endrin is highly toxic, with an acute oral LD_{50} in rats of 11 mg/kg for the pure compound.

The spatial configuration of endrin is

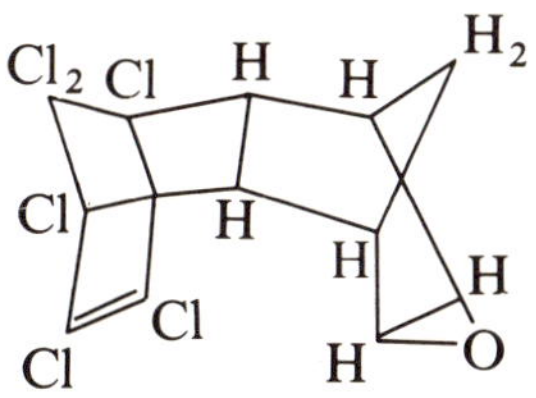

1,2,3,4,10,10-HEXACHLORO-6,7-EPOXY-1,4,4a,5,6,7,8,8a-OCTAHYDRO-
1,4-*endo-endo*-5,8-DIMETHANONAPHTHALENE

Being *endo-endo*, it has the boat formation.

TOXAPHENE

In a strict sense, both toxaphene and Strobane® are not cyclodienes but are chlorinated terpenes inasmuch as they are prepared by chlorinating naturally occurring terpenes. Toxaphene, empirical formula $C_{10}H_{10}Cl_8$, is prepared by chlorinating the bicyclic terpene camphene to contain 67–69% chlorine (Buntin, 1951; Parker and Beacher, 1947). It is an economically important insecticide, particularly for cotton. In 1971, 50 million pounds of toxaphene were produced, accounting for nearly 40% of all the chlorinated hydrocarbon insecticides produced in the United States in that year. One of the toxic components of toxaphene has recently been identified by Casida *et al.* (1974).

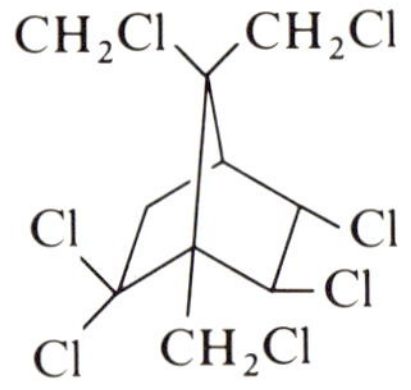

2,2,5-*endo*,6-*exo*,8,9,10-HEPTACHLOROBORNANE
(A toxic component of toxaphene)

Toxaphene is a mixture of isomers with a melting range of 65–70°C. It is a yellow wax and is soluble in most organic solvents and soluble in water at a level of 3 ppm (Gunther *et al.*, 1968). Toxaphene is stable except in the presence of alkali, sunlight, or heat above 155°C, all of which cause it to dehydrochlorinate (Metcalf, 1955).

The toxic nature of toxaphene is similar to that of chlordane. The dog is especially susceptible to toxaphene, with an LD_{50} of 20–30 mg/kg. Its acute oral LD_{50} in rats is 69 mg/kg.

STROBANE

A product similar to toxaphene is Strobane,® which is prepared by chlorinating camphene and pinene to contain 65% chlorine (Metcalf, 1955). It is called terpene polychlorinate and is a viscous straw-colored liquid with a mild aromatic odor. Strobane® is cheap to produce and has a residue effect. Its oral LD_{50} in rats is 200–250 mg/kg.

MIREX

Another analogue of these products is mirex, which is a white, crystalline, nonvolatile solid with a melting point of approximately 485°C.

Mirex is moderately toxic, with an acute oral LD_{50} in rats of 300–600 mg/kg. It is effective against fire ants, earwigs, slugs, snails, and wireworms.

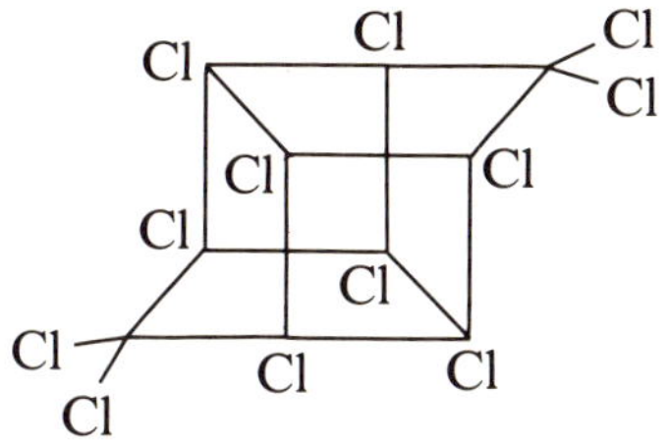

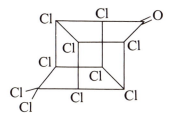

DODECACHLOROOCTAHYDRO-1,3,4-METHENO-2H-CYCLOBUTA[c,d]PENTALENE

KEPONE®

A product similar to mirex is Kepone,® which is a white to tan powder relatively soluble in water and some organic solvents. Its structure is

DECACHLOROOCTAHYDRO-1,3,4-METHENO-2H-CYCLOBUTA[c,d]PENTALEN-2-ONE

Kepone® is toxic, with an acute oral LD_{50} in rats of 95 mg/kg, and it is an acaricide in addition to being an insecticide. Both Kepone® and mirex are used as stomach poisons in the form of bait, and they control slugs, snails, and fire ants.

ENDOSULFAN

A related insecticide, also an acaricide, is endosulfan. It has the commercial name Thiodan.® It is a brownish crystalline solid with a setting point of 70–100°C.

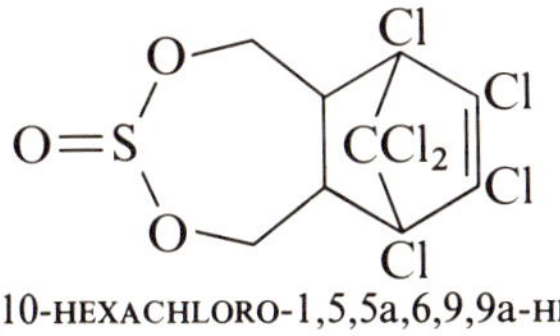

6,7,8,9,10,10-HEXACHLORO-1,5,5a,6,9,9a-HEXAHYDRO-
6,9-METHANO-2,4,3-BENZODIOXATHIEPIN 3-OXIDE

Endosulfan is a mixture of two isomers, one with a melting point of 106°C and the other 212°C. It is moderately soluble in most organic solvents and insoluble in water. Endosulfan is absorbed through the skin and has an acute oral LD_{50} in rats of 30–79 mg/kg. This compound caused a large-scale fish kill in the Rhine River by an accidental spilling.

TELODRIN® (ISOBENZAN)

Telodrin® has been developed as a soil insecticide for corn, alfalfa, clover, etc. It is known to be less stable than aldrin. It is solid, with a melting point of 248–257°C.

1,3,4,5,6,7,8,8-OCTACHLORO-3a,4,7,7a-
TETRAHYDRO-4,7-METHANOPHTHALAN

The acute oral LD_{50} of Telodrin® in rats is 4.8 mg/kg; it is readily absorbed through the skin.

3.3. ORGANOPHOSPHORUS INSECTICIDES

The organophosphorus insecticides (or, more commonly, the OP compounds) today account for approximately 30% of the registered synthetic insecticides and acaricides in the United States, and the number is steadily increasing. They are often referred to as organophosphates, because most of the member chemicals are some form of phosphates. The pioneering work on the organophosphates was started around 1934 by Gerhard Schrader in Germany. Thousands of OP compounds have been tested for insecticidal properties, and studies on these compounds have led to many discoveries about the biochemistry of the nervous system of both vertebrates and invertebrates. Whether they are true phosphates or other types of organophosphorus compounds, these chemicals possess the common characteristic that they are, or can become, excellent inhibitors of cholinesterases. In a strict sense, organophosphorus compounds that do not inhibit, or do not become inhibitors of, cholinesterases are not included in this group.

The OP group of insecticides includes such generally toxic compounds as parathion and TEPP and such selective compounds as malathion and ronnel. The OP insecticides are used as stomach and contact poisons, as fumigants, and as systemic insecticides for nearly every type of insect control.

Because the number of currently used organophosphorus insecticides is so large, description of their chemical properties will be restricted here to a few important ones which represent a group (or type) of OP compounds.

3.3.1. Pyrophosphates and Related Compounds

TEPP (TEPP)

Tepp or TEPP, a pyrophosphate, was the first widely marketed organophosphate insecticide (in 1944). It was synthesized by Clermont (1854), but apparently nothing was known of its toxic properties until Schrader in 1942 prepared this compound as a nicotinic substance. Schrader described it as hexaethyltetraphosphate (HETP), but now the main insecticidal component is known to be tetraethylpyrophosphate (TEPP) (Schrader, 1942).

The commercial preparation of HETP (Bladan®) contains 10–20 % TEPP. When a product contains 40 % or more tetraethylpyrophosphate, it is called tepp.

Tepp is a colorless, hydroscopic liquid, miscible with water but rapidly hydrolyzed to nontoxic diethylphosphoric acid. It is miscible with organic solvents and with aromatic oils but not with kerosene or paraffin.

Tepp is an extremely toxic compound, with an oral LD_{50} in male white rats of 1 to several mg/kg. It is not a systemic insecticide and is too toxic to be on the market today. The hydrolysis of TEPP accounts for its decomposition in sprays and in residues exposed to moisture. As a result of hydrolysis, free phosphoric acid is formed which is highly corrosive to metal containers. Tepp should therefore be stored in moisture-proof glass, stainless steel, or nickel containers. Its vapor pressure is on the order of 10^{-4} mm Hg (Schrader, 1963).

SCHRADAN (OMPA)

A pyrophosphate safe enough to use as a systemic insecticide is schradan or OMPA (for octamethylpyrophosphoramidate). It has an acute oral LD_{50} in rats of 5–55 mg/kg.

Schradan is of historical interest since it was the first OP compound to be studied as a systemic insecticide. The ability of plants to translocate this compound throughout their tissues, rendering them toxic to sucking insects, was first discovered by Kückenthal and Schrader in the early 1940s. Thus it kills sucking insects without killing their predators. Since systemic insecticides are polar, they do not penetrate the insect cuticle, which is permeable only to apolar compounds.

Its chemical formula is

$$(CH_3)_2N \overset{\overset{\displaystyle O}{\|}}{\underset{\diagup}{P}} -O- \overset{\overset{\displaystyle O}{\|}}{\underset{\diagdown}{P}} -N(CH_3)_2$$

$$(CH_3)_2N \qquad\qquad N(CH_3)_2$$

OCTAMETHYLPYROPHOSPHORAMIDATE

Schradan is a colorless, odorless liquid that is miscible with water and most organic solvents except petroleum hydrocarbons. Its boiling point is 118–122°C at 0.3 mm Hg, and its melting point is 20°C. It has a relatively high vapor pressure, 10^{-3} mm Hg at 20°C. It is stable in water.

3.3.2. Phosphorohalides and Cyanides

Phosphorohalides and cyanides are mostly known as "nerve gases," since they were first developed as chemical warfare agents and used in the form of mists at low concentrations (e.g., 1 ppm).

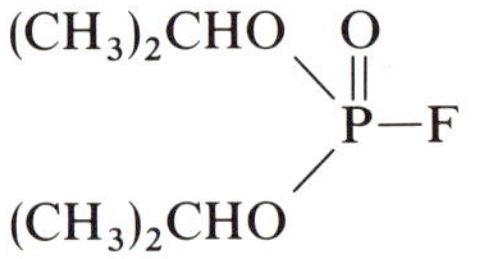

$$(CH_3)_2CHO \diagdown \overset{O}{\overset{\|}{P}}-F \qquad C_2H_5O \diagdown \overset{O}{\overset{\|}{P}}-CN$$
$$(CH_3)_2CHO \diagup \qquad\qquad (CH_3)_2N \diagup$$

DFP 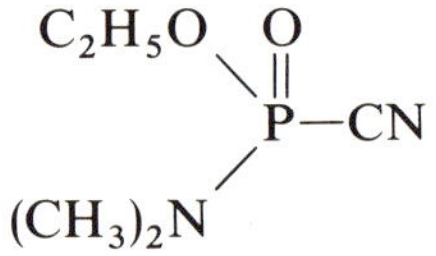   TABUN

DFP was made by a group of British scientists under B. C. Saunders at Cambridge. Tabun was made by the German group headed by Gerhard Schrader and is one of the most toxic compounds known. Neither compound is actually used as an insecticide, but a number of related compounds have been developed for insecticidal purposes.

DIMEFOX

The same German group also produced dimefox and other insecticidal phosphorohalides. Dimefox has a high vapor hazard (vapor pressure is 0.01 mm Hg at 20°C and 0.4 mm Hg at 30°C) and is more toxic to mammals than to insects.

The acute oral LD_{50} of dimefox in rats is 5 mg/kg. It is a colorless liquid with a boiling point of 67°C at 4 mm Hg. Dimefox is an inexpensive systemic insecticide and is highly soluble in water and acts as a powerful anticholinesterase. Its practical application should be limited to trunk or soil implantations. Dimefox is also called Pestox 14.®

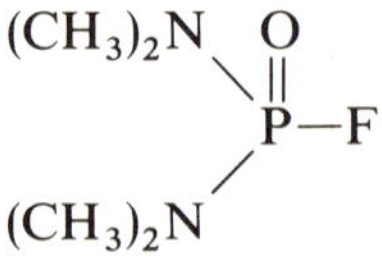

$$(CH_3)_2N \diagdown \overset{O}{\overset{\|}{P}}-F$$
$$(CH_3)_2N \diagup$$

N,N′-DIMETHYLDIAMIDOPHOSPHORYL FLUORIDE

MIPAFOX (ISOPESTOX®)

A similar compound to dimefox is mipafox, which is also called Iso-pestox® or Pestox 15.® It is less volatile (on the order of 10^{-3} mm Hg at 25°C) than dimefox. It is a white crystalline (melting point 60°C) and is slightly soluble in water (8%).

$$
\begin{array}{c}
(CH_3)_2CHNH \\
\diagdown \; \overset{O}{\underset{\parallel}{}} \\
P{-}F \\
\diagup \\
(CH_3)_2CHNH
\end{array}
$$

N,N'-DIISOPROPYLDIAMIDOPHOSPHORYL FLUORIDE

3.3.3. Dialkylarylphosphates, Phosphorothioates, and Phosphorodithioates

Trialkylphosphates (discussed in Section 3.3.4), dialkylarylphosphates, phosphorothioates, and phosphorodithioates form perhaps the most important present-day group of organophosphorus insecticides. Ordinarily, phosphorothioates or -dithioates are converted to corresponding phosphates or phosphorothionates in the animal or plant tissues and then become much more toxic ("activation"). Most of the following insecticides are marketed as the phosphorothioate (P=S) and are converted in the animal body to the phosphate (P=O), which is a more potent anticholinesterase. Generally speaking, the P=O analogues are less stable than P=S analogues and are susceptible to hydrolysis. Also, the P=O analogues may not penetrate well into the insect cuticle, which is more resistant to polar compounds (like phosphates) than to apolar analogues (P=S analogues are less polar than their P=O counterparts). For instance, paraoxon, a phosphate, is not a good mosquito larvicide; but parathion, a phosphorothioate, is, even though parathion is later converted to paraoxon in the insect's body.

PARATHION (THIOPHOS®)

The oxidation of parathion to paraoxon is a typical example of *in vivo* activation:

$$
\begin{array}{c}
C_2H_5O \quad S \\
\diagdown \; \overset{\parallel}{} \\
P{-}O{-}\!\!\bigcirc\!\!{-}NO_2 \\
\diagup \\
C_2H_5O
\end{array}
\xrightarrow{[O]}
\begin{array}{c}
C_2H_5O \quad O \\
\diagdown \; \overset{\parallel}{} \\
P{-}O{-}\!\!\bigcirc\!\!{-}NO_2 \\
\diagup \\
C_2H_5O
\end{array}
$$

PARATHION
O,O-DIETHYL O-p-NITROPHENYL
PHOSPHOROTHIOATE

PARAOXON
O,O-DIETHYL O-p-NITROPHENYL
PHOSPHATE

Parathion is highly toxic to mammals (acute oral LD_{50} in rats of 3.6 and 13 mg/kg for females and males, respectively), and its use has largely been superseded by less hazardous phosphates such as methylparathion and fenitrothion. Parathion is absorbed through the skin (acute dermal LD_{50} in rats of 6.8 and 21 mg/kg for females and males, respectively) and is also hazardous if inhaled (see Chapter 11). Paraoxon (or, more correctly, para-oxon) is a potent anticholinesterase with a pI_{50} in the range of 9–12. (I_{50} refers to the concentration which inhibits 50% of the enzymes; thus pI_{50} is the negative logarithm of I_{50}, like pH. A pI_{50} of 3, for instance, indicates a poor insecticide, while a pI_{50} of 9 indicates an excellent insecticide with a high degree of toxicity.) Johnsen and Hanstbarger (1966) note that parathion controls a variety of insects such as aphids, mites, beetles, Lepidoptera, leaf hoppers, leafminers, and other pests found on fruits, cotton, vegetables, and forage crops. It also controls several soil insects such as wireworms, rootworms, and symphilids.

Pure parathion is a colorless, almost odorless liquid, while the technical product is a dark brown liquid with a garlic odor. It is partially soluble in water (20–25 ppm, but paraoxon is much more soluble, 2400 ppm; Gunther *et al.*, 1968) and is completely miscible with a variety of alcohols, esters, ethers, and aromatic hydrocarbons. It is slowly hydrolyzed in water to form *p*-nitrophenol and diethyl orthothiophosphoric acid. The time for 50% hydrolysis is 120 days, but this occurs more rapidly in alkaline solution (Coates, 1949; Ketelaar, 1950; Peck, 1948; Topley, 1950). Thus parathion has a short residual life that is even shorter under alkaline conditions. It is not affected by oxygen or ultraviolet. The vapor pressure of parathion is 0.6×10^{-5} mm Hg at 20°C (Schrader, 1963).

Many analogues of parathion that have lower mammalian toxicity have been developed. These analogues are dimethyl instead of diethyl, as in parathion.

TABLE 3-3. Toxicities of Parathion and Methyl-parathion to the Rat

Rat LD_{50}		Parathion (mg/kg)	Methylparathion (mg/kg)
Oral	Male	13	14
	Female	3.6	24
Dermal	Male	21	67
	Female	6.8	67

From Schrader (1963).

METHYL PARATHION

Methylparathion is simply a dimethyl analogue of parathion, and it has an acute oral LD_{50} in rats of 14 mg/kg, which means it is still highly toxic. Its selectiveness appears to come from its relatively low dermal toxicity to mammals (Table 3-3). It has the same toxicity to insects as parathion and generally controls the same group of insects.

Methylparathion is a white crystalline compound with a melting point of 35–36°C.

O,O-DIMETHYL O-p-NITROPHENYL PHOSPHOROTHIOATE

The technical product is an amber liquid. Methylparathion is somewhat less stable than parathion, for it hydrolyzes more rapidly in alkali. Its volatility is similar to that of parathion (1×10^{-5} mm Hg at 20°C). It was the number one organophosphate insecticide produced in the United States in 1971 (Table 1-2).

CHLORTHION[®]

The addition of a chlorine atom markedly reduces the mammalian toxicity of the parathion analogue. Chlorthion[®] is also known as Bayer 22/190.[®]

O,O-DIMETHYL O-3-CHLORO-4-NITROPHENYL PHOSPHOROTHIOATE

Chlorthion[®] is equally toxic to insects but has an acute oral LD_{50} in rats of 880–980 mg/kg. Its chemical properties and solubility and hydrolysis are similar to those of parathion. Chlorthion[®] has a long residual life for a phosphate.

RONNEL

Ronnel is an excellent animal systemic insecticide. It is also known as fenchlorphos, Etrolene,[®] Dow-ET-57,[®] ET-14,[®] Korlan,[®] and Trolene.[®] It is crystalline and melts at 40–41°C.

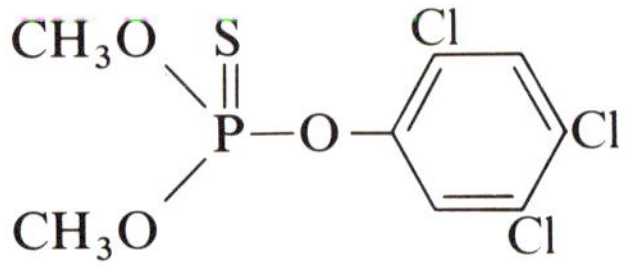

$$\text{CH}_3\text{O} \quad \text{S}$$

O,O-DIMETHYL O-2,4,5-TRICHLOROPHENYL PHOSPHOROTHIOATE

Ronnel is relatively soluble in water (1:250; Gunther *et al.*, 1968) and is soluble in most organic solvents. It has a mild mercaptan odor. It is a very safe insecticide, with an acute oral LD_{50} in rats of 1250 mg/kg for males and 2630 mg/kg for females.

It is administered orally and is effective against many ecto- and endoparasitic arthropods including cattle grubs, screw worms, and sucking lice. It is also used as a residual spray for controlling flies, fleas, and cockroaches. Its phytotoxicity limits its use as a forage and crop spray agent.

FENITROTHION (SUMITHION,® FOLITHION®)

Fenitrothion, which is also known as Sumithion,® is an insecticide with a moderate fumigant action (vapor pressure 5.4×10^{-5} mm Hg at 20°C; Schrader, 1963). It has an acute oral LD_{50} in rats of 250–670 mg/kg and therefore is a very selective insecticide. Kükenthal and Jung (1963) showed that it has a much longer residual effect (more than fourfold) than methylparathion against houseflies and mosquitoes.

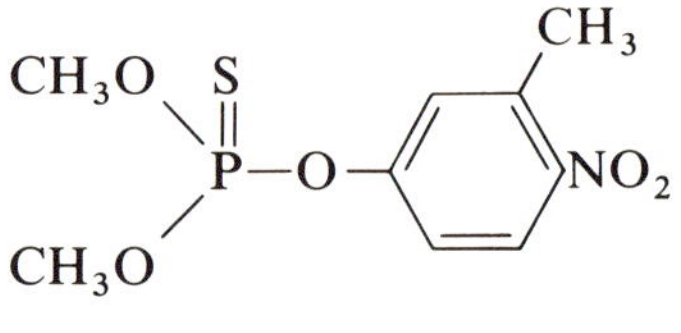

O,O-DIMETHYL O-(4-NITRO-*m*-TOLYL) PHOSPHOROTHIOATE

Because of its selectivity, it has been extensively used as a substitute for parathion and for DDT in forest pest control.

FENTHION

Fenthion, another parathion analogue, is also known as Lebaycid,® Baytex,® and Entex.® It is a systemic insecticide (water solubility 55 ppm; Gunther *et al.*, 1968) which gives excellent control of resistant flies, mosquitoes, ticks, and lice.

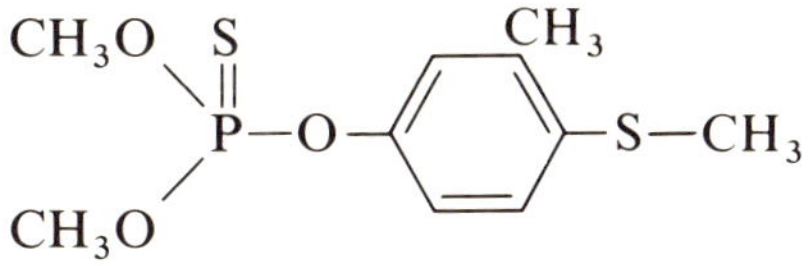

O,O-DIMETHYL O-[4-(METHYLTHIO)-m-TOLYL]PHOSPHOROTHIOATE

Fenthion has a long residual effect due to its low volatility (vapor pressure 3×10^{-5} mm Hg at 20°C). As compared to methylparathion, it is 3 and 20 times more resistant against acid and alkaline treatment, respectively (Schrader, 1963). Fenthion has an acute oral LD_{50} in rats of 178–310 mg/kg.

EPN

In all of the above parathion analogues, it is the "tail" part of the compound that is important in determining toxicity and selectivity. EPN, a phosphonate, is different, for it has a direct carbon–phosphorus link. EPN was introduced in 1949 and is one of the few phosphonates successfully marketed.

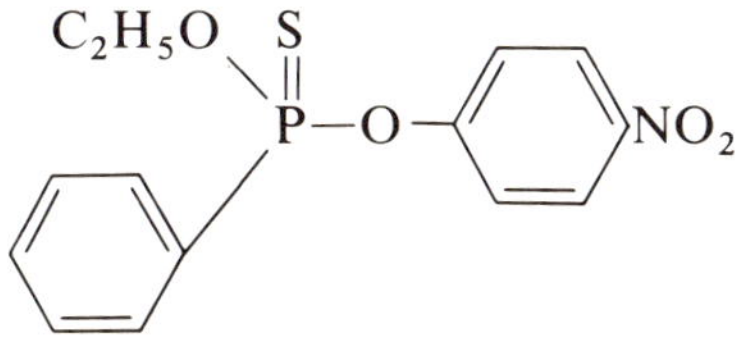

O-ETHYL O-p-NITROPHENYL PHENYLPHOSPHONOTHIOATE

EPN is a white crystalline solid with a melting point of 36°C. Technical EPN is a dark amber liquid. It is hydrolyzed in alkali to p-nitrophenol (hydrolysis constant at 37°C: $K = 13.8[OH^-]$ min^{-1}). It is less volatile than parathion.

EPN is relatively toxic to mammals and is an acaricide in addition to being an insecticide. Its acute oral LD_{50} in rats is 8–36 mg/kg. When applied topically to *Musca domestica*, it has an LD_{50} of 1.9 μg/g. It is an excellent insecticide and acaricide for orchard pests, including apple flea weevil, plum curculio, and coddling moth, and for some soil insects.

The following parathion analogues are diethyl, like parathion, but have greatly modified "tail" parts.

COUMAPHOS (CO-RAL®)

Coumaphos has an acute oral LD_{50} in rats of 90–110 mg/kg. However, its dermal toxicity is low (approximately 1/10). Its oxidation product, coroxon, is toxic, with a pI_{50} of 7.4 (versus 3.9 for coumaphos). It is used to treat cattle, poultry, horses, sheep, and dogs for control of cattle grubs, lice, horn flies, screw worms, mites, and ticks. Coumaphos is widely used today, and is known commercially as Co-ral,® Asuntol,® and Bayer 21/199.®

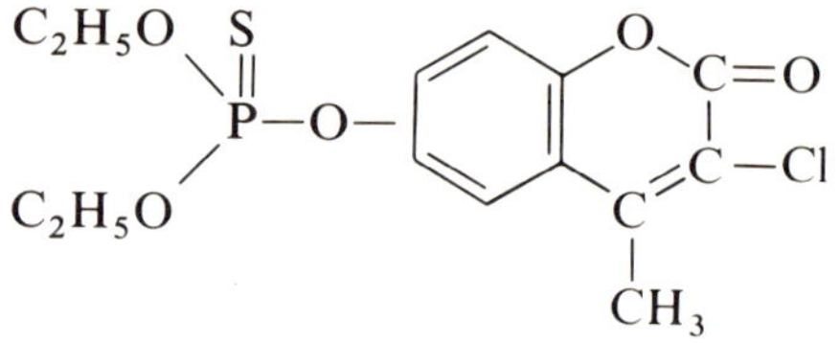

O-(3-CHLORO-4-METHYL-2-OXO-2H-1-BENZOPYRAN-7-YL)
O,O-DIETHYL PHOSPHOROTHIOATE

Coumaphos is a tan crystalline solid with a melting point of 90–92°C. It is soluble in organic solvents, less so in alcohols and ketones. It is almost insoluble in water (1.5 ppm) and is hydrolyzed in alkaline media.

DIAZINON

Diazinon is an important insecticide (number 3 in the United States) which is widely used today. It has a relatively low mammalian toxicity, with an acute oral and dermal LD_{50} in rats of 150–220 mg/kg and 500–900 mg/kg, respectively. Geese and ducks, however, are very susceptible to diazinon poisoning.

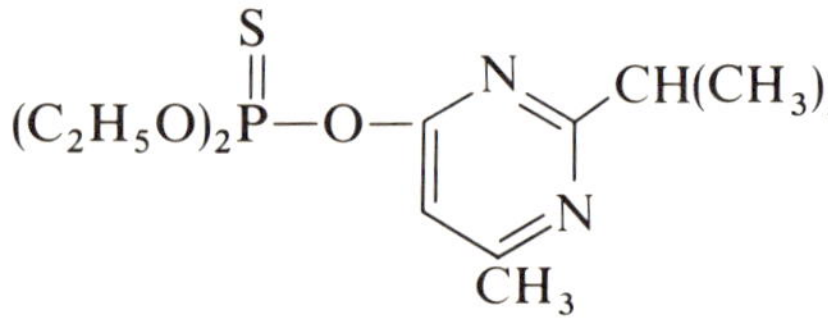

O,O-DIETHYL O-(2-ISOPROPYL-6-METHYL-4-PYRIMIDINYL)
PHOSPHOROTHIOATE

Diazinon is a colorless liquid with a boiling point of 83–84°C. The technical product is a brown liquid. It is sensitive to oxidation and heat, quickly degrading at temperatures above 100°C. It has a moderate volatility of 1.4×10^{-4} mm Hg at 20°C. Diazinon is relatively soluble in water (1:250; Gunther *et al.*, 1968) and is miscible with most organic solvents. It is relatively stable in dilute alkali but is slowly hydrolyzed in water or dilute acid. Diazoxon, its oxygen analogue, has an I_{50} (against fly-head cholinesterase) of 3.5×10^{-9} M, making it one of the most potent cholinesterase inhibitors.

Johnsen and Hanstbarger (1966) reported that diazinon has a wide spectrum of insect-killing power and can control various soil insects, DDT-resistant flies, household pests, and various vegetable and forage crop insects.

AZINPHOSMETHYL (GUTHION® OR GUSATHION)

Azinphosmethyl is a white solid with a melting point of 73–74°C. It is soluble in water to 0.003 % (Metcalf, 1971).

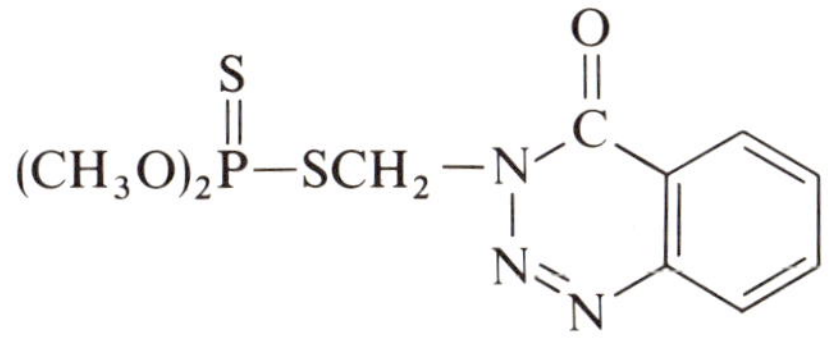

O,O-DIMETHYL S-4-OXO-1,2,3-BENZOTRIAZIN-
3($4H$)-YLMETHYL PHOSPHORODITHIOATE

Azinphosmethyl is an important broad-spectrum insecticide–acaricide with a fair residual property. Its acute oral LD$_{50}$ is 16.4 mg/kg. It penetrates the skin. Along with its ethyl analogue, azinphosmethyl is used against pests of fruits, vegetables, cotton, and ornamentals.

DURSBAN®

Dursban® is a residual organophosphate insecticide which has been effective in tests for controlling mosquitoes, ticks, soil and aerial insects on field crops, and household pests.

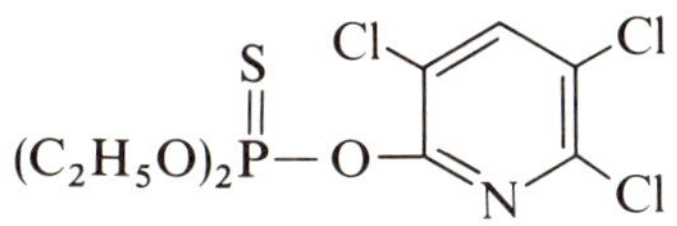

O,O-DIETHYL O-(3,5,6-TRICHLORO-2-PYRIDYL) PHOSPHOROTHIOATE

The toxicity of Dursban® to some aquatic fauna might limit its use in aquatic environments. It is a stomach and contact poison with a long residual life in the soil and a short one on foliage. Dursban® is moderately toxic to animals, with an acute oral LD$_{50}$ in rats of 97–276 mg/kg.

ABATE®

Abate® has an acute oral LD$_{50}$ in rats of 1000–3000 mg/kg and so is practically nontoxic to mammals. It also has a low hazard for birds and fish. Abate® is approved for control of various mosquito larvae (LC$_{50}$ 0.0004–0.01 ppm) on noncrop land. It is stable, e.g., no hydrolysis for weeks at pH 8, and is practically insoluble in water. It is a white crystalline which melts at 30–30.5°C.

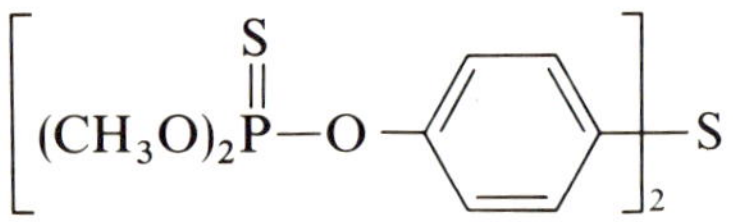

$$\left[(CH_3O)_2\overset{\overset{\displaystyle S}{\|}}{P}-O-\underset{}{\underset{}{\bigcirc}}- \right]_2 S$$

O,O,O',O'-TETRAMETHYL-*O,O'*-THIODI-*p*-PHENYLENE
PHOSPHOROTHIOATE

Technical Abate® is 90–95 % pure and is a brown viscous liquid.

3.3.4. Trialkylphosphates and Thiophosphates

TRICHLORFON (DIPTEREX®)

Trichlorfon is one of the few phosphonate organophosphate insecticides and has the trade name Dipterex.®

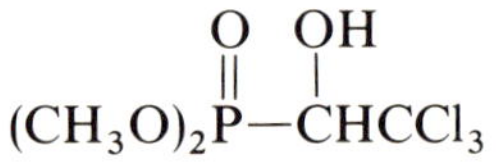

$$(CH_3O)_2\overset{\overset{\displaystyle O}{\|}}{P}-\overset{\overset{\displaystyle OH}{|}}{C}HCCl_3$$

DIMETHYL (2,2,2-TRICHLORO-1-HYDROXYETHYL)
PHOSPHONATE

Trichlorfon is a white to pale yellow crystalline powder with a melting point of 83–84°C. It is soluble in water, alcohols, and ketones; it is slightly soluble in toluene and ether. It is unstable to heat and is dehydrochlorinated in mild alkali to form the highly toxic insecticide dichlorvos (discussed next). Dehydrochlorination also occurs in the insect body; thus trichlorfon is a precursor of dichlorvos. It has a relatively low vapor pressure (7.8×10^{-6} mm Hg at 20°C).

Trichlorfon has systemic action and gives good control of endo- and ectoparasites of livestock. It is also effective against many species of flies, wasps, bugs, and beetles which attack plants. It has an acute oral LD_{50} in rats of 500 mg/kg.

DICHLORVOS (DDVP, VAPONA®)

Dichlorvos, which has the commercial names DDVP and Vapona,® is a very volatile insecticide (vapor pressure 1.2×10^{-2} mm Hg at 20°C) which gives rapid knockdown and kill of houseflies. It has been widely used in dry or liquid bait or in resin strands for fly control (fly strips) and is also used on livestock to control flies.

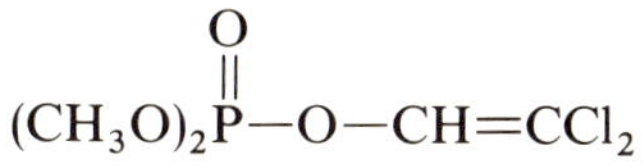

$$
\begin{array}{c}
\text{O} \\
\parallel \\
(CH_3O)_2P - O - CH = CCl_2
\end{array}
$$

O,O-DIMETHYL 2,2-DICHLOROVINYL PHOSPHATE

Dichlorvos is a colorless to amber liquid which is miscible with or soluble in most organic solvents and soluble in water to 1%. It has an acute oral LD_{50} in rats of 56–80 mg/kg.

PHOSPHAMIDON (DIMECRON®)

Phosphamidon is a systemic insecticide with an acute oral LD_{50} in rats of 27.8 mg/kg but a low dermal toxicity to rats of 125 mg/kg.

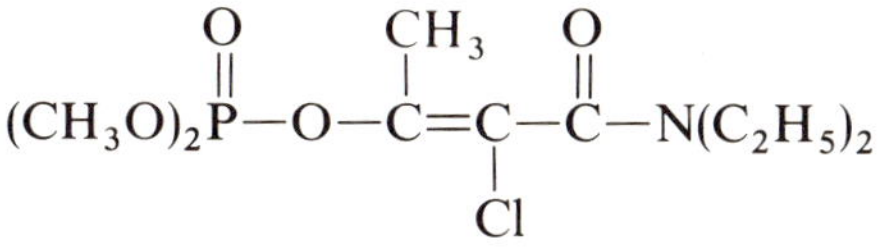

$$
\begin{array}{c}
\text{O} \qquad CH_3 \quad \text{O} \\
\parallel \qquad | \qquad \parallel \\
(CH_3O)_2P - O - C = C - C - N(C_2H_5)_2 \\
| \\
Cl
\end{array}
$$

DIMETHYL PHOSPHATE, ESTER WITH 2-CHLORO-N,N-
DIETHYL-3-HYDROXYCROTONAMIDE

Phosphamidon is not a residual insecticide–acaricide. It is incompatible with alkali. This compound is an oil which is soluble in water and most organic solvents. There is a wealth of information on this particular compound summarized in *Residue Review*, Vol. 37.

BIDRIN®

Bidrin® is a contact and systemic insecticide with an acute oral LD_{50} in rats of 22 mg/kg. It is water soluble and is used to control various crop pests.

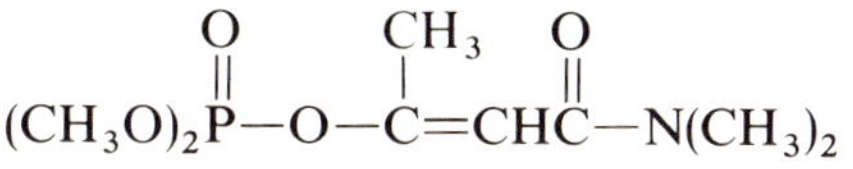

$$
\begin{array}{c}
\text{O} \qquad CH_3 \quad \text{O} \\
\parallel \qquad | \qquad \parallel \\
(CH_3O)_2P - O - C = CHC - N(CH_3)_2
\end{array}
$$

DIMETHYL PHOSPHATE, ESTER WITH *cis*-3-HYDROXY-
N,N-DIMETHYLCROTONAMIDE

Bidrin® is injected into elm trees with a Mauget injector to control the vector of Dutch elm disease.

NALED

Naled, commercially known as Dibrom,® is a quick-acting insecticide which has some fumigant action and kills by contact. It has an acute oral

LD_{50} in rats of 430 mg/kg. Naled is essentially a dibrominated dichlorvos. It has, however, the ability to act as a contact insecticide.

$$\underset{\text{1,2-DIBROMO-2,2-DICHLOROETHYL DIMETHYL PHOSPHATE}}{(CH_3O)_2\overset{\displaystyle O}{\overset{\displaystyle \|}{P}}-O-\overset{\overset{\displaystyle Br}{\displaystyle |}}{C}HCBrCl_2}$$

Naled is a straw-colored oil with a slightly pungent odor. Its melting point is 26°C. It is insoluble in water, slightly soluble in aliphatic solvents, and highly soluble in aromatic solvents. When it is mixed with various solvents, it may become corrosive to steel-containing metals.

Johnsen and Hanstbarger (1966) report that naled has a broad spectrum of insecticidal action. It has a relatively low hazard to fish and wildlife. Naled has a short residual life (hydrolyzed by 4% a day at pH 7) and can be used within 4 days of harvest without residue problems.

DEMETON (SYSTOX®)

Demeton, known commercially as Systox,® is a mixture of two isomers in the proportion of 65 parts thiono and 35 parts thiol. Demeton is a systemic insecticide which is highly toxic to mammals, with an acute oral LD_{50} in rats of 2–12 mg/kg. It is used mainly as a foliage spray and has a relatively long residual life within plants.

Technical demeton is a yellowish liquid with a boiling point of 134°C and a pronounced mercaptan odor (vapor pressure 5–18 × 10^{-4} mm Hg). The thiol isomer is a colorless oil and is ten to a hundred times more soluble in water than the thiono isomer, which is a colorless liquid (solubility of thiol 0.2%, thiono 0.02%). The thiol isomer is, however, more toxic than the thiono isomer, and it has an I_{50} almost ten times lower for a number of cholinesterases (Wirth, 1953).

$$\underset{\text{THIOL ISOMER}}{(C_2H_5O)_2\overset{\displaystyle O}{\overset{\displaystyle \|}{P}}-S-CH_2CH_2-S-CH_2CH_3}$$

$$\underset{\text{THIONO ISOMER}}{(C_2H_5O)_2\overset{\displaystyle S}{\overset{\displaystyle \|}{P}}-O-CH_2CH_2-S-CH_2CH_3}$$

O,O-DIETHYL *S*-(AND *O*-)2-(ETHYLTHIO) ETHYL PHOSPHOROTHIOATE

Both the isomers are readily hydrolyzed in either acid or alkali (hydrolysis constants for thiono and thiol isomers are 2.1 × 10^3 $[OH^-]\,min^{-1}$ and

$0.814\,[\text{OH}^-]\,\text{min}^{-1}$ at $25°\text{C}$, respectively). Both isomers are activated to the sulfones within the plant and become more toxic to the insect.

OXYDEMETONMETHYL (META-SYSTOX-R®)

Oxydemetonmethyl is the methyl analogue of the thiol isomer of demeton.

$$(\text{CH}_3\text{O})_2\overset{\displaystyle O}{\overset{\|}{\text{P}}}-\text{S}-\text{CH}_2\text{CH}_2-\text{S}-\text{C}_2\text{H}_5$$

S-2-(ETHYLSULFINYL) ETHYL O,O-DIMETHYL PHOSPHOROTHIOATE

Oxydemetonmethyl, which has the trade name Meta-Systox,® has an acute oral LD_{50} in rats of 65–75 mg/kg. Its water solubility is $1:300$ (Gunther *et al.*, 1968). It is a contact and systemic insecticide and miticide with a relatively long residual life. It is used mainly as a foliar spray or soil drench for sucking insects.

PHORATE (THIMET®)

Phorate is also closely related to demeton. It has the trade name Thimet® and has an acute oral LD_{50} against rats of 1–5 mg/kg. Its solubility in water is of the order of 50 ppm.

$$(\text{C}_2\text{H}_5\text{O})_2\overset{\displaystyle S}{\overset{\|}{\text{P}}}-\text{S}-\text{CH}_2-\text{S}-\text{CH}_2\text{CH}_3$$

O,O-DIETHYL S-(ETHYLTHIO) METHYL PHOSPHORODITHIOATE

Phorate is a systemic insecticide and miticide. It is used for cotton pests as well as for flea beetles, thrips, leafhoppers, and mites. It is not phytotoxic except for the mild toxicity to cabbage plants.

MALATHION

Malathion is one of the safest insecticides, for it has an acute oral LD_{50} in rats of 900–5800 mg/kg and is broken down by the mammalian liver. It is expected to become the number one organophosphate insecticide in the near future because of its safety features along with its compatibility with ultra-low-volume (ULV) spray methods. Malathion kills insects by contact or vapor action and also is a stomach poison (vapor pressure

1.25×10^{-4} mm Hg at 20°C). It is widely used for the control of many insects. Malathion's selectivity is due to the presence of the carboxyl groups, which are susceptible to mammalian hydrolysis. Introduction of a susceptible molecular group which would be selectively hydrolyzed by mammalian enzymes is an important consideration in designing new insecticides.

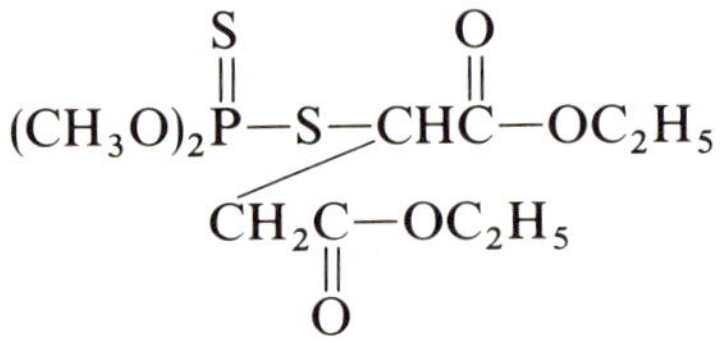

$$\begin{array}{c} \overset{S}{\overset{\|}{(CH_3O)_2P}}-S-\overset{\overset{O}{\|}}{CHC}-OC_2H_5 \\ \underset{\underset{O}{\|}}{CH_2C}-OC_2H_5 \end{array}$$

DIETHYL MERCAPTOSUCCINATE, *S*-ESTER WITH
O,O-DIMETHYL PHOSPHORODITHIOATE

Pure malathion is a clear colorless liquid (boiling point 156–157°C at 7 mm Hg), while the technical product is brown and has a garlic odor. It is miscible with most organic solvents and is slightly soluble in water (145 ppm; Johnson *et al.*, 1952). Malathion hydrolyzes at pH 5 or 8 and decomposes at excessive temperature. Because of its low mammalian toxicity, malathion is a particularly useful insecticide for household, home garden and greenhouse pests, besides being an excellent general insecticide.

DIMETHOATE

Dimethoate, also known as Rogor,® has a carboxyamide group through which it achieves selectivity. Mammals detoxify dimethoate with carboxyamidase, which splits amide bonds. It is a white solid, melting at 51–52°C.

$$\overset{S}{\overset{\|}{(CH_3O)_2P}}-SCH_2\overset{\overset{O}{\|}}{C}-NHCH_3$$

O,O-DIMETHYL *S*-(METHYL CARBAMOYLMETHYL) PHOSPHORODITHIOATE

Dimethoate is particularly toxic to houseflies, and no resistance to it has been reported. It is a systemic insecticide with a long residual life, and is used as a foliar spray to control sucking insects (e.g., aphids, red spider mites, and cherry flies). Dimethoate also kills insects by contact. It is soluble in water (2.5%; Gunther *et al.*, 1968) and is unstable in alkali. It has an acute oral LD_{50} in rats of 155–500 mg/kg.

3.4. CARBAMATE INSECTICIDES

The carbamates are the latest arrival in the field of anticholinesterase insecticides, and at present many new compounds are still in the process of being marketed. The insecticide index prepared by the Entomological Society of America (Kenaga and Allison, 1969) lists 21 carbamate insecticides. In general, carbamate insecticides are synthetic derivatives of physostigmine (commonly called eserine), which is the principal alkaloid of the plant *Physostigma venenosum* (calabar beans). Physostigmine was known to be an inhibitor of cholinesterase, and Stedman and Eason (1932) worked to develop synthetic analogues, such as prostigmine.

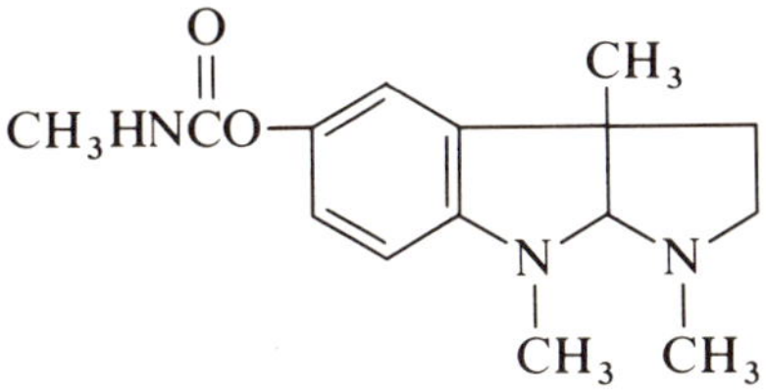

PHYSOSTIGMINE 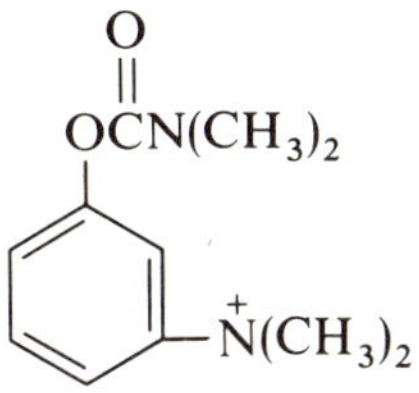   PROSTIGMINE

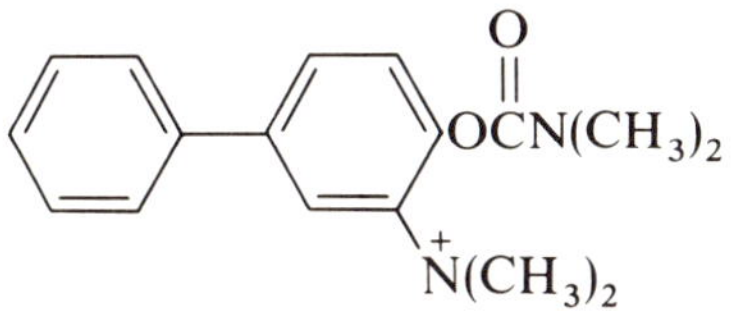

Nu 683

Although these and similar compounds are effective inhibitors of insect cholinesterase, they are unsuitable as insecticides, for they are water-soluble quaternary salts or amine hydrochlorides and hence too polar to penetrate the insect cuticle. Kolbezen *et al.* (1954) attributed their ineffectiveness as insecticides to their low lipoid solubility. O'Brien and Fisher (1958) considered that the ion-impermeable sheath of the insect nerve would also contribute to their ineffectiveness, for these drugs were ionized at physiological pH and would therefore be excluded from their site of action. Hence the modern carbamate insecticides have been modified by eliminating the polar moiety of physostigmine so that they can easily penetrate the insect cuticle as well as the nerve sheath. Of lipoid-soluble derivatives, Kolbezen *et al.* (1954) found the N-methylcarbamates of certain phenols to be highly toxic to houseflies and greenhouse thrips. Meanwhile, Gysin (1954) pursued another line of development and worked with dimethyl compounds, which led to the discovery of Dimetan,[®] pyrolan, and Isolan.[®]

The general carbamate structure is

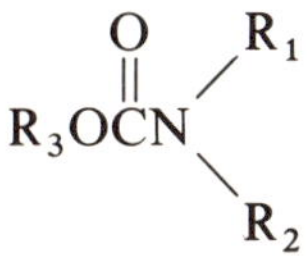

R_1 and R_2 are hydrogen, methyl, ethyl, propyl, or other short-chain alkyls, and R_3 is phenol, naphthalene, or other cyclic hydrocarbon rings.

3.4.1. Naphthylcarbamates

CARBARYL

Carbaryl, which was introduced in 1956 under the trade name Sevin,® is the most widely used carbamate today. It is a wide-spectrum insecticide which controls 100–150 species of insects but is virtually ineffective against houseflies, certain aphids, and spider mites. One of its major uses is for control of several cotton pests and forage pests such as insects that attack apples and pears. It is also effective against the gypsy moth.

Carbaryl has a low mammalian toxicity, with an acute oral LD_{50} in rats of 500–700 mg/kg. This compound is a reversible inhibitor of cholinesterase and is rapidly metabolized in mammals to 1-naphthol and other hydroxylation and conjugation products which are excreted in the urine.

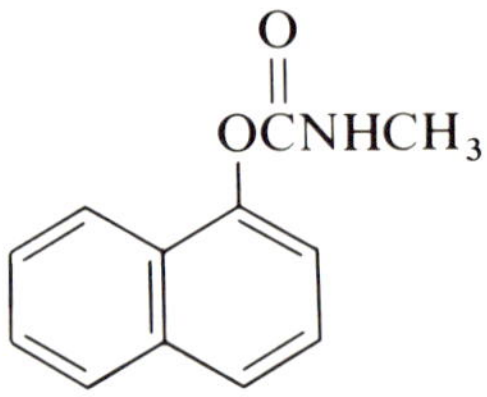

1-NAPHTHYL-*N*-METHYLCARBAMATE

Carbaryl is a white crystalline solid with a melting point of 142–143°C. It has a vapor pressure of less than 0.005 mm Hg at 26°C. Carbaryl is stable to heat, light, and hydrolysis except under alkaline conditions. It is soluble in most organic solvents and slightly soluble in water (about 40 ppm; Gunther *et al.*, 1968). Carbaryl has a short residual life and may be used right up to harvest.

Modifications of carbaryl have proved to be unsuccessful insecticides. The dimethyl analogue is only 1/10 as toxic as carbaryl, and the following analogues are nontoxic:

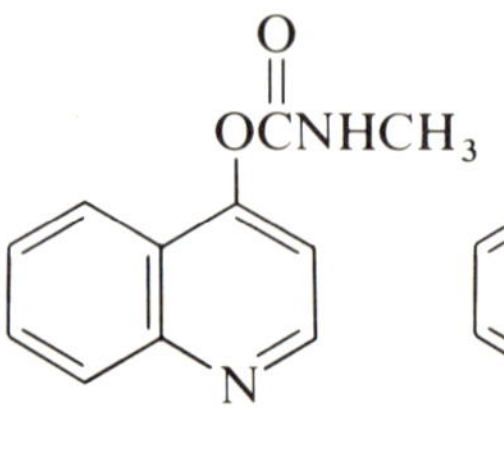

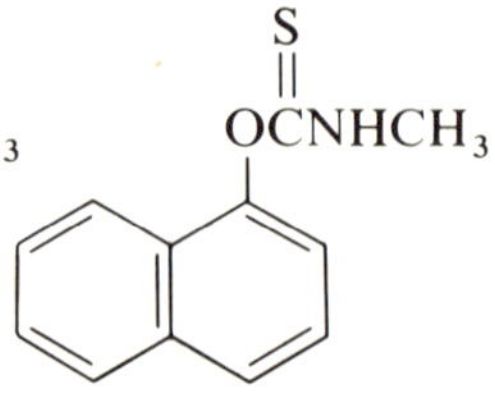

3.4.2. Phenylcarbamates

The basic structure of the phenylcarbamates is

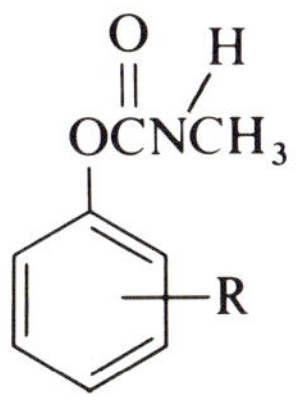

$$\overset{O}{\underset{\parallel}{}}\overset{H}{\underset{/}{}}$$
OCNCH$_3$

The position of R is important in determining toxicity for the following reason. Metcalf and Fukuto (1965*b*) explained that phenylcarbamates inhibit cholinesterase by competing with acetylcholine and interacting through Van der Waals forces at the esteratic and anionic sites of cholinesterase where normally acetylcholine interacts.

ZECTRAN,® MEXACARBATE

Zectran® is a contact and systemic insecticide which kills a relatively broad spectrum of insects and related pests, including some mollusks. It is highly toxic, with an acute oral LD$_{50}$ in rats of 15–63 mg/kg. Its toxicity can be enhanced by the addition of various pyrethrin synergists. It has a special effectiveness against Mexican bean beetles and southern army worms. Its usefulness for forest pest control is now being recognized.

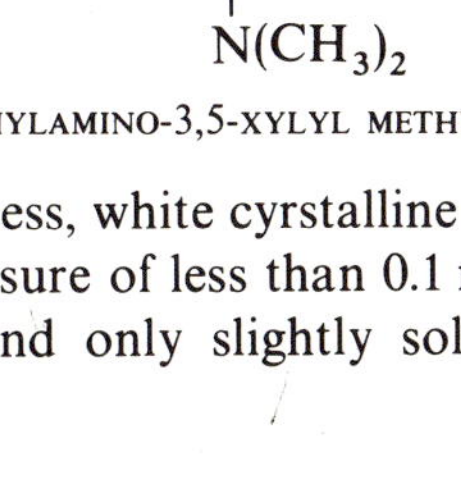

4-DIMETHYLAMINO-3,5-XYLYL METHYLCARBAMATE

Zectran® is an odorless, white cyrstalline solid with a melting point of 85°C. It has a vapor pressure of less than 0.1 mm at 139°C. It is soluble in most organic solvents and only slightly soluble in water (100 ppm, by estimate).

PROPOXUR, ARPROCARB (BAYGON® BAYER 39007)

Propoxur is a contact insecticide which shows systemic action with soil application. It has a rapid knockdown and has a long residual life.

Johnsen and Hanstbarger (1966) report that it is effectively used against cockroaches, flies, mosquitoes, chinch bugs, spiders, and sand flies.

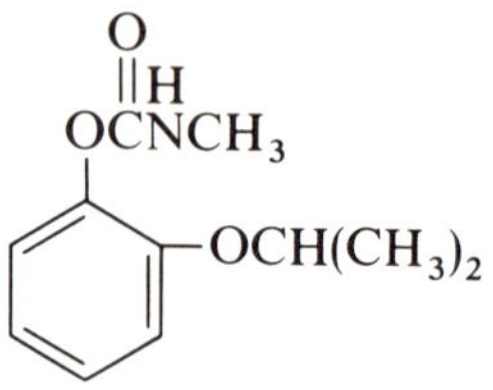

$$\overset{\displaystyle O}{\underset{\displaystyle OCNCH_3}{\|\,H}}$$

OCH(CH_3)_2

o-ISOPROPOXYPHENYL METHYLCARBAMATE

Propoxur has an LD_{50} in rats of 95–104 mg/kg. It is a promising insecticide for the control of adult mosquitoes.

BUX TEN,® *m*-(1-METHYLBUTYL)PHENYL METHYLCARBAMATE

Bux Ten® is actually a mixture of *m*-(1-methylbutyl)phenyl methylcarbamate and another compound, *m*-(1-ethylpropyl)phenyl methylcarbamate, in a 3:1 ratio.

BAYER 37344 (MESUROL® OR MERCAPTODIMETHUR)

Bayer 37344 is a crystalline material which melts at 121°C. It is virtually water insoluble.

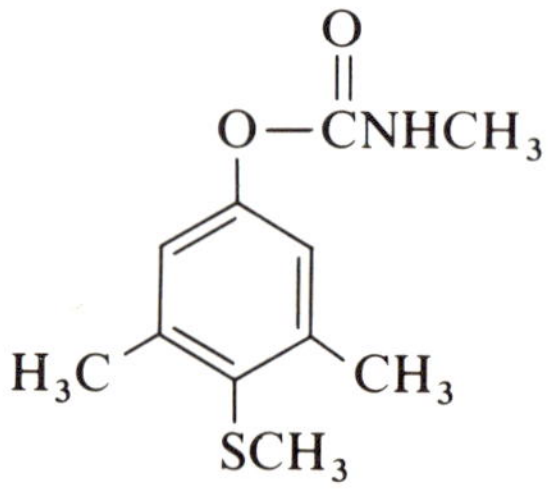

4-(METHYLTHIO)-3,5-XYLYL METHYLCARBAMATE

The acute oral LD_{50} of Bayer 37344 in rats is 130 mg/kg. It is a wide-spectrum insecticide and is effective against OP-resistant mites.

3.4.3. Heterocyclic Dimethylcarbamates

PYROLAN

Pyrolan (or G-22008) is an experimental insecticide. It is moderately toxic to mammals, with an oral LD_{50} in rats of 90 mg/kg. Pyrolan has some systemic action.

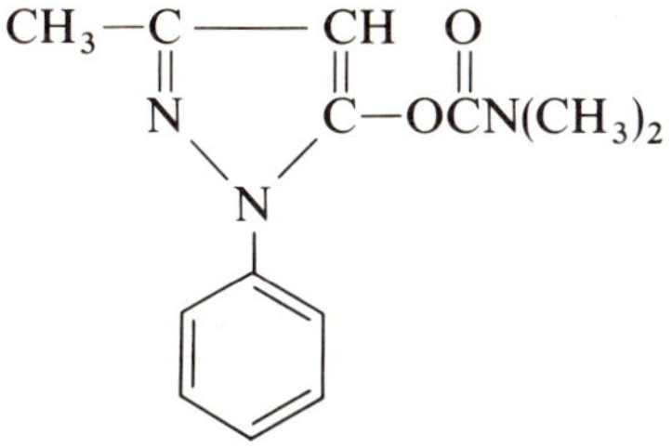

$$\text{CH}_3-\underset{\underset{\displaystyle N}{\|}}{C}\underline{\quad\quad}\underset{}{CH}\quad\overset{\displaystyle O}{\underset{\displaystyle C-OCN(CH_3)_2}{\|}}$$

N

3-METHYL-1-PHENYL-5-PYRAZOL DIMETHYLCARBAMATE

Pyrolan is a white crystalline material, has a melting point of 50°C, and is soluble in organic solvents and in water from 0.1 to 0.2%. It is said to be effective against aphids and the Oriental fruit fly in methyleugenol baits.

ISOLAN®

Isolan is also an experimental insecticide which is not commercially available. Gysin (1954) reports that it is more toxic to mammals than pyrolan and also has better systemic action.

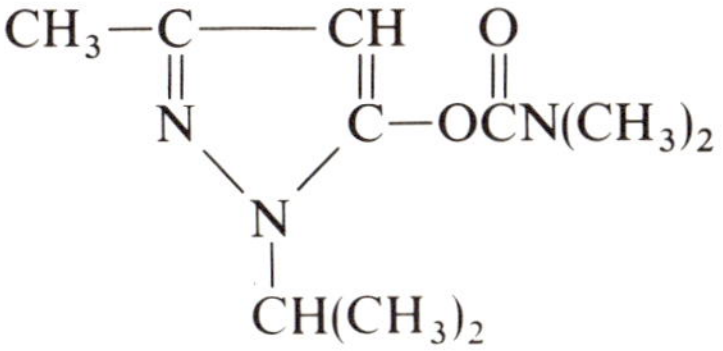

$$\text{CH}_3-\underset{\underset{\displaystyle N}{\|}}{C}\underline{\quad\quad}\underset{}{CH}\quad\overset{\displaystyle O}{\underset{\displaystyle C-OCN(CH_3)_2}{\|}}$$

N
|
CH(CH$_3$)$_2$

1-ISOPROPYL-3-METHYL-5-PYRAZOL DIMETHYLCARBAMATE

Isolan® and pyrolan have rapid knockdown action on houseflies (pyrethrin-like action) and are highly toxic by contact to aphids, thrips, bedbugs, and other small insects. They have a short residual life but have some systemic action and can be absorbed through leaves and roots of plants in quantities toxic to aphids and thrips.

DIMETILAN®

Dimetilan can be used to control flies in barns, stables, etc. A plastic fabric (fly band) is impregnated with dimetilan and then hung from the ceiling. It has a fairly high mammalian toxicity, with an acute oral LD_{50} in rats of 25–64 mg/kg.

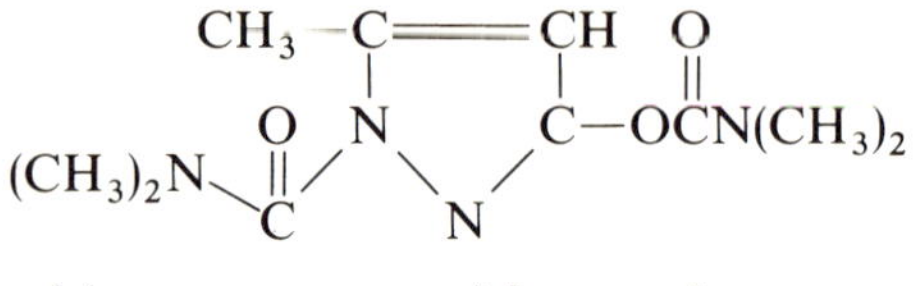

$$CH_3-C=\!=\!=CH\quad O$$
$$(CH_3)_2N \underset{C}{\overset{O}{\underset{\parallel}{}}} \underset{N}{\overset{N}{\diagup}} \underset{N}{\diagdown} C-O\overset{O}{\overset{\parallel}{C}}N(CH_3)_2$$

1-(DIMETHYLCARBAMOYL)-5-METHYL-3-PYRAZOL
DIMETHYLCARBAMATE

Dimetilan is a solid (melting point 68–71°C), with vapor pressure of 10^{-4} mm Hg at 20°C. Its water solubility is cited as 24%, and it has a systemic property.

3.4.4. Heterocyclic Methylcarbamates

MCA 600 (Mobam®)

MCA 600, which is now commercially available, is said to be effective in controlling leafhoppers, alfalfa weevils, and cotton insects.

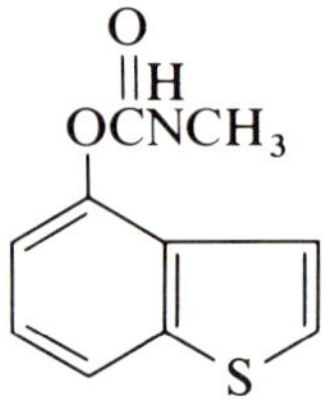

$$\overset{O}{\overset{\parallel}{\underset{}{OCNCH_3}}}H$$

4-BENZOTHIENYL-*N*-METHYLCARBAMATE

MCA 600 is a white crystalline solid with a melting point of 128°C. It has an acute oral LD_{50} in rats of 200 mg/kg.

Carbofuran (Furadan,® NIA 10242, ENT 27164)

Carbofuran, which ranks number two in the current (1971) list of carbamates produced in the United States, is highly toxic, with an acute oral LD_{50} in rats of 5 mg/kg. It is effective against soil insects in corn, cotton insects, and pests on potatoes. It is, however, not effective against spider mites. (See page 231 for formula drawing of carbofuran.)

3.4.5. Oximes

Temik® (Aldicarb, UC 21149)

Temik® is a new type of systemic carbamate insecticide, acaricide, and nematocide which is absorbed into the root system of plants and has a residual life of up to 10 weeks.

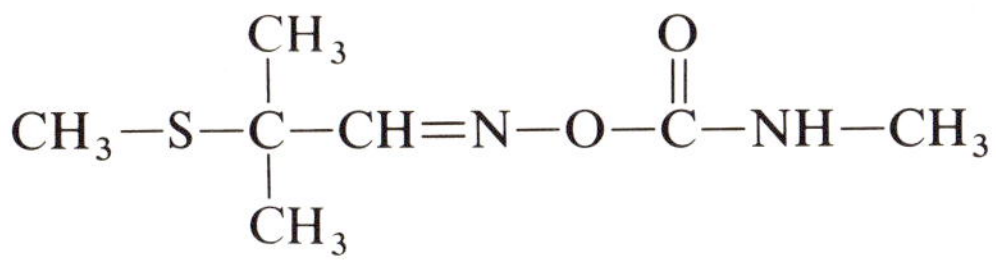

$$CH_3-S-\underset{\underset{CH_3}{|}}{\overset{\overset{CH_3}{|}}{C}}-CH=N-O-\overset{\overset{O}{\|}}{C}-NH-CH_3$$

2-METHYL-2-(METHYLTHIO)PROPIONALDEHYDE
O-(METHYLCARBAMOYL) OXIME

Temik® is highly toxic, with an acute oral LD_{50} in rats of 1–30 mg/kg. It is relatively soluble in water (6000 ppm in soil water; Gunther *et al.*, 1968).

3.5. THIOCYANATE INSECTICIDES

The organic thiocyanates were important insecticides before DDT and the chlorinated cyclodienes came into widespread use. In general, the thiocyanates produce rapid knockdown and paralysis of insects and have been especially used for fly sprays. No insects have developed resistance to these compounds. Thiocyanates are mild general poisons and are self-warning (i.e., if exposed, one feels ill immediately and thus moves away from the vicinity of the poison before receiving a fatal dose).

Methyl- and ethylthiocyanates have been used as fumigants. However, they are also toxic to plants, which limits their use somewhat.

LETHANE 60®

Many of the thiocyanates are known commercially as Lethanes and one of the most useful of these is Lethane 60,® 2-thiocyanatoethyl laurate. Its chemical formula is $C_{11}H_{23}COOCH_2CH_2SCN$. It is a clear amber liquid which is soluble in most organic solvents and insoluble in water.

It is moderately toxic, with an acute oral LD_{50} in rats of 500 mg/kg. It is harmful by skin absorption and if swallowed.

LETHANE 384®

Another thiocyanate in the Lethane series is Lethane 384,® 2-(2-butoxyethoxy) ethylthiocyanate. Its chemical formula is $C_4H_9OCH_2CH_2-OCH_2CH_2SCN$.

Lethane 384® is a clear brown oil, soluble in organic solvents and insoluble in water. It is toxic, with an acute oral LD_{50} in rats of 90 mg/kg. Like Lethane 60,® it is also harmful by skin absorption or if swallowed.

THANITE®

A thiocyanate with a rapid knockdown of houseflies is Thanite.® It is a clear amber liquid with an aromatic odor.

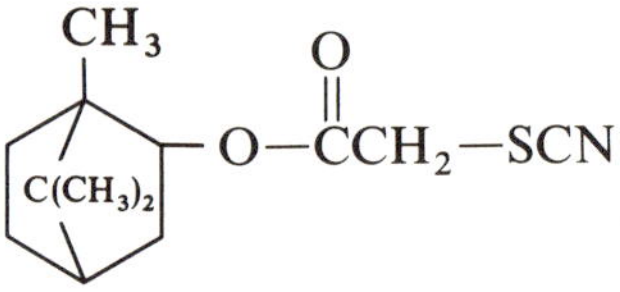

ISOBORNYL THIOCYANOACETATE

3.6. DINITROPHENOLS

The dinitrophenols were early insecticides and acaricides which were first used in the 1890s. They still have some usefulness today as dormant sprays and herbicides. These compounds are quite toxic to cells of all types, including plant cells. The dinitrophenols will kill mite eggs and have been used for this purpose as dormant sprays during the winter.

The basic structure of the dinitrophenol insecticides is

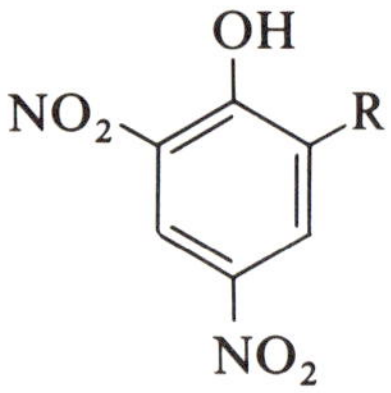

When R is CH_3, the compound is dinitrocresol, the most important dinitrophenol.

DINITROCRESOL (DNOC)

Dinitrocresol, also known as DNOC, is a yellow crystalline solid, melting point 85.8°C, with a slight, sharp odor. It has a vapor pressure of 5.2×10^{-5} mm Hg at 25°C and is slightly soluble in water (128 ppm at 15°C). The calcium, potassium, and ammonium salts of DNOC are freely soluble in water. Dinitrocresol also forms compounds with amines, phenols, and hydrocarbons.

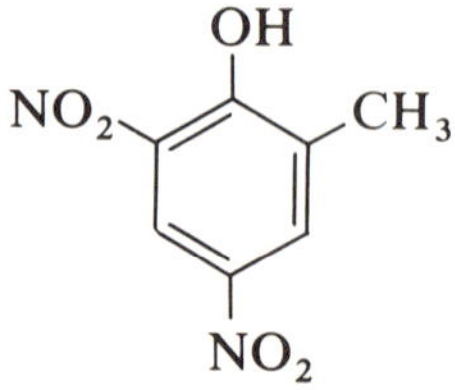

4,6-DINITRO-*o*-CRESOL

This insecticide, ovicide, and acaricide is marketed as Elgetol,[®] an aqueous solution containing 40% of the sodium salt of dinitrocresol (used as a dormant spray for orchards), and as Sinox,[®] a weed killer.

Dinitrocresol is highly toxic to mammals, with an acute oral LD_{50} in rats of 26–65 mg/kg. It is also toxic to plants.

DINITROCYCLOHEXYLPHENOL (DINEX[®])

Dinitrocyclohexylphenol, also known as DNCHP, DNOCHP, or Dinex,[®] is a yellow crystalline solid, melting point 106°C. It is an effective acaricide and is slightly soluble in water. Its amine salts are freely water soluble. Dinitrocyclohexylphenol is toxic, with an acute oral LD_{50} in rats of 65–333 mg/kg.

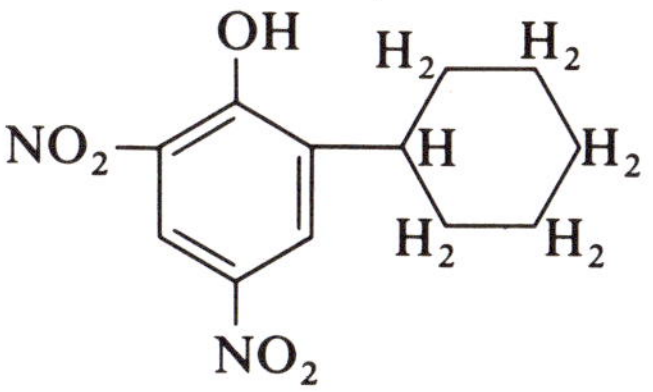

2-CYCLOHEXYL-4,6-DINITROPHENOL

Commercially, dinitrocyclohexylphenol is marketed as DNIII,[®] which is a 20% aqueous concentration of its dicyclohexylamine salt. DNIII[®] is less phytotoxic than DNCHP and is commonly used for foliage applications.

3.7. FLUOROACETATE DERIVATIVES

Fluoroacetate derivatives are characterized by their rigid structural requirement that only the compounds which can give fluoroacetic acid on activation in the animal or plant tissues are active. Their insecticidal actions are well defined in that fluoroacetate is further converted *in vivo* into fluorocitric acid, which inhibits aconitase (Peters, 1967) of the tricarboxylic acid cycle.

CH_2FCOOH FLUOROACETIC ACID

CH_2FCONH_2 FLUOROACETAMIDE

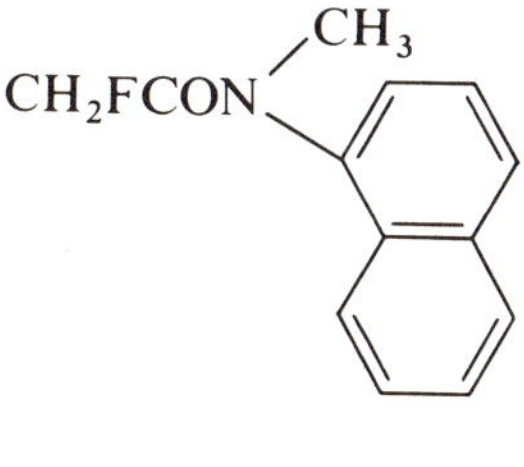

NISSOL,[®] 2-FLUORO-*N*-METHYL-*N*-(1-NAPHTHYL) ACETAMIDE

TABLE 3-4. Toxicities of Fluoroacetate Derivatives to the Mouse and the Housefly, Assayed by Injection (in LD_{50} mg/kg)

Compound	Mouse intraperitoneal	Housefly injection
Sodium fluoroacetate[a]	18	21
Fluoroacetamide[a]	85	9.5
Nissol[®][b]	200	14
p-Ethoxyfluoroacetanilide[a]	28	45
p-Chlorofluoroacetanilide	25	130

[a]From Matsumura and O'Brien (1963).
[b]From Johannsen and Knowles (1972).

Since these compounds attack the same biochemical processes in insects and mammals, it has been difficult to develop selective and thus useful fluoro-acetate derivatives, although Noguchi et al. (1968) have reported the discovery of one (Table 3-4). Fluoroacetate derivatives are effective as systemic insecticides and acaricides against aphids and mites. They are freely or relatively soluble in water.

3.8. ACARICIDAL CHEMICALS: SULFONATES, SULFONES, SULFIDES, AND NITROGEN-CONTAINING COMPOUNDS

There is a group of acaricidal compounds that normally contain two chlorinated benzene rings and are either sulfones, sulfonates, or sulfides (but never sulfates):

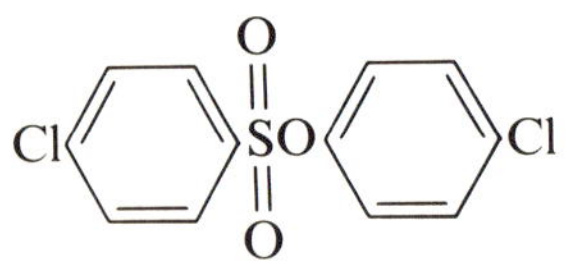

OVEX
(p-CHLOROPHENYL
p-CHLOROBENZENESULFONATE)

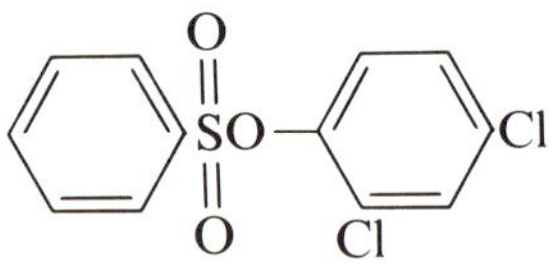

GENITE®

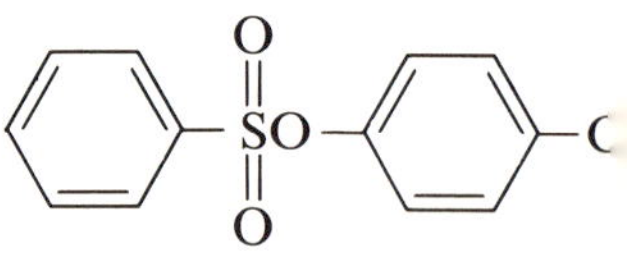

FENSON

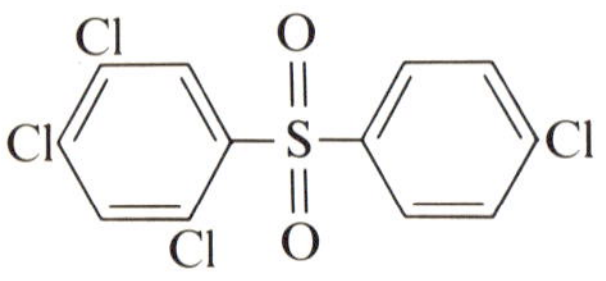

TETRADIFON
(TEDION®)

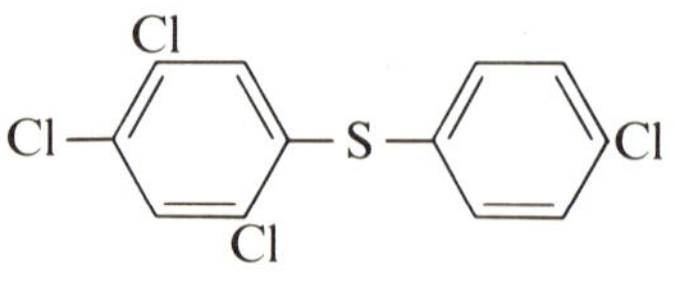

TETRASUL

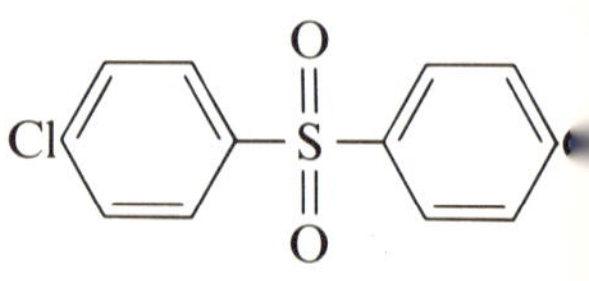

SULPHENONE®
(p-CHLOROPHENYL PHENYL SULFO

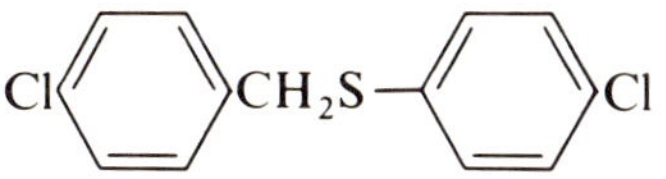

Cl⟨benzene⟩CH$_2$S—⟨benzene⟩Cl   ClCH$_2$CH$_2$OSOCHCH$_2$O—⟨benzene⟩C(CH$_3$)$_3$
$$\overset{O}{\overset{\|}{}}$$ over S, CH$_3$ below

CHLORBENSIDE 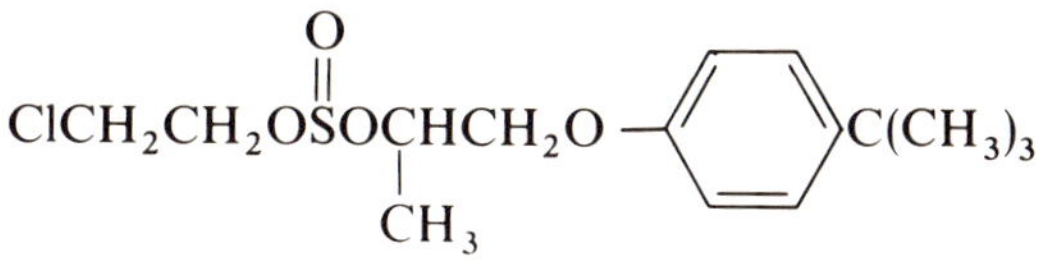   ARAMITE

OVEX (OVOTRAN®)

Ovex is a sulfonate ovicide which is too polar to be a good insecticide. It is also ineffective against adult mites and aphids. It is most effective against newly hatched larvae and eggs. It has a long-lasting residual effect and a low mammalian toxicity, with an acute oral LD_{50} in rats of 2050 mg/kg. This makes the compound useful as a spray for slow-growing crops such as fruit trees. Ovex has been used for the control of tetranychid mites.

Technical ovex is a flaky tan product with a melting point of 81°C. The pure compound is a colorless, crystalline solid with a melting point of 86°C. It is very stable but is hydrolyzed in alkali. Ovex is practically insoluble in water.

SULPHENONE®

A sulfone acaricide and ovicide is Sulphenone.® Pure Sulphenone® is a white solid with a melting point of 98°C. Because of its low phyto-toxicity to otherwise susceptible crops (e.g., some apple and pear varieties), it is used as a substitute acaricide. It is a short-life acardicide.

CHLORDIMEFORM, CHLORPHENAMIDINE, GALECRON,® FUNDAL®

Chlordimeform has been developed as an acaricidal compound, but it has been shown to have excellent properties in controlling some lepidopterous larvae in fields.

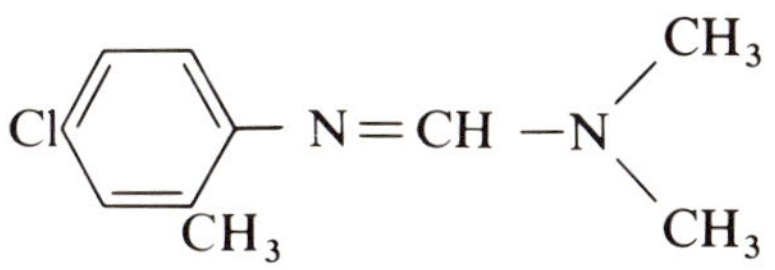

Cl⟨benzene, CH$_3$⟩— N=CH —N⟨CH$_3$, CH$_3$⟩

CHLORDIMEFORM

Its HCl salt is freely soluble in water. Although its exact volatility figure is not available, it is regarded as relatively volatile, since it kills mite eggs through vapor action.

3.9. FUMIGANTS

All fumigants are extremely volatile substances. Their usefulness is mainly in the area of control of stored-product insects and scale insects on citrus. Two factors play important roles in characterizing the fumigants: (1) flammability (2) self-warning properties. Table 3-5 summarizes the physical properties of some commonly used fumigants.

TABLE 3-5. Physical Properties of Commonly Used Fumigants

	Safe limit (ppm)	Boiling point (°C)	Relative weight (as air = 1)	Vapor pressure (mm Hg, 20°C)	Remarks (% in air flammability)
Chloropicrin	0.1	112	5.7	20	Corrosive, nonflammable
Methyl bromide	20	4	3.3	1420	Nonwarning, flammable (13.5)
Cyanide gas (HCN)	10	26	0.9	630	Flammable (6)
Ethylene dichloride	50	84	3.4	78	Highly flammable (6)
Ethylene oxide	50	11	1.5	1095	Explosive (3)
Carbon disulfide	20	46	2.6	314	Explosive (1)
Phosphine	0.3	−87.4	1.2	—	Flammable (2)

Mostly from Metcalf (1971).

METHYL BROMIDE (CH_3Br)

Methyl bromide is still the most widely used fumigant. It is a colorless, odorless liquid (boiling point 4.5°C). It is not self-warning since it has no odor in the gaseous form at room temperature. It is extremely toxic and flammable (safe vapor limit 17 ppm). It is stable and is heavier than air (relative weight against air 3.3). It is most widely used for insect pests in grain elevators and warehouses. It is also useful for soil pests including nematodes, although its high volatility somewhat restricts its use.

Methyl bromide fumigation of structures for the control of dry-wood termites has been used in the United States. Since it is not self-warning, it is extremely dangerous to applicators. Furthermore, the appearance of symptoms is late, making early detection of poisoning difficult.

ETHYLENE DICHLORIDE (CH_2ClCH_2Cl)

Ethylene dichloride is generally mixed with carbon tetrachloride and sold under a trade name, Dowfume.® It is used as a fumigant to control stored-product insects and soil insects. Its acute oral LD_{50} in rats is recorded as around 700–900 mg/kg.

CHLOROPICRIN (CCl_3NO_2)

Chloropicrin has been developed as a tear gas, but its usefulness as a fumigant was discovered in 1918 in France. It is self-warning because of its irritant effects, and is nonflammable. It is lethal at a concentration of 0.8 mg/liter for rats. Its main usefulness is for soil pests and to some extent for stored-product insects. It cannot be used for vegetables and other fresh crops and fruits.

PHOSTOXIN (PH_3, PHOSPHINE, HYDROGEN PHOSPHIDE)

Phostoxin is mainly composed of AlP powder, which slowly reacts with moisture in the air. It has approximately the same gas weight as air (1.18). It is generally regarded as nonflammable (spark point 100°C) and toxic (permissible limit for man 0.05 ppm). Because of its slow speed of activation (it generally takes 1–3 hr to become effective) and the ease of handling as tablets, it is considered a safe fumigant.

3.10. INORGANIC INSECTICIDES

Inorganic insecticides are relatively nonspecific, and since they are not too toxic to insects, large quantities (e.g., 10–100 lb/acre) are required to control insect pests in the field. Because of these limitations, inorganic insecticides have been gradually replaced by organic, particularly synthetic chemicals. Nevertheless, there are two groups of inorganic chemicals that are used today as insecticides: arsenicals and fluorides.

3.10.1. Arsenicals

Arsenicals are still widely utilized inorganic insecticides. The USDA (1969) data show that 9 million pounds of calcium and lead arsenate were produced in the United States in 1969.

The insecticidal activity of this group of compounds is generally directly related to the percentage content of metallic arsenic. These chemicals can be very phytotoxic, if the portion of water-soluble arsenic (H_3AsO_4) is high. Thus formulations which give high arsenic content with low water solubility have been developed. Two forms of arsenicals are important today: lead arsenate and calcium arsenate. Others are basic copper arsenate (BCA), $Cu(CuOH)AsO_4$; paris green, $(CH_2COO)_2Cu\cdot3Cu(AsO_2)_2$; magnesium arsenate, $Mg_3(AsO_4)_2\cdot MgO\cdot H_2O$; sodium arsenate, $NaAsO_2$; and zinc arsenate, $Zn_3(AsO_4)_2$, and zinc arsenite, $Zn_3(AsO_3)_2$. All arsenicals are stomach poisons to insects and leave water-insoluble residues on the top soil layers and on plant leaves.

Lead Arsenate, Basic Lead Arsenate

Lead arsenate is $PbHAsO_4$ and is sometimes referred to as "acid lead arsenate." It contains about 20% arsenic. It is 0.25% soluble in water, and it sometimes can cause phytotoxicity, particularly in the presence of alkali in water. Basic lead arsenate contains $Pb_4(PbOH)(AsO_4)_3\cdot H_2O$ and some $Pb_5(PbOH)_2(AsO_4)_4$. The commercial preparations contain 14% arsenic and are therefore less active than "acid lead arsenate." However, preparations of the latter are safer on delicate foliage. Altogether, lead arsenates are the least phytotoxic member of the arsenicals. Toxicities of these arsenicals are of the order of 100 mg/kg. The acid form is used more frequently, since it is more insecticidal. It is sprayed at 3–60 lb/acre, and is effective against chewing-type insects in orchards. The most popular formulation is dust containing 32% or more arsenic.

Calcium Arsenate [$Ca_3(AsO_4)_2$]

Calcium arsenate may contain up to 37% arsenic and therefore can be quite insecticidal. Commercial products (e.g., Kilmag®) contain about 25% arsenic and are considered to be in the safe basic form, $[Ca_3(AsO_4)_2]_3\cdot Ca(OH)_2$. Calcium arsenate is soluble in water to 0.4–0.5%. Generally, calcium arsenates are more apt to cause phytotoxicity than lead arsenates. The toxicity of the calcium arsenates is on the order of 35 mg/kg. They are formulated as 25% dusts, 15% bait, and 70% wettable powder and sprayed against the cotton boll weevil and insects in orchards and garden crops.

Sodium Arsenite ($NaAsO_2$)

Sodium arsenite is much more phytotoxic than the above arsenates and is not used on foliage. The arsenic content goes up to 44–57%, and it is quite insecticidal. The toxicity (LD_{50}) is quoted as 10 mg/kg. Because of its phytotoxicity, it is used against ticks, fleas, lice, ants, etc., in the form of bait under plants for leafy crop pests. These chemicals cause tremendous residue problems.

3.10.2. Inorganic Fluorides

Two types of inorganic fluorides are available as insecticides. They are sodium fluoride and a fluoride complex of aluminate or silicate. Generally, the degree of toxicity of these compounds is related to their fluorine content, but their phytotoxic action increases as their water solubilities increase, in analogy to arsenicals. They are also stomach poisons.

SODIUM FLUORIDE (NaF)

Sodium fluoride has been used as an insecticidal stomach poison since 1896. It contains 45.2 % fluorine and is soluble in water at 4.3 %. Its toxicity to man is cited as 75 mg/kg. It is formulated as powder (25–95 %) or bait and is used against cockroaches, silverfish, ticks, and lice. Because of phytotoxic problems, it is not used on plants. In the household and on premises where animals are kept, powders are spread at 10 oz/1000 ft^3.

CRYOLITE, SODIUM FLUOROALUMINATE (Na_3AlF_6)

Cryolite occurs naturally but can also be synthesized. It is relatively nontoxic to mammals. It contains 54 % fluorine and is soluble in water only to 0.06 %. Thus it is not too phytotoxic, although in damp climates some burning has been reported in corn. It is, however, soluble in dilute acid (and alkali), and some compatibility problems with other pesticide formulations have arisen. It is, however, a safe insecticide and causes few residual problems. Its residue tolerance limit has been set at 7 ppm. It is used on various berries, vegetables, and fruits.

SODIUM FLUOROSILICATE (Na_2SiF_6)

Sodium fluorosilicate contains over 60 % fluorine, but is highly phytotoxic since its water solubility is 0.65 %. Therefore, it cannot be applied on foliage. It is relatively nontoxic (LD_{50} 125 mg/kg) and has the same residual tolerance level as cryolite (for all fruits and vegetables). It is incompatible with alkaline material and is also known to accumulate in soil.

3.11. BOTANICAL INSECTICIDES

3.11.1. Nicotinoids

NICOTINE

Nicotine was first used as an insecticide in 1763, and the pure alkaloid was isolated in 1828 by Posselt and Reimann and synthesized in 1904 by Pictet and Rotschy. Nicotine alkaloid, nicotine sulfate, and other fixed nicotine compounds have been used as contact insecticides, fumigants, and stomach poisons. It appears in the market under the name Black Leaf 40,® which is an aqueous solution of the dibasic salt nicotine sulfate, containing 40 % nicotine. Nicotine is especially effective against aphids and other soft-bodied insects. It is highly toxic to mammals, with an acute oral LD_{50} in rats of 50–60 mg/kg. It is absorbed through the skin.

l-1-METHYL-2-(3-PYRIDYL)-PYRROLIDINE

Nicotine is a colorless, nearly odorless liquid with a boiling point of 247°C. On exposure to air, it darkens, becomes viscous, and develops a disagreeable odor. This aging does not affect its toxicity. It is volatile, with a vapor pressure of 0.042 mm Hg at 25°C and 0.12 mm Hg at 80°C. Nicotine is miscible with water below 60°C and is soluble in organic solvents. It readily forms salts with acids and dibasic salts with many metals and acids.

Commercially, nicotine is obtained from *Nicotiana tabacum* and *Nicotiana rustica*; it is found in the leaves of the former at a concentration of 2–5 % and in the latter at 5–14 %. Other parts of the plant have much smaller amounts of nicotine.

The nicotine is extracted from the plant by alkali and steam distillation or by extraction with benzene, trichloroethylene, or ether. Nicotine generally comprises 97 % of the alkaloid content of commercial tobacco. Anabasine and nornicotine are the only other alkaloids of insecticidal importance found in these plants.

NORNICOTINE

l-Nornicotine comprises 95 % of the alkaloid content of *Nicotiana sylvestris*, and *d*- and *dl*-nornicotine are found in the Australian plant *Duboisia hopwoodii* in variable amounts.

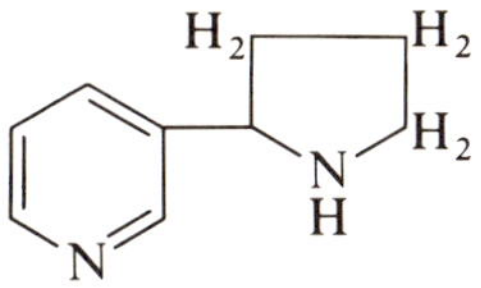

2-(3'-PYRIDYL)-PYRROLIDINE

In its pure form, nornicotine is a colorless, viscous liquid with a boiling point of 270–271°C. It is somewhat more stable than nicotine and does not darken as readily when exposed to air or have as pungent an odor as nicotine. Like nicotine, it is basic and readily forms salts. It is less volatile than nicotine.

Nornicotine is more effective against some insects than nicotine. Against mammals, nornicotine is equally as toxic as nicotine.

ANABASINE

Anabasine, which has also been known as neonicotine, is obtained from the new twigs of *Anabasis aphylla*, a small woody perennial that grows in Central Asia and North Africa. Anabasine is also found in tree tobacco in the southwestern United States. It is extracted by water or dilute acid or by steam distillation. Anabasine sulfate is used commercially as an insecticide in Russia.

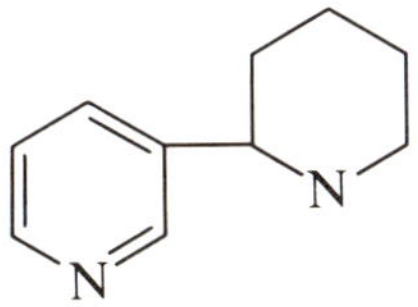

l-2-(3'-PYRIDYL)-PIPERIDINE

Anabasine is a colorless, somewhat viscous liquid which darkens rapidly on exposure to air. It has a boiling point of 280.9°C, and its volatility is similar to that of nicotine. It is soluble in water and in organic solvents. Like nicotine and nornicotine, it is basic and forms salts.

3.11.2. Rotenoids

ROTENONE

Plants containing rotenoids have been used as fish poisons for many centuries, and the active chemical ingredient was isolated in 1892 by Geoffroy and named *nicoulene*. The name *rotenone* was given in 1902 to an identical substance isolated from derris by Nagai Rotenone and allied substances are found in a large number of plants, all in the family Leguminosae. Economically important sources of rotenone are the plants *Derris elliptica* and *Derris malaccensis* from Malaya and the East Indies (where the dry product is called derris or tuba) and *Lonchocarpus utilis* and *Lonchocarpus urucu* from South America (where it is called timbo or cubé). Rotenone is used in the form of ground roots, resins, or as a crystalline material which is extracted by solvents, such as chloroform. Rotenone can be recrystallized further in alcohol. Commercially available rotenone-containing extracts vary considerably in the amount of rotenoids present, depending on the locality where produced and the botanical source. Although rotenone is considered the most active ingredient, the other extractives also possess appreciable toxicity.

Six rotenoids occur naturally, and rotenone, the most insecticidal, can act as either a contact or a stomach poison. All of the naturally occurring rotenoids apparently exist as levorotatory forms.

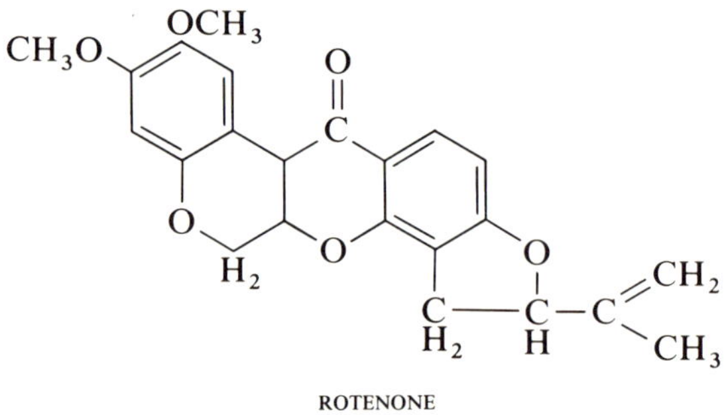

ROTENONE

Rotenone is white to yellowish white crystals, melting point 163°C, and is soluble in polar solvents and insoluble in water. It is readily detoxified by the action of air and light. Almost all toxicity may be lost after 2–3 days of summer exposure. Deterioration also takes place with heat. The chemical structure of rotenone was independently determined by Takei *et al.* (1932), Butenandt and McCartney (1932), and LaForge and Haller (1932).

The mammalian toxicity of rotenone varies greatly with the animal species, method of administration, and type of formulation. The acute oral LD_{50} of crystalline rotenone to rats is 132 mg/kg, to rabbits 3000 mg/kg, and to guinea pigs 60 mg/kg. When administered orally or intravenously in olive oil, it is more toxic. Derris powders were more toxic than would be accounted for by the rotenone content, indicating that the presence of other extractives had an effect on toxicity. Rotenone is very toxic to fish.

3.11.3. Pyrethroids

3.11.3a. Pyrethrum

Pyrethrum is found in the flowers of plants belonging to the family Compositae and the genus *Chrysanthemum*. The species which possess a high enough toxic content to be used for manufacture are *C. cinerariaefolium* and *C. coccineum*. Originally pyrethrum flowers came from Yugoslavia and Japan, but Kenya now supplies most of them. Kenya flowers contain an average of 1.3% pyrethrins, reaching 3% in selected strains; Japanese flowers contain 1% and Yugoslav flowers 0.7%.

Pyrethrum powders were first used around 1800, and by 1851 their use was world-wide. The ground flowers can be used as an insecticide, but this is very wasteful. Pyrethrum concentrates may be prepared from the ground flowers by extracting with petroleum ether, acetone, glacial acetic acid, ethylene dichloride, or methanol. Technical pyrethrum contains 20–30% of the toxic ingredient, and after filtration and reextraction the concentrate contains 90–100% pyrethrins.

Pyrethrum is essentially nontoxic to mammals (acute oral LD_{50} in rats about 1500 mg/kg) and is very fast-acting toward insects. It rapidly paralyzes insects, especially houseflies, and is commonly used in fly sprays and household insecticides. Often DDT or other insecticides are mixed with the pyrethrins because insects may later recover from pyrethrum alone.

The pyrethrum compounds found in pyrethrum flowers consist of four esters which are the combinations of two different alcohols with two different acids (Fujitani, 1909; Yamamoto, 1923; Staudinger and Ruzicka, 1924). The proportions may vary with the strain of flowers, conditions of culture, and method of extraction and concentration.

The alcohols are

PYRETHROLONE   CINEROLONE

The acids are

ACID I
CHRYSANTHEMUM MONOCAR-
BOXYLIC ACID

ACID II
CHRYSANTHEMUM DICARBOXYLIC
ACID MONOMETHYL ESTER

acid I + pyrethrolone = pyrethrin I

acid II + pyrethrolone = pyrethrin II

acid I + cinerolone = cinerin I

acid II + cinerolone = cinerin II

The method of attachment is as follows:

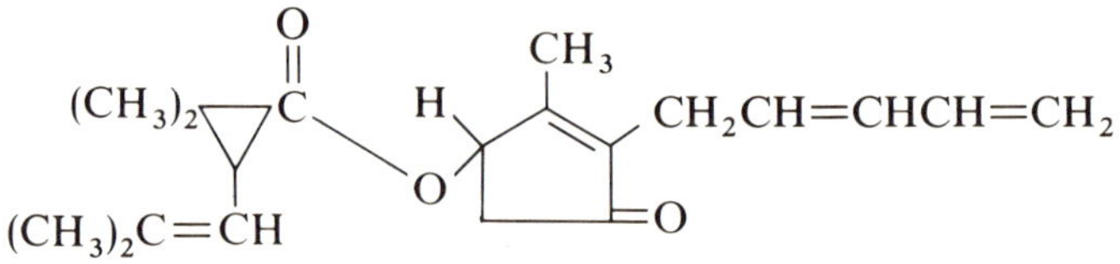

PYRETHRIN I

The pyrethrums are highly unstable in the presence of light, moisture, and air. Whole flowers decompose slower than ground flowers or dust. Stored powders lose about 20% of their potency in 1 year. The potency of the pyrethrums can best be preserved in sealed, lightproof containers kept at low temperatures. Various antioxidants have proved of value in preserving the insecticide.

The pyrethrin compounds are viscous liquids which are insoluble in water but soluble in organic solvents and oils. They are hydrolyzed in water, and the process is speeded by acid or alkali.

The pyrethrins are contact insecticides and have almost no stomach poison action because they are so readily hydrolyzed to nontoxic products. Their primary action is on the insect central nervous system since they produce such rapid paralysis.

3.11.3b. Synthetic Pyrethroids

Allethrin was the first synthetic pyrethrum analogue (Sanders and Taff, 1954). The synthetic pyrethrins are more stable than the natural pyrethrins due to the lower reactivity of the side-chains. Their toxic properties are similar to those of the natural pyrethrins, with some differences in relative toxicities according to the species of insect.

ALLETHRIN

Allethrin is cheap to produce and is effective against houseflies. The technical product is a mixture of the eight allethrin isomers, and its activity depends on the proportions of the isomers present. Allethrin is a clear brownish viscous liquid which is soluble in inert organic solvents and insoluble in water. It is more stable to heat and sunlight than the natural pyrethrins. It is moderately toxic, with an acute oral LD_{50} in rats of 680–920 mg/kg.

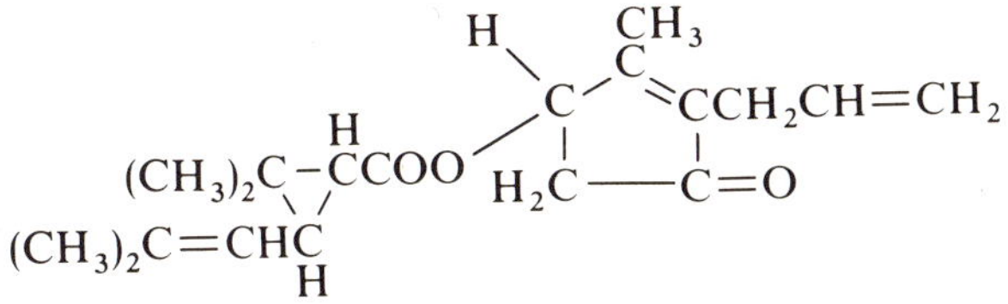

ALLETHRIN

PHTHALTHRIN

Phthalthrin is also a safe synthetic pyrethroid (mouse acute oral LD_{50} 1000 mg/kg). Its toxicity to houseflies is comparable to that of pyrethrin, with a modest increase in knockdown ability.

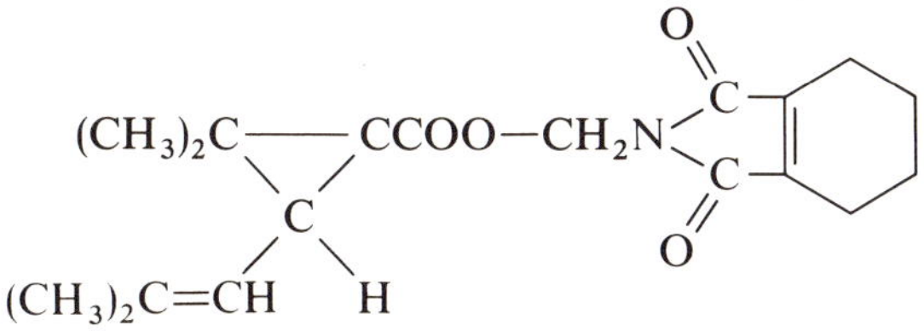

PHTHALTHRIN
N-(3,4,5,6-TETRAHYDROPHTHALIMIDO) METHYL *dl-cis-trans-*
CHRYSANTHEMATE

The following compounds have been synthesized by Elliot (1965, 1967):

(a)

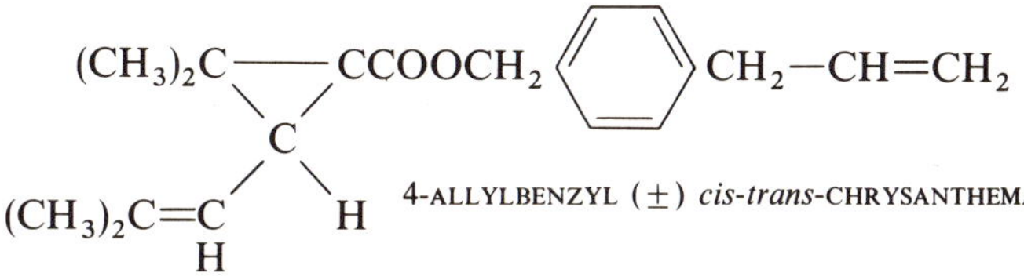

4-ALLYLBENZYL ($\pm$) *cis-trans-*CHRYSANTHEMATE

(b)

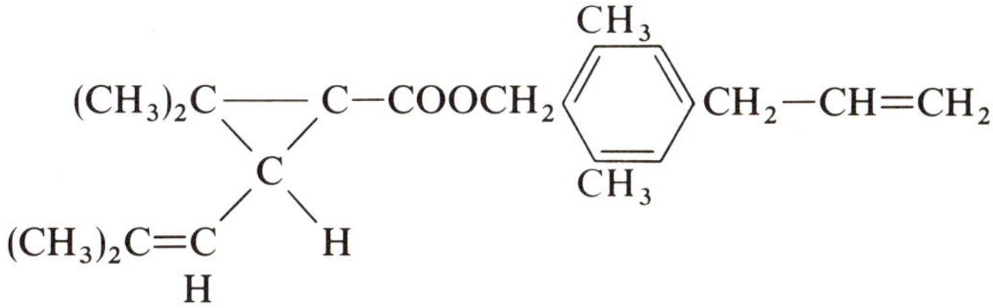

2,6-DIMETHYL-4-ALLYLBENZYL ($\pm$) *cis-trans-*CHRYSANTHEMATE

(c) 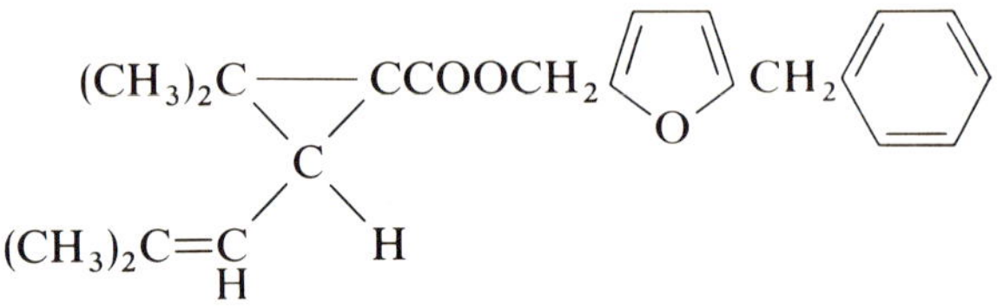 $(CH_3)_2C$————$CCOOCH_2$⟨furyl⟩CH_2⟨phenyl⟩

$(CH_3)_2C$=C H
          H

5-BENZYL-3-FURYL-METHYL (±) -*cis-trans*-CHRYSANTHEMATE

Their insecticidal potencies are rated as higher than those of pyrethrin and allethrin. The (+)-*trans* isomer of (c), for instance, is five times more effective than pyrethrin against houseflies.

3.12. SYNERGISTS

The toxicity of certain insecticides, notably pyrethrin, can be enormously increased by the addition of compounds which may not be insecticidal at all. These compounds are called synergists (e.g., pyrethrin synergists).

The majority of the synergists contain an active moiety, a methylenedioxyphenyl group.

SESAMIN

Sesamin is one of the active principals of sesame oil, which has long been known to synergize the action of pyrethrin (Eagleson, 1940).

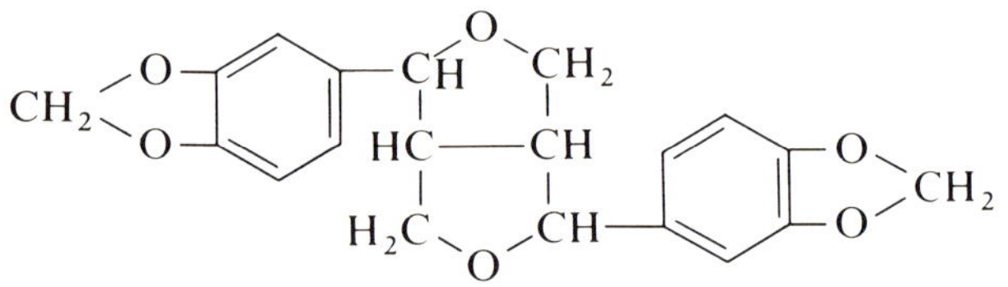

SESAMIN

It is a crystalline oleoresin, and its structure was determined by Haller *et al.* (1942).

SESAMOLIN

Sesamolin is another active principal of sesame oil.

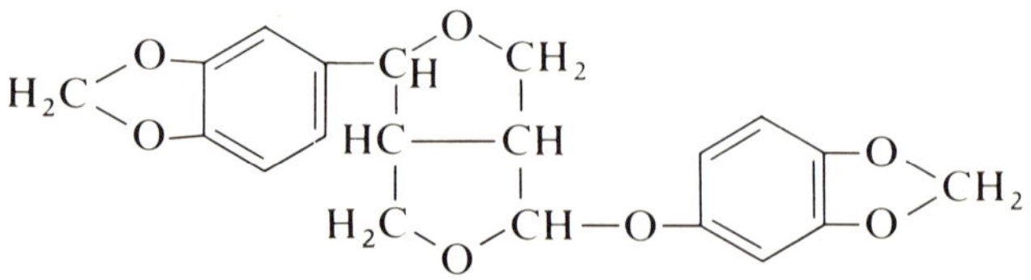

SESAMOLIN

It is a noncrystalline residue and is more effective than sesamin as a synergist with pyrethrin.

PIPERONYLBUTOXIDE

Perhaps the most widely used synthetic pyrethrin synergist is piperonylbutoxide.

α-[2-(2-BUTOXYETHOXY)ETHOXY]-4,5-METHYLENEDIOXY-2-
PROPYLTOLUENE

Its effectiveness is cited as around a tenfold increase in pyrethrin toxicity when it is mixed with pyrethrin in the ratio of 10:1. Its solubility in petroleum oils and in Freon® is acceptable enough for the preparation of aerosols. Its acute oral LD_{50} to rats is 7500–12,800 mg/kg.

SESAMEX (SESOXANE®)

Sesamex is a wide-spectrum synergist which acts not only with pyrethrin but also with carbamates, some organophosphates, and chlorinated hydrocarbon insecticides.

This synergist was originally found by Beroza (1956). Its synergistic action with allethrin is better than that of piperonylbutoxide.

ACETALDEHYDE 2-(2-ETHOXYETHOXY)ETHYL 3,4-METHYLENE-
DIOXYPHENYL ACETAL

It is very often used for *in vitro* studies to block the activity of mixed-function oxidases and is therefore of experimental interest as well.

3.13. REFERENCES

Arthur, B. W., and J. E. Casida (1959). *J. Econ. Entomol.* **52**:20.

Back, R. C. (1951). *Contrib. Boyce Thompson Inst.* **16**:451.

Balsom, E. W. (1947). *Trans. Faraday Soc.* **43**:54.

Beroza, M. (1956). *J. Agr. Food Chem.* **4**:53.

Brown, A. W. A. (1951). *Insect Control by Chemicals*. John Wiley & Sons, New York.

Brown, A. W. A., B. J. Wenner, and F. E. Park (1948). *Can. J. Res.* **D26**:188.

Buntin, G. (1951). U.S. Patent 2565471.

Busvine, J. R. (1964). *Nature* **174**:783.

Butenandt, A., and W. McCartney (1932). *Liebigs Ann.* **494**:17.

Casida, J. E., R. L. Holmstead, S. Khalifa, J. R. Knox, T. Ohsawa, K. J. Palmer, and R. Y. Wong (1974). *Science* **183**:520.

Clermont, A. (1854). *Ann. Chim. (Paris)* **91** :375.

Coates, H. (1949). *Ann. Appl. Biol.* **36**:156.

Daasch, L. W. (1947). *Ind. Eng. Chem. Anal. Ed.* **19**:779.

Davidow, B., and J. L. Radomski (1953). *J. Pharmacol. Exptl. Therap.* **107**:259, 266.

Eagleson, C. (1940). U.S. Patent 2202145.

Elliot, M. (1965). *Nature* **207**:938.

Elliot, M. (1967). *Nature* **213**:493.

Fujitani, J. (1909). *Arch. Exptl. Pathol. Pharmakol.* **61**:47.

Gaines, T. B. (1960). *Toxicol. Appl. Pharmacol.* **2**:88.

Geoffroy, E. (1892). *J. Pharm. Chim.* **26**:454.

Gunther, F. A., W. E. Westlake, and P. S. Jaglan (1968). *Residue Rev.* **20**:1. F. A. Gunther, ed. Springer-Verlag, New York.

Gysin, H. (1954). *Chimia* **8**:205, 221.

Haller, H. L., F. B. LaForge, and W. N. Sullivan (1942). *J. Org. Chem.* **7**:185.

Hayes, W. J. (1963). *Clinical Handbook of Economic Poisons*. Government Printing Office, Washington, D.C.

Johannsen, F. R., and C. O. Knowles (1972). *J. Econ. Entomol.* **65**:1754.

Johnsen, R. E., and W. M. Hanstbarger (1966). *Handbook of Insecticides*. Colorado State University, Fort Collins, Colo.

Johnson, G. V., J. Fletcher, K. G. Nolan, and J. Cassaday (1952). *J. Econ. Entomol.* **45**:279.

Kauer, K., R. DuVall, and F. Alquist (1947). *Ind. Eng. Chem. Ind. Ed.* **39**:1335.

Kenaga, E. E., and W. E. Allison (1969). *Bull. Entomol. Soc. Am.* **15**:85.

Ketelaar, J. A. A. (1950). *Rec. Trav. Chim.* **69**:649.

Kolbezen, M. J., R. L. Metcalf, and T. R. Fukuto (1954). *J. Agr. Food Chem.* **2**:864.

Kükenthal, H., and O. Jung (1963). Data obtained in Biologischen Institut der Farbenfabriken Bayer, Leverkusen. (As cited by Schrader, 1963, p. 295.)

LaForge, F. B., and H. L. Haller (1932). *J. Am. Chem. Soc.* **54**:810.

Maqsud-Nasiv, N. (1953). *J. Sci. Food Agr.* **4**:374.

Martin, H. (1964). *The Scientific Principles of Crop Protection*. St. Martin's Press, New York.

Matsui, M., and F. B. LaForge (1952). *J. Am. Chem. Soc.* **74**:2181.

Matsumura, F., and R. D. O'Brien (1963). *Biochem. Pharmacol.* **12**:1201.

Metcalf, R. L. (1947). *J. Econ. Entomol.* **40**:522.

Metcalf, R. L. (1948). *J. Econ. Entomol.* **41**:416.

Metcalf, R. L. (1955). *Organic Insecticides*. Interscience Publishers, New York.

Metcalf, R. L. (1971). In *Pesticides in the Environment*. R. White-Stevens, ed. Marcel Dekker, New York, p. 1.

Metcalf, R. L., and T. R. Fukuto (1965a). *J. Agr. Food Chem.* **13**:3.

Metcalf, R. L., and T. R. Fukuto (1965b). *J. Agr. Food Chem.* **13**:222.

Murphy, D. F., and C. E. Peet (1932). *J. Econ. Entomol.* **25**:123.

Nagai, K. (1902). *J. Tokyo Chem. Soc.* **23**:744.

Noguchi, T., Y. Hashimoto, and H. Miyata (1968). *Toxicol. Appl. Pharmacol.* **13**:189.

O'Brien, R. D., and R. W. Fisher (1958). *J. Econ. Entomol.* **51**:169.

Parker, W. I., and J. H. Beacher (1947). University of Delaware Bulletin 264, Tech. 36, 26 pp.

Peck, D. R. (1948). *Chem. Ind.* **1948**:526.

Peters, R. A. (1967). *Rec. Chem. Prog.* **28**:197.

Pictet, A., and A. Rotschy (1904). *Berichte* **37**:1225.

Richardson, C. H., and C. R. Smith (1926). *J. Agr. Res.* **33**:597.

Sanders, H. J., and A. W. Taff (1954). *Ind. Eng. Chem.* **46**:414.

Schneller, G., and G. Smith (1949). *Ind. Eng. Chem.* **41**:1027.

Schrader, G. (1942). German Patent 720577.

Schrader, G. (1963). *Die Entwicklung ueber insektizider Phosphorsäure-Ester.* Verlag Chemie, Weinheim.

Sherman, M. (1948). *J. Econ. Entomol.* **41**:575.

Slade, R. E. (1945). *Chem. Ind.* **1945**:314.

Soloway, S. B. (1963). Abstracts of the Fifth International Pesticides Congress, International Union of Pure and Applied Chemistry, London, p. 47.

Spencer, E. Y. (1961). *Can. J. Biochem. Physiol.* **39**:1790.

Staudinger, H., and L. Ruzicka (1924). *Helv. Chim. Acta* **7**:177.

Stedman, E., and L. Eason (1932). *Biochem. J.* **26**:2051.

Stephenson, O., and W. Waters (1946). *J. Chem. Soc.*, p. 339.

Takei, S., S. Miyajima, and M. Ono (1932). *Ber. Deutsch. Chem. Ges.* **65B**:1041.

Topley, B. (1950). *Chem. Ind.* **1950**:859.

USDA (1971). The Pesticide Review, 1970. Agricultural Stabilization and Conservation Service, Washington, D.C.

Velsicol Corp. (1964). *Tech. Bull. Entomol. Soc. Am.* **10**:99, 117.

Wirth, W. (1953). *Naunyn-Schmiedebergs Arch. Exptl. Pathol. Pharmakol.* **217**:144.

Yamamoto, R. (1923). *J. Chem. Soc. Japan* **44**:311.

Chapter 4

Modes of Action of Insecticides

4.1. INTRODUCTION

The modes of action of various drugs and poisons have fascinated mankind since the age of witchcraft-medicine. As stated before, insect toxicology is part of a much broader field, pharmacology and toxicology, in which studies of drug action are known as pharmacodynamics. Study of the actions of insecticides may therefore be called the pharmacodynamics of insecticides, or simply the pharmacology of insecticides. Though the history of the pharmacology of insecticides is just as brief as that of modern biochemistry and physiology, it is already growing into one of the most important disciplines of insect toxicology. Its existence is necessitated by (1) the need for therapeutic measures for accidental poisoning, (2) the demand for logical explanation of these toxic actions and of their subsequent side-effects in beneficial animals including man, and (3) the realization that it can provide a logical basis for developing even more useful compounds and can help in understanding the normal physiology and biochemistry of animals.

Often one group of insecticides is particularly active toward a certain kind of organism. The insecticide is then generally termed "selective." "Selective toxicity" generally refers to cases where mammals are less affected by a toxicant than insects and other pests. It is, however, equally appropriate to use the term in the cases where a beneficial species of insect is unharmed by an insecticide while pest species are killed. Intra- and interspecific differences in susceptibility to various kinds of insecticides are increasingly drawing more attention. Precise knowledge of both the modes of action of insecticides and the differences in responses of various organisms also helps in understanding the intricate differences in their physiology and biochemistry.

4.2. CLASSIFICATION OF INSECTICIDES BY THEIR ACTIONS

Brown (1951) has classified insecticides into five groups, based on mode of action: (1) physical poisons, (2) protoplasmic poisons, (3) respiratory poisons, (4) nerve poisons, and (5) poisons of a more general nature. Most modern insecticides are nerve poisons, and Brown's classification still covers almost all insecticides marketed today.

Another way of grouping the insecticides is to separate them into three groups according to mode of entry: (1) stomach poisons, (2) contact poisons, and (3) fumigants (Brown, 1951). This approach, though often very useful as a means to describe an insecticide to nonexperts, has some technical limitations such as the problem of having to classify a multipurpose insecticide as belonging to more than one category.

Table 4-1 indicates a rough grouping of insecticides by their mode of action. Of the five categories listed, metabolic inhibitors and neuroactive agents are the two major groups of modern insecticides. Strictly speaking,

TABLE 4-1. A Classification of Insecticides on the Basis of Their Mode of Action

Groups	Subgroup	Examples
Physical poisons[a]		Heavy mineral oils, inert dust
Protoplasmic poisons[a]		Heavy metals, e.g., Hg, acids
Metabolic inhibitors	Respiratory poisons[a]	HCN, CO, H_2S, rotenone, dinitro-phenols
	Inhibitors of mixed-function oxidase	Pyrethrin synergists
	Inhibitors of carbohydrate metabolism	Sodium fluoroacetate
	Inhibitors of amine metabolism	Chlordimeform
	Insect hormones	Juvenile hormone analogues
Neuroactive agents (nonmetabolic)	Anticholinesterases	Organophosphorus compounds, carbamates
	Effectors of ion permeability	DDT analogues, pyrethroids, cyclodiene compounds, BHC
	Agents for nerve receptors	Nicotine analogues
Stomach poisons		*Bacillus thuringiensis* toxin

[a]According to Brown (1951).

anticholinesterases and some of the other nerve poisons are also metabolic inhibitors, but they differ by specifically attacking the nervous system, or at least their actions on the nervous system constitute the major cause of insect (or mammal) death. It must be borne in mind, however, that the action of any insecticide could be multiple and that to trace its effect to a primary target is not always possible. The cause of death need not be congruent with the mode of action of an insecticide. For example, mosquito larvae exposed to an insecticide often die from lack of oxygen because they cannot reach the surface of the water, but the mode of action of the insecticide may be the inhibition of cholinesterase or any other immobilizing effect.

4.3. THE NERVOUS SYSTEM

The majority of the modern insecticides owe their toxicity to their ability to attack the nervous system as the primary target. The nervous system is one of the most susceptible and vulnerable portions of the body of highly developed organisms. The striking feature of insects is that they have such a well-developed central nervous system, almost comparable in organization to that of mammals. Poisoning the nervous system is the quickest and surest way of chemically upsetting the regular body mechanisms. Generally, the success of an insecticidal compound depends on the high degree of nerve development in the pest insects. By the same token, man, being the most highly developed organism, could be very susceptible to the same nerve poison. Fortunately, however, several useful compounds have been invented which can selectively attack insects without causing much damage to beneficial animals and humans. (There are also a number of compounds which can selectively attack the human nervous system, e.g., chemical warfare agents.)

In recent years, however, it has become apparent that in some cases the differences in susceptibility between mammals and insects are the result of inherent differences in their nervous systems. For instance, differences in susceptibility to nerve poisons between insects and mammals could partly be attributable to morphological differences in the pattern of nerve distribution: insects possess several nerve endings exposed without any protection, whereas mammals have relatively few of these vulnerable sites. The differences are of great interest not only from an economic point of view but also from the standpoint of comparative biochemistry. Historically, studies of the nervous system have been an amalgamation of invertebrate and vertebrate zoology, the basic organization of the nervous system being studied in mammals and its fundamental functions being studied in invertebrates (e.g., the giant axon of the squid and lobster).

The basic differences in the pattern of nerve organization between mammals and insects can best be demonstrated by first considering the fundamental structure of the mammalian nervous system and then pointing out the specific morphological and functional differences in the insect system. There are a number of excellent textbooks and review papers which will be of great help for readers who wish to study this subject further:

> J. E. Treherne, and J. W. L. Beament, eds. *The Physiology of the Insect Central Nervous System.* New York: Academic Press, 1965. 277 pp.
> J. W. L. Beament, J. E. Treherne, and V. B. Wigglesworth. *Advances in Insect Physiology,* Vol. 1. New York: Academic Press, 1963. 512 pp.
> R. D. O'Brien. *Insecticides.* New York: Academic Press, 1967. 332 pp.

Figures 4-1 and 4-2 indicate the general features of the mammalian and the insect nervous systems. Both have two important subdivisions in common: the central nervous system and the peripheral nervous system. The former consists of the brain and the spinal cord in mammals, and the brain and the central nerve cord in insects. The central nervous system, as its name

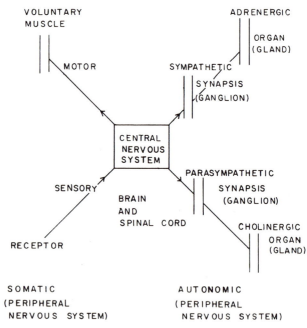

Fig. 4-1. Simplified diagram of mammalian nervous system. Modified from O'Brien (1960).

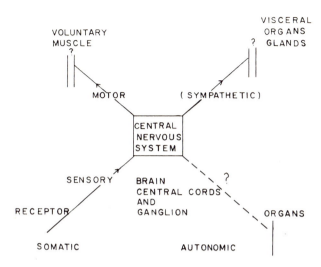

Fig. 4-2. Simplified diagram of insect nervous system. From O'Brien (1960).

implies, serves as the central integration system and is composed of millions of nerve cells which are connected one with another by junctions called synapses. The peripheral nervous system is composed of two subdivisions: the somatic system and the autonomic system.

4.3.1. Mammalian Nervous System

Since the mammalian nervous system is more complete with respect to these subdivisions than that of the insect, it will be examined first.

4.3.1a. The Somatic System

The somatic system handles that part of animal movement exemplified by a set of reactions to environmental stimulus and the muscle response that results. It is composed of incoming (afferent) and outgoing (efferent) pathways. As the means of detecting environmental changes, there are sensory receptors (eyes, ears, etc.) which are directly connected to the central nervous system. Generally the receptors are specialized organs, but in less specialized body areas nerve endings can serve the same purpose.

Conducted information can be either directly passed to the responding motor nerves (simple reflex arcs) or first analyzed by the brain. When the brain demands a response to the signal, the order is sent out to the voluntary muscles through the motor nerves. The efferent nerve fiber terminates at the neuromuscular junction, where the role of the nervous system ends when the message is passed on to the voluntary muscle. The message is conducted across the gap by a chemical mediator, acetylcholine; that is, the passage is controlled by the "cholinergic system." The role of acetylcholine in the system of nerve conduction will be discussed in later chapters.

4.3.1b. The Autonomic Nervous System

The autonomic nervous system controls the movements of the muscles of internal organs and stimulates various glands. The whole system is controlled by two subdivisions having opposite actions that are involuntary in nature; that is, the animal has no conscious control over them. These subdivisions are the sympathetic and parasympathetic systems.

The sympathetic system has characteristic large ganglia mostly chained together outside the central nervous system. The ganglia contain many synaptic gaps which are bridged by the cholinergic system. The axon connecting the central nervous system to a ganglion is often called the "preganglionic axon" and the one connecting the ganglion to the end organ the "postganglionic axon." At the terminal of the latter is a synaptic gap which is bridged to the organ or gland. The transmission of stimuli is mediated here by the release of adrenaline or noradrenaline.

The parasympathetic system basically resembles the sympathetic system except for the synaptic gap between the postganglionic axon and the organ, which is cholinergic. There are, however, a number of minor differences. The parasympathetic ganglia are often relatively small and are scattered in the body. They are usually located right beside the organs; therefore, the postganglionic axons are usually short. Generally speaking, the sympathetic stimulus accelerates the heartbeat and decreases the activity of the alimentary canal. The parasympathetic stimulus retards the heartbeat and increases the activity of the alimentary canal. Other important end results of sympathetic stimuli are dilation of the pupil and inhibition of salivary secretion.

4.3.1c. The Electric (Axonic) Conduction of Nerve Impulses

The nerve cells of these systems have a remarkable similarity to each other. Each cell has a cell body and a long extension, the axon, through which the stimulus is transmitted in the form of an electric wave (Fig. 4-3). An axon can be several feet long. At the end of the axon, the stimulus must be transmitted over the synaptic gap to the next cell. This is the place where a chemical mediator becomes necessary. Nerve conduction within the axon takes place by means of a series of changes in the membrane electric potential.

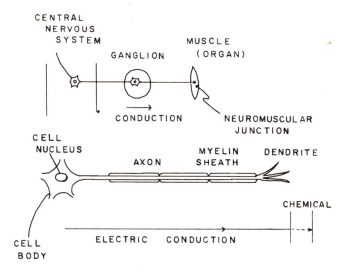

Fig. 4-3. A typical diagram of mammalian nerve cells.

The mechanism by which the nerve membrane performs the process of polarization and depolarization was extensively studied by Hodgkin (1951) and Hodgkin and Huxley (1952), who explained the meaning of ionic movements across the nerve membrane. As mentioned above, the axons usually maintain an electric potential difference across the membrane. It is known that normally the internal concentration of potassium ions far exceeds that outside. At the same time, the internal concentration of sodium ions is much lower than that outside. The potential can be generated by these differences in the ion concentrations across the membrane (Narahashi, 1962):

$$E_{Na} = \frac{RT}{F} \ln \frac{[Na]_{outside}}{[Na]_{inside}} \qquad E_K = \frac{RT}{F} \ln \frac{[K]_{outside}}{[K]_{inside}}$$

When the stimulus wave arrives at a particular portion of the membrane, depolarization in terms of inflowing of sodium ions and outflowing of potassium ions takes place (see Fig. 4-4).

The observed resting potential is usually lower than the value calculated by the above equation, since it does not include the movement of other ions. Suppose chlorine ions are taken into consideration (or any other ion). The equation can be transposed to

$$E_m = \frac{RT}{F} \ln \frac{PK[K]o + PNa[Na]o + PCl[Cl]i}{PK[K]i + PNa[Na]i + PCl[Cl]o}$$

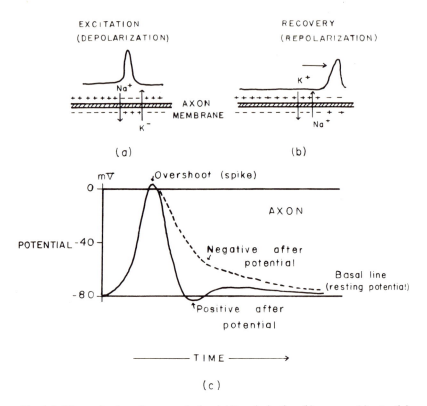

Fig. 4-4. The mechanism of nerve excitation. (a) Depolarization, (b) recovery, (c) potential change during excitation. Recorded with an oscilloscope from the internal and the external electrodes.

where E_m is the membrane potential and, for example, PK is the relative permeability of potassium ions across the axonic membrane.

Thus the excited membrane returns to its normal state by taking in potassium ions and ejecting sodium ions. The process is done against the gradient of ion concentrations and therefore requires energy. Hence the process is called "active ion transport." The mechanism of the ejection of sodium ions is popularly known as the "sodium pump." It is now generally accepted that this process is handled by a specialized enzyme, $(Na^+ + K^+)$-ATPase (Skou, 1965). On the other hand, the mechanisms for the ion permeability changes remain obscure.

The process of depolarization and polarization in the isolated crustacean nerve cord can be completely destroyed by the action of lecithinases A and C but not by trypsin or chymotrypsin, despite the fact that the latter treatments produce much more visible changes (as witnessed by electron microscopy) than the former ones. There are a number of reports available to indicate that ATP and other phosphates play a regulatory role in the ion transport of excitable membranes (see Abood, 1972). The role of phosphate transfer appears to be closely linked to that of Ca^{2+}, which is known to play a critical role in membrane excitation. The validity of the above model and related action theories is, however, yet to be confirmed. Therefore, only a brief account of the relationship between the nerve potentials and the ion movements has been presented here. More detailed information on current theories of the effects of various insecticides on the active transport mechanisms will be given later.

4.3.1d. The Chemical Transmission of Nerve Impulses

When a stimulus in the form of an electric conduction wave reaches the end of an axon, some means of transferring the message on to the next nerve cell or to an appropriate organ becomes necessary. The chemical mediators are generally there for this purpose.

Of the known chemical mediators, acetylcholine plays by far the most significant role in the process of chemical transmission:

$$CH_3-\overset{\overset{\displaystyle O}{\|}}{C}-O-CH_2-CH_2-\overset{\overset{\displaystyle CH_3}{|}}{\underset{\underset{\displaystyle CH_3}{|}}{N^+}}-CH_3$$

ACETYLCHOLINE

The acetylcholine–cholinesterase system—the cholinergic system—is the only known system for transferring the message across the neuromuscular junction of motor nerves, the parasympathetic neuroeffectors, and the sympathetic synapses; it generally does not operate at the final sympathetic neuroefferent junction. (An exception is the sympathetic innervation of the adrenals and the sweat glands, which is cholinergic.)

In the central nervous system, where numerous synapses exist, the cholinergic system plays an important part, as judged by the fact that powerful anticholinesterases are often good central nerve poisons. Two other important chemical mediators are adrenaline and noradrenaline:

ADRENALINE   NORADRENALINE

Noradrenaline is relatively more effective on the sympathetic neuro-effector junctions and is regarded as the actual, active substance involved in the transmission. There are other biogenic amines that play important roles in the central nervous system. Their significance will be discussed elsewhere (see Section 4.10).

4.3.2. Insect Nervous System

The insect nervous system is considerably simpler than that of mammals. However, in some instances, certain parts of it are highly specialized for adaptation to a complex mode of life.

As shown in Figs. 4-1 and 4-2, the basic differences between the insect nervous system and that of mammals are that in insects

1. No cholinergic systems are involved in the peripheral nervous system.
2. No distinct autonomic system exists. (There is an autonomic system which is mainly controlled by hormones.)
3. No ganglia are involved in the peripheral nervous systems. (The ganglia of insects correspond to the mammalian central nervous system.)
4. No chemical transmitter has been identified in the insect central nervous system other than acetylcholine, though it is believed that some active biogenic amines exist.

Besides the gross differences, the insect nervous system has the following morphological and histological differences:

1. There is no distinct myelination observable in the insect nervous system.
2. The insect nerve cords are covered with a tough fibrinous sheath, called simply the "nerve sheath," which limits the entry of many sub-stances. The sheath is said to be necessary to protect the nerve cord, and it seems to serve a similar purpose as the "blood–brain barrier" of the mammal.
3. The insect neuromuscular junction has no specific "end plates" as in the mammalian junction. The insect axon branches out into several "twigs" and innervates single muscle fibers, whereas the end plate of the mammalian system governs the whole group of muscle fibers.
4. The insect nervous system is supplied with tracheal systems which provide oxygen directly to the nerve cells through diffusion.

Otherwise, the nervous system of insects resembles that of mammals in many ways, and the basic similarities are more striking than the differences.

Nevertheless, variations among insect species are enormous, and so are the secondary variations of their neural organization. Interspecies differences in these patterns are, except for a few well-studied species (such as *Periplaneta americana*, the American cockroach), numerous and complex and beyond the scope of this chapter.

4.4. CHLORINATED HYDROCARBON INSECTICIDES

Chlorinated hydrocarbon nerve poisons are one of the most important insecticide groups used today. Chemically they are relatively unreactive stable compounds and are characterized by their long-lasting residual effects. Chlorinated hydrocarbon insecticides owe their biological activity mainly to their ability to upset the nervous system of living organisms, though some side-effects (such as effects on biological oxidation systems) have been noted in other parts of the body. Among them is the most studied compound, DDT.

4.4.1. DDT and Its Analogues

Tremendous efforts, probably unmatched by studies on any other insecticide, have been made to elucidate the action mechanism of DDT. This fact makes it the best example for discussing the mode of action of insecticides.

4.4.1a. Symptomatology and Characteristics

Perhaps the most unusual characteristic of DDT is that its insecticidal potency greatly increases with decrease in environmental temperature. A special term, "negative temperature correlation," was coined to describe this phenomenon.

DDT is relatively slow acting. The first sign of poisoning is usually uncoordinated movement of the test organism. This stage is followed by the very characteristic "DDT jitters": tremulousness of the entire body and limbs. According to Tobias and Kollros (1949), the sequence of symptoms in *P. americana* (the American cockroach) is (1) hyperextension of legs and uncoordinated movement; (2) general tremulousness; (3) ataxic gait, and hyperactivity resulting from external stimuli; (4) repeated falling on the back and righting efforts; (5) two separate leg movements, a high-frequency tremor and a slower flexion and extension; (6) disappearance of fast tremors, with the only symptoms left being isolated motions of body wall, tarsi, palpi, cerci, and antennae; and (7) complete stillness with the heartbeat being the only sign of life. Autotomically isolated legs of certain geometrid moths can continue to show the characteristic tremors (Wiesmann and Fenjves, 1944).

This phenomenon is interpreted as evidence of DDT's action on the motor nerves. The action of low doses of DDT on the nervous system seems to require an intact reflex arc; e.g., poisoning starts at a sensory cell at the campaniform sensilla, then reaches the central nervous synapses, and finally influences the motor neurons (Roeder and Weiant, 1946, 1948). At high concentrations, DDT can act directly on the motor nerves themselves.

The symptoms which follow DDT administration to the insect closely resemble those caused by curare, eserine, and nicotine. DDT itself can increase the amount of free acetylcholine (Tobias *et al.*, 1946; Lewis, 1953), although it can hardly be included in the group of cholinesterase inhibitors. It is generally considered that the release of acetylcholine *in vivo* (in the living body or tissue) is a result of DDT poisoning, or, more precisely, a result of the nervous overexcitement which is actually caused by some other interaction of DDT with the nervous system. A similar *in vivo* effect can be found in the respiratory responses of the insect toward DDT: the total oxygen consumption of the poisoned insect usually reaches severalfold more than the normal level. Buck and Keister (1949) found that insects die before all the reserves (such as glycogen, glucose, and lipids) have been depleted to starvation level. However, Ludwig (1946), as a result of an extensive study on the effects of DDT poisoning, suggested that the cause of death (but not necessarily the effect of DDT action) is starvation due to exhaustion.

It is possible that DDT interferes to some extent with the oxidation process in the nervous system. Saktor (1951) was able to show that DDT significantly inhibits cytochrome oxidase activity in the housefly. Morrison and Brown (1954) also reached a similar conclusion. Anderson *et al.* (1954) discovered, however, that DDE (a low-toxic metabolite of DDT) is a relatively potent inhibitor of succinoxidase and cytochrome oxidase of the housefly thorax compared to DDT; the reverse should be true if the primary target of DDT is these oxidative enzymes. Thus it is unlikely that the inhibitory effects of DDT on oxidation systems are the key to the process of actual DDT intoxication. Nor is any other evidence concerning the inhibitory effects of DDT on nonneural enzyme systems decisive enough to be of major importance (Merrill *et al.*, 1946; Ludwig, 1946; Rosedale, 1948).

4.4.1b. Action of DDT on Nerves and Theories of Its Mode of Action

Enough evidence has been accumulated to indicate that the primary target of DDT is indeed the nervous system. The characteristic "repetitive discharge" in the nerve impulse patterns has long been noted (Dresden, 1949; Roeder and Weiant, 1946, 1948; Yamasaki and Ishii, 1953). In essence, DDT causes the nerve fibers to produce this repetitive discharge (Fig. 4-5) in response to a single stimulus (Welsh and Gordon, 1947; Gordon and Welsh,

Fig, 4-5. Repetitive discharge caused by DDT treatment of the sensory neuron of *Periplaneta americana*. Redrawn from Yamasaki and Ishii (1954).

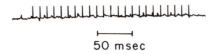

50 msec

1948). Welsh and Gordon pointed out that the phenomenon very closely resembles that of hypocalcemia (caused by reducing the level of Ca^{2+} or Mg^{2+} ions in the perfusing fluid), and they hypothesized that DDT interferes with surface recalcification which is needed to restore a normal resting potential after a single depolarization of the axonic membrane.

DDT has a strong affinity for cholesterol; this fact led several authors to postulate that it acts on cholesterol in the lipoid membrane and thereby reduces the membrane's permeability to Ca^{2+} ions (Laüger *et al.*, 1946; Welsh and Gordon, 1947). However, this hypothesis and the preceding one cannot unequivocally explain the mode of action of DDT: the first problem is that insect nervous systems do not produce the repetitive discharge in the absence of Ca^{2+} ions in the external medium. The second problem is that, as pointed out by Metcalf, there are a number of nontoxic DDT analogues which show high affinity for lipids, including cholesterol.

In addition to repetitive discharge, DDT-poisoned nerve cords show another characteristic response—"prolongation of afterpotential" (Yamasaki and Narahashi, 1952). When the negative afterpotential is increased to a certain level, a sudden burst of repetitive discharges can be provoked by a single stimulus. According to Narahashi and Yamasaki (1960), the period during which a single stimulus can produce a train of impulses (i.e., repetitive discharge) is relatively short, and finally the nerve reaches a point where it does not elicit multiple discharges. Although there is very little doubt as to the role of negative afterpotential in the direct process of producing the repetitive discharge in the insect nervous system, involvement of some other factors must be mentioned. The problem is that prolonged cathodal polarization, which should produce a similar effect as prolonged negative afterpotential (Fig. 4-6), does not induce repetitive discharge in normal axons. Therefore, it must be considered that DDT alters the axonic membrane in

Fig. 4-6. Single action potential recordings of the sensory neuron of *Periplaneta americana* before and after treatment with DDT. PAP, Positive afterpotential; NAP, negative afterpotential.

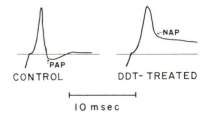

CONTROL   DDT- TREATED

10 msec

such a way that the membrane constants (e.g., conductance, resistance, and capacity) themselves become different from those in the normal membrane (Narahashi, 1963). By using intracellular microelectrode techniques, Yamasaki and Narahashi (1957) were successful in measuring the actual spike shape which is altered by the action of DDT. Comparing it with the spike shape produced by TEA (tetraethyl ammonium) treatment of cardiac muscle fibers and squid axons, and computing the relative ion permeabilities across the axonic membrane by the membrane potential equilibrium equation, Yamasaki and Narahashi concluded that the permeability of potassium ions across the nerve membranes must be greatly reduced by DDT. Studying the radioactive ion movement across the isolated nerve cord membrane, Matsumura and O'Brien (1966) demonstrated that nerve cords from DDT-poisoned *P. americana* discharged many more potassium ions than normal cords during a 20-min time interval without significantly altering the speed of uptake of potassium ions. The effect of DDT on the rate of sodium uptake and discharge was not so drastic as that on potassium.

Despite remarkable progress in the neurophysiology of DDT poisoning, studies on the biochemical aspects, namely the important questions as to how DDT can upset the normal functions of the nervous system (such as ion exchange across the axon membranes), are lagging far behind. In order to inhibit normal nerve function, DDT must first react with the nervous system in such a manner that the DDT molecule alters the nature of a nerve component(s) vital to normal nerve functions, or the DDT molecule, without altering any nerve component(s), must physically come in contact with the nerve membrane and thereby block the passage of important substances, such as ions.

Recently, it has been reported by two groups of researchers (Koch, 1969; Matsumura *et al.*, 1969; Matsumura and Patil, 1969) that DDT inhibits nerve ATPases. Although Mg-ATPase is affected by DDT, the extent of inhibition appears to be much more significant in $(Na^+ + K^+)$-ATPase, which is known to play an important role in the active transport of ions across the nerve membrane. The significance of this finding has not been critically examined at this stage. There are two major problems in attributing this biochemical effect of DDT to its mode of action: (1) DDT does not cause the same *in vivo* effects as ouabain, a well-known specific inhibitor of $(Na^+ + K^+)$-ATPase, and (2) there are many ATPases and ATP-involving systems present in any given nervous system. Indeed, Matsumura and Narahashi (1971) concluded that there is at least one other DDT-sensitive ATPase which behaves very much like $(Na^+ + K^+)$-ATPase does in the nerve cord of the American lobster (*Homarus americanus*). Subsequently, Doherty (1973) found an ATP-involving system that is highly sensitive to DDT and ionic environments in the axonic fraction of the same nerve material. Thus there are indications that one of the ATP-involving neural systems could be the DDT target. In the absence of the total elucidation of the mechanism of membrane excitation, the relationship between DDT poisoning and its action on these ATP-involving systems remains unconfirmed.

4.4.1c. Theories of Structure–Toxicity Relationships

The lack of evidence that DDT can react significantly with any well-defined enzyme system has led many researchers to suggest that the physical interference of DDT with membrane permeability is an important factor in DDT poisoning (see the excellent reviews by Kearns, 1956, and Martin, 1956).

Mullins (1954) proposed a theory that the activity of DDT comes from its physical shape, which is such that it can fit into the intermolecular spaces within the cylindrical lipoprotein lattice (Fig. 4-7). A foreign substance entering the interspace in the axonic membrane can temporarily hinder its permeability to ions and therefore have a narcotic effect. Permanent damage can be done if this substance is fixed in such a manner that some portion of it is firmly attached to the surrounding lipoprotein molecules and/or the orientation and the fixation of the substance bring some distortion of the interspace. The resulting effect will be ion leaks or blockage (depending on the type of distortion), which inevitably causes nerve disorders.

The orientation of the DDT molecule in Mullins' "interspace" is illustrated in Fig. 4-7. Here DDT enters the interspace from the trichlorocarbon group, which brings the attractive forces of the halogen atoms into a very favorable position to be effective. The di-(*p*-chlorophenyl) carbon group spreads enough to make a triangle so that the *p*-chlorophenyl groups constitute two stable "feet." In order to make this arrangement, the benzene rings should be able to rotate. Thus a compound like DDE, which possesses a double bond, cannot fit into the interspace (i.e., double bonds prevent rotation).

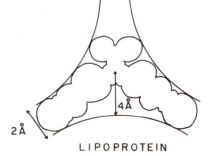

Fig. 4-7. "Intermolecular fit" of DDT. Each large circle indicates a lipoprotein molecule. From Mullins (1954).

2 Å

4 Å

LIPOPROTEIN

The concept of "fitting" in Mullins' hypothesis is a very interesting explanation of DDT action, though it is hardly appropriate to assume that such large lipoproteins (80 Å) and thus their interspaces exist in a membrane (100 Å). Holan (1969), for instance, has utilized this concept of fitting to explain the structure–activity relationships among cyclopropane analogues of DDT by proposing the fitting of DDT molecules in the Na and K channels.

From study of various DDT ring substituents, Riemschneider and Otto (1954) concluded that free rotation of the two phenyl groups and of the trichloro carbon group is the most important factor deciding the toxicity of DDT. In essence, in the 1,1-diphenyl-2,2,2-trichloroethane molecule (ring unsubstituted), free rotation of the two rings as well as of the trichloromethyl group is restricted only by the slight repulsion of the *o*-hydrogens of the phenyl groups, and *p,p*-substitution of chlorine atoms does not change the picture. However, if the *o*-position is occupied (e.g., *o,p'*-DDT) by a chlorine atom, rotation of the phenyl group becomes impossible. Indeed, *o,p'*-DDT has a relatively low insecticidal activity. The general order of insecticidal activity for various ring-substituted analogues is *p,p'-* > *m,p'-* > *o,p'-* > *m,o'-* > *o,o'*-DDT, Riemschneider and Otto claim that the order represents a complete parallelism with the freedom of rotation. The calculation of dipole moments (Wild, 1946) indicates that free rotation of a phenyl group in *o,p'*-DDT is impossible, despite the fact that the toxicity of this compound is not greatly below that of 1-(*p*-chlorophenyl)-1-phenyl-2,2,2-trichloroethane, in which free rotation is possible. Nevertheless, there is evidence indicating that, at least in several series of DDT analogues, free rotation of the phenyl group as well as of the trichloromethyl group is a very important factor in deciding toxicity (see also Martin, 1956). None of the ethylene-substituted analogues (DDE analogue I) show appreciable insecticidal activity though the toxicity can be restored by allowing free rotation between the 1-carbon and the 2-carbon (II):

(I)   DDE ANALOGUES   (II)

Brown and Rogers (1950) pointed out that in the DDT-type compounds the phenyl rings position themselves so that each ring will occupy two sides of a trihedral form. The tendency is more pronounced in a compound which has a carbon atom (e.g., the 2-atom-trichloro carbon) substituted with large atoms or bulky side-chains so that it hinders the rotation of ring groups.

Such conditions do exist in 1,1-*bis*(p-chlorophenol)-diphenyl-2,2-dimethyl-propane [i.e., CCl_3 replaced by $C(CH_3)_3$]:

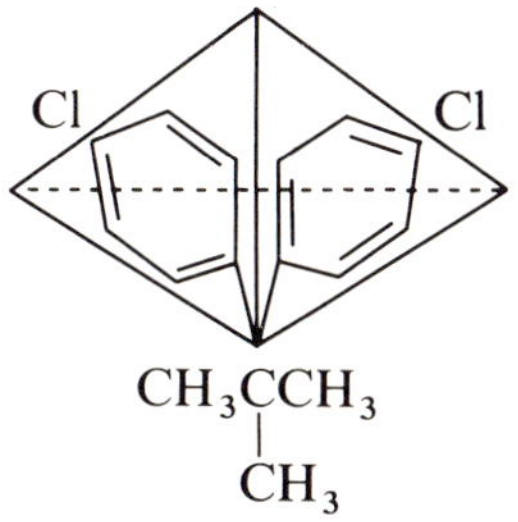

$$CH_3CCH_3$$
$$CH_3$$

Thus this hypothesis favorably explains the toxicity of a number of DDT-like compounds that have a bulky 2-carbon, e.g., 2-nitro-1,1-bis(p-chlorophenyl)-propane. However, the trouble with this hypothesis is that replacement of chlorine atoms with other substituents of equal or larger size does not always increase the toxicity of the resulting compound. For instance, replacement of chlorine atoms (atomic size 1.8 Å) with hydroxyl groups (1.7 Å) results in loss of activity. A similar replacement with methyl groups (2.0 Å) (i.e., $CClMe_2$ for CCl_3) also prevents insecticidal action against *Calendra granaria*, despite the fact that this compound fulfills the "trihedralization requirement" (Woodcock and Skerrett; see Woodcock, 1953). It is also surprising to note that a compound such as 1,1-bis(p-chlorophenyl)-2-nitro-propane, i.e.,

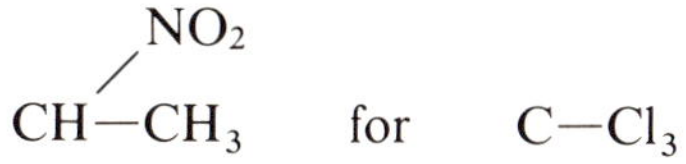

$$\overset{NO_2}{\underset{}{\diagup}}$$
$$CH{-}CH_3 \qquad \text{for} \qquad C{-}Cl_3$$

is highly insecticidal but 1,1-bis(p-chlorophenyl)-2,2-dichloro-2-nitroethane is not (Skerrett and Woodcock, 1952).

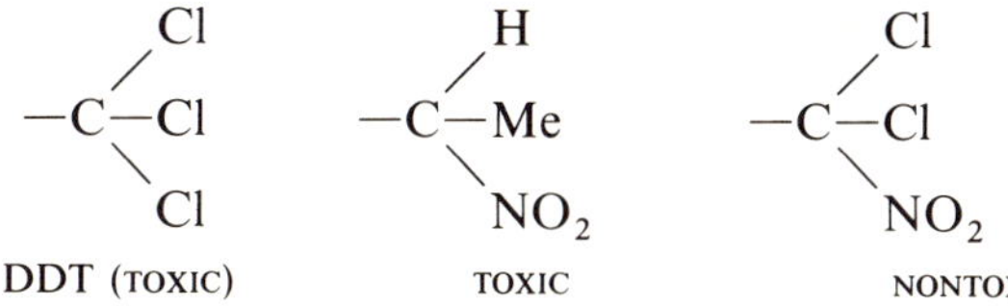

Cl	H	Cl
$-C-Cl$	$-C-Me$	$-C-Cl$
Cl	NO_2	NO_2
DDT (TOXIC)	TOXIC	NONTOXIC

Although at first sight these hypotheses (i.e., the "rotation" theory and the "trihedral configuration" theory) seem to conflict with each other, some reconciliation may be possible (Martin, 1956). If one can assume that the molecule must have a trihedral form to be active, the compounds which do not originally have such a rigid form should have the capacity for free rotation in order to assume the necessary spatial configuration at the site of action. Thus the requirement for DDT-like toxicity would be either a tri-hedral shape or the ability to form such a shape.

All the hypotheses discussed in this section have been concerned with the spatial requirements necessary for DDT analogues to have insecticidal potency. However, these might not be the only factors deciding the toxicity of DDT-like molecules. As mentioned before, replacement of chlorine atoms with hydroxy groups can result in a drastic reduction of toxicity because of increase in polarity, though the molecular shape may not be drastically altered. Obviously, lipid-solubility is one of the basic requirements for insecticidal activity. Such physicochemical and biochemical requirements must also be carefully studied in order to complete the picture of the "structure–toxicity relationship" of DDT analogues. Such approaches should be helpful in elucidating the mechanisms by which DDT-like molecules come in contact with the target and how they affect the nervous system.

Perkow (1956) noted that DDT, having five chlorine atoms, could assume a particular form of electron distribution: formation of two opposite electronegative ends (at the p,p'-chlorine positions) and one strong nucleophilic center (at the trichloro carbon). He concluded that such unbalanced electron distribution is necessary for the toxicity of DDT analogues (Fig. 4-8).

Gunther et al. (1954) discovered that the toxicity (in terms of $-\log LD_{50}$) of a series of DDT analogues was parallel with the sum of the logarithms of the Van der Waals attractive forces of the substituent groups. This theory, termed the "structural topography" theory, deals only with the primary action, i.e., the interaction of a DDT-like compound with (probably) some nerve proteins. If the shape and the size of the DDT-like molecules and the substituent groups control, for instance, molecular complementarity, Van der Waals attractive forces may govern the properties of "hold" or affinity of the compounds to the protein surface. The structural topography theory deals with very interesting aspects of molecular interactions, and it offers a new approach to the study of insecticide–nerve components interaction

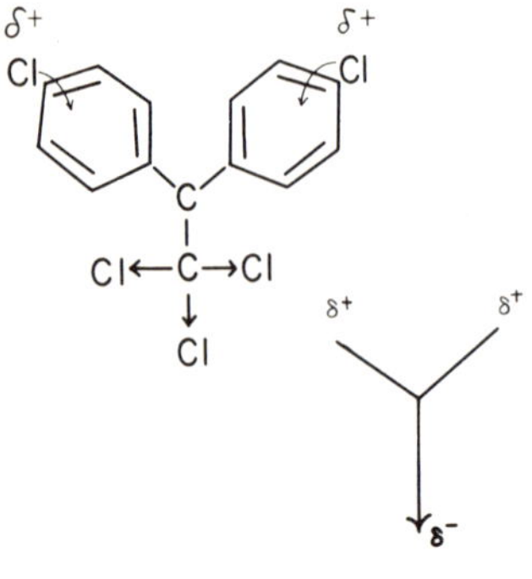

Fig. 4-8. Distribution of electrons within the DDT molecule.

problems. A question, however, remains as to the nature of the receptor protein and the biochemical implications of such a complex formation.

Laüger *et al.* (1944) have speculated that the effectiveness of DDT is due to the combination of the lipid-solubilizing ability of the trichloromethyl group and the toxic nature of the bis-(*p*-chlorophenyl)-methylene group. Laüger *et al.* (1946) also pointed out some similarities of the DDT molecule to the molecules of inhalation narcotics such as chloroform and methylene chloride. It is quite understandable that the liposolubility of DDT-like compounds can be one of the basic requirements for their toxicity, since the insect cuticle as well as the nervous system offers a strong penetration barrier against relatively polar substances. Nevertheless, this criterion cannot be the sole requirement for insecticidal activity, because there are many inactive compounds such as DDE which are highly liposoluble.

O'Brien and Matsumura (1964) have suggested that the process of DDT binding to the nerve components involves a reaction of "charge-transfer complex" formation and that this process induces the nervous disturbance, i.e., the symptoms of "nerve hyperexcitation." This hypothesis is based on three observations: (1) a number of chlorinated hydrocarbon insecticides are excellent electron acceptors as attested to by their extremely high sensitivity to the "electron-capture" device of the gas chromatographic detector, (2) many neurotropic and anticarcinogenic compounds (other than alkylating agents) are either good electron donors or acceptors, and (3) an artificial membrane composed of alternate layers of electron donors and acceptors forms a characteristic semipermeable membrane. A powerful electron acceptor or donor may easily upset this pattern of alternate layers and hence break up the normal function of the nerve membrane. Evidence in support of this hypothesis is that (1) DDT indeed forms tight complexes with nerve components, and (2) the formation makes the ultraviolet absorption spectrum shift. However, the theory has not been proven.

4.4.2. γ-BHC (Lindane)

4.4.2a. Symptomatology

The toxic principle of BHC is its γ-isomer, commonly known as lindane. Because of the specific toxicity of the γ-isomer as compared with other stereoisomers, the problem of the action mechanism of lindane has attracted the attention of many researchers and has led to many excellent research studies on structure–toxicity relationships.

Generally speaking, lindane is a more acute nerve poison than DDT. The symptoms of lindane poisoning in *P. americana* can be described as tremors, ataxia, convulsions, falling, and prostration (Savit *et al.*, 1946). Pasquier (1947) distinguished several phases of BHC poisoning in the desert locust (*Schistocerca*): prodromal phase (frequent abdomen-raising movements), typical phase (telescopic movements of the abdomen), choreoataxic phase (hyperexcitability, flying movements, and uncoordinated dance), clonic phase (falling on the back, tremors, etc.), and paralytic phase (paralysis). The tendency to cause a high rate of respiration in insects seems to be more pronounced in lindane than in DDT poisoning: 1 μg/roach of injected lindane produces oxygen consumption as high as that caused by DDT injection at 100 μg/roach in *Blattella germanica* (Harvey and Brown, 1951). Busvine (1954) observed that houseflies poisoned with lindane showed abnormal "fanning" movements with their wings, and this "fanning" symptom was not seen in DDT poisoning. The negative relationship between temperature and degree of poisoning is not as pronounced with lindane as with DDT (Guthrie, 1950).

Lindane appears to act as a stimulant to the mammalian central nervous system, while the β- and δ-isomers are depressant (McNamara and Krop, 1947). Applied to the leg of *P. americana*, lindane caused an increase in the nerve activity in the crural nerve after 1 hr. Repetitive spikes of two to four in a series appeared. They were definitely different from the repetitive discharges (trains) caused by DDT poisoning (Lalonde and Brown, 1954). The action of lindane seems to be more centralized than that of DDT, but like DDT it requires an intact reflex arc to produce a complete poisoning symptom (Bot, 1952). Summarizing briefly, lindane may be classified as belonging to a similar group of neurotoxicants as DDT though its mode of action is rather different from that of DDT. It is interesting to compare this conclusion with the observation that BHC-resistant insects are often not DDT-resistant (Busvine, 1954), although undoubtedly problems of different detoxification mechanisms for these two insecticides enter the picture. In general, however, there are more similarities between lindane poisoning and DDT poisoning than there are dissimilarities. For instance, Tobias *et al.* (1946) discovered that lindane causes an accumulation of acetylcholine in the ventral nerve cord of *P. americana* as DDT does. This may also be true in the mammalian nervous system, since lindane poisoning (like DDT poisoning) can be antagonized by atropine in mammals (McNamara and Krop, 1947).

4.4.2b. Theories of Structure–Toxicity Relationships

The most intriguing aspect of the chemistry of BHC is the large differences in biological activity found among the closely related isomers. This strongly suggests that a rigid spatial arrangement of the molecule is necessary for strong insecticidal activity. By using Stuart models, Mullins (1955)

TABLE 4-2. Molecular Configuration and Physiological Effects of the Isomers of BHC

Isomer	Melting point (°C)	Configuration[a]	Molecular diameters in plane of ring[b]		Molecular thickness		Physiological effect
β	297	*eeeee*	9.5	9.5	9.5	5.4	Inert or weak depressant
δ	130	*peeeee*	8.5	9.5	9.5	6.3	Strong depressant
α	157	*ppeee*	8.5	8.5	9.5	7.2	Weak excitant
γ	112	*pppee*	8.5	8.5	8.5	7.2	Strong excitant
ε	219	*peepee*	7.5	9.5	9.5	7.2	Not insecticidal
η	90	*peppee*	7.5	9.5	8.5	7.2	Not insecticidal
θ	124?	*pepeee*	8.5	9.5	8.5	6.3	Unknown

From Mullins (1955).

[a] *p* represents chlorine atoms distributed alternately above and below the plane of the ring; *e* represents chlorine atoms located approximately in the plane of the cyclohexane ring.
[b] Can be considered as Å units, insofar as the scale of the Stuart model is accurate. The three values for diameters in the plane of ring are at intervals of 60° each.

estimated the molecular diameters in the plane of the ring and discovered that only the γ-isomer has a value of less than 8.5 Å for all three maximum diameters (Table 4-2). According to Mullins, when lindane enters the interspace the membrane could be thrown out of equilibrium by the attractive force of chlorine atoms applied against the membrane constituents. Thereby it would become excited because of untimely ion leaks due to distortion of the lipoprotein molecules. (See 4.4.1*b* for the attractive force of chlorine atoms against the surrounding lipoprotein molecule.) As indicated before, Mullins' theory, though attractive, is very difficult to prove. Whether there is such an "interspace" existing in the nerve membrane is unknown. However, one cannot fail to note the strict relationship between the structural configuration and the toxicity of BHC isomers. The spatial as well as symmetrical (but not absolute symmetrical) requirements seem to be an absolute basis for lindane toxicity (see Fig. 4-9).

Practically nothing is known about the biochemical basis for insecticidal action of lindane. Nor is there any electrophysiological evidence to indicate the possible effect of lindane on the normal ion-transport mechanisms of the nervous system. There have been a number of attempts, very attractive by themselves, to account for the phenomena of lindane poisoning, but they have proven inadequate.

Slade (1945) was first to notice the similarity of lindane to B-vitamin *meso*-inositol. Kirkwood and Phillips (1946) presented the first evidence that lindane produces partially reversible inhibition of growth in a yeast strain which requires *meso*-inositol for growth. Although there have been a number of subsequent reports on the competitive properties of lindane against *meso*-inositol, unequivocal evidence contradicting this hypothesis has also appeared. Dresden and Krijgsman (1948) and Metcalf (1947) were unable to obtain any antidotal effects of *meso*-inositol against lindane poisoning. Moreover, it was later discovered that the structural resemblance of *meso*-inositol to lindane is rather superficial.

In brief, the mode of insecticidal action of lindane is far from clear: we simply know that lindane attacks the nervous system and that it requires a complete reflex arc to be effective. The structural requirement is rather strict, and the only toxic isomer is γ, which is bisymmetrical and compact. (In Table 4-2, none of the maximum plane diameters exceeds 8.5 Å as judged by the Stuart model.) Another hypothesis based on lindane's toxicity–structure relationships with other structural isomers has been proposed by Soloway (1965). According to him, the presence of two electronegative centers (Fig. 4-10) positioned evenly across the plane of symmetry (or plane that is closest to symmetry, if it is not absolutely symmetrical) is a very important factor in addition to the general molecular shape in deciding the insecticidal potency of all chlorinated cyclodiene compounds and lindane. Since all chlorine atoms are electrophilic (inductive effect), the center carbon atoms of the three equatorial and the opposing three axial chlorines should become electronegative. (In Fig. 4-10, they are indicated by carbons 2 and 5.) This tendency at carbon 5 is aided by the fact that two opposing equatorial

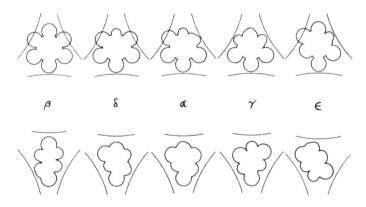

β δ α γ ϵ

Fig. 4-9. Stuart models showing the fit of various isomers of BHC into a membrane interspace for plane (upper row) and one end-on orientation. The interspace is the same model employed for the action of DDT (see Fig. 4-7). While all end-on orientations are possible, only the γ-isomer (lindane) can get into the interspace in a plane orientation.

ELECTRONEGATIVE CENTER

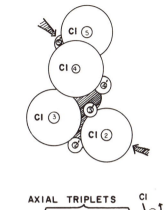

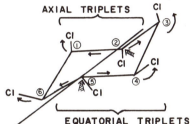

Fig. 4-10. Molecular models of lindane indicating the general molecular shape (upper Courtauld molecular model) and schematic positions (lower) of two electronegative centers. Redrawn from Soloway (1965).

chlorine atoms (i.e., carbons 6 and 4) are situated close to those two axial chlorine atoms (1 and 3) which take the *cis* position relative to chlorines 6 and 4, respectively (Fig. 4-11). Since lindane does not have any particular functional group, it would appear that physical interaction involving such electronegative centers takes place between lindane and some biological site in the nervous system that normally carries out an important physiological role. Soloway's hypothesis also involves the general topography and the shape of the lindane molecule. This portion of his theory can best be explained in the following section on the mode of action of cyclodiene insecticides. The mode of action of lindane is, however, known to be different from that of either DDT or cyclodiene insecticides, such as dieldrin or heptachlor.

4.4.3. Cyclodiene Insecticides

4.4.3a. Symptomatology

The cyclodiene insecticides include chlordane, heptachlor, aldrin, dieldrin, isodrin, endrin, and endosulfan. The characteristic chlorinated endomethylene bridge structure distinguishes this group of compounds from

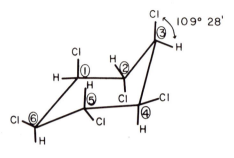

γ-isomer: Cl atoms are arranged in the
order of a a a e e e (① ② ③ ④ ⑤ ⑥)

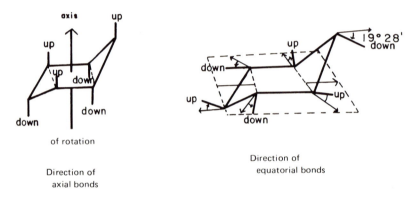

of rotation

Direction of
axial bonds

Direction of
equatorial bonds

Fig. 4-11. Schemes of the molecular structure of lindane and its bond orientations.

other chlorinated hydrocarbon insecticides. Symptomatologically, this group of compounds has a "positive temperature correlation" (i.e., toxicity is enhanced by high external temperatures) and therefore is considered to belong to an entirely different series of insecticides than DDT and the BHC analogues. On the whole, however, the superficial resemblances of dieldrin poisoning and that caused by DDT are striking. In *P. americana*, chlordane first acts as a depressant. At later stages of intoxication, the insect often responds to external stimulus with bursts of violent tremors. Electrophysiologically, the cyclodiene insecticides produced a profound excitation symptom in the crural nerve of *P. americana* (Lalonde and Brown, 1954). Trains of "repetitive discharges" were also observed for the nerve cords of roaches treated with aldrin and dieldrin. An important observation was that there were always time lags (latent periods) between application and the appearance of symptoms. The latent period was 2 hr for dieldrin and 4 hr for aldrin. Heptachlor, α-chlordane, and β-chlordane gave similar results, with latent periods of 3, 3, and 8 hr, respectively. Honeybees became highly agitated on contact with chlordane, and the state of uncoordinated actions occurred in 4 hr (Eckert, 1948). As mentioned previously, Busvine (1954) described the "fanning" movements of chlordane- and lindane-poisoned flies that distinguished these two compounds from DDT analogues.

The effect of cyclodiene compounds on the rate of respiration is much more pronounced than with DDT, but less so than with lindane (Harvey and Brown, 1951). Here again, the presence of the latent period for cyclodiene compounds is plainly observable (Fig. 4-12). The latent periods measured by Harvey and Brown for *B. germanica* injected with 10 μg each of cyclodiene compounds were 4–8 hr for α- and β-chlordane, approximately 1.5 hr for dieldrin, and 2.5 hr for aldrin. Heptachlor and toxaphene caused a very gradual increase in the rate of respiration after 15–45 min, reaching a peak around 1–1.5 hr. With all of these compounds, the resulting hyperactivity correlated well with the sudden increase in respiratory activity. Orser and Brown (1951) studied the heartbeat of chlordane- and toxaphene-treated *P. americana* and observed the irregularity of heart pulsation in both normal and decapitated insects. The heart stopped at the diastole stage with chlordane and at the systole stage with toxaphene.

Using cyclodiene-resistant and -susceptible housefly strains, Yamasaki and Narahashi (1958) discovered that the latent periods for nerve excitation

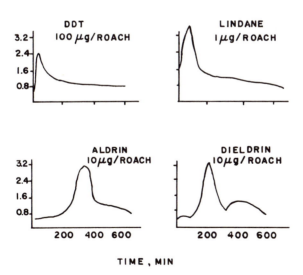

Fig. 4-12. Rate of oxygen consumption of *Blattella* injected with DDT, lindane, aldrin, and dieldrin. From Harvey and Brown (1951).

after the application of dieldrin were different: the resistant individuals had a latent period two to three times longer than the susceptible ones. This finding can be regarded as an important piece of evidence suggesting the primary target of dieldrin to be the central nervous system.

As for the site of action in the nervous system, Telford and Matsumura (1970) were able to determine that 50–60 % of all dieldrin entering the nervous system of *B. germanica* was bound to the axonic membrane. These workers also showed that the total amount of dieldrin picked up by the nerve cords of the dieldrin-resistant cockroaches was much less than that of their susceptible counterparts, confirming an earlier *in vitro* observation by Matsumura and Hayashi (1969). Wang and Matsumura (1970) studied the neurotoxicity-toxicity relationship among several cyclodiene insecticides and concluded that the latent *in vivo* toxicities correlated well with the latent neurotoxicities. Of these compounds, dieldrin showed an outstanding latency, while its metabolite, 6,7-aldrin-*trans*-diol was found to elicit much more intense neurotoxic symptoms in a much shorter time. In a more detailed study, Wang *et al.* (1971) also concluded that 6,7-aldrin-*trans*-diol is the compound most toxic to the nerve cord, particularly at the metathoracic ganglion of *P. americana*, among five derivatives of dieldrin tested.

Practically nothing is known about the biochemical mechanisms by which cyclodiene insecticides cause neurotoxicities. Certainly ATPases, particularly $(Na^+ + K^+)$-ATPases, have been found to be partially inhibited by various cyclodiene insecticides (Koch, 1969, 1970; Koch *et al.*, 1969; Chu and Cutkomp, 1971). The lack of similarities in symptoms of neurotoxicity between ouabain and these chlorinated hydrocarbon insecticides makes it difficult to explain their toxicities solely by their action on $(Na^+ + K^+)$-ATPases. Moreover, the relationships between *in vitro* inhibitory properties and *in vivo* toxicities are not always clear. For example, Chu and Cutkomp (1971) indicate that a $(Na^+ + K^+)$-ATPase from the coxal muscle of *P. americana* was 25 times more sensitive to γ-chlordane than another $(Na^+ + K^+)$-ATPase from the nerve cord.

As for the effects of cyclodiene insecticides on nerve function, electrophysiologists have not actually pinpointed the outcome of cyclodiene poisoning. Wang *et al.* (1971) indicated that sensory neurons in the cockroach leg were stimulated by direct application of dieldrin to produce trains of impulses, in essence confirming the findings of Lalonde and Brown (1954). Thus it is possible that the mechanism of dieldrin poisoning also involves eventual changes in ion permeability at axonic membrane levels. Hayashi and Matsumura (1967) studied the effects of dieldrin on the rate of ion flux in the cockroach nervous system and found that dieldrin affects the rate of exchange of Ca^{2+} ion. The overall effects of dieldrin on other ion-transport systems were otherwise similar to those of DDT.

4.4.3b. Theories of Structure–Toxicity Relationships

Although Martin (1946) suggested that a general relationship might exist between toxicity and the possible dehydrochlorination of chlorinated hydrocarbon insecticides, there seems to be no apparent correlation. Busvine observed that the end view of the hexachlorocyclopentadienes resembles that of the γ-isomer of hexachlorocyclohexane (lindane), and considered that the chlorine pentagon may represent the "toxaphore" (i.e., toxic molecular moiety) of this group of compounds. Unfortunately, there are a number of compounds which possess the chlorine pentagons and yet are nontoxic, while some other related compounds are toxic without the pentagon.

The structure–toxicity relationships of this group of compounds seem to be very complicated, and no definite rule other than trivial structural similarities of toxic compounds is available. One of the most intriguing phenomena is the difference in toxicity between the α- and β-isomers of chlordane, with α-chlordane being only one-tenth as toxic as β-chlordane (Soloway, 1965).

α-CHLORDANE   β-CHLORDANE

In addition, heptachlor is generally agreed to be more toxic than β-chlordane and enneachlor is approximately half as toxic as heptachlor (Cristol, 1950).

HEPTACHLOR   ENNEACHLOR

It is very tempting to consider that this order of toxicities is directly related to each compound's ability to form a toxic derivative of chlordane, i.e., oxychlordane. However, as will be shown in Chapter 5, it is actually easier to form oxychlordane from α-chlordane (*trans-*). Thus the reason for the high toxicity of the β-isomer must be sought in its structural characteristics or in other metabolic factors such as faster degradation of α-chlordane through the dehydrochlorination process.

Soloway (1965) wrote an interesting review summarizing the work on the structure–toxicity relationship of numerous cyclodiene insecticides. The essence of his theory, described in Section 4.4.2b, is that all active chlorinated cyclodiene insecticides should have two prominent "electronegative" centers positioned evenly across the plane of symmetry. A number of key compounds played important roles in the derivation of this conclusion. The first group of such compounds consists of α- and β-chlordane, where β-chlordane has both of the chlorines in the 1,2-positions *cis* while α-chlordane has the corresponding chlorines *trans*:

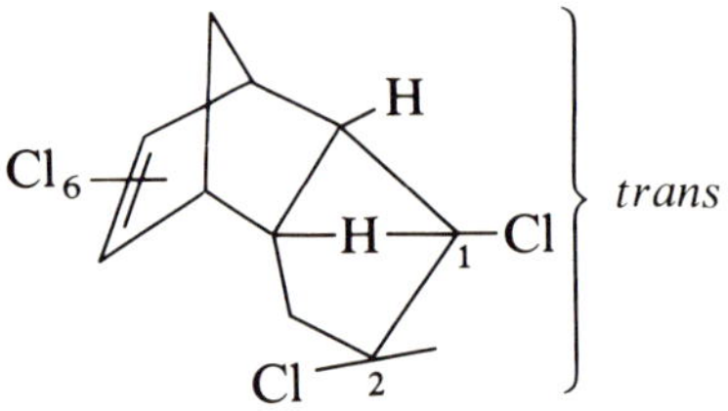

α-CHLORDANE
($\frac{1}{10}$ TOXIC AS β)

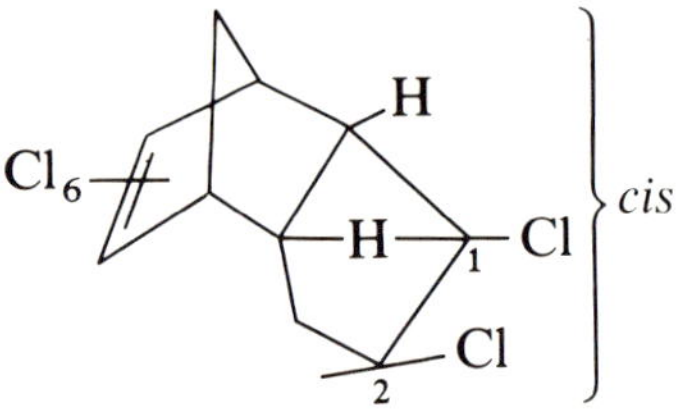

β-CHLORDANE
(TOXIC)

In β-chlordane, therefore, the effect of the chlorine atoms is doubled while in the α-isomer the effect is greatly reduced.

In another example, heptachlor epoxide, which is the actual form of heptachlor stored by animals (Davidow *et al.*, 1951; Davidow and Radomski, 1953), was compared with its stereoisomer (compound I):

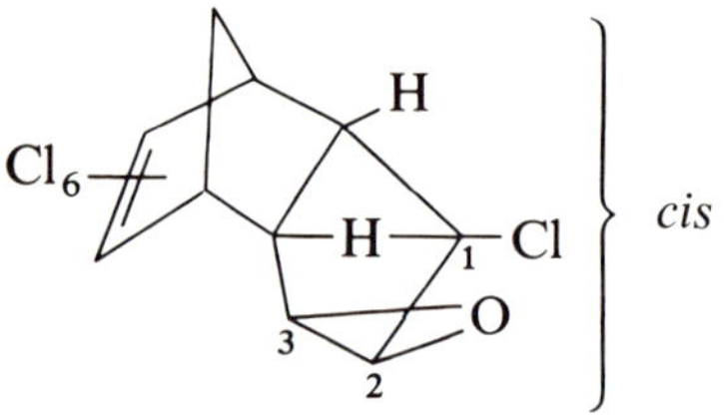

HEPTACHLOR
EPOXIDE
(TOXIC)

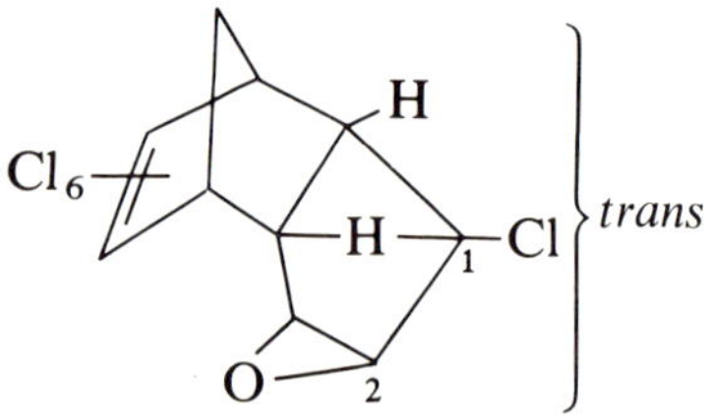

COMPOUND I
(NONTOXIC)

Heptachlor epoxide, stereochemically, is presumed to have both the epoxy oxygen and the chlorine in the 1-position *cis* to the angular hydrogens, while compound I has the epoxy oxygen *trans*.

A similar situation exists among the hexachlorodimethanonaphthalenes; among three dieldrin analogues, only compound II (*endo* epoxyisomer of dieldrin) is notably nontoxic:

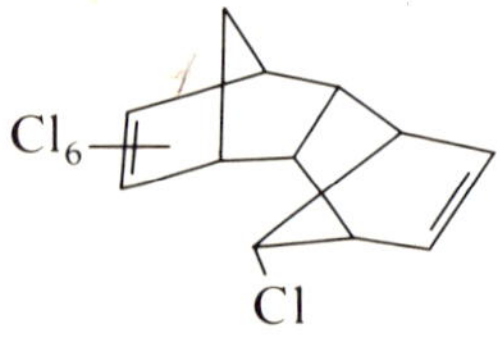

9a-CHLOROALDRIN
(TOXIC)

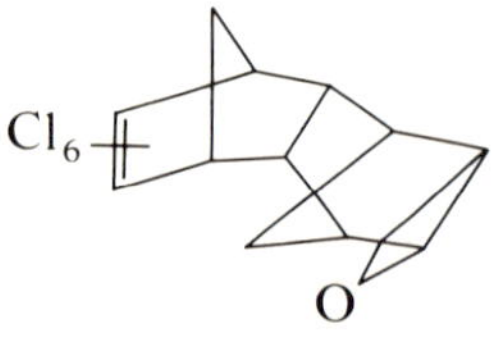

DIELDRIN
(TOXIC)

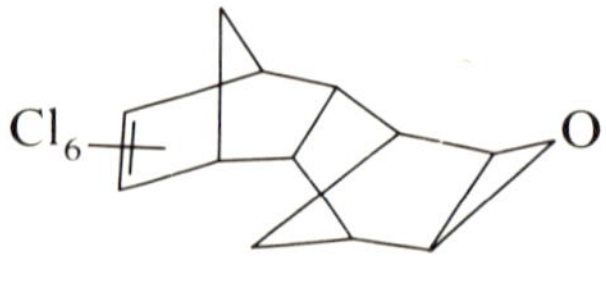

COMPOUND II
(NONTOXIC)

It can be seen that in these two highly insecticidal analogues (9a-chloro-aldrin and dieldrin) the positions of chlorine and oxygen, two electrophilic substituents, are the same. Obviously, these electronegative centers should not be too far away from the other electronegative center (e.g., the center of the hexachloronorbornene moiety), as is the case with compound II. This can be illustrated by drawing the molecules in a two-dimensional manner (the drawings are simplified here for easier understanding):

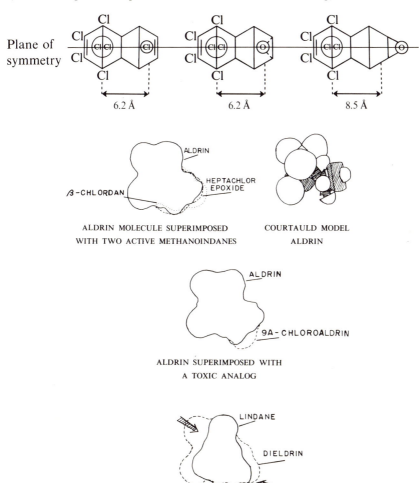

Fig. 4-13. Comparison of two-dimensional views of various chlorinated hydrocarbons. Note that two centers of electronegativity match well in all cases. From Soloway (1965).

One could draw similar two-dimensional views of those hexachloro-methanoindanes (aldrin analogues) in question (Fig. 4-13). Moreover, when three-dimensional models are positioned in such a manner that the hexa-chloronorbornene nucleus comes at the upper left corner (see Fig. 4-13) all toxic molecules have remarkably similar outlines. It is interesting to note that the molecular model of lindane can also be positioned so that two electronegative centers fit well into the same position as with other active cyclodienes.

These considerations of molecular topography vs. toxicity not only indicate the chemical requirements for toxic properties of these groups of chlorinated hydrocarbon insecticides but also are helpful to those who study the same problem from the biochemical end. Since all these compounds have two electron-rich sites positioned opposite each other along the line of symmetry, it can be postulated that they should fit into a particular bio-logical site in the nervous system in such a way as to block its normal physiological function. Physical interaction as a basis for the toxic action of these chlorinated hydrocarbon insecticides (in terms of binding or complex formation between the insecticide and a nerve component) is a likely possibility.

4.5. NATURALLY OCCURRING BOTANICAL INSECTICIDES

4.5.1. Pyrethroids and Synergists

Pyrethrum is a well-known insecticide because of its quick knockdown action or, more precisely, induction of temporary paralysis. This unique action does not necessarily represent the killing property: the amount required to produce mortality is usually much greater than that for paralysis. Usually the action of pyrethroids, paralytic or lethal, can be effectively augmented by various "synergists." The primary target of the pyrethrin group of insecticides is considered to be the nervous system, as judged by their quick action and the results of symptomatological studies. The symp-toms of pyrethrin poisoning follow the typical pattern of nerve poisoning: (1) excitation, (2) convulsions, (3) paralysis, and (4) death. The effects of pyrethrins on the insect nervous system closely resemble those of DDT (Welsh and Gordon, 1947; Yamasaki and Ishii, 1952), but are apparently much less persistent. Regular, rhythmic, and spontaneous nerve discharges have been observed in insect and crustacean nerve–muscle preparations poisoned with pyrethrins. The primary target of pyrethrins seems to be the ganglia of the insect central nervous system (Roy *et al.*, 1943; Krüger, 1931), although some pyrethrin-poisoning effect can be observed in isolated legs.

Pyrethrins injected into the abdomen or a spiracle produce progressive symptoms in the leg innervated by the ganglion nearest the point of injection. A leg which is severed, however, continues to contract or relax for about $\frac{1}{2}$ hr or more.

Electrophysiologically, pyrethrins cause repetitive discharges and conduction block (Welsh and Gordon, 1947; Yamasaki and Ishii, 1952). As with DDT, the negative afterpotential of a single spike is increased by the application of allethrin (a synthetic pyrethrin analogue) to the giant axon of *P. americana*, but unlike with DDT the appearance of repetitive discharges is not augmented by low-temperature treatments. Instead, there is a critical temperature below which no repetitive discharge is elicited (Narahashi, 1961, 1962). The large negative afterpotential probably is not caused by a low potassium or a high sodium concentration inside the axonic membrane, as attested to by the actual ineffectiveness of a high external concentration of K^+ ions or a low external concentration of Na^+ ions. Narahashi (1963) considers that the large negative afterpotential is caused by an accumulation of some unknown depolarizing substance inside or outside the membrane and that metabolic reactions are involved in this process of allethrin poisoning. In more recent work, Narahashi (1971) studied the effects of allethrin on ion permeability with the aid of a voltage-clamp technique. He found that (1) the peak amplitude of the sodium conductance is somewhat suppressed, (2) the sodium inactivation mechanism is greatly prolonged, and (3) potassium conductance is slightly suppressed. The net effect is suppression of the peak height of the action potential and increase and prolongation of the negative afterpotential.

The effects of temperature on allethrin poisoning are complex. The ability of allethrin to initiate repetitive afterdischarges increases as the temperature is increased. However, the nerve-blocking action, which most likely is responsible for the characteristic paralysis produced by pyrethroid insecticides, is enhanced at low temperatures (negative correlation) in both the cockroach (nerve cord) and squid (axon) nerves. Lowering the temperature apparently decreases the level of sodium conductance available to the axonic membrane.

4.5.2. Nicotinoids

Nicotine, in the form of tobacco extract, has been used as an insecticide since the mid-eighteenth century and still ranks as an excellent contact insecticide. As a whole, symptoms of poisoning by nicotine follow the general sequence of (1) excitation, (2) convulsion, (3) paralysis, and (4) death. Nicotine stimulates the rate of heatbeat at low concentrations and decreases it at high concentrations. The onset of symptoms from nicotine poisoning in honeybees is described as (1) period of inactivity, (2) sequence of paralysis beginning with the hind legs, then the wings and other legs, (3) the insect falls on its back, and (4) total paralysis (McIndoo, 1916). This sequence for nicotine poisoning seems to occur in a number of other insects, although in some cases a violent excitation period may precede paralysis (Raffy and Portier, 1931).

The most conspicuous feature of nicotine poisoning in insects is its selective inhibitory actions on nerve ganglia and synapses. Application of nicotine to the ganglion of an isolated leg reflex arc of *P. americana* produces violent tremors of the leg which disappear on disconnection of the leg from the ganglia; nicotine exerts no effect on the isolated leg (Yeager and Munson, 1942).

There is now a biochemical explanation of why and how nicotine blocks the synapses. Richards and Cutkomp (1945) found that nicotine has no *in vitro* inhibitory effect on cholinesterase activity. DFP and eserine, both well-established anticholinesterase agents, do not compete with nicotine, possibly because of the structural difference in the actual target sites in the synapses (Wiersma and Schallek, 1947). In mammalian tissues, nicotine is known to act as acetylcholine does at the neuromuscular junction and at the ganglion. It therefore causes stimulation of voluntary muscles and of the sympathetic innervated smooth muscles and glands. This action may be compared with that of muscarine, which acts as acetylcholine does, but only at the junction of the parasympathetic axon with the organ (cholinergic junction). The balance of evidence therefore seems to indicate that nicotine can attack some of the cholinergic junctions and all neuromuscular junctions of mammals by acetylcholine-like actions.

Yamamoto *et al.* (1962) studied the structure–toxicity relationship of nicotine-like compounds and concluded that

1. The essential skeleton for high toxicity is

2. The basicity of the compounds is very important; i.e., pK_a values for the nonpyridinyl nitrogen moiety (pK_{a1}) should be higher than 5.8, preferably 7.4–9.0 (see Table 4-3).
3. The minimum distance between the two nitrogens is about 4.2 Å, and this distance is in approximate agreement with that of the two functional groups of the acetylcholine molecule.

TABLE 4-3. Toxicity and Basicity of Nicotinoids

Compound	Relative basicity[a]		Relative toxicity[b]
	$pK_{a2'}$	$pK_{a1'}$	
Nornicotine (*l*)	3.3	9.0	1–2
5'-Methylnornicotine			
(*dl-cis, trans*)	3.4	8.9	0.94–1.00
Anabasin (*d, l*)	3.1	8.7	10
Nicotine (*l*)	3.1	7.9	1.00
Dihydronicotyrine (*d, l*)	2.9	7.4	1.27–3
Anabaseine	2.8	6.7	0.32–0.61
5'-Methylmyosmine (*d, l*)	2.5	5.6	0.10–0.47
Myosmine	2.8	5.5	0.03–0.27
Nicotirine	2.4	4.7	0.08–0.10
2-(3-Pyridyl)-2-oxazoline	2.5	3.0	0.01

From Yamamoto *et al.* (1962).
[a]$pK_{a2'}$ for pyridyl nitrogen and $pK_{a1'}$ for the other nitrogen.
[b]Test insects are *Aphis rumicis, Musca domestica, Periplaneta americana* and *Oncopeltus fasciatus.*

On this basis, Yamamoto *et al.* postulated that nicotine attacks acetylcholine receptors, which resemble cholinesterases (Fig. 4-14). Inhibition takes place mainly at the anionic site by the competitive binding of the high-basic nitrogen atom, which is normally ionized at physiological pH. At the same time, the pyridinyl nitrogen assists the action of the high-basic nitrogen by affecting the electronic status of the receptor protein.

There is no reason to believe that nicotine inhibits cholinesterase either *in vivo* or *in vitro*, though its acetylcholine-like activity has been documented.

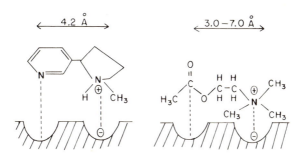

Ch-ACh receptor   Ch-ACh receptor

Fig. 4-14. Similarities between acetylcholine and nicotine binding to the acetylcholine receptor.

For instance, Roeder and Roeder (1939) observed that perfusion of an isolated nerve cord of *P. americana* with nicotine at 1×10^{-5} M caused an appreciable increase in nerve activity, while Richards and Cutkomp (1945) found that nicotine had no *in vitro* inhibitory effect on cholinesterase activity in bee brains. O'Connor *et al.* (1965) discovered that nicotine and succinylcholine are the two most active neuromuscular blocking agents in *P. americana*, while many other anticholinesterases are virtually ineffective. Since nicotine has no effect on isolated legs, these data should not be regarded as direct evidence for the presence of acetylcholine receptors in the insect neuromuscular junctions.

In conclusion, it appears that nicotine is very selective in choosing the binding sites in both insects and mammals. The balance of evidence indicates that nicotine selectively attacks the neuromuscular junction of mammals and the synaptic ganglion of insects (Flattum and Sternburg, 1970a) without severely attacking other known acetylcholine receptors or the acetylcholinesterase, which is known to be similar in protein structure to the acetylcholine receptors. However, nicotine can mimic the action of acetylcholine *in vivo* under the proper conditions. Therefore, it is likely that the receptors susceptible to nicotine cannot distinguish between acetylcholine and nicotine. In actual *in vivo* poisoning, nicotine is also known to cause the release of synaptically active material in the abdominal nerve cord of *P. americana* (Flattum and Sternburg, 1970b). Thus the cause of poisoning could be complex.

4.5.3. Rotenoids

Rotenone represents a curious example of a toxicant which is a metabolic inhibitor as well as a nerve poison. Its mode of action and resulting symptoms are distinctly different from those of any other compound which has been mentioned in this chapter. Essentially, rotenone is a slow-acting inhibitor and its depressing effect is more pronounced than its stimulating action. In some cases, the mouthparts of insects can be paralyzed so that they die of starvation. The poisoning of insects by rotenone is usually a slow process: (1) inactivity, locomotive instability, and refusal to eat; (2) knockdown; (3) paralysis; and (4) slow death. The rates of heartbeat and respiration are depressed (Harvey and Brown, 1951). In insects, the ultimate cause of death is most likely failure of respiratory functions since it is not uncommon to observe the heart beating even after all other biological signs of life are gone.

Rotenone is an excellent fish poison, acting through the gills. It is also a potent poison for mammals; the mammalian oral LD_{50} is of the order of

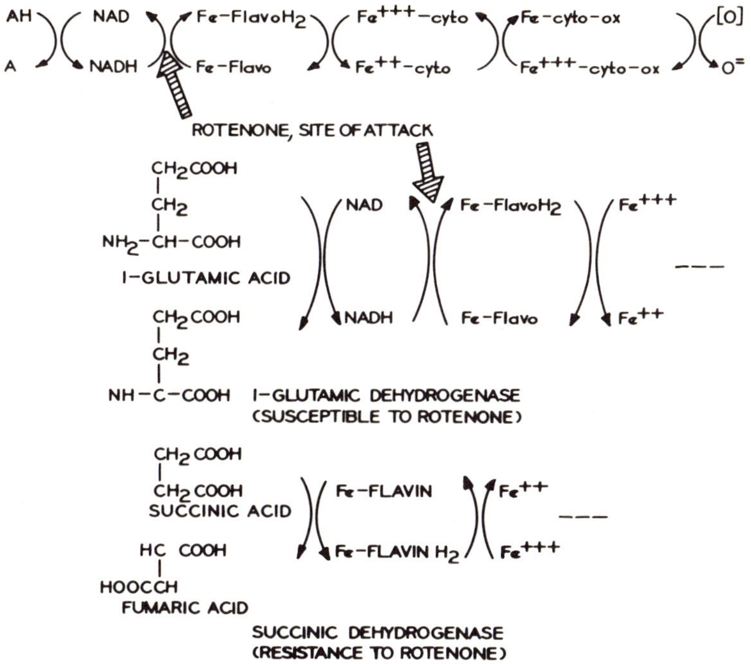

Fig. 4-15. Scheme of the action site of rotenone.

10–30 mg/kg (Lightbody and Mathews, 1936). Its paralytic effect is more pronounced in vertebrates than in insects.

It has been shown that the phenomena which accompany rotenone poisoning mainly derive from the ability of rotenone to inhibit respiratory metabolism (Fukami, 1956, 1961; Lindahl and Öberg, 1961). More precisely, rotenone owes its inhibitory potency to its ability to interfere with the electron-transport process between reduced diphosphopyridine nucleotide (DPNH or NADH) and cytochrome *b* (Fig. 4-15). Fukami *et al.* (1959) surveyed the structure–toxicity relationship of the rotenone derivatives and examined the relationship between inhibition of respiratory metabolism and block of nerve conduction. The results clearly suggested that inhibition of respiratory metabolism is one of the major causes of nerve conduction block.

4.6. ORGANOFLUORINE COMPOUNDS: FLUOROACETATE AND ITS ANALOGUES

Fluoroacetate is one of the few poisons for which the biochemical mode of action in vertebrates is precisely known (see Peters, 1957). Fluoroacetate

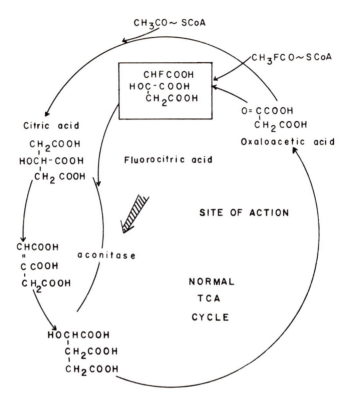

Fig. 4-16. Scheme of the mechanism of fluoroacetate inhibition of
the TCA cycle.

is a quick-acting poison, causing symptoms 20–60 min after administration
of a lethal dose to mammals. The LD_{50} for humans is estimated to be 5 mg/kg.
Repeated vomiting is the first symptom, followed by tonic convulsions
(Gajdusek and Luther, 1950). The cause of death appears to be heart failure.

Fluoroacetate is nontoxic to all enzymes tested *in vitro*. The toxicity
of fluoroacetate *in vivo* comes from the fact that it is converted to a potent
enzyme inhibitor, fluorocitric acid, in the living body just as ordinary pyruvic
acid is incorporated into the TCA cycle to form citric acid (Fig. 4-16).
Because of the structural similarity of fluorocitric acid to citric acid, the
former competes with the latter for the target enzyme, aconitase, and thereby
blocks the enzyme action. It is clear that the result of fluoroacetate poisoning
is the accumulation of citric acid in vertebrates and insects (Peters, 1957;
Miura *et al.*, 1961; Matsumura and O'Brien, 1963), but the cause of death
is not certain. Buildup of a large dose of citric acid at vital areas could
indirectly cause a state of hypocalcemia, in addition to the disruption of the

TCA cycle due to the inhibition of aconitase. All fluoroacetate derivatives tested so far owe their toxicity to the eventual metabolic formation of fluoroacetic acid in the animal body. Thus the basis for the selectivity of fluoroacetamide appears to be the ease with which insect systems hydrolyze the parent compound to yield fluoroacetic acid, in comparison to the mammalian carboxyamidase which distinguishes fluoroacetamide from acetamide and thus produces less fluoroacetic acid (Matsumura and O'Brien, 1963).

The same principle appears to apply to the more selective and recent compound, Nissol®, 2-fluoro-N-methyl-N-(1-naphthyl)acetamide (Noguchi *et al.*, 1968; Johannsen and Knowles, 1972), which also causes accumulation of citric acid in insects and mammals.

Sodium fluoroacetate is known to have good insecticidal action against mustard beetles, aphids, and cabbage white caterpillars, but because of its hazard it has not found commercial use as an insecticide. However, fluoroacetamide, a less hazardous analogue, is used as a commercial insecticide in Britain and Japan.

4.7. ANTICHOLINESTERASES: ORGANOPHOSPHORUS AND CARBAMATE INSECTICIDES

Anticholinesterases constitute a major portion of the modern synthethic insecticides. The list prepared for the Entomological Society of America (1963 revision, Kenaga, 1966) includes 103 organophosphorus and 13 carbamate insecticides versus 19 DDT analogues and 19 chlorinated aryl hydrocarbons representing the chlorinated hydrocarbon insecticides, and the number of valuable anticholinesterase insecticides is steadily increasing. A number of excellent books and review papers are available concerning the mode of action of organophosphorus and carbamate insecticides; therefore, only a brief description of the minimum basic principles will be given in this chapter. Readers who are interested in more detailed information should refer to the following sources:

J. E. Casida. "Mode of Action of Carbamates." *Ann. Rev. Entomol. 8:* 39–58, 1963.

D. F. Heath, *Organophosphorus Poisons.* New York: Pergamon Press, 1961. 403 pp.

R. D. O'Brien. *Toxic Phosphorus Esters.* New York: Academic Press, 1960. 434 pp.

R. D. O'Brien. "Organophosphates and Carbamates." *Metabolic Inhibitors 2:* 205–241, 1963.

4.7.1. Organophosphorus Compounds

4.7.1a. Symptoms

The usual symptoms of organophosphorus poisoning in mammals are defecation, urination, lacrimation, muscular twitching, and muscular weakness. In severe cases, prostration and clonic, sometimes tonic, convulsions follow. Light symptoms are usually parasympathetic in nature (e.g., bradycardia, salivation, and myosis), though a number of symptoms due to stimulation of organs innervated by complex systems (e.g., acidosis and hyperglycemia) are also observed.

The therapeutic methods used for cases of organophosphorus poisoning are:

1. Atropinization to block acetylcholine action at the parasympathetic nerve endings and in the central nervous system.
2. Curarization to block acetylcholine action at the neuromuscular junction.
3. Application of hexamethonium for ganglia protection.
4. Application of a reactivator of phosphorylated acetylcholinesterases (such as 2-PAM) to regenerate the enzyme.
5. Application of appropriate first-aid techniques—especially artificial respiration, which has reversed many otherwise fatal cases.

Symptoms in insects roughly follow the general pattern of nerve poisoning: (1) restlessness, (2) hyperexcitability, (3) tremors and convulsions, and (4) paralysis. The ultimate cause of death in insects is usually difficult to prove, with the exception of a few insect species which have a particularly vulnerable mode of life (e.g., respiration failure in mosquito larvae). The ability of organophosphates to poison insects is now regarded to be the result of their anticholinesterase activity, but actual proof is still needed. The strongest support for the anticholinesterase theory of organophosphate poisoning comes from analogy with mammalian poisoning and the overall adequateness of the theory. Mengle and Casida (1958) showed that for 12 out of 17 organophosphates tested, the time point for the maximum inhibition of cholinesterase in flies coincided with that for onset of death. However, Hopf and Taylor (1958) reported that the cholinesterase of locust nerve can be totally inhibited by an organophosphorus compound, $(C_2H_5O)(C_2H_5)$-$P(O)SCH_2CH_2N^+(C_2H_5)_3$, and yet the insect will not be affected. There seems to be little chance that inhibition of any other esterases, such as common aliesterases, can cause actual poisoning of insects (Stegwee, 1959). Many other supporting or contradicting data for the theory of anticholinesterase activity of organophosphates in insect poisoning could be described, but the balance of evidence, at present, strongly supports the view that cholinesterase inhibition in mammals is analogous to what happens in insects. The reader is referred to the references at the beginning of this chapter for detailed discussions of this matter.

Although the majority of toxic organophosphates are anticholinergic, at least one group of organophosphates, bicyclic phosphates, do not behave in this manner (Bellet and Casida, 1973) and yet they are highly toxic; e.g., one of them (Compound I below) is 33 times more toxic than parathion, LD_{50} being 0.18 mg/kg against mice. They propose that the toxicity of these bicyclic phosphates is due to their structural similarity to cyclic AMP.

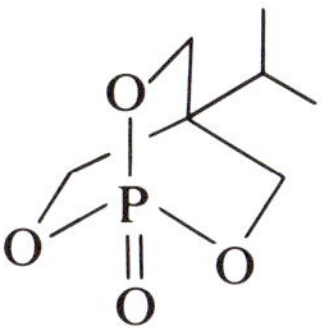

4-ISOPROPYL BICYCLIC PHOSPHATE (I) 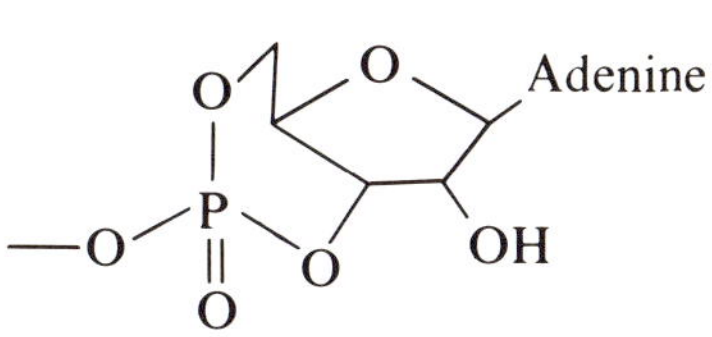   CYCLIC AMP (II)

4.7.1b. Inhibition

The most conspicuous feature of all organophosphorus (OP) compounds is their structural complementarity with the target enzyme molecule, cholinesterase. In essence, OP compounds mimic the gross molecular shape of the natural substrate of cholinesterase, acetylcholine. Cholinesterase is perhaps one of the most studied enzymes in biological systems. It is known to have two active sites, the esteratic site and the anionic site* (Wilson and Bergmann, 1950):

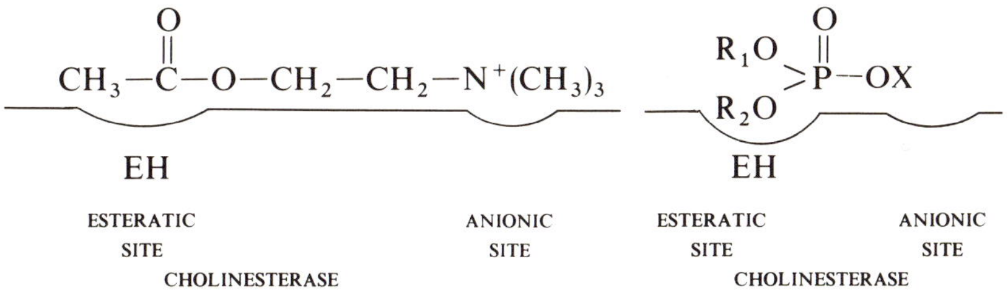

EH		EH	
ESTERATIC SITE	ANIONIC SITE	ESTERATIC SITE	ANIONIC SITE
CHOLINESTERASE		CHOLINESTERASE	

The process of OP inhibition of cholinesterase is essentially analogous to the early stage of acetylcholine hydrolysis. The first step is coulombic binding of the acyl carbon or phosphorus atom to the esteratic site and, in the case of acetylcholine, to that of the cationic nitrogen moiety. In some

* In the structures shown, E indicates enzyme and EH its hydrogen form.

cases, there can be an analogous cationic moiety which acts as the cationic nitrogen atom of the acetylcholine molecule in the molecule of OP compounds.

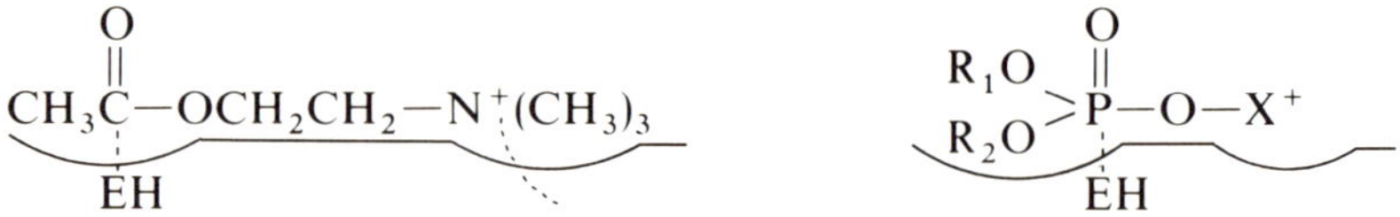

STEP 1

In step 2, a hydrogen atom is transferred from the enzyme (esteratic site) to the choline part of acetylcholine (or O—X part for the OP compound) by hydrogen bond formation and then by cleavage and rearrangement of the C—O—C or P—O—X part of the substrate. The processes are called acylation and phosphorylation, respectively.

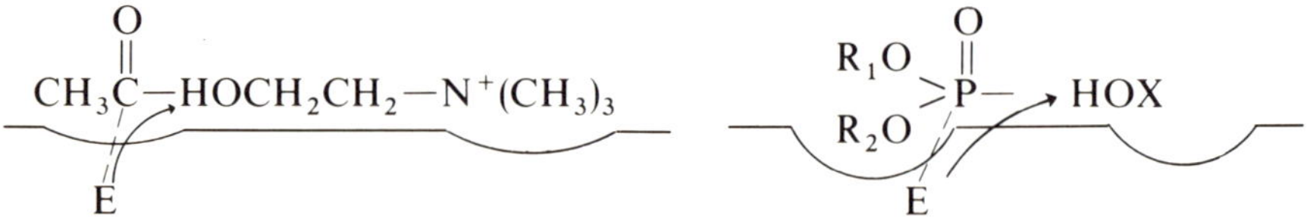

ACYLATION   PHOSPHORYLATION

STEP 2

In step 3, the acetylated or phosphorylated enzymes can undergo further rearrangement by the action of ambient water to form a hydroxylated complex.

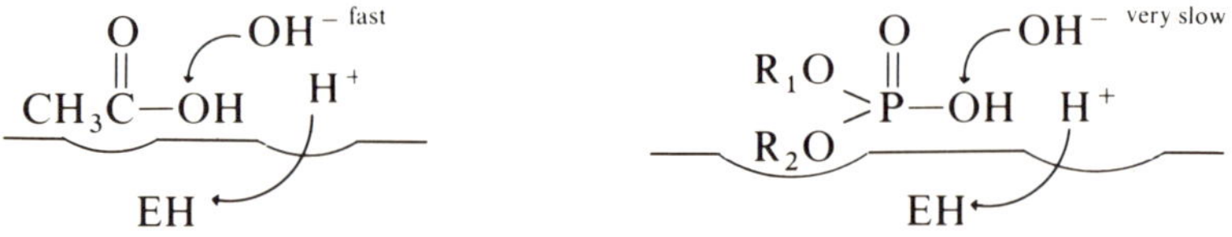

STEPS 3 AND 4

The last stage, step 4, is the only step where the analogy between acetylcholine and OP compounds almost fails. The process of deacylation from the enzyme (sometimes called enzyme recovery) occurs very rapidly, whereas dephosphorylation takes place at an extremely slow pace. It is this particular step that makes OP compounds powerful cholinesterase inhibitors.

The above processes can also be expressed in the forms of equations:

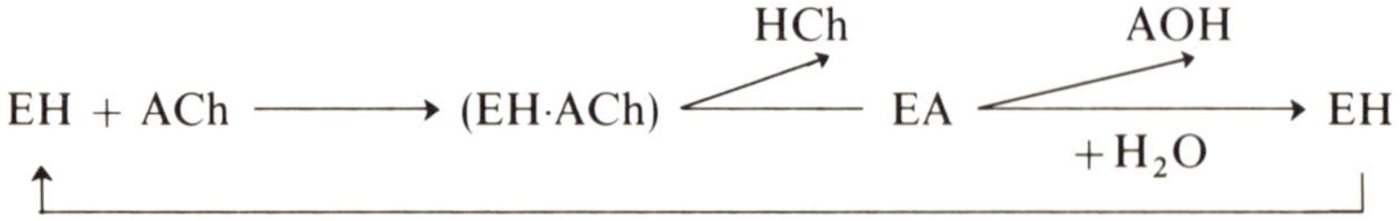

or

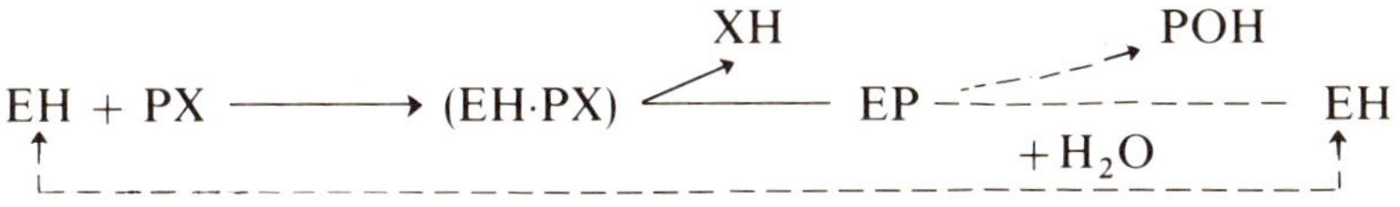

$$EH + PX \longrightarrow (EH \cdot PX) \xleftarrow{\quad XH \quad} EP \dashrightarrow \xrightarrow{\quad POH \quad}_{+H_2O} EH$$

Phosphorylated enzymes can be regarded as a kind of phosphate ester and can slowly be hydrolyzed by water (or other nucleophilic agents):

$$EP + H_2O \xrightarrow{\text{slow}} EH + POH$$

The reaction ultimately yields the original enzyme, and therefore is called reactivation. This reactivation process involves only the dialkylphosphate moiety of the organophosphate molecule, and therefore the rates of reaction vary with the basic alkyl groups on the phosphorus:

$$(MeO)_2 PO \cdot E > (EtO)_2 PO \cdot E > (iPrO)_2 PO \cdot E > (iPrNH)_2 PO \cdot E$$

The rate of cholinesterase inhibition, however, is very dependent on the first part of the reaction with the phosphate, where the initial affinity of the phosphate for the enzyme becomes very important.

$$EH + PX \underset{k_2}{\overset{k_1}{\rightleftharpoons}} (EHPX) \xrightarrow[k_3]{\quad XH \quad} EP$$

STEP 1   STEP 2

In step 1, the affinity can be expressed as

$$K_a = k_2/k_1$$

and it is controlled by the nature of the physiochemical properties of the phosphorus atom moiety. In order to inhibit the cholinesterase molecule, the phosphate molecule must make an electrophilic attack on the active site (possibly serine) of the enzyme. Therefore, those substituents which make P a better electrophilic reagent should improve its anticholinesterase activity (O'Brien, 1960):

$$(RO)_2 \overset{\overset{\displaystyle O_{\delta+}}{\|}}{P} - O \quad X$$

IMPROVED ANTICHOLINESTERASE

$$(RO)_2 \overset{\overset{\displaystyle O_{\delta-}}{\|}}{P} - O \quad Y$$

WORSENED ANTICHOLINESTERASE

Indeed, the degree of electrophilicity of the X substituents as exemplified by Hammett's σ constant (Hammett, 1940) for aromatic compounds closely follows that of anticholinesterase activity (Fig. 4-17). In usual cases of

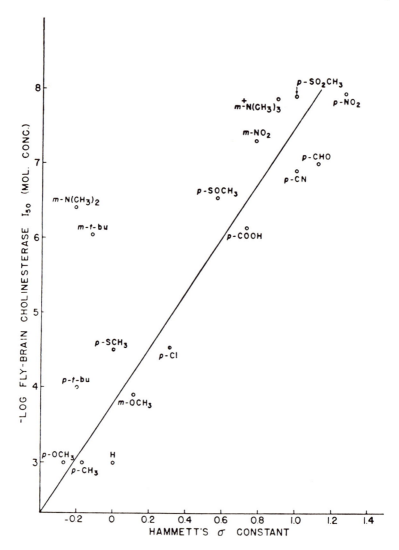

Fig. 4-17. Relationship between anticholinesterase activity and Hammet's σ constants for substituted diethylphenylphosphates. From Fukuto and Metcalf (1965).

organophosphate inhibition, the rate of complex formation (step 1) does not depend on k_3, and therefore the equilibrium of EH, PX, and EHPX is close to

$$\frac{[EH][PX]}{[EHPX]} = \frac{k_2}{k_1} = K_a \qquad \text{affinity constant} \qquad (1)$$

The rate of inhibition is, however, expressed as

$$\frac{d[EP]}{dt} = k_3[EHPX] = \frac{k_1 k_3}{k_2}[EH][PX]$$

or

$$\frac{d[EP]}{dt} = k_i[EH][PX] \qquad (2)$$

The original concentration of enzyme $[E_0H]$ is given by

$$[E_0H] = [EH] + [EHPX] + [EP]$$

Assuming that $[EH] \gg [EHPX]$ and that the reaction goes directly to form $[EP]$ (e.g., from enzyme and inhibitor $[EP]$ is formed directly; Aldridge's assumption),

$$[EH] = [E_0H] - [EP]$$

Therefore,

$$\frac{d[EH]}{dt} = -\frac{d[EP]}{dt} \qquad (3)$$

Substituting in equation (2),

$$-\frac{d[EH]}{dt} = k_i[EH][PX] \qquad (4)$$

Which is a typical bimolecular equation:

$$-\frac{1}{[EH]}d[EH] = k_i[PX]\,dt$$

By integrating $t = 0$ to t and the enzyme level a (the original level at $t = 0$) to $a - x$ (level reached after time t):

$$-\int_{a-x}^{a}\frac{1}{[EH]}d[EH] = k_i[PX]\int_{t}^{0}dt$$

Since,

$$\log_n\left(\frac{a}{a-x}\right) = k_i[PX]t$$

therefore,

$$k_i = \frac{1}{[PX]t}\log_n\left(\frac{a}{a-x}\right)$$

or

$$k_i = \frac{2.303}{ti} \log_{10}\left(\frac{a}{a-x}\right) \tag{5}$$

where a is the initial enzyme activity and x is the activity of the enzyme after incubation for t minutes. Here k_i is the bimolecular rate constant. The ratio $a/(a-x)$ could be rewritten as $100/P$, where P is the percent enzyme activity remaining at time t. Thus

$$k_i = \frac{2.303}{t[PX]} \log_{10}\left(\frac{100}{P}\right)$$

or

$$k_i = \frac{2.303}{t[PX]}(2 - \log_{10} P)$$

or by substituting i, a more popular term for the initial inhibitor concentration, for $[PX]$ and rearranging the equation,

$$\log_{10} P = 2 - \frac{k_i i}{2.303} t \tag{6}$$

Thus it is customary to record percent inhibition $(100 - P)$ of the enzyme at various time intervals and calculate the enzyme remaining uninhibited (P) and plot the values on a log scale vs. time on a linear scale, e.g.,

$$2 - \log_{10} P = \frac{k_i i}{2.303} t \tag{7}$$

Indeed, equation (7) agrees well with the experimental data, as shown by Aldridge and Davison (1952) (Fig. 4-18).

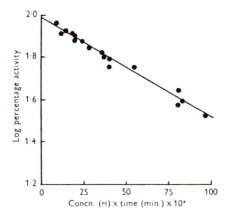

Fig. 4-18. Graph of all results for cholinesterase inhibition (i.e., at several inhibitor concentrations) by purified diethyl-*p*-chlorophenylphosphate. Bimolecular rate constant derived from slope of line is 1.1×10^2 ($M^{-1} \cdot min^{-1}$). From Aldridge and Davison (1952).

The bimolecular rate constant is one way of expressing the potency of organophosphate poisons, provided that pseudo-first-order kinetics are obeyed, but another convenient way is to express it by the inhibitor concentration which causes 50% inhibition of the target enzyme. The median inhibition concentration is known as I_{50}, and, as one can easily see, the lower the value the more potent the inhibitor is. The advantage of using I_{50} is that it can be expressed independently of the enzyme concentrations, which are, in usual biological preparations, difficult to estimate. There is a convenient equation to interconvert the "bimolecular constants" and I_{50} (O'Brien, 1960) where t is the incubation time: Also, the negative logarithm of I_{50},

$$k_i = 0.695/I_{50}t$$

the pI_{50} (like pHs for hydrogen concentrations) may prove to be quite useful (e.g., pI_{50} of 1×10^{-5} M is 5 and of 1×10^{-6} is 6). The above solution of the problem has one difficulty: the assumption that $[EH] \gg [EP]$ might not always lead to a correct answer. Main (1964) points out that for

$$\text{EH} + \text{PX} \underset{k_2}{\overset{k_1}{\rightleftharpoons}} (\text{EHPX}) \overset{k_3}{\longrightarrow} \text{EP}$$

$$e \qquad i \qquad\qquad r \qquad\qquad q$$

the relation of e and i vs. the first-step product r should be expressed by

$$k_1(e - r - q)i = k_2 r \tag{8}$$

and

$$r = \frac{(e - q)i}{i + K_a} \tag{9}$$

The rate of irreversible inhibition is expressed as

$$dq/dt = k_3 r \tag{10}$$

$$dq/dt = \frac{i}{(i + K_a)} k_3(e - q) \tag{11}$$

One can integrate the above equation between q_1 and q_2 and t_1 and t_2 and observe that $\log(e - q_1)(e - q_2) = \log(v_1/v_2)$ for the time interval, Δt, is

$$\frac{1}{i} = \frac{\Delta t}{2.303 \, \Delta \log v} \frac{k_3}{K_a} - \frac{1}{K_a} \tag{12}$$

since

$$k_3/K_a = k_i \tag{13}$$

$$\frac{1}{i} = \left(\frac{\Delta t}{2.303 \, \Delta \log v}\right) k_i - \frac{1}{K_a} \tag{14}$$

Therefore, when $1/i$ is plotted against $\Delta t/2.303 \, \Delta \log v$, a straight line is expected.

Moreover, the slope should be $k_i = (k_3/K_a)$, the intercept of the ordinate is expected to be $-1/K_a$, and the intercept of the abscissa should give $1/k_3$.

Main (1964) was indeed able to find such an enzyme–inhibitor relationship in malaoxon inhibition of human cholinesterase, where the phosphorylation step was slow enough to be a limiting factor (Fig. 4-19). Typical K_a and k_i values in the malaoxon case are

$$K_a = 7.7 \times 10^{-4} \, \text{M}$$

$$k_3 = 8.7 \, \text{min}^{-1}$$

$$k_i = 1.42 \times 10^5 \, \text{M}^{-1} \cdot \text{min}^{-1}$$

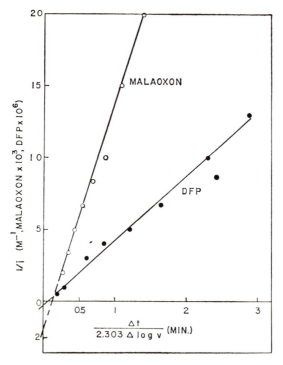

Fig. 4-19. Enzyme–inhibitor relationship in malaoxon.
From Main (1964).

whereas the same experiment with DFP gave the corresponding values of

$$K_a = 1 \times 10^{-5} \text{ M}_1 \quad \text{(high affinity)}$$

$$k_3 = 30 \text{ min}$$

$$k_i = 3 \times 10^6 \text{ M}^{-1} \cdot \text{min}^{-1}$$

In ordinary cases, however, Aldridge's assumption could well be utilized

$$\text{E} + \text{I} \longrightarrow \text{EI}$$
$$\quad e \quad i \qquad q$$

since in the case of a highly potent organophosphate it is not uncommon that K_a is much larger than the inhibitor concentration i. In place of equation (11), therefore, the simplified equation can be written as

$$dq/dt = k_i(e - q)i \tag{15}$$

and therefore

$$k_i = \frac{2.303 \, \Delta \log v}{\Delta t} \cdot \frac{1}{i} \tag{16}$$

Thus k_i can be expressed in exactly the same manner as in equation (5).

4.7.1c. Recovery

Previously it was stated that the phosphorylated enzyme recovers very slowly. However, the rate of recovery is relatively fast for dimethylphosphorylated cholinesterase (e.g., 50% recovery of a dimethylphosphorylated erythrocyte cholinesterase takes about 80 min, while that for the same enzyme diethylphosphorylated is of the order of 500 min; Aldridge and Davison, 1952). The reversal of isopropylphosphorylcholinesterase is almost negligible.

Reactivation is temperature dependent, as is the case with the other chemical reactions involving hydrolysis. The nature of the enzyme is also important. The rate of recovery for one type of enzyme (e.g., pseudocholinesterase) could vary from species to species (Davison, 1955) and from one organ to another even within the same organism (Davison, 1953).

The process of recovery can also be induced by artificial means. The reversal of inhibition involves hydrolysis of the phosphorylated enzyme. Thus it is natural that strong nucleophilic agents can accelerate this recovery process. Also, the addition of a quaternary (or tertiary) nitrogen group at a particular distance from the nucleophilic moiety to mimic acetylcholine greatly enhances the compound's ability to assist the hydrolysis process of the phosphorylated enzyme. Thus it is natural that strong nucleophilic agents can accelerate this recovery process. Also, the addition of a quaternary (or tertiary) nitrogen group at a particular distance from the nucleophilic moiety to mimic acetylcholine greatly enhances the compound's ability to assist the hydrolysis process of the phosphorylated enzyme:

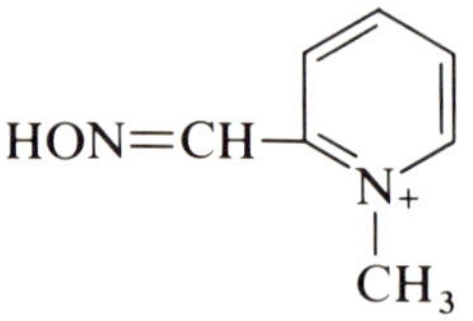

$$HON=CH-\underset{\underset{CH_3}{\underset{|}{N+}}}{\bigcirc} \qquad \underset{\underset{CH_2-CH_2-CH_2}{\underset{|}{\underset{N+}{\bigcirc}}}}{\overset{HON=CH}{\bigcirc}} \qquad \underset{\underset{N+}{\underset{|}{}}}{\overset{CH=NOH}{\bigcirc}}$$

PYRIDINE-2-ALDOXIMO-   1,3-BIS(N-PYRIDINIUM-4-
METHIODIDE   ALDOXIME) PROPANE
(2-PAM)   (4-PAM)

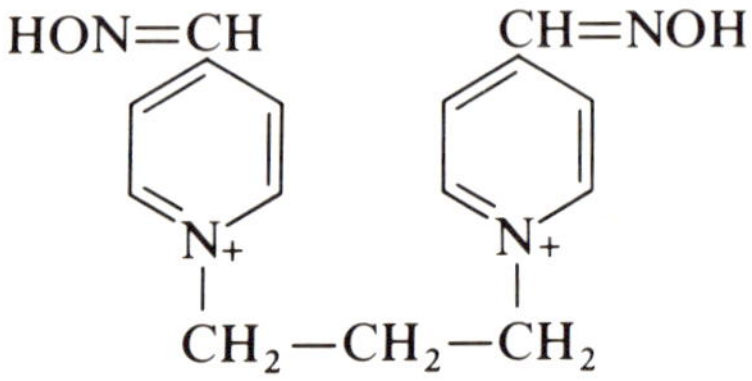

These compounds are extremely important from a therapeutic point of view. Their reactivities are extremely high: 80% recovery of inhibited enzyme at 10^{-5} M may take less than a minute. The reaction is also temperature and pH dependent.

4.7.1d. Aging

One of the characteristics of organophosphate inhibition of cholin-esterases is that the rate of recovery (induced by reactivating agents) becomes less and less as the time of the inhibitor–enzyme contact becomes longer (Hobbiger, 1955). The phenomenon is often referred to as "aging." This process is also temperature and pH dependent, being markedly accelerated at lower pHs and/or higher temperatures. The most likely explanation of aging is that phosphorylated cholinesterase can change its nature as time passes. Berends *et al.* (1959) suggested the following change for the DFP-inhibited pseudocholinesterase:

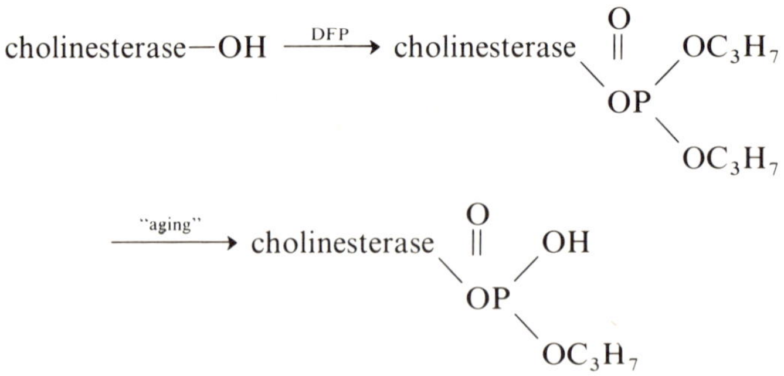

$$cholinesterase-OH \xrightarrow{\text{DFP}} cholinesterase \underset{OP}{\overset{O}{\underset{\diagdown}{\overset{\parallel}{\diagup}}}} \begin{array}{c} OC_3H_7 \\ \diagup \\ \diagdown \\ OC_3H_7 \end{array}$$

$$\xrightarrow{\text{"aging"}} cholinesterase \underset{OP}{\overset{O}{\underset{\diagdown}{\overset{\parallel}{\diagup}}}} \begin{array}{c} OH \\ \diagup \\ \diagdown \\ OC_3H_7 \end{array}$$

4.7.2. Carbamates

4.7.2a. Action

The basic difference between medical and insecticidal carbamates is that the former compounds always possess either a quaternary or a basic nitrogen group which can attack the anionic site of cholinesterase, whereas the latter compounds are never basic (i.e., ionization reduces the ability of the compounds to penetrate the insect cuticle as well as the nerve sheath). Carbamates are as potent cholinesterase inhibitors as organophosphates are, and they behave in an almost identical manner in biological systems, with certain differences which will be discussed below. A number of excellent reviews on medically useful carbamates (e.g., Stempel and Aeschlimann, 1956) and on insecticidal carbamates (Gysin, 1954; Casida, 1963; O'Brien, 1963) are available for reference.

Certain carbamates (e.g., eserine and prostigmine) are very selective inhibitors of cholinesterase and not of aliesterase. At the discriminating concentration, 1×10^{-6} M, eserine can inhibit cholinesterase (both pseudo- and acetylcholinesterase) with little or no effect on aliesterase (Myers and Mendel, 1949). Yet some carbamates are potent inhibitors of aliesterase. The selectivity of carbamates is sometimes more pronounced against the cholinesterases of different species. For instance, cholinesterases from the frog brain and from *Planaria* (a flatworm) are relatively insensitive toward eserine (Hawkins and Mendel, 1946). Even among human enzymes, Nu-1250

Nu-1250

at 1×10^{-6} M inhibits 96% of the erythrocyte cholinesterase, while it inhibits only 16% of the serum cholinesterase (Hawkins and Mendel, 1949). The carbamates do not inhibit or carbamylate chymotrypsin, which is readily phosphorylated by the organophosphates (Casida *et al.*, 1960).

The second important difference between the action mechanism of carbamates and that of organophosphates is that the primary mode of cholinesterase inhibition by carbamates is apparently reversible. There are two factors contributing to this reversibility: first, the step 1 reaction is reversible, and, second, the step 3 reaction is easy for carbamates, yielding the original enzyme which results in the appearance of recovery. The first factor can be attested to by the phenomenon

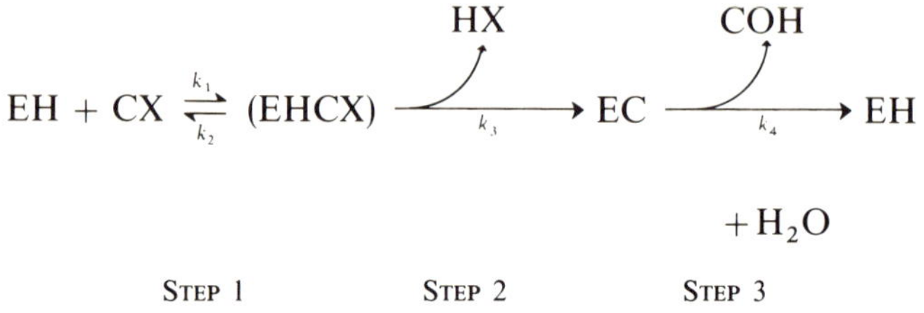

$$EH + CX \underset{k_2}{\overset{k_1}{\rightleftharpoons}} (EHCX) \xrightarrow[k_3]{\text{HX}} EC \xrightarrow[k_4]{\text{COH}} EH + H_2O$$

STEP 1   STEP 2   STEP 3

that the step 1 reaction is reversed by the addition of acetylcholine at high (e.g., 10^{-3} to 10^{-4} M) concentrations. At any given time, carbamate inhibition therefore proceeds on a steady-state basis: both EHCX and EC represent the inhibited enzyme. The overall reaction progresses steadily so that eventually all the inhibitor can be hydrolyzed. Thus the steady state lasts only for a limited time period. The decarbamylation process (step 3) can be speeded up by the addition of hydroxylamine to the inhibited cholinesterase, as is the case with phosphorylated cholinesterases, but unlike the latter cases, 2-PAM does not show recovery activity on the carbamylated enzymes. The decarbamylation process is entirely dependent on the N-alkyl substituents, and the order of lability is known to be $-C(O)CH_3 \gg -C(O)NH_2 > -C(O)NHCH_3 > -C(O)N(CH_3)_2 > -P(O)(OR)_2$. The magnitude of the decarbamylation constant k_4 in min^{-1} is on the order of 4×10^{-1} for $-C(O)NH_2$, 1.8×10^{-2} for $-C(O)NHCH_3$, and 2.6×10^{-2} for $-C(O)N-(CH_3)_2$ (see Table 4-4).

Table 4-4. Various Reaction Constants^a for Three Representative Carbamates

	k_i ($M^{-1} \cdot min^{-1}$)	K_a (M)	k_3 (min^{-1})	k_4 (min^{-1})
Methylcarbamate				
Phenyl^b	5.4×10^2	2.9×10^{-3}	1.6	1.8×10^{-2}
Naphthyl (carbaryl)	1.3×10^5	1.1×10^{-5}	1.3	1.8×10^{-2}
Dimethylcarbamate				
1-Isopropyl-3-methyl 5-Pyrozolyl (Isolan®)	1.6×10^5	8.0×10^{-6}	1.3	2.6×10^{-2}

From O'Brien (1967).

^{a}Note that low K_a means a good inhibitor and high K_a a bad inhibitor.
^{b}Nontoxic carbamate.

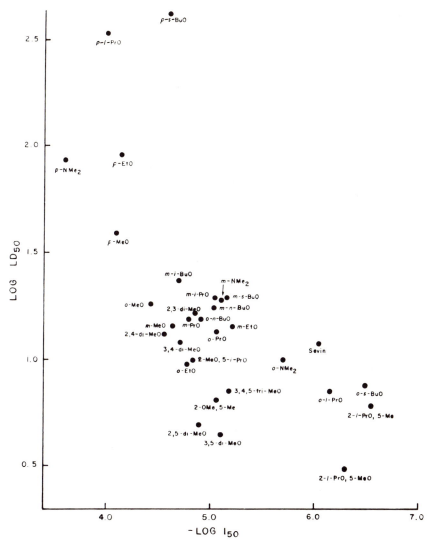

Fig. 4-20. Plot of $-\log I_{50}$ for fly-head cholinesterase inhibition vs. $\log LD_{50}$ for houseflies for carbamates synergized with piperonylbutoxide. From Metcalf and Fukuto (1965).

An exhaustive study made by Metcalf and Fukuto (1965) on the structure–toxicity relationship among various phenyl *N*-methylcarbamates indicated that (1) the inhibition constant $(-\log I_{50})$ for fly-head cholinesterase is roughly related to the compound's *in vivo* toxicity when synergists are used (Fig. 4-20); (2) there is a second molecular center that acts as a

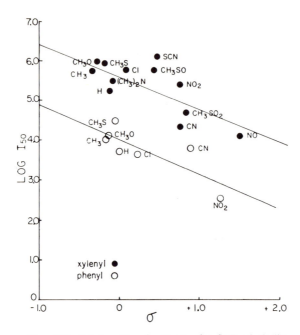

Fig. 4-21. Relationship of $-\log I_{50}$ for fly-head cholinesterase inhibition to σ values for p-substituted phenyl and xylenyl N-methylcarbamates. From Metcalf and Fukuto (1965).

supplementary fit to cholinesterase at the anionic site, and, moreover, the optimum distance between C=O and the second center is 5 Å in magnitude; (3) unlike organophosphates, addition of electron-accepting substituents to the phenyl ring results in a reduction of inhibitory potency, and, moreover, addition of electron-donating substituents enhances its inhibitory power (Figs. 4-20 and 4-21). According to O'Brien (1967), the initial step of complex formation (step 1) is favored by nucleophilic substitution, while step 2 is enhanced by electrophilic substitution. In carbamate inhibition, the important process appears to be step 1. The data shown in Table 4-4 clearly indicate that the K_a (i.e., affinity) of the carbamate is the most important factor in determining its toxicity. This "molecular-fitting" consideration of Metcalf and Fukuto (1965) also indicates the importance of the initial reaction, step 1, for carbamates. The fact that electrophilic substitutions lessen carbamate potency simply indicates the importance of the precise electrophilic nature of the essential moiety of carbamates in achieving the step 1 reaction, which is the actual limiting factor for carbamates but not for organophosphates.

4.7.2b. Symptoms

Carbamates are powerful anticholinesterase agents, and act in mammals as parasympathomimetic drugs. Their action appears to be the result of local accumulation of acetylcholine. Clinically they are used as (1) drugs for topical administration as in miotic glaucoma, (2) stimulants to intestinal peristalsis, (3) drugs for treatment of urinary retention and for myasthenia gravis and muscle spasms. Unlike organophosphates, carbamates do not seem to cause delayed ataxia and demyelination of nerves in man and chickens, possibly because the process of decarbamylation takes place with relative ease and thus has little after effect on the nervous system. It has never been reported that any of the carbamate insecticides have carcinogenic action; anxiety has arisen from the fact that many analogues of C_2H_5OC-$(O)NH_2$ show such activity.

The symptoms in insects are primarily those of poisoning of the central nervous system (which is understandable, since the insect neuromuscular junction is not cholinergic). The action of the carbamate may be blocked by application of nicotine, atropine, or barbituric acid to the ganglion (Tobias *et al.*, 1946; Wiesmann and Kocher, 1951). The rate of beating of the isolated heart of *P. americana* was markedly accelerated by *o*-, *m*-, and *p*-isopropylphenyl *N*-methylcarbamate at a threshold concentration of 5×10^{-7} M (Metcalf and Fukuto, 1965). The same preparation had already been shown to be sensitive to acetylcholine (at 10^{-9} M) and to organophosphates such as DDVP and paraoxon (at 1×10^{-7} M). The latter compounds caused only a change in the rate of beating, whereas administration of carbamates resulted in an erratic beat with incomplete systolic and diastolic movement, leading often to complete arrest (Metcalf *et al.*, 1964). The results could be interpreted to mean that these carbamates have a direct action on the acetylcholine receptors as a result of their well-recognized inhibitory action on cholinesterase.

4.8. INHIBITORS OF RESPIRATORY ENZYMES

4.8.1. Inhibitors of the Electron Transport System

There are a number of insecticidal compounds which affect the respiratory system of animals. The case of rotenone, a specific inhibitor, has already been described, and while not many insecticidal compounds can match rotenone in their specificity, a considerable number of insecticidal compounds can be called inhibitors of respiratory enzymes. An insecticidal antibiotic, piericidin A, for instance, is known to inhibit the process of electron transport between NADH dehydrogenase and ubiquinone (Tamura and Takahashi, 1971) just as rotenone does (Fukami, 1956, 1961). The mechanisms of inhibition in terms of specific binding of these compounds to electron transport particles have been extensively studied. The site appears to be a protein–lipid complex and a portion of NADH dehydrogenase, which must play an important role in the process of electron transfer. While conformational requirements for the inhibitory property appear to be strict, the binding does not involve a rigid bonding, inasmuch as the inhibitors can be reextracted in their original forms by the use of organic solvents.

4.8.2. Inhibitors of Oxidative Phosphorylation

Many acaricides and insecticides are known to affect the process of oxidative phosphorylation. Williamson and Metcalf (1967) studied several salicylanilide derivatives and found that they are excellent uncouplers. Two compounds, 5-Cl-3-(p-Cl-phenyl)-2′-Cl-5′-CF$_3$-salicylanilide and 5-Cl-3-t-butyl-2′-Cl-4′-NO$_2$-salicylanilide (I), had pI_{50} values (for Pi-ATP exchange) of 9.08 and 9.14 (i.e., in the range of 10^{-9} M) against a housefly mitochondrial system. They are claimed to be the highest recorded pI_{50} values.

5-Cl-3-t-BUTYL-2′-Cl-4′-NO$_2$-SALICYLANILIDE (I)

In terms of their effects on housefly mitochondrial ATPase activities, these active analogues were 1000–10,000 times more potent (i.e., in their preparation ATPase activity caused stimulation that was increased approximately fourfold by 10^{-4} M dinitrophenol) than dinitrophenol.

Ilivicky and Casida (1969) studied the similar effects of substituted 2,4-dinitrophenols, 2-trifluoromethylbenzimidazoles, salicylanilides, carbonyl cyanide phenylhydrazones, and other compounds. They found that, aside from compound I above, carbonylcyanide p-trifluoromethoxyphenylhydrazone and three other compounds are excellent uncouplers, particularly against insect and mouse-brain mitochondrial systems.

4.9. INHIBITORS OF MIXED-FUNCTION OXIDASES

Inhibitors of mixed-function oxidases have been developed as synergists for pyrethrin insecticides, and generally are methylenedioxyphenyl (or benzodioxole) derivatives. The characteristics of these synergists in relation to the microsomal mixed-function oxidases (MFO) will be described in Chapter 5. While it is certain that these methylenedioxyphenyl compounds (MDP) attack MFO, the mechanisms causing such phenomena are not clear. In some instances, MDP compounds appear to act as alternative substrates (competitive inhibitors), while in other cases the relationship appears to be complex, i.e., interaction with the MFO system through allosteric modification, acylation or other types of reactions to change the nature of the MFO (Wilkinson, 1971; Casida, 1969, 1970). Since in all cases studied such MDP–MFO interactions involve some changes in P_{450} spectra (Matthews *et al.*, 1970; Philpot and Hodgson, 1972), the site of action appears to be at least near the cytochrome P_{450} portion of the MFO system. The complicating matter is that such interactions generally involve more than one type of spectral change, and these are also influenced by the conditions of the reaction (e.g., by the presence and absence of oxygen, and by the methods of reduction of P_{450}).

4.10. CHLORDIMEFORM AND ITS ANALOGUES

Chlordimeform (Galecron®) was mainly developed as an acaricide but later was found to have excellent insecticidal activity, particularly against some lepidopterous insects in the field. Studies in our laboratory (Beeman and Matsumura, 1973) indicated that chlordimeform causes some sympathomimetic symptoms, notably dilation of the pupil in the rat. Its effects can be classified into two distinct types: initial hyperexcitation including jumping at high doses (greater than 100–200 mg/kg), and subsequent tranquilization generally involving low posturing due to spreading of the lower limbs. (At low doses, this latter phase occurs at the beginning.) The poisoned animal generally remains motionless in an open space and shows locomotive difficulty. It can react to sudden external stimuli, but indicates a lack of self-stimulation. As a result of *in vivo* studies, it became apparent that chlordimeform induced accumulation in the rat brain of serotonin (5-hydroxytryptamine) and norepinephrine, two important biogenic amines. *In vitro* studies on rat liver monoamine oxidase revealed that chlordimeform analogues are monoamine oxidase inhibitors. This *in vitro* finding has been confirmed by Aziz and Knowles (1973). Both groups indicated that the I_{50} values for the monoamine oxidase inhibition by active chlordimeform analogues were of the order of 10^{-5} M.

Abo-Khatwa and Hollingworth (1972), on the other hand, found that chlordimeform is an inhibitor of oxidative phosphorylation and that it also stimulates mitochondrial ATPase activities. The concentrations found to inhibit 50% of the oxidative phosphorylation activities of mitochondria from *B. germanica* were of the order of 10^{-1} to 10^{-2} M. These workers found, however, that chlordimeform could stimulate mitochondrial ATPase at 10^{-5} M.

The role of this uncoupling activity of chlordimeform in the total process of chlordimeform poisoning is not certain. As the above workers have pointed out, the symptoms of chlordimeform poisoning at LD_{50} levels are very different from those for DNP, a well-acknowledged uncoupling agent. The overall symptoms of chlordimeform poisoning in *B. germanica* indicate extensive involvement of the nervous system.

Studies in our laboratory indicate that the threshold concentration for chlordimeform toxicity to the nervous system of *P. americana* is of the order of 10^{-5} to 10^{-4} M and that its effects on the isolated cockroach heart can be seen at 10^{-5} M levels. The overall neurosymptoms (electrophysiological) agree well with the *in vivo* poisoning symptoms (Beeman, 1974). The fact that chlordimeform directly stimulates isolated cockroach heart indicates, in addition, that chlordimeform could act directly on some receptors in insects at relatively low concentrations.

4.11. REFERENCES

Abo-Khatwa, N., and R. M. Hollingworth (1972). *Life Sci.* **11**:1181 (Part II).

Abood, L. G. (1972). In *Basic Neurochemistry*. R. W. Albers, ed. Little, Brown, Boston, p. 223.

Aldridge, W. N., and A. N. Davison (1952). *Biochem. J.* **51**:62.

Anderson, A., R. B. March, and R. L. Metcalf (1954). *Ann. Entomol. Soc. Am.* **47**:597.

Aziz, S. A., and C. O. Knowles (1973). *Nature* **242**:417.

Beeman, R. W. (1974). Studies on the mode of action of chlordimeform. Master of Science thesis, University of Wisconsin.

Beeman, R. W., and F. Matsumura (1973). *Nature* **242**:273.

Bellet, E. M., and J. E. Casida (1973). *Science* **182**:1135.

Berends, F., C. H. Posthumus, I. van der Sluys, and F. A. Deierkauf (1959). *Biochim. Biophys. Acta* **34**:576.

Bot, J. (1952). *Doc. Med. Geog. Trop.* **4**:57.

Brown, A. W. A. (1951). *Insect Control by Chemicals*. John Wiley & Sons, New York.

Brown, H. D., and E. F. Rogers (1950). *J. Am. Chem. Soc.* **72**:1864.

Buck, J. B., and M. L. Keister (1949). *Biol. Bull.* **97**:64.

Busvine, J. (1954). *Nature* **174**:783.

Casida, J. E. (1963). *Ann. Rev. Entomol.* **8**:39.

Casida, J. E. (1969). In *Microsomes and Drug Oxidations*. A. H. Conney, G. J. Cosmides, J. R. Gillette, R. W. Estabrook, J. R. Fuuts, and G. J. Mannering, eds. Academic Press, New York, pp. 517–530.

Casida, J. E. (1970). *J. Agr. Food Chem.* **18**:753.

Casida, J. E., K. B. Augustinsson, and G. Jonsson (1960). *J. Econ. Entomol.* **53**:205.

Chu, Y. C., and L. K. Cutkomp (1971). *J. Econ. Entomol.* **64**:559.

Cristol, S. (1950). *Advan. Chem. Ser.* **50**:184.

Davidow, B., and J. Radomski (1953). *J. Pharmacol. Exptl. Therap.* **107**:259.

Davidow, B., E. Hagan, and J. Radomski (1951). *Fed. Proc.* **10**:291.

Davison, A. N. (1953). *Biochem. J.* **54**:583.

Davison, A. N. (1955). *Biochem.* **60**:339.

Doherty, J. (1973). ATP related systems in lobster nerve. PhD. thesis, University of Wisconsin.

Dresden, D. (1949). *Physiological Investigations into the Action of DDT.* Drukkerij en Uitgeverij G. W. van der Wiel, Arnhem.

Dresden, D., and B. Krijgsman (1948). *Bull. Entomol. Res.* **38**:575.

Eckert, J. E. (1948). *J. Econ. Entomol.* **41**:487.

Flattum, R. F., and J. G. Sternburg (1970*a*). *J. Econ. Entomol.* **63**:62.

Flattum, R. F., and J. G. Sternburg (1970*b*). *J. Econ. Entomol.* **63**:67.

Fukami, J. (1956). *Botyu-Kagaku* **21**:122.

Fukami, J. (1961). *Bull. Natl. Inst. Agr. Sci. Ser. C* **13**:33.

Fukami, J., T. Nakatsugawa, and T. Narahashi (1959). *Japan. J. Appl. Entomol. Zool.* **3**:259.

Fukuto, T. R., and R. L. Metcalf (1956). *J. Agr. Food Chem.* **4**:930.

Gajdusek, D. C., and G. Luther (1950). *Am. J. of Diseases of Child.* **79**:310.

Gordon, H. T., and J. H. Welsh (1948). *J. Cell. Comp. Physiol.* **31**:395.

Gunther, F. A., R. C. Blinn, G. E. Carman, and R. L. Metcalf (1954). *Arch. Biochem. Biophys.* **50**:504.

Guthrie, F. (1950). *J. Econ. Entomol.* **43**:549.

Gysin, H. (1954). *Chimia* **8**:205.

Hammett, L. P. (1940). *Physical Organic Chemistry.* McGraw-Hill, New York.

Harvey, G. T., and A. W. A. Brown (1951). *Can. J. Zool.* **29**:42.

Hawkins, R. D., and B. Mendel (1946). *J. Cell. Comp. Physiol.* **27**:69.

Hawkins, R. D., and B. Mendel (1949). *Biochem. J.* **44**:260.

Hayashi, M., and F. Matsumura (1967). *J. Agr. Food Chem.* **15**:622.

Hobbiger, F. (1955). *Brit. J. Pharmacol.* **10**:356.

Hodgkin, A. L. (1951). *Biol. Rev.* **26**:339.

Hodgkin, A. L., and A. F. Huxley (1952). *J. Physiol.* **116**:449.

Holan, G. (1969). *Nature* **221**:1025.

Hopf, H. S., and R. T. Taylor (1958). *Nature* **182**:1381.

Ilivicky, J., and J. E. Casida (1969). *Biochem. Pharmacol.* **18**:1389.

Johannsen, F. R., and C. O. Knowles (1972). *J. Econ. Entomol.* **65**:1754.

Kearns, C. W. (1956). *Ann. Rev. Entomol.* **1**:123.

Kenaga, E. E. (1966). *Bull. Entomol. Soc. Am.* **12**(2):117.

Kirkwood, S., and P. Phillips (1946). *J. Biol. Chem.* **163**:251.

Koch, R. B. (1969). *J. Neurochem.* **16**:269.

Koch, R. B. (1970). *Chem.-Biol. Interactions* **1**:199.

Koch, R. B., L. K. Cutkomp, and F. M. Do (1969). *Life Sci.* **8**:289 (Part II).

Krüger, F. (1931). *Z. Angew. Entomol.* **18**:344.

Lalonde, D., and A. W. A. Brown (1954). *Can. J. Zool.* **32**:74.

Laüger, P., H. Martin, and P. Müller (1944). *Helv. Chim. Acta* **27**:892.

Laüger, P., R. Pulver, C. Montigel, and H. Wild (1946). Mechanism of intoxication by DDT insecticides in insects and warm-blooded animals. Address reprinted by Geigy Co., New York.

Lewis, W. (1953). *Nature* **172**:1004.

Lightbody, H., and J. Mathews (1936). *Ind. Eng. Chem.* **28**:809.

Lindahl, P. E., and K. E. Öberg (1961). *Exptl. Cell. Res.* **23**:228.

Ludwig, D. (1946). *Ann. Entomol. Soc. Am.* **39**:496.

Main, A. R. (1964). *Science* **144**:992.

Martin, H. (1946). *J. Soc. Chem. Ind. (London)* **65**:402.

Martin, H. (1956). *Ann. Rev. Entomol.* **1**:149.

Martin, H. (1964). *The Scientific Principles of Crop Protection.* St. Martin's Press, New York.

Matsumura, F., and M. Hayashi (1969). *J. Agr. Food Chem.* **17**:231.

Matsumura, F., and R. D. O'Brien (1963). *Biochem. Pharmacol.* **12**:1201.

Matsumura, F., and R. D. O'Brien (1966). *J. Agr. Food. Chem.* **14**:39.

Matsumura, F., and T. Narahashi (1971). *Biochem. Pharmacol.* **20**:825.

Matsumura, F., and K. C. Patil (1969). *Science* **166**:121.

Matsumura, F., T. A. Bratkowski, and K. C. Patil (1969). *Bull. Environ. Contam. Toxicol.* **4**:262.

Matthews, H. B., M. Skrinjaric-Spoljan, and J. E. Casida (1970). *Life Sci.* **9**:1039 (Part II).

McIndoo, N. (1916). *J. Agr. Res.* **7**:89.

McNamara, B. P., and S. Krop (1947). *Pharmacological Effect of Lindane and Isomers.* Chem. Corps, U.S. Army Med. Div., Rept. 125.

Mengle, D. C., and J. E. Casida (1958). *J. Econ. Entomol.* **51**:750.

Merrill, R., J. Savit, and J. Tobias (1946). *J. Cell. Comp. Physiol.* **28**:465.

Metcalf, R. L. (1947). *J. Econ. Entomol.* **40**:522.

Metcalf, R. L., and T. R. Fukuto (1965). *J. Agr. Food Chem.* **13**:220.

Metcalf, R. L., M. Y. Winton, and T. R. Fukuto (1964). *J. Insect. Physiol.* **10**:353.

Miura, K., T. Uchiyama, and K. Honda (1961). *Agr. Biol. Chem. (Tokyo)* **25**:83.

Morrison, P. E., and A. W. A. Brown (1954). *J. Econ. Entomol.* **47**:723.

Mullins, L. J. (1954). *Chem. Rev.* **54**:289.

Mullins, L. J. (1955). *Science* **122**:118.

Myers, D. K., and B. Mendel (1949). *Proc. Soc. Exptl. Biol. Med.* **71**:357.

Narahashi, T. (1961). *Physiologist* **4**:80.

Narahashi, T. (1962). *J. Cell. Comp. Physiol.* **59**:61.

Narahashi, T. (1963). *Advan. Insect Physiol.* **1**:175.

Narahashi, T. (1971). *Bull. World Health Org.* **44**:337.

Noguchi, T., H. Miyata, T. Mori, Y. Hashimoto, and S. Kosaka (1968). *Pharmacometrics* **2**:376.

O'Brien, R. D. (1960). *Toxic Phosphorus Esters.* Academic Press, New York, 434 pp.

O'Brien, R. D. (1963). *Metabolic Inhibitors.* Academic Press, New York, Vol. 2, p. 205.

O'Brien, R. D. (1967). *Insecticides: Action and Metabolism.* Academic Press, New York, 332 pp.

O'Brien, R. D., and F. Matsumura (1964). *Science* **146**:657.

O'Connor, A., R. D. O'Brien, and M. M. Saltpeters (1965). *J. Insect Physiol.* **11**:1351.

Orser, W., and A. W. A. Brown (1951). *Can. J. Zool.* **29**:54.

Pasquier, R. (1947). *Bull. Semestr. Off. Natl. Anti-Acridien Algeria* **4**:5.

Perkow, W. (1956). *Z. Naturforsch.* **11b**:38.

Peters, R. A. (1957). *Advan. Enzymol.* **18**:113.

Philpot, R. M., and E. Hodgson (1972). *Mol. Pharmacol.* **8**:204.

Raffy, A., and P. Portier (1931). *Compt. Rend. Soc. Biol.* **108**:1062.

Richards, A., and L. Cutkomp (1945). *J. Cell. Comp. Physiol.* **26**:57.

Riemschneider, R., and H. D. Otto (1954). *Z. Naturforsch.* **9b**:95.

Roeder, K., and S. Roeder (1939). *J. Cell. Comp. Physiol.* **14**:1.

Roeder, K. D., and E. A. Weiant (1946). *Science* **103**:304.

Roeder, K. D., and E. A. Weiant (1948). *J. Cell. Comp. Physiol.* **32**:175.

Rosedale, J. (1948). *J. Entomol. Soc. S. Afr.* **11**:34.

Roy, D., S. Ghosh, and R. Chopra (1943). *Ann. Appl. Biol.* **30**:42.

Saktor, B. (1951). *J. Econ. Entomol.* **43**:838.

Savit, J., J. Kollros, and J. Tobias (1946). *Proc. Soc. Exptl. Biol. Med.* **62**:44.

Skerrett, E. J., and D. Woodcock (1952). *J. Chem. Soc.* **3308**:12.

Skou, J. C. (1965). *Physiol. Rev.* **45**:597.

Slade, R. (1945). *Chem. Ind.* **1945**:314.

Soloway, S. B. (1965). *Adv. Pest Control Res.* **6**:85.

Stegwee, D. (1959). *Nature* **184**:1253.

Stempel, A., and J. A. Aeschlimann (1956). *Medical Chemistry*, Vol. III. F. F. Bliche and R. H. Cox, eds. John Wiley & Sons, New York, p. 238.

Tamura, S., and N. Takahashi (1971). In *Naturally Occurring Insecticides*. M. Jacobson, and D. G. Crosby, eds. Marcel Dekker, New York.

Telford, N., and F. Matsumura (1970). *J. Econ. Entomol.* **63**:795.

Tobias, J. M., and J. J. Kollros (1949). *Biol. Bull.* **91**:247.

Tobias J. M., J. J. Kollros, and J. Savit (1946). *J. Cell. Comp. Physiol.* **28**:159.

Wang, C. M., and F. Matsumura (1970). *J. Econ. Entomol.* **63**:1731.

Wang, C. M., T. Narahashi, and M. Yamada (1971). *Pesticide Biochem. Physiol.* **1**:84.

Welsh, J. H., and H. T. Gordon (1947). *J. Cell. Comp. Physiol.* **30**:147.

Wiersma, C. A. G., and W. Schallek (1947). *Science* **106**:421.

Wiesmann, R., and P. Fenjves (1944). *Mitt. Schweiz. Entomol. Ges.* **19**:179.

Wiesmann, R., and C. Kocher (1951). *Z. Angew. Entomol.* **33**:297.

Wild, H. (1946). *Helv. Chim. Acta* **29**:497.

Wilkinson, C. F. (1971). *Bull. World Health Org.* **44**:171.

Williamson, R. L., and R. L. Metcalf (1967). *Science* **158**:1694.

Wilson, I. B., and F. Bergmann (1950). *J. Biol. Chem.* **185**:479.

Woodcock, D. (1953). *Science and Fruit*, No. 256. University of Bristol, Bristol, England, p. 308.

Yamamoto, I., H. Kamimura, R. Yamamoto, S. Sakai, and M. Goda (1962). *Agr. Biol. Chem.* **26**:709.

Yamasaki, T., and T. Ishii (1952). *J. Appl. Entomol. (Tokyo)* **7**:157.

Yamasaki, T., and T. Ishii (1953). *J. Appl. Entomol. (Tokyo)* **9**:87.

Yamasaki, T., and T. Ishii (1954). *Botyu-Kagaku* **19**:1.

Yamasaki, T., and T. Narahashi (1957). *Botyu-Kagaku* **22**:296.

Yamasaki, T., and T. Narahashi (1958). *Botyu-Kagaku* **23**:146.

Yeager, J., and S. Munson (1942). *J. Agr. Res.* **64**:307.

Chapter 5

Metabolism of Insecticides by Animals and Plants

5.1. GENERAL TYPES OF METABOLIC ACTIVITIES

Organic compounds which do not belong in the category of fat, carbohydrate, protein, vitamin, steroid, or mineral are considered to be foreign to the body. Such foreign compounds are utilized on a vast scale by man as drugs, pesticides for agricultural use, additives to foods and fabrics, etc. It is important to know the fate of these chemicals within the animal body because organisms very often do eliminate them through various defense mechanisms.

Formerly the metabolism of all foreign organic compounds, i.e., xenobiotics, was called "detoxification," and indeed since organisms do convert many compounds into less toxic or harmless products the term is still applicable. However, it has been found that the body converts some compounds into more toxic substances, and so the term "activation" is often used to describe this phenomenon. The type of change that occurs depends on the chemical structure of the compound, but other factors such as species of animal, method of administration, and diet may also be involved. Some compounds that are very polar or are insoluble in both water and lipids are not metabolized by the body and are excreted unchanged. According to Williams (1959), the four types of chemical changes that occur are oxidation, reduction, hydrolysis, and synthesis. Each group of reactions can be subdivided into a number of different metabolic activities depending on the type of substrate involved.

165

There are two major steps of detoxification of xenobiotics (Williams, 1959; Parke and Williams, 1969): the primary phase involving oxidative, hydrolytic, and other enzymatic processes to produce polar end-products (nonsynthetic process), and the secondary phase producing water-soluble conjugates (synthetic process). In the nonsynthetic metabolism of insecticides in animals, three major types of enzyme systems are dominant throughout:

1. *Hydrolases*, which split the insecticide substrates through hydrolysis. Examples are carboxylesterases, amidases, phosphatases, and A-type esterases (O'Brien, 1960).
2. *Glutathione S-transferases*, which are characterized by dependency on reduced glutathione (GSH) for their actions. Examples are DDT-dehydrochlorinase (Kearns, 1956), BHC-degrading enzymes (Ishida and Dahm, 1965; Clark *et al.*, 1969), the methylparathion demethylation system (Fukami and Shishido, 1966), and other dealkylating and perhaps dearylating systems.
3. *Microsomal oxidases*, which are characterized by the requirement of NADPH, microsomes, and oxygen *in vitro* for degradation of their substrates. They are also characterized by their sensitivities toward methylenedioxyphenyl derivatives (pyrethrin synergists) such as sesamex and piperonylbutoxide.

The majority of insecticides are (1) chlorine-containing aromatics; (2) chlorine-containing cyclic hydrocarbons; (3) organophosphorus esters (including thioesters), amidates, and halides; (4) carbamic esters and oximes, (5) olefinic, cyclic, and heterocyclic natural products; and (6) short alkyl halides (fumigants). Also, any of these compounds may have a number of N-, S-, and O-containing alkyl, aromatic, and heterocyclic compounds as side-chains.

By far the most important biochemical reactions involving the initial stages of insecticide metabolism are the abovementioned NADPH-requiring general oxidation system and the hydrolysis of esters. In addition, metabolism of compounds containing halogens, particularly chlorine, must be considered in this chapter. The conjugation mechanisms (synthetic metabolism) will also be described to give an overall view of the general patterns of metabolic activities in biological systems. Actually, this particular field has been well reviewed, and there are a number of excellent articles covering the subject. Examples are

D. L. Bull. "Metabolism of Organophosphorus Insecticides in Animals and Plants." *Residue Rev. 43*:1, 1972.
J. E. Casida. "Mixed-Function Oxidase Involvement in the Biochemistry of Insecticide Synergists." *J. Agr. Food Chem. 18*:753, 1970.

H. W. Dorough. "Metabolism of Insecticidal Methylcarbamates in Animals." *J. Agr. Food Chem. 18* : 1015, 1970.

E. Hodgson and F. W. Plapp. "Biochemical Characteristics of Insect Microsomes." *J. Agr. Food Chem. 18* : 1048, 1970.

J. B. Knaak. "Biological and Nonbiological Modifications of Carbamates." *Bull. World Health Org. 44* : 121, 1971.

C. O. Knowles, S. Ahmad, and S. P. Shrivastava. "Chemistry and Selectivity of Acaricides." In *Insecticides—Pesticide Chemistry I.* A. S. Tahori, ed. London : Gordon and Breach Science Publishers, 1972.

R. Kuhr. "Metabolism of Carbamate Insecticide Chemicals in Plants and Insects." *J. Agr. Food Chem. 18* : 1023, 1970.

S. Matsunaka. "Metabolism of Pesticides in Higher Plants." In *Environmental Toxicology of Pesticides.* F. Matsumura, G. M. Boush, and T. Misato, eds. New York : Academic Press, 1972.

R. E. Menzer and W. C. Dauterman. "Metabolism of Some Organophosphorus Insecticides." *J. Agr. Food Chem. 18* : 1031, 1970.

C. M. Menzie. *Metabolism of Pesticides.* Bureau of Sport Fisheries and Wildlife, Special Scientific Report—Wildlife No. 127, 1969.

A. W. J. Schlagbauer and B. G. L. Schlagbauer. "The Metabolism of Carbamate Insecticides—A Literature Analysis. II." *Residue Rev. 42* : 85, 1972.

B. G. L. Schlagbauer and A. W. J. Schlagbauer. "The Metabolism of Carbamate Insecticides—A Literature Analysis. I." *Residue Rev. 42* : 1, 1972.

5.2. PRIMARY METABOLIC PROCESSES

5.2.1. Oxidation Through Mixed-Function Oxidase Systems

The NADPH-requiring general oxidation system, commonly referred to as the "mixed-function oxidase system" or MFO, is located in the microsomal portions of various tissues, particularly the liver (Brodie *et al.*, 1958; Gillette *et al.*, 1969). It is characterized by (1) requiring NADPH as a cofactor, (2) involving an electron transport system with cytochrome P_{450}, and (3) being capable of oxidizing many different kinds of substrates (i.e., substrate nonspecificity).

The complete mechanism by which NADPH facilitates this electron transport system has not been elucidated, but the general scheme of such a system is known (Fig. 5-1). In short, it involves mainly cyctochrome P_{450} along with cytochrome b_5, which requires NADH. The final process of oxidation (or hydroxylation) of drugs and pesticides involves reduced

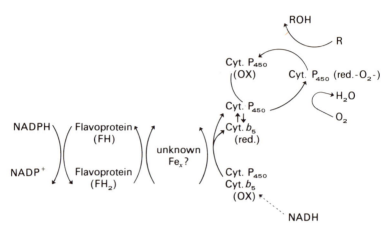

Fig. 5-1. Schematic diagram indicating the electron transport system involved in the NADPH-requiring general oxidation system in liver microsomes. From Kamin and Masters (1968).

cytochrome P_{450}–oxygen complex, which, on oxidation of the substrate, becomes a stable oxidized form itself. Omura and Saito (1964) isolated and described this important cytochrome P_{450} by reacting it with CO: cytochrome b_5 is insensitive to CO. (Cytochrome P_{450} is so named because in the reduced state it exhibits an absorption band with the peak at 450 nm, while in the oxidized state the peak position shifts to 420–422 nm.)

Within the microsomal particles, NADPH-requiring oxidation activity generally concentrates in smooth microsomes (i.e., smooth endoplasmic reticulum), although in some cases the activity can be evenly distributed among smooth and rough fractions (Gram and Fouts, 1968).

The reactions catalyzed by this system include (1) deamination, (2) demethylation, (3) dealkylation, (4) aromatic ring hydroxylation, (5) alkyl and N-hydroxylation, (6) cleavage of ester bonds, (7) epoxidation, (8) oxidation of sulfides to sulfoxides and sulfones, (9) conversion of phosphorothioates to phosphoates, (10) conversion of methylenedioxyphenyls to catechols, and (11) oxidation of alcohols and aldehydes to acids. There are a number of excellent review papers available on this subject by Casida (1969, 1970), Gillette *et al.* (1969), Hodgson (1968), Hodgson and Plapp (1970), and Wilkinson (1968).

One of the most important characteristics of the mixed-function oxidase systems is that they are sensitive to chemicals which contain a methylenedioxyphenyl moiety (MDP). The mechanism of MDP inhibition is mainly a competitive one. For instance, Ray (1967) indicates that in housefly microsomes inhibition of naphthalene hydroxylation by sesamex is competitive, while epoxidation of aldrin to dieldrin is inhibited by a noncompetitive reaction (with lower K_m) and by a competitive reaction of sesamex. The inhibitory action of methylenedioxyphenyl compounds on microsomal oxidation systems appears to be universal among animal species and therefore should be of great importance in understanding the action mechanisms of these oxidative systems.

Wilkinson *et al.* (1966) demonstrated that the structural requirements for MDP compounds are strict: replacement of one oxygen atom by anything other than one sulfur atom greatly reduces the activity.

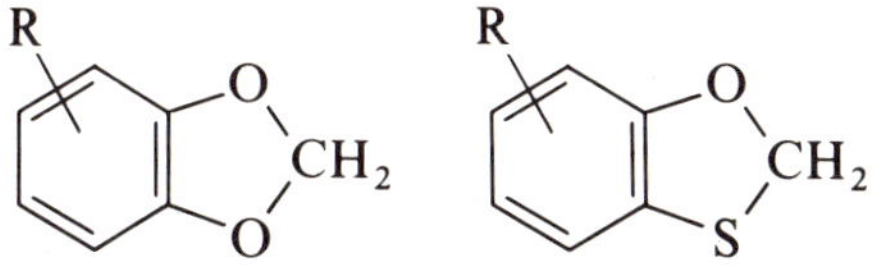

ACTIVE   FAIRLY ACTIVE

In view of the reduction in potency of MDP that results from the replacement of methylene hydrogens by deuterium atoms, it is generally considered that the first step of MDP action is formation of the benzodioxolium ion and release of a hydrogen (Hennessy, 1965):

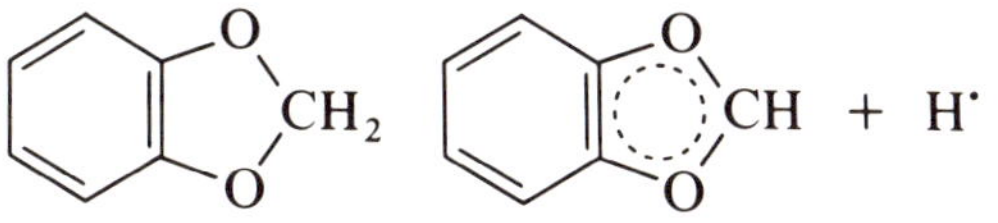

1,3-BENZODIOXOLE   BENZODIOXOLIUM ION

5.2.2. Reduction

Reduction is much less common than oxidation, but for some compounds it is a general biochemical reaction. Reduction reactions known to occur in systems of higher animals are (1) reductive dehalogenation, where a halogen atom is replaced by a hydrogen atom, such as in the DDT → TDE reaction; (2) reduction of nitro groups to hydroxylamine and to amines, e.g., parathion to aminoparathion; (3) reduction of pentavalent arsenics to trivalent arsenics; and probably (4) formation of a double bond from an epoxy ring. Other reductive-type reactions that could take place (Williams, 1959) are (1) reduction of aldehydes to alcohols and ketones to secondary alcohols, (2) saturation of double bonds, and (3) sulfhydryl formation from disulfides. However, the frequency of such reactions on insecticidal compounds is rather rare.

Generally speaking, there appear to be two types of reductive reactions that can take place on insecticide substrates—NADPH (or NADH in some cases) dependent and independent—but the distinction between them is by no means clear. Perhaps the best example of the former is the so-called nitroreductase reaction, which converts aromatic nitro groups to corresponding amines. The system is located in the soluble and microsomal fractions of the mammalian liver and kidney and is NADPH or NADH dependent, as well as O_2 and CO sensitive. Although the latter factor is indicative of P_{450} involvement, its relationship to this cytochrome system is not certain. An example of the NADPH-independent system is the reductive dechlorination reaction on DDT, which yields TDE (DDD) as the major end-product. This reaction can be produced even in the absence of an enzymatic system; here only the presence of reduced iron–porphyrin complexes appears to be necessary (Miskus *et al.*, 1965), although enzymatic systems are much more efficient in producing the same end result (Walker, 1969). Walker has shown that this reductive dechlorination reaction can also be processed by a NADPH-stimulated, CO- and O_2-susceptible microsomal system in the pigeon liver, indicating that the difference between the NADPH-independent and -dependent systems is not easy to establish.

To cite a few examples of reduction of insecticidal chemicals, incubation of 2-ketodihydrochlordene with pig liver or housefly microsomes and NADPH can result in the formation of 1-hydroxydihydrochlordene (Brooks and Harrison, 1967*a, b*). Aminoparathion is known to form from parathion as a result of reductive conversion of the *p*-nitro group to *p*-amino in the presence of NADPH in the housefly (Lichtenstein and Fuhremann, 1971). SD-8447 [2-chloro-1-(2,4,5-trichlorophenyl)vinyl dimethylphosphate] is first hydrolyzed in rats to form a ketone (I), which is reduced to alcohols (Whetstone *et al.*, 1966):

1-(2,4,5-TRICHLOROPHENYL)ETHANOL (II) AND 2,4,5-TRICHLOROETHANEDIOL (III)

5.2.3. Hydrolytic Processes: Metabolism of Esters and Ethers

A large number of insecticides are esters or ethers of one form or another, e.g., phosphorus esters and carbamic esters. In addition, many ester and ether groups exist among various side-chains of insecticides such as malathion, methoxychlor, and arprocarb. These side-chain esters often play important roles in determining the species differences in susceptibility to the chemicals.

Esterases are hydrolases which split ester compounds with the addition of water to yield alcohols and acids. They usually do not require any co-factors for their actions but are often stimulated by divalent cations. The pH of the reaction medium often plays an important role, for both H^+ and OH^- ions exert a profound influence on the process of hydrolysis.

Esterases can be subdivided into three groups: A-type esterases, which are resistant to organophosphates and degrade them as substrates; B-type esterases, which are susceptible to organophosphate inhibition; and C-type esterases, which are resistant to organophosphates but do not degrade them. There are two types of esterases that are important in metabolizing insecticidal chemicals—carboxylesterases and phosphatases. The former are actually B-type esterases and are also called aliesterases. They are thus inhibited by various organophosphates. Carboxylesterases, however, play vital roles in degrading several organophosphorus compounds, particularly carboxylic esters of phosphorothiolates. The best example is malathion degradation:

$$
\begin{array}{c}
\text{desmethylase} \quad \text{carboxylesterase (to produce } \alpha\text{-acid)} \\
\downarrow \qquad\qquad\qquad \downarrow \\
\text{CH}_3\text{O} \quad \text{S} \\
\diagdown \; \| \\
\text{P—S—CHCOOC}_2\text{H}_5 \\
\diagup \qquad\qquad | \\
\text{CH}_3\text{O} \quad \uparrow \quad \text{CH}_2\text{COOC}_2\text{H}_5 \\
| \qquad\qquad\qquad \uparrow \\
\text{phosphatase} \qquad \text{carboxylesterase (to produce} \\
\beta\text{-acid)}
\end{array}
$$

It has been generally acknowledged that carboxylesterase (sometimes referred to as carboxyesterase) is responsible for the selective toxicity of this compound, favoring mammals over insects (Krueger and O'Brien, 1959) and resistant insects over susceptible ones (Matsumura and Brown, 1961; Matsumura and Hogendijk, 1964a, b). A purified preparation of this enzyme from the rat liver has all the characteristics of a B-type esterase (Main and Braid, 1962). Since there are many B-type esterases in any given animal system, it is not possible to conclude that definite carboxylesterase systems for insecticide substrates are present. However, all insecticidal chemicals that contain one or more carboxylic acid esters are likely to be hydrolyzed by one or more B-type esterases, inasmuch as the substrate specificities of these carboxylesterases are generally broad.

It is not certain, however, whether these B-type esterases are the ones that hydrolyze insecticidal carbamic esters. Hydrolysis of carbamates generally plays a less significant role in the initial step of degradation, and so no critical examination of this possibility is available. Nevertheless, these carbamates are known to show high affinities for aliesterases, and carbamylated enzymes can slowly recover. In fact, carbamates can be regarded as bad substrates for esterases, even including cholinesterases. It is therefore likely that aliesterases also serve as rather inefficient degradative hydrolases of insecticidal carbamic esters.

Some other types of insecticidal chemicals which are susceptible to hydrolytic attack are amides, ethers, phosphorohalides, and thioethers. Dimethoate, being an amide, is mainly metabolized through amidases (e.g., Dauterman *et al.*, 1959; Krueger *et al.*, 1960), although it is not certain whether such an enzymatic action can be accomplished by esterases, particularly by B-type esterases.

It is certain that there are a number of A-type esterases that actively metabolize insecticidal esters. In the case of organophosphates, these esterases are first phosphorylated at esteratic sites. By definition, if dephosphorylation takes place with relative ease the esterase can be called an A-type esterase. Thus true phosphates, as opposed to thiophosphates, must be degraded by this route. In the past, enzymes which degrade thiophosphates and phosphates were collectively called phosphatases. It is certain that true phosphates, having high affinities for esterases, are preferred substrates for esterases, although their involvement in the hydrolysis of phosphorothionates is far from clear (see Section 5.4.2), except that phosphorothionates are known to be degraded by GSH- or NADPH-independent hydrolytic systems in animals.

Another hydrolytic system which has recently been studied is that involving hydrate epoxy rings (e.g., Matthews and Matsumura, 1969; Brooks and Harrison, 1969). The system apparently is different from the common esterases in that it is located mainly in the microsomal fraction. Brooks (1972) treats this system as analogous to hydration processes known to react with benzene oxides and naphthalene-1,2-oxide to yield corresponding *trans*-dihydrodiols, e.g., the "epoxide hydrase" found in rabbit liver microsomes.

Other hydrolytic systems available are lipases; phosphatases such as acid and alkaline phosphatases, ATPases and phospholipases; peptidases; and proteinases. Their actions against insecticide derivatives are not well known, except in obvious cases such as proteinase actions against bacterial protein toxins (e.g., that of *Bacillus thuringiensis*) and peptidase actions against polypeptide antibiotics.

5.2.4. Glutathione-Mediated Metabolism

Glutathione is known to be involved in many metabolic reactions with xenobiotics. Insecticidal chemicals and their metabolic products are also known to be catalyzed by glutathione-mediated systems. There appear to be two types of such systems. In the first, glutathione is utilized in a purely catalytic manner; in the second, glutathione is consumed by direct binding to the substrate, at least at the initial stage of the reaction, and is not regenerated.

5.2.4a. Glutathione-Catalyzed Metabolism

The catalytic-mediated reactions could also involve direct binding to the substrate, but the important difference is that glutathione levels do not change at the end of the reaction, e.g., DDT-dehydrochlorinase (Lipke and Kearns, 1960; Lipke and Chalkley, 1962):

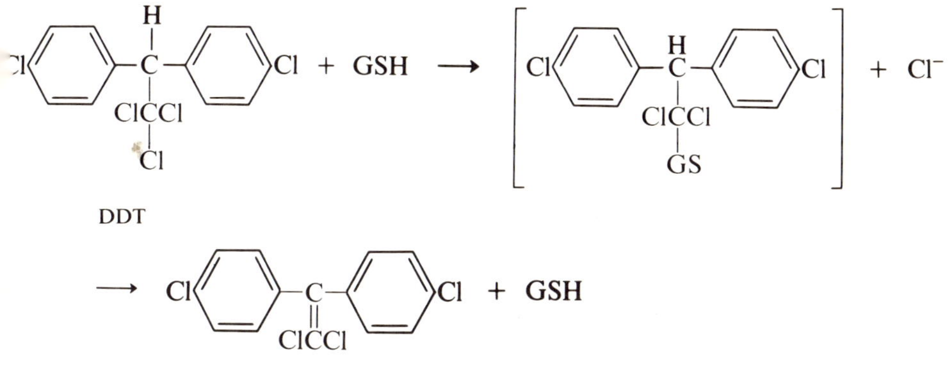

DDT

DDE

In this scheme, which is purely hypothetical since the intermediate shown above has never been isolated, GSH is regenerated at the end of each reaction.

This enzyme system can be inhibited by *p*-chloromercuribenzoate or iodoacetate and by competitive substrates such as DDD or DDMS, 1-chloro-bis(*p*-chlorophenyl)ethane. The first inhibitor probably binds with the —SH group of GSH, and/or the enzyme, and the second inhibitor could be competing for GSH, since the inhibition takes place at approximately the same levels for iodoacetate and GSH, in addition to the fact that it can be reversed by the addition of larger amounts of GSH. Thus there is a possibility that this enzyme system is related to a GSH-alkyltransferase type process. However, iodoacetate is also a —SH inhibitor, and there is no example of GSH being regenerated in the course of S-alkyltransferase actions.

5.2.4b. Glutathione S-Transferases

In the glutathione S-transferase reaction, glutathione is expected to bind with the substrate either by replacing a labile moiety or by directly attaching to the substrate molecule (Table 5-1). Boyland and Chasseaud (1969) classified the GSH S-transferases of mammals into five different enzyme systems. These enzymes are characterized by being localized in soluble fractions and requiring GSH.

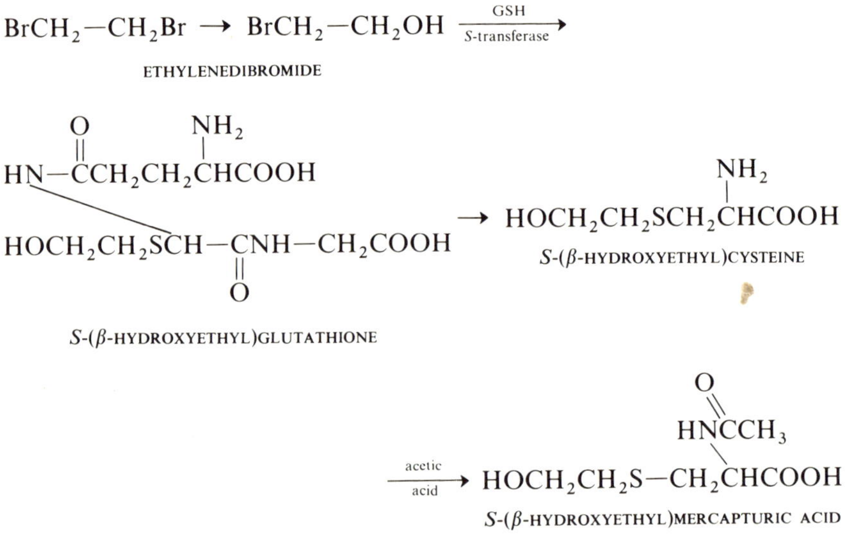

$$BrCH_2{-}CH_2Br \rightarrow BrCH_2{-}CH_2OH \xrightarrow[\text{S-transferase}]{\text{GSH}}$$

ETHYLENEDIBROMIDE

$$\rightarrow HOCH_2CH_2SCH_2CHCOOH$$

S-(β-HYDROXYETHYL)CYSTEINE

S-(β-HYDROXYETHYL)GLUTATHIONE

$$\xrightarrow[\text{acid}]{\text{acetic}} HOCH_2CH_2S{-}CH_2CHCOOH$$

S-(β-HYDROXYETHYL)MERCAPTURIC ACID

In most cases, glutathione conjugates thus formed are expected to go through further transformations to finally yield more excretable compounds, such as mercapturic acids. The metabolic process involving ethylene dibromide illustrates the generalized pattern of such a transformation.

Whatever the final form of the metabolic products, the initial process in the reactions that are catalyzed by GSH S-transferases is formation of glutathione complexes such as shown in Table 5-1. There appear to be enormous variations in the activity of transferases among mammalian

TABLE 5-1. Known Glutathione S-Transferases

Enzyme	Substrate	Reaction products
	Type A, GSH replacing labile moiety	
GSH S-alkyltransferase	CH_3I	$CH_3 \cdot SG + H^+ + I^-$
GSH S-aralkyltransferase	Benzylchloride	$Ph \cdot CH_2SG + H^+ + Cl^-$
GSH S-aryltransferase	1,2-Dichloro-4-nitrobenzene	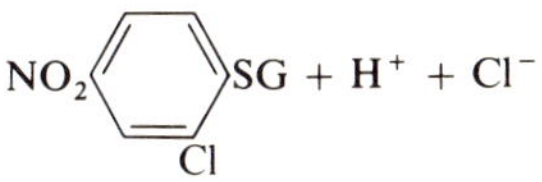
	Type B, GSH direct addition	
GSH S-epoxidetransferase	2,3-Epoxy-propyl-phenylether	$Ph \cdot O \cdot CH_2CH(SG)CH_2OH$
GSH S-alkene(cis-ester) transferase	Diethylmaleate	$C_2H_5O_2CCH(SG) \cdot CH_2 \cdot CO_2C_2H_5$

From Boyland and Chasseaud (1969).

species examined so far. In rats, the order of transferase activity is alkene (7.5 μmoles/min/g liver) > alkyl (3.0) > aralkyl (2.3) > epoxide (2.1) > aryl (0.5), while in pigeons alkyl is the predominant enzyme. In mice, alkene S-transferase activities are outstanding; 12.2 μmoles/min/g liver was recorded, the highest activity noted among nine species examined.

It is likely that the same GSH S-alkyltransferase is acting on the O-alkyl moiety of insecticidal organophosphates. Hollingworth (1969) studied demethylation mechanisms on methylparaoxon and fenitrothion.

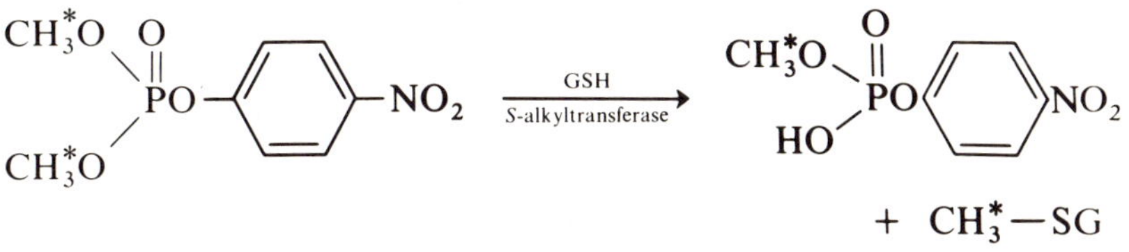

As a result of *in vitro* reaction with methylparaoxon in mice, S-methyl-glutathione was formed along with desmethyl methylparaoxon. The system can be antagonized by methyl iodide, which is a standard substrate used for GSH S-alkyltransferase. Also, in fenitrothion-treated mice, glutathione was depleted. The system is activated only by glutathione and not by cysteine of S-(N-ethylsuccinimido)glutathione. It is located in the soluble (super-natant) fraction of mouse liver. The system reacts well with dimethyl-phosphates but only slowly with diethyl and dipropyl analogues. Thus it is likely that it is the GSH S-alkyltransferase itself that de-O-methylates these dimethylphosphates. Examples of such demethylation reactions are to be found in the studies of Fukami and Shishido (1966) for methylparathion and methylparaoxon, of Hutson *et al.* (1968) for the methyl analogue of chlorfevinphos, of Morello *et al.* (1968) for mevinphos, of Stenersen (1969) for bromophos, and of Miyata and Matsumura (1971) for famphur and dichlorvos.

GSH S-aryltransferase is also likely to act on aromatic pesticide derivatives, although the relationship has not yet been established.

The most likely case of S-aryltransferase action has been observed with parathion, which probably degrades in the housefly according to the following scheme (Dahm, 1970; Oppenoorth *et al.*, 1972):

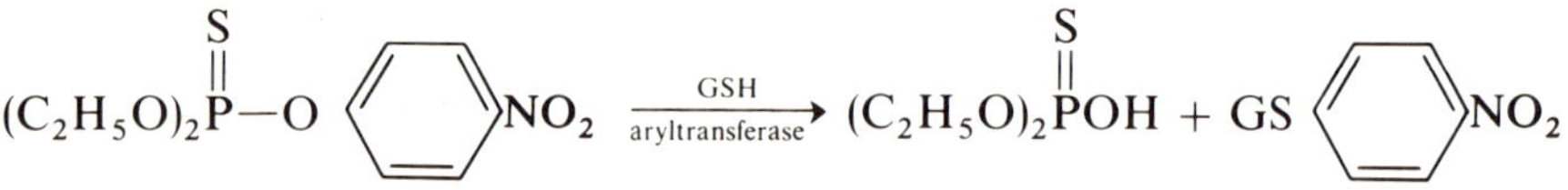

$$(C_2H_5O)_2\overset{\overset{\displaystyle S}{\|}}{P}-O \overset{}{\underset{}{\bigcirc}} NO_2 \xrightarrow[\text{aryltransferase}]{\text{GSH}} (C_2H_5O)_2\overset{\overset{\displaystyle S}{\|}}{P}OH + GS \overset{}{\underset{}{\bigcirc}} NO_2$$

The case of the GSH-stimulated dearylation system for diazinon in the housefly is better understood. Lewis (1969) and Lewis and Sawicki (1971) reported increased production of diethylphosphorothioic acid from the soluble fraction of the housefly in the presence of GSH. Yang *et al.* (1971) also found in a multiresistant housefly strain an increase in diethylphosphorothioic acid and diethylphosphoric acid produced from diazinon on addition of GSH. Shishido *et al.* (1972*b*) finally identified S-(2-isopropyl-4-methyl-6-pyrimidinyl)glutathione along with diethylphosphorothioic acid as the reaction product of diazinon and the supernatant fractions from rat liver and cockroach fat body:

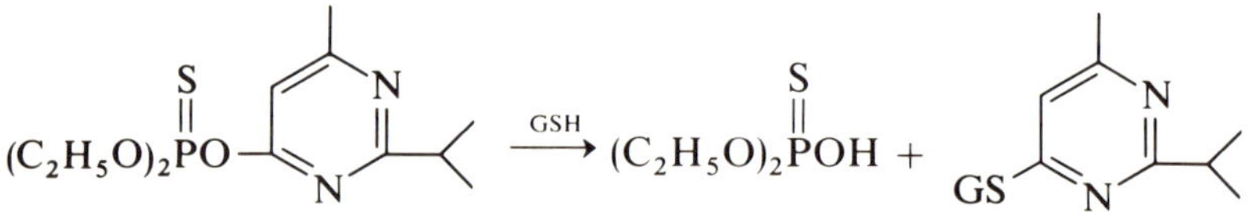

$$(C_2H_5O)_2\overset{\overset{\displaystyle S}{\|}}{P}O \xrightarrow{\text{GSH}} (C_2H_5O)_2\overset{\overset{\displaystyle S}{\|}}{P}OH + \quad GS$$

The system is specific for GSH, and either oxidation of GSH or replacement with cysteine, thioglycolic acid, or 2-mercaptoethanol eliminated the stimulatory effects of GSH. Probably another example of S-aryltransferase activities can be found in the enzymatic system that metabolizes BHC in the presence of GSH (Clark *et al.*, 1969). Confirmation is lacking that any of these enzyme systems are indeed identical to the enzymes defined as S-aryltransferases. Nor is there any evidence that other GSH transferases actively operate against any pesticidal chemicals.

5.3. CONJUGATION SYSTEMS: SECONDARY METABOLIC PROCESSES

The synthetic reactions, some of which are also called conjugation processes, are limited in number as compared with other metabolic activities toward xenobiotics. The occurrence of metabolic synthetic activity is governed by the presence of certain chemical centers (e.g., —OH) and the biochemical capability of the animal species. For instance, some compounds which do not have such a group may acquire it after oxidation or reduction. Some of the synthetic reactions are limited to a few classes of animals, and others occur in almost all species.

With the synthetic mechanisms, part of the molecule synthesized is provided by the animal, and this moiety is called the conjugating agent. When the dose of foreign material is not excessive, the conjugating agent may be provided from waste material or from the tissues without harm to the animal. The most important aspect of such conjugation activities by organisms is that the foreign compound, often apolar hydrophobic material, becomes water soluble and hence can be excreted through the bile–feces or the kidney–urine system.

5.3.1. Glucuronide or Glucuronic Acid Conjugation

β-Glucuronide formation is the most common synthetic process, occurring in practically all species of animals. However, it does not occur in insects and may not occur extensively in cats. Uridine diphosphate glucuronate (UDPGA) and β-glucuronic transferase are required to accomplish the tasks of conjugation in mammals. In insects, β-glucosides appear to be formed instead of the glucuronides (Smith, 1962). Compounds which can form glucuronides are those that contain hydroxyl, carboxyl, amino, or sulfhydryl groups or that can form them by oxidation, reduction, or some other process.

Hydroxyl groups in almost any compound are capable of conjugating with glucuronic acid in the body (e.g., 4-hydroxycarbaryl and naphthol). Carboxyl groups are capable of conjugating with glucuronic acid. Conjugation has been shown to occur with aromatic acids, aliphatic acids in which β-oxidation is hindered in some way, and phenyl-substituted acetic acids (e.g., 2,6-dichlorobenzoic acid). Aliphatic and aromatic amino groups are capable of conjugating with glucuronic acid (e.g., Banol® and carbaryl for *N*-glucuronide formation). The compounds formed are presumably *N*-glycosides. The sulfhydryl group (—SH), as the analogue of the —OH group, should also be capable of conjugating with glucuronic acid *in vivo*. This has been observed with thiophenol (C_6H_5SH), which forms the corresponding thioglucosiduronic acid.

There are a number of insecticidal chemicals which are known to form glucuronides. By far the best known group of chemicals that mainly go through this route of conjugation are the carbamates. For example, Banol[®] is 88.8 % metabolized by this route, carbaryl 82.6%, carbofuran 91.0%, Meobal[®] 85.0%, and Mobam[®] 78.0% according to the tabulation made by Mehendale and Dorough (1972). Some others are dieldrin (Matthews and Matsumura, 1969), dichloro- and trichlorophenols which can form BHC, etc., hydroxy-biphenyls, and chlorfevinphos (Hutson *et al.*, 1967). Most of these are substituted phenols, cyclic hydroxy compounds, or benzol derivatives. It is thought that the enzymatic system involved in the glucuronide formation is located in the liver (or other detoxification organs), mainly in microsomal fractions. The specificity for the cofactor is strict, as uridine diphosphoglucuronic acid (UDPGA) is known to be the sole agent catalyzing the reaction.

5.3.2. Sulfate Conjugation or Ethereal Sulfate Synthesis

Sulfate conjugation is a common reaction of phenol and naphthols in most species of animals including insects, and consists of the combination of a phenol with sulfate to form an acid ester of sulfuric acid or an aryl sulfuric acid, $RHSO_4$ or $RO \cdot SO_2 \cdot OH$. Both are easily hydrolyzed by acid and therefore are readily distinguished from other conjugates. There are a few examples of sulfate conjugation in insecticidal chemicals. Sulfonyl diphenyl or sulfinyldiphenol, which forms as a result of metabolic action on Abate,[®] can form both glucuronides and sulfates in bean plants (Blinn, 1968). 1-Naphthol, which is the hydrolytic product of carbaryl, also forms both types of conjugates (e.g., Knaak *et al.*, 1965; Leeling and Casida, 1966). Although sulfate formation is less important than glucuronide formation, sulfates also can form from carbofuran (Dorough, 1968). Similarly, chlorine-substituted phenols from any chlorinated pesticide are expected to go through sulfate conjugation (e.g., BHC). In all cases, sulfate conjugation in insecticidal compounds can be found with substituted phenols and naphthols only. It is probable that other insecticidal hydroxy compounds besides phenols can undergo this reaction since the sulfuric esters of steroid alcohols and carbohydrates are natural products (Williams, 1959). Judging by the examples of other xenobiotics, there is a remote possibility that aromatic amines and thiophenols also conjugate with sulfate to form sulfamates and thiosulfates, $Ar \cdot NH \cdot SO_3H$ and $AR \cdot S \cdot SO_3H$. For example, 6-amino-4-nitro-*o*-cresol is excreted in locusts through this route (Kikal and Smith, 1959).

5.3.3. Glycine Conjugation or Hippuric Acid Synthesis

Conjugation with glycine in the animal body is a reaction of the carboxyl group only. Only a few examples of glycine conjugation are known in insects (Smith, 1962). It is most commonly observed with aromatic acids, where the glycine conjugates of which are sometimes known as hippuric acids. Acids which undergo glycine conjugation are (1) aromatic acids, (2) substituted acetic acids, (3) β-substituted acrylic acids, and (4) certain naturally occurring steroid acids.

For instance, 2,4-dimethylbenzoic acid (II), a major metabolic product of dimethrin (I), is known to conjugate with glycine to form 2,4-dimethyl-hippuric acid (III):

Piperonylic acid from piperonal can also form a glycine conjugate (Kamienski and Casida, 1968):

PIPERONYLIC ACID   ITS GLYCINE CONJUGATE

The herbicide dichlorobenil (2,6-dichlorobenzenethioacetamide) gives off 2,6-dichloro-3-hydroxybenzoic acid, which forms a hippurate conjugate (Wit and van Genderen, 1966).

5.3.1. Cysteine Conjugation or Mercapturic Acid Synthesis

The types of compounds undergoing cysteine conjugation are few in number. The mercapturic acid synthesis involves addition of an L-acetylcysteyl residue to an aromatic ring (i.e., the benzoyl residue). In many cases, this is achieved by an initial reaction with glutathione followed by hydrolysis and acetylation. Another type of cysteine conjugation can occur on compounds with high affinities for the —SH group, such as organic mercury compounds and organohalides.

Formation of mercapturic acid may consist of the following general reactions. With aromatic hydrocarbons, mercapturic acid formation takes place with replacement of nuclear hydrogen by the acetylcysteyl residue:

$$R-C_6H_5 \longrightarrow R-C_6H_4-SCH_2\underset{\underset{NHCOCH_3}{|}}{C}HCOOH$$

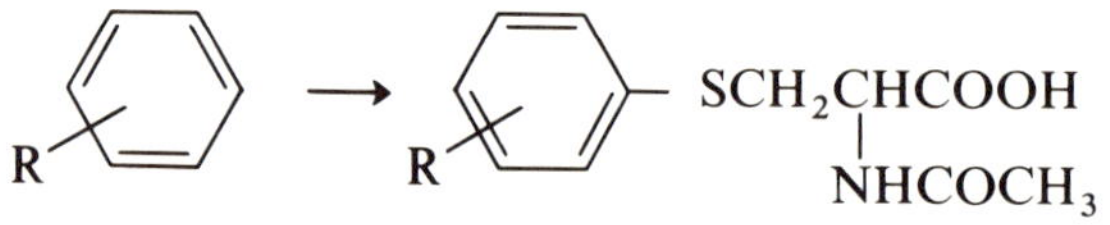

PHENYLMERCAPTURIC ACID

With halogenated aromatic hydrocarbons, the mercapturic acid is formed by acetylcysteyl dehalogenation:

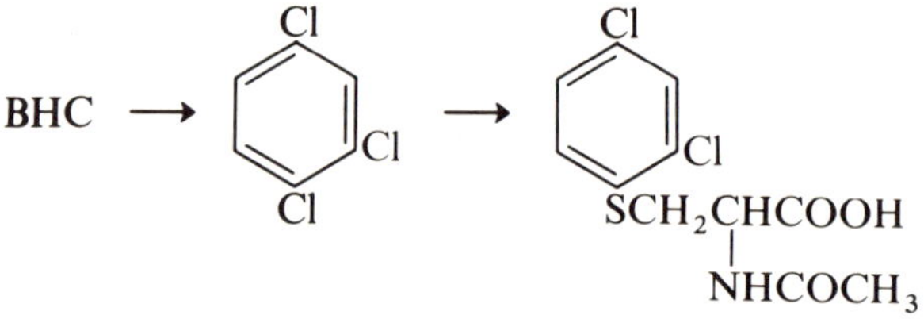

$$BHC \longrightarrow \text{1,2,4-TRICHLOROBENZENE} \longrightarrow \text{2,4-DICHLOROMERCAPTURIC ACID}$$

1,2,4-TRICHLOROBENZENE   2,4-DICHLOROMERCAPTURIC ACID

With halogenated nitrobenzenes, mercapturic acids are formed either by replacement of halogen as above or by replacement of the nitro group (i.e., acetylcysteyl denitration):

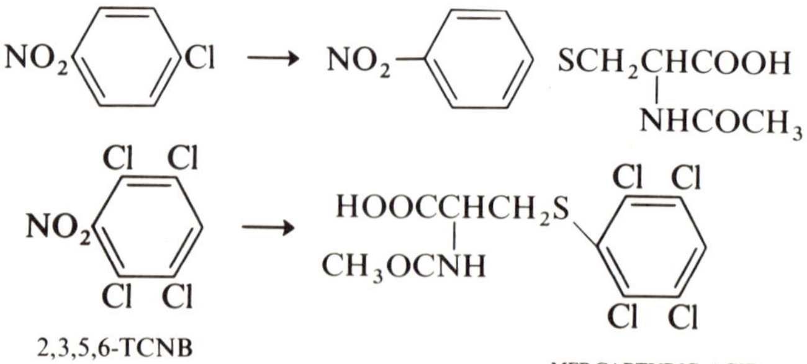

$$NO_2-C_6H_4-Cl \longrightarrow NO_2-C_6H_4-SCH_2\underset{\underset{NHCOCH_3}{|}}{C}HCOOH$$

2,3,5,6-TCNB

MERCAPTURIC ACID

For instance, in rabbits 2,3,5,6-TCNB is converted to a mercapturic acid and excreted along with a sulfate and a glucuronide (Bray *et al.*, 1953). The insecticidal compounds known to form cysteine conjugates are BHC (through trichlorobenzenes and dichlorobenzenes as illustrated above) (Grover and Sims, 1965) and ethylene dibromide (Nachtomi *et al.*, 1966).

Possible involvement of cysteine conjugation reactions in γ-BHC in insects will be discussed later.

5.3.5. Histidine, Lysine, and Glutamine Conjugation

Histidine is known to catalyze the decomposition process of methyl bromide by taking its methyl group and thereby releasing inorganic bromide:

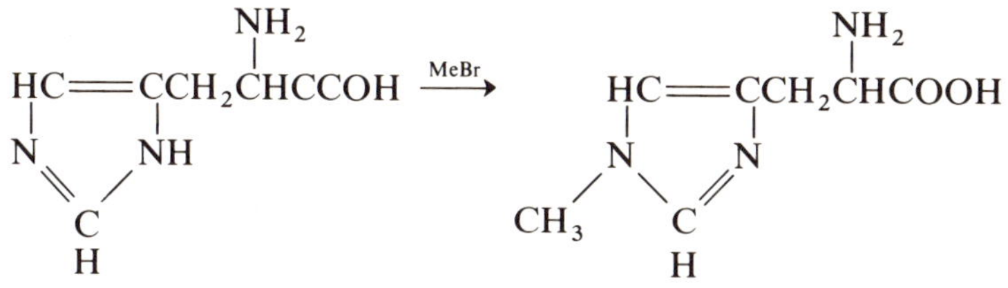

HISTIDINE   3-*N*-METHYLHISTIDINE

In addition, 1-*N*-methylhistidine and 1,3-*N,N'*-methylhistidine are known to form. In the fumigation of wheat, its protein (gluten) fraction accounts for most of the decomposition activity. Lysine is also utilized to demethylate methyl bromide (Bridges, 1955; Winteringham *et al.*, 1955). Glutamine conjugation has not been reported for insecticidal derivatives. It is known to take place with phenylacetic acid and indolyl-3-acetic acid in mammals.

5.3.6. Glucoside Formation

Glucosides are formed in plants and insects, and the process involves carbohydrates. Glucoside formation was originally observed by Smith (1955*a, b*) in certain species of Orthoptera, Coleoptera, Lepidoptera, and Hemiptera. It also occurs in Thysanura. Cockroaches require uridine diphosphate glucose (UDGP) to form glucoside, while mammalian systems require uridine diphosphate glucuronate (UDPGA) to form similar conjugates. In other cases, the conjugation products in insect and mammalian systems can be very different. For instance, piperonic acid from methylenedioxyphenyl compounds is known to form a glycine conjugate in mammals, whereas in insects the major conjugation product is a glucoside (Esaac and Cassida, 1968, 1969). Glucoside conjugation systems are apparently present in plants. For example, Blinn (1968) found β-glucosides among the metabolic products from Abate® in bean leaves:

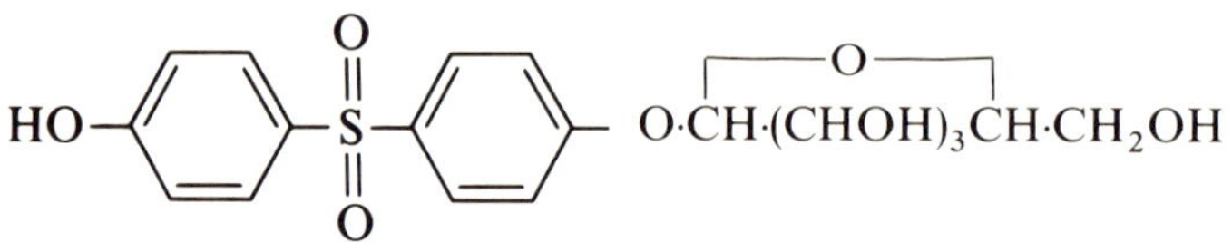

$$HO-\langle \rangle-\overset{\overset{O}{\|}}{\underset{\underset{O}{\|}}{S}}-\langle \rangle-O\cdot CH\cdot(CHOH)_3CH\cdot CH_2OH$$

SULFONYL DIPHENOL-β-GLUCOSIDE

5.3.7. Cyanide–Thiocyanate Detoxification

Sublethal doses of hydrocyanic acid or inorganic cyanides can be converted to thiocyanates, which are then excreted in the urine:

$$CN^- \rightarrow CNS^- \rightarrow (RCN \rightarrow RCNS)$$

Thiocyanates are also formed from organic cyanides, provided that they are converted *in vivo* to the cyanide ion. This conversion process can be interpreted as a true detoxification process since the thiocyanates are much less toxic than cyanides (Williams, 1959).

5.3.8. Methylation

Methylation is a normal metabolic process of some natural primary, secondary, and tertiary amines. For instance, the pyridinyl nitrogen of nicotine can be methylated to form isomethonium ion, which can be excreted in urine (McKennis *et al.*, 1963; Turnbull *et al.*, 1960):

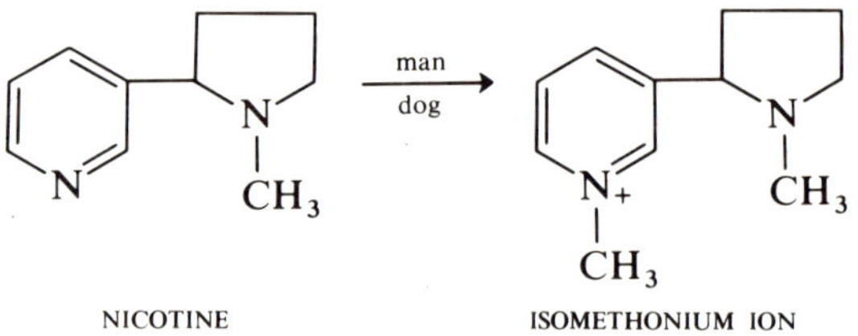

NICOTINE   ISOMETHONIUM ION

It may also occur with some foreign organic substances such as carbamates. The pyridines and quinolines are methylated *in vivo* at the heterocyclic nitrogen atom to yield highly ionized quaternary metal ammonium derivatives which are less toxic than their precursors and are readily excreted. Methylation may also occur with the phenolic hydroxyl group. The methylation of the —SH group of homocysteine is a normal metabolic process, and it should also occur with foreign sulfur compounds. This methylation process has not been found to take place in insect systems (Gilmour, 1959).

Although it is not an enzymatic process, the sulfide bond of Meta-Systox[®] is known to be methylated (Niessen *et al.*, 1963):

$$(CH_3O)_2P(O)SCH_2CH_2-S-CH_2CH_3 \rightarrow$$

META-SYSTOX[®]

$$CH_3O \diagdown$$
$$\diagup P(O)SCH_2CH_2-SCH_2CH_3$$
$$-O$$

$$+ (CH_3O)_2P(O)SCH_2CH_2 \overset{+}{S}CH_2CH_3$$
$$\qquad\qquad\qquad\qquad\quad |$$
S-METHYL META-SYSTOX[®] CH_3

5.3.9. Acetylation and Coupling with CoA

Acetylation is mainly a reaction against foreign amines and is a simple modification of normal reaction processes, e.g., conversion of choline to acetylcholine and acetylation of the —SH group of coenzyme A. Acetylation of foreign amines is usually regarded as a general reaction of aromatic amines and some unnatural aromatic acids such as α-amino-α-phenylbutyric acid and phenylcysteine. There are isolated instances of acetylation of other amino groups. Williams (1959) notes that the possible existence of additional forms of metabolic acylation of foreign compounds should be considered, since "active formyl" groups and acyl esters of coenzyme A other than acetyl have been detected in the tissues. Dichlone (2,3-dichloro-1,4-naphtho-quinone) for instance, has been suggested (Owens and Novotny, 1958) to form —S—CoA complex at the 2-position:

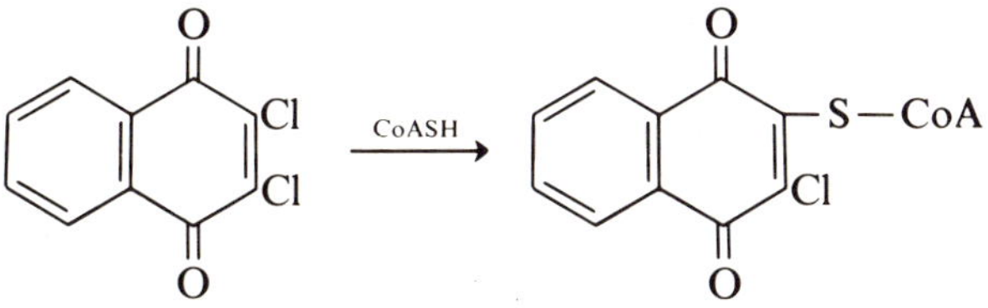

DICHLONE   —S—CoA COMPLEX

All monofluoroacetic acid derivatives are expected to form CoA derivatives before being incorporated into the TCA cycle to become fluorocitrate:

$$CH_2FCOOH \xrightarrow{\text{CoASH}} CH_2FCO-S-CoA$$
FLUOROACETATE   ACTIVATED —S—CoA FORM

5.3.10. Glutathione Conjugations

Although glutathione-involving metabolic transformations of insecticidal compounds are numerous and play very important roles in degradation, they have been mainly regarded as primary processes. The reason for the problem in classification is that the end-products of such processes are not the glutathione conjugates.

The question of whether glutathione conjugation systems are primary or secondary processes is purely academic. The former view is supported by the fact that intact glutathione complexes are not excreted by higher animals. The latter view is also tenable, since in plants glutathione conjugates such as that of atrazine (Lamoureux *et al.*, 1969) are stored as detoxified products:

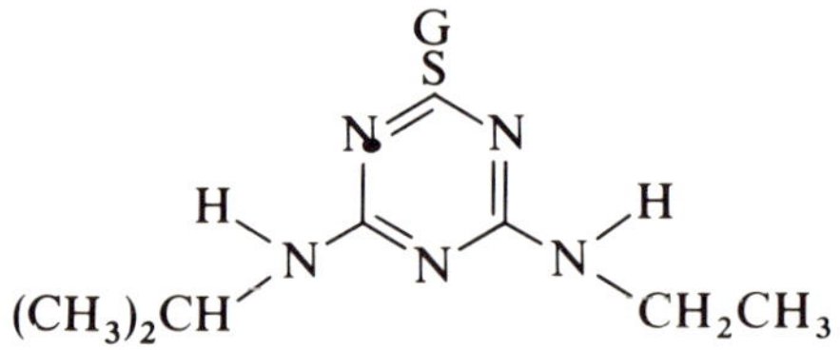

GLUTATHIONE CONJUGATE OF ATRAZINE

Because of the importance of this system in the primary processes, the details of these reactions were explained in Section 5.2.4.

5.3.11. Other Examples and Possibilities of Conjugations

Conjugation reactions with various uronic acids for aromatic compounds are expected to occur in plants. In mammals, ribosides can form instead of glucosides. Schayer (1956), for instance, observed the conversion of imidazole-4-acetic acid into a ribose derivative, particularly in rats, mice, and rabbits. Also, in insects metals are known to be removed from tissues in the form of insoluble sulfides. In birds, another possibility is ornithuric acid synthesis, which is equivalent to hippuric acid synthesis. Aromatic acids are known to react in this fashion in birds.

5.4. METABOLIC REACTIONS CHARACTERISTIC OF EACH GROUP OF INSECTICIDAL CHEMICALS

In the preceding section, general enzymatic systems which metabolize insecticidal chemicals have been explained. There are, however, many metabolic reactions which are pertinent to or very characteristic of certain groups of insecticides. Moreover, it seems advisable to study all metabolic reactions that occur on one or on a series of related chemicals in order to understand the sequence and thus the significance of individual reactions. That is, metabolic changes of insecticidal chemicals are generally judged according to whether they represent activation or detoxification, and in either case the initial reaction is usually the most critical step. For example, oxidation of phosphorothionates to phosphates ($P=S$ to $P=O$) or epoxidation of aldrin, heptachlor, etc., to dieldrin and heptachlor epoxide is carried out by the same "mixed-function oxidase," resulting in group-specific reactions. The same enzyme system can ring-hydroxylate various phenyl and naphthyl N-methylcarbamates, which may result in "activation" or "detoxification" of these carbamates. Chemical-group-specific reactions thus are indeed helpful in getting a general idea of the significance of metabolic activities.

Another reason why group-specific metabolism is of interest is that there are many common metabolic reactions that act on common or similar molecular fragments. For instance, once the initial metabolic process causes cleavage of organophosphorus compounds, the remaining phosphorus moieties often give very similar molecules, such as dimethylphosphoric acid, dimethylthiophosphoric acid, diethylphosphoric acid, and diethylthiophosphoric acid. This is important since the majority of researchers do not determine the fate of these small molecular fragments once the original molecule goes through the initial cleavage or disintegration process.

5.4.1. Chlorinated Hydrocarbon Insecticides

5.4.1a. Dehydrochlorination

The most characteristic reaction with the chlorinated hydrocarbon insecticides is dehydrochlorination. In this reaction, a chlorine atom is removed from the initial molecule along with a hydrogen from the adjacent carbon. The system requires GSH as a cofactor (Kearns, 1956; Lipke and Kearns, 1960; Dinamarca *et al.*, 1969). The resulting end-product is an olefinic compound. The dehydrochlorination system is present in many insect and vertebrate species. However, its activities in plants appear to be rather weak (Klein, 1972).

BHC apparently goes through a series of dehydrochlorination steps to yield various chlorinated olefinic and aromatic substances. The metabolism of BHC isomers has been studied extensively by Bradbury and Standen (1955), Oppenoorth (1956), and Sternburg and Kearns (1956). The last workers found that an initial dehydrochlorination product 1,3,4,5,6-pentachloro-1-cyclohexene (PCCH), is a metabolic intermediate of γ-BHC in the housefly.

Bradbury and Standen (1959, 1960), however, pointed out that less than 3 % of γ-BHC is converted to this compound in houseflies. Another minor metabolic product has been identified as 1,2,4-trichlorobenzene (Bradbury and Standen, 1958). The major metabolite appears as a single spot on paper chromatographic systems and is water soluble. On alkaline hydrolysis, this major metabolite gives all six possible isomers of dichlorothiophenol, i.e., 2,3-, 2,4-, 2,5-, 2,6-, 3,4-, and 3,5-dichlorothiophenol. Ishida and Dahm (1965a, b) partially purified the housefly enzyme system which efficiently degrades α-, γ-, and δ-isomers of BHC in addition to γ- and δ-PCCH into water-soluble metabolites. The order of metabolic susceptibility among isomers was δ-PCCH $\gg$ γ-PCCH $>$ α-BHC. β-BHC was not degradable by this system. Among the insect and mammalian systems tested, houseflies were by far the most active organisms in degrading BHC.

Clark *et al.* (1969) found that GSH plays an important role in BHC degradation in houseflies and grass grubs. PCCH did not appear to be the intermediate for the production of dichlorophenylhydrogen sulfides. Reed and Forgash (1970), on the other hand, studied organic soluble products in houseflies and found 1,2,4- and 1,2,3-trichlorobenzene and isopentachloro-cyclohexene. These metabolites could be formed through dehydrochlorina-tion processes, although another metabolite, pentachlorobenzene, could not.

One note of caution is that these dehydrochlorination systems reacting with BHC isomers have never been critically compared to DDT-dehydro-chlorinase. Chances are that the housefly enzymes are not DDT-dehydro-chlorinase, since DDT resistance due to this enzyme does not usually extend to BHC resistance in this species. There are probably many more substrates for various dehydrochlorination systems such as TDE, some cyclodiene analogues, and toxaphene, but details of reaction mechanisms have not been worked out. There is one case in which a dehydrochlorinase is believed to act on an insecticidal chemical other than chlorinated hydrocarbon insecticides —i.e., the dehydrochlorination process by which trichlorfon produces di-chlorvos, which actually represents activation (Metcalf *et al.*, 1959).

5.4.1b. Reductive and Hydrolytic Dechlorination

Reductive dechlorination, so called because it requires hydrogen atoms to replace chlorine, is a very common reaction in the microbial world, but this method of degradation must also be widely distributed among animal and plant species as judged by the appearance of TDE (DDD) in them as a result of DDT administration. The major route of DDT degrada-tion in mammals appears to be DDT $\rightarrow$ DDD $\rightarrow$ (DDMU)* $\rightarrow$ DDMS $\rightarrow$ (DDNU) $\rightarrow$ DDOH $\rightarrow$ DDA (e.g., Peterson and Robison, 1964; Datta, 1970). The system requires a series of dechlorination and/or dehydrochlorination steps followed by hydrogenation steps. Since DDE is very stable and is regarded as a terminal product, the initial dechlorination step must play a very important role in deciding the overall rate of degradation of DDT in any animal. *In vivo*, there is always the question of contributions by intestinal microbial fauna and flora in the formation of DDD from DDT (Peterson and Robison, 1964), and such a possibility cannot be easily dismissed.

There are a number of examples of hydrolytic dechlorination of cyclodiene insecticides (Brooks, 1969). In these reactions, chlorine atoms are replaced by —OH groups. Though such reactions can take place by the action of oxidative hydroxylation systems, hydrolytic dechlorination reactions are characterized by their great susceptibility to pH changes and insensitivity to NADPH and methylenedioxyphenyl inhibitors such as sesamex. This type of hydrolytic reaction can occur only with the chlorinated chemicals that are already unstable. A good example can be found in *N*-chlorotriazines such as atrazine, which is easily hydrolyzed to yield hydroxyatrazine (Armstrong and Harris, 1965).

5.4.1c. Oxidative Reactions

There are basically three types of oxidative attacks on chlorinated insecticides. They are epoxidation of a double bond, hydroxylation of either direct or nonsubstituted carbons or replacement of chlorine atoms, and formation of aldehydes, ketones, and acids from alcohols.

The epoxidation process is widely distributed and occurs with various cyclodiene insecticides. Originally it was assumed that only double bonds were attacked, such as aldrin to dieldrin, isodrin to endrin, and heptachlor to heptachlor epoxide:

HEPTACHLOR   HEPTACHLOR EPOXIDE

Schwemmer *et al.* (1970) and Lawrence *et al.* (1970) independently showed that epoxidation can take place on a saturated ring of chlordane, possibly

---

* The common names for DDT metabolites are not in accordance with the official method of nomenclature, i.e., DDMU stands for 2,2-dichlorodiphenyl-1-monochlorinated unsaturated ethane, DDMS for 2,2-dichlorodiphenyl-1-monochlorinated saturated ethane, and DDNU for 2,2-dichlorodiphenyl-1-nonechlorinated unsaturated ethane.

through a dehydrogenation step:

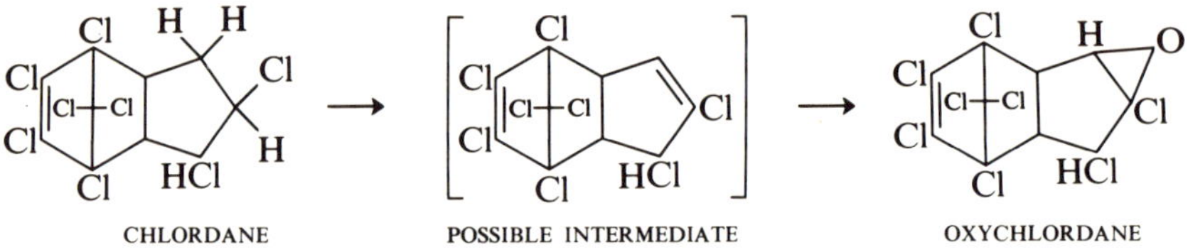

CHLORDANE   POSSIBLE INTERMEDIATE   OXYCHLORDANE

Examples of epoxidation in chlorinated insecticides other than cyclo-diene insecticides are rare. Probably an epoxy ring forms in the process of BHC metabolism to give two different trichlorophenols as follows (Grover and Sims, 1965):

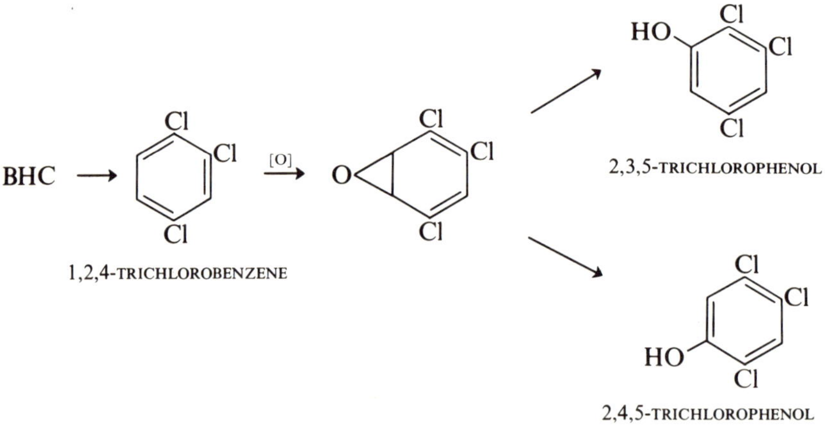

2,3,5-TRICHLOROPHENOL

BHC →

1,2,4-TRICHLOROBENZENE

2,4,5-TRICHLOROPHENOL

Hydroxylation can take place either on aromatic rings or on alkyl carbons. DDT is hydroxylated in *Drosophila melanogaster* (Tsukamoto, 1959) and in the German cockroach, *Blattella germanica* (Menzel *et al.*, 1961), probably to form dicofol:

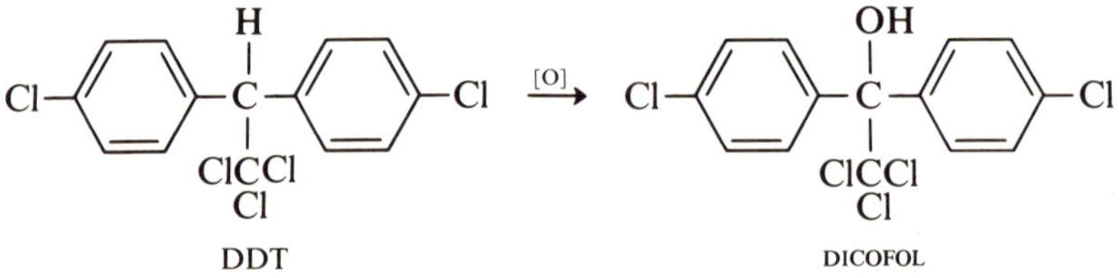

DDT   DICOFOL

There are a number of reports indicating that hydroxylation processes are important in the process of metabolism of cyclodiene insecticides.

Richardson *et al.* (1968) reported one major urinary and one major fecal metabolite to form from dieldrin *in vivo* in the rat. The urinary metabolite was identified as 2-ketodieldrin by these workers and by Damico *et al.* (1968). Feil *et al.* (1970) later identified the fecal metabolite as 9-hydroxydieldrin. Matthews and Matsumura (1969) studied the metabolic activities *in vivo* and *in vitro* and found three pathways to produce 2-ketodieldrin, 9-OH dieldrin, and *trans*-aldrindiol: each system appeared to be operating independently of the others. Brooks and Harrison (1967) showed that hydroxylation reactions by pig liver microsomes are most prevalent among dihydrochlordene derivatives. Baldwin *et al.* (1970) similarly found that 9-hydroxyendrin forms from endrin in the rat.

5.4.1d. *Metabolism of Aliphatic Chlorinated Hydrocarbons: Fumigants and Metabolic Fragments*

Some of the short-chain fumigants such as chloroform and methyl chloride are volatile and may be eliminated from the body unchanged in the expired air. A number of chlorine-containing insecticides give up halogens or a halogenated moiety of the molecule, although the majority do have relatively high molecular weights and therefore are not volatile enough to be eliminated through the respiratory processes. Sterner (1949) has expressed the opinion that the toxicity of these fumigants is due to the undegraded molecule rather than to metabolic decomposition products. Several of these compounds are quite stable *in vivo*, and a high percentage of unchanged material can often be recovered from fatally poisoned animals. In the tissues, they are expected to form various conjugates with various amino acids, e.g., ethylene dichloride (Morrison and Munro, 1965):

$$ClCH_2CH_2Cl \rightarrow \quad \underset{HOOC}{\overset{H_2N}{\diagdown}}CHCH_2S-CH_2CH_2SCH_2CH\underset{COOH}{\overset{NH_2}{\diagup}}$$

ETHYLENEDICHLORIDE $\qquad$ *S,S'*-ETHYLENE-BIS-CYSTEINE

The expected metabolic change which halogenated hydrocarbons undergo *in vivo* is a hydrolytic dehalogenation with the formation of oxidized hydrocarbon and halogen ions. This is an enzymatic reaction and occurs in the liver, kidney, and spleen:

$$RCH_2Cl + H_2O \rightarrow RCH_2OH + Cl^- + H^+$$

$$RCHCl_2 + H_2O \rightarrow RCHO + 2Cl^- + 2H^+$$

$$RCCl_3 + 2H_2O \rightarrow RCOOH + 3Cl^- + 3H^+$$

Brominated compounds are expected to go through similar hydrolytic reactions. Ethylene dibromide gives 2-bromoethanol as the metabolic product in the body.

The α-halogen-substituted aldehydes and alcohols form additional compounds at the carbonyl group with nucleophilic reagents due to the presence of the electron-attracting halogen groups. Therefore, they readily add water to form dihydroxy compounds or hydrates. The chlorinated aldehydes are metabolized mainly by reduction to the corresponding alcohols and are oxidized only with difficulty. They are mainly excreted as glucuronides of the corresponding alcohol. In the case of chloral hydrate, the metabolic routes are

$$glucuronic\ conjugation$$

$$\nearrow$$

$$CCl_3CH(OH)_2 \ \rightarrow \ CCl_3CHO \ \rightarrow \ CCl_3CH_2OH \ \rightarrow \ CCl_3CH_2OC_6H_9O_6$$

$$\searrow \qquad\qquad \nearrow$$

$$CCl_3COOH \ \rightarrow \ protein\ bound$$

$$CH_3CClHCOOH \xleftarrow[\text{dechlorination}]{\text{plants}} CH_3CCl_2COOH \ (dalapon) \xrightarrow{\text{cow}} glycerides$$

$$CH_3CCl(OH)COOH$$

The resulting chlorinated acids are bound to either proteins or glycerides. The speed of dechlorination, either reductive or hydrolytic, is expected to be slow (Kutschinski, 1961).

5.4.1e. Metabolism of Halogenated Aromatic Compounds

Many insecticidal chemicals including organophosphates and carbamates give chlorinated aromatic compounds as metabolic products. The majority of them are chlorinated benzenes and phenols. Investigation into the metabolism of halogenated aromatic hydrocarbons has generally centered on the formation of mercapturic acids and the consequent relationship to cysteine and sulfur metabolism. However, in many cases mercapturic acid formation is not an important reaction; the chemical is mainly metabolized by hydroxylation. Monohalogen benzenes form appreciable amounts of mercapturic acids, but tri-, tetra-, penta-, and hexachlorobenzenes form little or no mercapturic acids.

In general, halogen-substituted benzenes are metabolized in the animal body to phenols, catechols, and mercapturic acids. The output of these metabolites depends on the number and orientation of the halogen atoms and on the species of animal. As the number of substituted chlorine atoms increases, formation of mercapturic acids and catechol decreases and the polychlorinated benzenes are metabolized mainly by oxidation to monohydric phenols. The higher chlorinated benzenes are not readily absorbed into excretory systems so that total metabolites in the urine tend to decrease as chlorine content increases.

On the other hand, halogenated toluene derivatives do not form mercapturic acids. They are metabolized mainly (if not entirely) by oxidation of the methyl group to yield the corresponding halogenated benzoic acids, which may be excreted in part as halogenated hippuric acids.

p-Dichlorobenzene is widely used against moths and is metabolized to phenols and does not form mercapturic acid products. The *p*-isomer is more toxic than the *o*-isomer. This could be due to differences in metabolism, since the *o*-compound is a mercapturic acid former while the *p*-compound is not. The *m*-isomer behaves in a similar manner to the *o*-compound. The trichlorobenzenes are metabolized by oxidation to phenols, which are excreted in the conjugated form. Small amounts of mercapturic acid products are also present. This occurs by direct replacement of one of the chlorines with the —SH group of the cysteine molecule.

Chlorinated naphthalenes are widely used as insecticides, and the higher polychlorinated naphthalenes are nonflammable waxlike substances useful in electrical insulation. Very little is known about the metabolic fate of these compounds, but present data suggest that toxicity is related to their metabolic fate. It appears that metabolites consist of conjugated glucuronic acids and small amounts of ethereal sulfates and mercapturic acids. Cornish and Block (1959) suggest that a high degree of chlorination of naphthalene may interfere with the formation of 1,2-dihydro-1,2-diols, which they regard as the normal mechanism of naphthalene metabolism.

5.4.1f. DDT Metabolism

Insects. In insects, the usual route of metabolism of DDT is through DDE (Fig. 5-2). However, there are some additional routes with as many as seven metabolites in addition to DDE.

Sternberg *et al.* (1950) showed that DDT-resistant houseflies detoxify DDT mainly to its noninsecticidal metabolite DDE. The ability to change absorbed DDT to DDE appears to be a major factor in the survival of poisoned flies. The rate of dehydrohalogenation of DDT to DDE varies a great deal among fly strains and among individual flies. Unidentified metabolites of DDT have also been found in houseflies. Kimura and Brown (1964) found that DDT-resistant *Aedes aegypti* strains produced more DDE from DDT than did susceptible individuals.

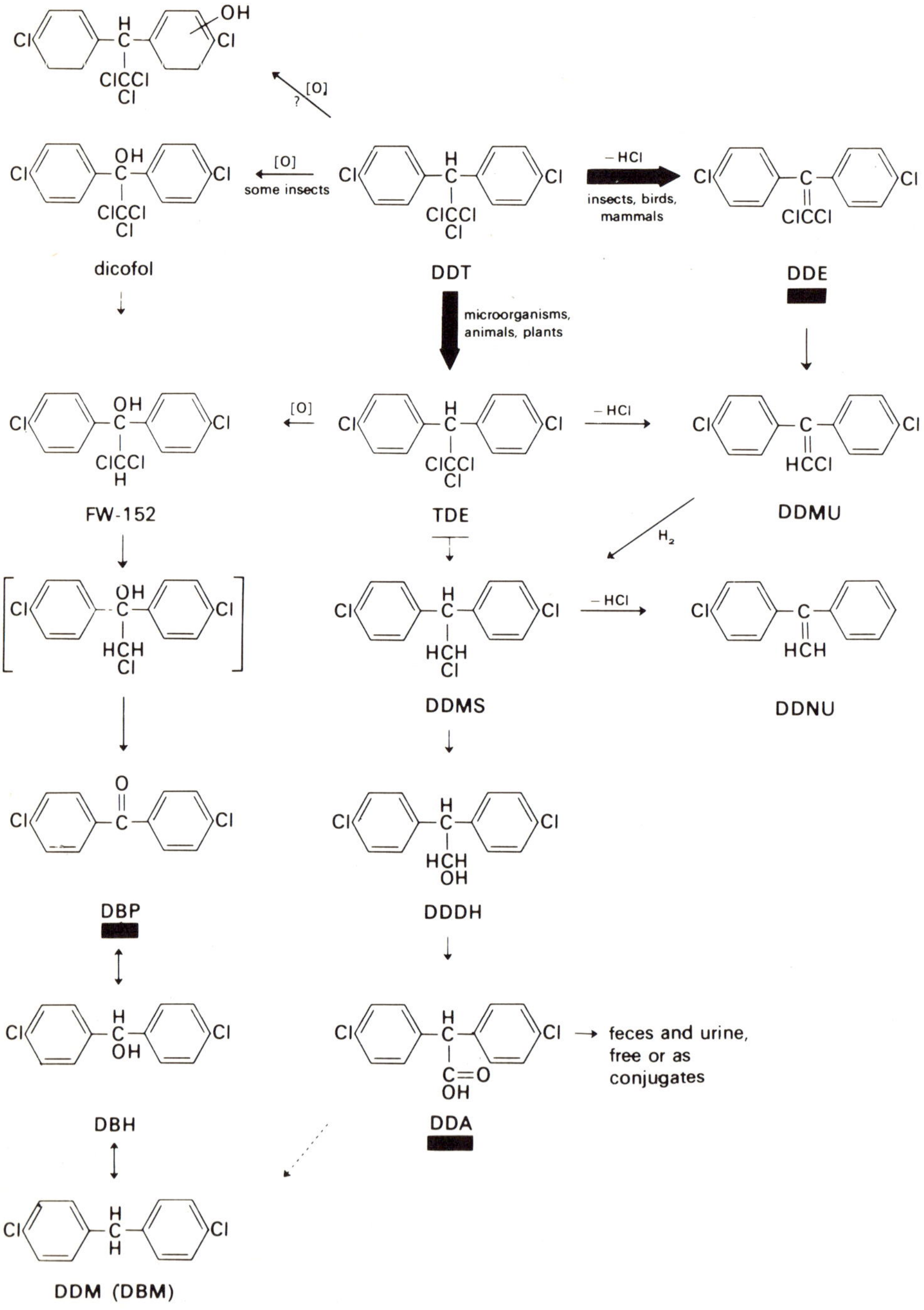

Fig. 5-2. General pattern of DDT metabolism in animals and plants.

DDT-dehydrochlorinase, a glutathione-dependent enzyme, was isolated by Lipke and Kearns (1959, 1960), Miyake *et al.* (1957), and Sternberg *et al.* (1954). This enzyme system catalyzed the degradation of *p,p'*-DDT to *p,p'*-DDE or the degradation of *p,p'*-DDD (TDE) to its corresponding ethylene (TDEE). The *o,p'*-isomer of DDT was not attacked by DDT-dehydrochlorinase, suggesting that the *p,p*-orientation is necessary for enzyme action. Metcalf (1955) cites the activity of this enzyme to analogues of DDT in the order DDT > DBrDT > DDD > methoxychlor > DFDT. In general, DDT resistance of fly strains is correlated with the activity of DDT-dehydrochlorinase, although other resistance mechanisms are known to exist.

In many studies, portions of the DDT applied cannot be accounted for as DDT, DDE, or other known metabolites producing the Schechter–Haller reaction. Butts *et al.* (1953) injected ^{14}C-DDT labeled at the 2-carbon into *Periplaneta americana*. As much as 55% of the total DDT injected was converted to an unknown metabolite, and aqueous extraction indicated that 43% was in the aqueous phase. It could not be extracted by ether until treated by sulfuric acid, suggesting that the metabolite was a conjugate between a DDT derivative and a carbohydrate. No evidence of excretion of $^{14}CO_2$ was obtained.

Grasshoppers show a natural tolerance to DDT. Sternberg and Kearns (1952) showed that their tolerance depends on metabolism in the cuticle and gut to DDE and on the rapid passage of ingested DDT through the gut without appreciable absorption. These factors prevent DDT from reaching its site of action in the nervous system. A small injected dose of DDT is fatal to these insects, but they can withstand large oral or dermal doses. DDE was the only metabolite identified, but unidentified metabolites were also present.

Although the usual route of metabolism of DDT by insects appears to be through DDE, there are some exceptions. Perry and Buckner (1958) found that DDT-resistant human body lice metabolize DDT to what probably is *p*-chlorobenzoic acid. Also, Tsukamoto (1959, 1960) found that *Drosophila melanogaster* metabolized DDT to its hydroxy analogue, dicofol (Kelthane®). The tobacco hornworm and *Blattella germanica* degrade TDE to its hydroxy analogue (Gatterdam *et al.*, 1964).

Mammals. Several groups have isolated and identified DDE, DDA, and 4,4-dichlorobenzophenone as metabolites in mammals fed or exposed to DDT. DDE and DDA appear in the feces and bile, probably in amide linkage. The principal water-soluble metabolite is DDA.

Studies on DDT metabolism in rats indicated DDD to be an intermediate (Datta *et al.*, 1964; Finley and Pillmore, 1963). The presence of a phenolic compound among the metabolites was suggested by Morello

(1965). It appears certain from various data that DDA is the final major metabolite of DDT in the rat.

As for the form of DDT metabolites that are excreted in feces and urine, Jensen *et al.* (1957) concluded that the fecal complex was apparently a DDA complex, but it differed from the biliary complexes. Pinto *et al.* (1965) studied both fecal and urinary metabolites of DDT. The fecal material gave one major and two minor DDA conjugates. The urinary metabolites also gave three distinct peaks. The major fecal conjugate was purified and found to be a DDA conjugate containing DDA, serine, and aspartic acid in equimolar proportions. Other DDT metabolites found in rats by Pinto *et al.* (1965) were *p,p'*-dichlorobenzhydrol (DBH), *p,p'*-dichlorodiphenylmethane (DDM or DBM), *p,p'*-dichlorobenzophenone (DBP), and DDE. Peterson and Robison (1964) found large amounts of DDD and DDMU in rats fed DDT via stomach tubes. They subsequently fed DDMU to rats and obtained DDMS. Thus these authors concluded that the major scheme of DDD degradation is DDD → DDMU → DDMS → DDNU → DDOH → DDA.

Monkeys fed DDT excreted DDA that was probably produced without DDE as an intermediate metabolite (Durham *et al.*, 1963). DDT has an affinity for fatty materials and is stored in mammalian fat and excreted in milk. Fat storage in rats takes place at intakes as low as 1 ppm. The fat storage is composed of a mixture of DDT and DDE (Lang *et al.*, 1951).

Wildlife. Judging by the levels of DDE detected in wild avian species, DDE production appears to be one of the most prevailing biochemical activities on DDT. Bailey *et al.* (1969) found in pigeons (*Columba liva*) that DDE is produced from DDT in relatively large quantities. They also found that DDE is not metabolized by the bird, while DDD (TDE) gives rise to small amounts of DDE and is metabolized rapidly to DDMU. Thus in this species dehydrochlorination activity is strong. The rate-limiting factor is further degradation of DDMU, which requires hydrogenation steps. They concluded that the final degradation product of DDMU is DDM (DBM), which can also form DDMU after prolonged exposure to light. Abou-Donia and Menzel (1968), on the other hand, found DDE, DDD, DDMU, DDMS, DDNU, DDOH, DDA, DDM, and DBP in chicks administered DDT either at the egg stage or after hatching. Essentially the same metabolites were detected when they substituted *p,p'*-DDD for DDT, indicating that the major metabolic route is through this compound. Treatment of chicks with DDE, interestingly, gave a small amount of DBP as the metabolic product.

It is generally acknowledged that fish do not have efficient detoxification mechanisms; thus there has been some discussion as to whether the low levels of DDT metabolism observed in a few fish species are due to microbial action. Wedemyer (1968), however, showed that the liver of the rainbow trout could convert DDT to DDE, indicating that at least some of the activities come from the fish themselves. Cherrington *et al.* (1969) studied DDT degradation *in vitro* in the intestinal contents of Atlantic salmon and found DDE and DDD. At this stage, it appears that fish metabolism of DDT is not extensive, and moreover that at least some part of such metabolic activity comes from microbial action. Only a few metabolites have been described in fish.

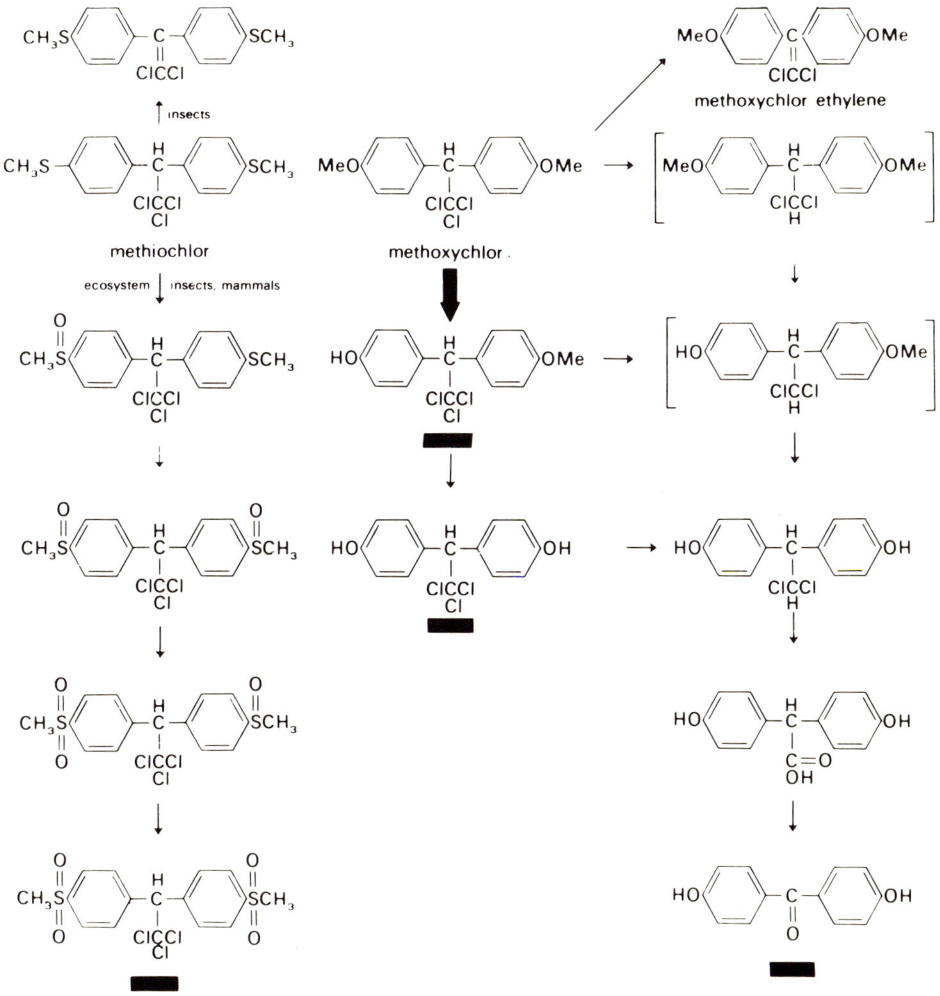

Fig. 5-3. Metabolic pattern of methoxychlor and methiochlor in the mouse and housefly and in an ecosystem. From Kapoor *et al.* (1970).

5.4.1g. Methoxychlor Metabolism

Methoxychlor is a safe insecticide to use since it is rapidly destroyed in the mammalian body and is not accumulated in fat or excreted in milk. Methoxychlor is not excreted in urine either intact or as its acetic acid derivative that is found in feces. Kapoor *et al.* (1970) studied the metabolism of DDT, methoxychlor, and methiochlor in mice and insects and in a model ecosystem and found that the pattern of metabolism for the latter two compounds is different from that for DDT in that modification of the *p*-substituent plays a large role. The general pattern is illustrated in Fig. 5-3 to demonstrate this point.

Methoxychlor is slowly dehydrochlorinated in alkali to 2,2-bis(*p*-methoxyphenyl)-1,1-dichloroethylene. This reaction proceeds slower than with DDT. Heavy metals catalyze this reaction.

5.4.1h. BHC Metabolism

Mammals. At present, there appear to be two schools of thought regarding BHC metabolism. One group follows the traditional line of thinking that γ-BHC first forms γ-PCCH (γ-2,3,4,5,6-pentachlorocyclohex-1-ene), which undergoes various changes (route A, Fig. 5-4). The second group claims that γ-PCCH is not an intermediate, but γ-BHC is immediately hydroxylated by an efficient system to yield various chlorophenols (route B, Fig. 5-4).

The first group feels that in mammals the metabolism of γ-BHC proceeds via γ-PCCH to 1,2,4-trichlorobenzene (1,2,4-TB) and then to phenols, sulfates, and glucuronic acid conjugates which are excreted in the urine. In studies with dogs, rabbits, and rats (Coper *et al.*, 1951; van Asperen and Oppenoorth, 1954; San Antonio, 1959; Broniz *et al.*, 1962; Koransky *et al.*, 1964), BHC was rapidly broken down to 1,2,4-trichlorobenzene and other unidentified compounds. In studies by Grover and Sims (1965), γ-BHC and γ-PCCH were converted by rats to 2,3,5- and 2,4,5-trichlorophenol and excreted in the urine as free phenols, sulfates, and glucuronic acid conjugates and 2,4-dichloromercapturic acid. Since Ishida and Dahm (1965a, b) found that rat liver microsomes in the presence of $NADPH_2$ and O_2 metabolized γ- and δ-PCCH, it appears certain that there is an efficient oxidative system to metabolize PCCH.

When it comes to γ-BHC itself, however, the claim for the γ-PCCH pathway is not certain. Chadwick and Freal (1972a) studied the urinary metabolites of γ-BHC and found no PCCH or chlorinated benzenes. In the neutral fraction from urine, they found a large quantity of 2,3,4,5,6-penta-chloro-2-cyclohexen-1-ol (PCCOL). From the acidic fraction, they recovered various chlorinated phenols, particularly 2,3,4,6-tetrachloro, 2,4,6-trichloro

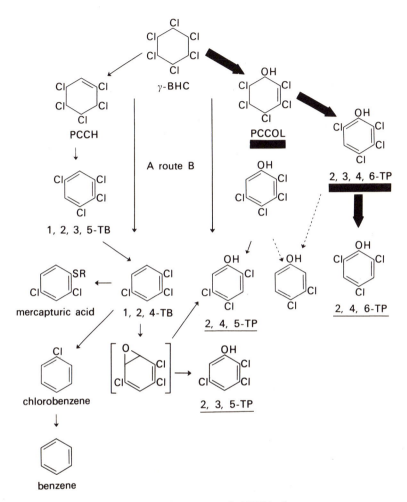

Fig. 5-4. Metabolic patterns of γ-BHC in the rat.

> 2,3,4,5-tetrachloro, 2,4,5-trichloro > 2,3,5-trichloro > 3,4-dichloro. They actually isolated PCCOL and established its identity through mass, NMR, infrared, and ultraviolet spectroscopy. Thus it appears certain that PCCOL is the immediate metabolic product from γ-BHC. The same authors (1972b) repeated the work with DDT-pretreated rats and found that DDT enhances the 2,3,4,5-TP and 2,4,5-TP route appreciably, while the production of 2,3,4,6-TP also increases somewhat.

Two explanations can be offered to reconcile this discrepancy. The first possibility is that γ-PCCH, or similar intermediates such as iso-PCCH isolated by Reed and Forgash (1968), can be converted into PCCOL through an oxidative attack. The second explanation is that only hydroxylated metabolic products can be excreted into the urine by rats and that PCCH and tetrachloro- and trichlorobenzenes are retained in the body. Chadwick and Freal (1972*a*) studied only the urinary products that could be extracted from the acidified (pH 2) urine into benzene.

Insects. Oppenoorth (1954) demonstrated that houseflies metabolize γ-BHC and that BHC-resistant flies degrade it faster than susceptible ones. In other studies on the fate of ^{14}C-γ-BHC in houseflies, pentachloro-cyclohexene and some water-soluble metabolites have been found. No respired gases contained ^{14}C. Most of the radioactivity was extracted by carbon tetrachloride, but an unextracted residual activity remained in both resistant and susceptible flies (Bradbury and Nield, 1953; Bradbury and Standen, 1955, 1958; Bradbury, 1957; Bridges, 1959; Sternberg and Kearns, 1956).

Ishida and Dahm (1965) purified an active enzyme fraction from the housefly that converted γ-BHC to unidentified water-soluble metabolites. Studies by Bradbury and Standen (1959) and Sims and Grover (1965) indicated that the first step in the metabolism of γ-BHC by flies was the removal of one chlorine atom, and the formation of a C—S bond, followed by the loss of other chlorine atoms to form dichlorothiophenol.

Work by Bogdarina (1957) and San Antonio (1959) has indicated that pentachlorocyclohexane is a metabolite of BHC in plants.

5.4.1i. Aldrin and Dieldrin Metabolism

Many studies have shown that aldrin is converted to its epoxide ana-logue dieldrin by mammals, soil microorganisms, plants, and insects. For example, Brooks and Harrison (1963) and Giannotti (1958) demonstrated that houseflies and *P. americana* epoxidized isodrin to endrin. Wong and Terriere (1965) showed that the epoxidation of aldrin, isodrin, and heptachlor was performed by rat liver microsomes. Furthermore, they found that female rats converted these compounds less rapidly than male rats. Although aldrin/dieldrin is considered to be a very stable compound in animals, mammalian species, in particular, degrade it rather rapidly (Fig. 5-5).

Ludwig *et al.* (1964) fed ^{14}C-aldrin to rats and found that the active material excreted in the urine and feces consisted of aldrin, dieldrin, and considerable amounts (up to 75% in the feces and 95% in the urine) of a mixture of unidentified hydrophilic metabolites. This group also showed that a saturation level does occur and that most (99.5%) of the administered insecticide eventually is excreted either as the insecticide itself or as a meta-bolic product and is not stored in the rat tissue.

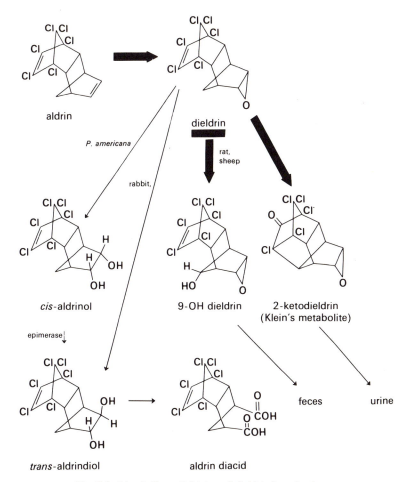

Fig. 5-5. Metabolism of aldrin and dieldrin in animals.

Korte and Arent (1965) and Ludwig and Korte (1965) isolated six metabolites of dieldrin and identified the main one as one of the two enantiomorphic isomers of 6,7-*trans*-dihydroxydihydroaldrin. This compound as tested *in vivo* has an acute oral toxicity to mammals of about one-twelfth to one-sixth that of dieldrin. After intravenous administration of this metabolite to rats, a more hydrophilic compound was found which was more likely the corresponding diacid.

Cueto and Hayes (1962) and Fletcher (1959) found at least two unidentified, neutral, polar, chlorinated metabolites of dieldrin in the urine of men occupationally exposed to dieldrin.

Although there were many studies indicating the presence of hydrophilic metabolites of dieldrin in the urine and feces of rats and some other mammalian species, it was not until 1968 that any serious efforts were made to unequivocally isolate and identify them. Richardson *et al.* (1968) and Damico *et al.* (1968) identified the major urinary metabolite as 2-ketodieldrin, and Matthews and Matsumura (1969) confirmed this identification. Although Richardson *et al.* (1968) tentatively proposed the major fecal product of dieldrin to be 4a- or 5-hydroxydieldrin, Feil *et al.* (1970) finally concluded that it is 9-hydroxydieldrin. Since this metabolite gives a carbonyl group to form a ketone on oxidation, it cannot be a tertiary alcohol as postulated before. In the rat, excretion of 9-OH dieldrin through the fecal route is much more important than that through the urinary route (Matthews *et al.*, 1971). The structures of these two major metabolic products of dieldrin have been confirmed in a detailed study made by McKinney *et al.* (1972).

With respect to insects, Sellers and Guthrie (1972) found a hydroxydieldrin in the housefly. Nelson and Matsumura (1973) also found at least two monohydroxylated dieldrins in the tissues of houseflies treated with dieldrin, one of which behaved identically to 9-OH dieldrin in various chromatographic systems. In *B. germanica*, the major metabolic product was *trans*-aldrindiol, and there were two other monohydroxydieldrins. The major metabolite was not 9-OH dieldrin, since it could not be oxidized to a ketone by the oxidation procedure successful for 9-OH dieldrin. Apparently, it was either 4a- or 5-hydroxydieldrin. In *P. americana*, the major metabolic product was *cis*-aldrindiol.

5.4.1j. Isodrin and Endrin Metabolism

As with aldrin, isodrin is metabolically converted to endrin (e.g., Brooks and Harrison, 1963), and since isodrin is not an economically important compound, endrin is the one that has been studied by scientists.

In mammalian metabolism of endrin, the key argument has been whether Δ-ketoendrin can be the major metabolic product. Δ-Ketoendrin is a thermal isomerization product of endrin (Phillips *et al.*, 1962) and is also formed as a result of microbial action. Klein *et al.* (1968) reported that Δ-ketoendrin is the metabolite, but Richardson (1965) could not find it using a GLC analysis technique in the tissues of rats fed endrin.

Baldwin *et al.* (1970) isolated two major fecal metabolites and one urinary metabolite from rats fed endrin (Fig. 5-6). Their infrared, mass, and NMR spectroscopic data indicate that the major fecal metabolite is 9-OH endrin, which can be oxidized to 9-ketoendrin, the major urinary metabolite. The other fecal metabolite appears to be endrin hydroxylated at a carbon atom other than 9. While there are some insect metabolism data, it seems

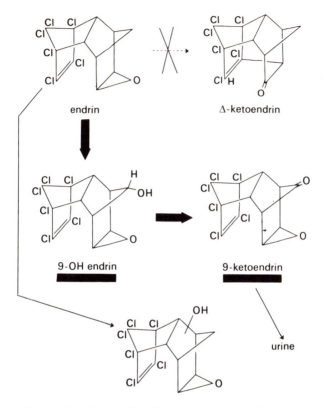

Fig. 5-6. Metabolism of endrin in the rat. Drawn from the data
of Baldwin *et al.* (1970).

wise not to try to draw any definite conclusion from them at this stage,
in view of the mammalian data.

5.4.1k. Metabolism of Heptachlor and Chlordane Analogues

Heptachlor is metabolized to heptachlor epoxide by mammals, insects,
plants, and soil microorganisms. When cows are grazed on heptachlor-
treated pasture, only heptachlor epoxide is found in the body tissue and
milk (Stoddard *et al.*, 1954; Ely *et al.*, 1955; Bache *et al.*, 1960; Rusoff *et al.*,
1962, 1963). Wong and Terriere (1965) showed that, in the rat, epoxidation
of heptachlor occurred in the microsomal part of the liver, but they did not
study further metabolism. Klein *et al.* (1968) intravenously injected rats and
rabbits with heptachlor and found the epoxide and another hydrophilic
compound. They compared this hydrophilic compound with an authentic

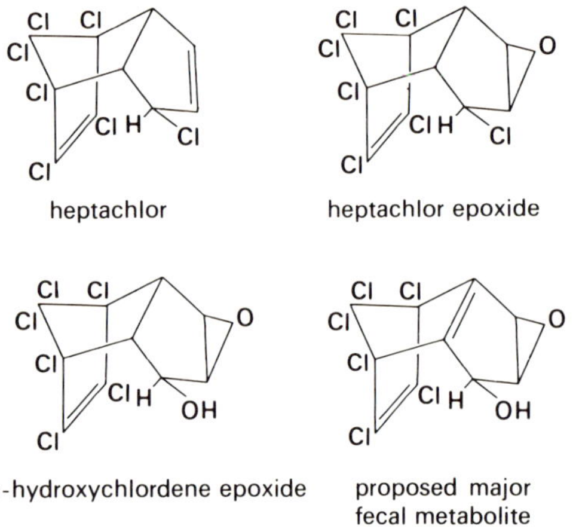

heptachlor   heptachlor epoxide

1-exo-hydroxychlordene epoxide   proposed major
                                                    fecal metabolite

Fig. 5.7. Heptachlor metabolic analogues. From Matsumura
and Nelson (1971).

1-exo-hydroxychlordene epoxide by thin-layer, paper, and gas chromato-graphic methods and concluded that the two were identical.

Matsumura and Nelson (1971) fed heptachlor epoxide to rats and also collected a hydrophilic metabolite from feces. However, they noted some difference in its thin-layer chromatographic behavior from that of 1-exo-hydroxychlordene epoxide. NMR, infrared, and mass spectroscopic analyses of this metabolite indicated that it had one more double bond than the reference compound. They proposed the structure shown in Fig. 5-7 for the major metabolic product.

Little is known about the metabolism of chlordane. Stohlman and Smith (1950) fed chlordane to rabbits and found that it was partially excreted in the urine as one or more chlorine-containing acidic degradation products soluble above pH 8.6. Studies with rats by Poonawalla and Korte (1964) showed that when α-chlordane-^{14}C was injected intravenously, 29% of the total radioactivity was excreted within 60 hr in the feces and only 1% in the urine. Hydrophilic metabolites accounted for 75% of the radioactivity in the feces. Hoffman and Lindquist (1952) showed that resistant houseflies can metabolize chlordane, but no metabolites have been identified.

Perhaps the most concrete discovery regarding chlordane metabolism is that both α- and β-chlordane can be oxidized to form oxychlordane (Schwemmer et al., 1970; Lawrence et al., 1970) (see Fig. 5-8). This is startling since such conversion requires an initial dehydrogenation reaction. Street and Blau (1972) indeed found such an intermediate by incubating α-chlordane (trans-isomer) in the presence of NAD with rat liver homogenates from both sexes. They concluded that this initial dehydrogenation reaction is the

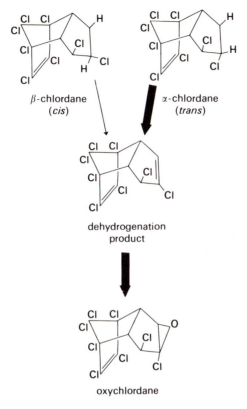

β-chlordane
(*cis*)

α-chlordane
(*trans*)

dehydrogenation
product

oxychlordane

Fig. 5-8. Metabolic formation of oxychlordane from
α- and β-chlordane. From Street and Blau (1972).

rate-limiting factor and that it is easier to dehydrogenate α-chlordane than β-chlordane (*cis*-isomer); the net result was a much higher accumulation of oxychlordane in the fat of the rats fed α-chlordane than in those fed β-chlordane. Polen *et al.* (1971) chemically characterized oxychlordane and found that it does not form in plants and soil treated with technical chlordane. Their studies also indicate that accumulation of oxychlordane is more pronounced in animals (pigs) fed α-chlordane than in those fed β-chlordane. The oxychlordane/α-chlordane ratio was on the order of 10–20 in the rat (Street and Blau, 1972) and 18 in the pig (Polen *et al.*, 1971).

As for dihydrochlordene analogues, two papers by Brooks and Harrison (1967*a, b*) are particularly pertinent. Dihydrochlordene can be hydroxylated in housefly and in pig microsomal preparations (Fig. 5-9). The major route of hydroxylation is at the 1- and the 2-carbon from the *exo* direction (= *syn*) in about equal proportions (Brooks and Harrison, 1967*a*). In the three analogues of dihydroheptachlor studied the most likely site of hydroxylation is at the 1-carbon from the *exo* direction, if such an attack is possible (Brooks and Harrison, 1967*b*). The hydroxylation at the 2-position is preferably done from the *exo* direction, although an *endo* (= *anti*) hydroxylation is possible either to form a diol of the metabolite or to replace the chlorine atom already positioned *endo* by an —OH group.

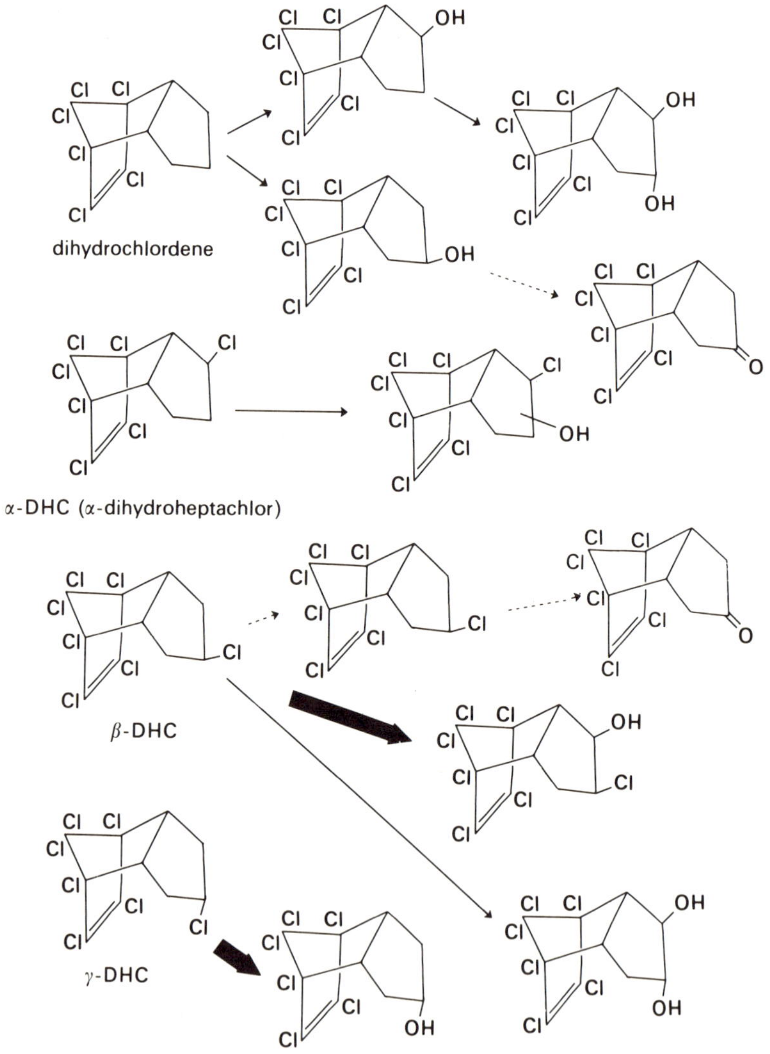

Fig. 5-9. Metabolism of dihydrochlordenes by pig and housefly microsomal preparations in the presence of NADPH.

5.4.11. Endosulfan Metabolism

Barnes and Ware (1965) studied the metabolism of ^{14}C-labeled endosulfan in houseflies and identified endosulfan sulfate as an oxidized metabolite. Cassil and Drummond (1965) also reported endosulfan sulfate formation in leafy plants such as spinach, celery, and alfalfa and in tree foliage. Ware (1967) noted that endosulfan resembles heptachlor and aldrin in that it undergoes oxidation to form a primary insecticidal metabolite, endosulfan sulfate, which is the terminal residue. Apparently this oxidation system is widely present in the environment. Deema *et al.* (1966) found the chief metabolite of endosulfan in mice to be the sulfate, which was stored in fat and excreted in feces. Another metabolite found in the urine appeared to be endosulfan diol. This metabolite is excreted when mice are given endosulfan as the sulfate, ether, or diol and therefore is the major urinary metabolite (see Fig. 5-10).

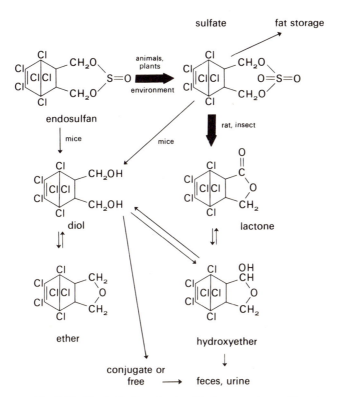

Fig. 5-10. Metabolic fate of endosulfan in rats, mice, and insects.

In contrast, in the rat the major metabolites found in the feces and urine are endosulfan hydroxyether and lactone. These two compounds are apparently interconvertible, since rats given either can excrete both (Terranova and Ware, 1963; Schuphan *et al.*, 1968). Endosulfan hydroxyether and lactone are also excreted by rats given endosulfan ether and diol, indicating that all four compounds are interconvertible, with the lactone and hydroxyether being the two preferred forms.

5.4.2. Metabolic Reactions Specific to Organophosphates

5.4.2a. General Reactions

The following types of reactions occur in the metabolism of organophosphate compounds. Reactions (1–5) are oxidative changes through the action of the mixed-function oxidase, and (6) is isomerization. Any of these reactions, along with reactions 11, can yield more toxic substances than the original insecticides (i.e., "activation").

1. Conversion of phosphorothionates to phosphates:

$$(RO)_3P(S) \xrightarrow{+O_2} (RO)_3P(O) + SO_4^{2-}$$

In most compounds, two of the R groups are the same, hence the formula representing them can be $(RO)_2PSOR'$. The enzyme which catalyzes this reaction is the mixed-function oxidase. This process is often called desulfuration. An example is the conversion of parathion to paraoxon (Nakatsugawa and Dahm, 1965).

2. Oxidative demethylation of *N*-methylated phosphoramides, $(R_2N)_3$ PO; phosphorodiamic acid derivatives, $(R_2N)_2PO \cdot OR$; and phosphorodiamic halides, $(R_2N)_2PO \cdot X$, where X is halogen or other alkylamides, $R_1NR_2R_3$:

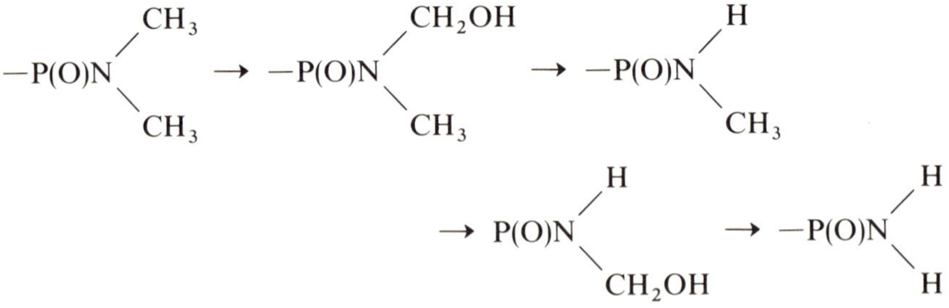

$$-P(O)N \begin{array}{c} CH_3 \\ CH_3 \end{array} \rightarrow -P(O)N \begin{array}{c} CH_2OH \\ CH_3 \end{array} \rightarrow -P(O)N \begin{array}{c} H \\ CH_3 \end{array}$$

$$\rightarrow P(O)N \begin{array}{c} H \\ CH_2OH \end{array} \rightarrow -P(O)N \begin{array}{c} H \\ H \end{array}$$

These *N*-dealkylation reactions are also carried out by the mixed-function oxidases in plants and animals. Examples are the metabolism of schradan (O'Brien, 1960), Bidrin® (Menzer and Casida, 1965), and phosphamidon (Bull *et al.*, 1967).

3. Oxidation of thioethers to sulfoxides and sulfones:

$$(RO)_2P(O)\cdot OR'\cdot S\cdot R'' \rightarrow (RO)_2P(O)\cdot OR'\cdot SO\cdot R''$$

$$\rightarrow (RO)_2P(O)\cdot OR'\cdot SO_2\cdot R''$$

An example is the metabolism of Systox® and Iso-Systox® and their isomers (Metcalf *et al.*, 1957).

4. Oxidative degradation of the P—O-aryl and possibly other ether and ester bonds. Parathion, for instance, can be degraded into diethylphosphorothioate and *p*-nitrophenol by the action of the mixed-function oxidase system (Nakatsugawa and Dahm, 1967). Welling *et al.* (1974) suggest that malaoxon, but not malathion, can be degraded by an oxidative mechanism to yield β-monocarboxylic acid of malaoxon in the houseflies, particularly in the resistant strains.

5. Hydroxylation of side-chains. The isopropyl moiety of diazinon can be hydroxylated at two different positions (Mücke *et al.*, 1970). Hydroxylation of the aromatic ring in triarylphosphates may occur with compounds such as tri-*o*-cresyl phosphate (Eto *et al.*, 1962).

6. Isomerization of phosphorothioates:

$$\begin{array}{c} R_1O \quad S \\ \diagdown \, \| \\ P-O- \\ \diagup \\ R_2O \end{array} \rightarrow \begin{array}{c} R_1S \quad O \\ \diagdown \, \| \\ P-O- \\ \diagup \\ R_2O \end{array} \quad \text{(e.g., BAY 25141)}$$

$$\begin{array}{c} S \\ \| \\ (RO)_2P-O- \end{array} \rightarrow \begin{array}{c} O \\ \| \\ (RO)_2P-S- \end{array} \quad \text{(e.g., SYSTOX}^®)$$

7. Hydrolysis of orthophosphates and phosphorohalidic acids through esterases:

$$(RO)_2P(O)OR' \rightarrow (RO)(OH)P(O)\cdot OR' + ROH$$

$$(RO_2)_2P(O)\cdot OR' \rightarrow (RO_2)_2P(O)\cdot OH + R'OH$$

$$(RO_2)_2P(O)\cdot X \rightarrow (RO_2)_2P(O)\cdot OH + HX$$

The metabolism of DFP is an example of this reaction (Mounter *et al.*, 1955).

8. Hydrolysis of carboxylic esters (or amide for dimethoate) of organophosphorus compounds (e.g., Krueger and O'Brien, 1959; Krueger

et al., 1960; Dauterman *et al.*, 1959):

$$(RO)_2P(O) \cdot OR'COOR'' \rightarrow (RO)_2P(O) \cdot OR'COOH + R''OH$$

The detoxification of malathion and dimethoate is an example.

9. Glutathione-mediated dealkylation (Fukami and Shishido, 1966; Hollingworth, 1969) and dearylation (Dahm, 1970; Oppenoorth *et al.*, 1972; Shishido *et al.*, 1972*b*).

10. Reduction of the nitrophenol moiety:

$$(RO)_2P(S)OPh \cdot NO_2 \xrightarrow{H_2} (RO)_2P(S)OPh \cdot NH_2$$

Parathion derivatives can be reduced to form nontoxic amines.

11. Formation of "toxic" analogues through minor changes on the side-chain, e.g.,

$$(CH_3O)_2P(O)OCH_2-CCl_3 \xrightarrow{-HCl} (CH_3O)_2P(O)\overset{\overset{\displaystyle OH}{\displaystyle |}}{C}=CCl_2$$

$$(CH_3O)_2P(O)OCHBr-CCl_2Br \xrightarrow{-2Br} (CH_3O)_2P(O)\overset{\overset{\displaystyle OH}{\displaystyle |}}{C}=CCl_2$$

$$(CH_3O)_2P(O)\overset{\overset{\displaystyle }{\displaystyle |}}{\underset{\underset{\displaystyle OAc}{\displaystyle |}}{C}}=CCl_2 \xrightarrow{+H_2O} (CH_3O)_2P(O)\overset{\overset{\displaystyle OH}{\displaystyle |}}{C}=CCl_2$$

Three different reaction mechanisms (dehydrochlorination, debromination, and hydrolysis) are involved in the above "activation" of three insecticidal precursors to produce dichlorvos (Metcalf *et al.*, 1959).

5.4.2b. Enzyme Systems Involved in the Degradation of Organophosphates

Among the biochemical systems which degrade organophosphates, two different groups of enzyme systems are important enough to be called major degradation mechanisms. They are the oxidative systems described in the preceding section and the hydrolytic systems.

As for the esterases, three basic types are known to occur: A, B, and C. Both A- and B-type enzymes play important roles in degrading organophosphates. In rat liver, for instance, four enzymes degrade various organophosphates:

1. An esterase degrading diethylphosphates and phosphorothioates which is stimulated by Ca^{2+} (Kojima and O'Brien, 1968).
2. A DFPase which is stimulated by Mn^{2+} and Co^{2+} (Mounter *et al.*, 1955).
3. A malathion carboxylesterase (Main and Braid, 1962).
4. A soluble enzyme which degrades dichlorvos and is characterized by being activated by Mn^{2+} (Hodgson and Casida, 1962).

The last enzyme could be identical to (2), and (1) could also be related to serum esterase, often called "paraoxonase."

In addition, there are glutathione-stimulated, apparently hydrolytic, enzymatic activities. It is not known whether such activities come from GSH S-transferase at this stage.

GSH S-transferases are defined enzymatic systems where the alkyl or aryl group of the substrate is transferred to glutathione. There are a number of reports now to show that O-dealkylation by this mechanism is common among organophosphate compounds (Morello *et al.*, 1968; Stenersen, 1969; Lewis, 1969; Miyata and Matsumura, 1972). Dahm (1970), Oppenoorth *et al.* (1972), and Shishido *et al.* (1972b), on the other hand, presented evidence for O-aryl transfer by the soluble GSH-dependent enzyme.

Individual examples of metabolic patterns for organophosphates will be presented in the following discussion.

5.4.2c. Schradan and Dimefox Metabolism

The metabolism of schradan was the center of attention in the 1950s since it was known to be converted into a much more toxic anticholinesterase compound *in vitro* and *in vivo* in plants, yeast bacteria, and animals. The magnitude of "activation" is enormous in that there is a 6000-fold increase in toxicity over the original compound (e.g., DuBois *et al.*, 1950; Gardiner and Kilby, 1952; Kilby, 1953; Casida and Stahmann, 1953). Moreover, schradan is an effective plant systemic insecticide. The question has arisen as to whether this activation process represents a beneficial reaction in making the insecticide more toxic to insects or a nonbeneficial step (Metcalf, 1955) because this activated material apparently degrades faster (Hartley, 1951). The activation process has been logically concluded to be oxidative, and the liver (mammals) and fat body (O'Brien and Spencer, 1953) have been found to be the sites of oxidation.

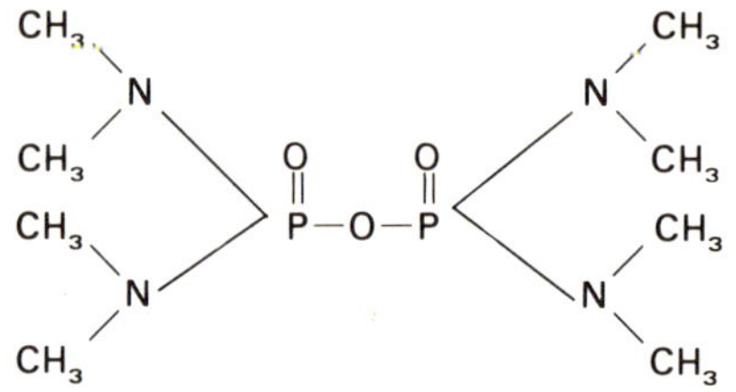

Not so toxic

schradan

↓

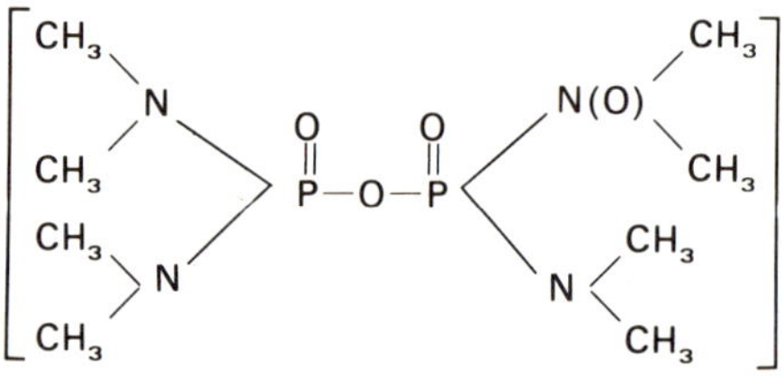

schradan N-oxide

↓

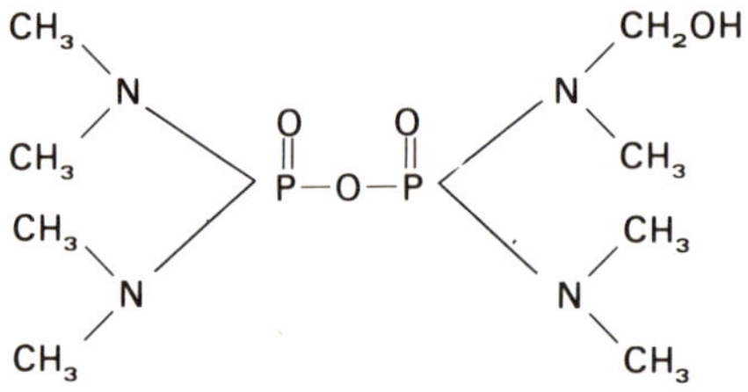

Toxic

hydroxymethylschradan

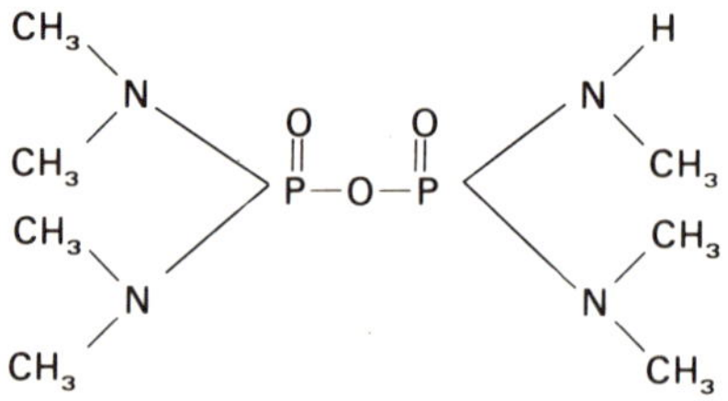

Toxic

N-desmethylschradan

Fig. 5-11. Activation and degradation of schradan through oxidative *N*-demethylation processes.

As for the identity of the active substance, several scientists originally favored the structure of *N*-oxide (Hartley, 1954; Casida *et al.*, 1954; etc.). Spencer *et al.* (1957) studied the infrared spectrum of the purified chloroform solution of the active substance and decided that it must be a methylol (*N*-hydroxylmethyl) derivative. Judging by other examples of stepwise oxidative *N*-demethylation reactions, they concluded that $-N(CH_3)$ (CH_2OH) is an unstable intermediate of *N*-desmethylschradan (see Fig. 5-11).

As for the role of plant oxidation, Metcalf (1955) concluded that the oxidative conversion in the plant is nonbeneficial from the standpoint of insecticidal action, since insects themselves can activate the insecticide (Tonelli and March, 1954). Moreover, schradan itself penetrates better into insect systems. Schradan has a relatively short half-life in plants, disappearing in several weeks (Heath *et al.*, 1952) despite more or less theoretical predictions that it should last for a few years (Hartley, 1951; Heath and Casapieri, 1951).

By analogy, it is reasonable to assume that dimefox is also oxidized, and indeed Aldridge and Barnes (1952) found that dimefox is converted into a more potent anticholinesterase inhibitor in rat liver. Subsequently, Arthur and Casida (1958) found the analogous activated material, which must be *N*-hydroxymethyldimefox.

5.4.2d. Trichlorfon, Dichlorvos, and Naled Metabolism

The major argument about the metabolism of trichlorfon has been whether this compound's insecticidal activity actually comes from its dehydrochlorinated and more potent analogue, dichlorvos. To be sure, nobody argues against trichlorfon being converted to dichlorvos. For instance, in pigs dichlorvos appeared in the blood as a result of subcutaneous injection of trichlorfon *in vivo* (Schwarz and Dedek, 1965). But the question remains.

The argument of Metcalf *et al.* (1959) is that trichlorfon's anticholinesterase activity increases under mild alkaline conditions, which promote dehydrochlorination, and remains poor under acidic conditions, an observation supported by Miyamoto (1959). O'Brien (1960), on the other hand, contends that trichlorfon has a direct effect on cholinesterase and that its *in vivo* action can be attributed to this effect. In fact, Arthur and Casida (1957) could not find dichlorvos in trichlorfon-treated houseflies. The truth may lie somewhere in between, although the dichlorvos theory is somewhat favored in that the conversion takes place so easily even at pH 7. Also supporting the dichlorvos theory is the fact that naled (Dibrom®), because of its low anticholinesterase activity and instability, must be converted *in vivo* into enough dichlorvos to cause toxicity. Indeed, there are many cases where detection of an "activated" substance is not easy because of low levels of the compound (e.g., paraoxon) at a given time even in the vital organs.

Trichlorfon is metabolized quickly. In cows, 65% of administered ^{32}P-trichlorfon was excreted in the urine in 12 hr. The urinary metabolites were analyzed and found to contain a small amount (17%) of dimethylphosphate (Robbins *et al.*, 1956). Hassan and Zayed (1965) and Hassan *et al.* (1965) showed that in rats the *O*-demethylation process is an important degradation mechanism. In brain homogenates, trichlorfon is rapidly metabolized to desmethyltrichlorfon, monomethylphosphate, and 2,2,2-trichlor-1-hydroxyethylphosphoric acid. *In vivo* studies by Hassan and Zayed (1965) showed recovery of 60% of the ^{14}C in urine and in the expired air after administration of ^{14}C-trichlorfon. Similarly with ^{32}P-trichlorfon, Hassan *et al.* (1965) found 75–85% of the radioactivity in the urine in 48 hr. Mono- and dimethylphosphate were the major metabolites in the urine. With respect to insects, Zayed and Hassan (1965) and Hassan *et al.* (1965) found that the cotton leafworm (*Prodenia lituria*), which is quite resistant to trichlorfon, degraded it to desmethyltrichlorfon, a glucuronic conjugate of 2,2,2-trichloro-1-hydroxyethylphosphoric acid (about 68%), monomethylphosphone acid (5%), and dimethylphosphoric acid (27%), and $^{14}CO_2$. Similarly, in the cotton plant ^{32}P-trichlorfon is metabolized into monomethylphosphate, dimethylphosphate, and phosphoric acid (Mostafa *et al.*, 1965).

Basically the metabolic pattern of dichlorvos in the rat is similar to that of trichlorfon (Hodgson and Casida, 1962). The hydrolytic enzyme responsible for P—O cleavage is mainly in the soluble fraction and is activated by Ca^{2+}. Under experimental conditions using rat liver homogenate, 68% is recovered as dimethylphosphate, 11% as desmethyldichlorvos, and 21% as monomethylphosphate. While demethylation activity is relatively strong, Miyata and Matsumura (1972) could observe only a modest increase with GSH. The same authors (1971) could obtain high degrees of stimulation of dichlorvos degradation by the addition of GSH to *P. americana* homogenates, suggesting that at least in some animals this demethylation process is carried out by the GSH *S*-alkyltransferase.

Casida *et al.* (1962) compared metabolism of ^{32}P-dichlorvos and ^{32}P-naled (Dibrom®). They found an essentially identical rate of appearance of ^{32}P in the blood for these two insecticides. Also, the rates of excreted mono- and dimethylphosphates (69 and 71%, respectively) and desmethyldichlorvos were about 29.8 and 29%, respectively, as judged by the percent recovery of the radioactivity excreted 0.5–1 hr after administration. Inorganic phosphate excretion was low: 1.3% for the dichlorvos-fed cow and less than 0.2% for the naled-fed one (Fig. 5-12).

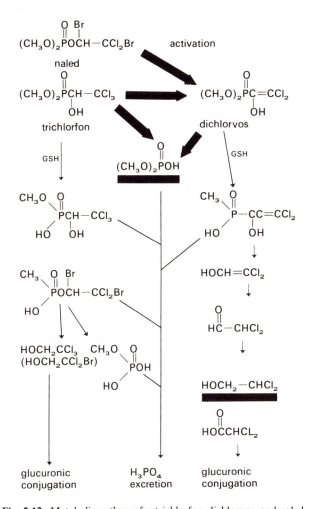

Fig. 5-12. Metabolic pathway for trichlorfon, dichlorvos, and naled.

The hydrolytic demethylation process in the soluble fraction of rat liver (Hodgson and Casida, 1962) is interesting because demethylation is now generally acknowledged to be mediated by GSH *S*-alkyltransferase or at least certainly not by the microsomal NADPH-dependent system. However, Miyata and Matsumura (1972) have disputed the existence of a GSH-mediated system (in this particular reaction) since they could obtain only a slight increase in the total hydrolytic products (i.e., products from hydrolysis at both P—O—vinyl and Me—O—P sites) from the addition of

GSH to the 20,000 g supernatant of rat liver. The rate of metabolism of dichlorvos was 11 nmoles/mg/hr for the control and 12 nmoles/mg/hr in the presence of GSH. Under the same experimental conditions, metabolism of famphur increased from 1.1 to 7.8 nmoles/mg/hr.

5.4.2e. Bidrin, Azodrin, Phosphamidon, Mevinphos, and Ciodrin Metabolism

Bidrin,[®] azodrin, phosphamidon, mevinphos, and Ciodrin[®] are essentially very similar organophosphates. In terms of metabolic behavior, they should be divided into the two major groups of amides and esters. The major metabolic route for amides is oxidative N-dealkylation in the

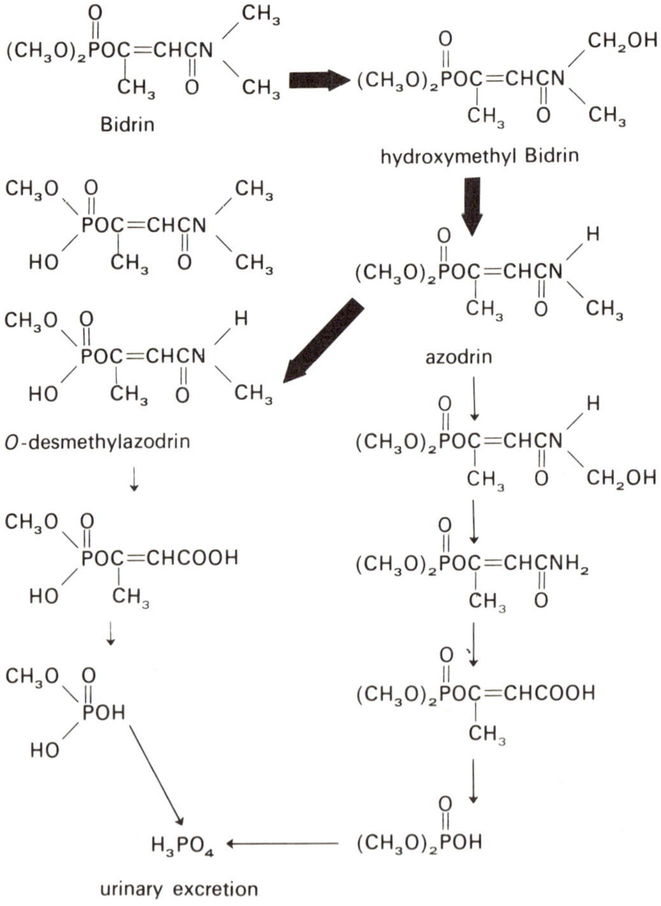

Fig. 5-13. Metabolic fate of Bidrin[®] and azodrin in mammals, insects, and plants.

same fashion as the reactions already described for schradan, while esters are simply hydrolyzed to yield corresponding acids.

Metabolism of Bidrin® has been studied in rats, plants, and insects (Bull and Lindquist, 1964; Menzer and Casida, 1965; Hall and Sun, 1965) and in soil (Corey, 1965). All the patterns of metabolism appear to be similar. The N-dealkylation system is oxidative inasmuch as it is inhibited by sesamex. Within 6 hr, over 50% of Bidrin® administered to the rat is excreted, mostly in the form of hydrolysis products (approximately 35% of the total or 70% of the excreted radioactivity). The most predominant oxidative N-dealkylation product is azodrin (5–8% in 6 hr; Fig. 5-13).

Phosphamidon is essentially a diethylamide analogue of Bidrin® with a chlorine at the 2-carbon. There are two factors that make the metabolic pattern of this compound slightly different from that of Bidrin® and azodrin (Fig. 5-14). First, the extra chlorine atom is labile and therefore can be

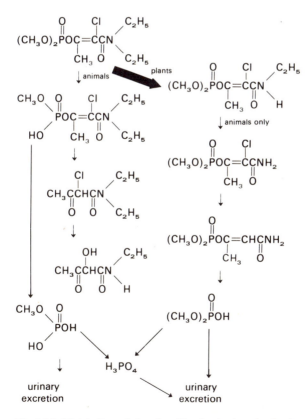

Fig. 5-14. Metabolism of phosphamidon in plants and animals.

subjected to hydrolytic dehalogenation, which could be either a chemical or a biochemical reaction, or to reductive dechlorination, which is enzymatic. Second, the fact that phosphamidon is a diethyl, and not a dimethyl, amide could make this oxidative N-dealkylation process a little more difficult, particularly for plant systems. The whole subject has been ably discussed by Geissbühler *et al.* (1971).

Since only the oxidatively N-dealkylated products of phosphamidon are expected to have any anticholinesterase activity, the above question becomes rather crucial from a toxicological standpoint. Two phenomena are pertinent here: first, nobody has yet observed a N-hydroxyethyl intermediate of phosphamidon (e.g., see Jacques and Bein, 1960; Anliker *et al.*, 1961; Menzer and Dauterman, 1970), although N-desethylphosphamidon does indeed form. Second, plants appear to be capable of carrying out the reaction only to N-desethylphosphamidon, while animals are capable of forming the corresponding amide (Clemons and Menzer, 1968).

Both mevinphos (Phosdrin®) and Ciodrin® are carboxylic esters of Bidrin® acid. Casida *et al.* (1958) studied mevinphos metabolism in cows and found that dimethylphosphate is the major urinary metabolite. Other hydrolysis products are found in the feces. Chamberlain (1964*a*, *b*) found that ^{32}P-Ciodrin® is quickly metabolized in the lactating cow and eliminated, mostly in the urine, in 2 days. Among the metabolites, 61–90 % is dimethylphosphoric acid. Three hours after administration, 11 % of the metabolites are in the form of the carboxylesterase product.

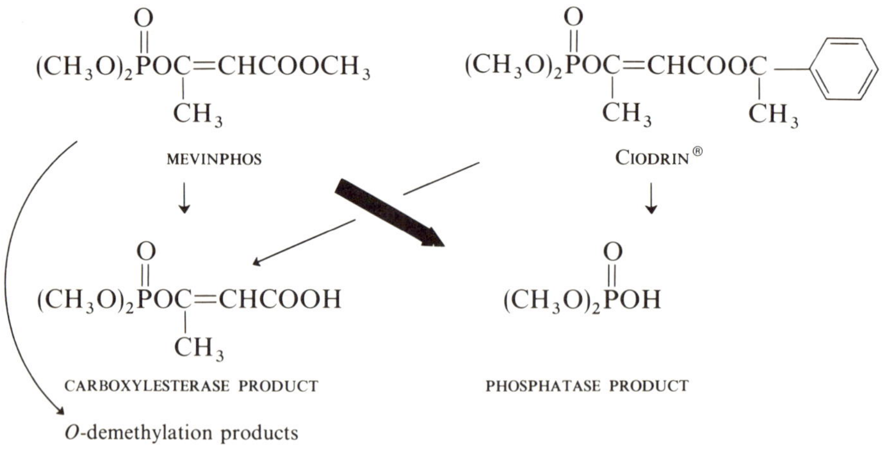

5.4.2f. Metabolism of Parathion and Analogues

Since parathion was one of the first thiophosphates to be used as an insecticide, the interest of scientists originally centered around the initial

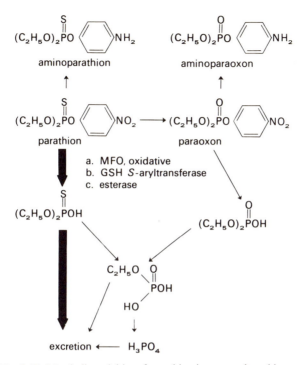

Fig. 5-15. Metabolic activities of parathion in mammals and insects.

oxidation process forming the corresponding phosphate, paraoxon, which is the actual anticholinergic agent (Fig. 5-15). These intensive studies (e.g., Diggle and Gage, 1951; Metcalf and March, 1953a, b) led to the conclusion that all thiophosphates must be first converted *in vivo* to become toxic. Thus in the late 1950s various animal tissues were studied to determine the enzymatic basis of such desulfuration processes. While several cofactors, ions, and other conditions have been tested and claimed to promote such a reaction process, it is now firmly established that this reaction is carried out by the mixed-function oxidase (MFO). Thus Nakatsugawa and Dahm (1962) showed that the system requires NADPH and oxygen using *P. americana* as the test animal. The same authors (1967) using ^{35}S-parathion demonstrated that ^{35}S is removed by the microsomes of the rat, the rabbit, and the cockroach fat body by formation of a tight protein complex. The system requires NADPH and O_2 in addition to having characteristic sensitivity toward a MFO inhibitor, SKF 525-A, establishing firmly that the system desulfurating parathion is MFO.

In mammals in general, the major metabolic products are formed from P—O bond cleavage (e.g., Miyamoto *et al.*, 1963a, b; Miyamoto, 1964a, b;

TABLE 5-2. Parathion Reduction Activities by Liver Homogenates from Various Species

	Number of animals	Aminoparathion (μmoles formed/ 100 mg/30 min)
Rat	6	1.256
Mouse	6	1.191
Guinea pig	9	0.455
Chicken	4	0.518
English sparrow	8	0.196
Bullhead	6	1.431
Sucker	3	1.364
Flounder	5	0.839
Sculpin	5	0.544
Largemouth bass	4	0.913
Sunfish	6	0.954
Bluegill	4	1.093
Alewife	4	0.850

From Hitchcock and Murphy (1967).

Vardanis and Crawford, 1964), e.g., mono- and diethylphosphoric acids, diethylthiophosphoric acid, and phosphoric acid. As expected, such phosphate cleavages leave *p*-nitrophenol, which is excreted in the urine and not in the bile.

Since reductive activities are known to be high in the microbial world, it was not surprising that incubation of parathion with bovine rumen fluid resulted in the formation of aminoparathion and aminoparaoxon, among other metabolic products (Cook *et al.*, 1957; Pankaskie *et al.*, 1952; Ahmed *et al.*, 1958). However, it is now apparent that such nitroreductase activity can also be found in the animals themselves. Hitchcock and Murphy (1967) made comparative studies on the formation of aminoparathion in many mammalian, avian, and fish species (Table 5-2). Variation among species appears to be considerable. The system requires NADPH, as in the case with the housefly enzyme (Lichtenstein and Fuhremann, 1971).

As for the P—O cleaving enzymes, there appear to be three systems working at the parathion molecule independently. They are (1) oxidative cleavage, (2) esterase action, and (3) GSH *S*-transferase reactions. Cleaving the P—O bond of parathion itself has been the interest of scientists. Matsumura and Hogendijk (1964a) found that in the housefly parathion is directly converted into mostly diethylthiophosphoric acid. Particularly with *in vitro* preparations, this was almost the only metabolic product. Since the reaction could be performed in the absence of added cofactors, these workers assumed a hydrolytic action. The system was found to be more active against parathion than paraoxon, which is unusual for A-type esterases. Accordingly they labeled the system as "thionase," in contrast to "oxonase," which should preferably degrade paraoxon, such as in the system found by Main (1960). Nakatsugawa and Dahm (1965) reexamined the enzymatic properties of such a parathion P—O cleavage reaction. They found the same end-product, and made the surprising discovery that this process is activated by NADPH. Subsequent studies with microsomal preparations of rat liver and cockroach fat body unequivocally established that it is MFO that splits the P—O bond of parathion to give diethylthiophosphoric acid and *p*-nitrophenol. However, this was not the end of the story. Previously, Matsumura and Hogendijk (1964*a*) had tried to purify the "thionase" by conventional biochemical methods. While the activity of the homogenate can be explained by the small amounts of endogenous NADPH or NADPH-generating systems generally present in such preparations, explanation of the elution of apparently soluble protein from a DEAE column was difficult if one assumed the microsomal origin alone. It was found by two different groups that addition of reduced glutathione to the soluble fraction from housefly homogenate can stimulate the P—O cleavage activities (Dahm, 1970; Oppenoorth *et al.*, 1972). Thus it now appears that parathion is degraded by a GSH *S*-aryltransferase in a similar fashion as diazinon (Shishido *et al.*, 1972*b*). The only thing that is not certain at this stage is the action of esterases on parathion itself. While all investigators have noted some degree of P—O splitting of parathion using unfortified homogenate or supernatant sources, such observations cannot be used as evidence for esterase actions. Kojima and O'Brien (1968) showed that the partially cleaned esterase fraction (e.g., washed mitochondria) of rat liver, which hydrolyzes paraoxon to diethylphosphoric acid, can also degrade parathion, but to a lesser degree. In view of the possibility of enzyme contamination, no definite conclusion can be reached on this particular point at present.

Paraoxon, on the other hand, has been acknowledged to be degraded through esterase action; since it is a phosphate with a strong inhibitory action on cholinesterases, it is expected to have high affinities for other esterases, making this substrate available for esterases. The general route of hydrolytic attack is at the P—O nitrophenol bond, liberating diethylphosphate. The enzyme is sometimes called "paraoxonase" and is often present in mammalian systems as soluble serum enzyme (Main, 1960). Another enzyme which hydrolytically desethylates paraoxon has been found by Nolan and O'Brien (1970). They incubated ^{3}H-ethylparaoxon with housefly systems and obtained ^{3}H-ethanol and monoethylparaoxon as the reaction products. The detection of ^{3}H-ethanol is important, since it indicates that neither GSH (product expected is *S*-ethylglutathione) nor an oxidative system (product should be ^{3}H-aldehyde) is involved in such a reaction.

In this field, we have witnessed that small changes in the molecules often cause an enormous difference in toxicity to certain animal species. Methylparathion and fenitrothion are good examples of such minor molecular alterations that brought spectacular reductions in mammalian toxicity without reducing the insect toxicity. The patterns of metabolic transformation of these two compounds are quite different from that of parathion in that they are largely demethylated initially. Indeed, it is with these two compounds that Fukami and Shishido (1963) originally found the effect of GSH on the dealkylating activities of organophosphates, leading to the discovery that GSH S-transferases also actively degrade insecticidal chemicals. In their later paper (1966), these workers studied the enzymatic aspect and concluded that the enzyme system is omnipresent, with highest activity in the liver. As a result of a comparison between the rat liver system and the midgut and fat body of horn beetle larvae, they concluded that the rat liver has approximately 5.5 times more demethylating activity than the insect system. Hollingworth (1969) confirmed this enzyme to be a GSH S-alkyl-transferase by establishing the presence of S-methylglutathion as the result of a reaction between GSH and methyl-labeled ^{14}C-methylparaoxon where the radioactivity was transferred to glutathione. It must be mentioned, however, that other enzyme systems also attack these insecticides. Several workers (Miyamoto *et al.*, 1963*a, b*; Miyamoto, 1964*a, b*; Vardanis and Crawford, 1964) detected phosphoric, thiophosphoric, dimethylthiophosphoric, and dimethylphosphoric acids in mammals exposed to fenitrothion, indicating that parathion-type degradation occurs. The question here is the relative activity compared to parathion.

Demethylation systems appear to be present in plants (Miyamoto and Sato, 1965*a, b*), as desmethylfenitrothion has been found in rice plants exposed to fenitrothion, along with dimethylphosphorothioic acid. Such a system is also expected to play an important role in degrading all methylparathion analogues. For instance, Plapp and Casida (1958*b*) found that about 50 % of the urinary metabolite of ronnel in the rat was an O-demethylated product. When the phosphate analogue of ronnel was fed to rats, again about one-half of the metabolites found in the urine were O-demethylation products. Other prominent metabolites of ronnel were dimethylphosphate and dimethylthiophosphate.

5.4.2g. Diazinon, Chlorfevinphos, and Coumaphos Metabolism

Diazinon is a diethylthiophosphate. Thus Matsumura and Hogendijk (1964*a*) found that, like parathion, diazinon is largely degraded to diethylthiophosphoric acid. Indeed, *in vivo* Robbins *et al.* (1957) found that cows given diazinon produced diethylthiophosphoric acid as the major metabolic product. Two factors make diazinon somewhat different from parathion. First, diazinon can be oxidized at the side-chain (the reverse of nitroreductase action), and, second, it has been proven that GSH attaches to the pyrimidyl ring and thereby facilitates GSH S-aryltransferase action (Shishido *et al.*, 1972*a*, *b*). Hydroxylation activities on the isopropyl group of diazinon have been reported by Pardue *et al.* (1970), Mücke *et al.* (1970), and Shishido *et al.* (1972*b*).

On the other hand, Lewis (1969) and Yang *et al.* (1971) found that the system responsible for the P—O split for both diazinon and diazoxon is MFO in the housefly. This is interesting because Nakatsugawa and Dahm (1967) found that the same system in the rat and the cockroach is specific to parathion and not to paraoxon; Lewis (1969) and Yang *et al.* (1971) however, mention that diazinon is the preferred substrate. In the absence of GSH but in the presence of NADPH, Shishido *et al.* (1972*b*) found that the proportions of these metabolic products formed in the rat liver microsome preparation incubated with diazinon were 14.8% for F, 7.2% G, 6.7% diazoxon (B), and 4.7% A (see Fig. 5-16). It is interesting to

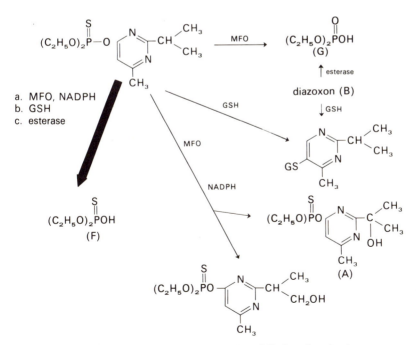

Fig. 5-16. Enzymatic degradation of diazinon in animals.

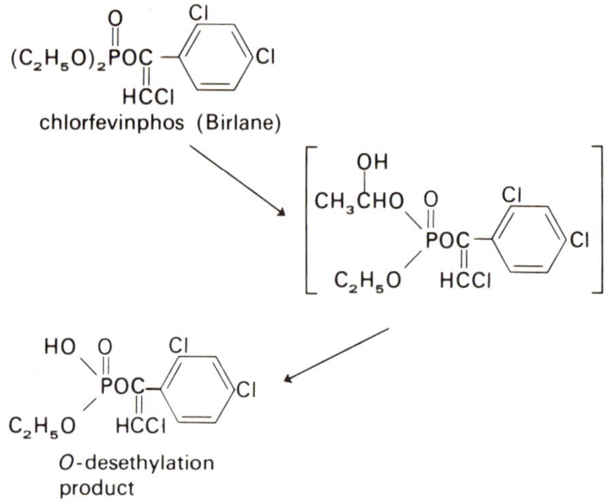

Fig. 5-17. Oxidative O-desethylation of chlorfevinphos.

note that Lewis (1969) states that diazoxon is oxidatively desethylated by the microsomal preparation of resistant houseflies, since there are other organophosphates which are degraded by such means.

Chlorfevinphos has been shown to be O-desethylated through the action of the MFO system by Donninger *et al.* (1967). They postulate that there is an unstable 1-hydroxyethyl intermediate (Fig. 5-17).

Coumaphos is, on the other hand, likely to be degraded by esterase action, although this compound is also mainly desethylated (Lindquist *et al.*, 1958; O'Brien and Wolfe, 1959; Vickery and Arthur, 1960). Krueger *et al.* (1959) showed that in the cow, rat, and goat coumaphos is attacked by esterases at two places: one to desethylate and the other to form diethylthio-phosphoric acid.

O'Brien (1960) proposes that the selectivity of this animal systemic insecticide comes from the hydrolytic desethylation ability of mammals. It would be interesting to study the possibility of nonhydrolytic degradation of coumaphos in view of recent knowledge on other enzyme systems. Certainly oxidative O-dealkylation is a possibility with the oxon analogue at least, and the possibility of GSH S-alkyltransferase action on such a compound is not remote. There are other examples of hydrolytic dealkylation (Hodgson and Casida, 1962; Nolan and O'Brien, 1970), and thus such a reaction mechanism is also possible for coumaphos.

5.4.2h. Demeton, Phorate, and Disulfoton:
Oxidation of Thioether Side-Groups

The group of chemicals that includes demeton, phorate, and disulfoton is characterized by the presence of a R—S—R group that is oxidized to form the corresponding sulfoxide and sulfone. The original discovery of such oxidative changes was made by the group of scientists at the Riverside Campus, University of California (the Citrus Experimental Station), namely Metcalf *et al.* (1954, 1955) and Fukuto *et al.* (1954, 1955) (Fig. 5-18).

Another important point is the toxicity and metabolic differences between two isomers of demeton (thiol and thiono isomers; see Section 3.3.4 in Chapter 3). Although the rate of metabolism of demeton appears to be highest in mammals, followed by insects and plants, the basic pattern of metabolism is the same. In the scheme illustrated in Fig. 5-18, X and Y are either S or O (Schrader, 1963). For instance, with thiol isomer X = O and Y = S, and with thiono isomer X = S and Y = O. The thiono isomer can be oxidatively desulfurated to give the corresponding phosphate, which is metabolized to the sulfoxide and the sulfone and then to diethylphosphoric acid. In the case of oxydemetonmethyl, oxidation of the thioether similarly precedes the final hydrolysis to give dimethylphosphate (Niessen *et al.*, 1963).

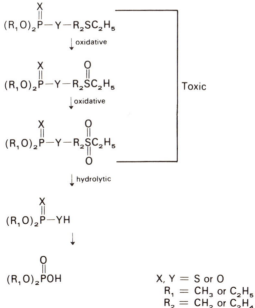

Fig. 5-18. Generalized scheme for oxidative metabolism of demeton, phorate, and disulfoton.

In the case of phorate (Thimet®) and disulfoton (Di-Syston®), both X and Y are S. R_1 is ethyl for both insecticides, and $R_2 = CH_2$ for phorate and C_2H_4 for disulfoton. Bowman and Casida (1958) found the sulfoxide of dithioate and the sulfoxide and the sulfone of thiolate in rat liver. Thus oxidation of X(S) to O takes place after the initial oxidation to sulfoxide. In the urine, they found diethylphosphorothioic acid (80%), diethylphosphoric acid (17%), and diethylphosphorodithioic acid (3%). The metabolic patterns in insects and plants appear to be similar. In all cases, the major degradation product is either diethylphosphoric acid or diethylphosphorothioic acid.

The metabolic pathways for disulfoton in animals and insects (Bull, 1965) and in plants (Metcalf *et al.*, 1957; Bull, 1965) are very much alike and also similar to those for phorate. The initial step of oxidation is formation of the sulfoxide analogue, which can be either desulfurated or oxidized to the sulfone. Similar hydrolytic products are also found.

Other insecticides which possess thioether side-chains and are liable to sulfoxide and sulfone formation are fenthion (Francis and Barnes, 1963; Niessen *et al.*, 1962; Fikudo *et al.*, 1962; Brady and Arthur, 1961), famphur (O'Brien *et al.*, 1965), and Abate® (Blinn, 1968). In all cases, oxidation of thioethers eventually to sulfones is accomplished, whether they are aromatic or alkyl thioethers.

5.4.2i. Metabolism of Malathion, Acethion, and Dimethoate: Degradation Through Hydrolytic Cleavage of Side-Chains

Malathion appears to degrade mainly through hydrolytic pathways (March *et al.*, 1956a; Knaak and O'Brien, 1960; Mattson and Sedlak, 1960; Seume and O'Brien, 1960) (Fig. 5-19). The importance of the hydrolytic degradation at its carboxyesters in animals has been determined by March *et al.* (1956b), Cook and Yip (1968), Wells *et al.* (1958), Weidhaas (1959), O'Brien *et al.* (1961), and Pasarela *et al.* (1962). These authors found that malathion mono- and dicarboxylic acids formed as a result of malathion incubation with animal systems. In plants, Rowlands (1964, 1965) showed that malathion carboxylic acids are also formed.

It was O'Brien (e.g., 1960) who actually promoted the view that this strong carboxylesterase action by some of the animal systems can explain malathion's selectivity. The view is now supported by experimental evidence on malathion-resistant insects, which invariably show high carboxylesterase activities compared to their susceptible counterparts (Matsumura and Brown, 1961, 1963; Matsumura and Hogendijk, 1964b; Townsend and Busvine, 1969). Malathion resistance is usually specific, extending only to

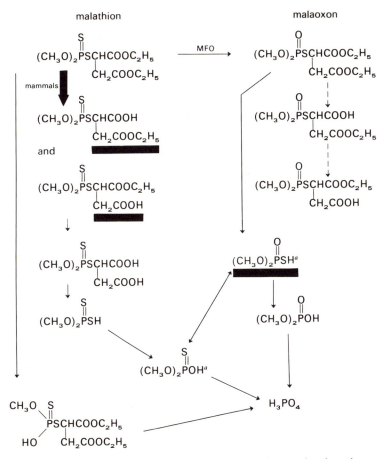

Fig. 5-19. Metabolic fate of malathion. [a]Indistinguishable from each other, they probably isomerize either way according to pH.

other carboxylesters (Dauterman and Matsumura, 1962), and therefore O'Brien's view appears to be certain. That such carboxylesterases are of B-type origin (i.e., actually aliesterase) has been shown by Main and Braid (1962), who purified the enzyme to study its properties.

It is interesting to note that there is also some variation among the tissues and species from which carboxylesterases are obtained. Table 5-3 indicates ways by which they hydrolyze either at the α-position or the β-position.

As for the hydrolytic attack on P—O and P—S bonds (often referred to as "phosphatase" attack), many of the above scientists have observed that

TABLE 5-3. Ratio of α- and β-Malathion[a] Monoacid Formation in Several Animal Species

Enzyme sources	Ratio α/β
Pure horse liver aliesterase	0.1
Rat liver microsomes[b]	0.07
Beef liver acetone powder	2.5
Pig pancreas acetone powder	1.0
Pig kidney acetone powder	1.0
Pig liver esterases (partially purified)	2.0
Housefly, whole-fly homogenate	3.5–5.0
Tribolium castaneum, resistant whole-beetle homogenate	1.6

From Welling and Blaakmeer (1971).

[a]
$$(CH_3O)_2 \overset{\displaystyle S}{\overset{\|}{P}}{-}S{-}\underset{\displaystyle CH_2COO\overset{\beta}{-}C_2H_5}{\overset{|}{CH_2COO}}\overset{\alpha}{-}C_2H_5$$

$$-S{-}\underset{\displaystyle CH_2COOC_2H_5}{\overset{|}{CHCOOH}}$$

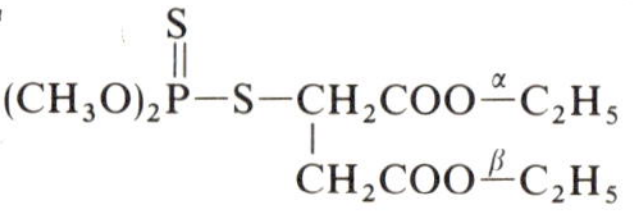

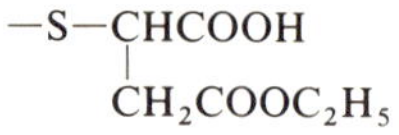

α-acid

$$-S{-}\underset{\displaystyle CH_2COOH}{\overset{|}{CHCOOC_2H_5}}$$

β-acid

[b]This is contrary to what Chen *et al.* (1969) found in the urine of rats: they detected only α-acid.

various phosphoric, thiophosphoric, and dithiophosphoric acid derivatives do form in the absence of GSH or NADPH *in vitro*. Miyata and Matsumura (1971), however, observed GSH stimulation of two of the five partially purified soluble cockroach enzyme preparations in the rate of desmethyl-malathion production. Thus it is more than likely that some of the activities so far ascribed to simple esterases can be mediated either by a GSH-stimulated system or by a GSH S-alkyltransferase. It is interesting to note that Welling *et al.* (1974) feel that in the housefly malaoxon's carboxylic ester bond can be degraded via oxidation. They based their conclusion on the susceptibility of this degradation system to sesamex. The system is also partially inhibited by malathion which cannot act as a substrate. More work is needed to confirm their findings as oxidative degradation to carboxylic ester bonds has not been known.

The pattern of degradation is similar in acethion, which is an S-ethyl-acetate substitute for the S-diethylsuccinate moiety of malathion. Perry (1960) has shown that acethion degrades very rapidly, and O'Brien *et al.*

(1958) and Krueger *et al.* (1960) have demonstrated that in mammals and insects acethion (carboxylic) acid is the major metabolic product:

$$(C_2H_5O)_2\overset{\overset{\displaystyle S}{\|}}{P}-SCH_2COOC_2H_5 \xrightarrow{\text{carboxylesterase}} (C_2H_5O)_2\overset{\overset{\displaystyle S}{\|}}{P}-SCH_2COOH$$

ACETHION   ACETHION ACID

As with malathion, O'Brien (1960) has proposed that this carboxylester hydrolysis could account for this compound's selectivity favoring mammalian species.

Dimethoate, on the other hand, is a carboxyamide instead of a carboxylester. It is generally considered that carboxylesterases cannot hydrolyze carboxyamide linkages and that there are separate amidases, such as the arylamidase of the mammalian systems, to handle amides. Nevertheless, it has been suggested that the selectivity of dimethoate also comes from the superior hydrolytic ability of mammalian systems compared to corresponding insect systems (Dauterman *et al.*, 1959; Krueger *et al.*, 1960; Uchida *et al.*, 1964). Indeed, Uchida and O'Brien (1967) could obtain a good correlation between the carboxyamidase activities and *in vivo* toxicities of dimethoate in six vertebrate species. Rats given dimethoate (Dauterman *et al.*, 1960; Krueger *et al.*, 1960; Uchida *et al.*, 1964; Plapp and Casida, 1958a; Sanderson and Edson, 1964) excreted dimethylphosphoric acid, dimethylthiophosphoric acid, and other minor metabolites in the urine. Dauterman *et al.* (1960) were able to detect dimethoate acid (i.e., carboxyamidase product) in a steer treated with dimethoate, along with other metabolites such as desmethyldimethoate, dimethylphosphoric acid, dimethylphosphorothioic acid, and dimethylphosphorodithioic acid. Indeed, these dimethylphosphoric acid derivatives could be obtained by administering dimethoate acid to the rat, suggesting that dimethoate acid is rather susceptible to further hydrolytic attacks. In insects (Bull *et al.*, 1963; Sanderson and Edson, 1964), accumulation of the toxic oxygen analogue (thiolate) is noticeable; otherwise, the pattern of metabolism appears to be similar:

$$(CH_3O)_2\overset{\overset{\displaystyle S}{\|}}{P}-S-CH_2\overset{\overset{\displaystyle O}{\|}}{C}NHCH_3 \xrightarrow{\text{amidase}} (CH_3)_2\overset{\overset{\displaystyle S}{\|}}{P}-SCH_2COOH$$

DIMETHOATE   DIMETHOATE ACID

hydrolysis ⟶ phosphoric acid derivatives

Uchida *et al.* (1964) and Uchida and O'Brien (1967) found that this carboxyamidase is located mainly in the liver microsomal fraction. It is not mediated by the oxidative enzyme, however. Its identity and properties have not been studied in detail.

5.4.3. Metabolism of Carbamate Insecticides

5.4.3a. General Patterns

There are a number of excellent review articles on the metabolism of carbamate insecticides. Examples are those by Casida (1963, 1970), Dorough (1970), Kuhr (1970), Schlagbauer and Schlagbauer (1972*a, b*), and Knaak (1971).

The major characteristic of the carbamates is that oxidative degradation activities through the mixed-function oxidase systems really dominate their metabolism. The reasons for this are not entirely clear. One possible explanation is that because of the chemical nature of these insecticidal carbamic esters, hydrolytic products, such as naphthols and phenols, are often less polar than the parent compounds. This view is in agreement with the generally held view that metabolism of xenobiotics in animals is geared up to making them more polar so that they can be excreted (Williams, 1959). Another obvious reason is that aromatic compounds predominate in the group, and animal systems are prepared to oxidatively handle these apolar chemicals inasmuch as decomposition of aromatic rings is not possible. Also an important consideration is that by design insecticidal carbamic esters are made utilizing rather poorly electrophilic moieties (e.g., substituted phenols and naphthols) to couple with *N*-methyl or dimethyl carbamic acids. By contrast, all insecticidal organophosphates are the ones which possess strongly electrophilic substituents and thus are naturally susceptible to hydrolytic attack (O'Brien, 1967). For example, paraoxon is also an aromatic insecticide except that *p*-nitrophenol is a strongly electrophilic substitution, and accordingly it is susceptible to enzymatic as well as non-enzymatic hydrolytic reactions.

Whatever the oxidative metabolic activities for carbamate insecticides, they certainly cause a variety of changes which can influence the toxicity and the residual characteristics of these chemicals. In general, most oxidative changes cannot be regarded as either "detoxification" or "activation" processes. The toxicities of these oxidized compounds can be more or less than those of the parent chemicals, but usually they do not become totally nontoxic. In Table 5-4 are listed the metabolic products known from Oonnithan and Casida's (1968) monumental work to be anticholinesterase agents equal or superior in potency to their parent compounds. Thus the question of activation vs. detoxification can be judged only by the stability, availability, and frequency of appearance of each metabolite in the animal system, in addition to its potential to inhibit cholinesterase *in vitro*.

In this regard, the potential for conjugation reactions in these metabolic products becomes exceedingly important. Indeed, conjugation reactions

TABLE 5-4. Known Toxic Metabolites of Carbamate Insecticides[a]

Compound[b]	Minimum detectable level[c] (μg)	Structure
Carbaryl 5-Hydroxycarbaryl I	0.2 0.1	I 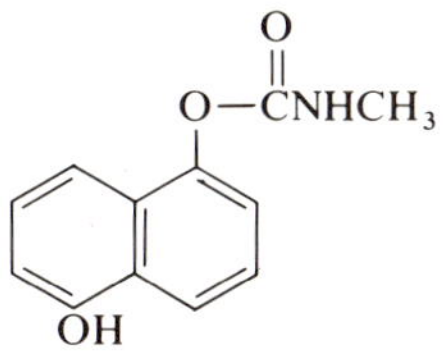
Propoxur 5-Hydroxypropoxur II	0.5 0.06	II
UC-1085 3-(1-Hydroxy-1-methylethyl)phenyl methyl carbamate III	0.004 0.004	III
Mesurol® 4-Methylsulfinyl-3,5-xylyl methylcarbamate IV 4-Methylsulfonyl-3,5-xylyl methylcarbamate V	0.1 0.01 0.1	IV, V
Metacil 4-Aminometacil VI 4-Dimethylaminometacil VII 4-Methylaminometacil VIII	1.0 0.1 1.0 0.1	VI, VII, VIII

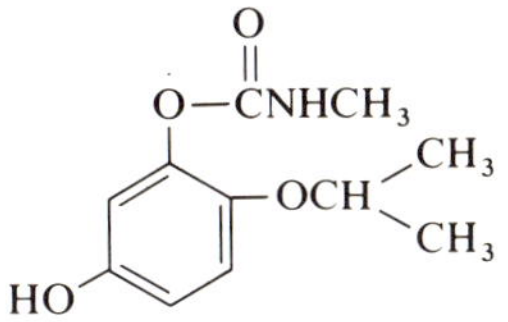

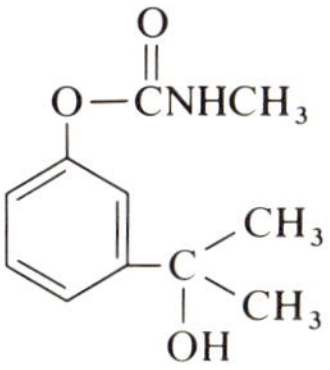

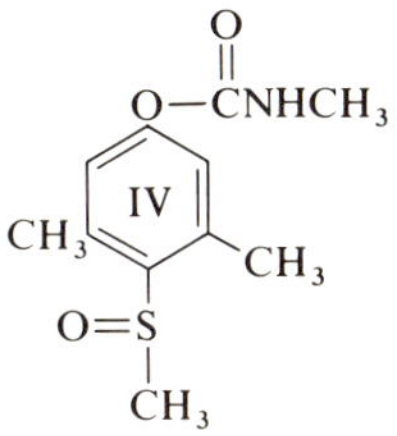

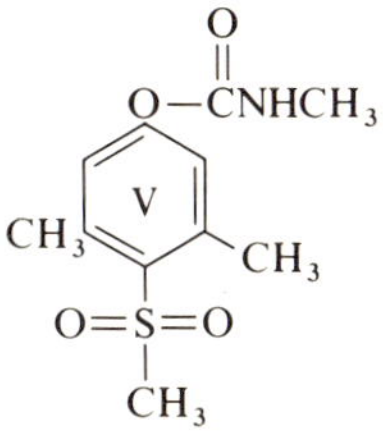

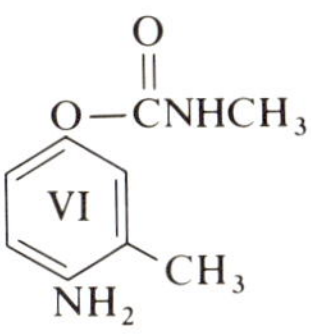

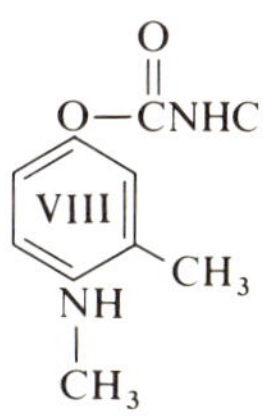

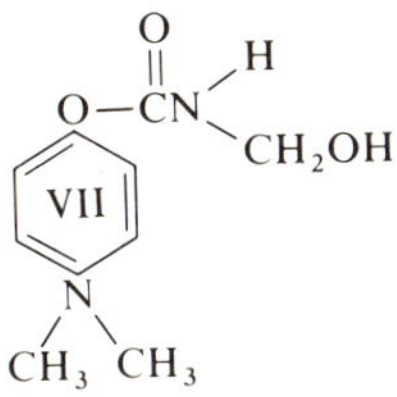

Compound[b]	Minimum detectable level[c] (μg)	Structure
Zectran®	0.02	
4-Amino Zectran® (as VI)	0.004	
4-Methylamino Zectran® (as VIII)	0.004	
Dimetilan	0.2	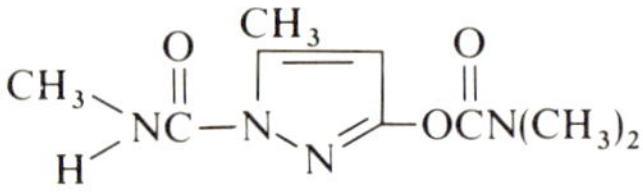
2-Methylcarbamoyl-3-methyl-5-		
pyrazolyl methylcarbamate IX	0.2	

From Oonnithan and Casida (1968).

[a]Only the metabolites that are actually found and are equal or superior in potency to their parent carbamates as inhibitors of cholinesterase are listed.
[b]Banol®, HRS-1422, and Isolan® did not produce metabolites that were more potent cholinesterase inhibitors than the original compounds.
[c]Assayed as the minimum quantity required to inhibit acetylcholinesterase.

do occur frequently for the metabolic products of carbamates and are considered to be the second most important type of reactions. Hydrolytic reactions come third in the order of importance for carbamates. However, the point to remember is that these two reactions are clearly detoxification processes.

5.4.3b. Oxidative Metabolism of Carbamates

Basically the oxidative reactions can be classified into two groups: (1) ring hydroxylation, and sometimes further oxidation to ketones, and (2) oxidation of side-chains. Ring hydroxylation was originally reported by Dorough et al. (1963) and Dorough and Casida (1964) on carbaryl. Leeling and Casida (1966) further identified 4-hydroxy-, 5-hydroxy, and 5,6-dihydro-dihydroxycarbaryl. Thus ring positions 4 and 5 are susceptible to hydroxylation, and probably the 5–6 positions are susceptible to epoxidation. Similarly, various positions on benzene rings are also susceptible to hydroxylation, depending on the nature of other ring substitutions. In the case of propoxur, hydroxylation occurs at the 5-position, and with UC-10854 it is at the 4-position (Oonnithan and Casida, 1966, 1968):

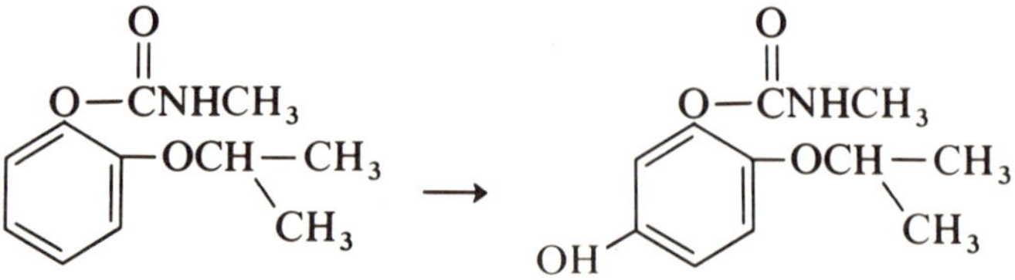

PROPOXUR

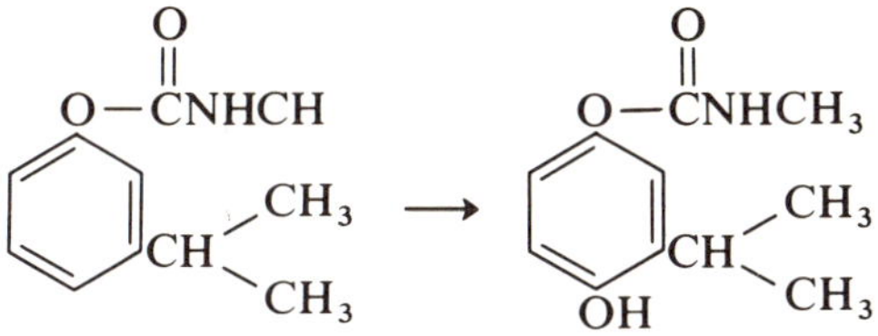

$$
\begin{array}{ccc}
\text{UC-10854} & \longrightarrow &
\end{array}
$$

UC-10854

Ring hydroxylation can also take place with nonaromatic rings. The position of hydroxylation for carbofuran is at the 3-position (Dorough, 1968; Metcalf *et al.*, 1968):

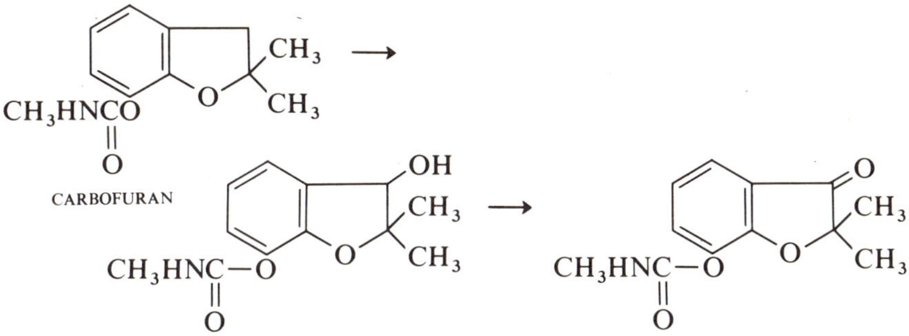

CARBOFURAN $\longrightarrow$

3-HYDROXYCARBOFURAN $\longrightarrow$ 3-KETOCARBOFURAN

Oxidation of side-chains can occur in many forms. First, various aliphatic ring substitutions can be hydroxylated. Methyl groups can become hydroxymethyls (e.g., Meobal,® Miyamoto, 1970), and isopropyl moieties can become 1-hydroxyisopropyl groups (e.g., HRS-1422; Oonnithan and Casida, 1968):

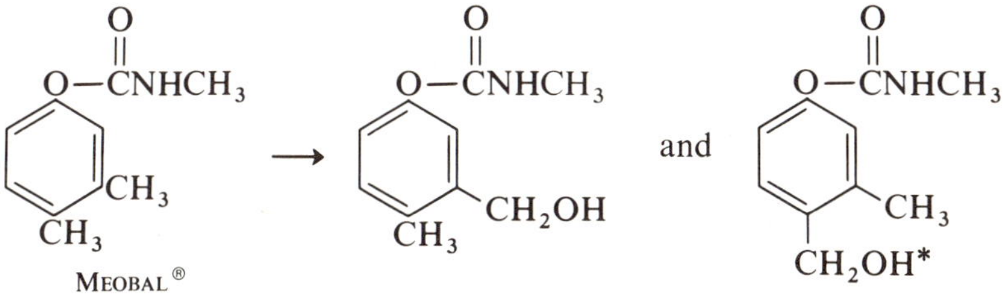

MEOBAL® $\longrightarrow$ and

CH_2OH*

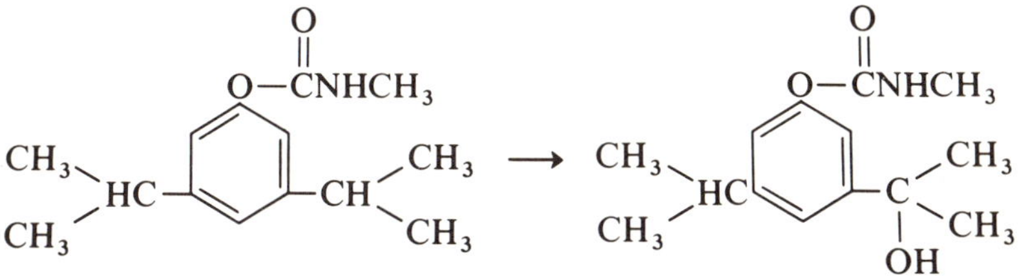

HRS-1422 $\longrightarrow$

* Only this moiety can be oxidized farther to an acid.

Another important type of oxidative attack on carbamate insecticides is N-demethylation. This process is extremely important for carbamates, since every carbamate possesses an N-alkyl moiety and some even have extra N-alkyl moieties (e.g., Zectran®). As discussed previously, the process of oxidative N-dealkylation actually takes place by stepwise reactions (R = CH₃ or H for insecticidal carbamates):

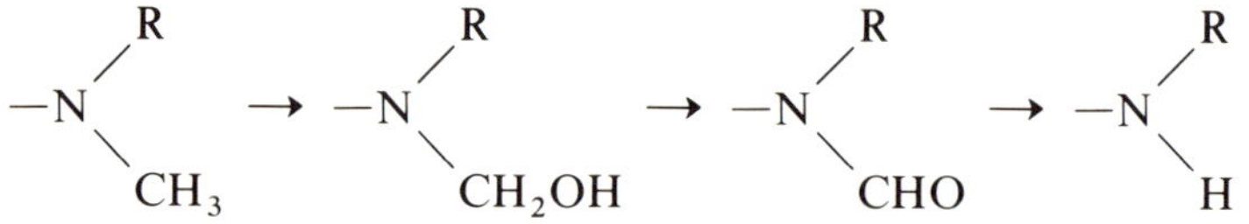

$$-N\begin{smallmatrix}R\\CH_3\end{smallmatrix} \rightarrow -N\begin{smallmatrix}R\\CH_2OH\end{smallmatrix} \rightarrow -N\begin{smallmatrix}R\\CHO\end{smallmatrix} \rightarrow -N\begin{smallmatrix}R\\H\end{smallmatrix}$$

In the case of Zectran,® both types of reactions have been reported to occur (Oonnithan and Casida, 1968):

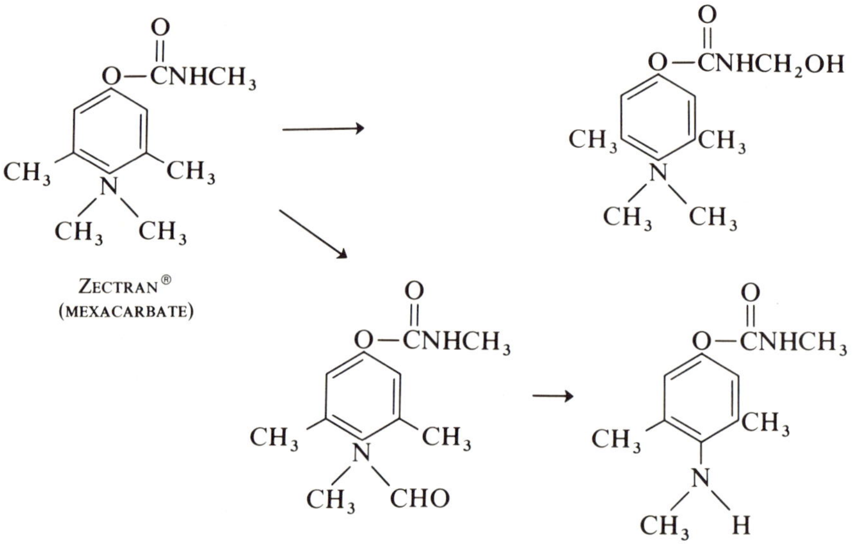

ZECTRAN®
(MEXACARBATE)

The N-hydroxymethyl intermediate at the carbamyl moiety has been observed in, e.g., Banol® (Baron and Doherty, 1967), propoxur (Oonnithan and Casida, 1968), and Bux® (Sutherland et al., 1970). Since nobody has yet found a completely N-dealkylated carbamyl metabolite, i.e., ROC(O) NH₂, such intermediates must be labile and are probably hydrolytically degraded to corresponding phenols by the action of the arylamidase, at least in mammals.

Another site of oxidative attack is the thioether bonding, if the molecule contains such a side-chain. Knaak et al. (1966), Andrawes et al. (1967), and Dorough et al. (1970) showed that Temik® (aldicarb) can be oxidized with sulfoxide and sulfone in the rat and in the cow (found in the milk). Robbins et al. (1970) found sulfoxides of Mobam,® 4-benzothienyl sulfate-1-

oxide and 4-hydroxybenzothiophene-1-oxide, in the milk and urine of a cow and goats fed Mobam.®

Two hydrolytic routes are available: (1) hydrolysis of the carbamyl moiety and (2) hydrolysis of side-chains, if there are any. The ease with which carbamate insecticides are enzymatically hydrolyzed depends largely on the chemical nature of the insecticide itself, and to a lesser extent on differences in species and tissues.

As for the relative importance of the hydrolytic processes vs. oxidative ones, there are a wide variety of data showing species, tissue, and timing differences. To elaborate a little further, the ratio is largely dependent on the nature of the chemicals and the species. For instance, nobody has really detected a substantial amount of carbofuran hydrolysis products in any animals and plants, whereas with carbaryl there are numerous hydrolysis data showing hydrolysis products amounting to 20–80 % of the total metabolic products in various systems, depending on the species. Even within one individual the ratio is expected to vary. For instance, Casida and Augustinsson (1959) found that carbaryl, on incubation with plasma exclusively, was converted into 1-naphthol. Knowing that the same compound is extensively oxidized in the liver (Dorough *et al.*, 1963), one must conclude that the source of the enzymes is very important in deciding the ratio in *in vitro* experiments. Let us consider the species difference even further. In the case of propoxur, Dawson *et al.* (1964) found only the hydrolysis product 2-isopropoxyphenol and its conjugate in human urine. In contrast, Kuhr and Casida (1967) could not find any hydrolysis product of propoxur in bean plants! The general "rule of thumb" appears to be that the order of ratios for hydrolytic processes in comparison to oxidative reactions is as follows (the figures in parentheses give the percentages for propoxur):

man > rats and other mammals > insects > plants

(100 %)   (20–40 %)   (1–20 %)   (0 %)

As for the variations due to chemical differences within one species, Schlagbauer and Schlagbauer (1972b) prepared a convenient table of percent hydrolysis figures for various compounds in the rat: 23 % for carbofuran, 31 % for propoxur, 39–65 % for carbaryl, 49 % for HRS-1422, 53 % for UC-10854, 53–62 % for aldicarb, 76 % for Zectran,® and 99 % for Mobam.® So, surprisingly, hydrolysis plays a rather major role in the metabolism of many carbamates as far as the rat is concerned.

One might get the impression that oxidative reactions completely predominate in carbamate metabolism because (1) they produce far more varieties of metabolites, (2) *in vitro* data, particularly the liver microsomal data with NADPH, have been extensively quoted, and (3) often only solvent-extractable metabolites have been reported, while the bulk of hydrolysis products and conjugates remain in aqueous phases. In the final analysis, however, both systems are important. Hydrolysis might be aided by initial oxidation of the molecule, as well. Neglecting one system and overemphasizing the other could result in a very erroneous concept of the total metabolism of carbamates.

5.4.3c. Carbaryl Metabolism

Earlier metabolic studies on carbaryl indicated that the hydrolysis product, 1-naphthol, is a major metabolic product. Casida and Augustinsson (1959), for instance, incubated carbaryl with the plasma of 15 different animals and identified 1-naphthol as the metabolic product. Carpenter *et al.* (1961) found that the rat metabolized carbaryl and excreted 31% of the original amount as conjugates of 1-naphthol in the feces. Eldefrawi and Hoskins (1961) treated *B. germanica* and found 1-naphthol in the excreta along with six other metabolites in 24 hr.

Dorough *et al.* (1963) and Dorough and Casida (1964) made an important discovery that various hydroxylation products are formed as a result of carbaryl incubation with the rat liver microsomal system. They identified 1-naphthol-*N*-hydroxymethylcarbamate and two ring hydroxylation products, 4-hydroxy- and 5-hydroxycarbaryl, along with a dihydroxydihydro analogue, 1-naphthol, and several other unidentified metabolic products. Leeling and Casida (1966) further identified 5,6-dihydroxydihydrocarbaryl and 5,6-dihydroxydihydro-1-naphthol by using similar enzyme preparations. Actually, this was the first indication that carbamate insecticides are extensively metabolized by oxidative systems, i.e., by the mixed-function oxidases, in plants and animals (Fig. 5-20).

Other discoveries followed soon thereafter. Knaak *et al.* (1965) studied the urinary metabolites of carbaryl by using ^{14}C-naphthyl-labeled and *N*-methyl-^{14}C-labeled carbaryl in rats and guinea pigs. Of the ^{14}C-naphthyl-labeled carbaryl, 80–90% was excreted in the urine as anionic metabolites within 24 hr. The major metabolic products they found were sulfate and glucuronic acid conjugates of 1-naphthol (39%) and 4-hydroxycarbaryl (17.5%). Another major conjugate was tentatively identified as the glucuronide of 1-naphthylmethylimidocarbonate (26%).

The patterns of carbaryl metabolism in insects (e.g., Dorough and Casida, 1964; Terriere *et al.*, 1961; Kuhr, 1970) and in plants (bean plants; Kuhr and Casida, 1967) are very similar except for the nature of the conjugation. In plants, the conjugation products are glucosides (or other sugar conjugates) of 4- and 5-hydroxycarbaryl, 5,6-dihydroxydihydrocarbaryl, and 1-naphthol. In insects, the nature of the conjugates has not been determined. Smith (1955*b*) suggested glucose conjugates for phenolic metabolites in 15 insect species, while Terriere *et al.* (1961) indicated glucuronide and sulfate conjugation for 1-naphthol in the housefly.

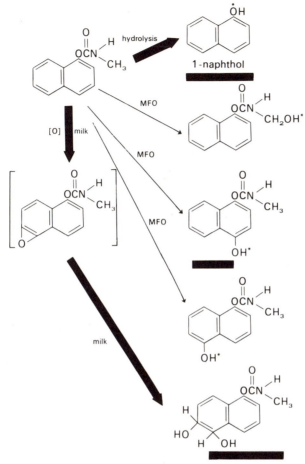

Fig. 5-20. Metabolism of carbaryl in animals and plants. The positions of conjugation are indicated by asterisks.

There is variation among different species and tissues in the quantity of these metabolites produced. In some cases, there is quite a departure from the general pattern of carbaryl metabolism. Knaak and Sullivan (1967), for instance, could not find any of the major rat urinary metabolites in the feces and urine of dogs treated with carbaryl. Dogs appear to directly conjugate carbaryl, and 35% of ^{14}C-naphthyl labeled metabolites are excreted in the feces as opposed to 40% in the urine. On the other hand, the major chloroform-extractable metabolite appearing in cow's milk is 5,6-dihydroxydihydrocarbaryl (Dorough, 1967; Baron *et al.*, 1969). In insects, Ku and Bishop (1967) report that 1-naphthol and 1-naphthol conjugates are the major metabolic products. Thus a complete generalization appears to be a little risky.

5.4.3d. Substituted Phenylcarbamate

Propoxur can be metabolized through hydrolysis or hydroxylation. The relative proportion of these two competing activities appears to be largely species specific. Oonnithan and Casida (1968) incubated propoxur with rat liver and microsomal preparation in the presence of NADPH and found 5-hydroxypropoxur, N-hydroxymethylpropoxur (both oxidation products), 2-isopropoxyphenol, and 1-phenyl-2-hydroxymethylcarbamate (both hydrolysis products). In houseflies, the isopropoxy moiety is apparently metabolized, according to Metcalf *et al.* (1967), after being liberated from propoxur, since 2-isopropoxy-1,3-^{14}C-labeled propoxur is converted eventually to

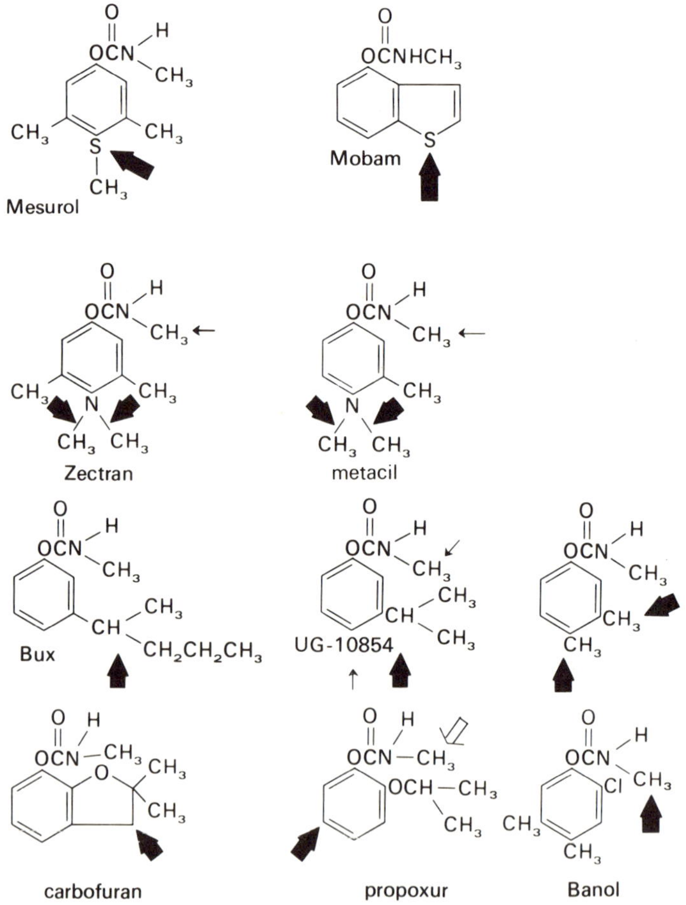

Fig. 5-21. Site of oxidative attack on various substituted phenyl methylcarbamates. Thick black arrows indicate strong reactions, white arrows intermediate ones, and thin arrows weak ones.

$^{14}CO_2$. However, Shrivastava *et al.* (1969) found the metabolic pattern of propoxur in the housefly to be similar to that in the rat liver: identical metabolites were found. According to their studies, hydrolytic processes play a very minor role. They found the major metabolic product to be 5-hydroxypropoxur (Fig. 5-21).

Obviously, species-specific variation does influence the relative importance of the systems. In man, so far only the hydrolysis product has been detected from the metabolism of propoxur. Dawson *et al.* (1964) could recover only 2-isopropoxyphenol and its glucuronide in the urine of humans fed propoxur. In bean plants, the glucose conjugate of 2-hydroxyphenyl-methylcarbamate is the major metabolic product, although there are nine other metabolites including 4-hydroxy- (or 5-hydroxy), *N*-hydroxy propoxur (Kuhr and Casida, 1967).

For alkyl-substituted phenyl *N*-methylcarbamates, hydroxylation on side chains takes place instead of ring hydroxylation. Thus Bux® is hydroxylated at the 1-carbon position of the 1-methylbutane substitute (Sutherland *et al.*, 1970). HRS-1442 (3,5-diisopropylphenyl *N*-methylcarbamate) and UC-10854 (3-isopropyl methylcarbamate) are also hydroxylated at the 1-position (Oonnithan and Casida, 1966, 1968). Both Moebal® (3,4-dimethylphenyl methylcarbamate) (Miyamoto, 1970) and Landrin® (3,4,5-trimethyl- and 2,3,5-trimethylphenyl methylcarbamate) (Slade and Casida, 1970) are metabolized through a series of hydroxylations at the methyl substitutes. One can probably add carbofuran to this group of chemicals, since it is a cyclic alkyl-ether substituted phenyl *N*-methylcarbamate. It is hydroxylated at the 3-position (Dorough, 1968; Metcalf *et al.*, 1968).

For ring-substituted dialkyl amines and thioethers, oxidation of this moiety occurs rather readily, as discussed in Section 5.4.2. As expected, both metacil (4-dimethylamino-3-cresyl methylcarbamate) and Zectran® go through a series of oxidative *N*-demethylations to finally yield amino analogues of these insecticides (Oonnithan and Casida, 1968; Williams *et al.*, 1964*a, b*). Mobam® (4-benzothieny methylcarbamate) and Mesurol® (4-methylthio-3,5-xylyl methylcarbamate) can be oxidized at the thioether bonds to the respective sulfoxides, and in the case of the latter to the sulfone (Robbins *et al.*, 1969; Oonnithan and Casida, 1968). Oxidative *N*-dealkylation apparently is such a common reaction that it is just about the only one that has been observed to take place among the metabolic changes occurring in all pyrazolyl dimethylcarbamates, such as dimetilan (Zubairi and Casida, 1965; Eberle and Gunther, 1965) and Isolan® and pyrolan (Hodgson and Casida, 1961; Eberle and Gunther, 1965).

It is interesting to note that so far there is no report that ring chlorine-substituted phenyl methylcarbamates can be hydroxylated at any position around the ring. In the case of Banol® (Baron and Doherty, 1967), the only place that was hydroxylated was the *N*-methyl moiety.

To summarize, there are two types of reactions that can attack ring-substituted phenyl methylcarbamates: hydrolysis through esterases and oxidation. The relative ease of hydroxylation apparently is as follows: thioether oxidation, *N*-dealkylation > hydroxylation of alkyl side-chains (particularly at the carbon next to the ring) > ring hydroxylation for unsubstituted compounds or those monosubstituted with other groups than cited above ≥ hydroxylation of the carbamyl *N*-methyl moiety.

5.4.3e. Oxime Insecticide: Aldicarb Metabolism

Aldicarb or Temik® is an oxime insecticide. Its metabolic fate is different than that of the carbamate insecticides because it possesses an extra hydrolyzable bond. Since it is also a thioether, oxidation of aldicarb to a sulfoxide analogue is expected. Hydrolysis, which follows the oxidative reaction,

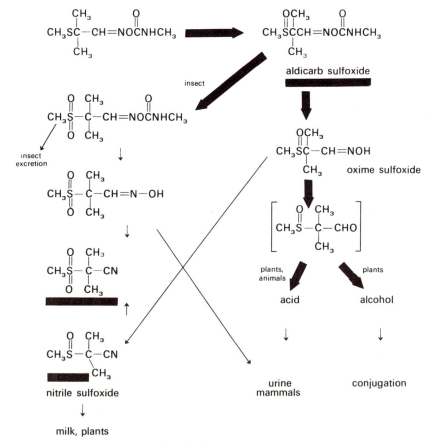

Fig. 5-22. Metabolism of aldicarb by mammals, insects, and plants.

probably plays the second most important role in mammalian systems (Knaak *et al.*, 1966; Andrawes *et al.*, 1967). To a lesser degree, further oxidation to the sulfone is possible. In all cases, the hydrolyzed products (oximes) becomes propionaldehyde analogues, which are further oxidized to acids. Dorough and Ivie (1968) found that while the bulk of aldicarb metabolites are eliminated in the urine in the lactating cow (83 % in 24 hr), a small portion is found in milk. These authors made an interesting observation that there are two nitrile fragments in addition to the sulfoxides and sulfones of aldicarb and its oximes (Fig. 5-22). The same nitrile metabolites were found in cotton plants along with other metabolites mentioned above (Metcalf *et al.*, 1966; Coppedge *et al.*, 1967; Bartley *et al.*, 1970). Another difference in the metabolic pattern in this plant was that the main metabolites were an alcohol and its glycoside.

Metcalf *et al.* (1966) also found an interesting metabolic difference in the housefly: apparently the hydrolytic reaction does not take place. The resulting oxidation end-product, aldicarb sulfone, is excreted.

5.4.4. Metabolism of Botanical and Other Insecticides

Basically, the naturally occurring botanical insecticides are a collection of unrelated compounds. Therefore, grouping them together with respect to metabolic reactions is for convenience only.

5.4.4a. Nicotine Metabolism

Nicotine is the basic principle of tobacco smoke, and as such its metabolism has been studied in man and animals. In essence, in all animals so far tested (e.g., Bowman *et al.*, 1959; McKennis *et al.*, 1962, 1963; Self *et al.*, 1964; Turner, 1969) cotinine has been shown to be the major metabolic product. Even in plants (mustard greens), Gunther *et al.* (1959) found cotinine to be the principal metabolite of nicotine. The route of cotinine formation is apparently through the immediate hydroxylation product, 2-hydroxynicotine. Hucker *et al.* (1959), using rabbit liver microsomes with NADPH, showed that 2-hydroxynicotine as well as cotinine is formed *in vitro*. Cotinine is further metabolized to desmethylcotinine. Judging by the results of *in vivo* work, the overall metabolic pattern of nicotine can be summarized as shown in Fig. 5-23. There is a possibility that *N*-demethylation can also occur at an early stage of degradation to produce *N*-desmethylnicotine and then *N*-desmethylcotinine. Though it takes place, this route will be a minor one, since cotinine accumulation is dominant in all organisms.

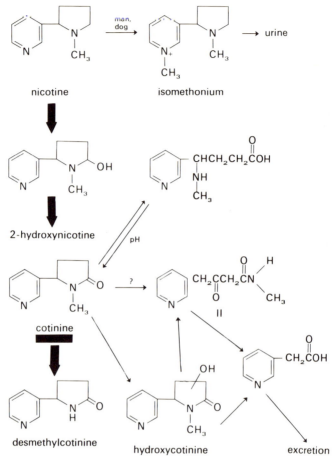

Fig. 5-23. Metabolism of nicotine in animals. Cotinine is also a principal metabolite in insects.

An interesting sidelight is that an immediate conjugation system, pyridyl *N*-methylation, is available in man and dogs but not in insects. This compound, isomethonium, is polar enough by itself and can be excreted in the urine.

From a comparative biochemical standpoint, it is interesting to note that the hamster apparently has six times as much ability as the rat to produce cotinine from nicotine (Harke *et al.*, 1970).

5.4.4b. Rotenone Metabolism

Metabolism of rotenone has been studied by Casida's group (Fukami *et al.*, 1967, 1969; Yamamoto and Casida, 1967; Yamamoto *et al.*, 1969). The positions of hydroxylation are

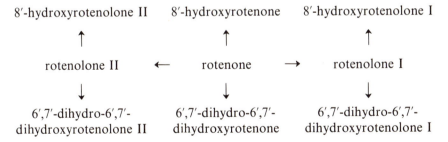

ROTENONE

Rotenolones I and II are distinguished by the direction of hydroxylation:

ROTENOLONE I   ROTENOLONE II

In either case, all three compounds, rotenone, rotenolone I, and rotenolone II, are hydroxylated either at the 8'-position or at the 6'- and 7'-positions:

8'-hydroxyrotenolone II   8'-hydroxyrotenone   8'-hydroxyrotenolone I

↑   ↑   ↑

rotenolone II   ←   rotenone   →   rotenolone I

↓   ↓   ↓

6',7'-dihydro-6',7'-   6',7'-dihydro-6',7'-   6',7'-dihydro-6',7'-
dihydroxyrotenolone II   dihydroxyrotenone   dihydroxyrotenolone I

There are other uncharacterized polar metabolites. Also, some of these metabolites appear to be similarly toxic to mice as rotenone.

5.4.4c. Pyrethrin I, Allethrin, Dimethrin, and Phthalthrin Metabolism

Chang and Kearns (1964) studied pyrethroid metabolism in houseflies and found that hydrolysis of the ester linkage is not a major detoxification mechanism. Because of the effective synergistic actions of various methylenedioxyphenyl compounds, it has been suspected that pyrethrins must be metabolized largely through oxidative means.

Yamamoto and Casida (1966) found that ^{14}C-allethrin was converted to 13 metabolites by housefly-abdomen homogenate fortified with $NADPH_2$. Each of these metabolites had the ester group intact, and the major product also appeared to be acidic based on its chromatographic position. The authors found that the major metabolite of ^{14}C-allethrin is O-desmethyl-allethrin II [3-(3-allyl-2-methyl-4-oxo-2-cyclopenten-1-yl) chrysanthemum dicarboxylate].

^{14}C-Pyrethrin was converted *in vitro* by the housefly-abdomen–$NADPH_2$ system to at least ten metabolites, and the major metabolites differed only slightly in chromatographic positions from those derived from ^{14}C-allethrin. On saponification, the major metabolite of ^{14}C-pyrethrin I yielded a product which cochromatographed with chrysanthemum dicarboxylic acid, and on reaction with diazomethane it cochromatographed with pyrethrin II.

The major metabolites of dimethrin and phthalthrin yielded chrysanthemum dicarboxylic acid on saponification. These findings suggest that pyrethrin I, dimethrin, and phthalthrin are metabolized by the *in vitro* enzyme systems by oxidation of the analogous methyl group for allethrin and that other modifications of the molecule are not involved in forming the major metabolites.

The relative ease of oxidation of the 1-methyl group in the isobutenyl moiety of the pyrethroids is similar to known pathways for mammalian metabolism of terpenes, such as citral, dihydromyrcene, geranical, and geranic acid (Williams, 1959).

Preliminary studies by Yamamoto and Casida indicate that allethrin is also converted in part to O-desmethylallethrin II *in vivo*. This reaction involves a detoxification because allethrin is greater than thirtyfold more toxic than O-desmethylallethrin II when these compounds are injected into houseflies.

The same group has published three more papers on the metabolic fate of pyrethroids: Yamamoto and Casida (1968) and Yamamoto *et al.* (1969) for houseflies and Yamamoto *et al.* (1971) for rats. The principal metabolites of pyrethrin I are the same as the ones for pyrethrin II. In all cases, the metabolites have a carboxyl group on the isobutenyl group of the acid moiety. Four metabolites identified in the last paper had $-CH_2-CH=CHOH-CH_2-OH$ (*cis*), $-CH_2CHOH-CH=CHCH_2OH$ (*trans*), $-CH_2-CH=CH-CH(CH_2OH)$ O-conjugate (*cis*), and $-CH_2-CH=CH-CH=CH_2$ for their alcohol side-chains. Similarly, allethrin metabolites were found to be diacid esters of alcohols containing $-CH_2CHOH-CH_2OH$, $-CHOH-CH=CH_2$, and $-CH_2-CH=CH_2$ as their side-chains.

One metabolite had both end methyl groups of the isobutenyl oxidized to COOH and CH_2OH with the unchanged $-CH_2-CH=CH_2$ side-chain.

As for acaricidal compounds, Sen Gupta and Knowles (1970) and Ahmad and Knowles (1970) studied degradation of formetanate in rats and rat liver preparations. The metabolic products were *m*[(methylamino)-methylene]amino-phenyl methylcarbamate (demethylformetanate), *m*-form-aminophenyl-*N*-methylcarbamate, *m*-formaminophenol, *m*-acetamidophenol, and *m*-aminophenol. Sen Gupta and Knowles (1970) and Knowles and Sen Gupta (1970) studied the metabolism of chlordimeform Galecron® in the dog and goat and in the rat, respectively. Major urinary metabolites of chlordimeform were *N*-formyl-4-chloro-*o*-toluidine, 4-chloro-*o*-toluidine, 5-chloroanthranilic acid, and *N*-formyl-5-chloroanthranilic acid. Bowker and Casida (1969) studied the metabolism of Fenazaflor® (5,6-dichloro-1-phenoxycarbonyl-2-trifluoromethylbenzimidazole) and related compounds in mammals, insects, and plants. Both hydrolytic and oxidative metabolic routes were found.

Terranova and Crystal (1970) studied the metabolic fate of the chemosterilant *N,N'*-tetramethylene *bis*(1-aziridinecarboxamide) in the black fly and the screw worm fly. They found the sterilant to be rather persistent; about 25% was left after 504 hr on *P. regina*. Jones (1970) studied the metabolism of HMPA and found that in the rat and the mouse HMPA lost methyl groups to yield several metabolites including *N*-formyl (pentamethyl) phosphoramide.

5.5. REFERENCES

Abou-Donia, M. B., and D. B. Menzel (1968). *Biochem. Pharmacol.* **17**:2143.

Ahmad, S., and C. O. Knowles (1970). *J. Econ. Entomol.* **63**:1690.

Ahmed, M. K., J. E. Casida, and R. E. Nichlos (1958). *J. Agr. Food Chem.* **6**:740.

Aldridge, W. N., and J. M. Barnes (1952). *Nature* **169**:345.

Andrawes, N. R., H. W. Dorough, and D. A. Lindquist (1967). *J. Econ. Entomol.* **60**:979.

Anliker, R., E. Beriger, M. Geiger, and K. Schmid (1961). *Helv. Chim. Acta* **44**:1622.

Armstrong, D. E., and R. F. Harris (1965). *Agron. Abst.* **57**:81.

Arthur, B. W., and J. E. Casida (1957). *J. Agr. Food Chem.* **5**:186.

Arthur, B. W., and J. Casida (1958). *J. Econ. Entomol.* **51**:49.

Bache, C. A., G. G. Gyrisco, S. N. Fertig, E. W. Huddleston, D. J. Lisk, F. H. Fox, G. W. Trimberger, and R. F. Holland (1960). *J. Agr. Food Chem.* **8**:408.

Bailey, S., P. J. Banyan, B. D. Rennison, and A. Taylor (1969). *Toxicol. Appl. Pharmacol.* **14**:13.

Baldwin, M. K., J. Robinson, and D. V. Parke (1970). *J. Agr. Food Chem.* **18**:1117.

Barnes, W. W., and G. W. Ware (1965). *J. Econ. Entomol.* **58**:286.

Baron, R. L., and J. D. Doherty (1967). *J. Agr. Food Chem.* **15**:830.

Baron, R. L., J. A. Sphon, J. T. Chen, E. Lustig, J. D. Doherty, E. A. Hansen, and S. M. Kolbye (1969). *J. Agr. Food Chem.* **17**:883.

Bartley, W. L., N. R. Andrawes, E. L. Chancey, W. P. Bagley, and H. W. Spurr (1970). *J. Agr. Food Chem.* **18**:446.

Boyland, E., and L. F. Chasseaud (1969). *Biochem. J.* **115**:985.

Blinn, R. C. (1968). *J. Agr. Food Chem.* **16**:441.

Bogdarina, A. A. (1957). *Transl. Plant Physiol. A.I.B.S.* (*Fiziologiya Rasteny*) **4**:254.

Bowker, D. M., and J. E. Casida (1969). *J. Agr. Food Chem.* **17**:956.

Bowman, J. S., and J. E. Casida (1958). *J. Econ. Entomol.* **51**:838.

Bowman, E. R., L. B. Turnbull, and H. McKennis, Jr. (1959). *J. Pharmacol. Exptl.Therap.* **127**:92.

Bradbury, F. R. (1957). *J. Sci. Food Agr.* **8**:90.

Bradbury, F. R., and P. Nield (1953). *Nature* **172**:1052.

Bradbury, F. R., and H. Standen (1955). *J. Sci. Food Agr.* **6**:90.

Bradbury, F. R., and H. Standen (1958). *J. Sci. Food Agr.* **9**:203.

Bradbury, F. R., and H. Standen (1959). *Nature* **183**:983.

Bradbury, F. R., and H. Standen (1960). *J. Sci. Food Agr.* **11**:92.

Brady, U. E., and B. W. Arthur (1961). *J. Econ. Entomol.* **54**:1232.

Bray, M. G., Z. Hybs, S. P. James, and W. V. Thorpe (1953). *Biochem. J.* **53**:266.

Bridges, R. G. (1955). *J. Sci. Food Agr.* **6**:261.

Bridges, R. G. (1959). *Nature* **184**:1337.

Brodie, B. B., J. R. Gillette, and B. N. LaDu (1958). *Ann. Rev. Biochem.* **27**:427.

Broniz, H., Z. Bidzinski, and J. Lenicka (1962). *Medycyna Pracy* **13**:449.

Brooks, G. (1969). *Residue Rev.* **27**:81. F. A. Gunther, ed. Springer-Verlag, New York.

Brooks, G. T. (1972). In *Environmental Quality and Safety.* F. Coulston, and F. Korte, eds. Georg Thieme Verlag, Stuttgart, p.106.

Brooks, G. T., and A. Harrison (1963). *Biochem. J.* **87**:5.

Brooks, G. T., and A. Harrison (1967a). *Life Sci.* **6**:681.

Brooks, G. T., and A. Harrison (1967b). *Life Sci.* **6**:1439.

Brooks, G. T., and A. Harrison (1969). *Bull. Environ. Contam. Toxicol.* **4**:352.

Brown, A. W. A. (1971). In *Pesticides in the Environment*, Vol. 1, Part II. R. White-Stevens, ed. Marcel-Dekker, New York, p. 457.

Bull, D. L. (1965). *J. Econ. Entomol.* **58**:249.

Bull, D. L., and D. A. Lindquist (1964). *J. Agr. Food Chem.* **12**:310.

Bull, D. L., D. A. Lindquist, and J. Hacskaylo (1963). *J. Econ. Entomol.* **56**:129.

Bull, D. L., D. A. Lindquist, and R. R. Grabbe (1967). *J. Econ. Entomol.* **60**:332.

Butts, J. S., S. C. Chang, B. E. Christensen, and C. H. Wang (1953). *Science* **117**:699.

Carpenter, C. P., C. S. Weil, P. E. Palm, M. W. Woodside, J. H. Nair, III, and H. F. Smyth, Jr. (1961). *J. Agr. Food Chem.* **9**:30.

Casida, J. E. (1963). *Ann. Rev. Entomol.* **8**:39.

Casida, J. E. (1969). *Microsomes and Drug Oxidations.* J. R. Gillette, A. H. Conney, G. J. Cosmides, R. W. Estabrook, J. R. Fouts, and G. J. Mannering, eds. Academic Press, New York, pp. 517–530.

Casida, J. E. (1970). *J. Agr. Food Chem.* **18**:753.

Casida, J. E., and K. B. Augustinsson (1959). *Biochim. Biophys. Acta* **36**:411.

Casida, J. E., and M. A. Stahmann (1953). *J. Agr. Food Chem.* **1**:883.

Casida, J. E., R. K. Chapman, M. A. Stahmann, and T. C. Allen (1954). *J. Econ. Entomol.* **47**:64.

Casida, J. E., P. E. Gatterdam, J. B. Knaak, R. D. Lence, and R. P. Niedermeier (1958). *J. Agr. Food Chem.* **6**:658.

Casida, J. E., L. McBride, and R. P. Niedermeier (1962). *J. Agr. Food Chem.* **10**:370.

Cassil, C. C., and P. E. Drummond (1965). *J. Econ. Entomol.* **58**:356.

Chadwick, R. W., and J. J. Freal (1972a). *Bull. Environ. Contam. Toxicol.* **7**:137.

Chadwick, R. W., and J. J. Freal (1972b). *Food Cosmet. Toxicol.* **10**:789.

Chamberlain, W. F. (1964a). *J. Econ. Entomol.* **57**:119.
Chamberlain, W. F. (1964b). *J. Econ. Entomol.* **57**:329.
Chang, S. C., and C. W. Kearns (1964). *J. Econ. Entomol.* **57**:397.
Chen, P. R., W. P. Tucker, and W. C. Dauterman (1969). *J. Agr. Food Chem.* **17**:86.
Cherrington, A. D., U. Paim, and O. T. Page (1969). *J. Fish. Res. Board Can.* **26**:47.
Clark, A. G., S. Murphy, and J. N. Smith (1969). *Biochem. J.* **113**:89.
Clemons, G. D., and R. E. Menzer (1968). *J. Agr. Food Chem.* **16**:312.
Cook, J. W., and G. Yip (1958). *J. Assoc. Offic. Agr. Chemists* **41**:407.
Cook, J. W., J. R. Blake, and M. W. Williams (1957). *J. Assoc. Offic. Agr. Chemists* **40**:664.
Coper, H., H. Herken, and J. Klempaw (1951). *Arch. Exptl. Pathol. Pharmakol.* **212**:463.
Coppedge, J. R., D. A. Lindquist, D. L. Bull, and H. W. Dorough (1967). *J. Agr. Food Chem.* **15**:902.
Corey, R. A. (1965). *J. Econ. Entomol.* **58**:112.
Cornish, H. H., and W. D. Block (1959). *Fed. Proc.* **18**:207.
Cueto, C., and W. Hayes, Jr. (1962). *J. Agr. Food Chem.* **10**:366.
Dahm, P. A. (1970). Some aspects of metabolism of parathion and diazinon. In *Biochemical Toxicology of Insecticides*. R. D. O'Brien and I. Yamamo, eds. Academic Press, New York, pp. 51–63.
Damico, J. N., J.-Y. T. Chen, C. T. Costello, and E. A. Haenni (1968). *J. Assoc. Offic. Anal. Chemists* **51**:48.
Datta, P. R., E. P. Laug, and A. K. Klein (1964). *Science* **145**:1052.
Dauterman, W. C., and F. Matsumura (1962). *Science* **138**:694.
Dauterman, W. C., J. E. Casida, J. B. Knaak, and T. Kawalczyk (1959). *J. Agr. Food Chem.* **7**:188.
Datta, P. R. (1970). *Industrial Medicine* **39**:4, 49.
Dauterman, W. C., G. B. Viado, J. E. Casida, and R. D. O'Brien (1960). *J. Agr. Food Chem.* **8**:115.
Dawson, J. A., D. F. Heath, J. A. Rose, E. M. Thain, and J. B. Ward (1964). *Bull. World Health Org.* **30**:127.
Deema, P., E. Thompson, and G. W. Ware (1966). *J. Econ. Entomol.* **59**:546.
Diggle, W. M., and J. C. Gage (1951). *Biochem. J.* **49**:491.
Dinamarca, M. L., I. Saavedra, and E. Valdes (1969). *Comp. Biochem. Physiol.* **31**:269.
Donninger, C., H. D. Hutson, and B. A. Pickering (1967). *Biochem. J.* **102**:26.
Dorough, H. W. (1967). *J. Agr. Food Chem.* **15**:261.
Dorough, H. W. (1968). *J. Agr. Food Chem.* **16**:319.
Dorough, H. W. (1970). *J. Agr. Food Chem.* **18**:1015.
Dorough, H. W., and J. E. Casida (1964). *J. Agr. Food Chem.* **12**:294.
Dorough, H. W., and G. W. Ivie (1968). *J. Agr. Food Chem.* **16**:460.
Dorough, H. W., N. C. Leeling, and J. E. Casida (1963). *Science* **140**:170.
Dorough, H. W., R. B. Davis, and G. W. Ivie (1970). *J. Agr. Food Chem.* **18**:135.
DuBois, K. P., J. Doull, and J. M. Coon (1950). *J. Pharmacol. Exptl. Therap.* **99**:376.
Durham, W. F., P. Ortega, and W. J. Hayes (1963). *Arch. Internat. Pharmacodyn. Therap.* **141**:111.
Eberle, D. O., and F. A. Gunther (1965). *J. Assoc. Offic. Agr. Chemists* **48**:927.
Eldefrawi, M. E., and W. H. Hoskins (1961). *J. Econ. Entomol.* **54**:401.
Ely, R. E., L. A. Moore, P. E. Hubanks, R. H. Carter, and R. W. Poos (1955). *J. Dairy Sci.* **38**:669.
Esaac, E. G., and J. E. Casida (1968). *J. Insect Physiol.* **14**:913.
Esaac, E. G., and J. E. Casida (1969). *J. Agr. Food Chem.* **17**:539.
Eto, M., J. E. Casida, and T. Eto (1962). *Biochem. Pharmacol.* **11**:337.

Feil, V. J., R. D. Hedde, R. G. Zaylskie, and C. H. Zachrison (1970). *J. Agr. Food Chem.* **18**:120.

Fikudo, H., T. Masuda, and Y. Miyahara (1962). *Japan. J. Appl. Entomol. Zool.* **6**:230.

Finley, R. B., Jr., and R. E. Pillmore (1963). *Am. Inst. Biol. Sci. Bull.* **13**:41.

Fletcher, T. (1959). World Health Organization Symposium on Pesticides, Brazaville, Republic of Congo. WHO/PA/40.60, p. 103.

Francis, J. I., and J. M. Barnes (1963). *Bull. World Health Org.* **29**:205.

Fukami, J., and T. Shishido (1963). *Botyu-Kagaku* **28**:77.

Fukami, J., and T. Shishido (1966). *J. Econ. Entomol.* **59**:1338.

Fukami, J., I. Yamamoto, and J. E. Casida (1967). *Science* **155**:713.

Fukami, J., T. Shishido, K. Fukunaga, and J. E. Casida (1969). *J. Agr. Food Chem.* **17**:1217.

Fukuto, T. R., R. L. Metcalf, R. B. March, and M. G. Maxon (1954). *J. Am. Chem. Soc.* **76**:5103.

Fukuto, T. R., R. L. Metcalf, R. B. March, and M. G. Maxon (1955). *J. Econ. Entomol.* **48**:347.

Gardiner, J., and B. Kilby (1952). *Biochem. J.* **51**:78.

Gatterdam, P., R. De, F. Guthrie, and T. Bowery (1964). *J. Econ. Entomol.* **57**:258.

Geissbühler, H., G. Voss, and R. Anliker (1971). *Residue Rev.* **37**:39. F. A. Gunther, and J. D. Gunther, eds. Springer-Verlag, New York.

Giannotti, O. (1958). *Sao Paulo Inst. Biol.* **25**:253.

Gillette, J. R., A. C. Conney, G. J. Cosmides, R. W. Estabrook, J. R. Fouts, and G. J. Mannering, eds. (1969). *Microsomes and Drug Oxidations.* Academic Press, New York, 547 pp.

Gilmour, D. (1959). *Biochemistry of Insects.* Academic Press, New York.

Gram, T. E., and J. R. Fouts (1968). *The Enzymatic Oxidation of Toxicants.* E. Hodgson, ed. North Carolina State University, Raleigh, N.C., p. 47.

Grover, P., and P. Sims (1965). *Biochem.* **96**:521.

Gunther, F. A., R. C. Blinn, E. Benjamini, W. R. Kinkade, and L. D. Anderson (1959). *J. Agr. Food Chem.* **7**:330.

Hall, W. E., and Y. P. Sun (1965). *J. Econ. Entomol.* **58**:845.

Harke, H. P., B. Frahm, C. Schultz, and W. Dontenwill (1970). *Biochem. Pharmacol.* **19**:495.

Hartley, G. (1951). 15th International Chemistry Congress, New York.

Hartley, G. S. (1954). *Chem. Ind.* **1954**:529.

Hassan, A., and S. M. A. D. Zayed (1965). *Can. J. Biochem.* **43**:1271.

Hassan, A., S. M. A. D. Zayed, and F. M. Abdel-Hamid (1965). *Biochem. Pharmacol.* **14**:1577.

Heath, D., and P. Casapieri (1951). *Trans. Faraday Soc.* **47**:1093.

Heath, D., D. Love, and M. Llewellyn (1952). *J. Sci. Food Agr.* **3**:60, 69.

Hennessy, D. J. (1965). *J. Agr. Food Chem.* **13**:218.

Hitchcock, M., and S. D. Murphy (1967). *Biochem. Pharmacol.* **16**:1801.

Hodgson, E., ed. (1968). *The Enzymatic Oxidation of Toxicants.* Proceedings of a Conference held at North Carolina State University, Raleigh, N.C., 229 pp.

Hodgson, E., and J. E. Casida (1961). *Biochem. Pharmacol.* **8**:179.

Hodgson, E., and J. E. Casida (1962). *J. Agr. Food Chem.* **10**:208.

Hodgson, E., and F. W. Plapp (1970). *J. Agr. Food Chem.* **18**:1048.

Hoffman, R., and A. Lindquist (1952). *J. Econ. Entomol.* **45**:233.

Hollingworth, R. M. (1969). *J. Agr. Food Chem.* **17**:987.

Hucker, H. B., J. R. Gillette, and B. B. Brodie (1959). *Nature* **183**:47.

Hutson, D. H., D. A. A. Akintonwa, and D. H. Hathway (1967). *Biochem. J.* **102**:133.

Hutson, D. H., B. A. Pickering, and C. Donninger (1968). *Biochem. J.* **106**:20.

Ishida, M., and P. Dahm (1965a). *J. Econ. Entomol.* **58**:602.

Ishida, M., and P. Dahm (1965b). *J. Econ. Entomol.* **58**:383.

Jaques, R., and M. J. Bein (1960). *J. Arch.Toxikol.* **18**:316.

Jensen, J., C. Cueto, W. Dale, C. Rothe, G. Pearce, and A. Mattson (1957). *J. Agr. Food Chem.* 5:919.

Jones, A. R. (1970). *Biochem. Pharmacol.* 19:603.

Kamienski, F. X., and J. E. Casida (1968). Abstracts of the 155th American Chemical Society Meeting, San Francisco, April, Am. Chem. Soc., Washington, D.C.

Kamin, H., and B. S. S. Masters (1968). In *The Enzymatic Oxidation of Toxicants*. E. Hodgson, ed. North Carolina State University, Raleigh, N.C., p. 5.

Kapoor, I. P., R. L. Metcalf, R. F. Nystrom, and G. K. Sangha (1970). *J. Agr. Food Chem.* 18:1145.

Kearns, C. E. (1956). *Ann. Rev. Entomol.* 1:123.

Kikal, T., and J. N. Smith (1959). *Biochem. J.* 71:48.

Kilby, B. (1953). *Chem. Ind.* 1953:856.

Kimura, T., and A. W. A. Brown (1964). *J. Econ. Entomol.* 57:710.

Klein, W. (1972). In *Environmental Quality and Safety*. F. Coulston, and F. Korte, eds. Georg Thieme Verlag, Stuttgart, p. 164.

Klein, W., W. Muller, and F. Korte (1968). *Liebigs Ann. Chim.* 713:180.

Knaak, J. B. (1971). *Bull. World Health Org.* 44:121.

Knaak, J. B., and R. D. O'Brien (1960). *J. Agr. Food Chem.* 8:198.

Knaak, J. B., and L. J. Sullivan (1967). *J. Agr. Food Chem.* 15:1125.

Knaak, J. B., M. J. Tallant, W. J. Bartley, and L. J. Sullivan (1965). *J. Agr. Food Chem.* 13:537.

Knaak, J. B., M. J. Tallant, and L. J. Sullivan (1966). *J. Agr. Food Chem.* 14:573.

Knowles, C. O., and A. K. Sen Gupta (1970). *J. Econ. Entomol.* 63:856.

Kojima, K., and R. D. O'Brien (1968). *J. Agr. Food Chem.* 16:574.

Koransky, W., J. Portig, H. Vohland, and I. Klempar (1964). *Arch. Exptl. Pathol. Pharmakol.* 247:49.

Korte, F., and H. Arent (1965). *Life Sci.* 4:2017.

Krueger, H. R., and R. D. O'Brien (1959). *J. Econ. Entomol.* 52:1063.

Krueger, H. R., J. E. Casida, and R. P. Niedermeier (1959). *J. Agr. Food Chem.* 7:182.

Krueger, H. R., R. D. O'Brien, and W. C. Dauterman (1960). *J. Econ. Entomol.* 53:25.

Ku, T. Y., and J. L. Bishop (1967). *J. Econ. Entomol.* 60:1328.

Kuhr, R. T. (1970). *J. Agr. Food Chem.* 18:1023.

Kuhr, R. T., and J. E. Casida (1967). *J. Agr. Food Chem.* 15:814.

Kutschinski, A. H. (1961). *J. Agr. Food Chem.* 9:365.

Lamoureux, G. L., R. H. Shimabukuro, H. R. Swanson, and D. S. Frear (1969). Abstract, Agriculture and Food Chemistry Division, 157th American Chemical Society Meeting, April 14–18, Minneapolis.

Lang, E., F. Kunze, and C. Pickett (1951). *Arch. Ind. Hyg. Occup. Med.* 3:245.

Lawrence, J. H., R. P. Barron, J.-Y. T. Chen, P. Lombardo, and W. R. Benson (1970). *J. Assoc. Offic. Anal. Chemists* 53:261.

Leeling, N. C., and J. E. Casida (1966). *J. Agr. Food Chem.* 14:281.

Lewis, J. B. (1969). *Nature* 224:917.

Lewis, J. B., and R. M. Sawicki (1971). *Pesticide Biochem. Physiol.* 1:275.

Lichtenstein, E. P., and T. W. Fuhremann (1971). *Science* 172:589.

Lindquist, D. A., E. C. Burns, C. P. Pant, and P. A. Dahm (1958). *J. Econ. Entomol.* 51:204.

Lipke, H., and J. Chalkley (1962). *Biochem.* 85:109.

Lipke, H., and C. Kearns (1959). *J. Biol. Chem.* 234:2123.

Lipke, H., and C. Kearns (1960). *Adv. Pest Control Res.* 3:253.

Ludwig, G., and F. Korte (1965). *Life Sci.* 4:2027.

Ludwig, G., J. Wies, and F. Korte (1964). *Life Sci.* 3:123.

Main, A. R. (1960). *Biochem. J.* 74:10.

Main, A. R., and P. E. Braid (1962). *Biochem. J.* **84**:255.
March, R. B., T. R. Fukuto, R. L. Metcalf, and M. G. Maxon (1956*a*). *J. Econ. Entomol.* **49**:185.
March, R. B., R. L. Metcalf, T. R. Fukuto, and F. A. Gunther (1956*b*). *J. Econ. Entomol.* **49**:679.
Matsumura, F., and A. W. A. Brown (1961). *J. Econ. Entomol.* **54**:1176.
Matsumura, F., and A. W. A. Brown (1963). *J. Econ. Entomol.* **56**:381.
Matsumura, F., and C. J. Hogendijk (1964*a*). *J. Agr. Food Chem.* **12**:447.
Matsumura, F., and C. J. Hogendijk (1964*b*). *Entomol. Exptl. Appl.* **7**:179.
Matsumura, F., and J. O. Nelson (1971). *Bull. Environ. Contam. Toxicol.* **5**:489.
Matthews, H. B., and F. Matsumura (1969). *J. Agr. Food Chem.* **17**:845.
Matthews, H. B., J. D. McKinney, and G. W. Lucier (1971). *J. Agr. Food Chem.* **19**:1244.
Mattson, A. M., and V. A. Sedlak (1960). *J. Agr. Food Chem.* **8**:107.
McKennis, H., Jr., L. B. Turnbull, S. L. Schwartz, E. Tamaki, and E. R. Bowman (1962). *J. Biol. Chem.* **237**:541.
McKennis, H., Jr., L. B. Turnbull, and E. R. Bowman (1963). *J. Biol. Chem.* **238**:719.
McKinney, J. D., H. B. Matthews, and L. Fishbein (1972). *J. Agr. Food Chem.* **20**:597.
Mehendale, H. H., and H. W. Dorough (1972). In *Insecticide–Pesticide Chemistry*, Vol. 1. A. S. Tahori, ed. Gordon and Breach, London, p. 15.
Menzel, D. B., S. M. Smith, R. Miskus, and W. M. Hoskins (1961). *J. Econ. Entomol.* **54**:9.
Menzer, R. E., and J. E. Casida (1965). *J. Agr. Food Chem.* **13**:102.
Menzer, R. E., and W. C. Dauterman (1970). *J. Agr. Food Chem.* **18**:1031.
Menzie, C. M. (1969). *Metabolism of Pesticides*. Special Scientific Report, Wildlife No. 125, Department of the Interior, Washington, D.C.
Metcalf, R. L. (1955). *Organic Insecticides*. Interscience Publishers, New York.
Metcalf, R. L., and R. B. March (1953*a*). *Science* **117**:527.
Metcalf, R. L., and R. B. March (1953*b*). *Ann. Entomol. Soc. Am.* **46**:63.
Metcalf, R. L., R. B. March, T. R. Fukuto, and M. G. Maxon (1954). *J. Econ. Entomol.* **47**:1045.
Metcalf, R. L., R. B. March, T. R. Fukuto, and M. G. Maxon (1955). *J. Econ. Entomol.* **48**:364.
Metcalf, R. L., T. R. Fukuto, and R. B. March (1957). *J. Econ. Entomol.* **50**:338.
Metcalf, R. L., T. R. Fukuto, and R. B. March (1959). *J. Econ. Entomol.* **52**:44.
Metcalf, R. L., T. R. Fukuto, C. Collins, K. Borck, J. Burk, H. T. Reynolds, and M. F. Osman (1966). *J. Agr. Food Chem.* **14**:579.
Metcalf, R. L., M. F. Osman, and T. R. Fukuto (1967). *J. Econ. Entomol.* **60**:445.
Metcalf, R. L., T. R. Fukuto, C. Collins, K. Borck, S. A. El-Aziz, R. Munoz, and C. C. Cassil (1968). *J. Agr. Food Chem.* **16**:300.
Miskus, R., D. Blair, and J. Casida (1965). *J. Agr. Food Chem.* **13**:481.
Miyake, S., C. Kearns, and H. Lipke (1957). *J. Econ. Entomol.* **50**:359.
Miyamoto, J. (1959). *Botyu-Kagaku* **24**:130.
Miyamoto, J. (1964*a*). *Agr. Biol. Chem.* **28**:411.
Miyamoto, J. (1964*b*). *Agr. Biol. Chem.* **28**:422.
Miyamoto, J. (1970). In *Biochemical Toxicology of Insecticides*. R. D. O'Brien, and I. Yamamoto, eds. Academic Press, New York, p. 115.
Miyamoto, J., and Y. Sato (1965*a*). *Botyu-Kagaku* **30**:45.
Miyamoto, J., and Y. Sato (1965*b*). *Botyu-Kagaku* **30**:49.
Miyamoto, J., Y. Sato, T. Kadota, and A. Fujinami (1963*a*). *Agr. Biol. Chem.* **27**:669.
Miyamoto, J., Y. Sato, T. Kadota, A. Fujinami, and M. Endo (1963*b*). *Agr. Biol. Chem.* **27**:381.
Miyata, T., and F. Matsumura (1971). *Pesticide Biochem. Physiol.* **1**:267.
Miyata, T., and F. Matsumura (1972). *J. Agr. Food Chem.* **20**:30.

Morello, A. (1965). *Can. J. Biochem. Physiol.* **43**:1289.

Morello, A., A. Vardanis, and E. Y. Spencer (1968). *Can. J. Biochem.* **46**:885.

Morrison, A. B., and I. C. Munro (1965). *Can. J. Biochem.* **43**:33.

Mostafa, I. Y., A. Massan, and S. M. A. D. Zayed (1965). *Z. Naturforsch.* **20b**:67.

Mounter, L. A., L. T. H. Dien, and A. Chanutin (1955). *J. Biol. Chem.* **215**:691.

Mücke, W., K. O. Alt, and O. Esser (1970). *J. Agr. Food Chem.* **18**:208.

Nachtomi, E., E. Alumot, and A. Bondi (1966). *Israel J. Chem.* **4**:329.

Nakatsugawa, T., and P. A. Dahm (1962). *J. Econ. Entomol.* **55**:594.

Nakatsugawa, T., and P. A. Dahm (1965). *J. Econ. Entomol.* **58**:500.

Nakatsugawa, T., and P. A. Dahm (1967). *Biochem. Pharmacol.* **16**:25.

Nelson, O., and F. Matsumura (1973). *Arch. Environ. Toxicol.* **1**:224.

Niessen, H., H. Tietz, and H. Frehse (1962). *Pflanzenschutz-Nachr.* **15**:125.

Niessen, H., H. Tietz, G. Hecht, and G. Kimmerle (1963). *Arch. Toxikol.* **20**:44.

Nolan, J., and R. D. O'Brien (1970). *J. Agr. Food Chem.* **18**:802.

O'Brien, R. D. (1960). *Toxic Phosphorus Esters.* Academic Press, New York.

O'Brien, R. D. (1967). *Insecticides: Action and Metabolism.* Academic Press, New York, 332 pp.

O'Brien, R. D., and E. Y. Spencer (1953). *J. Agr. Food Chem.* **1**:946.

O'Brien, R. D., and L. S. Wolfe (1959). *J. Econ. Entomol.* **52**:692.

O'Brien, R. D., S. D. Thorn, and R. W. Fisher (1958). *J. Econ. Entomol.* **51**:714.

O'Brien, R. D., W. C. Dauterman, and R. P. Niedermeier (1961). *J. Agr. Food Chem.* **9**:40.

O'Brien, R. D., E. C. Kimmel, and P. R. Sferra (1965). *J. Agr. Food Chem.* **13**:366.

Omura, T., and R. Sato (1964). *J. Biol. Chem.* **239**:2379.

Oonnithan, E. S., and J. E. Casida (1966). *Bull. Environ. Contam. Toxicol.* **1**:59.

Oonnithan, E. S., and J. E. Casida (1968). *J. Agr. Food Chem.* **16**:28.

Oppenoorth, F. J. (1954). *Nature,* **173**:1000.

Oppenoorth, F. J. (1956). *Arch. Neerl. Zool.* **12**:1.

Oppenoorth, F. J., V. Rupes, S. ElBashir, N. W. H. Houx, and S. Voerman (1972). *Pesticide Biochem. Physiol.* **2**:262.

Owens, R. M., and H. M. Novotny (1958). *Contributions Boyle Thompson Inst.* **19**:464.

Pankaskie, J., F. Fountaine, and P. Dahm (1952). *J. Econ. Entomol.* **45**:51.

Pardue, J. R., E. A. Hansen, R. P. Barron, and J.-Y. T. Chen (1970). *J. Agr. Food Chem.* **18**:405.

Parke, D. V., and R. T. Williams (1969). *Brit. Med. Bull.* **25**:256.

Pasarela, N. R., R. G. Brown, and C. B. Shaffer (1962). *J. Agr. Food Chem.* **10**:7.

Perry, A., and A. Buckner (1958). *Am. J. Trop. Med. Hyg.* **7**:620.

Perry, A. S. (1960). *J. Agr. Food Chem.* **8**:266.

Peterson, J. E., and W. H. Robison (1964). *Toxicology and Applied Pharmacology* **6**:321.

Phillips, D. D., G. E. Pillard, and S. B. Soloway (1962). *J. Agr. Food Chem.* **10**:217.

Pinto, J., M. Camien, and M. Dunn (1965). *J. Biol. Chem.* **240**:2148.

Plapp, F. W., and J. E. Casida (1958*a*). *J. Econ. Entomol.* **51**:800.

Plapp, F. W., and J. E. Casida (1958*b*). *J. Agr. Food Chem.* **6**:662.

Polen, P. B., M. Hester, and J. Benziger (1971). *Bull. Environ. Contam. Toxicol.* **5**:521.

Poonawalla, N., and F. Korte (1964). *Life Sci.* **3**:1497.

Ray, J. W. (1967). *Biochem. Pharmacol.* **16**:99.

Reed, W. T., and A. J. Forgash (1968). *Science,* **160**:1232.

Reed, W. T., and A. J. Forgash (1970). *J. Agr. Food Chem.* **18**:475.

Richardson, A. (1965). Tunstall Laboratory, Shell Research Ltd. (Indirectly cited from Baldwin *et al.*, 1970.)

Richardson, A., M. Baldwin, and J. Robinson (1968). *J. Chem. Ind.* **18**:588.

Robbins, J. D., J. E. Bakke, C. Fjelstul, G. O. Alberts, and V. J. Feil (1969). Abstracts of the 157th American Chemical Society Meeting, Minneapolis, April.

Robbins, J. D., J. E. Bakke, and V. J. Feil (1970). *J. Agr. Food Chem.* **18**:130.

Robbins, W. E., T. L. Hopkins, and G. W. Eddy (1956). *J. Econ. Entomol.* **49**:801.

Robbins, W. E., T. L. Hopkins, and G. W. Eddy (1957). *J. Agr. Food Chem.* **5**:509.

Rowlands, D. G. (1964). *J. Sci. Food Agr.* **15**:824.

Rowlands, D. G. (1965). *J. Sci. Food Agr.* **16**:325.

Rusoff, L., W. Waters, J. Ghosson, J. Frye, Jr., L. Newsom, E. Burns, W. Barthel, and R. Murphy (1962). *J. Agr. Food Chem.* **10**:377.

Rusoff, L., R. Temple, R. Meyers, L. Newsom, E. Burns, W. Barthel, C. Corley, and A. Allsman (1963). *J. Agr. Food Chem.* **11**:289.

San Antonio, J. (1959). *J. Agr. Food Chem.* **7**:322.

Sanderson, D. M., and E. F. Edson (1964). *Brit. J. Ind. Med.* **21**:52.

Schayer, R. W. (1956). *Brit. J. Pharmacol.* **11**:472.

Schlagbauer, A. W. J., and B. G. L. Schlagbauer (1972a). *Residue Rev.* **42**:85.

Schlagbauer, B. G. L., and A. W. J. Schlagbauer (1972b). *Residue Rev.* **42**:1.

Schrader, G. (1963). *Die Entwicklung neuer insektizider Phosphorsaure-Ester.* Verlag Chemie, Weinheim.

Schuphan, I., K. Ballschmiter, and G. Tolo (1968). *Z. Naturforsch.* **23b**:701.

Schwarz, H., and W. Dedek (1965). *Zentralbl. Veterinaermed.* **12b**:653.

Schwemmer, B., W. P. Cochrane, and P. B. Polen (1970). *Science* **169**:1087.

Self, L. S., F. E. Guthrie, and E. Hodgson (1964). *Nature* **204**:300.

Sellers, L. G., and F. E. Guthrie (1972).

Sen Gupta, A. K., and C. O. Knowles (1970). *J. Econ. Entomol.* **63**:10.

Seume, F. W., and R. D. O'Brien (1960). *J. Agr. Food Chem.* **8**:36.

Shishido, T., K. Usui, M. Sato, and J. Fukami (1972a). *Pesticide Biochem. Physiol.* **2**:51.

Shishido, T., K. Usui, and J. Fukami (1972b). *Pesticide Biochem. Physiol.* **2**:27.

Shrivastava, S. P., M. Tsukamoto, and J. E. Casida (1969). *J. Econ. Entomol.* **62**:483.

Sims, P., and P. Grover (1965). *Biochem. J.* **95**:156.

Slade, M., and J. E. Casida (1970). *J. Agr. Food Chem.* **18**:467.

Smith, J. N. (1955a). *Biol. Rev.* **30**:455.

Smith, J. N. (1955b). *Biochem. J.* **60**:436.

Smith, J. N. (1962). *Ann. Rev. Entomol.* **7**:465.

Spencer, E. Y., R. D. O'Brien, and R. W. White (1957). *J. Agr. Food Chem.* **5**:123.

Stenersen, J. (1969). *J. Econ. Entomol.* **62**:1043.

Sternberg, J., and C. Kearns (1952). *J. Econ. Entomol.* **45**:497.

Sternberg, J., and C. Kearns (1956). *J. Econ. Entomol.* **49**:548.

Sternberg, J., C. Kearns, and W. Bruce (1950). *J. Econ. Entomol.* **43**:214.

Sternberg, J., C. Kearns, and H. Moorefield (1954). *J. Agr. Food Chem.* **2**:1125.

Sterner, J. H. (1949). *Ind. Hyg. Toxicol.* **2**:783. F. A. Patty, ed. Interscience Publishers, New York.

Stoddard, G., G. Bateman, J. Shupe, J. Harris, H. Bahler, L. Harris, and D. Greenwood (1954). Proceedings of the Annual Meeting, Western Division of the American Dairy Association Progress Report, Vol. 35, p. 295. Am. Dairy Assoc., Chicago, Ill.

Stohlman, E., and M. Smith (1950). *Adv. Chem. Ser.* **1**:228.

Street, J. C., and S. E. Blau (1972). *J. Agr. Food Chem.* **20**:395.

Sutherland, G. L., J. W. Cook, and R. L. Baron (1970). *J. Assoc. Off. Anal. Chemists* **53**:993.

Terranova, A., and G. W. Ware (1963). *J. Econ. Entomol.* **56**:596.

Terranova, A. C., and M. M. Crystal (1970). *J. Econ. Entomol.* **63**:455.

Terriere, L. C., R. B. Boose, and W. T. Roubal (1961). *Biochem. J.* **79**:620.

Tonelli, P., and R. March (1954). *J. Econ. Entomol.* **47**:902.

Townsend, M. G., and J. R. Busvine (1969). *Entomol. Exptl. Appl.* **12**:243.

Tsukamoto, M. (1959). *Botyu-Kagaku* **24**:141.

Tsukamoto, M. (1960). *Botyu-Kagaku* **25**:156.

Turnbull, L. B., E. R. Bowman, and H. McKennis, Jr. (1960). *Fed. Proc.* **19**:268.

Turner, D. M. (1969). *Biochem. J.* **115**:889.

Uchida, T., and R. D. O'Brien (1967). *Toxicol. Appl. Pharmacol.* **10**:89.

Uchida, T., W. C. Dauterman, and R. D. O'Brien (1964). *J. Agr. Food Chem.* **12**:48.

van Asperen, K., and F. J. Oppenoorth (1954). *Nature*, **173**:1000.

Vardanis, A., and L. Crawford (1964). *J. Econ. Entomol.* **57**:136.

Vickery, D. S., and B. W. Arthur (1960). *J. Econ. Entomol.* **53**:1037.

Walker, C. H. (1969). *Life Sci.* **8**:1111 (Part II).

Ware, G. W. (1967). Research Circular 151. Ohio Agricultural Research Development Center, Wooster, O.

Wedemyer, G. (1968). *Life Sci.* **7**:219.

Weidhaas, D. E. (1959). *J. Assoc. Offic. Agr. Chemists* **42**:445.

Welling, W., and P. T. Blaakmeer (1971). In *Pesticide Chemistry*, Vol. 2. A. S. Tahori, ed. Gordon and Breach, London, p. 61.

Welling, M., A. W. De Vries, and S. Voeman (1974). *Pesticide Biochem. Physiol.* **4**:31.

Wells, A. L., Z. Stelmach, G. E. Guyer, and E. J. Benne (1958). *Mich. State Univ. Agr. Expt. Sta. Quart. Bull.* **40**:786.

Whetstone, R. R., D. D. Phillips, Y. P. Sun, and L. F. Ward, Jr. (1966). *J. Agr. Food Chem.* **14**:352.

Wilkinson, C. F. (1968). In *The Enzymatic Oxidation of Toxicants*. E. Hodgson, ed. North Carolina State University, Raleigh, N.C., p. 113.

Wilkinson, C. F., R. L. Metcalf, and T. R. Fukuto (1966). *J. Agr. Food Chem.* **14**:73.

Williams, E., R. W. Meikle, and C. T. Redemann (1964*a*). *J. Agr. Food Chem.* **12**:457.

Williams, E., R. W. Meikle, and C. T. Redemann (1964*b*). *J. Agr. Food Chem.* **12**:453.

Williams, R. T. (1959). *Detoxication Mechanisms*, 2nd ed. John Wiley & Sons, New York, 796 pp. (See pp. 520–521.)

Winteringham, F. P. W., A. Harrison, and P. M. Bridges (1955). *Biochem. J.* **61**:357.

Wit, J. G., and H. van Genderen (1966). *Biochem. J.* **101**:698.

Wong, D., and L. Terriere (1965). *Biochem. Pharmacol.* **14**:375.

Yamamoto, I., and J. E. Casida (1966). *J. Econ. Entomol.* **59**:1542.

Yamamoto, I., and J. E. Casida (1967). Abstracts of the 153rd Meeting of the American Chemical Society, Miami.

Yamamoto, I., and J. E. Casida (1968). *Agr. Biol. Chem.* **32**:1382.

Yamamoto, I., E. C. Kimmel, and J. E. Casida (1969). *J. Agr. Food Chem.* **17**:1227.

Yamamoto, I., M. Elliott, and J. E. Casida (1971). *Bull. World Health Org.* **44**:347.

Yang, R. S., H. E. Hodgson, and W. C. Dauterman (1971). *J. Agr. Food Chem.* **19**:10, 14.

Zayed, S. M. A. D., and A. Hassan (1965). *Can. J. Biochem.* **43**:1257.

Zubairi, M. Y., and J. E. Casida (1965). *J. Econ. Entomol.* **58**:403.

Chapter 6

Entry of Insecticides into Animal Systems

6.1. PENETRATION OF INSECTICIDES THROUGH THE INSECT CUTICLE

The body surface of insects consists of a hard skin known as the cuticle. At first glance it appears that this barrier should be very effective against the entry of insecticides, but there are two main reasons why it is not: (1) the insect exposes a far greater surface area relative to its volume than the mammal, and (2) the insect cuticle is hydrophobic (i.e., lipophilic) so that it can resist desiccation and drowning. Most modern insecticides are apolar and therefore easily penetrate the insect cuticle. Mammalian skin, by contrast, is relatively resistant to the entry of insecticides. As a result, it is common that the acute oral toxicity of an insecticide is much higher in mammals than the contact toxicity, whereas the contact and oral toxicities are almost equal in insects. Though in insects cuticular penetration is usually the major pathway, in some instances insecticides do enter by way of the mouth, repiratory system, and other vulnerable places such as the antennae, eyes, and tarsi.

6.1.1. Morphology of the Insect Cuticle

The insect cuticle is 60–70 μm thick and is composed of the epicuticle, the exocuticle, and the endocuticle. Beneath the cuticle are the epidermis and basal membrane. The cuticle is secreted by the hypodermal cells and at first is clear, soft, pliable, and moist. In most insects, the cuticle quickly becomes hard and dark. This is the process of sclerotization and results from tanning of the protein by quinones. The sclerotized layer is the exocuticle, and the remainder is the endocuticle. The exocuticle and the endocuticle together make up the procuticle.

It is well recognized that the texture of the outermost layer of the cuticle (epicuticle) is very important in determining insecticide penetration. The epicuticle typically consists of a thin layer of tanned protein impregnated on its outer surface with lipoid or wax. It may be overlaid with grease (blattids) or with a heavy layer of wax (*Dytiscus*, coccids, psyllids) or with a thin layer of cement. The rest of the epicuticle beneath the outermost wax layer is cuticuline, which is protein mixed with polymerized phenols to form a hard crust. The cuticuline is about 3 μm thick. The epicuticle is resistant to strong acids (which dissolve the remainder of the cuticle). It is impermeable to water but is dissolved by strong alkali and is penetrated by fat solvents such as acetone and chloroform. The function of the epicuticle is to prevent the passage of moisture outward through the cuticle, thus protecting the insect from desiccation.

The endocuticle is 70% water and contains crystallites of chitin, arranged parallel to the cuticle surface but fully rotatable in that plane, and imbedded in a matrix of arthropodin (water-soluble protein). The exocuticle, which is sclerotized, is dark, hard, dry, and rigid. It is insoluble in water. The cuticle of most insects is perforated by pore canals which reach from the hypodermal cells to the exocuticle. They are generally overlaid by the epicuticle, although in some insects (*Periplaneta*) they extend through the epicuticle. There are as many as 300 pore canals per hypodermal cell in *Periplaneta*; however, in other insects they are sparse, and they are completely absent from the cuticle of mosquito larvae.

Dermal glands are found in the cuticle of many insects which secrete wax or grease (Blattidae), glandular incrustations (Coleoptera), or odorous material (Lepidoptera).

In many insects, especially sclerotized active forms such as flies, bees, and wasps, the cuticle is set with hair sensilla. They are inserted in little circular membranes, the areolar membranes, which are extremely thin to allow free movement of the sensilla.

6.1.2. Insect Cuticles as Membranes

The insect cuticle is well constructed to prevent desiccation of the body. A drop of water about the size of an average insect would evaporate in a few minutes. The cuticle is characterized not only by its great resistance to the loss of water by evaporation but also by its power to absorb water. Pal (1951) observed that drops of water placed on the leg of a cockroach appeared in the trachea below. Lees (1946, 1947) has shown that many species of ticks can take up water from the air even when it is far below saturation and that intake occurs through the cuticle against the concentration gradient of water, although the speed of penetration would be comparatively slow.

The cement and wax layers, which restrict the outward permeation of water, also oppose the entry of insecticides to a certain extent, but entry is greatly accelerated after these layers have been injured by abrasion (Wigglesworth, 1945). It also seems generally true that the presence of wax solvents facilitates the passage of toxic substances into insects (O'Kane *et al.*, 1940).

In considering the rate of penetration of any material through the insect cuticle, two clearly distinguishable factors may be relevant: (1) the nature of the cuticle as a membrane and (2) the nature of the material, particularly its polarity. Let us assume that the insect cuticle is a uniform thin membrane through which material diffuses. The rate of diffusion of material in a homogeneous medium at a certain temperature and pressure is represented by

$$\partial c / \partial t = a^2 (\partial^2 c / \partial X^2 + \partial^2 c / \partial Y^2 + \partial^2 c / \partial Z^2) \tag{1}$$

where t is time, c is the concentration of the material, a is the diffusivity, and X, Y, and Z are the three-dimensional axes. Diffusion through the cuticle can be interpreted as a one-dimensional problem:

$$dc / dt = a^2 \cdot \partial^2 c / \partial X^2 \tag{2}$$

Here dc/dt is the diffusion speed in terms of concentration change, and $\partial^2 c / \partial X^2$ represents the changing rate of concentration gradient against the thickness, X, of the membrane. Now let us assume that the original concentration is very high and constant, and therefore the changing rate of the concentration gradient can be decided only by the material in the diffusing cell (i.e., outside the membrane). Then the equation can be written as

$$dc / dt = K(C_0 - C_i) \tag{3}$$

where C_0 is the originally applied concentration or the original concentration of the material in the diffusing cell, C_i is the final concentration in the diffusion cell (i.e., inside the membrane) and K is a constant (Davison and Danielli, 1952). Since the diffusivity constant can be determined as a function of area and size of the diffusion cell, the equation may be written as

$$dc / dt = \frac{PA}{V} (C_0 - C_i) \tag{4}$$

where P is the permeability constant, V is the volume of the diffusion cell, and A is the area of contact between the material and the membrane. By rearranging equation (4), one obtains

$$\int_0^{C_i} \frac{dc}{C_0 - C_i} = \int_0^t \frac{PA}{V}\, dt \tag{5}$$

$$\ln \frac{C_0}{C_0 - C_i} = \frac{PA}{V} t \tag{6}$$

or

$$\frac{C_0}{C_0 - C_i} = e^{PAt/V} \tag{7}$$

The actual amount penetrated can therefore be expressed as

$$S = \frac{C_i}{V} = \frac{C_0}{V}(1 - e^{-PAt/V}) \tag{8}$$

According to equation (8), the rate of penetration must be proportional to the concentration of the material administered and exponential to the penetration time, or, according to equation (6), the logarithm of percent remaining material outside the membrane should be negatively proportional to the time. These two different criteria for testing the general applicability of the law of diffusion have been employed in the past by Treherne (1957) and Matsumura (1963) has used the first criterion and Olson and O'Brien (1963) have used the second. These authors have noted that the observed rate of diffusion is in fair to good accordance with the theoretical rate, with very little variation among replicates. Matsumura (1963) measured the rate of penetration of topically applied P^{32}-malathion through the cockroach pronotum, which had previously been found to show a homogeneous response based on the results of a radioautographic test (Matsumura, 1959), and found that the rate of penetration at low concentrations tended to exceed that predicted by theory, whereas at high concentrations the rate distinctly fell away from theoretical linearity (Fig. 6-1). The data of several workers (Sternberg *et al.*, 1950; Lindquist and Dahm, 1956; Matsumura, 1963), plotted as log % recovery of unpenetrated insecticides vs. time, also show breaks in the curve, with sudden slowing of penetration. Such a biphasic relationship can be explained by the assumption that the insecticide is quickly absorbed by the cuticle, and then a second, slower penetration function (i.e., diffusion into the body) takes over. By recovering malathion absorbed in the cuticle itself from either separated pronotum or by hot water extraction and subtracting the absorbed values from the rest of the penetra-

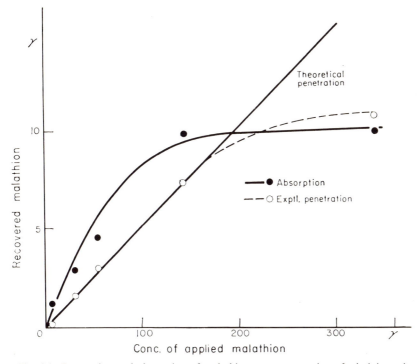

Fig. 6-1. Penetration and absorption of malathion vs. concentration of administered malathion: concentrations are shown as μg malathion per droplet of acetone (0.02 mm^3), which was topically applied on the pronotum of *Periplaneta americana*.

tion values, Matsumura (1963) obtained a straight line. However, Olson and O'Brien (1963) believe that the rapid early penetration is due to abrasion of the cuticle and that, as a whole, simple first-order kinetics accounts for the rate of cuticular penetration. An excellent example of how biphasic penetration is related to absorption of insecticide by the cuticular material is provided by Richards and Cutkomp (1946), who showed that DDT is first tightly absorbed by the cuticular chitin, which then facilitates the total uptake of DDT by the insect.

6.1.3. Factors Influencing the Rate of Penetration: Chemical Nature of Insecticides

Another factor which affects the rate of penetration is the polarity of the compound itself. It has been generally considered that the rate of penetration

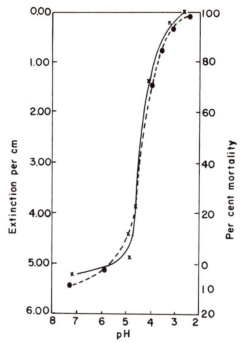

Fig. 6-2. Relation between contact toxicity and
dissociation of DNOC. Dotted line, dissociation
measured photometrically; solid line, percentage
mortality of *Ephestia* eggs. From Dierick (1943).

is directly related to the liposolubility of the compound since the outer layers
of the cuticle are highly apolar. A clear-cut example of the toxicity–dissociation relationship of dinitro-*o*-cresol (DNOC) has been provided by Dierick
(1943): DNOC, being highly undissociated at pH 2, can be very toxic in acidic
conditions, but almost nontoxic at neutrality where it is highly dissociated.
He also showed that the percent mortality for the insects tested closely
followed the line of dissociation for DNOC. At pH 2 it killed all the insects
tested, and at pH 5 it killed none. At pH 5, DNOC is 50% dissociated and
50% undissociated. Below pH 5, DNOC is in its apolar, and hence toxic,
form (Fig. 6-2).

Nicotine is dissociated (and polar) at acidic pH and undissociated
(apolar) at alkaline pH and is hence more toxic when in a basic solution.
However, Ellisor (1936) showed that molecular nicotine and ionized nicotine
paralyze and kill *Periplaneta* at equal rates when injected into the body
fluids, thus demonstrating that differences in toxicity are determined by the

rate of cuticular penetration. Iljinskaya (1946) showed that the related alkaloid anabasine applied as a contact spray in alkaline solution, where it is a free base, is twice as toxic to *Aphis* as an equivalent acidic solution of anabasine sulfate.

There is a danger, however, in conclusions derived from data where toxicity was the only reliable criterion for assessing the rate of penetration. The toxicity of insecticides can also depend on other factors such as the route of transport to the nervous system, the distance to the target (nervous system), and the permeability of the insect nerve sheath, which is more permeable to highly liposoluble compounds (O'Brien and Fisher, 1958).

By using radiotracer techniques, Olson and O'Brien (1963) tested the rate of penetration of two electrolytes and four insecticides (two organophosphate and two chlorinated hydrocarbon insecticides) through the pronotum of *Periplaneta americana* and came to the conclusion that permeability is directly related to polarity (Table 6-1). In other words, the

TABLE 6-1. Half-Times of Disappearance of Solutes from the Surface and Partition Coefficients[a]

Solute	Half-time (min)	Partition coefficient olive oil/H_2O
K_2HPO_4	9	—
H_3PO_4	16	0.1
Dimethoate	27	0.34
Paraoxon	55	4.06
Dieldrin	320	64.0
DDT	1584	316.0

[a]Acetone was the solvent in the half-time determinations. The partition coefficients were determined at 25–27°C; the value for K_2HPO_4 was too small to measure with this technique. From Olson and O'Brien (1963).

polar compounds penetrated much faster than the less polar (lipophilic) compounds.* This finding was completely opposite to the generally accepted

*The criterion for penetration that these authors used was somewhat different from that of previous authors. They measured penetration in terms of the remaining external insecticide which could be washed out with 1 ml of water and then with 1 ml of acetone. Therefore, the amount of insecticide which stayed inside the cuticle itself was included in the total penetration.

theory of insecticide penetration (see the opening of this chapter). Olson and O'Brien decided that the cause for the increase in penetration with polarity is that with the technique of topical application the solute is generally directly introduced into the lipoidal part of the cuticle, so that it does not have to penetrate the lipids; rather, it is mixed with the lipids. Thereafter the solute must penetrate relatively polar material, such as procuticle (consisting largely of relatively polar chitin and protein), and so highly apolar compounds tend to stay with the epicuticular wax. Clearly, then, the mode of penetration depends on not only the nature of the compound but also on the method by which the compound is introduced to the surface of the cuticle.

Indeed, Olson and O'Brien (1963) discovered that phosphoric acid, when applied in a water droplet, hardly penetrated until the water had completely evaporated, whereas the same applied in acetone penetrated extremely fast. Likewise, dimethoate did not appreciably penetrate into the cuticle when applied with water. Hence the results obtained from topical application with solvent cannot be directly compared to those from contact or topical application with aqueous solution, for the former method tends to disturb the external barrier and greatly alter the rate of penetration. In nature, however, the insecticide must be first picked up by the insect to be fully effective: here lipophilicity of the compound becomes much more important than the simple rate of penetration through the cuticle. Thus for extremely apolar and lipophilic insecticides the initial efficiency of pickup is high, whereas the rate of infusion from the cuticular wax into the more aqueous tissue components is slow, as shown by Gerolt (1969) with dieldrin. In this study, Gerolt found that transfer of dieldrin from the cuticle to the hemolymph is extremely slow, whereas the horizontal diffusion rate within the cuticle is faster. This fact led him to conclude that for dieldrin tracheal entrance is more important than cuticular penetration. Benezet and Forgash (1972a), on the other hand, found that topically applied malathion, a much more polar compound than dieldrin, can be found in the hemolymph within 15 sec after application. Using ^{3}H-malathion, Ahmed and Gardiner (1968) reached a similar conclusion that the circulatory system plays an important role in the distribution of malathion in the desert locust.

6.1.4. Factors Influencing Penetration of Insecticides: Effects of Cuticular Components

Treherne (1957) has shown that the rate of penetration by solutes of an excised cuticle, with water on both sides of it, is proportional to the water solubility of the solutes. However, although this may be true in the specific case where the cuticle is water saturated and is held in water containing electrolytes, it may not have general applicability to the mode of penetration of most types of insecticides, which are usually topically applied, or applied by contact, to the surface of the cuticle where the insecticide–epicuticular wax system is the external component and the insecticide–body fluid is the internal component. Once the insecticide is picked up by the cuticular lipid, it must be gradually dispersed from the point of contact (Benezet and Forgash, 1972*a*) until it reaches an equilibrium dictated by the partitioning properties of the compound vs. various body tissue components and by the initial concentration of the insecticide. That these cuticular components play an important role is apparent. For polar insecticides, the presence of lipids should increase the rate of penetration of the insecticide already there (Benezet and Forgash, 1972*b*), and for apolar lipophilic insecticides the opposite is expected (Gerolt, 1969).

On the other hand, cuticular lipids may not be the only materials that influence the rate of pickup and penetration of insecticides; for instance, Richards and Cutkomp (1946) demonstrated that the presence of chitin is the major factor determining the easy pickup of DDT by insects and other arthropods, since this substance has a strong affinity for chitin.

In conclusion, there appear to be at least three major factors that play important roles in deciding the actual amount of insecticide penetrating the insect body: lipid solubility (or polarity), affinity for the cuticular components other than lipids (e.g., protein and chitin), and solubility in hemolymph. In considering the overall picture of insecticide pickup and penetration, it is important to clearly define the "limiting factor" which controls the actual amount of insecticide entering the body. For instance, with residually sprayed insecticides the lipid solubility becomes very important, because initial pickup by the insect is the rate-limiting factor. Assuming, however, that the insecticide is already on the insect (direct spray or topically applied), its solubility in the hemolymph (aqueous) becomes important, because this factor determines the rate and amount of penetration through the cuticular lipid layer to an aqueous body fluid. The data obtained by Treherne (1957) and by Olson and O'Brien (1963) indicate this tendency. Absorption into the cuticular components could play an intermediate role as a factor in aiding the initial pickup, and at the same time the insecticide in the cuticle would act as both a reservoir and the mediator between the lipid and the aqueous layers of the cuticle.

6.1.5. Effect of Carriers and Solvents

6.1.5a. Oil

The addition of oil to insecticide preparations often increases toxicity. Tattersfield *et al.* (1947) found DDT in benzene solution to be four to six times more toxic to *Macrosiphum* than in aqueous suspension. Addition of oil to rotenone greatly increases its toxicity. In many cases, oil solutions of chlordane are toxic to insects which are relatively unaffected by dusts or suspensions of this insecticide.

The oil solvents used as insecticide carriers are relatively apolar, are insoluble in water, do not dissociate, and lack reactive groups. Lipophilic insecticides readily dissolve in such compounds. Hurst (1943) found that the effect of kerosene (or chloroform or its vapor) is to remove lipid from its normal structural combination with protein in the lipoprotein mosaic of the cuticle. Treatment of the cuticle with apolar carriers (solvents) allows polar compounds to penetrate, provided they are not too highly dissociated. Thus kerosene aids the penetration of alcohols, ketones, fatty acids, amines, and phenols.

There is a threefold effect of oil as a carrier: (1) It gives the insecticide a chance to attach to the insect (an apolar compound in solvent attaches to organic material). (2) It breaks the epicuticular wax by dissolving it or else carries the insecticide through it. (3) It disrupts the internal protein organization of the cuticle. Also, the lighter the oil, the more insecticidal it is as a carrier.

One important phenomenon is that water-insoluble crystalline insecticides are virtually ineffective when they are inserted in the insect body cavity. Fisher (1952) observed this in the housefly by using crystalline DDT inserted inside the mesonotum. Gerolt (1969) confirmed this phenomenon in the housefly by administering either crystalline dieldrin or filter paper impregnated with dieldrin in the abdominal cavity. Three organophosphate insecticides, dichlorvos, chlorfevinphos (Birlane®), and methylparathion, were not as toxic when they were applied internally as solvent-free deposits on filter paper as when they were topically applied externally with acetone. Thus solubilization of these water-insoluble insecticides by the cuticular lipids, sometimes aided by solvents, appears to be a very important step in their penetration. This observation is also supported by Benezet and Forgash (1972b), who showed that elimination of the cuticular lipids drastically reduces the rate of malathion penetration. Thus evenly dissolved insecticides, either in cuticular lipids or in solvents such as acetone, can disperse in the body with ease. Gerolt has shown that the amount of acetone is extremely important in determining the effectiveness of injected dieldrin against houseflies: the toxicity was drastically reduced when the volume of acetone was decreased from 0.3 to 0.03 μl.

On the other hand, there is evidence that too much externally applied apolar oil has detrimental effects on penetration, particularly with very lipophilic insecticides (e.g., Gerolt, 1969), in contrast to the effects of solvents or polar low boiling point carrier oils. For instance, Wigglesworth (1942) showed that the action of pyrethrins could be speeded up by adding light

TABLE 6-2. Rate of Penetration of Pyrethrins in Combination with Various Oil Carriers into *Rhodnius*

Oil	Boiling point (°C)	Time to paralysis
White spirit	150–200	2 hr
Odorless distillate	200–260	4 hr
Fraction, high b.p.	260–360	6 hr
Fraction, high b.p.	320	6–28 hr
Olive, castor, or sesame oil	—	$1\frac{1}{2}$–3 days

From Wigglesworth (1942).

(low boiling point) oils to the preparations (Table 6-2), while vegetable oils had the opposite effect. It is probable that apolar oils with high boiling points and vegetable oils act as a reservoir for apolar insecticides, thus causing an actual slowing down of the rate of diffusion from the cuticle to the rest of the tissues. Opposite effects are expected for polar insecticides. Ahmed and Gardiner (1967) report that dilute malathion in mineral oil is more toxic than concentrated malathion topically applied on the desert locust.

6.1.5b. Detergents

Detergents form a "bridge" between lipophilic substances and water-soluble substances and thus enhance penetration of insecticides. Richards and Korda (1948) found that detergents disrupt not only the lipoid layer of the epicuticle but also the protein layers of the endocuticle. The properties rendering a detergent most effective are (1) enough liposolubility to penetrate and emulsify the epicuticular wax, (2) sufficient solubility in water (i.e., not excessively lipophilic), and (3) ability to penetrate the outer cement layer of the epicuticle.

6.1.5c. Dusts

Controlling insects with dusts such as soot, ashes, or road dust is an ancient and effective practice. Now activated charcoal, the oxides and carbonates of magnesium and calcium, the siliceous minerals, and finely powdered alumina are used. The mode of action of dusts involves withdrawal of body water and lipids from the insect, i.e., desiccation through abrasion of the cuticular surface. With hygroscopic materials (soot, activated charcoal), moisture is directly adsorbed onto the dust particles. With nonhygroscopic materials (pyrophyllite, alumina), the cuticle is lacerated, allowing moisture to be lost to the surrounding air. Dusts are ineffective at 100% relative humidity, and their efficiency increases with increasing dryness of the air.

Beament (1945) found that inert dusts can mop up by crystalline action the outer layers of wax or grease on the surface of the epicuticle. Silica aerogel absorbs lipids, hence disrupting the wax layer. Benezet and Forgash (1972b) found that diminution of the lipids by powder (DriDi or Florisil, both silicic acid) without causing abrasion in the housefly drastically reduced the permeability of subsequently applied malathion with acetone. Severe mechanical abrasion of the cuticle, on the other hand, caused an increase in penetration of the same compound in *P. americana* (Matsumura, 1963). The inner layers of the epicuticle are also disrupted mechanically by the insect's own movements, which rub the abrasive dust around. Abrasion facilitates the entry of contact insecticides, and dust preparations of many of these are common.

6.2. ROUTES OF INSECTICIDE ENTRY INTO INSECTS

Because of the large proportion of cuticular surface in relation to the rest of the body surface area, the penetration of insecticides through the cuticle is undoubtedly one of the most important factors in the total picture of insecticide entry. But insects do possess several vulnerable parts exposed to the outside through which insecticides may enter, e.g., the tracheal system, the mouth, or any other exposed sensory organ. Roy *et al.* (1943) observed that most pyrethrum enters by way of the spiracles instead of passing through the cuticle. In general, the site of entry is largely dependent on the type of insecticide; for instance, if the insecticide has high vapor pressure it tends to enter through the spiracles or antennae. Whenever insecticides are very lipophilic (e.g., DDT and dieldrin), vertical penetration through hemolymph appears to be limited (Lewis, 1965; Gerolt, 1969). In such cases, entry through sensory organs, the tracheal system, and the intersegmental membrane (Quraishi and Poonawalla, 1969) may be very important. If the insecticide is given with food material (e.g., in bait), it naturally enters through the mouth. Very polar chemicals such as arsenicals and inorganic fluorides are known to enter through this route. Even for general cuticular penetration the site of entry may vary: Matsumura (1959) made a general survey of malathion pickup by *P. americana* from a glass surface, and, as shown in Table 6-3, the highest amount was picked up by the legs, although in terms of specific activity (i.e., amount of insecticide/ weight of organ) the tracheal system showed the highest tendency to collect malathion. However, an experiment in which the spiracles of the cockroaches were closed with rubber cement (Table 6-3) showed that malathion in the trachea actually plays a very minor role in the total pickup process. Appar-

TABLE 6-3. Penetration and Distribution[a] of Malathion in a Live *Periplaneta americana* Specimen

Organs	Malathion recovered (counts/3 min)	Weight of organs (mg)	Specific activity (counts/3 min/mg)
Tracheal system[b]	272[c]	3.8	71.5
Cuticle exclusive of head	3,938	284.8	13.8
Head and antennae	947	53.2	17.8
Legs	11,700	283.1	41.4
Intestines and fat tissues[b]	1,173	255.2	4.6

From Matsumura (1959).

[a]For 3 hr at 22.5°C. ^{32}P-Malathion, 40 mg, was evenly applied on the inner surface of a large battery jar (12-pint). Specific activity 42/3 min/mg. The insect was in knockdown state.
[b]As much as possible was collected, but it does not represent the total tissue.
[c]When the spiracles were closed with rubber cement, the radioactivity decreased by 10–30 %.

ently the route via the legs is most important in this type of insecticide application.

O'Kane *et al.* (1933) studied the rate of response of *Blattella, Blatta,* and *Periplaneta* produced by the application of nicotine or kerosene at various sites, and reached the conclusion that the most rapid reaction was obtained on application to the ventral cervical region. It appears that the effectiveness of nicotine is almost entirely dependent on how close it is applied to the major ganglion of the insect. In *Tenebrio* larvae, the quickest response in terms of average time for the first appearance of convulsion induced by nicotine was produced by application at the ventrum pro- and mesonota. It seems likely that the thoracic and head regions are vulnerable points for various contact insecticides. For flies, adult mosquitoes, and honeybees, the tarsi, on which the insect walks, are one of the most important points of entry for DDT owing to the presence of tarsal chemoreceptors immediately beneath the thin cuticle (Hayes and Liu, 1947). Residual insecticides are often ineffective against insects which lack tarsal chemoreceptors (e.g., the beetle *Epilachna* and the bug *Oncopeltus*) (Hayes and Liu, 1947 ; Sarkaria and Datton, 1949). In agreement with these observations, Gerolt (1969) found that the relative knockdown time of the housefly is shortest (i.e., the insecticide is most toxic) when dieldrin is given at the fore tibia and mesopleuron (Table 6-4).

It is generally believed that increased thickness and sclerotization of the insect cuticle work against insecticide penetration. In fact, Matsumura

TABLE 6-4. Differences in Relative Knockdown Time After Treatment with Dieldrin at Various Loci of Application on the Housefly

Locus of application	Relative knockdown time (min)
Dorsally to fourth abdominal segment	100
Scutellum	105
Dorsally between compound eyes	85
Labellum	41
Fore tibia	32
Mesopleuron	33

From Gerolt (1969).

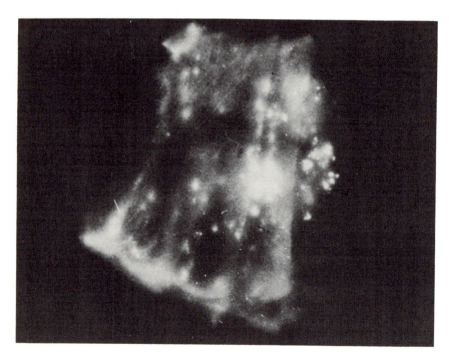

Fig. 6-3. Radioautographic visualization of penetration of ^{32}P-malathion through a *Periplaneta americana* adult. The abdominal sterna is at the bottom of the illustration. The photograph was taken from the inside of the cuticle after dissection and desiccation. Light areas represent ^{32}P-malathion. Note that intersegmental areas are particularly loaded with malathion. From Matsumura (1959).

(1959) found that newly molted *P. americana* adults picked up three to four times more malathion than the darkly tanned ones, although the most malathion was actually recovered in the cuticle itself. However, Lewis (1965) found no such difference in the case of DDT against houseflies. It is possible that the polarity difference plays some role in this discrepancy. As for the thickness, Klinger (1936) reported that the susceptibility of four species of lepidopterous larvae to pyrethrins decreased as the thickness of the larval cuticle increased. This general rule should not be extended beyond comparisons within closely related (taxonomically or ecologically) groups of insects. For instance, even among the dipterous larvae, the impermeable cuticle of *Corethrun plumicornis* is only 2 μm thick, while the highly permeable cuticle of *Chironomus plumosus* is 7 μm thick (Alexandrov, 1935). Among the different parts of the body of an individual insect, certain thin cuticular parts seem to provide quite vulnerable targets for insecticides. Indeed, insecticides seem to creep into the intersegmental space and penetrate the intersegmental membrane (Matsumura, 1959; Quraishi and Poonawalla, 1969) (see Fig. 6-3),

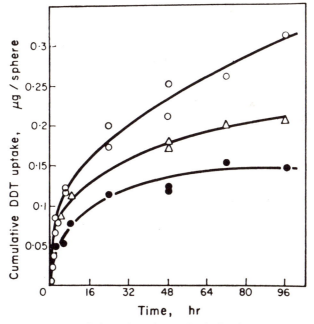

Fig. 6-4. Rates of absorption of DDT (topically given as spheres of lanoline containing DDT) through two different regions of *Phormia terraenovae*: Open circles, application to antennae of living flies; filled circles, same to dead flies; open triangles, application to genae of living flies. From Lewis (1965).

while the thick cuticle, such as the pronotum or tergum, remains resistant to penetration. Lewis (1965) found a rough correlation between the surface area of thin cuticles including the area of cross-section of ducts and the rate of DDT penetration in the housefly. The most susceptible areas for DDT penetration were covered by thin, porous cuticle having numerous duct openings (see Fig. 6-4).

Insecticides affect the physiological state and the behavior of insects, and the question is whether the rate of penetration is increased or decreased by such changes. Armstrong *et al.* (1951) conducted an interesting experiment on the rate of penetration of BHC isomers in *Calendra granaria*. Table 6-5 shows that the γ-isomer is the most toxic because it is apolar enough to penetrate the exterior cuticle and also water soluble enough to be carried by

TABLE 6-5. Absorption[a] and Penetration of the Isomers of BHC in *Calendra granaria*

	BHC isomers			
	α	β	γ	δ
	μg/g			
Recovered amount in				
Exterior wax	12	3	60	102
Interior (penetration)	4	4	43[b]	8
	g BHC/100 g			
Solubility in				
Cuticular wax	1.7	0.4	8.7	14.8
Petroleum oil	1.5	0.7	5.2	13.4

[a]Sorption from a filter paper during the 7–12 hr testing period. The total of exterior wax plus interior represents total absorption.
[b]Possibly helped by onset of symptoms.

From Armstrong *et al.* (1951).

the hemolymph throughout the interior of the insect to its site of action. The δ-isomer penetrates the exterior cuticle more than the γ-isomer, but it is too lipophilic to be carried throughout the interior body. The α- and β-isomers are more water soluble than the γ-isomer, but they are too polar to penetrate the exterior cuticle. When the exterior cuticle is disrupted (by abrasion), larger amounts of the α- and β-isomers are found in the interior body, but abrasion also facilitates the penetration of the γ-isomer and even larger amounts of it are found there. When the insect is dead, there is no circulation to carry the insecticide throughout the interior body, which explains the low amounts of all the isomers found there. The amounts found in the exterior cuticle of the dead insect are practically the same as with the live insect since circulation is not involved in the outer layers. Lewis (1965) noted that antennal penetration of DDT is also less in dead flies than in live ones (Fig. 6-4).

6.3. ROUTES OF INSECTICIDE ENTRY INTO HIGHER ANIMALS

6.3.1. Penetration of Insecticides Through Mammalian Skin

As a rule, mammalian skin appears to offer a more resistant barrier to apolar insecticides (such as DDT) than insect cuticle. Table 6-6 suggests that DDT penetrates into some insects' bodies as though there were no

TABLE 6-6. Variation in the Toxicity of DDT to Mammals and Insects as Affected by the Mode of Administration

	Toxic dose (mg/kg)	
Mammal	Cutaneous	Oral
Rat	3000	400
Rabbit	300–2820	300
Insect	Topical	Injection
Periplaneta americana	10	7
Popillia japonica	93	162

From Negherbon (1959).

barrier at all, while an approximately tenfold difference exists between the toxicities of injected and topically applied DDT in mammalian species. It is not, however, easy to establish any reliable basis for explaining the general tendencies of insecticide penetration through mammalian skin, because of a present lack of information.

O'Brien and Dannelley (1965) studied the pattern of insecticide penetration through the skin of female rats using malathion, DDT, dieldrin, carbaryl, and famphur (Table 6-7). Only DDT appeared to give a straightforward

TABLE 6-7. Relationship Between the Rate of Insecticide Penetration Through Rat Skin and the Partition Coefficients of the Penetrating Insecticides

Insecticide	Half-time (hr)	Partition coefficient[a] olive oil/H_2O
Carbaryl	14.5	64.5
Famphur	19	174
Malathion	5.5	413
DDT	26	932
Dieldrin	3.5	1805

From O'Brien and Dannelley (1965).

[a]Obtained by the following calculation:

$$P = R\left(1 - v\frac{L}{\alpha}\right)\left(1 - \frac{L}{\alpha}\right)$$

where P is the partition coefficient, L is the amount of the insecticide originally in oil, α is the amount left in oil after twice partitioning with water, and R is the volume ratio of water to oil. Note that there is no apparent correlation between half-time and partition coefficient.

pattern of penetration over a 24-hr observation period. Malathion had a biphasic pattern of penetration, a fast phase (initial 2 hr) and a slow phase (2–24 hr). The half-time penetration values* were 5.5 hr for malathion, 26 hr for DDT, 3.5 hr for dieldrin, 14.5 hr for carbaryl, and 19 hr for famphur (all approximate). The values would have been higher if the experiments had been made on unshaven skin. Although malathion (relatively polar) penetrated much faster than DDT (apolar), it appears that polarity alone cannot explain the pattern of penetration; for instance, the difference in rate of penetration between DDT and dieldrin was much more than one would expect from the difference in their polarities.

Feldmann and Maibach (1970), on the other hand, studied penetration of insecticides through human skin by topically applying six radioactive insecticides and measuring the amounts of radioactivity appearing in the urine for 5 days. The total percentages of radioactivities recovered in the

---

*Strictly speaking, it is not quite accurate to use half-time penetration values for dieldrin, carbaryl, and famphur, because their penetration does not follow simple first-order kinetics.

urine during this time were 73.9% for carbaryl, 9.7% for parathion, 9.3% for lindane, 8.2% for malathion, 7.7% for dieldrin, and 7.6% for aldrin. By this criterion, carbaryl is most systemic against human skin, while the remaining five insecticides have only modest penetration. Again there is no

TABLE 6-8. Comparison of Acute Oral and Acute Dermal Toxicities of Various Insecticidal Compounds to Female Rats[a]

	Oral	Dermal	Dermal/oral ratio
Chlorinated hydrocarbons			
Aldrin	60	98	1.63
Chlordane	430	530	1.23
DDT	118	2510	21.27
Dieldrin	46	60	1.30
Endrin	7.5	15	2.0
Heptachlor	162	250	1.54
Isodrin	7.0	23	3.29
Kelthane® (dicofol)	1000	1000	1.00
Lindane	91	900	9.89
Toxaphene	80	780	9.75
Organophosphates			
Chlorthion®	980	4100	4.18
DDVP (dichlorvos)	56	75	1.34
Delnav®	23	63	2.74
Demeton	2.5	8.2	3.28
Diazinon	76	455	5.99
Dicapthon	330	1250	3.79
Dipterex®	560	>2000	>3.57
EPN	7.7	25	3.25
Azinphosmethyl (Guthion®)	11	220	20.00
Malathion	1000	>4444	>4.44
Methylparathion	24	67	2.79
Parathion	3.6	6.8	1.89
Schradan	42	44	1.05
Thimet®	1.1	2.5	2.27
Trithion®	10.0	27	2.70
Others			
Isolan®	13	6.2	0.48
Carbaryl	500	>4000	>8.0
Calcium arsenate	298	>2400	>8.05
Lead arsenate	1050	>2400	>2.29
Nicotine sulfate	83	285	3.43

From Gaines (1960).

[a]Stomach tube with peanut oil vs. dermal tests with xylene applied on 3.0 by 4.5-cm rectangles of shaven backs.

clear-cut relationship between the polarity of the compound and the rate of penetration. This type of *in vivo* test cannot be thoroughly reliable for determining the systemic property of these compounds, since the rates of metabolism and excretion are different for each compound. In this particular case, carbaryl is known to be quickly hydrolyzed or hydroxylated and excreted in urine as conjugates, while dieldrin, for instance, is stored in fat or slowly metabolized. Moreover, a significant portion of the dieldrin metabolites are excreted in the feces rather than in the urine.

Perhaps under the circumstances the most reasonable compromise is to study the difference between acute oral toxicity and acute dermal toxicity for animals of the same sex. Of course, subacute and chronic tests would give different data for cumulative poisons than for noncumulative ones. The data presented in Table 6-8 indicate that there is a great variety in the ratio of dermal vs. oral toxicity. The compounds which have high ratios are DDT, lindane, toxaphene, azinphosmethyl, carbaryl, and calcium of arsenate, indicating that these compounds do not readily penetrate the rat skin. (These are ratios of approximately 8 or more.) The ones which do not show large differences are aldrin, chlordane, dieldrin, dichlorvos, dicofol, heptachlor, parathion, schradan, and Isolan,[®] each with a ratio of less than 2, meaning that these compounds penetrate the rat skin with relative ease. According to the ratios in Table 6-8, the order of penetration for the four compounds O'Brien and Dannelley (1965) studied is dieldrin > malathion > carbaryl > DDT, which agrees well with their observation.

6.3.2. Other Routes of Entry into Higher Animals

Insecticides are known to enter animal bodies through oral ingestion as residues in food and drinking water, and high doses are occasionally ingested accidentally or in cases of suicide and homicide. The extent and the meaning of continuous intake of residual insecticides will be discussed in Chapter 11. Oral poisoning is expected to be significant for those insecticides which are not systemic nor volatile. For instance, both calcium arsenate and lead arsenate are expected to get into animal systems through this route.

Another very important route of entry is through the respiratory system. Inhalation toxicity is expected to be important for volatile insecticides and for insecticides formulated as fine mists and powders. At the temperatures at which insecticides are normally used, compounds which have vapor pressures above 10^{-3} mm Hg are considered to be volatile enough to cause inhalation toxicity in the form of true vapor. For example, the high vapor pressure of dichlorvos (DDVP) enables this compound to be used as a vapor-action insecticide to control flying insects such as mosquitoes and flies when it is formulated as Vapona® strips or any other vapor-releasing device (Zavon and Kindel, 1966). All of the fumigant insecticides on the market today have higher vapor pressures than dichlorvos. Insecticidal compounds vaporized by the use of heating devices, e.g., the "lindane vaporizer," can also cause inhalation problems. Such devices are intended to control flying insects in an enclosed room. The inhalation problems from this usage have been discussed by West (1967).

The other type of inhalation hazard, fine powder or mist formation, is mostly caused by the spraying of insecticides or by manufacturing and formulating processes. It is known that particles larger than 10μm in diameter are effectively eliminated before reaching the lungs by the naso-pharynx barrier. Particles of 3μm or less, optimally $1–3 \mu$m (Gage, 1968), are likely to be deposited in the alveolar regions.

Assessment of the systemic action of these insecticides or insecticidal formulations is difficult. Arbitrarily, the Federal Hazardous Substances Act defines as an inhalation toxicant "any substance that produces death within 14 days in one-half of a group of white rats . . . when inhaled continuously for a period of 1 hour or less at an atmospheric concentration of more than 200 ppm but not more than 20,000 ppm . . . or more than 2 mg but not more than 200 mg per liter of mist or dust." Recommended threshold limit values for the air concentration of various insecticides have been listed in Chapter 2 and in Chapter 3 (Section 3.9). Such toxicity data, however, do not indicate the rate of penetration of insecticides via inhalation routes, because of the differences in the inherent toxicities of these compounds.

Nevertheless, there is evidence to show that the rate of insecticide penetration through the alveolar tissue far exceeds that through the skin; alveolar tissue is directly supplied with blood vessels and is lined with a thin, moist membrane which does not pose as formidable a barrier for insecticidal chemicals. Indeed, from comparative studies on dermal and inhalation toxicities of parathion to human volunteers, Hartwell *et al.* (1964) concluded that parathion is ten times more toxic by the inhalation route than by the dermal route.

6.4. REFERENCES

Ahmed, H., and B. G. Gardiner (1967). *Nature* **214**:1338.
Ahmed, H., and B. G. Gardiner (1968). *Bull. Entomol. Res.* **57**:651.
Alexandrov, W. J., (1935). *Acta Zool.* **16**:1.
Armstrong, G., F. Bradburg, and H. Standen (1951). *Ann. Appl. Biol.* **38**:555.
Beament, J. W. L. (1945). *J. Exptl. Biol.* **21**:115.
Benezet, H. J., and A. J. Forgash (1972*a*). *J. Econ. Entomol.* **65**:53.
Benezet, H. J., and A. J. Forgash (1972*b*). *J. Econ. Entomol.* **65**:895.

Davison, H., and J. F. Danielli (1952). *The Permeability of Natural Membranes.* Cambridge University Press, Cambridge.

Dierick, G. F. E. M. (1943). *Tijdschr. Plantenziekt.* **49**:22.

Ellisor, L. O. (1936). *Iowa State Coll. J. Sci.* **11**:51.

Feldman, R. J., and H. I. Maibach (1970). *J. Invest. Dermatol.* **54**:435.

Fisher, R. W. (1952). *Can. J. Zool.* **30**:254.

Gage, J. C. (1968). *Brit. J. Ind. Med.* **25**:304.

Gaines, T. B. (1960). *Toxicol. Appl. Pharmacol.* **2**:88.

Gerolt, P. (1969). *J. Insect Physiol.* **15**:563.

Hartwell, W. V., G. R. Hayes, Jr., and A. J. Fundses (1964). *Arch. Environ. Health* **8**:820.

Hayes, W. P., and Y.-S. Liu (1947). *Ann. Entomol. Soc. Am.* **40**:401.

Hurst, H. (1940). *Nature* **145**:388.

Hurst, H. (1943). *Trans. Faraday Soc.* **39**:390 and *Nature* **152**:292.

Iljinskaya, M. I. (1946). *Compt. Rend. Acad. Sci. U.S.S.R.* **51**:557.

Klinger, H. (1936). *Arb. Physiol. Angew. Entomol.* **3**:49 and 115.

Lees, A. D. (1946). *Parasitology* **37**:1.

Lees, A. D. (1947). *J. Exptl. Biol.* **23**:291.

Lewis, C. T. (1965). *J. Insect Physiol.* **11**:683.

Lindquist, D. A., and P. A. Dahm (1956). *J. Econ. Entomol.* **49**:579.

Matsumura, F. (1959). The permeability of insect cuticle. M.S. thesis, University of Alberta, Edmonton, Canada.

Matsumura, F. (1963). *J. Insect Physiol.* **9**:207.

Negherbon, W. O. (1959). *Handbook of Toxicology.* Vol. 3: *Insecticides.* Saunders, Philadelphia.

O'Brien, R. D., and C. E. Dannelley (1965). *J. Agr. Food Chem.* **13**:245.

O'Brien, R. D., and R. W. Fisher (1958). *J. Econ. Entomol.* **51**:169.

O'Kane, W. C., G. L. Walker, H. G. Guy, and O. J. Smith (1933). *Tech. Bull. New Hampshire Agr. Expt. Station* **54**:1.

O'Kane, W. C., L. C. Glover, R. L. Blickle, and B. M. Parker (1940). *Tech. Bull. N.H. Agr. Expt. Sta.* **74**:1.

Olson, W. D., and R. D. O'Brien (1963). *J. Insect Physiol.* **9**:777.

Pal, R. (1951). *Bull. Entomol. Res.* **51**:121.

Quraishi, M. S., and Z. T. Poonawalla (1969). *J. Econ. Entomol.* **62**:988.

Richards, A. G., and L. K. Cutkomp (1946). *Biol. Bull.* **90**:97.

Richards, A. G., and F. M. Korda (1948). *Ann. Entomol. Soc. Am.* **43**:49.

Roy, D. N., S. M. Ghosh, and R. N. Chopra (1943). *Ann. Appl. Biol.* **30**:42.

Sarkaria, D. S., and R. L. Datton (1949). *Trans. Entomol. Soc. Am..* **175**:71.

Sternberg, J., C. W. Kearius, and W. N. Bruce (1950). *J. Econ. Entomol.* **43**:214.

Tattersfield, F., C. Potter, and E. M. Gillhem (1947). *Bull. Entomol. Res.* **37**:497.

Treherne, J. E. (1957). *J. Insect Physiol.* **1**:178.

Webb, J. E., and R. A. Green (1946). *J. Exptl. Biol.* **22**:8.

West, I. (1967). *Arch. Environ. Health* **15**:97.

Wigglesworth, V. B. (1942). *Bull. Entomol. Res.* **33**:205.

Wigglesworth, V. B. (1945). *J. Exptl. Biol.* **21**:9.

Zavon, M. R., and E. A. Kindel, Jr. (1966). In *Organic Pesticides in the Environment.* Advances in Chemistry Series 60, American Chemical Society, Washington, D.C., p. 177.

Chapter 7

Dynamics of Insecticide Movement in the Animal Body

7.1. TOTAL INTAKE–ELIMINATION DYNAMICS

It is known that insecticidal compounds do not keep accumulating when continuous doses are given to an animal. That is, in chronic administration studies it has been observed that the levels of insecticides in various tissues rise to certain heights and then generally show some decline, even though the same daily doses are still maintained (e.g., Hayes, 1965; Robinson, 1969, 1970; Ludwig *et al.*, 1964; Kaul *et al.*, 1970; Korte, 1970). One such example is shown in Fig. 7-1.

The central scheme developed by Robinson and his associates (Robinson, 1967, 1969; Hunter *et al.*, 1969; Robinson *et al.*, 1969; Walker *et al.*, 1969) is that both accumulation and elimination processes follow concentration-dependent first-order kinetics. The speed of elimination of insecticides from the body after termination of continuous exposure to an insecticide can be expressed as

$$\frac{dc}{dt} = kc \qquad (1)$$

where c is the concentration of the insecticide and k is the first-order reaction constant. Integration of equation (1) gives

$$C = C_0 e^{-kt} \qquad (2)$$

where C is the concentration of the insecticide in a body tissue (compartment) and C_0 is the insecticide concentration at the start of the termination

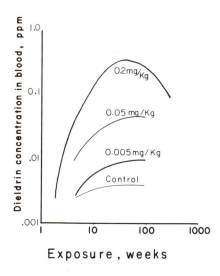

Fig. 7-1. Concentration of dieldrin in the blood of dogs given daily doses of dieldrin for 2–5 years. From Robinson (1969).

experiment. Similarly, the rate at which the insecticide accumulates as a result of continuous administration to the animal can be expressed as

$$\frac{dc}{dt} = \alpha - kt \tag{3}$$

where α is a constant representing a maximum ($=$ initial) velocity. Integration of (3) gives

$$C = C_\infty(1 - e^{-kt}) \tag{4}$$

where C_∞ is the concentration of the insecticide at $t = \infty$ (i.e., the maximum or saturation level in the experimental conditions).

In writing these equations, a generalized "compartments" concept has been employed. That is, an animal is considered to be composed of an infinite set of compartments, each of which functions as a unit in relation to insecticide level changes. Each tissue, such as the liver or the adipose tissue, can be thought of as a compartment, and the assumption has been made that the concentration of insecticide within a compartment is constant and changes occur simultaneously at all points in a compartment. In practice, a convenient residue-monitoring tissue, body constituent, or product has been adopted by researchers. Blood is the most frequently studied one, but fat and products such as excreta and eggs (Brown et al., 1965) have also been used for monitoring. Any of these materials can be regarded as representing a hypothetical "compartment."

In the preceding equations, the whole animal can be considered to be one "compartment" as long as the system behaves as one unit. The mathematical explanation of such reasoning is that the whole body consists of many "compartments" which behave similarly; that is, from equation (4)

$$C = C_\infty - \sum_{i=1}^{n+1} C_{i\infty}\, e^{-k_i t} \tag{4a}$$

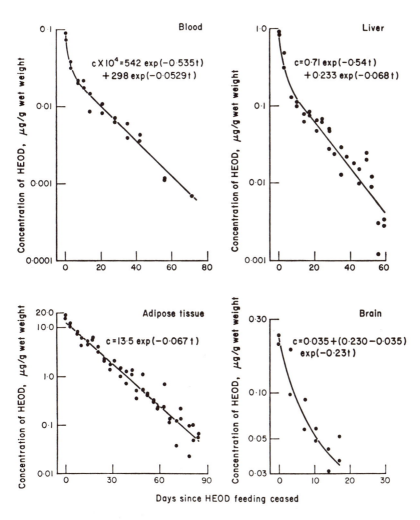

Fig. 7-2. Decline in dieldrin concentration in the blood, liver, adipose tissue, and brain of rats fed 10 ppm dieldrin in their diet for 8 weeks followed by a basic diet without added dieldrin for the periods indicated. Solid circles denote values for individual rats. From Robinson *et al.* (1969).

where k_i is the reaction constant for the compartment i, there are n number of compartments ($i = 1$ to $i = n + 1$) in the body, and $C_{i\infty}$ is the concentration in i of the insecticide at $t = \infty$. As long as each compartment follows the relationship in equation (4), the summation of these reactions will follow first-order kinetics. Similarly for elimination reactions, from equation (2)

$$C = \sum_{i=1}^{n+1} C_{i0}\, e^{-k_i t} \tag{2a}$$

where C_{i0} is the concentration of the insecticide at $t = 0$ in compartment i. An example of a (2a)-type reaction is shown in Fig. 7-2, from Robinson *et al.* (1969). For each of the four tissues, a plot of the logarithm of dieldrin concentration vs. time on a linear scale is presented. The tissues having a biphasic relationship (e.g., blood and liver) instead of a simple linear relationship (e.g., adipose tissue) are assumed to consist of two subcompartments. Thus the equation here is

$$C = A\, e^{-k_a t} + B\, e^{-k_b t} \tag{5}$$

by analogy to the preceding equations. Why should there be two subcompartments in blood and liver, and not in adipose tissue? Robinson *et al.* believe the reason to be that biotransformation of dieldrin takes place in the former tissues but not in the latter one, which represents a mere passive storage compartment:

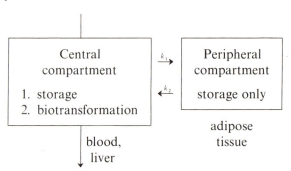

While it is purely speculative to consider the second subcompartment as metabolic, the general tendency of the adipose tissue elimination curve seems to indicate a single-compartment reaction, while there must be two or more subcompartments operating independently in the other organs.

There are other groups of researchers who have treated their data on a different mathematical basis. McCully *et al.* (1966) fed DDT-contaminated forage to beef steers for 83 days and measured the decline of residues of

p,p'-DDT, DDD, and p,p'-DDE in the fat for 331 days thereafter. Their data, when plotted on a log vs. log graph, gave straight lines, indicating a power function. Witt *et al.* (1966*b*) summarized data from the experiments of seven different groups on DDT residue relationships between the feed and the milk fat and found that all these data conveniently fit into a log vs. log relationship (see Fig. 7-7). Bruce *et al.* (1965) also treated their data on milk fat residues of heptachlor epoxide, DDT, and dieldrin on the basis of power function relationships, but it is not clear whether these authors were using mathematical or pharmacokinetic models. However, such power function relationships can be observed in the elimination of radioisotopes from bone tissue (Norris *et al.*, 1958), as Robinson *et al.* (1969) point out.

It is unfortunate that there are no comparable common data in these two sets of experiments to provide a reasonable explanation for the differences. At present, however, Robinson's model seems the more logical one, since there are other environmental (see Chapter 8) and pharmacological data which fit this type of relationship.

Zatz (1972), using Robinson's model, attempted to mathematically predict the degrees of accumulation of dieldrin and DDT. The degree of accumulation, R, is defined as the level that can be reached by repeated dosing with an insecticide divided by the level that is attained by a single administration at the same dose (i.e., if a single administration gives 1 ppm and repeated doses finally reach 100 ppm in the fat, the degree of accumulation is 100).

$$R = \frac{1}{1 - e^{kt}} \tag{6}$$

where R is the degree of accumulation, t is the time interval between doses (i.e., once-daily dosing would give $t = 1$), and k is the elimination constant, which can be estimated from half-life figures by the following equation:

$$K = \frac{0.693}{t_{1/2}} \tag{7}$$

where $t_{1/2}$ is the half-life of the compound in a given animal system. Thus assuming the biological half-life of dieldrin and DDT to be 369 and 115 days, respectively, and t to be $\frac{1}{3}$, meaning that the animal takes these insecticides 3 times a day in its feed, the degree of accumulation would be 1600 for dieldrin and 500 for DDT. Since such equations for predicting the degree of accumulation are of obvious value to environmental toxicologists, more work to test their validity is needed.

7.2. TRANSPORT OF INSECTICIDES BY BLOOD AND BODY FLUID

The major vehicle by which insecticides are distributed throughout the body is blood for vertebrate species and body fluid-hemolymph for invertebrates. Thus generally the method of application that best gets the chemical to the bloodstream elicits the fastest toxic symptoms. Witt *et al.* (1966*a*), for instance, made an attempt to measure the rate of disappearance of DDT from blood by injecting a DDT preparation directly into the jugular vein of cows. They found that the DDT concentration rapidly declined almost to equilibrium by 24 hr. The highest level of DDT reached in blood by use of this method was about 30 ppm on an extractable lipid basis, and the equilibrium level was 0.1 ppm. Assuming a two-compartment process, these workers assessed the half-life value to be on the order of 60–80 min for the fast phase and 8 hr for the slower phase. Hathway *et al.* (1967) found that 5 min after an intravenous dose of dieldrin in rabbits less than 0.5 % remained in circulation. Kaul *et al.* (1970) similarly administered ^{14}C-β-dihydroheptachlor and found that the radioactivity in various organs reached a maximum in 3–5 hr. Thereafter, the rate of decrease in radioactivity was different in each organ. After 24 hr 60 % and after 48 hr 70 % of the radioactivity was excreted, mainly in the form of hydrophilic metabolites. These two examples clearly indicate that transfer from blood to other tissues is rather rapid, although there are not enough data on other insecticides to indicate whether there are specific differences among compounds in the speed of transport by blood.

Since the majority of insecticides are not readily soluble in aqueous solutions, the mechanisms by which the blood carries them have aroused scientific curiosity. Moss and Hathway (1964) conducted a detailed study on the behavior of two cyclodiene insecticides, Telodrin® and dieldrin, in the blood and found that their solubility in rabbit serum is 4000 times greater than their solubility in water. In the blood they were found mainly in the erythrocytes (particularly in the erythrocyte contents as opposed to the membrane) and plasma and not in the leukocytes, platelets, or stroma. From centrifugal and electrophoretic examination, Moss and Hathway concluded that the insecticides mainly bind with hemoglobin and an unknown constituent in rabbit erythrocyte contents, with albumin and α-globulin in rabbit serum, and with pre- and postalbumin in rat serum. The ratio of distribution between the plasma and the cells was 37:19 for Telodrin® and 38:20 for dieldrin. Ultracentrifugal separation of Telodrin®-containing rabbit serum caused flotation of a proportion of Telodrin®-bound components with high-density lipoprotein, which has similar electrophoretic properties to those of α_1-globulin.

Morgan *et al.* (1972) studied the distribution of *p,p'*-DDT, *p,p'*-DDE, and dieldrin among the blood constituents of human workers with a long history of pesticide exposure and intake. They found that the percentages of *p,p'*-DDT and *p,p'*-DDE in the erythrocytes were 12.5% and 13.4%, respectively, much lower than the 39.8% found for dieldrin, which was distributed roughly according to the volume of erythrocytes in relation to the plasma. They also found that plasma albumin and, secondarily, the smaller globulins are the principal plasma protein constituents associated with blood-borne *p,p'*-DDT and *p,p'*-DDE, and that they occur primarily in the triglyceride-rich low-density and very-low-density lipoproteins.

While direct association of these lipophilic insecticides with triglycerides has been proposed by several workers, Rose (1971) could not find a direct relationship between the levels of free fatty acid and dieldrin in the blood under several different conditions of free fatty acid contents. Since Morgan *et al.* (1972) observed that protein-free serum is virtually devoid of pesticides, it is reasonable to assume that the bulk of lipophilic insecticides are carried in the blood in protein- and lipoprotein-associated form.

7.3. DISTRIBUTION AND REDISTRIBUTION WITHIN THE ANIMAL BODY

7.3.1. Distribution After Acute Administration

Whatever the route of entry, insecticides in the body of an animal are initially carried by the blood system to be distributed to various tissues. It is generally agreed that the initial distribution is rapid and that the pattern of distribution roughly corresponds to the speed and pattern of blood circulation. Kaul *et al.* (1970), for instance, found that the radioactivities in various tissues after intravenous injection of ^{14}C-β-dihydroheptachlor in male rats reached maxima in 3 hr, except in intestinal organs (maximum at 5–7 hr), liver (maximum at 12 hr), and gut contents (maximum at 16 hr). The radioactivity declined thereafter in all organs, indicating redistribution of the insecticide and its metabolites into gastrointestinal organs. Heath (1962) and Heath and Vandekar (1964) studied the pattern of distribution of dieldrin in rats and mice and found that dieldrin disappears quickly from the blood and is distributed among all tissues within the first few minutes. The tissues receiving the highest per gram amounts of dieldrin were brain, liver, lungs, kidneys, and heart. In about 80 min the concentration of the insecticide in most organs, notably the brain, declined, and the dieldrin was redistributed to the gastrointestinal system and fatty tissues.

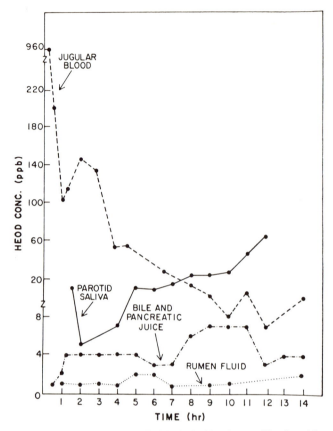

Fig. 7-3. Concentration of dieldrin in blood, parotid saliva, bile
and pancreatic juice, and rumen fluid after injection of 20 mg of
dieldrin into the jugular vein of a goat. From Cook (1970).

Cook (1970) studied the fate of dieldrin (20–50 mg) injected into the
jugular vein of goats, sheep, and calves. He found that dieldrin levels in the
blood declined over a 12-hr period and that the levels in the bile and pancreatic juice and the rumen fluid increased during this period (Fig. 7-3).
An interesting observation is that the dieldrin level in parotid saliva (and
probably in all saliva from the various glands) drastically increased in the
first hour, declined in the next hour, and showed a steady increase thereafter.
Thus dieldrin reenters the gastrointestinal system from the blood via saliva,
bile, and pancreatic juice. In calves, the only ruminants with separate bile and
pancreatic ducts among those tested, Cook (1970) estimated that the concentrations of dieldrin in bile and pancreatic juice over a 24-hr period were

10.6 and 2.6 ppb, respectively. Considering the even larger flow volume for bile juice (0.57 ml for bile and 0.32 ml for the pancreas in 24 hr), dieldrin entry via the bile route appears to be of great importance, in agreement with other workers' observations (Heath and Vandekar, 1964; Morsdorf et al., 1963). Davison (1970) studied the overall pattern of distribution of dieldrin and its metabolites in sheep 24 hr after administration of low doses of dieldrin. It is apparent from his data (Table 7-1) that the bulk of dieldrin and its metabolites end up in the carcass, rumen, and other gastrointestinal regions in 24 hr.

TABLE 7-1. Mean Distribution of Dieldrin in a Short-Term Experiment[a] and in a Long-Term 32-Week Feeding Experiment

	Short-term radiometry		Chronic data: dieldrin mean concentration after 32 weeks[d] (ppm)	Percent dry matter	Percent fat in dry matter
	Percent recovery of radioactivity[b]	Extractable with petroleum ether[c] (%)			
Adipose tissue	1.1	102.3	126	68.1	88.4
Heart	0.4	102.3	107	30.8	49.4
Muscle	0.1	105.4	104	25.3	18.2
Brain	0.1	100.9	20.5	22.5	35.5
Spinal cord	0.05	97.5	18.9	32.3	57.9
Carcass	55.5	91.8	110	36.7	39.8
Liver	3.0	78.5	323	29.4	11.4
Kidneys	0.1	64.3	80	20.7	15.8
Rumen contents	17.4	63.9	—	—	—
Gastrointestinal contents	7.0	66.1	—	—	—
Feces	3.2	61.7	—	—	—
Urine	4.4		—	—	—
Total	92.35				

Data taken from Davison (1970).

[a]Radioactivity in sheep 24 hr after administration of 50–100 μCi ^{14}C-dieldrin specific activity 25–190 μCi/mg. The sheep were fed 2 mg of nonradioactive dieldrin for 4–32 weeks prior to the radioactive tests.
[b]Expressed as percent of the total ^{14}C-dieldrin originally administered as judged by the results of combustion assays.
[c]Lower recoveries probably indicate the presence of more hydrophilic metabolites.
[d]Result of concurrently conducted nonradioactive tests at 0.5 mg dieldrin/kg/day. Other tissue values were 98.6 ppm for bones and 65.9 ppm for adrenals. Expressed as ppm in fat on dry tissue basis.

One of the most convenient ways to visualize the pattern of distribution of insecticides is to use whole-body radioautography. Bäckström *et al.* (1965) have examined the pattern of distribution of ^{14}C-DDT in pregnant mice by such a method. While the general tendency of redistribution in the whole body appears to indicate the presence of a rather rapid active transport system carrying the bulk of insecticides toward organs related to excretory processes, the remaining portion of insecticides, representing only a small part of the total, may redistribute much more slowly. Woolley and Runnells (1967) studied changes in DDT concentrations in the brain and the spinal cord of the rat following intubation with 100 mg/kg of *p,p'*-DDT. The level of DDT decreased drastically in the liver as compared to the brain and spinal cord. The levels in the liver at 6, 12, and 24 hr after administration were 216, 165, and 43 ppm, respectively, while those in the whole brain were 22.0, 22.2, and 11.8 ppm and those in the spinal cord were 24.5, 27.2, and 25.1 ppm. Hunter *et al.* (1960) showed that 8 days after dosing rats with 4.3 mg dieldrin the level in the adipose tissue increased to 34% of the total present in the body at that time.

Apparently the reason behind the slow buildup of lipophilic insecticides in adipose tissue and other fatty tissues is that these portions of the body have an unusually poor supply of blood, in addition to the phenomena of (1) very slow establishment of equilibrium in the extracellular water in these tissues (Jones, 1950) and (2) slow elimination from these tissues.

7.3.2. Distribution After Chronic Dosing and Dynamics of Redistribution

It is apparent from the foregoing discussion that in acute dosing lipophilic insecticides gradually settle into fatty tissues after a certain period of time for redistribution and that in continuous daily dosing the general pattern is gradual accumulation in the fat approaching a plateau after some months. Thus for chronic administration the final pattern of accumulation at equilibrium is expected to be different from the acute dosing distribution.

Generally speaking, in a steady state or equilibrium the level of insecticide in blood is related to its concentrations in other organs and tissues. The relationship between blood concentration and fat concentration of pesticides has been extensively investigated in the attempt to find a readily available substitute for fat for residue analyses. In brief, the relationship is a rather reliable one for dieldrin (Hunter and Robinson, 1967), fair for DDT, and poor for endrin (Richardson *et al.*, 1967) and other chlorinated hydrocarbon insecticides. Such equilibria are difficult to attain when the systems are very slow to respond to changes in the blood levels or are sensitive to

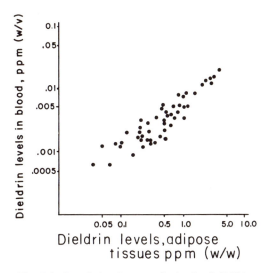

Fig. 7-4. Correlation between the levels of dieldrin in the blood and those in the adipose tissues of human volunteers fed constant doses of dieldrin for 18 months. From Hunter and Robinson (1967).

fluctuations in the physiological and biochemical states of the animal. Also, for easily degradable chemicals establishment of equilibrium against the active organs could be difficult. An example of good correlation between blood and adipose tissue levels of dieldrin is shown in Fig. 7-4. The ratio between the concentration of dieldrin in the adipose tissue and that in the blood appears to be on the order of 140 for humans (Hunter and Robinson, 1967) and 180 for dogs (Keane and Zavon, 1969). Richardson *et al.* (1967) obtained good correlations between dieldrin levels in blood and those in the liver, kidneys, lungs, and fat but not in the combinations blood–pancreas, blood–spleen, and blood–muscles in dogs fed 0.1 ppm of dieldrin for 128 days. Davison (1970) found good correlations in the combinations blood–bone, blood–brain, blood–adipose tissue, blood–heart, blood–muscle, and blood–spinal cord but not in blood–kidney and blood–liver in sheep fed 0.5–4 mg/kg/day of dieldrin for 32 weeks. The balance of evidence supports a good correlation between blood levels of dieldrin and those of relatively nonactive tissues.

A typical end result of such distribution levels in various organs is summarized in Table 7-1. Because of its large total volume and percent fat content, the adipose tissue accumulates by far the highest proportion of the total body burden of dieldrin. However, when one examines the data on a ppm in fat basis, it becomes apparent that the levels of dieldrin in various tissues are fairly constant. The notable exceptions are those in the brain and spinal cord, and in the liver. That the nervous tissues, including the brain, show much lower levels of chlorinated hydrocarbon residues than expected from their fat content has been observed by many researchers (see human residue data, Chapter 11, Fig. 11-9). The reason for such a phenomenon is not known, since this tissue is adequately supplied with blood and is rich in lipoprotein and other lipid matter known to show high affinity for these insecticides.

The levels of chlorinated hydrocarbon insecticides in the liver tend to fluctuate, and hence the relatively high dieldrin level in the liver fat is not as consistent as the low level in the brain. Generally, however, the liver fat, contains high levels of chlorinated hydrocarbon insecticides, probably because these chemicals are mainly detoxified in this organ and the rate of detoxification is rather slow, causing a backup of residues when the total body burden becomes higher than the maximum rate of turnover.

7.4. FACTORS AFFECTING STORAGE AND RELEASE

Storage of insecticides in the body tissues has been well recognized through residue analyses of human tissues in various communities. The tendency toward storage is naturally high with stable and lipophilic insecticides, i.e., chlorinated hydrocarbons. Residues in man will be discussed in detail in Chapter 11; however, it should be mentioned here that there are sex, race, age, and regional specific factors that determine the levels of pesticides. For instance, human males accumulate more pesticide residues than human females in any community. The logical conclusion is that there must be some hormonal, biochemical, and physiological body conditions which favor accumulation. The sexual difference in rate of accumulation is evident in experimental animals, also. In rats, it is the female that accumulates more endrin. Klevay (1971) has shown that the liver of male rats eliminates endrin via the bile system two to 12 times faster than that of female rats. Accordingly, the level of storage in a certain organ is most likely decided by the rate of elimination and discharge in that part of the body.

It is well known that chlorinated hydrocarbon residues are released into the bloodstream from the adipose tissue and other sources during starvation. For example, Dale *et al.* (1962) showed that DDT can be mobilized in rats by starvation. Heath and Vandekar (1964) demonstrated that starvation of dieldrin-treated rats caused an increase in bile secretion and fecal excretion of dieldrin and its metabolites. Ecobichon and Saschenbrecker (1969) found that food deprivation in cockerels caused massive mobilization of DDT into the plasma. This in turn resulted in complete redistribution of DDT. The tissues which showed consistent increases on starvation at all DDT concentrations tested were liver > brain > heart, whereas the ones which showed consistent declines were kidney > skeletal muscle. Increased residues in the blood in starved avian species have been also observed by Donaldson *et al.* (1968). In studying the effect of various forms of stress, Brown (1970) found that the blood concentration of DDT is increased by exercise, cold treatments, and starvation in the male rat given 5 mg/kg of DDT orally for 14 days.

Rose (1971) studied the relationship between mobilization of free fatty acid into the bloodstream and the levels of dieldrin in the blood. He used short-term stress such as cold exposure, swimming, and administration of a chemical inducer ("fat-mobilizing factor") and long-term stress by starvation to change the level of free fatty acid in the blood of white rats. In all cases, he could observe increased free fatty acid levels in the blood, but dieldrin levels did not increase according to the fatty acid changes. For example, swimming caused an increase of fatty acid from 155 to 403 μmoles/liter plasma, while the level of radioactive dieldrin remained unchanged at 2105 dpm (disintegrations per minute) per milliliter of blood for controls and 2016 dpm/ml for dieldrin-treated rats. Similarly, the fat-mobilizing factor, an extract of pig pituitary, caused an increase in fatty acid from 253 to 457 μmoles/liter, while dieldrin levels remained constant at 186 dpm/ml for the controls and 181 dpm/ml for the treated rats. Starvation induced a small increase in dieldrin—125 dpm/ml for controls and 178 dpm/ml for fasted rats—while the fatty acid levels were 266 μmoles/liter for controls and 655 μmoles/liter for fasted rats. However, this dieldrin increase can be explained by the decrease in size of the adipose tissue alone, without considering the release of fatty acid. The undeniable message here is that free fatty acid levels have nothing to do with the levels of dieldrin in blood.

However, Rose (1971) also demonstrated that treatments with an anesthetic ether could drastically increase dieldrin levels in the blood of rats and rabbits. This finding confirms an earlier observation that anesthesia treatments result in an increase in blood dieldrin (Hathway *et al.*, 1967). The phenomenon is not caused by the change in blood–adipose tissue partitioning characteristics, nor is it related to fatty acid mobilization by ether. It does appear to be pertinent to the observation by many workers that barbiturates and other chlorinated hydrocarbon insecticides can cause reduction of dieldrin storage in tissues. It was originally reported by Street (1964) that DDT, simultaneously fed to rats, has the effect of significantly reducing the level of dieldrin storage. Subsequent studies by Street *et al.* (1966a) showed that four drugs, tobutamide, aminopyrine, heptabarbital, and phenylbutazone, fed to female rats for 10 days could also reduce the levels of dieldrin storage. Of these, heptabarbital and aminopyrine, which are sedatives and analgesic agents, showed high potencies. On the basis of these observations, Street and Blau (1966) proposed that antagonism of dieldrin storage by these chemicals is due to induction of dieldrin-metabolizing liver microsomal enzymes.

Street *et al.* (1966*b*), however, could not suppress this DDT effect with ethionine (up to 250 mg/kg/day in the diet) and actinomycin D (up to 25 μg/day for 10 days), acknowledged inhibitors of protein synthesis which have the potential to suppress enzyme induction activities in the liver. This finding is in direct conflict with the "induction" theory. Furthermore, the DDT effect could not be observed in sheep, hens (Street *et al.*, 1966*b*), or guinea pigs (Wagstaff and Street, 1971) despite the fact that DDT at doses employed should have caused marked induction effects.

Using liver slices *in situ*, Matsumura and Wang (1968) showed that release of ^{14}C-dieldrin from the liver tissue of male rats was accelerated by phenobarbital, DDT, and even electric shocks. At the same time, DDT and phenobarbital administered *in vivo* reduced the rate of uptake of dieldrin by the liver slices. The *in situ* experiments were conducted in a short time period, up to 90 min, and hence the results are not likely related to the induction phenomena. In this connection it is interesting to note that Street (1968) himself points out the possibility that such effects can be caused by the drug displacement phenomenon, where one drug effectively displaces another by competing for the same binding site. This type of competitive binding is relatively well known; for instance, methylenedioxyphenyl synergists are acknowledged to bind with the same liver microsomal action sites as the drug- and pesticide-degrading system, mostly as competitive inhibitors, and thus to increase the effectiveness of the drugs and pesticides. In such cases, competition will favor the chemical which shows higher affinity for the binding site.

Examination of available data shows that in female rats the order of accumulation in the adipose tissue appears to be DDT > dieldrin > methoxychlor; the actual data show 100 ppm accumulation of DDT/DDE after doses of 5 ppm DDT daily in the diet for 10 weeks vs. 67 ppm and 32.2 ppm, respectively, for dieldrin and methoxychlor after daily doses of 10 ppm each for 10 weeks (Street and Blau, 1966). Thus at comparable doses DDT can reduce the storage of dieldrin and methoxychlor, and dieldrin can reduce the storage of methoxychlor, but no reduction is expected in other combinations. In guinea pigs (Wagstaff and Street, 1971), the order of accumulation is not the same: dieldrin (542 ppm) ≫ DDT (21 ppm) ≥ lindane (12 ppm) after feeding of 25 ppm of each insecticide per day for 14 days. Assuming that the degree of accumulation is a good index of affinity for the binding site, dieldrin can reduce DDT storage but DDT cannot influence dieldrin storage in guinea pigs, which is precisely opposite to the case in the rat! Indeed, comparatively small amounts of dieldrin (1 ppm) are enough to cause a reduction of much larger amounts of DDT (50 ppm) in guinea pigs, in accordance with the large difference in their degrees of accumulation in this species. The competition of DDT and dieldrin for the same binding sites is a likely possibility since these two insecticides show very similar binding patterns, in the rat brain at least (Matsumura and Hayashi, 1969) (Table 7-2).

TABLE 7-2. Distribution of Five Insecticides Among Various Components of the Central Nervous System of the Rat and Cockroach[a]

	Dieldrin	DDT	Lindane	Phthalthrin	Nicotine
Rat brain[b]					
Supernatant[c]	1.00	1.12	4.98	3.74	18.26
Cell membrane	1.92	2.49	1.02	1.64	1.55
Myelin fragment	6.05	6.17	5.45	10.25	2.58
Synaptic complexes	4.60	5.73	1.81	0.16	3.62
Mitochondria	1.23	1.16	0.73	0.71	1.90
Debris and nucleus	8.32	8.12	8.00	7.52	1.97
Cockroach nerve cord, ganglia[b]					
Supernatant[c]	144.6	209.0	382.3	512.5	568.5
Cell membrane	151.2	104.8	59.6	62.1	10.4
Cell membrane	99.6	40.2	6.1	9.7	4.0
Synaptic complexes	58.3	40.2	1.6	4.6	0.1
Synaptic complexes–					
mitochondria	27.5	45.6	0.6	4.9	0.5
Sheath, nucleus debris	21.0	42.2	0.5	3.9	1.4

From Matsumura and Hayashi (1969).

[a]The nerve homogenates were incubated *in vitro* with 10^{-5} M of insecticide. Expressed as nmoles of insecticide bound per mg protein.
[b]There is a difference in the amount of neural matter: 100 mg of neural matter/ml for the rat brain and 10 mg/ml for the roach nerve.
[c]Including free insecticide.

The balance of evidence, therefore, supports the view that competitive displacement plays an important role in these drug and insecticide interactions. All these chemicals are also membrane-interacting agents, and molecular interactions at the membrane level are expected to take place. Thus it is not surprising that both surface-activating agents and electric shock trigger a change in storage levels of pesticides, for these treatments are expected to alter the membrane properties. While in the case of long-term treatments there must be some induction effects, such processes appear to be relatively less important.

7.5. PENETRATION AND DISTRIBUTION INTO VITAL ORGANS AND TISSUES

7.5.1. Nervous System

The most common site of insecticide attack is the nervous system, and thus the penetration and subsequent distribution of insecticides therein have been investigated by several researchers.

That the brain tends to accumulate lower levels of insecticide residues in mammalian systems has already been shown (e.g., Davison, 1970; Yoshioka *et al.*, 1972; see also Fig. 11-9). Thus the question is whether there is some sort of "blood–brain barrier" against insecticidal chemicals. The "blood–brain barrier" is believed to line the capillaries of the cerebral vessels. After passing this barrier, a drug or an insecticide must still penetrate the membrane of the brain. A variety of drugs are known to penetrate from the blood into the cerebrospinal fluid by simple diffusion at rates parallel to their lipid–water partition coefficients (e.g., Albert, 1968). In other words, highly lipophilic compounds, such as chlorinated hydrocarbon insecticides, are expected to penetrate into the brain very rapidly. The more polar the insecticide, the slower the rate of penetration is going to be. Exceptions are the chemicals which are indistinguishable from those ions which are actively transported into the central nervous system. A general observation is that the "barrier" is effective only against very polar insecticides, of which there are relatively few.

Insect nerves, on the contrary, appear to pose a much more formidable barrier to polar compounds. To cite examples, acetylcholine, physostigmine or eserine, prostigmine, and tubocurarine are virtually ineffective against insect nerves, while they are highly active against mammalian nerves. That this difference is not due to an inherent susceptibility difference is apparent from the fact that apolar carbamate derivatives are highly toxic to both insects and mammals. Hence one wonders just where a polar toxic compound becomes insecticidal. Of the above compounds, prostigmine, which is almost completely ionized at *p*H 7, is nontoxic to insects, while eserine, which is expected to have a fraction (0.091; O'Brien and Fisher, 1958) in the un-ionized state, is slightly toxic (in terms of LD_{50}, it is about one-thirtieth as toxic to insects as to the mouse). As discussed previously, the insect CNS is more resistant to polar chemicals because it is covered with the ion-impermeable nerve sheath (Hoyle, 1953) To avoid penetration problems, most neuroactive insecticides are designed to be rather apolar and lipo-philic. Accordingly, it is not expected that the mammalian CNS would show any resistance against them, and neuroactive insecticides are by far the most numerous of the chemicals in use at present.

Then why should the CNS tend to accumulate less chlorinated hydro-carbon residues? The answer should be found in the processes of partitioning of pesticides with respect to the blood rather than in the blood–brain barrier. In fact, the level of residues in the brain is known to rise within a few minutes after intravenous administration and reach a maximum within a few hours along with the levels in other tissues, such as the adipose tissue (e.g., Kaul *et al.*, 1970), but finally to attain a somewhat lower equilibrium than in other tissues. It is possible that the virtual absence of neutral lipids in the brain somehow influences the final residue levels, but much more information is needed to confirm such a view.

Woolley and Runnells (1967) found that within the CNS of the DDT-poisoned rat, distribution is fairly uniform 24 hr after administration of DDT both on a total fresh tissue basis and on a gram lipid extract basis. Bäckström *et al.* (1965), on the other hand, observed that DDT shows a preference for the gray matter, whereas dieldrin seems to concentrate in white matter. Matsumura and Hayashi (1969) studied the pattern of distri-bution among various subcellular components of the CNS *in vitro* (Table 7-2). In general, the amounts of insecticides that remained in the supernatant were directly related to their water solubility. In the case of lipophilic insecticides, the pattern of binding was rather similar: most of the insecticides were bound to membranes, particularly in nuclear myelin fragments, synaptic membranes, and cell membranes. Telford and Matsumura (1971) used a radioautographic method for electron microscopic preparations and observed that in the nerve cord of *Blattella germanica* 47.6% of silver grains representing dieldrin were on the axonic membrane, 14.3% on the glial membrane, 32.6% on the axoplasm, 3.3% on the nerve sheath, 1.3% in extracellular space, and 0.9% on other sites. Thus it is likely that the bulk of these insecticides are bound to the membrane and that the degree of binding is related to their lipid–water partition coefficients.

7.5.2. Transfer to Fetus and Reproductive Organs

The placenta also contains a barrier system, referred to as the "placental barrier." Its selectiveness is mainly for active transport of amino acids, glucose, vitamins, and inorganic ions. Other polar compounds are trans-ferred very slowly. Compared to the "blood–brain barrier," the "placental barrier" is less selective and efficient; thus even polar chemicals can eventually find their way to the fetus. Also, the fetus does not have an efficient elimination mechanism for polar chemicals, as the brain does, which aggravates its tendency to be a rather inefficient barrier mechanism. Lipid-soluble insecticides generally have no problem in reaching the fetus. Since these lipid–soluble chemicals are also expected to diffuse out quickly, their final accumulation must be decided by partitioning against the blood of the mother. On the other hand, polar chemicals or polar metabolites of insecticides are expected to reach the fetus slowly, but once there to have a very slow rate of elimination.

Finnegan *et al.* (1949) first demonstrated the placental passage of DDT in dogs. Bäckström *et al.* (1965) confirmed the observation in pregnant mice by using a whole-body radioautographic technique. Both DDT and dieldrin accumulated in the fetus, particularly in the liver, fat, and intestines. The levels of accumulation in the fetus appeared to be no different or slightly less than those in the brain and the heart of the mother. High amounts of DDT and dieldrin were found in ovarian corpora lutea. Other tissues which showed high levels of DDT and dieldrin accumulation were the liver, placenta, and mammary glands.

As for more polar insecticides, Fischer and Plunger (1965) were unable to detect any trace of parathion in an 8-month fetus from a mother who had taken a fatal dose of this insecticide. However, Fish (1966) reports that the cholinesterase of rat embryos was inhibited when their mothers were treated with parathion, methylparathion, or DFP by intraperitoneal injections, indicating eventual transfer into the fetus. Both Ackermann and Engst (1970) and Villeneuve *et al.* (1972) made more complete attempts to quantitatively study the extent of organophosphate transfer. The former group found that methylparathion, bromophos, and Imidan® were present in various tissues of rat embryos. Methylparaoxon, the toxic metabolite of methylparathion, was also found. The latter group studied the relationship between the amounts of ^{14}C-parathion in the blood of the embryo and of the mother, and made an attempt to relate the data to the levels of plasma cholinesterase. They found that parathion levels in the fetal plasma were much lower than in the brain. Plasma cholinesterase activities were also indicative of lower levels of fetal exposure to paraoxon (Table 7-3). Thus it is apparent that some sort of placental barrier must exist for relatively polar insecticidal chemicals in the rat.

With respect to DDT accumulation, Dedek and Schmidt (1972) studied the rate of transfer to mouse embryos by using pregnant mice injected with 0.5 mg/day of DDT intraperitoneally. They found that the levels of DDT and its metabolites in fetal blood were about 50 % of those in maternal blood and that the fetal liver and brain contained only slightly lower amounts than those in the maternal organs. Starvation stress caused an increase in

TABLE 7-3. Parathion Contents of Maternal and Fetal Plasma and Amniotic Fluid[a]

Sample	Time (min)	Parathion[b] (ng/ml)	Approximate plasma ChE inhibition[c] (%)
Maternal plasma	10	593	41
	20	131	43
	30	110	44
	60	44.5	40
	120	35.5	41
	240	30.5	24
Fetal plasma	10	3.85	21
	30	1.10	24
	60	0.65	23
Amniotic fluid	60	ND[d]	—
	120	ND	—
	240	ND	—

From Villeneuve *et al.* (1972).

[a]Intravenous injection through jugular vein at 0.1 mg/kg.
[b]Metabolites are not included.
[c]Graphically obtained from their Fig. 1; only approximations.
[d]Not detected.

the fetal concentrations of DDT and its metabolites. Huber (1965) fed pregnant mice 40 ppm of Kepone® in the diet and found that on the average 5 ppm accumulated in seven embryos weighing 0.3 g. This is a comparable level to that in the maternal brains but is considerably lower than the levels found in their livers (45 ppm) and fat (13 ppm). Hathway *et al.* (1967), on the other hand, studied the mechanisms of transport of dieldrin from the rabbit mother to the blastocyst and to the fetus. They found that free blastocysts pick up dieldrin rather rapidly from the mother's blood (Fig. 7-5), but after implantation the rate of pickup measurably slows down. During the second half of pregnancy, passage of dieldrin to the fetus is limited to the transplacental mode, as attested to by the complete lack of dieldrin in allantoic and amniotic fluid. As with the blastocysts, dieldrin concentration in the whole fetus reached the same level as that in the mother's blood in 40–50 min, but then declined to approximately one-third that in maternal blood. These workers also injected ^{14}C-dieldrin into the fetus and observed the appearance of dieldrin in the maternal blood to prove the two-way traffic

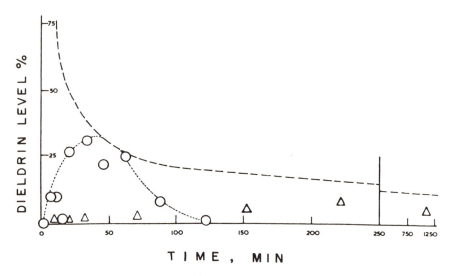

Fig. 7-5. Time courses for uptake of ^{14}C-dieldrin by 6-day rabbit blastocysts (circles) and 9-day blastocyst fluid (triangles) in relation to the maternal blood level (broken line) of dieldrin. All values are expressed as percent of the original (5 min after injection) dieldrin level in the maternal blood. From Hathway *et al.* (1967).

pattern for dieldrin; however, they did not assess relative speeds of these two transport processes.

Thus it appears likely that the "placental barrier" is not too efficient for highly liposoluble compounds, while for polar compounds its effectiveness becomes highly significant, as ratios of 100:1 in the case of parathion indicate. However, because the fetus is expected to excrete polar compounds slowly and is much more susceptible to toxicants than the adult animal, and because the difference in cholinesterase inhibition is only modest, the effectiveness of the barrier against polar compounds cannot be regarded as entirely satisfactory.

As for the localization of insecticides in reproductive organs, indirect evidence that they cause either functional or morphological changes may be relevant. For instance, Burlington and Lindeman (1950) showed that DDT strongly inhibited testicular and other secondary sex growth in cockerels. Deichmann *et al.* (1971) observed subnormal reproductive performance in dogs fed DDT for an extended period. Mandzhgalodze and Vashakidze (1972) state that carbaryl at 1/200, 1/500, and 1/1000 LD_{50} doses in rats exerted gonadotoxic effects and that it interfered with the estrus cycle. Gellert *et al.* (1972) showed that *o,p'*-DDT injected intraperitoneally for 27 days delayed vaginal opening and increased the weight of ovaries and uteri in the rat. This evidence, however, cannot be regarded as direct proof of insecticide action on these organs, as insecticides are known to interfere either with hormone-producing organs or with the liver, where through induction they can cause marked effects on hormonal levels. Direct evidence of localization of insecticides in individual organs is needed.

In this respect, Bäckström *et al.* (1965) observed a high concentration of DDT and dieldrin in ovarian corpora lutea and mammary glands by using a radioautographic technique. Smith *et al.* (1972*a, b*) studied the distribution of ^{3}H-DDT in male mice after a single oral dose of DDT and found that high levels of radioactivity were transferred into male reproductive organs within a short period, such as 1–4 hr. Although epididymal fat was the fattiest tissue, other tissues appeared to collect more DDT-R. The concentration of DDT-R in the seminal plasma increased with time, but after 12 days settled down at a low level. In contrast, a high level of radioactivity was found after 12 days in the anterior prostate gland, indicating the actual site of accumulation in the long run.

TABLE 7-4. Accumulation of Kepone® Residues in Gonads as Compared to Other Organs and Tissues in Mice Fed 40 ppm Daily in the Diet

Days on 40 ppm	Level of Kepone® (ppm)					
	Liver	Brain	Kidney	Fat	Muscle	Gonads[a]
Females						
5[b]	45	3	7	13	—	—
15[c]	39	20	30	27	5	—
30	78	26	27	25	20	25
30	64	30	32	33	15	25
90	67	29	27	29	15	13
90	66	30	39	34	16	—
150	168	26	74	81	23	26
150	90	61	37	49	19	—
500	120	75	88	84	26	20
500	96	55	35	29	11	18
Males						
100	60	25	25	41	10	26
300	113	65	49	22	8	17

From Huber (1965).

[a]Uterus and ovaries for females.
[b]Kepone® feeding at 40 ppm for 6 days produced 5 ppm in seven embryos, 12 ppm in four placentas, 3 ppm in a 6-day-old suckling mouse.
[c]Kepone® levels in the brain and liver of a 15-day-old suckling mouse were 36 and 47 ppm, respectively.

Huber (1965) assayed the levels of Kepone® residues in the gonads of male and female mice by feeding them a daily diet containing 40 ppm for an extended period. His data (Table 7-4) indicate that the levels of Kepone® in the gonads remain constant and slightly lower than those found in the brain, or at levels rather comparable to those in muscle.

7.6. ELIMINATION OF INSECTICIDES: EXCRETION AND SECRETION

7.6.1. Biliary and Urinary Excretion

Excretion of foreign compounds (xenobiotics) may be through two major systems, the renal and the hepatic. Other minor systems available are exhalation from the lungs for volatile compounds, secretion in sweat and saliva, and elimination to nonvital tissues such as hair and nails, but for insecticidal compounds these minor systems are not expected to play any significant role. An exception is secretion of pesticide residues in milk, which will be explained later. For the two main excretory processes, it is necessary that the insecticidal compounds enter the bloodstream. The liver is supplied with the portal vein and the kidneys with the renal artery, through which xenobiotics are transported by blood to reach them. Since the blood can flow through these organs via hepatic veins and renal veins, the material not taken up by one organ can be further transported to the other. The material taken up by these organs is first metabolically altered (or, to a limited extent, taken up as the original compound) and then either excreted via the bile duct into the intestine to be eliminated in feces (hepatic system) or excreted in the urine (renal system). In other limited cases, poorly soluble ingested material, such as chlorinated dibenzo-*p*-dioxins, may go through the digestive system without being absorbed into the blood and then appear in the feces in the original form (Norback *et al.*, 1973).

In the liver–bile system, the insecticidal derivatives emerge from the liver with the bile through two hepatic ducts which merge into a "common hepatic duct" to enter the gall bladder for bile storage (Fig. 7-6). At intervals, bile leaves the gall bladder by the bile duct to be discharged into the second part of the duodenum, which is directly connected with the small intestine. Some compounds are absorbed from the small intestine to be returned to the portal vein; thus the probability of recycling through this route exists.

The rate and the route of excretion of insecticidal derivatives (i.e., the insecticidal compound and its *in vivo* metabolic products) are largely determined by (1) the rate of uptake by the liver and the kidneys from the

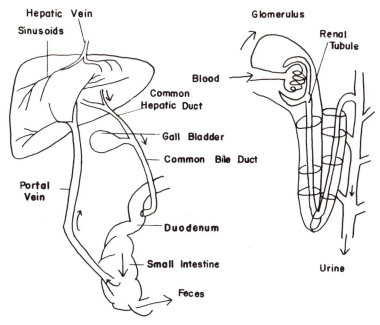

Fig. 7-6. Schematic diagrams of the biliary and urinary excretion mechanisms.
Mainly after Millard *et al.* (1961) and Albert (1968).

blood and (2) the ease with which these organs can dispose of the materials
via the respective excretory systems. There are large pores in the membranes
of the hepatic parenchymal cells, and molecules as large as inulin (molecular
weight 5000) pass rather easily into the liver system. The general consensus
is that this facility is designed to pass the bilirubin complex into the hepatic
system from the blood (Segre, 1972), for it has been shown that bilirubin is
bound to albumin in the plasma and the liver takes up the whole complex as
such. Lipid solubility is an important factor in the hepatic uptake of drugs.
Thus most insecticidal compounds are expected to be picked up by the liver
rather readily : numerous examples cited in Section 7.3 on liver accumulation
of lipophilic insecticides support this view. It is not without reason that DDT
and other extremely liposoluble insecticides concentrate in the liver at the
early stages of poisoning or whenever the system is overloaded. However,
when in equilibrium (such as the case with DDT residues in the general
public) the liver residues tend to settle at levels expected from its fat content.
If the uptake process does not prove to be a critical one, the process of bilary
excretion should. In brief, the major purpose of hepatic degradation activities
against xenobiotics is to convert apolar and lipophilic compounds into

polar, water-soluble, and therefore excretable material. Compounds of extreme polarity either are not absorbed by the hepatic system but rather excreted by the urinary system or are excreted through the hepatic system unchanged.

Williams (1967) has also shown that the process of biliary excretion is related to the size of the compound: the larger the compound the faster the rate of excretion. Indeed, compounds with a molecular weight of less than 150 have a hard time getting through the system. Williams is one of the investigators who considers this to be related to the normal function of biliary excretion; i.e., it is designed to pass bilirubin glucuronide, bile salts, and other conjugate complexes. Bilirubin cannot be excreted as such, but various compounds including insecticidal chemicals have been reported to be excreted through the biliary system without being conjugated (Preisig, 1972). The prerequisites for excretion thus appear to be polarity and molecular size.

The process of uptake by the kidney can be divided into two functional processes in the nephron itself: (1) the initial pickup through the glomerulus and (2) the reabsorption process through the renal tubules. Blood flows into the glomerulus, which has a quite porous membrane that does not pass particulate matter or most of the protein. Thus protein and particulate-bound pesticides are excluded here. The human kidneys are said to produce 185 liters of glomerular filtrate each day, but only 1.5 liters of urine is produced, the rest being reabsorbed through the nonporous membrane of the renal tubules (Albert, 1968). Passage of liposoluble material through either membrane appears to be simply in accordance with the law of diffusion in either direction, while for certain ions the process is carried out through active transport mechanisms. Thus it is possible for a liposoluble substance to be first picked up by the glomerulus and then reabsorbed by the renal tubules. Whatever the mechanism of selective passage, renal pickup and excretion are more effective for polar substances which cannot be picked up by the hepatic system. Thus the "renal barrier" is the more serious one for insecticidal chemicals, which are generally too apolar to be excreted by this route. Only polar metabolites are excreted by this system, although often very small amounts of the liposoluble original insecticides (e.g., Matthews *et al.*, 1971) are excreted; the phenomenon is indicative of the somewhat perfunctory nature of this absorption–reabsorption system, depending only on concentration gradients. That is, some degree of slipup is expected at peak congestion periods.

The biliary–urinary excretion ratio is generally determined by the nature of the compound and by the species. Williams (1967) concludes from a large number of examples that rats, hens, dogs, and cats generally excrete xenobiotics in the bile, while guinea pigs and rabbits tend to excrete in the urine. However, the ratio differences are expected to be more pronounced among the same species of animals fed different types of chemicals. For instance, Baron and Doherty (1967) found that only 6% of the radioactivity was found in the urine (and the remainder in the feces) in the rat fed ring-labeled Banol,® 6-chloro-3,4-dimethylphenyl-N-methylcarbamate. Matthews *et al.* (1971) studied the metabolism and excretion of ^{14}C-dieldrin in male and female rats and found that in both sexes approximately ten times as much radioactivity was found in the feces as in the urine. The difference in metabolic products found in this study is very intriguing. In the feces of male rats, 80–93% of the extractable material was in the form of 9-hydroxy-dieldrin (F-1) along with 1–2% dieldrin (except on the first day, when 7% for dieldrin was recorded). In the urine, more than 70% on the first day, 90% on the third day, and after the sixth day more than 97% of the extractable radioactivity was in the form of Klein's metabolite (U-1, sometimes referred to as pentachloroketone). The amount of dieldrin in the urine declined from 20% on the first 2 days to almost zero. Thus, for dieldrin, there is a clear division of metabolic pathways. Matthews *et al.* also found that Klein's metabolite along with dieldrin was distributed all around the body, particularly in the kidneys, lungs, heart, and liver, but there was no F-1 anywhere except in the gastrointestinal tract. This led them to conclude that Klein's metabolite is produced by the liver, as suggested by Matthews and Matsumura (1969), but is reabsorbed in the small intestine probably as a result of deconjugation of its conjugated precursor and thus distributed widely, since it is not readily picked up by the hepatic system and is only slowly excreted from the urinary system because of its low solubility.

7.6.2. Secretion in Milk

It has been well documented that chlorinated hydrocarbon insecticides are readily incorporated into milk of ruminants (Shepherd *et al.*, 1949; Ely *et al.*, 1957; Zweig *et al.*, 1961) and humans (Tanabe, 1972; Quinby *et al.*, 1965; see also Chapter 11). The levels of residues in milk appear to be species specific inasmuch as human milk appears to contain at least twice as much insecticide residues as that of cows (e.g., Eagan *et al.*, 1965; Quinby *et al.*, 1965). The general relationship between the levels of DDT in the feed and the appearance of DDT residues in cow's milk has been summarized by Witt *et al.* (1966b). According to the general data (Fig. 7-7), the rule of thumb for the relationship is approximately 1:1. At lower levels of DDT, the relationship is slightly in favor of the milk fat residue: 0.5 ppm of DDT residues in the feed could cause, perhaps, 1 ppm of contamination in the milk fat.

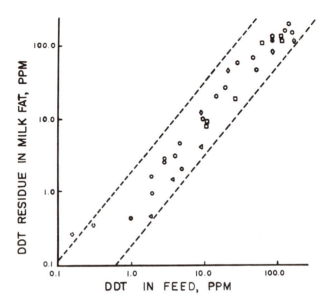

Fig. 7-7. DDT residues in milk fat vs. DDT residues in feed.
From Witt *et al.* (1966*b*).

To study the relative degrees of residue accumulation in milk among chlorinated hydrocarbon insecticides, Gannon *et al.* (1959*a*) fed Holstein cows a daily ration of insecticide-containing feed for 16 weeks at varying doses. The rates of accumulation in milk for the insecticides were aldrin (excreted as dieldrin) > dieldrin > heptachlor (excreted as epoxide) ≫ methoxychlor. How do the levels of insecticide residues in milk compare to those in other tissues? Gannon *et al.* found that the residues in milk were on the whole about one-tenth those in the fat of the cows: the fat-to-milk ratios were 9.23, 11.29, 16.11, 9.27, and 9.85 for aldrin, dieldrin, DDT, heptachlor, and methoxychlor, respectively (calculated on a ppm basis, e.g., 10 ppm in fat and 1 ppm in whole milk gives the ratio of 10). In another experiment, Gannon *et al.* studied dieldrin excretion in milk in detail and found that the fat-to-milk ratios varied from 12 to 18 for cows fed 0.1–2.25 ppm of dieldrin in the diet for 12 weeks. Since this particular milk contained 4% fat, the body fats–milk fat ratio would become 0.48–0.72, indicating that milk fat tends to concentrate somewhat higher levels of dieldrin than the body fats in general (approximately twice as much for the above insecticides). On the other hand, they observed that the residue level of dieldrin in milk declined about twice as fast as that in the body fats when the animals were taken off spiked feeds. There are not enough data to show how other groups of insecticides behave. However, it is expected that more polar and more degradable chemicals are not likely to accumulate in milk. The data for Guthion® (Everett *et al.*, 1966), for instance, show that the feed-to-milk ratio is of the order of 200:1 (or on a milk fat basis 10:1) instead of 1:1 for DDT and dieldrin.

Wilson and Cook (1972) compared the proportion of dieldrin excreted in milk to that in feces and urine. Cows were orally given 0.1 mg/kg of dieldrin for 6 weeks, and urine, feces, and milk were collected. At the end of the experiment, the animals were sacrificed. Altogether, 42.77 % of administered dieldrin was recovered: 31.6 % in feces, 8.0 % (at the end of 2 weeks) in fat on the day of biopsy, 3.1 % in milk, 0.07 % in body fluid, and 0 % in urine. The proportion of dieldrin in milk was rather low compared to that found in feces. The rest of the unaccounted for portion (57 %) must have been excreted as metabolic products in feces and urine, and so the actual proportion of excretion in milk would be even smaller. Witt *et al.* (1966b) studied the effect of the route of administration on the rate of appearance of DDT in the cow's milk (Table 7-5). It is apparent from the data that intravenous

TABLE 7-5. Differences in the Rate of DDT Residue Appearance in Cow's Milk by the Mode of DDT Administration

Method of application	Consecutive days dosed	Mean maximum response[a] (ppm in milk fat)			
		DDT	DDE	DDD	Total DDT–R
Intratracheal	1	0.68	0.06	0.0	0.74
	6	1.54	0.01	0.08	1.64
Rumen, capsular[b]	1	0.59	0.0	0.28	0.87
	6	0.49	0.08	1.18	1.75
Rumen, aged residue[c]	1	0.38	0.90	0.56	1.84
Intravenous	1	3.00	0.12	0.12	3.24
	6	7.60	0.25	0.68	8.53

From Witt *et al.* (1966b).

[a] Average of three cows; net gain over baseline.
[b] An oil solution in capsules. Expected to be washed down to lower alimentary tracts quickly.
[c] Actually, DDT in acetone was put in the feed prior to feeding.

injections cause the highest degree of total DDT-R in milk. This is understandable, since these residues are transported to the mammary gland via the bloodstream. It is also evident from the data that major metabolic activities are found in the rumen, since the two methods which allowed DDT to initially enter the rumen via the oral route gave the highest proportion of metabolites.

It is important to stress here that metabolic products are also produced in milk. While there are not sufficient data to show how well relatively polar metabolites can be excreted through this route, several carbamate metabolites that have been found in urine have also been detected in milk. For example, Robbins et al. (1970) found 4-benzothienyl sulfate-1-oxide in the milk of a cow and goats as the result of oral capsule administration of Mobam®. This metabolite has also been found to be one of the major urinary metabolites in goats and cows. However, unlike in urine, this metabolite was the only significant residue found in milk, indicating some selectivity of the milk excretion system.

7.7. REFERENCES

Ackermann, H., and R. Engst (1970). *Arch. Toxicol.* **26**:17.

Albert, A. (1968). *Selective Toxicity*, 4th ed. Butler & Tanner, London, p. 79.

Bäckström, J., E. Hansson, and S. Uliberg (1965). *Toxicol. Appl. Pharmacol.* **7**:90.

Baron, R. L., and J. D. Doherty (1967). *J. Agr. Food Chem.* **15**:830.

Brown, J. R. (1970). *Toxicol. Appl. Pharmacol.* **17**:504.

Brown, V. K., A. Richardson, J. Robinson, and D. E. Stevenson (1965). *Food Cosmet. Toxicol.* **3**:675.

Bruce, W. N., R. P. Link, and G. C. Decker (1965). *J. Agr. Food Chem.* **13**:63.

Burlington, H., and V. F. Lindeman (1950). *Proc. Soc. Exptl. Biol. Med.* **74**:48.

Cook, R. M. (1970). *J. Agr. Food Chem.* **18**:434.

Dale, W. E., T. B. Gaines, and W. J. Hayes (1962). *Toxicol. Appl. Pharmacol.* **4**:89.

Davison, K. L. (1970). *J. Agr. Food Chem.* **18**:1156.

Dedek, W., and R. Schmidt (1972). *Pharmazie* **27**:294.

Deichmann, W. B., W. E. MacDonald, A. G. Beasley, and D. Cubit (1971). *Ind. Med.* **40**:10.

Donaldson, W. E., T. J. Sheets, and M. D. Jackson (1968). *Poultry Sci.* **47**:237.

Eagan, H., R. Goulding, J. Roburn, and J. O'G. Tatton (1965). *Brit. Med. J.* **11**:66.

Ecobichon, D. J., and P. W. Saschenbrecker (1969). *Toxicol. Appl. Pharmacol.* **15**:420.

Ely, R. E., L. A. Moore, R. H. Carter, and B. A. App (1957). *J. Econ. Entomol.* **50**:348.

Everett, L. J., C. A. Anderson, and D. MacDougall (1966). *J. Agr. Food Chem.* **14**:47.

Finnegan, J. K., H. B. Haag, and P. S. Larson (1949). *Proc. Soc. Exptl. Biol. Med.* **72**:357.

Fischer, R., and C. Plunger (1965). *Mitt. Arch. Toxicol.* **21**:101.

Fish, S. A. (1966). *Am. J. Obstet. Gynecol.* **96**:1148.

Gannon, N., R. P. Link, and G. C. Decker (1959a). *J. Agr. Food Chem.* **7**:829.

Gannon, N., R. P. Link, and G. C. Decker (1959b). *J. Agr. Food Chem.* **7**:824.

Gellert, R. J., W. L. Heinrichs, and R. S. Swerdloff (1972). *Endocrinology* **91**:1095.

Hathway, D. E., J. A. Moss, J. A. Rose, and D. J. M. Williams (1967). *Europ. J. Pharmacol.* **1**:167.

Hayes, W. J., Jr. (1965). *Ann. Rev. Pharmacol.* **5**:27.

Heath, D. F. (1962). In *Radioisotopes and Radiation in Entomology*. International Atomic Energy Agency, Vienna.

Heath, D. F., and M. Vandekar (1964). *Brit. J. Ind. Med.* **21**:269.

Hoyle, G. (1953). *J. Exptl. Biol.* **30**:121.

Huber, J. J. (1965). *Toxicol. Appl. Pharmacol.* **7**:516.

Hunter, C. G., and J. Robinson (1967). *Arch. Environ. Health* **15**:620.

Hunter, C. G., A. Rosen, R. T. Williams, J. G. Reynolds, and A. N. Worden (1960). *Meded. Landbouwhogesch. Wageningen* **25**:1296.

Hunter, C. G., J. Robinson, and M. Roberts (1969). *Arch. Environ. Health* **18**:12.

Jones, H. B. (1950). In *Medical Physics*, Vol. 2. O. Glaser, ed. Yearbook Publishers, Chicago, p. 855.

Kaul, R., W. Klein, and F. Korte (1970). *Tetrahedron* **26**:99.

Keane, W. T., and M. R. Zavon (1969). *Bull. Environ. Contam. Toxicol.* **4**:1.

Klevay, L. (1971). *Proc. Soc. Exptl. Biol. Med.* **136**:878.

Korte, F. (1970). *Biochem. J.* **118**:45p.

Ludwig, G., J. Weis, and F. Korte (1964). *Life Sci.* **3**:123.

Mandzhgalodze, R. N., and V. I. Vashakidze (1972). *Soobshch. Acad. Nauk Gruz. SSR* **65**(2): 485. (Indirectly cited from *Health Aspects of Pesticides*, pp. 73–1196.)

Matsumura, F., and M. Hayashi (1969). *Residue Rev.* **25**:265. F. A. Gunther, ed. Springer-Verlag, New York.

Matsumura, F., and C. M. Wang (1968). *Bull. Environ. Contam. Toxicol.* **3**:203.

Matthews, H. B., and F. Matsumura (1969). *J. Agr. Food Chem.* **17**:845.

Matthews, H. B., J. D. McKinney, and G. W. Lucier (1971). *J. Agr. Food Chem.* **19**:1244.

McCully, K. A., D. C. Villeneuve, W. P. McKinley, W. E. Phillips, and M. Hidiroglou (1966). *J. Assoc. Offic. Anal. Chemists* **49**:966.

Millard, N. D., B. G. King, and M. J. Showers (1961). *Human Anatomy and Physiology*, 4th ed. W. B. Saunders, Philadelphia.

Morgan, D. P., C. C. Roan, and E. H. Paschal (1972). *Bull. Environ. Contam. Toxicol.* **8**:321.

Morsdorf, K., G. Ludwig, J. Vogel, and F. Korte (1963). *Med. Exptl.* **8**:90.

Moss, J. A., and D. E. Hathway (1964). *Biochem. J.* **91**:384.

Norback, D. H., J. H. Engblom, and J. R. Allen (1973). Chlorinated dibenzo-*p*-dioxin distribution within rat tissues and subfractions of the liver. In *Environmental Health Perspectives*, *Issue* **5**:233, 1973. National Institute of Environmental Health Science, Research Triangle Park, N.C.

Norris, W. P., S. A. Tyler, and A. M. Brues (1958). *Science* **128**:456.

O'Brien, R. D., and R. W. Fisher (1958). *J. Econ. Entomol.* **51**:169.

Preisig, R. (1972). Evaluation of the action of foreign compounds on binary excretions. In *Liver and Drugs*. F. Orlandi, and A. M. Jezequel, eds. Academic Press, New York, p. 107.

Quinby, G. E., J. F. Armstrong, and W. F. Durham (1965). *Nature* **207**:726.

Richardson, L. A., J. R. Lane, W. S. Gardner, J. T. Peeler, and J. E. Campbell (1967). *Bull. Environ. Contam. Toxicol.* **2**:207.

Robbins, J. D., J. E. Bakke, and V. J. Feil (1970). *J. Agr. Food Chem.* **18**:130.

Robinson, J. (1967). *Nature* **215**:33.

Robinson, J. (1969). *Can. Med. Assoc. J.* **100**:180.

Robinson, J. (1970). *Ann. Rev. Pharmacol.* **10**:353.

Robinson, J., M. Robert, M. Baldwin, and A. I. T. Walker (1969). *Food Cosmet. Toxicol.* **7**:317.

Rose, J. A. (1971). The effect of fat mobilization on the release of dieldrin from adipose tissue. M.S. thesis, University of Strathclyde, Glasgow.

Segre, G. (1972). Kinetics of drugs in the hepatobiliary system. In *Liver and Drugs*. F. Orlandi, and A. M. Jezequel, eds. Academic Press, New York, p. 85.

Shepherd, J. B., L. A. Moore, R. H. Carter, and F. W. Poos (1949). *J. Dairy Sci.* **32**:549.

Smith, M. T., J. A. Thomas, C. G. Smith, M. G. Mawhinney, and J. J. McPhillips (1972*a*). *Toxicol. Appl. Pharmacol.* **22**:327.

Smith, M. T., J. A. Thomas, C. G. Smith, M. G. Mawhinney, and J. W. Lloyd (1972*b*). *Toxicol. Appl. Pharmacol.* **23**:159.

Street, J. C. (1964). *Science* **146**:1580.

Street, J. C. (1968). In *Enzymatic Oxidation of Toxicants.* E. Hodgson, ed. Proceeding of a Conference held at North Carolina State University at Raleigh, N.C., p. 197.

Street, J. C., and A. D. Blau (1966). *Toxicol. Appl. Pharmacol.* **8**:497.

Street, J. C., M. Wang, and A. D. Blau (1966*a*). *Bull. Environ. Contam. Toxicol.* **1**:6.

Street, J. C., R. W. Chadwick, M. Wang, and R. L. Phillips (1966*b*). *J. Agr. Food Chem.* **14**:545.

Tanabe, H. (1972). In *Environmental Toxicology of Pesticides.* F. Matsumura, G. M. Boush, and T. Misato, eds. Academic Press, New York, p. 239.

Telford, J. N., and F. Matsumura (1971). *J. Econ. Entomol.* **64**:230.

Villeneuve, D. C., R. F. Willes, J. B. Lacroix, and W. E. J. Phillips (1972). *Toxicol. Appl. Pharmacol.* **21**:542.

Wagstaff, D. J., and J. C. Street (1971). *Bull. Environ. Contam. Toxicol.* **6**:273.

Walker, A. J., D. E. Stevenson, J. Robinson, E. Thorpe, and M. Roberts (1969). *Toxicol. Appl. Pharmacol.* **15**:345.

Williams, R. T. (1967). In *Drug Responses in Man.* G. Wolstennolme, and R. Porter, eds. Little, Brown, Boston, p. 71.

Wilson, K. A., and R. M. Cook (1972). *J. Agr. Food Chem.* **20**:391.

Witt, J. M., W. H. Brown, G. I. Shaw, L. S. Maynard, L. M. Sullivan, F. M. Whiting, and J. W. Stull (1966*a*). *Bull. Environ. Contam. Toxicol.* **1**:187.

Witt, J. M., F. M. Whiting, and W. H. Brown (1966*b*). In *Organic Pesticides in the Environment.* Advances in Chemistry Series 60, American Chemical Society, Washington, D.C., p. 99.

Woolley, D. E., and A. L. Runnells (1967). *Toxicol. Appl. Pharmacol.* **11**:389.

Yoshioka, Y., A. Hara, H. Nawa, H. Yoshioka, and I. Iwasaki (1972). *Igaku No Ayumi (Progr. Med. Sci.)* **80**(13):811.

Zatz, J. L. (1972). *J. Pharm. Sci.* **61**:948.

Zweig, G., L. M. Smith, S. A. People, and R. Cox (1961). *J. Agr. Food Chem.* **9**:481.

Chapter 8

Movement of Insecticides in the Environment

8.1. INTRODUCTION

The major source of environmental contamination by pesticides is the deposits resulting from application of these chemicals to control agricultural pests and pests causing public health problems. Other sources of contamination could be pesticides used in urban areas and industrial wastes. Misplacement and leakage during the transportation, distribution, or storage processes related to the manufacture of pesticides could result in environmental contamination. In any case, the pesticides enter the environment. They are absorbed by various constituents of the environment and transported to other places mainly by water and air movements. They are picked up by various biological systems and are, at the same time, chemically or biochemically transformed to other nontoxic or toxic compounds in the environment.

What concerns us most is the biological effects of these pesticide derivatives which, at various stages of environmental alteration, can come in contact with many forms of biological systems. Two aspects of the problem appear to be particularly important: (1) effects on man and domestic animals, which may take up pesticides mostly through contaminated foods and feeds, and (2) effects on wildlife, where certain species can be severely affected by absorption or accumulation of pesticides through the food web so that the balance of the ecosystem in nature is directly or indirectly disturbed.

8.2. RESIDUES OF INSECTICIDES

8.2.1. General

There is no question about the presence of insecticide residues in the environment and in food commodities. Thus the FDA determines the levels of residues which can be tolerated in marketed food, called "tolerance levels." Some examples are shown in Table 8-1.

According to Gunther and Blinn (1956), the term *residue* should not be regarded as synonymous with *deposit*. The use of *deposit* should be limited to the insecticidal chemical as initially laid down on the solid surface by the treatment; *residue* should refer to the chemical, regardless of locality, found on or in a substance, with the implication of aging by time lapse or alteration or both.

Residues of insecticides may sometimes decompose or disappear from the deposit site at a constant rate. Factors influencing these reactions are, for instance, enzymatic degradation and evaporation and washing out (weathering). If the residues are measured in terms of concentration (e.g., parts per million, or ppm), other factors such as growth of the subject plants may contribute to the disappearance rates.

Like many other chemical reactions, the initial rate of insecticide disappearance follows the law of "first-order kinetics": the rate of disappearance is related to the amount deposited. Thus a plot of the log amount of insecticide left at the initial deposit site against time (usually days) should be a straight line.

In nature, however, quite often disappearance takes place in two steps: (1) an initial phase of fast disappearance and (2) a subsequent phase of slow decrease in the amounts of residues (Fig. 8-1). The initial fast phase is often called "dissipation" and the second phase "persistence." The reasons for such two-step reactions could be multiple. One of the major causes appears to be that insecticides can be absorbed or translocated to sites where they escape the initial vigorous weathering effects. Since the translocated insecticides are stored, reacted, or degraded at various rates (depending on the type of reactions), first-order kinetics is no longer followed. Although a semilog graph of the phase may also be a straight line as before, this should be regarded as purely coincidental, since it is the end result of many complex reactions.

Nevertheless, the use of such measurements as "half-life" values or 90–100% disappearance values to express the approximate rates of insecticide residue disappearance has certain advantages, for these values are largely determined by the initial fast phase of dissipation through which the major portion of the initial deposit is often eliminated. Also, such values can serve as a general quick reference to the rate of insecticide disappearance.

TABLE 8-1. Extracts from "Official F.D.A. Tolerance 1969" (National Agricultural Chemicals Association, 1970)

	Insecticide	Tolerance (ppm)	Remarks
Apples, pears, and quinces	Aldrin (dieldrin)	0	Total residue result of aldrin application
	BHC	5	
	DDT	7	
	Malathion	8	
	Parathion (and/or methyl-parathion)	1	
Corn (fresh)	BHC	5	
	DDT	3.5	Determined on kernels plus cobs only
	Methoxychlor	14	
	Chlordane	0.3	
Corn (grain)	Aldrin (dieldrin)	0	
	Dieldrin	0	
	Hydrogen cyanide	100	Post-harvest fumigation
	Malathion	8	Post-harvest application
Wheat	Allethrin	2	Post-harvest
	Carbaryl	0	
	Heptachlor	0	
	Methoxychlor	2	Storage treatment
	Parathion	1	
	Methyl bromide	50	Inorganic bromide calculated as Br
	Toxaphene	5	
Meat (fat)	Lindane	7	Cattle, goats, horses, and sheep
	DDT	7	
	Methoxychlor	3	
	Toxaphene	7	
Meat	Malathion	4	
Meat and milk	Heptachlor	0	
Milk	DDT (DDD, DDE)	0.05[a]	
	Dioxathion	0	
	Malathion	0	(0.5 ppm in milk fat reflecting negligible residues in milk)
	Methoxychlor	0	

[a]These tolerances are not established to provide for residues from the purposeful use of DDT, DDD, or DDE on dairy cattle, in dairy barns, or on crops intended to be used for feeding dairy cattle.

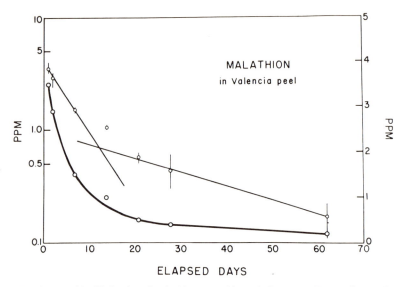

Fig. 8-1. Residual behavior of malathion on and in peel of oranges. (From Blinn *et al.*, 1959). ⚬—⚬ logarithmic coordinate; ⚬—⚬ linear coordinate.

8.2.2. Terrestrial Plants Including Agricultural Crops

Generally speaking, terrestrial plants do not bioaccumulate very apolar insecticides, such as chlorinated hydrocarbons, beyond the levels found in soil. The physicochemical characteristics both of the compounds and of the plants play important roles in determining the rate of pickup by the plants. For instance, there are a number of systemic insecticides which can easily be picked up by plants. These are generally compounds with high polarity (relatively speaking) and water solubility. To cause residue problems, there must be an unfortunate combination of persistence, translocation to edible (or usable) parts of the plant, storage, and accumulation. Since most of the systemic insecticides marketed today are degradable, and since these polar chemicals do not cause eventual bioaccumulation in other ecosystems, only a few key factors which are pertinent to the context of this chapter will be considered here.

Despite their low water solubilities, some of the chlorinated hydrocarbon insecticides do migrate into plant tissues. For instance, BHC and, to a lesser extent, aldrin, are picked up by plants. The transport mechanism of such apolar compounds is not really known, although it is clear that lindane is translocated through the root system. Lichtenstein and Schulz (1960*b*) compared the rate of appearance of lindane in the aerial parts of pea plants (after soil application of lindane) in uncovered pots and in pots covered with aluminum foil so that none of the soil surface was exposed except for $\frac{1}{16} - \frac{1}{18}$-inch holes through which plants were allowed to grow. They found that the pea vines from the covered pots contained 10% more lindane than the controls, but the soil of the covered pots also contained 10% more lindane than that of the uncovered ones. Under identical experimental conditions, only one-hundredth as much aldrin (as measured as aldrin plus dieldrin in the plant) and a trace of heptachlor were picked up by the plant. DDT did not translocate into the aerial parts at all. These studies indicate the importance of the chemical nature of the insecticide. Another factor influencing the rate of pickup is the nature of the soil. It is clear that soils with minimum sorptivity (i.e., clean sand) would cause the highest plant pickup of a given insecticide; in fact, these studies were conducted with sand.

Penetration of pesticides into plant leaves has been ably reviewed by Hull (1970). In general, semilipophilic compounds tend to penetrate into leaves because of the nature of the outer cuticle in most plants; compounds whose polarity is increased on entry into plant tissue usually penetrate best. This phenomenon clearly indicates the presence of a lipophilic barrier at the outer surface and a hydrophilic transport mechanism inside. Thus the polar–apolar characteristics of the compound determine the fate of the pesticide in terms of plant-mediated translocation. High temperature and atmospheric humidity favor penetration.

A survey of the literature indicates that high accumulation of chlorinated hydrocarbon insecticides is found whenever the soil samples are nonsorptive (e.g., Lichtenstein, 1959; Lichtenstein *et al.*, 1967a) or the plants studied are root crops (e.g., Lichtenstein *et al.*, 1967a; Bruce *et al.*, 1966). Apparently, roots tend to accumulate chlorinated hydrocarbon insecticides mostly in their peels. For instance, Lichtenstein (1965) examined five varieties of carrots for their absorption rates of aldrin and heptachlor in the field and found that one white variety, White Belgium, contained 50% of the residues in the pulp, while all others contained 70–86% of the residues in the peels. Other portions of plant tissues that generally tend to accumulate insecticide derivatives are the oily parts. Soybeans, for instance, accumulate five times as much heptachlor as oat seed under similar conditions (Bruce *et al.*, 1966), and pesticides tend to accumulate in the oily peels of citrus fruits, as shown in Fig. 8-1 (Blinn *et al.* 1959).

In reality, however, other factors such as method and frequency of pesticide application, processing, and size are also important in deciding the levels of pesticide residues in marketed agricultural commodities (Table 8-2). Thus empirical approaches to monitoring pesticide residues are still the only way to insure the safety of these plant materials for food consumption.

TABLE 8-2. DDT Residues Found in Agricultural
Plant Commodities in the United States

Food commodities	Duggan (1968) (ppm)	Corneliussen (1970) (ppm)
Peas and beans	0.01	0.0204
Leaf vegetables	0.025	0.0214
Fruit vegetables	0.048	0.0398
Root vegetables	0.008	0.0002
Fruits	0.012	0.0094
Grains and cereals	0.005	0.005

Surface or "effective" residues (the portion of the insecticide left from the initial deposit) can be eliminated, at least in part, by washing, brushing, scraping, or any other means to remove the surface layer of the plant surface. For instance, Westlake and Butler (1953) note that malathion residues can be greatly reduced by simply washing fruits and vegetables. Thompson and Van Middelem (1955) cite a number of examples where 10–90 % of the residues of various insecticides can be removed by washing. Interestingly enough, this treatment removed not only water-soluble insecticides but also lipophilic ones; for instance, when mustard greens were treated with nicotine (1 hr), parathion (24 hr), or toxaphene (24 hr), washing removed approximately 90 % of the deposit or residue from all samples. At least in a few cases, the insecticides remaining on the surface are known to be degraded or activated by the action of ultraviolet rays or other external factors. For instance, aldrin and dieldrin can be converted on the plant surfaces to corresponding bridged forms which are at least equally toxic as their parent compounds (Rosen *et al.*, 1966).

Because a number of persistent insecticides are lipophilic and tend to accumulate in plant oils and waxes, they often stay within the skin, peel, or outer surface of the plants after penetration. As already mentioned, a major portion of chlorinated hydrocarbon insecticides can be recovered in the skin of various carrot varieties (Lichtenstein *et al.*, 1965). Possibly, removal of skins or peels could eliminate a large part of these insecticides.

Insecticides which actually enter plant tissues can be degraded, activated (to become more toxic), or conjugated. Such metabolic activities can even be observed in stored products such as grains (Rowlands, 1971). It is not likely, however, that the insecticides stored directly in the wax or oil of plant tissues are further metabolized (activation and degradation of insecticides by plant tissues were described in Chapter 5). Table 8-3 gives half-life values of various insecticides on and in major crops (Gunther and Jeppson, 1960).

TABLE 8-3. Half-Life Values for Persistent Insecticides

Insecticides	Half-life in or on crop[a] (days)
DDT	Alfalfa (5–7), citrus fruit (38–50), clover (10–14), lettuce (2–3), peach fruit (8–11), peach leaves (14–18)
Dieldrin	Apple leaves (3–6), clover (5–6), citrus fruits (44–60), peach leaves (6–7)
EPN	Apple leaves (3–5), peach fruit (8–10)
Malathion	Apple fruit (2–3), lettuce (3–4), onions (1–2), citrus fruit (17–32), peach fruit (4–6), stored wheat (150–190)
Parathion	Alfalfa (2), apple fruit (3–6), apple leaves (1–3), citrus fruit (61–78), lettuce (1–3), peach fruit (3–7), pear fruit (2), tomatoes (3–7)
Pyrethrins	Stored wheat (28)

From Gunther and Jeppson (1960).

[a]The half-life of a residue is the time required for half of the residue to react or otherwise dissipate.

8.2.3. Soils

The pesticides used to control agricultural pests (e.g., insects, weeds, and fungi) are normally directed to plants, although the major portion is deposited on the surface of the soil. A number of external factors can play important roles in deciding the fate of these deposited insecticides : (1) absorption by the soil particles as well as by the organic matter in the soil, (2) leaching and washing off by water, (3) evaporation into the air, including mechanical transport by water vapor, (4) degradation and/or activation by soil micro-organisms, (5) physiological decomposition or activation catalyzed by soil conditions or soil constituents, (6) photodecomposition, and (7) translocation through biological systems (including the plants) to other environments. Among all these processes, only those belonging to categories (4), (5), (6), and to some extent (7) can actually reduce the total amount of pesticides in the environment. Other processes merely serve for translocating the pesticides.

The single most important factor deciding overall persistence is the nature of the insecticidal compound itself. Characteristics such as water solubility and polarity, which in turn influence the lipophilic nature of the compound, as well as volatility and chemical reactivity–stability are especially important. The overall persistence figures for each insecticidal compound have been summarized by Edwards (1966) and Kearney *et al.* (1969) and are listed in Tables 8-4 and 8-5. Thus through volatilization aldrin disappears much faster than dieldrin. Arsenicals tend to accumulate in the top few

TABLE 8-4. General Persistence of Chlorinated Hydrocarbon Insecticides in Soils

	95% disappearance[a] (years)	75–100% disappearance[b] (years)
Aldrin	1–6 (3)	3
Chlordane	3–5 (4)	5
DDT	4–30 (10)	4
Dieldrin	5–25 (8)	3
Heptachlor	3–5 ($3\frac{1}{2}$)	2
Lindane	3–10 ($6\frac{1}{2}$)	3
Telodrin®	2–4 (4)	—

[a]From Edwards (1966). Averages given in parentheses.
[b]From Kearney et al. (1969).

TABLE 8-5. General Persistence of Pesticides in Soils Expressed as Time Required for Disappearance of 75–100% of the Original Amount

Pesticide	Time
Organophosphates	
Diazinon	12 weeks
Malathion and parathion	1 week
Herbicides	
Propazine and picloraur	18 months
Simazin	12 months
Atrazine and monuron	10 months
2,4,5-T	5 months
MCPA	3 months
2,4-D	1 month
Dalapon and ClPC	8 weeks
1-PC, EPTC	4 weeks

From Kearney et al. (1969).

inches of soil. Accumulation of residues is particularly high in soils beneath orchard trees and in cotton fields. In the Pacific Northwest of the United States, as much as 1400 lb of arsenic trioxide per acre was found. The deciding factor here is the poor solubility and low volatility of the compound. A number of chlorinated hydrocarbon insecticides such as DDT, BHC, chlordane, dieldrin, and heptachlor are generally stable and remain in the soil for a long time.

Aside from their chemical characteristics, there are several important factors that influence the fate of insecticides in soil. They are soil type, moisture, temperature, mode of cultivation, cover crop practices, and application and formulation of the insecticides. The influence of water, air, sunlight, and microbial action on the fate of pesticide residues in the soil will be discussed in Chapter 9. Discussion here will be limited to absorption, leaching, and overall geographical influences.

That various insecticides and some herbicides can be strongly bound to soil particles has become increasingly evident. This can be seen in the phenomenon that many pesticides persist only in the top layer of soils and do not penetrate vertically into the lower layers (Edwards, 1966; Lichtenstein *et al.,* 1967*b*). The materials to which various pesticides attach have not been clearly defined. Two types of materials can be assumed to be present: inorganic and organic soil components. Also, it should be possible to differentiate between specific and nonspecific types of absorption. Lichtenstein and Schulz (1960*a*) showed that the amounts of aldrin and dieldrin recovered after treatment with aldrin were directly related to the amount of organic matter found in the soil (Table 8-6). Clearly, the organic matter can absorb

TABLE 8-6. Residues of Aldrin and Dieldrin Found in a Quartz Sand and Three Soil Types 56 Days After Treatment (10 ppm) with Aldrin

Soil types	Organic matter (%)	Residue (as % of initial aldrin recovery)		
		Total	Aldrin	Dieldrin
Quartz sand	0.0	1.2	0.7 ± 0.3	0.5 ± 0.1
Plainfield sand	0.8	61.4	57.9 ± 0.7	3.5 ± 0.0
Carrington loam	4.1	70.2	55.0 ± 0.8	15.2 ± 0.5
Muck	40.0	75.6	65.1 ± 1.4	10.5 ± 0.2

From Lichtenstein and Schulz (1960*a*).

aldrin (and dieldrin) and thereby decrease the chance of disappearance of aldrin through the action of various factors.

Carter and Stringer (1970) studied the factors influencing the initial rate of penetration into soil and found that both soil type and the moisture content play very important roles in determining the depth of penetration of heptachlor, aldrin, γ-chlordane, and dieldrin. They also noted that clay content (vs. sand) is important in that the insecticides did not migrate in soils with high clay and silt composition. Studies of the effects of soil constituents other than organic matter on the persistence and fate of insecticidal compounds are scarce. Generally speaking, insecticides persist longer in acidic soil than in alkaline soil. This is particularly true with organophosphorous compounds but is also applicable to chlorinated hydrocarbon insecticides to some extent. The mineral content of soil plays some role in determining the fate of insecticides by either catalyzing decomposition (Fowkes et al., 1960) or influencing sorption behavior (Gallaher and Evans, 1961).

It is clear that external conditions such as temperature and cultivation greatly affect the persistence of insecticides (Lichtenstein, 1965). Insecticides disappear much faster at high temperatures. Cover crops tend to increase the persistence, probably by lowering the rate of evaporation from the soil. Cultivation of soils may have several effects on the fate of insecticides, by allowing more uniform exposure to various soil constituents and air and by burying a portion much deeper than the level expected by surface application. Lichtenstein and Schulz (1961) studied the effect of daily discing to a depth of 5 inches of a loam soil which had been treated with DDT and aldrin and found that the practice caused a 38% reduction of aldrin residues and a 25% reduction of DDT residues. On the other hand, insecticide residues on the surface of the soil are likely to disappear much faster than the ones thoroughly buried in soil by one initial cultivating (Lichtenstein et al., 1962).

Air movements over the insecticide-containing soil are also important. Spencer and Cliath (1969) demonstrated that the rate of dieldrin disappearance from soil is a function of air movement through soils, in accordance with the observation by Lichtenstein's group that repeated cultivation (aeration) helps evaporation of the pesticide.

As for the effects of formulation, both Lichtenstein (1965) and Edwards (1970) agree that insecticides in granular form persist longer than the ones in emulsion form. The latter formulation, however, persists longer than miscible liquids and wettable powders in terms of longevity of residues in soil.

Soils perhaps serve as the largest environmental reservoir of pesticides, and, similarly, pesticides in water would be present in the form of soil-bound particles at the bottom.

The preliminary report by Ware (1971) is interesting. It notes that in Arizona, where the use of DDT was banned in 1969, the levels of DDT in crops and milk declined within 2 years, while the average residues in soils remained constant at about 2 ppm. In this case, DDT was apparently bound to soil components not available to plants. The factors controlling the rate of release of DDT to biological systems are very important.

8.3. MOVEMENT OF RESIDUES IN THE ENVIRONMENT

8.3.1. Water- and Air-Mediated Transport

In the environment, water can provide the major means of transporting various substances from one source to another. The bulk of pesticide residues are generally confined to the upper 1–2 inches of soils, and although vertical transport of pesticides through soil by water is limited (Lichtenstein *et al.* 1967*b*), water can wash away the soil particles that contain pesticide residues (Eye, 1966; Lichtenstein *et al.*, 1967*b*). For instance, Walker (1967) found large amounts of chlorinated hydrocarbons bound in soil particles. Lichtenstein *et al.* (1967*b*) found, however, that small amounts of various insecticides can be truly eluted through percolation (Table 8-7) with water through soils.

TABLE 8-7. **Insecticide Residues in Water After Percolation Through Columns of Carrington Silt Loam Soil**[a]

	Water solubility (ppm)	Original compound	Major metabolites (ppm)
Parathion	24	1	0.05 (paraoxon)
Guthion®	34	1	0.1 (oxygen analogue)
Di-Syston®	25	0.5	1.0 (sulfoxide), 0.2 (sulfone)
Carbaryl	40	1	0.1 (1-naphthol)
Lindane	6.6	1	—
Aldrin	0.01	nil[b]	—
DDT	0.04	nil[b]	—

From Lichtenstein *et al.* (1967*b*).

[a]Insecticide soil treatment: 50 ppm.
[b]Detection limit 1 μg.

Only insecticides with water solubility above 5 ppm were eluted under the experimental conditions. Some insecticides can stay in water as a suspension (higher concentration than their true solubility suggests), although the presence of soil particles in water accelerates the process of precipitation. Apparently, compounds with relatively high water solubilities stay suspended longer in water than ones with low solubilities. Bailey and Hannum (1967) found an inverse relationship between the sediment particle size of a lake bottom and pesticide concentration. Apparently, the bulk of hydrophobic insecticides are bound to the mud particles and eventually precipitate with them. The abundance of such particles around lake regions with "soft" bottoms could easily explain such a phenomenon.

Most of the agricultural and urban areas in the United States lie in plains remote from mountains. The rivers serving these areas are generally slow flowing and cover great areas. The Mississippi River system, for instance, drains at least one-third of the land mass of the United States. Riseborough *et al.* (1968) calculate that roughly 10,000 kg/year of pesticides are transported by this system to the Gulf of Mexico. This figure alone might not seem impressive, but consideration of the additional amount of pesticides adhering tightly to soil particles that are leached out into the river reveals that the total pesticides present in the system may well be much higher.

A characteristic of U.S. waterways is the abundance of lakes, particularly in the northern portions. In general, water in lakes exchanges very slowly. The Great Lakes in particular are susceptible to pollution because of their extremely slow rate of discharge into the Atlantic Ocean and because of the presence of many urban (and some agricultural) areas around them. (The rate of total water exchange in Lake Michigan is said to be 30.8 years, and it takes only 5 tons of DDT to give 1 ppt of DDT pollution; Rainey, 1967.) The level of DDT in Lake Michigan is on the order of 1 ng/liter (1 ppt), and it seriously affects the hatching of coho salmon eggs. Larger coho salmon from Lake Michigan are relatively resistant to the effect of DDT and contain 3–4 ppm (in 2- to 4-lb fish) to 12–13 ppm (in 11- to 12-lb fish) (Reinert, 1969).

The levels of pesticides in water systems do change gradually from year to year. Lichtenberg *et al.* (1970) summarized the results of synoptic surveys on chlorinated hydrocarbon insecticides in surface waters of the United States during 1964 through 1968. They showed that the occurrence of all compounds tested reached a peak in 1966 and declined sharply in 1967 and 1968 except for BHC, which showed only a slight decline. Dieldrin remained the most serious pollutant in U.S. surface waters. Zabik *et al.* (1971) studied the effect of urban and agricultural pesticide use on residue levels in a river located in south-central Michigan and concluded that the concentrations of pesticides in bottom samples give a good indication of long-term contamination, whereas levels in the suspended matter indicate the amounts of pesticides being carried on a short-term basis. The most intriguing phenomenon is that the major source of DDT contamination appears to be waste water treatment plants (urban discharges) rather than agricultural areas.

Another important route of transportation is through the air. This includes carrier transport with water particles and dust in addition to true evaporation of pesticides through volatilization. The return of pesticides to the earth is mainly through rain and to a lesser extent, through falling dust.

In field conditions, 50 % of DDT applied to the surface of soil was found to disappear in 16–20 weeks (Wheatley, 1965) and 60 % of DDT applied to an apple orchard could not be accounted for (Stringer and Pickard, 1968). Lloyd-Jones (1971) estimated the loss of DDT through true volatilization to be on the order of 2 lb acre/year in summer and 0.3 lb acre/year for non-summer months from laboratory experiments and DDT volatility figures.

Degree of volatilization is directly related to the amount of pesticide available to the outside. Harris and Lichtenstein (1961) found that volatilization of aldrin from soils containing high concentrations of organic matter was much less than that from soils with low concentrations. It appears that organic matter can absorb aldrin and thereby decrease the available residue portion. Thus an increase in the amount of insecticide applied on a fixed amount of soil invariably increases the rate of evaporation.

Other factors affecting the process of pesticide evaporation are soil moisture, relative humidity of the air (particularly effective against dry soil samples), soil temperature, and air movement over the soil samples. Because of the phenomenon that pesticides can be evaporated along with water vapor, Acree *et al.* (1963) proposed the term "co-distillation," implying that these chemicals can distill away from the soil surface along with water vapor. Indeed DDT and water can co-distill at 100°C as Bowman *et al.* (1959) demonstrated. However, it is doubtful whether presticides can actually "co-distill" at temperatures far below the boiling point of water as there should be no bulk-flow of water vapor at these temperatures, as pointed out by Hartley (1969) and Hamaker (1972). Spencer and Cliath (1973) and Spencer *et al.* (1973) explain the phenomenon by citing two important factors: First, water facilitates the movement of pesticides toward the surface of the soil through the "wick effect." Second, water can also facilitate the release of pesticides from soil constituents by replacing them at the sorption site or inactivating the adsorption material in the soil (i.e., "desorption phenomenon"). It is apparent from the above discussion that the term "co-distillation" does not correctly describe the event, and therefore it cannot be recommended for future use.

In addition to the effect of water the volatility of the insecticidal compound itself plays an important role. Acree *et al.* (1963), for instance, found that the volatile compounds aldrin, heptachlor, phorate, lindane, heptachlor epoxide, and dieldrin appreciably disappeared from soil through evaporation, while DDT, carbaryl, malathion, and parathion did *not*.

As for transport in dust, Cohen and Pinkerton (1966) collected samples of dust-rain brought to Ohio by a mammoth dust storm (as well documented by the Air Pollution Division, U.S. Public Health Service) originating in the Texas–Oklahoma–New Mexico area. Their data showed large amounts of chlorinated hydrocarbon insecticides and an organophosphate (ronnel) in dust. DDT was found in concentrations of 3–90 ppm during the application season of June and July (1965). Since the average monthly dustfall in Cincinnati, Ohio, is about 15 tons per square mile, this gives a figure of 45–270 g of DDT falling per square mile.

The dominant mechanism for the removal of pesticides from the atmosphere is probably rainfall. Tarrant and Tatton (1968) reported that the rainwater in Great Britain contained an astonishing amount of pesticidal residues (Table 8-8). It is interesting to note that the pesticide concentrations

TABLE 8-8. **Average Concentration (ppt) of Chlorinated Hydrocarbon Residues in Rainwater in Great Britain**[a]

Location	α-BHC	γ-BHC	Dieldrin	DDE	DDT	TDE
London	29	59	16	25	61	7
Camborne	5	43	6	28	53	34
Lerwick (island)	24	121	11	20	46	7

From Tarrant and Tatton (1968).

[a]Camborne is located at the southwest tip (Cornwall) of Great Britain, and Lerwick is on an island off the northern tip of Scotland.

were not very different among the locations tested, indicating the presence of some force of equalization (transport).

8.3.2. Long-Range Transport

DDT and other persistent pesticides are known to be transported for surprisingly long distances, but studies of long-range transport on a transcontinental or global scale are not very numerous. An often-cited paper by Peterle (1969) indicates that melt waters of Antarctic snow contain DDT. This pesticide must somehow be transported into such a remote place.

As discussed above, one of the major forces for direct transport of pesticides on a global scale appears to be that by dust in the air. Radioactive nuclear fallout and industrial pollutants are examples of other contaminants transported by airborne dust (Hurtig, 1972); the tracer studies on strontium-90 in connection with latitudinal fallout from the atmosphere are of particular interest (Appleby, 1969; cited by Hurtig, 1972). For insecticide studies, we must rely on limited amounts of information at present to assess their movement.

Riseborough *et al.* (1968) collected dust over Barbados and estimated that the total pesticide content in the dust was 41 ppb, which according to their calculations should yield 600 kg/year of pesticides in the area of the Atlantic Ocean between the equator and 30° north.

In addition to air transport, pesticides can be spread by major rivers and by ocean currents. As described already, large quantities of pesticides are discharged into the Gulf of Mexico by the Mississippi River system—roughly 10,000 kg/year not counting the amount adhering to leached soil particles. The large accumulation of pesticide residues in the Mississippi delta attests to this fact. It is not clear how much of the pesticides thus washed into the Gulf would be carried farther on by ocean currents. Burnett (1971) studied distribution of DDT residues along coastal California by using an indicator organism, the common surf-zone sand crab, *Emerita analoga*, which is a nonmigrating common particulate filter-feeder species in that area. He found that the animals near the Los Angeles County sewer outfall contained over 45 times more DDT (mostly in the form of DDE) derivatives than animals near major agricultural drainage areas. He attributed this to the nearby Montrose Chemical plant, which was the sole DDT manufacturer in the United States at that time. The sediments near the outfall, by his estimate, contained over 100 metric tons of DDT, acting as one of the largest known reservoirs of this type. Despite the presence of such a large reservoir in the area, Burnett found a relatively modest spread of DDT contamination in the crab samples. With 100 ppb of total DDT contamination as the base value, the total spread ranged from approximately 33°22′ N to 35°42′ N. The farthest transport of DDT along the shore was 3573 km. It is likely that the higher residues toward the north represented the work of ocean currents. This finding is also supported by evidence that DDE/DDT ratios were lower on the northern side of the peak contamination area; that is, "fresh" contaminants are expected to have a lower DDE/DDT ratio than older sewage samples. While biological transport cannot be neglected from biological and health standpoints, its relative contribution to the global movement of pesticides (in terms of quantity) is small.

It is not clear just how much of atmospheric transport is actually due to dust-mediated transport and how much comes from true pesticide vapors. A detailed analysis of air over nine U.S. cities in 1967–1968 indicated that the concentration of DDT in the air is $1–150 \times 10^{-9}$ g/m^3 in the winter. Thus the presence of DDT, at least in the atmosphere, is certain (Stanley *et al.*, 1971). Woodwell *et al.* (1971) estimate that the equilibrium of DDT at 20°C would be 3×10^{-6} g/m^3 or 2 ppb by weight. The total saturation capacity of the atmosphere would thus equal the amount of DDT produced to date (10^{12} g) without considering the portion that could be carried by dust in the air.

There appears to be more than enough capacity in the earth's atmosphere to accommodate all available pesticide residues.

As mentioned before, rainfall probably plays the most important role in recycling pesticides back to the earth, mostly to the ocean. Basing their judgment on the figures obtained by Tarrant and Tatton (1968) and by Wheatley and Hardman (1965), Woodwell *et al.* (1971) estimated that all the DDT residues produced to date could be washed down in 3.3 years, calculating the average DDT concentration in rainwater as 60 ppt. Since DDT concentration in rainwater varies appreciably throughout the year, this figure can be considered as the upper limit; in any event, the rate of turnover is rather fast since it is on the order of a few years. Thus Woodwell *et al.* (1971) proposed that the worldwide pattern of DDT movement is evaporation from the land, dispersion through the atmosphere, rainfall into the ocean, settling into the mixed layer (less than 75–100 m) of the ocean, and finally elimination through sedimentation to the abyss of the ocean. Only a fraction (less than one-thirtieth of 1 year's production of DDT during the mid-1960s, or approximately 3×10^9 g) of potentially available DDT has been estimated to be present in the atmosphere and mixed layers of the ocean. The reason for such nonavailability of DDT to the biota is not known. Presumably, the bulk of DDT is tied up either in soil and soil sediments or in other nonavailable sites such as oil sediments (Hartung and Klingler, 1970), in addition to the force of degradation in the environment.

In summary, atmospheric transport of persistent pesticide residues appears to play the most important role in the long-range dispersion of pesticides. Dust-mediated transport serves as the direct means of dispersion in addition to its role in increasing the total holding capacity of the atmosphere. Even without considering dust-microcrystal mediated input, the atmosphere offers more than enough capacity to accommodate pesticides in the form of vaporized gaseous molecules. Rainfall facilitates the final phase of global transport of pesticides into the ocean. As the amount of total available pesticides (e.g., DDT) diminishes, the holding ability of soil aquatic sediments would become the limiting factor for the global distribution of pesticides.

8.3.3. Mathematical Models for Insecticide Movement

Woodwell *et al.* (1971) proposed a model to explain the dynamics of DDT accumulation in a biosphere. It mainly explains the worldwide movement of this compound in relation to accumulation in the atmosphere and in the mixed layers of the ocean. A set of first-order rate equations to give estimates of DDT loads in various reservoirs (biospheres) as a function of time has been proposed:

$$\frac{dN_i}{dt} = R_i(t) - \sum_{j=1}^{m} \frac{N_i}{T_{ij}^{(l)}} + \sum_{j=1}^{m} \frac{N_i}{T_{ij}^{(g)}} \tag{1}$$

where $R_i(t)$ is the rate at which newly produced DDT is introduced into the ith reservoir and N_i is the amount of DDT in the ith reservoir. The sums represent losses from $(-\sum)$ and gains to $(+\sum)$ the ith reservoir; $T_{ij}^{(l)}$ is the time constant for DDT loss from the ith to the jth reservoir, and $T_{ij}^{(g)}$ represents inputs from the jth to the ith reservoir. The major assumption here is that the increase (apart from the newly introduced DDT, R_i) and decrease of DDT in one reservoir follow first-order kinetics (i.e., are concentration dependent) from the adjacent reservoir.

Robinson (1967) arrived at a similar conclusion that transfer of chlorinated hydrocarbon insecticides from one trophic level to another is concentration (of the next trophic level) dependent. For instance, the rate of loss of an insecticide from the ith trophic level can be expressed as

$$\frac{dx_{ijk}}{dt} = - m(x_{ijk}) \tag{2}$$

where x_{ijk} is the average concentration of the insecticide in the whole body of one individual (i.e., individual k of the jth species in the ith trophic level). According to Robinson, the total insecticide present in a trophic level (e.g., the ith level) can be expressed as

$$\sum_{jk} m_{ijk} x_{ijk} \tag{3}$$

where m_{ijk} is the mass (body weight) of individual k.

The transfer of the insecticide from the $(i-1)$th level to the ith level can be expressed as

$$\text{Amount of transfer} = \left(\sum_{jk} \Delta m_{(i-1)jk} - \sum_{jk} \Delta m_{ijk} \right) x_{(i-1)jk} \tag{4}$$

Thus both increase and loss of the insecticide appear to be concentration dependent: the loss depends on the concentration of the ith level and the increase on that of the $(i-1)$th level.

Assuming that $\sum_{jk} \Delta m_{(i-1)jk} \gg \sum_{jk} \Delta m_{ijk}$ and that the former is a function of $\sum_{jk} m_{(i-1)jk}$,

$$f \sum_{jk} m_{(i-1)jk} \tag{5}$$

Thus

$$\frac{dx_{ijk}}{dt} = \frac{f \sum m_{(i-1)jk}}{m_{ijk}} x_{(i-1)jk} + \frac{f \sum m_{ijk}}{m_{ijk}} x_{ijk} \tag{6}$$

This becomes analogous to what Woodwell *et al.* (1971) have proposed.

Eberhardt *et al.* (1971) similarly propose that transfer of pesticides can be expressed in terms of

$$\frac{dy_2(t)}{dt} = c_2 y_1(t) - \mu_2 y_2(t) \qquad [y_i(0) = 0] \tag{7}$$

where $y_2(t)$ represents DDT concentration in a given "compartment" (i.e., mathematically analogous to trophic level or biosphere as above), c_2 is a "transfer coefficient," μ_2 is a rate of loss, and $y_1(t)$ is the concentration of the source of DDT taken in (i.e., $i - 1$ equivalent) by the given compartment. Equation (7) is then very similar to equation (6) in its general concept.

By applying ^{36}Cl-labeled DDT (on phenyl rings) in a granular form at a rate of 0.2 lb DDT per acre (220 g/ha) at the southwestern edge of Lake Erie, these workers examined the validity of their model in ten plant species, six invertebrate species, two fish species, and a tadpole species. The concentration of DDT in filtered water was assumed to follow the equation

$$\frac{dy}{dt} = \lambda e^{-kt} - \mu y \tag{8}$$

where λe^{-kt} represents release from granules and μy losses from water due to all causes (but not metabolic, since the ^{36}Cl-phenyl moiety is expected to be present in all DDT metabolites).

They found that in the tadpole and most of the plant species there were two phases of uptake; in the fast compartment, uptake occurred in 24–72 hr, and in the slow compartment, which followed equation (7), it occurred after 14 days. They assumed that the fast compartment was essentially depleted of DDT by 14 days after application. Examination of the field-collected samples of DDT accumulation revealed that the relationship held well in five species. For two species which did not fit into the equation, the tadpole and the narrow-leaf pondweed, these workers noted that these organisms showed an unusually high proportion of the fast compartment, if the fast-compartment data are not neglected. If this fast-compartment portion is subtracted from each accumulation curve, the relationship appears to become compatible with the equation. Then since both fish species showed only the slow compartment (peak time on the order of 30 days), it was assumed that the fast compartment found in tadpoles and narrow-leaf pondweed was due mostly to input from direct contact with DDT at spray time, while the slow compartment found in fish was probably due to the effects of the food chain.

While there are still a number of problems, these studies at least fit to a general degree with the actual field data. Thus it is possible that these equations of first-order kinetics—i.e., equations (1), (6), and (7)—roughly correspond to what takes place in nature, at least in the case of food-chain-oriented accumulation of DDT and related compounds.

8.4. REFERENCES

Acree, F., M. Beroza, and M. C. Bowman (1963). *J. Agr. Food Chem.* **11**:278.

Appleby, W. G. (1969). Pesticide residues in our environment. Paper presented to IUCN, Delhi, Nov. 24–Dec. 1. (Indirectly cited from Hurtig, 1972.)

Bailey, T. E., and J. R. Hannum (1967). *Journal of the Sanitary Engineering Division*, Proceedings of the American Society of Civil Engineers, Ann Arbor, Mich. Vol. 93 No. SA5 p. 27.

Blinn, R. C., G. E. Carman, W. H. Ewart, and F. A. Gunther (1959). *J. Econ. Entomol.* **52**:42.

Bowman, M. C., F. Acree, C. H. Schmidt, and M. Beroza (1959). *J. Econ. Entomol.* **52**:1038.

Bruce, W. H., G. C. Decker, and J. G. Wilson (1966). *J. Econ. Entomol.* **59**:179.

Burnett, R. (1971). *Science* **174**:606.

Carter, F. L., and C. A. Stringer (1970). *Bull. Environ. Contam. Toxicol.* **5**:422.

Cohen, J. M., and C. Pinkerton (1966). Organic pesticides in the environment, *Adv. Chem. Ser.* **60**:163.

Corneliussen, P E. (1970). *Pest. Monit. J.* **4**:89.

Duggan, R. E. (1968). *Pest Monit.* **J.2**:2.

Eberhardt, L. L., R. L. Meeks, and T. J. Peterle (1971). *Nature* **230**:60.

Edwards, C. A. (1966). *Residue Rev.* **13**:83.

Edwards, C. A. (1970). Persistent pesticides in the environment. In *Critical Reviews in Environmental Control*, Vol. 1. Chemical Rubber Co., Cleveland, p. 7.

Eye, J. D. (1966). Aqueous transport of dieldrin in soils. Ph.D. thesis, University of Cincinatti.

Fowkes, F. M., H. A. Benes, L. B. Ryland, W. M. Sawyer, K. D. Detling, E. S. Loettler, F. B. Folckemer, M. R. Johnson, and Y. P. Sun (1960). *J. Agr. Food Chem.* **8**:203.

Gallaher, P. J., and L. Evans (1961). *New Zealand Agr. Res.* **4**:466.

Gunther, F. A., and R. C. Blinn (1956). *Ann. Rev. Entomol.* **1**:167.

Gunther, F. A., and L. R. Jeppson (1960). *Modern Insecticides and World Food Production.* John Wiley & Sons, New York, p. 82.

Hamaker, J. W. (1972). In *Organic Chemicals in the Soil Environment.* Dekker, New York, pp. 341–397.

Harris, C. R., and E. P. Lichtenstein (1961). *J. Econ. Entomol.* **54**:1038.

Hartley, G. S. (1969). In *Pesticide Formulation Research, Physical and Colloidal Chemical Aspects. Advances in Chemistry Series* **86**:115.

Hartung, R., and G. W. Klingler (1970). *Environ. Sci. Technol.* **4**:407.

Hull, H. M. (1970). *Residue Rev.* **31**:1. F. A. Gunther, and J. D. Gunther, eds. Springer-Verlag, New York.

Hurtig, B. (1972). In *Environmental Toxicology of Pesticides.* F. Matsumura, G. M. Boush, and T. Misato. Academic Press, New York, p. 257.

Kearney, P. C., J. R. Plimmer, and C. S. Helling (1969). *Encycl. Chem. Technol.* **18**:515.

Lichtenburg, J. J., J. W. Eichelberger, R. C. Dressman, and J. E. Longbottom (1970). *Pesticides Monitoring J.* **4**:71.

Lichtenstein, E. P. (1959). *J. Agr. Food Chem.* **7**:430.

Lichtenstein, E. P. (1965). *Research in Pesticides.* Academic Press, New York, pp. 199–203.

Lichtenstein, E. P., and K. R. Schulz (1960a). *J. Econ. Entomol.* **53**:192.

Lichtenstein, E. P., and K. R. Schulz (1960b). *J. Agr. Food Chem.* **8**:452.

Lichtenstein, E. P., and K. R. Schulz (1961). *J. Econ. Entomol.* **54**:517.

Lichtenstein, E. P., C. H. Muller, G. R. Myrdal, and K. R. Schulz (1962). *J. Econ. Entomol.* **55**:215.

Lichtenstein, E. P., G. R. Myrdal, and K. R. Schulz (1965). *J. Agr. Food Chem.* **13**:126.

Lichtenstein, E. P., T. W. Fuhremann, N. E. A. Scopes, and R. F. Skrentny (1967a). *J. Agr. Food Chem.* **15**:864.

Lichtenstein, E. P., T. W. Fuhremann, K. R. Schulz, and R. F. Skrentny (1967b). *J. Econ. Entomol.* **60**:1714.

Lloyd-Jones, C. P. (1971). *Nature* **229**:65.

National Agricultural Chemicals Association (1970). Official F.D.A. Tolerance 1969, *NAC News Pesticide Rev.* **28**: No. 3 (February).

Peterle, T. J. (1969). *Nature* **244**:620.

Rainey, R. E. (1967). *Science* **155**:1242.

Reinert, R. E. (1969). *Limnos* **2**:3.

Riseborough, R. W., R. J. Huggett, J. J. Griffin, and E. D. Goldberg (1968). *Science* **159**:1233.

Robinson, R. (1967). *Nature* **215**:33.

Rosen, J. D., D. J. Sutherland, and G. R. Lipton (1966). *Bull. Environ. Contam. Toxicol.* **1**:133.

Rowlands, D. G. (1971). *Residue Rev.* **35**:91. F. A. Gunther, and J. D. Gunther, eds. Springer-Verlag, New York.

Spencer, W. F., and M. M. Cliath (1969). *Environ. Sci. Technol.* **3**:670.

Spencer, W. F., and M. M. Cliath (1973). *J. Environ. Quality* **2**:284.

Spencer, W. F., W. J. Farmer, and M. M. Cliath (1973). *Residue Review* **49**:1, F. A. Gunther, ed. Springer-Verlag, New York.

Stanley, C. W., J. E. Barney, II, M. R. Helton, and A. R. Yobs (1971). *Environ. Sci. Technol.* **5**:430.

Stringer, A., and J. A. Pickard (1968). *Annual Report of the Long Ashton Research Station.* University of Bristol, p. 80.

Tarrant, K. R., and J. O. G. Tatton (1968). *Nature* **219**:725.

Thompson, B. D., and C. H. Van Middelem (1955). *Proc. Am. Soc. Hort. Sci.* **65**:357.

Walker, K. C. (1967). Monitoring river water and sediments. Paper presented at Symposium on the Science and Technology of Residual Insecticides with Special Reference to Aldrin and Dieldrin, Nov. 3, St. Charles, Ill., USDA, Washington, D.C.

Ware, G. (1971). *Pesticides Monitoring J.* **5**:276.

Westlake, W. E., and L. I. Butler (1953). *J. Econ. Entomol.* **46**:850.

Wheatley, G. A. (1965). *Ann. Appl. Biol.* **55**:325.

Wheatley, G. A., and J. A. Hardman (1965). *Nature* **207**:486.

Woodwell, G. M., P. P. Craig, and H. A. Johnson (1971). *Science* **174**:1101.

Zabik, M. J., B. E. Pape, and J. W. Bedford (1971). *Pesticides Monitoring J.* **5**:301.

Chapter 9

Environmental Alteration of Insecticide Residues

9.1. CHARACTERISTICS OF ENVIRONMENTAL ALTERATION

From the viewpoint of environmental toxicologists, the disappearance of insecticide residues in their original form at a given location does not mean the end of the problem. Rather, the problem starts there. The question is whether such disappearance means actual degradation of the hazardous chemical or whether it is actually a signal that the insecticide has been translocated for, say, bioconcentration into some ecosystem or converted into more dangerous chemicals.

In Chapter 8, this particular aspect was neglected for the sake of simplicity, but it must be pointed out that food residues are mostly in the form of environmentally altered chemicals—e.g., DDE, dieldrin, and heptachlor epoxide derived from the original insecticides DDT, aldrin, and heptachlor in the environment. With this in mind we can then ask, what are the key problems in the environmental reactions on pesticidal chemicals? First let us recall our basic position that pesticide pollution is a biological problem, and the problem cannot be assessed without due consideration of biological effects. Any environmental change which results in the creation of biologically active, but not necessarily toxic, substances is of concern. Also, from the information given in the previous chapters we know that alteration products having a long residual life are what actually cause problems.

Degradation of pesticide residues in the environment can be affected by various factors. The most important factor contributing to the speed of degradation appears to be the chemical nature of the pesticide itself. It is difficult to find the physical or biological means to degrade already persistent pesticide residues (e.g., DDE). On the other hand, if the pesticide is labile (e.g., malathion), many physical conditions, such as high and low pH and high temperature, and biological processes can be found to degrade it. The emphasis of this chapter is on the stable and persistent pesticides which cause environmental problems by accumulating in various ecosystems because of their chemical stability and often accompanying affinity for organic matter.

For a given compound, there are naturally many factors influencing its residual characteristics. In Chapter 8, factors influencing the transport of pesticide residues were discussed—e.g., absorption by soil constituents, leaching by rainwater, pickup by plants and animals, evaporation either directly or with water vapor, and dust-mediated transport by wind—all of which can reduce the level of the pesticide at the site. When we think of the environment as a whole (i.e., as one enclosed system), the total amount of the pesticide is not decreased by these processes. The processes that play important roles in actually decreasing the total amount of the pesticide residues are the ones mediated by microorganisms, animals, plants, and sunlight. Other factors in addition to extreme pH and heat, such as catalytic agents in the soil and soil enzymes, are important in degrading relatively unstable pesticides, say, simazin and atrazine (e.g., soil pH ; Hamilton and Moreland, 1962 ; Armstrong *et al.*, 1967) and malathion (pH and possibly soil enzymes ; Tiedje and Alexander, 1967 ; Getzin and Rosefield, 1968). These factors could also be important for further degradation of unstable metabolic intermediates once other systems have attacked the original molecules and made them more vulnerable to environmental weathering factors. For the stable pesticides, only under extreme conditions (such as high temperature and high alkalinity against BHC) would such physical factors have any significant effect on their metabolic fate (e.g., Yoshida and Castro, 1970 ; Bradbury, 1963).

9.2. CHARACTERISTICS OF MICROBIAL METABOLISM

The major groups of soil microorganisms are actinomycetes, fungi, and bacteria (Alexander, 1965). Generally speaking, soils are rich with microbial fauna and flora, while oceanic and atmospheric environments are both almost devoid of microbial activities. Estuarine and many freshwater environments perhaps rank second to soil in their microbial populations.

There are a number of typical microbial metabolic pathways: β-oxidation, ether cleavage, ester and amide hydrolysis, oxidation of alcohols and aldehydes, dealkylation, hydroxylation, hydrohalogenation, epoxidation, reductive dehalogenation, N-dealkylation, etc. The most noticeable degradative characteristics of microorganisms are their reductive systems. Examples are the reductive dechlorination of DDT (to form TDE; Plimmer *et al.*, 1968); production of aldrin from dieldrin (Matsumura *et al.*, 1968), in contrast to the reverse process of epoxidation of aldrin which commonly occurs; and conversion of parathion to aminoparathion (Menzel *et al.*, 1961). The ability to enzymatically cleave the aromatic ring itself is limited to microorganisms.

Soil is regarded as generally nutrient deficient for microbial growth, which requires carbon. Thus a unique characteristic of microorganisms is their utilization of pesticidal chemicals as a carbon source. In a few cases, pesticidal chemicals have been found to serve as the sole carbon source for microbial cultures, indicating that at least some of the pesticide substrates are true energy sources. Of course, higher animals and plants also metabolize pesticidal chemicals to CO_2, but here the metabolism is incidental to energy production.

Microorganisms almost completely lack efficient conjugation systems to convert xenobiotics into excretable forms, in contrast to higher animals. So far there is no record of, for instance, the formation of sulfate glucoside or glucuronide among microbial metabolites of pesticides despite their frequent occurrence in animals and plants. The only synthetic activities clearly demonstrated in microorganisms are methylation and acetylation. Cserjesi and Johnson (1972), for instance, found that pentachlorophenol (PCP) is methylated at the OH position in liquid cultures of *Trichoderma virgatum*. Since this methylated metabolite is more stable and less biologically active, it is likely that this process indicates a genuine conjugation system. On the other hand, acetylation is a very common conjugation process for aromatic amines in microorganisms. The reaction is particularly known to produce halogen-substituted anilines, which are frequent metabolic products of various herbicides (Tweedy *et al.*, 1970):

Another important consideration is that microorganisms generally lack well-defined mixed-function oxidase systems, although similar oxidation –hydroxylation processes catalyzed by cytochrome P_{450} and NADPH do exist in highly developed microorganisms.

9.3. METABOLISM OF CHLORINATED HYDRO-CARBON INSECTICIDES BY MICROORGANISMS

9.3.1. DDT Analogues

The major microbial metabolic step with respect to DDT analogues is the reductive dechlorination reaction, but dehydrochlorination and oxidation reactions are also observed. For instance, DDT is mainly degraded by the former route (Fig. 9-1) to yield a series of dechlorinated analogues, TDE (Finley and Pillmore, 1963; for review, see Pfister, 1972), DDMS, and DDNS (Matsumura *et al.*, 1971a). The enzymatic basis of the microbial reductive dechlorination reaction has been studied by French and Hoopin-garner (1970), who incubated DDT *in vitro* with various subcellular fractions of *Escherichia coli* and found that TDE production was highest in the system where both the membrane and the cytoplasm were present. Therefore, the enzyme is apparently located in the cell membrane, with cytoplasmic factors that promote it. Among various cofactors investigated, FAD was found to be the only compound that showed stimulatory properties. This pheno-menon can be observed only under anaerobic conditions, indicating that prior reduction of FAD is required for the dechlorination reaction. When-ever Krebs cycle metabolic activities were promoted by exogenously added NAD, NADH, pyruvate, or malate, the production of TDE drastically declined.

Anderson and Lichtenstein (1971) studied the effects of nutritional factors on the rate of DDT metabolism in a pure strain of fungus, *Mucor alternans*. This fungal strain degraded DDT into unidentified water-soluble metabolites under aerobic conditions in the dark. The rate of degradation was maximal when glucose and ammonium nitrate were present in the medium as compared to seven other carbon and eight other nitrogen sources. It is unfortunate that the identity of these metabolic products was not established since it appears that the rate of degradation is rather vigorous, e.g., 30 % degradation at 1 ppm DDT in 5 days.

The two other enzymatic reactions that influence the metabolism of DDT are, as just mentioned, oxidation and dehydrochlorination. Under aerobic conditions, the oxidative system works to hydroxylate DDT and TDE at the 2-carbon to yield dicofol and FW-152 (Matsumura and Boush,

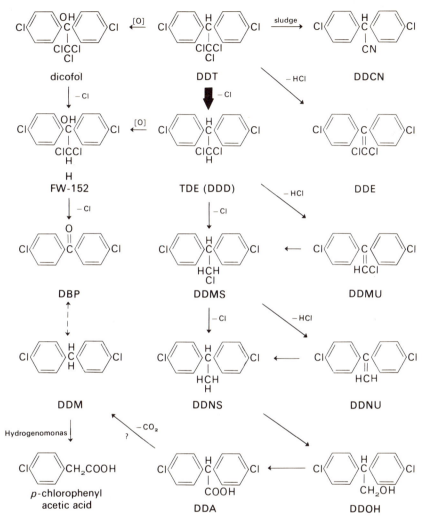

Fig. 9-1. Metabolic conversion of DDT by microorganisms and in the environment.

1968). This is usually a moderate side-reaction. Dehydrochlorination yields DDE (Keil and Priester, 1969), DDMU, and DDNU (Matsumura *et al.*, 1971a). BHC can also be degraded by this route to yield γ-PCCH.

Albone *et al.* (1972) and Jensen *et al.* (1972) have described the discovery of *p,p'*-DDCN in sewage sludges. The former workers incubated radiolabeled and nonradioactive DDT in a vessel with anaerobic sewage sludge from a water treatment plant in Bristol, England, for 88 days at 37°C. Minced beef was added to the system to promote microbial activities. The latter workers incubated DDT with sludge at 20°C for up to 8 days under nitrogen atmosphere and found that by 24 hr approximately 60% of DDT was consumed, and apparently no further degradation took place. At 24 hr of incubation, the percentages of metabolic products formed were approximately 28 for DDD, 8 for DDCN, 5 for DDMU, and 2 for DBP. They reasoned that DDCN formed directly from DDT in a similar chemical reaction as the formation of benzonitrile from α-trichlorotoluene and ammonium chloride at high temperature. Jensen *et al.* (1972) were actually able to detect DDCN in the natural sediment from a lake in Sweden. The toxicological significance of DDCN formation in the environment is not apparent, but the authors point out that it is known as a biocide for soil pests.

Cleavage of aromatic rings to form aliphatic carbons or ring-opening of cyclic hydrocarbons is not a common metabolic reaction. Focht and Alexander (1971) demonstrated that the chlorine atoms at *p,p'*-positions provide stability to DDT, inasmuch as its analogues without these chlorine atoms are degraded quickly by a strain of *Hydrogenomonas* to yield phenol and benzoic acid. This particular strain has been grown on diphenylmethane but has been found to aerobically decompose *p,p'*-dichlorodiphenylmethane (DDM) as well as 1,1-diphenyl-2,2,2-trichloroethane (DTE), but not DDT. Since phenyl- and *p*-chlorophenylacetic acid are formed from DDM and DTE, these authors assumed that ring cleavage through a catechol intermediate in one of the rings must have taken place.

In agreement with the general observation that the initial stage of DDT degradation in microorganisms is reductive dechlorination, Alexander (1972) found that this same strain of *Hydrogenomonas* can degrade DDT under anaerobic conditions. Thus it is apparent that even within the same species the metabolic patterns can be influenced by environmental conditions. In anaerobic environments a series of dechlorination processes will take place, eventually yielding DDMS, DDNS (Matsumura *et al.*, 1971*a*), and probably DDM, which could also form as a result of hydrogenation of DBP or decarboxylation of DDA (Fig. 9-1). These analogues with fewer chlorines on the 2-carbon could presumably be degraded further through oxidative reactions such as the one shown by Focht and Alexander (1971). By the same token, substitution of these chlorine atoms with various alkoxy and alkylthio groups greatly enhances the probability of biodegradation (Kapoor *et al.*, 1970; Mendel *et al.*, 1967).

9.3.2. BHC

BHC has long been known to disappear in the soil, particularly under anaerobic conditions. Both microbial actions and soil alkalinity could

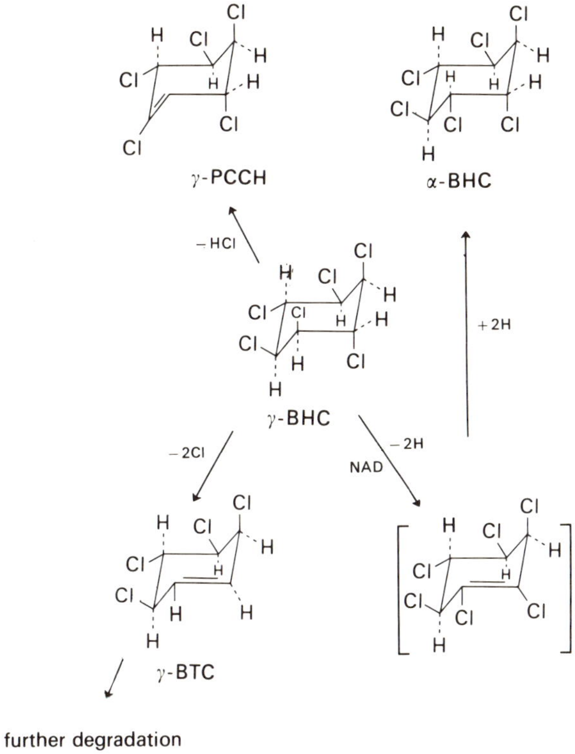

Fig. 9-2. Metabolism of γ-BHC by microorganisms.

play important roles in degrading BHC (Bradbury, 1963). The microbial dehydrochlorination process (Yule *et al.*, 1967) appears to be rather minor in importance (Fig. 9-2). Exceptions could be the unicellular algae (Sweeney, 1969), which appear to degrade lindane exclusively to γ-PCCH. The major metabolic product of such a process on γ-BHC is γ-tetrachloro-1-cyclo-hexene (γ-BTC) (Tsukano and Kobayashi, 1972). Other dehydrochlorination products such as 1,2,3,5-tetrachlorobenzene (Allan, 1955) could form as the result of further microbial attack. It is generally acknowledged that BHC can be degraded much more easily under anaerobic conditions than aerobic ones (Raghu and MacRae, 1966; Yoshida and Castro, 1970; Sethunathan *et al.*, 1969); for instance, BHC disappears rather rapidly in submerged paddy fields. However, Sethunathan *et al.* (1969) found that the major metabolic product produced by *Clostridium* sp. isolated from rice fields was not PCCH. Judging by the chromatographic data presented by these workers, it was apparently γ-BTC. Studies in our own laboratory (Benezet and Matsumura, 1973*a*) confirmed that *γ*-BTC is the major metabolic product, which is further degraded to more polar materials.

The most important aspect of BHC degradation in the environment is the difference in the rate of disappearance among the BHC isomers. Japanese scientists (Goto, 1971) found that *α*- and *β*-BHC accumulated in soils, rice straws, milk, and other agricultural commodities after extensive use of BHC. Benezet and Matsumura (1973*a*) found that microorganisms are capable of producing *α*-BHC from pure *γ*-BHC, both in the laboratory and in aquatic sediments. Among microorganisms tested, a strain of *Pseudomonas putida* was found to be particularly active. This isomerization reaction is catalyzed by NAD and to a lesser extent by FAD. The production of *γ*-BTC has been observed always to accompany that of *α*-BHC. Thus it is most likely that the presence of high levels of *α*- and *β*-BHC is the result both of isomerization reactions by microorganisms and of selective degradation of *γ*-BHC, leaving other constituents of technical BHC.

9.3.3. Cyclodiene Insecticides

This group of cyclodiene insecticides includes dieldrin, aldrin, heptachlor, and chlordane. Also, for the sake of convenience, chlorinated terpenes such as toxaphene and Strobane® are included. Generally speaking, these insecticides are stable and not too many microorganisms are capable of degrading them, although toxaphene and endrin are known to be degraded relatively fast. Matsumura and Boush (1967) found that only ten isolates out of 600 soil microbial cultures were capable of degrading dieldrin. The chlorine-containing ring is particularly stable, and the major microbial attack appears to be on the nonchlorinated rings. The only exception to this rule is the finding by Bixby *et al.* (1971) that a strain of *Trichoderma koningi* slowly degraded dieldrin labeled on the 1,2,3,4,10-carbons (i.e., chlorinated carbons) into CO_2.

The most thoroughly studied microbial metabolic process for cyclodienes is the epoxidation reaction, which converts aldrin to dieldrin, heptachlor to heptachlor epoxide, and isodrin to endrin (Kiigemagi *et al.*, 1958; Lichtenstein and Schulz, 1960). This is an exceedingly common oxidation process in many biological systems. As mentioned previously, microorganisms also appear to be capable of reducing dieldrin to form aldrin (Matsumura *et al.*, 1968).

Another characteristic reaction associated with cyclodiene metabolism is rearrangement, such as intramolecular bridge formation (e.g., photodieldrin and photoaldrin formation, Fig. 9-3) (Rosen *et al.*, 1966; Robinson *et al.*, 1966) and rearrangement of the epoxy ring to form ketones, aldehydes, and alcohols (Fig. 9-4) (Matsumura *et al.*, 1968, 1971*b*).

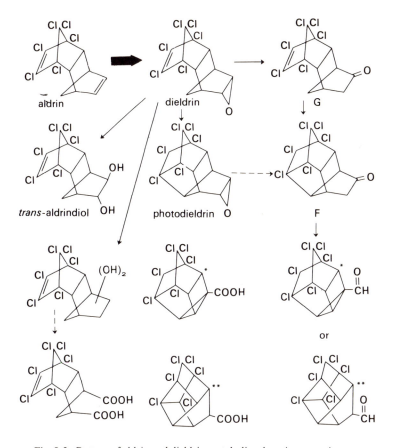

Fig. 9-3. Pattern of aldrin and dieldrin metabolism by microorganisms.
*These are only tentatively proposed structures. Alternative structures (indicated by **) have been proposed by Rosen (1969).

For both dieldrin and endrin, the major reaction product is ketones, which are formed as a result of isomerization of the epoxy ring. Judging by the difference in the general degradation rates of these two compounds, it is likely that the ketoendrin can be formed more easily by microorganisms. Certain microorganisms are capable of forming photodieldrin from dieldrin (Matsumura *et al.*, 1970), and such an isomerization reaction appears to be very common in the case of endrin. For instance, Patil *et al.* (1970) examined 20 common soil isolates of microorganisms for their metabolic abilities against endrin, aldrin, DDT, γ-BHC, and propoxur. They found that all of them could degrade DDT and endrin, while only 13 degraded aldrin and none degraded either γ-BHC or propoxur. All of these isolates were capable of forming ketoendrin from endrin (Matsumura *et al.*, 1971*b*).

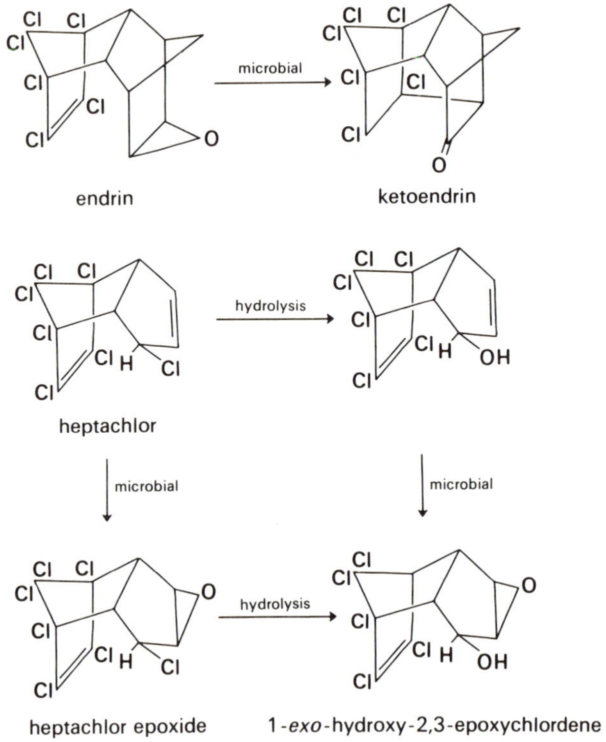

Fig. 9-4. Metabolic conversion of endrin and heptachlor.

The significance of such rearrangement reactions is not known. However, it is unlikely that the microorganisms are able to derive any energy from them. The rearrangements could, at least, give the microorganisms the chance to degrade the compounds further, since most yield more polar compounds that the parent molecules. Indeed, Korte and his associates (Korte, 1970) have shown that photodieldrin can be degraded relatively rapidly by *Aspergillus flavus* and *Penicillium notatum* to two hydrophilic metabolites. Preliminary studies in our laboratory also show that photodieldrin is degraded much faster than dieldrin by many microorganisms. Another important process of cyclodiene degradation is the hydrolysis reaction of the epoxy ring to yield dihydroxy analogues (Wedemyer, 1968; Matsumura *et al.*, 1969). It must be remembered that none of these stable metabolic products has been shown to be less toxic than the parent molecules against wildlife. According to Batterton *et al.* (1971), photodieldrin, metabolites F and G, photoaldrin, and ketoendrin are about as toxic as dieldrin to two species of blue-green algae. Wang and Matsumura (1970) and Wang *et al.* (1971) found that *trans*-aldrindiol and photodieldrin are more neurotoxic to the nervous system of cockroaches than the parent compounds.

It is somewhat surprising that so far no hydroxylation products have been found among the metabolites of cyclodiene insecticides, as is the case for mammals. Several microbial metabolites containing hydroxyl moieties have been shown to form as the result of hydrolytic process. Miles *et al.* (1969) concluded that formation of 1-hydroxychlordene from heptachlor is due to chemical hydrolysis, since this compound appeared in medium that contained no microorganisms (Fig. 9-4). Telodrin® is metabolized to a hydrophilic metabolite by *Aspergillus niger*, *A. flavus*, and *P. notatum.* Although this metabolite has not been identified, it is most likely a hydroxy product or a derivative of it, as judged by its similarity to one of the Telodrin® metabolites in mosquito larvae (Korte *et al.*, 1962; Korte and Stiasni, 1964). The opening of epoxy rings to form various diols (e.g., *trans*-aldrindiol from dieldrin) (Wedemyer, 1968; Matsumura and Boush, 1967) is also likely to be caused by a hydrolytic enzyme, as in mammalian systems.

Since the oxidative systems that form epoxy rings from aldrin, heptachlor, etc., are omnipresent in the microbial world, this lack of oxidative hydroxylation processes on cyclodiene compounds must be judged to indicate that microbial oxidase systems in general are not suited for cyclic hydrocarbons. Since many of the cyclodienes lack vulnerable sites for dehydrochlorination and hydroxylation reactions, the major metabolic reactions that appear to occur on these compounds are isomerization and hydrolysis. Ketones, aldehydes, and alcohols are formed by these reactions mostly by alteration of the structure of the nonchlorinated (i.e., the position forming the epoxy ring) sides of the molecules.

9.4. METABOLISM OF ORGANOPHOSPHATE AND CARBAMATE INSECTICIDES BY MICROORGANISMS

The most common reaction mechanism used by microorganisms to degrade organophosphate and carbamate insecticides appears to be the hydrolysis process through esterases (Matsumura and Boush, 1971). The oxidative processes important in higher animals for degrading these compounds are less frequently observed. This is probably because of the lack of defined mixed-function oxidase systems in microorganisms. It is sometimes difficult to distinguish the enzymatic hydrolysis process from physical processes such as *p*H and catalytic reactions for labile chemicals.

Another interesting possibility is that soil enzymes, in the absence of microbial action, can degrade pesticides. Soil enzymes are extracellular enzymes, found outside living soil organisms. They could be exoenzymes produced by living microbes as well as enzymes released on the death of soil organisms (Skujins, 1966). Getzin and Rosefield (1968), for instance, found that sterilization of soil samples by autoclaving destroyed 90 % of the degradation activity, while a γ-radiation treatment (4 mrads, at 250,000 rads/hr) hardly reduced it. The authors were able to extract a fraction with 0.2 N NaOH that could actively degrade malathion.

As for the true enzymatic hydrolysis system, Matsumura and Boush (1966) demonstrated that malathion can be hydrolyzed by an enzyme preparation from *Trichoderma viride*. The system is prepared by an acetone powder extraction procedure and is sensitive to DFP at 10^{-6} M, indicating its enzymatic origin. Boush and Matsumura (1967) also observed degradation of dichlorvos, diazinon, parathion, DFP, and carbaryl by cultures of *Pseudomonas melophthora*. All organophosphate insecticides were converted into water-soluble products, and carbaryl was mainly metabolized into 1-naphthol in the absence of added cofactors. The last observation agrees with the finding of Bollag and Liu (1971) that 1-naphthol is the major conversion product of carbaryl by soil microorganisms. Although phorate (Thimet®) has two vulnerable sites for oxidative attacks (i.e., P=S and the sulfide bond), it is degraded only through hydrolytic processes in *Pseudomonas fluorescens* and *Thiobacillus thiooxidans* (Ahmed and Casida, 1958). This is surprising since these organisms are known to utilize sulfur as an energy source. In any event, 60–75 % of phorate was hydrolyzed in 8 days by these organisms.

The evidence for hydrolytic degradation of diazinon is circumstantial. Bro-Rasmussen *et al.* (1968) studied the effects of various soil environmental factors on the rate of diazinon degradation and concluded that microorganisms play by far the most important role in determining its overall disappearance rate. Konrad *et al.* (1967) showed that hydrolysis is the major mechanism of diazinon degradation in soil. Laanio *et al.* (1972) found a hydrolysis product, 2-isopropyl-4-methyl-6-hydroxypyrimidine, to be the most abundant product in rice paddy water and soil that had been treated with ^{14}C-diazinon, particularly in the several weeks immediately following the application.

Zayed *et al.* (1965) found that trichlorfon is degraded through hydrolytic processes in *Aspergillus niger*, *Penicillium notatum*, and a *Fusarium* species. The metabolic products found were O-methyl-2,2,2-trichloro-1-hydroxy-ethylphosphonic acid and probably 2,2,2-trichloro-1-hydroxyphosphonic acid.

Fenitrothion (Sumithion®) is degraded through hydrolytic processes by *Bacillus subtilis*, although in this particular bacterium reductive metabolism of the compound is much more pronounced (Miyamoto *et al.*, 1966). The hydrolytic metabolic products identified by Miyamoto *et al.* (1966) were desmethylfenitrothion, didesmethylfenitrothion, dimethylphosphorothioate, and, to a lesser extent, desmethylaminofenitrothion. Interestingly enough, no fenitroxon (Sumioxon®) was observed, indicating that oxidative desulfuration does not take place under the culture conditions, in this species or that fenitroxon is quickly degraded by esterases, although the absence of fenitroxon metabolites makes the latter probability unlikely.

Zectran® (mexacarbate) is mostly degraded through hydrolytic processes to 4-dimethylamino-3,5-xylenol by microorganisms (Benezet and Matsumura, 1974). Among 12 microbial isolates tested, nine of them produced this hydrolysis product, while in five isolates 4-*N*-demethylation was the major metabolic process.

Oxidative metabolism is also known to take place on these insecticidal chemicals, but its extent appears to be limited compared to that in herbicide degradation (e.g., see Kearney *et al.*, 1967; Bollag, 1972). This is rather surprising, since the insecticidal carbamic and phosphorus esters, particularly the former compounds, are known to undergo extensive oxidative reactions in higher animals and plants. The general lack of microbial oxidative reactions against these insecticides is emphasized by the fact that there is no confirmed report on conversion of P=S to P=O (i.e., desulfuration) among phosphorothioates so far examined; this reaction is omnipresent among other biological systems. Although it might mean that "oxons" are formed and then further degraded by hydrolytic enzymes, the conspicuous absence of "oxons" among microbial metabolites of thiophosphates suggests a basic difference between microbial metabolism and that of other organisms. Since epoxidation of cyclodiene insecticides is well known among microbial species, this phenomenon should not be regarded as indicating a general lack of any oxidative system.

While there are only a few examples of metabolic studies with respect to oxidative reactions, there are apparently two such processes that commonly occur in microbial systems. One is the oxidation of thioethers to sulfoxides and sulfones, and the other is *N*-dealkylation of alkylamines. Coppedge *et al.* (1967) compared degradation of Temik® in cotton plants and in soils. In soils, they found Temik® sulfoxide, sulfone, nitrile sulfoxide, and oxime sulfoxide, indicating that oxidation of the thioether bond had taken place along with hydrolysis and further conversion of the oxime to the nitrile.

Ahmed and Casida (1958) found that a yeast species, *Torulopsis utilis*, and a green alga, *Chlorella pyrenoidosa*, could oxidize phorate (Thimet®) to the corresponding sulfoxide.

Williams *et al.* (1972) examined the residues of fensulfothion, *O,O*-diethyl *O*-[*p*-(methylsylfinyl)phenyl] phosphorothioate and found its sulfone

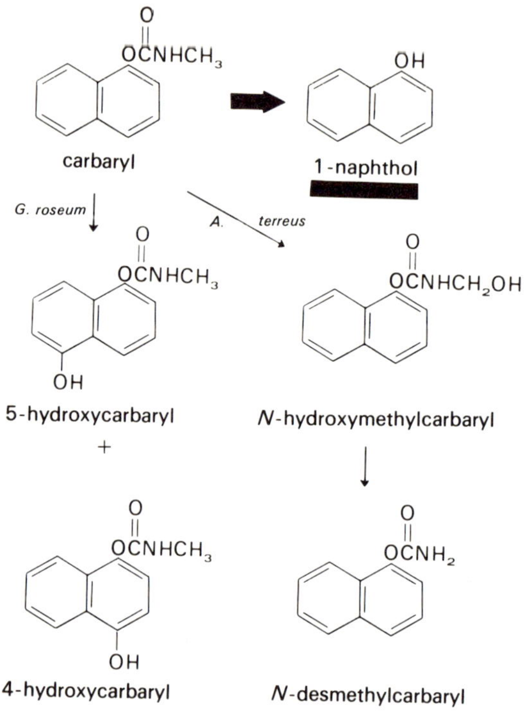

Fig. 9-5. Metabolism of carbaryl by microorganisms. Mainly from Liu and Bollag (1971*a, b*)

analogue, in muck soil in which carrots were grown. Such reactions of sulfur oxidation are not surprising, because of the numerous microorganisms which are specialized in deriving energy from these particular processes.

With respect to the *N*-dealkylation reaction, Liu and Bollag (1971*a*) isolated a soil fungus, *Gliocladium roseum*, from soil which had been treated with carbaryl for 4 weeks. This fungus was found to metabolize carbaryl to three major metabolites: *N*-hydroxymethylcarbaryl and 4- and 5-hydroxy-carbaryl (Fig. 9-5). After 7 days of incubation, they found 7500 cpm of radioactive carbaryl remaining out of 12,000 cpm, in contrast to 410, 360, and 696 cpm of the corresponding metabolites in the growth medium. They noted that the fungus further metabolized these intermediates. Liu and Bollag (1971*b*) demonstrated that *N*-desmethylcarbaryl forms along with *N*-hydroxymethylcarbaryl in *Aspergillus terreus* and to a lesser extent in *A. flavus*. As mentioned previously, Benezet and Matsumura (1973*b*) found that Zectran® (mexacarbate) is largely degraded through hydrolytic and 4-*N*-dealkylation processes by several microorganisms (Fig. 9-6). A few microorganism species exclusively utilize the latter process. One of the bacterium species, HF-3, in particular produces no hydrolytic metabolite, in contrast to *Trichoderma viride*, which exclusively produces the hydrolysis product, 4-dimethylamino-3,5-xylenol. There is a good probability that the *N*-demethylation process is related to energy production by microorganisms, since both cultures produced *N*-dealkylation products in larger quantities in the absence of an added carbon source than under the standard growing conditions with added mannitol.

The only hydroxylation reactions so far known for these insecticidal chemicals are the ones observed by Liu and Bollag (1971*a*, *b*) on carbaryl. Since there are a number of reports of microbial hydroxylation reactions on herbicidal chemicals, it is likely that a greater incidence of such oxidative reactions will be reported in the near future. The overall impression at this stage is that hydroxylation reactions occur less frequently in the microbial world.

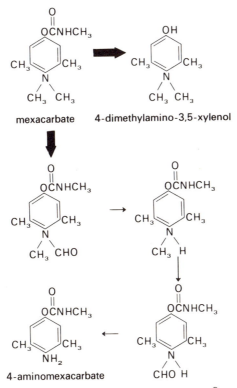

Fig. 9.6. Metabolism of mexacarbate (Zectran®) by microorganisms. From Benezet and Matsumura (1974).

Reductive reactions are, on the other hand, known to occur frequently in microbial species. Thus fenitrothion and EPN are converted to amino-fenitrothion and amino-EPN by *Bacillus subtilis* (Miyamoto *et al.*, 1966), and the major microbial metabolic product of parathion is aminoparathion (Cook, 1957; Ahmed *et al.*, 1958).

9.5. METABOLISM OF ACARICIDAL COMPOUNDS BY MICROORGANISMS

Metabolism of Chlorobenzilate® and chloropropylate was studied by Miyazaki *et al.* (1969, 1970) by using a yeast species, *Rhodotorula gracilis*. The basic pattern of metabolism is through hydrolysis followed by de-carboxylation and dehydrogenation to give DBP (4,4'-dichlorobenzo-phenone):

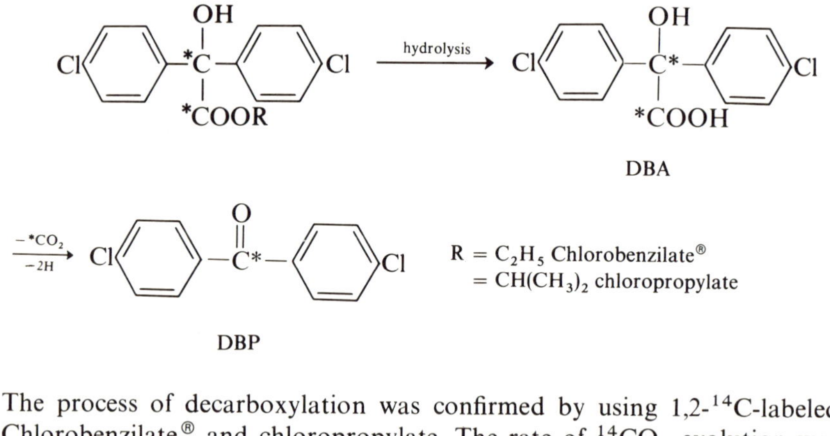

The process of decarboxylation was confirmed by using 1,2-^{14}C-labeled Chlorobenzilate® and chloropropylate. The rate of $^{14}CO_2$ evolution was found to parallel the rate of yeast growth. In the case of chloropropylate, the initial hydrolysis to yield DBA (4,4'-dichlorobenzilic acid) is the rate-limiting factor. The decarboxylation process is stimulated when citric acid is used as the main carbon source and is inhibited by the presence of α-ketoglutaric acid. On the basis of the chemical similarity of citric acid to DBA and this observation, Miyazaki *et al.* suggested that these two sub-strates could be handled by the same decarboxylation–dehydrogenation system. Further *in vitro* studies are needed to confirm or deny such a possibility.

Metabolism of chlordimeform (Galecron® or Fundal®) was studied by Johnson and Knowles (1970) in bacteria, actinomycetes, and fungi:

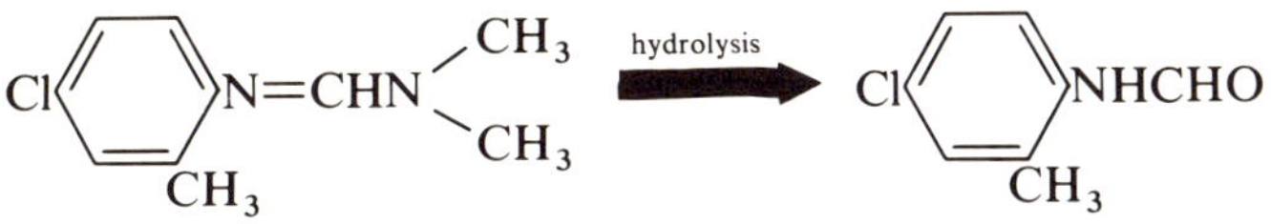

CHLORDIMEFORM

N-FORMYL-4-CHLORO-o-TOLUIDINE

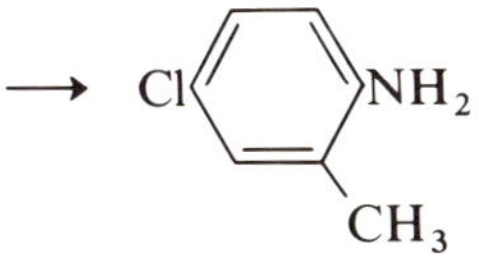

4-CHLORO-o-TOLUIDINE

All organisms produced N-formyl-4-chloro-o-toluidine; approximately 80–95 % of chlordimeform was metabolized in 48 hr at 28°C, and between 10 and 88 % of the radioactivity was found as this metabolite. *Streptomyces griseus* and *Serratia marcesens* in particular metabolized it further to 4-chloro-o-toluidine. An important observation is that N-demethylation, which usually takes place in animals and plants (e.g., Knowles, 1970) to yield a toxic N-demethylchlordimeform, does not occur in microorganisms. Assuming that this N-demethylation is an oxidative process, the finding of Johnson and Knowles (1970) is consistent with the general observation that hydrolytic processes are more pronounced in microbial systems than oxidative ones.

In a similar study, Arurkar and Knowles (1970) studied degradation of Formetanate® in river-bottom sediment (pH 8.0):

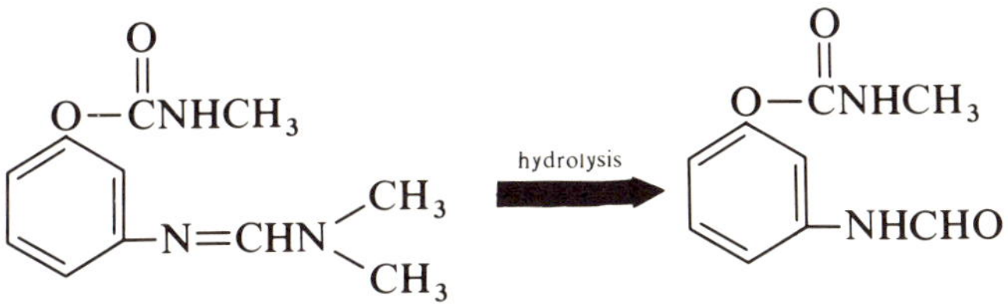

FORMETANATE®

Hydrolytic processes dominated here, also, except that Formetanate® is so susceptible to alkaline hydrolysis that the role of microbial degradation was not clear.

9.6. DEGRADATION BY SUNLIGHT AND OTHER PHYSICAL FACTORS

Among the physical factors known to influence the residual fate of pesticides in nature (e.g., light, air, surfaces, moisture, and pH), sunlight, particularly the ultraviolet portion of sunlight, appears to make the most significant contribution. The sunlight reaching the surface of the earth does not have any ultraviolet component below 280 nm, because the atmosphere effectively eliminates such short-wave ultraviolet rays (Koller, 1965). It is possible, therefore, that artificial ultraviolet radiation, by using an intense mercury lamp (253.7 nm), for instance, can create degradation products which are not produced by the action of natural sunlight. Theoretically, compounds which do not show absorption in any given range of wavelength are not supposed to go through photochemical reactions, and yet we know that even such a compound as dieldrin can be affected by sunlight. Several factors may contribute to cause this phenomenon.

One of the most important factors affecting the rate of sunlight degradation of pesticides and other organic chemicals is the presence of photosensitizers, compounds that facilitate the transfer of the energy of light into the receptor chemicals. In the past, photolytic research work has been carried out in the presence and absence of photosensitizers, although no significant qualitative differences have been found in the metabolic (photolytic) routes. It is known that various photosensitizers also facilitate photolysis of pesticidal compounds. Rosen and Carey (1968) and Rosen et al. (1969, 1970) used both benzophenone and riboflavin-5'-phosphate (FMN) as sensitizers for their studies on photodecomposition of pesticides. Ivie and Casida (1970, 1971a) found rotenone and other pesticides and nonpesticides to be sensitizers for degradation of various insecticides. In addition to rotenone, good photosensitizers were some aromatic amines, anthraquinone (which showed the broadest spectrum), and benzophenone. Insecticidal combinations which synergistically acted as sensitizers were: Abate®–dieldrin, carbyne (12E)–Sumithion,® phenothiazine–DDT, and rotenone–dieldrin. The same authors (1971b) also reported that substituted 4-chromanones and rotenone–sensitized dieldrin converted to photodieldrin on bean leaves. Chlorophyll from spinach chloroplasts plus rotenone acted as a sensitizer for organophosphates, carbamates, pyrethroids, and dinitrophenol insecticides.

Another important factor which affects the nature of photochemical reactions is the medium in which the reacting substance is dissolved or suspended. In laboratory tests, the effect of solvents becomes an important factor in deciding either the speed of the reaction or the nature of the reaction products. The medium and the solvent can influence the outcome of photochemical reactions in two different ways: (1) they can be photosensitizers, and (2) they can act as the reaction partner to the pesticide molecule energized by light. The difference may be best expressed in the following equations:

$$A + B \xrightarrow[hv]{light} A + B^* \longrightarrow (AB) \longrightarrow A' + B' \tag{1}$$

$$A + B \xrightarrow[hv]{light} A^* + B \longrightarrow A' + B' \tag{2}$$

where A is a pesticide molecule and B is the substance which promotes photochemical reactions. In equation (1), B is the photosensitizer and therefore is energized by light first. It in turn reacts with the pesticide to form the photochemical reaction product A'. In equation (2), B is the reacting agent (e.g., solvent or medium) for the pesticide that has been energized by light to produce the product A'. An example of a photonucleophilic reaction is the displacement of a chlorine atom on a benzene ring with —OH in aqueous solution containing nucleophilic agents (Crosby *et al.*, 1972):

$$Cl\langle\hspace{-0.5em}\bigcirc\hspace{-0.5em}\rangle OCH_2COOH \xrightarrow[light]{H_2O} HO\langle\hspace{-0.5em}\bigcirc\hspace{-0.5em}\rangle OCH_2COOH \tag{3}$$

$$\xrightarrow[KCN]{\begin{subarray}{c} light \\ H_2O \end{subarray}} CN\langle\hspace{-0.5em}\bigcirc\hspace{-0.5em}\rangle OCH_2COOH \tag{4}$$

In the range of sunlight irradiation (above 280 nm), the H—OH bond is not expected to break up to provide the nucleophil needed for reaction (3). The most likely source of the nucleophilic agent in this case is the hydroxide ion (OH−), which can be replaced by the cyanide ion (CN−) to yield the corresponding *p*-cyanophenol product (equation 4). There are other examples of solvent-aided photochemical reactions. For instance, Mazzocchi and Rao (1972) found that methanol acted as a hydrogen donor in photolysis of monuron:

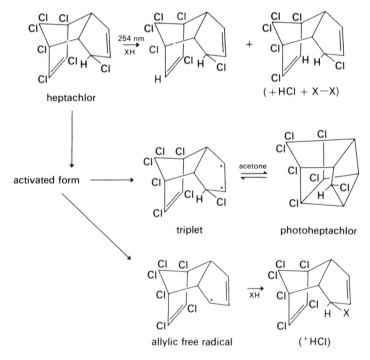

A similar case with heptachlor involving cyclohexane as a hydrogen donor for the released chlorine ion has been studied in detail by McGuire *et al.* (1972). In this case, short-wave ultraviolet (maximum 254 nm) irradiation causes dechlorination at the 5- or 6-position in cyclohexane, which acts

Fig. 9-7. Effects of ultraviolet irradiation and solvents on photochemical reactions on heptachlor. Here acetone acts as a photosensitizer and cyclohexane (designated as XH) as a hydrogen donor, either to the chlorine ion released or to the energized form of heptachlor itself. From McGuire *et al.* (1972).

as a hydrogen donor to heptachlor and the chloride ion released from hepta-chlor. This process appears to involve an activated form of heptachlor, which probably is in a singlet state. On ultraviolet irradiation at 300 nm, heptachlor is exclusively converted into photoheptachlor in acetone. The reaction proceeds via a triplet state ("triplet" in Fig. 9-7), a highly energized state compared to the singlet state, and acetone here acts as a "sensitizer." McGuire *et al.* found that the photochemical reaction proceeded much faster when cyclohexane was added to acetone at 300 nm. This process leads to the formation of 1-cyclohexane substitute of heptachlor. These workers call the process "solvent substitution," since the solvent here simply substitutes for the chlorine in the following fashion:

$$\text{heptachlor} \xrightarrow{hv} \text{heptachlor allylic free radical} + \text{Cl}'$$

$$\text{Cl}' + \text{XH} \rightarrow \text{HCl} + \text{X}'$$

$$\text{X}' + \text{allylic free radical} \rightarrow \text{heptachlor 1-cyclohexane substitute}$$

Thus solvents (or media) can act as either a "sensitizer substitute" or a "donor" to promote photochemical reactions.

9.6.1. Chlorinated Hydrocarbons

There appear to be two types of major reaction processes catalyzed by ultraviolet radiation affecting the chlorinated hydrocarbon compounds: the intramolecular rearrangement process to form isomers of the original compound and the dechlorination process. The former process is best exemplified by the cage formation reaction of dieldrin (Rosen *et al.*, 1966; Robinson *et al.*, 1966). Rosen and Sutherland (1967) further confirmed that a similar photochemical reaction can take place with aldrin to form caged photoaldrin. These reaction products can be found in nature. For instance, Robinson *et al.* (1966) and Korte (1967) and his associates found photodieldrin on the surface of plant leaves. Lichtenstein *et al.* (1970) found photodieldrin in soil samples which had previously been treated with a large amount of aldrin in the field.

For endrin photolysis reactions, Rosen *et al.* (1966) found that keto-endrin (Fig. 9-8D) forms as a result of ultraviolet irradiation. The same rearrangement product has already been reported to form from endrin as a result of thermal decomposition (Phillips *et al.*, 1962). Zabik *et al.* (1971) confirmed this photolytic pathway for endrin to show that ketoendrin, caged endrin aldehyde, and alcohol are the major photolytic products of endrin.

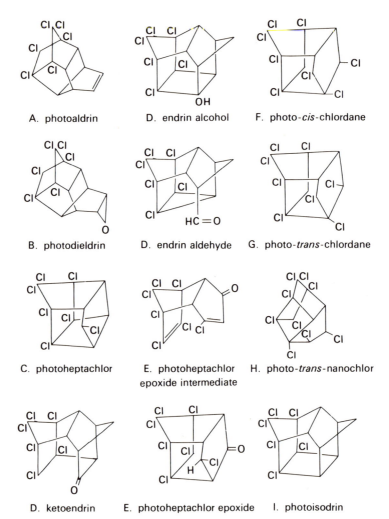

Fig. 9-8. Examples of photoisomerization products among chlorinated hydrocarbon insecticides.

In the presence of a photosensitizer, benzophenone, heptachlor and isodrin photochemically convert into corresponding caged isomers (Fig. 9-8C, I) (Rosen *et al.*, 1969).

Heptachlor epoxide is also known to form a half-cage isomer (Fischler and Korte, 1969; Benson *et al.*, 1971). While there has been some discussion about its chemical identity, it now appears that the product has the molecular structure shown in Fig. 9-8E (Ivie *et al.*, 1972). This photochemical product is severalfold more toxic than heptachlor epoxide to houseflies and mice.

Similar half-cage isomers are also known to form from *cis*-chlordane, γ-chlordane (Fischler and Korte, 1969; Benson *et al.*, 1971; Vollner *et al.*, 1969), *trans*-chlordane, and *trans*-nonachlor (Ivie *et al.*, 1972). Apparently, the orientation of the chlorine atoms on the cyclopentane ring is important in the formation of these half-cage isomers: the chlorine atom oriented toward the center (i.e., *endo* position) prevents the bridge formation at that carbon (Benson *et al.*, 1971).

The second major photolytic reaction, the photodechlorination process, was originally reported by Henderson and Crosby (1967) and by Rosen (1967). The former authors irradiated dieldrin with short-wave

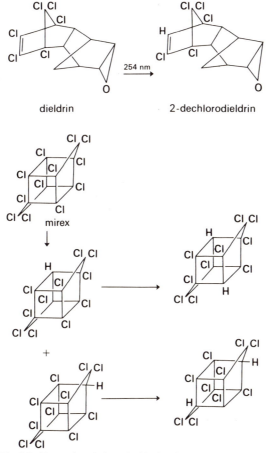

Fig. 9-9. Examples of photodechlorination processes among chlorinated hydrocarbon insecticides.

ultraviolet (ranging less than 290 nm) and found that one of the olefinic chlorines was removed from the original compound, as judged by the results of NMR (nuclear magnetic resonance) and mass spectroscopic analyses. Anderson *et al.* (1968) proposed that the photodechlorination takes place via a triplet state. McGuire *et al.* (1970) showed that the dechlorination process could also take place on heptachlor when it is directly irradiated with short-wave ultraviolet (253.7 nm) in hexane or cyclohexane, while irradiation of the same compound with a relatively long-wave source (300-nm peak) in acetone yielded only the caged isomer of Rosen *et al.* (1969). Benson (1971) irradiated solid and dissolved (in ethyl acetate) dieldrin and confirmed the production of both caged (photodieldrin) and dechlorinated dieldrin isomers. On further irradiation of photodieldrin, two photolytic products were formed, one of which appeared to be a chlorohydrin of photodieldrin.

As for photodegradation of mirex, Alley *et al.* (1972) irradiated pure mirex in cyclohexane or isooctane and obtained several monohydro and dihydro derivatives, indicating that mirex in hydrocarbon solvents mainly degrades via dechlorination processes. The structures of these dechlorination products proposed by Alley *et al.* are shown in Fig. 9-9.

Whenever the pesticide molecule possesses vulnerable moieties, naturally other types of photochemical reactions than the ones cited occur involving structural changes on these sites. For instance, endosulfan possesses a sulfone ring which is susceptible to chemical and biological attack. Indeed, Archer *et al.* (1972) found that it is photochemically degraded with relative ease to yield mainly endosulfandiol.

9.6.2. Aromatic Pesticides

Four basic types of photochemical reactions may take place when aromatic pesticides are exposed to ultraviolet: ring substitution, hydrolysis (in aqueous solution), oxidation, and polymerization. Since many workers have chosen herbicidal compounds for such photodegradation studies, we shall discuss several examples of these.

The most common form of the nucleophilic ring substitution reaction for the chlorinated aromatics is the replacement of a ring chlorine by a hydroxyl group. An example is the photolytic degradation of 2,4-D in aqueous solution. The final reaction product of 2,4-D is 1,2,4-benzenetriol, indicating both cleavage of the ether group and replacement of ring chlorines by hydroxyl (Crosby and Tutass, 1966). A similar reaction can take place with cyclic triazine herbicides, where the chlorine atom on the triazine ring is known to be vulnerable (Pape and Zabik, 1970).

The hydrolytic photodecomposition process is also a common reaction in aqueous solution. In addition to the example of 2,4-D, Pape *et al.* (1970) demonstrated that an *N*-methylphenylcarbamate (C-8353, Ciba and Geigy Corporation) can be decarbamylated to yield the corresponding substituted phenol. Crosby and Tutass (1966) showed that carbaryl was photochemically decomposed to methyl isocyanate and 1-naphthol. Exposure of parathion to sunlight gave rise to *p*-nitrophenol (Hasegawa, 1959), indicating a hydrolytic splitting of the P—O linkage of parathion. An example of nonhydrolytic cleavage of an organophosphate, however, can be found in the photolytic degradation reaction of Guthion® (azinphosmethyl) to *O,O*-dimethyl-*S*-4-oxo-1,2,3-benzotriazin on irradiation in hexane, indicating utilization of two hydrogens from the solvent to degrade the parent molecule (Kurihara *et al.*, 1966).

Oxidative and reductive photochemical reactions are also important degradation processes of pesticides. Thus chlorobenzoic acid gives rise to benzaldehyde on ultraviolet irradiation (Crosby, 1966). Exposure of parathion to sunlight and ultraviolet radiation results in the production of paraoxon and other oxo analogues of parathion (Koivistoinen, 1963; Payton, 1953; Sandi, 1958; Hasegawa, 1959). 4,4'-Dichlorobenzophenone is also formed from DDT by the aid of ultraviolet; the reaction is partly carried out by an oxidative photochemical reaction (i.e., in the presence of oxygen; Fleck, 1949).

Polymerization of substituted aromatics by the aid of ultraviolet irradiation sometimes takes place along with other initial structural changes of the pesticide derivatives. Thus DDT loses two chlorine atoms first and then forms a dimer (Fig. 9-10) in the absence of air.

A combination of such complex oxidation, ring substitution, and polymerization reactions can produce many unexpected photolytic products. An excellent example of such an interaction of different photochemical reactions on a single pesticidal compound can be found in the work by Munakata and Kuwahara (1969) on the degradation of pentachlorophenol under sunlight. Various monomers of chlorinated quinones, hydroxyquinones, and phenols are originally formed. These intermediates and the parent compounds react with each other to form ether linkages to yield both dimers and trimers.

Despite the complexity of the degradation patterns of pesticides by environmental factors, several basic principles can be pointed out to facilitate understanding. It is important to consider (1) whether such degradation products are stable enough to become "terminal residues," (2) whether such terminal residues have high affinity for biological materials (either direct or through secondary means such as a food chain) to cause biological magnification phenomena, and (3) whether these products are harmful to some forms of biological systems.

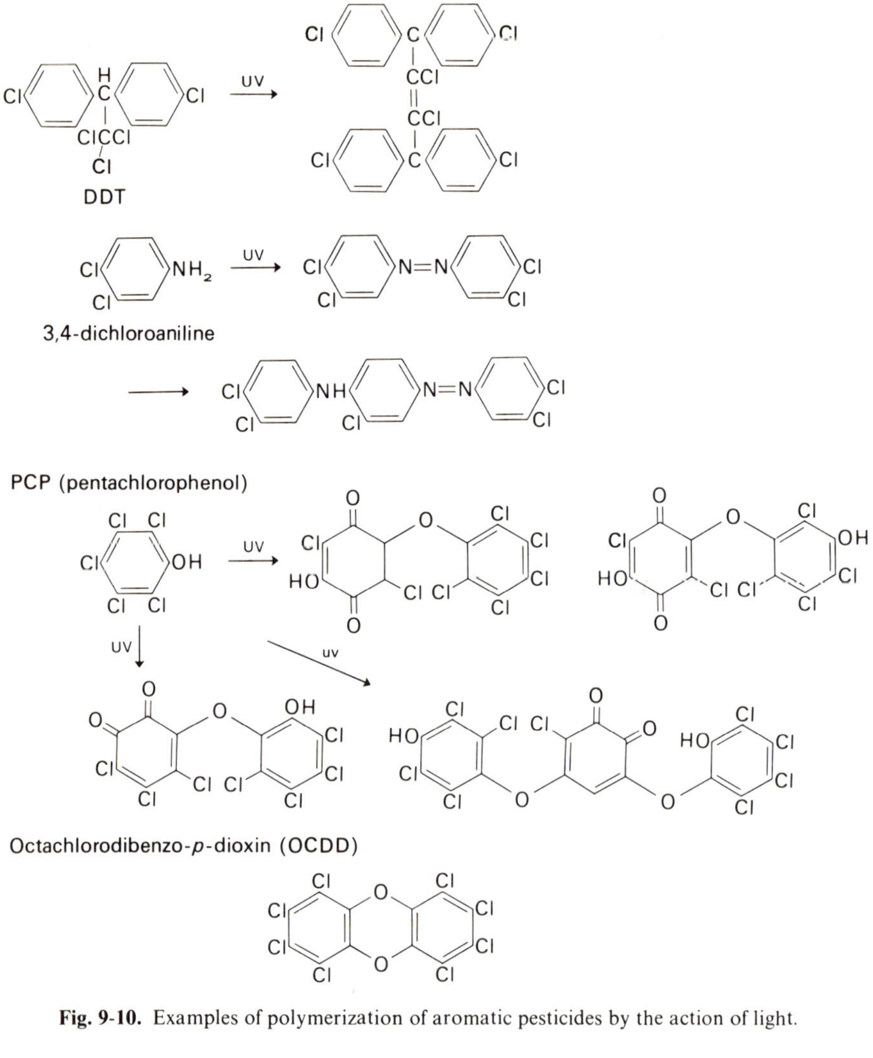

Fig. 9-10. Examples of polymerization of aromatic pesticides by the action of light.

Pesticides, in general, are low-volume and highly biologically active environmental contaminants. Unlike fertilizers, industrial wastes, etc., pesticides do not physically alter environments as such in quantity. Their potential hazards, therefore, must be judged in terms of their biological effects, and studies of metabolic and physical degradation activities in the environment should be primarily concerned with the detection of potential "terminal residues."

It is now generally acknowledged that the toxic contaminant of 2,4,5-T is 3,4,7,8-tetrachlorodibenzo-*p*-dioxin. This compound exists in crude 2,4,5-T preparations at levels of 0.1 to several ppm. Its oral LD_{50} in rats is in the neighborhood of $\mu g/kg$. Studies made by the United States Department of Agriculture (Kearney, 1970) and by us (Matsumura and Benezet, 1973) indicate that it is much more stable to microbial attack than 2,4,5-T. Along with the suggestion that octachlorodibenzo-*p*-dioxin (OCDD, the last formula at the bottom of Fig. 9-10) may form by photolysis from pentachlorophenol (Crosby and Wong, 1970), it becomes probable that chlorinated dibenzodioxins are the toxic terminal residues of both chlorinated phenoxy and phenol pesticides. Another group of toxic material, chlorinated dibenzofurans, have been reported to form from certain components of PCBs (Crosby *et al.*, 1973):

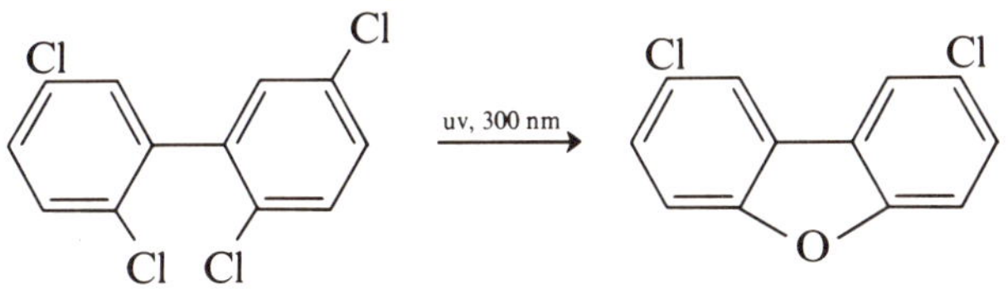

Accumulation of abnormal amounts of β-BHC in the environment, particularly in biological systems, also serves as a good example of a buildup of unexpected terminal residues. Metabolic studies in the laboratory should be followed, therefore, with quantitative survey work in the field as well as with toxicological evaluations of the effects of such environmental contaminants on various biological systems.

9.7. REFERENCES

Ahmed, M. K., and J. E. Casida (1958). *J. Econ. Entomol.* **51**:59.

Ahmed, M. K., J. E. Casida, and R. E. Nichols (1958). *J. Agr. Food Chem.* **6**:740.

Albone, E. S., G. Eglinton, N. C. Evans, and M. M. Rhead (1972). *Nature* **240**:420.

Alexander, M. (1965). *Soil Sci. Soc. Am. Proc.* **29**:1.

Alexander, M. (1972). In *Environmental Toxicology of Pesticides*. F. Matsumura, G. M. Boush, and T. Misato, eds. Academic Press, New York, p. 365.

Allan, J. (1955). *Nature* **175**:1131.

Alley, E. G., D. A. Dollar, B. R. Layton, and J. P. Minyard, Jr. (1972). *J. Agr. Food Chem.* **21**:139.

Anderson, C. M., J. B. Bremmer, I. W. McCay, and R. N. Warrener (1968). *Tetrahedron Letters* **33**:1255.

Anderson, J. P. E., and E. P. Lichtenstein (1971). *Can. J. Microbiol.* **17**:1291.

Archer, T. E., I. K. Nazer, and D. G. Crosby (1972). *J. Agr. Food Chem.* **20**:954.

Armstrong, D. E., G. Chesters, and R. F. Harris (1967). *Soil Sci. Soc. Am. Proc.* **31**:61.

Arurkar, S., and C. O. Knowles (1970). *Bull. Environ. Contam. Toxicol.* **5**:324.

Batterton, J. C., G. M. Boush, and F. Matsumura (1971). *Bull. Environ. Contam. Toxicol.* **6**:589.

Benezet, H. J., and F. Matsumura (1973). *Nature* **243**:480.

Benezet, H. J., and F. Matsumura (1974). *J. Agr. Food Chem.* In press.

Benson, W. R. (1971). *J. Agr. Food Chem.* **19**:66.

Benson, W. R., P. Lombardo, I. J. Egry, R. D. Ross, Jr., R. P. Barron, D. W. Mastbrook, and E. A. Hansen (1971). *J. Agr. Food Chem.* **19**:857.

Bixby, M., G. M. Boush, and F. Matsumura (1971). *Bull. Environ. Contam. Toxicol.* **6**:491.

Bollag, J.-M. (1972). *Crit. Rev. Microbiol.* **2**(1):35. A. I. Laskin and H. Lechevalier, eds. CRC Press, Cleveland.

Bollag, J.-M., and S.-Y. Liu (1971). *Soil. Biol. Biochem.* **3**:337.

Boush, G. M., and F. Matsumura (1967). *J. Econ. Entomol.* **60**:918.

Bradbury, F. R. (1963). *Ann. Appl. Biol.* **52**:361.

Bro-Rasmussen, F., E. Noddegaard, and K. Voldum-Claussen (1968). *J. Sci. Food Agr.* **19**:278.

Cook, J. W. (1957). *J. Agr. Food Chem.* **5**:859.

Coppedge, J. R., D. A. Lindquist, D. L. Bull, and H. W. Dorough (1967). *J. Agr. Food Chem.* **15**:902.

Crosby, D. G. (1966). In *Abstracts of the 152nd Meeting of the American Chemical Society*, New York. American Chemical Society, Washington, D.C.

Crosby, D. G. (1969). *Residue Rev.* **25**:1.

Crosby D. G., and H. P. Tutass (1966). *J. Agr. Food Chem.* **14**:596.

Crosby, D. G., and A. S. Wong (1970). In *160th Meeting of the American Chemical Society*, Chicago. American Chemical Society, Washington, D.C.

Crosby, D. G., K. W. Moilanen, M. Nakagawa, and A. S. Wong (1972). In *Environmental Toxicology of Pesticides*. F. Matsumura, G. M. Boush, and T. Misato, eds. Academic Press, New York, p. 423.

Crosby, D. G., K. W. Moilanen, and A. S. Wong (1973). Environmental generation and degradation of dibenzodioxins and dibenzofurans, *Environ. Health Perspect.* **5**:259.

Cserjesi, A. J., and E. L. Johnson (1972). *Can. J. Microbiol.* **18**:45.

Dalton, R. L., A. W. Evans, and R. C. Rhodes (1965). *Southern Weeds Conf.* **18**:72.

Finley, R. B., and R. E. Pillmore (1963). *Am. Inst. Biol. Sci. Bull.* **13**:41.

Fischler, H.-M., and F. Korte (1969). *Tetrahedron Letters* **34**:2967.

Fleck, E. E. (1949). *J. Am. Chem. Soc.* **71**:1034.

Focht, D. D., and M. Alexander (1971). *J. Agr. Food Chem.* **19**:20.

French, A. L., and R. A. Hoopingarner (1970). *J. Econ. Entomol.* **63**:756.

Getzin, L. W., and I. Rosefield (1968). *J. Agr. Food Chem.* **16**:598.

Goto, M. (1972). *Environmental Safety and Quality*, Vol. 1, p. 211. F. Coulson and F. Korte, eds. Georg Thieme Verlag, Berlin.

Hamilton, R. H., and D. E. Moreland (1962). *Science* **135**:373.

Hasegawa, T. (1959). *Fukuoka Acta Med.* **50**:1900.

Henderson, G. L., and D. G. Crosby (1967). *J. Agr. Food Chem.* **15**:888.

Ivie, G. W., and J. E. Casida (1970). *Science* **167**:1620.

Ivie, G. W., and J. E. Casida (1971a). *J. Agr. Food Chem.* **19**:405.

Ivie, G. W., and J. E. Casida (1971b). *J. Agr. Food Chem.* **19**:410.

Ivie, G. W., J. R. Knox, S. Khalifa, I. Yamamoto, and J. E. Casida (1972). *Bull. Environ. Contam. Toxicol.* **7**:376.

Jensen, S., R. Goethe, and M.-O. Kindstedt (1972). *Nature* **240**:421.

Johnson, B. T., and C. O. Knowles (1970). *Bull. Environ. Contam. Toxicol.* **5**:158.

Kapoor, I. P., R. L. Metcalf, R. F. Nystrom, and G. K. Sangha (1970). *J. Agr. Food Chem.* **18**:1145.

Kearney, P. C. (1970). Private communication to author.

Kearney, P. C., and T. J. Sheets (1965). *J. Agr. Food Chem.* **13**:369.

Kearney, P. C., D. D. Kaufman, and M. Alexander (1967). In *Soil Biochemistry*, Vol. 1. A. D. McLaren and G. H. Peterson, eds. Marcel Dekker, New York, p. 318.

Keil, J. E., and L. E. Priester (1969). *Bull. Environ. Contam. Toxicol.* **4**:169.

Kiigemagi, U., H. E. Morrison, J. E. Roberts, and W. B. Bollen (1958). *J. Econ. Entomol.* **51**:198.

Knowles, C. O. (1970). *J. Agr. Food Chem.* **18**:1038.

Koivistoinen, P. (1963). *Acta Agr. Scand.* **12**:285.

Koller, L. R. (1965). *Ultraviolet Radiation*, 2nd ed. John Wiley & Sons, New York.

Konrad, J. G., D. E. Armstrong, and G. Chesters (1967). *Agron. J.* **59**:591.

Korte, F. (1967). In *Proceedings of the Commission of Terminal Residues and of the Commission of Residue Analysis*, International Union of Pure and Applied Chemistry, Vienna, August. IUPAC, Vienna, Austria.

Korte, F. (1970). *J. Assoc. Offic. Anal. Chemists* **35**:988.

Korte, F., and M. Stiasni (1964). *Ann. Chem.* **673**:146.

Korte, F., G. Ludwig, and J. Vogel (1962). *Ann. Chem.* **656**:35.

Kurihara, N. H., D. G. Crosby, and H. F. Beckman (1966). In *Abstracts of the 152nd Meeting of the American Chemical Society*, New York. American Chemical Society, Washington, D.C.

Laanio, T. L., G. Dupuis, and H. O. Esser (1972). *J. Agr. Food Chem.* **20**:1213.

Lichtenstein, E. P., and K. R. Schulz (1960). *J. Econ. Entomol.* **5**:192.

Lichtenstein, E. P., K. R. Schulz, T. W. Fuhremann, and T. T. Liang (1970). *J. Agr. Food Chem.* **18**:100.

Liu, S.-Y., and J.-M. Bollag (1971*a*). *J. Agr. Food Chem.* **19**:487.

Liu, S.-Y., and J.-M. Bollag (1971*b*). *Pesticide Biochem. Physiol.* **1**:366.

Matsumura, F., and G. M. Boush (1966). *Science* **153**:1278.

Matsumura, F., and G. M. Boush (1967). *Science* **156**:959.

Matsumura, F., and G. M. Boush (1968). *J. Econ. Entomol.* **61**:610.

Matsumura, F., and G. M. Boush (1971). In *Soil Biochemistry*, Vol. II. A. D. McLaren and J. J. Skujins, eds. Marcel Dekker, New York, p. 320.

Matsumura, F., G. M. Boush, and A. Tai (1968). *Nature* **219**:965.

Matsumura, F., T. A. Bratkowski, and K. C. Patil (1969). *Bull. Environ. Contam. Toxicol.* **4**:262.

Matsumura, F., K. C. Patil, and G. M. Boush (1970). *Science* **170**:1206.

Matsumura, F., K. C. Patil, and G. M. Boush (1971*a*). *Nature* **230**:325.

Matsumura, F., V. G. Khanvilkar, K. C. Patil, and G. M. Boush (1971*b*). *J. Agr. Food Chem.* **19**:27.

Matsumura, F., and H. J. Benezet (1973). *Envir. Health Perspect.* **5**:253.

Mazzocchi, P. H., and M. P. Rao (1972). *J. Agr. Food Chem.* **20**:957.

McGuire, R. R., M. J. Zabik, R. D. Schuetz, and R. D. Flotard (1970). *J. Agr. Food Chem.* **18**:319.

McGuire, R. R., M. J. Zabik, R. D. Schuetz, and R. D. Flotard (1972). *J. Agr. Food Chem.* **20**:856.

Mendel, D. B., A. K. Klein, J. T. Chen, and M. S. Walton (1967). *J. Assoc. Offic. Agr. Chemists* **50**:897.

Menzel, D. B., S. M. Smith, R. Miskus, and W. M. Hoskins (1961). *J. Econ. Entomol.* **54**:9.

Miles, J. R., C. M. Tu, and C. R. Harris (1969). *J. Econ. Entomol.* **62**:1334.

Miyamoto, J., K. Kitagawa, and Y. Sato (1966). *Japan. J. Exptl. Med.* **36**:211.

Miyazaki, S., G. M. Boush, and F. Matsumura (1969). *Appl. Microbiol.* **18**:972.

Miyazaki, S., G. M. Boush, and F. Matsumura (1970). *J. Agr. Food Chem.* **18**:87.

Munakata, K., and M. Kuwahara (1969). *Residue Rev.* **25**:13.

Pape, B. E., and M. J. Zabik (1970). *J. Agr. Food Chem.* **18**:202.

Pape, B. E., M. F. Para, and M. J. Zabik (1970). *J. Agr. Food Chem.* **18**:490.

Patil, K. C., F. Matsumura, and G. M. Boush (1970). *Appl. Microbiol.* **19**:879.

Payton, J. (1953). *Nature* **171**:355.

Pfister, R. M. (1972). *Crit. Rev. Microbiol.* **2**:1. CRC Press, Cleveland.

Phillips, D. D., G. E. Pollard, and S. B. Soloway (1962). *J. Agr. Food Chem.* **10**:217.

Plimmer, J. R., and P. C. Kearney (1969). In *158th Meeting of the American Chemical Society*, New York. American Chemical Society, Washington, D.C.

Plimmer, J. R., P. C. Kearney, and D. W. Vonendt (1968). *J. Agr. Food Chem.* **16**:594.

Raghu, K., and I. C. MacRae (1966). *Science* **154**:264.

Robinson, J., A. Richardson, B. Bush, and K. E. Elgar (1966). *Bull. Environ. Contam. Toxicol.* **1**:127.

Rosen, J. D. (1967). *Chem. Commun.* **1967**:189.

Rosen, J. D. (1969). Private communication to the author.

Rosen, J. D., and W. F. Carey (1968). *J. Agr. Food Chem.* **16**:536.

Rosen, J. D., and D. J. Sutherland (1967). *Bull. Environ. Contam. Toxicol.* **2**:1.

Rosen, J. D., D. J. Sutherland, and G. R. Lipton (1966). *Bull. Environ. Contam. Toxicol.* **1**:133.

Rosen, J. D., D. J. Sutherland, and M. A. Q. Khan (1969). *J. Agr. Food Chem.* **17**:404.

Rosen, J. D., M. Siewierski, and G. Winnett (1970). *J. Agr. Food Chem.* **18**:494.

Sandi, E. (1958). *Nature* **181**:499.

Sethunathan, N., E. M. Bautista, and T. Yoshida (1969). *Can. J. Microbiol.* **15**:1349.

Skujins, J. J. (1966). *Soil Biochemistry*. A. D. McLaren, and G. H. Peterson, eds. Marcel Dekker, New York, p. 371–414.

Smith, J. W., and T. J. Sheets (1966). *Weed Sol. Am. Abst.* **6**:39.

Sweeney, R. A. (1969). In *Proceedings of the 12th Conference on Great Lakes Research*, p. 98. International Association for Great Lakes Research, U.S. Bureau of Commercial Fisheries, P.O. Box 640, Ann Arbor, Michigan.

Tiedje, J. M., and M. Alexander (1967). In *Abstracts of the American Society of Agronomy Annual Meeting*, Washington, D.C. p. 94. American Society of Agronomy, Madison, Wisconsin.

Tsukano, U., and A. Kobayashi (1972). *Agr. Biol. Chem.* **36**:116.

Tweedy, B. G., C. Loeppky, and J. A. Ross (1970). *Science* **168**:482.

Vollner, L., W. Klein, and F. Korte (1969). *Tetrahedron Letters* **34**:2967.

Wang, C. M., and F. Matsumura (1970). *J. Econ. Entomol.* **63**:1731.

Wang, C. M., T. Narahashi, and M. Yamada (1971). *Pesticide Biochem. Physiol.* **1**:84.

Wedemyer, G. (1968). *Appl. Microbiol.* **16**: 661.

Williams, I. H., M. J. Brown, and G. Finlayson (1972). *J. Agr. Food Chem.* **20**:1219.

Yoshida, T., and T. F. Castro (1970). *Soil Sci. Soc. Am. Proc.* **34**:440.

Yule, W. N., M. Chiba, and H. W. Morley (1967). *J. Agr. Food Chem.* **15**:1000.

Zabik, M. J., R. D. Schuetz, W. L. Burton, and B. E. Pape (1971). *J. Agr. Food Chem.* **19**:308.

Zayed, S. M. A. D., I. Y. Mostafa, and A. Hassan (1965). *Arch. Mikrobiol.* (*Berlin*) **51**:118.

Chapter 10

Effects of Pesticides on Wildlife

There are two important factors which must be considered before the impact of any pesticidal material on various ecosystems can be critically discussed. These are (1) the current levels of contamination in the ecosystems and (2) the susceptibility of biological material to the pesticides. The contamination factor should be considered in relation to the dynamics of pesticide movement (e.g., accumulation through the food chain). The susceptibility data should eventually include such subtle effects as changes in egg shell thickness and courting and nesting behavior, irritability, and chronic effects to be complete. One of the most important parameters for assessing the sublethal effects of pesticides appears to be to what degree they affect the reproductive processes.

10.1. GENERAL SURVEY OF RESIDUE LEVELS IN VARIOUS ECOSYSTEMS

Since pesticides are not initially sprayed uniformly everywhere, and since the process of biological concentration takes place in nature, no general statement can be made as to the uniform levels of pesticide contamination in any country. According to the USDA survey (1968), 75% of all the insecticides used in this country are applied to less than 2% of the land and, furthermore, 75% of the total land area of the United States has never been sprayed with pesticides.

Perhaps the most convenient criteria for measuring the levels of available pesticides in the area are the amounts of pesticides in water and, to a lesser extent, air. (Pesticides irreversibly bound in soils, for instance, are not considered available.) Table 10-1 shows that the average concentrations

TABLE 10-1. Mean Concentrations (expressed as ppt[a]) of Chlorinated Hydrocarbon Insecticides Found in U.S. River Systems

Location	Reference	Number of sites	DDT analogues	γ-BHC	Dieldrin	Endrin
U.S. major river basins	Breidenbach *et al.*					
	(1967)	99	8.2	trace	6.9	2.41
	Weaver *et al.* (1965)	97	9.3	—	7.5	5.5
	Green *et al.* (1967)	109	8.3	2.2	5.9	3.6
Mississippi delta	USDA (1966)	10	112	28.0	10.0	541
California rivers	Keith and Hunt (1966)	82	0.62	0.01	—	—
U.S. western streams	Brown and Nishioka					
	(1967)	11	10.3	2.8	2.3	1.4
British rivers	Lowden *et al.* (1969)	7	1.6	18.7	3.3	—
Recommended maxima for U.S. drinking water	Nicholson (1969)	—	42	56	17	1

From Edwards (1970).

[a]Same as ng/liter.

of pesticides are relatively constant throughout the major rivers, except in the Mississippi delta, where the concentrations of DDT analogues, endrin, and γ-BHC increase by tenfold. The data for British rivers reveal that the level of γ-BHC in relation to other insecticides is high, reflecting its heavier usage in Britain than in the United States. Comparable figures cited by Metcalf (1964) are DDT 0–87, DDE 0–18, dieldrin 0–118, and endrin 0–94 ppt (parts per trillion) in water from 100 locations in the United States.

Studies on the levels of pesticides in the air have been less extensive. West (1964) collected 18 samples of ambient air over California cities and found DDT in 16 of them. Air samples from the national air-sampling network contained 0.0002–0.34 μg of DDT per 1000 m^3. Nevertheless, evidence indicates that high levels of chlorinated hydrocarbon insecticides accumulate in dust particles, which could be returned to the earth with rainwater (Table 10-2). The levels of chlorinated hydrocarbon insecticides in rainwater appear to be higher than in other water samples, both in the United States and in Britain. The level of γ-BHC in rainwater is higher in Britain,

TABLE 10-2. Pesticide Levels Found in Dust and Rainwater Collected in Ohio[a]

Pesticide	Dust[b] (ppm)	Rainwater[c] (ppt)
DDT	0.6	190
DDE	0.2	18
γ-BHC	—	25
Dieldrin	0.003	
Chlordane	0.5	
Heptachlor epoxide	0.04	
Ronnel	0.2	
2,4,5-T	0.04	
Total organic chlorine	1.34	240

From Cohen and Pinkerton (1966).

[a]Monthly dustfall in Cincinnati, Ohio, is about 15 tons per square mile.
[b]Expressed per air-dried weight of dust.
[c]British rainwater contained 1–79 (DDT analogues), 29–100 (γ-BHC), and 9–28 (dieldrin) ppt of pesticides (Tarrant and Tatton, 1968; Wheatley and Hardman, 1965).

in accordance with the data from all other water samples from that country.

Since these media (water, air, dust, etc.) are mobile forces in the environment, the pesticide levels in them are of great concern. Indeed, Cohen and Pinkerton (1966) suggest that the major mode of atmospheric transportation of chlorinated insecticides in the United States is by dust particles.

While determining pesticide levels in water is the most convenient method for assessing the extent of contamination in any particular area, there are several problems in using such data as a direct index. The initial problem is the low water solubility of many pesticides. The water solubility of DDT is cited as around 1.2 ppb (parts per billion) or 1200 ppt (see Tables 10-1 and 10-2, which show values found in water approaching 200 ppt). Therefore, even though a much higher amount of DDT may be present at a site, the water sample cannot show a higher concentration value than 1.2 ppb. Also, pesticides with low water solubility tend to adhere to particles and other solid matter in the water, so that measurement of pesticide levels in water is greatly influenced by removal of particulate matter. Tiny particles tend to stay in water after filtration and can give rise to wrong figures for the pesticide levels. An equally great problem is how to extract accurate amounts of pesticide from a large volume of water (Lamar *et al.*, 1966; Hylin, 1971).

Zahik *et al.* (1971) have used as a criterion for suspended matter that retained by a 5-μm filter ; according to them, this represents short-term contamination, while more long-term contamination is revealed in the bottom sediment (using a core sample 6 cm in diameter and 12 cm long).

However, the effects of locality cannot be neglected. Studies by Reinert (1970) clearly demonstrate that all fish species so far tested from Lake Michigan contain more insecticides (DDT derivatives and dieldrin) than those from Lake Superior (Table 10-3). While the levels of insecticides in

TABLE 10-3. Average Concentration (ppm on whole fish basis) of DDT (1965–1968) and Dieldrin (1967–1968) in Two Fish Species from Individual Great Lakes

	DDT		Dieldrin	
	Alewife	American smelt	Alewife	American smelt
Lake Michigan	3.89	2.31	0.11	0.06
Lake Ontario	1.99	1.58	0.06	0.10
Lake Huron	2.44	0.75	0.05	0.04
Lake Erie	1.59	1.06	0.14	0.04
Lake Superior	0.72	0.32	0.05	0.02

From Reinert (1970).

water particles and sediments alone do not determine the final level of contamination in biological systems, local differences in contamination clearly support the view that a concentration-dependent accumulation phenomenon does take place in any given ecosystem.

Another way to assess contamination is to measure the pesticide levels in indicator organisms or in wildlife and plants. Indicator organisms* are ones which tend to accumulate pesticides from the ambient medium. Two factors are generally considered important in selecting an indicator organism : (1) the organism shows a tendency toward a high degree of biological concentration, and (2) the degree of concentration of pesticides is proportional to the level of pesticides in the ambient medium. In the case of water samples, oysters and other shellfish which filter water to obtain food are generally regarded to be useful indicator organisms. In parts per trillion ranges, mollusks do not show any ill effects (unlike shrimp and crabs). Fish accumulate residues gradually (much slower than mollusks), but their pesticide

*This term is also sometimes used to refer to organisms that are susceptible to pesticides. Such organisms are used for bioassay purposes for various pesticides.

levels change when fat is mobilized. Oysters and other shellfish not only accumulate chlorinated hydrocarbon pesticides rapidly but also discharge them when they are returned to pesticide-free media (Butler, 1967*a,b*). More than 150 estuarine stations have been established to monitor pesticide levels on the coasts of the United States by using indicator organisms. Hammerstrom *et al.* (1967), for instance, analyzed 270 oyster samples from estuarine areas of Louisiana and Alabama, along with water and sediment samples, and concluded that the number of oyster samples containing pesticides matched generally well with the water and sediment samples. The data summarized in Table 10-4 show a general relationship between DDT in the environment and the amount accumulated in oysters.

TABLE 10-4. Levels of DDT Analogues[a] in Oyster, Water, and Bottom Sediment Samples

Location	Oyster (ppm)	Water (ppb)	Sediment (ppm)
Alabama, Mobile Bay area[b]	0.68	1	0.012
	0.29	1	0.013
	0.51	1	0.010
	0.32	1	0.007
East coast[c]	151.0	10.0[d]	—
	30.0	1.0	—
	7.0	0.1	—
Pacific coast[c]	20.0	1.0	—

[a]DDT(*p, p'*), DDD (TDE), and DDE.
[b]From Hammerstron *et al.* (1967).
[c]From USDI (1964).
[d]This value exceeds the solubility limit of DDT in water (i.e., 1.2 ppb). It is probable that DDT attached to suspended matter is included in this figure.

Other species could become important indicator organisms if they are abundant locally, or important in the energy cycle, or have a wide range of distribution. Burnett (1971), for instance, used a common surf-zone sand crab when studying the California coast. Earthworms are good indicator organisms for soil residue assessment since they engorge a large quantity of soil to obtain nutrition. Wheatley and Hardman (1968) studied the relationship between the amount of pesticides taken up by earthworms and the levels in the soil (Fig. 10-1). While interspecies differences were obvious (e.g., *Allolobophora chlorotica* accumulated more pesticides than other species), a general proportionality did exist between the levels of pesticides

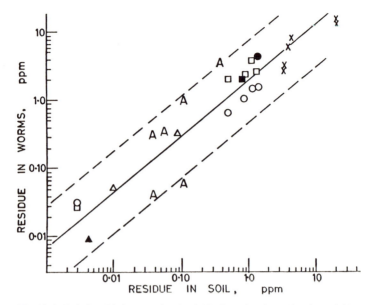

Fig. 10-1. Relationship between levels of chlorinated hydrocarbon insecticides in the soil and in earthworms. From Wheatley and Hardman (1968); indirectly cited from Edwards (1970). Comparable U.S. figures (designated by "A") have been added for comparison. From the data of Davis (1968).

in the soil and in the earthworms. The choice of earthworms as an indicator is also excellent in view of their importance in the food chain for bird populations. However, the relationship between the amount of pesticide intake and the amount appearing in certain tissues or products is better understood in aquatic animals (e.g., see Fig. 10-2).

Use of the residue data from wild animals and plants has been the most direct method of assessing pesticide contamination. For instance, if fish from one region are found to contain more pesticides than ones of the same species from another area, this indicates at least that higher levels of pesticides are available in the former area. Such data also provide an overall picture of the contamination, which is certainly useful as a diagnostic tool.

It is abundantly clear that simple residue data on soil, water, and sediments alone cannot provide indications of the level of pesticides in the wildlife. Even if the fish from Lake Erie show a lower level of DDT than those from Lake Michigan (Table 10-3), it does not necessarily mean that the DDT concentration in Lake Erie is lower than that in Lake Michigan. Water that contains particles high in organic matter tends to make lipophilic pesticides less available to fish. Bottom mud of high organic content readily

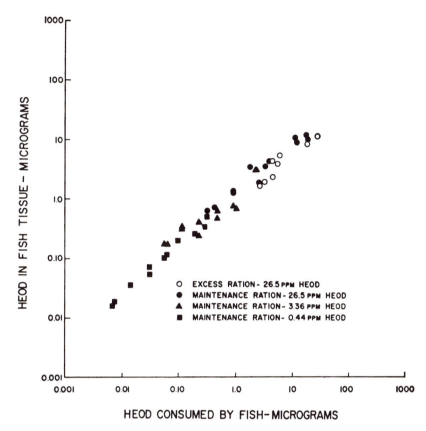

Fig. 10-2. Relationship between dieldrin in tissues of fish (Reticulate sculpins) and the amount of dieldrin consumed through the ration of dieldrin-containing tubificid worms (food). Altogether four different concentrations of dieldrin were given. From Chadwick and Brocksen (1969).

binds pesticides, in contrast to sandy or rocky bottom materials (Cope, 1966). It is not uncommon that fish from clean lakes contain more DDT than ones from lakes with higher degrees of eutrophication.

Another important consideration is the species-specific differences in the rate of insecticide pickup. The data summarized in Table 10-5 demonstrate that in earthworms the species-specific difference is much more pronounced than the difference among insecticides. For instance, *Lumbricus terrestris* accumulates each insecticide at a level slightly higher or lower than that found in soil, while *Allolobophora chlorotica* consistently accumulates much higher levels of all the insecticides.

TABLE 10-5. Examples of Chlorinated Hydrocarbon Residues in ppm Found in Soil Invertebrates in Agricultural Areas in Great Britain and the United States

				DDT-R			
	Aldrin	Dieldrin	Lindane	p, p'-DDT	o, p'-DDT	p, p'-DDE	p, p'-TDE
Great Britain[a]							
Soil	0.72	0.64	0.004	0.63	0.14	0.17	—
Earthworms							
Lumbricus terrestris	0.053	1.6	0.0064	0.54	0.068	0.49	—
Allolobophora longa	0.28	2.2	0.0060	0.77	0.19	0.38	—
A. calignosa	0.52	3.8	0.011	1.5	0.35	0.65	—
A. chlorotica	0.98	4.6	0.013	2.9	0.72	1.0	—
A. rosea	0.64	3.9	0.017	1.6	0.30	0.70	—
Octolasion cyaneum	0.84	2.4	0.0076	0.67	0.19	0.38	—
United States[b]							
Soil	0.03	0.1	—	0.3	0.04	0.1	0.03
Earthworms	0.02	0.5	0.05	2.1	0.2	0.9	0.2
Beetle sp.[c]	0	0.08	0	0	0	1.4	0

[a]From Wheatley and Hardman (1968).
[b]From Davis (1968).
[c]Average of two species, *Harpalus* sp. and *Agronum* sp. Insignificant residues are listed as 0.

Direct residue data from animals and plants can be utilized to estimate current levels of actually available pesticides entering an ecosystem from the surrounding environment by monitoring chronological changes. Residue data from soil, water, and sediment can be used to elucidate the "partition" behavior of any pesticide with respect to the surrounding environment and the biological systems. It is possible that the kinetics of such "partitioning" in an ecosystem can be empirically developed for each pesticide to complete the background knowledge on levels of contamination in relation to their effects on wildlife.

10.2. HAZARDS TO WILDLIFE

10.2.1. Acute Toxicity

The most convenient and direct parameter for measuring pesticide effects on wildlife is the direct, acute LD_{50} value. Basic acute toxicity data can

TABLE 10-6. Summary of Acute Oral LD_{50} in mg/kg Values of Representative Insecticides for Several Wildlife Species

	Chlorinated hydrocarbons			Organophosphates			Carbamates		
	DDT	Dieldrin	Endrin	Abate®	Dursban®	Parathion	Carbaryl	Propoxur	Zectran®
Mallards ♀	2240	381	5.64	80–100	75.6	1.90	2179	11.9	3.0
Pheasants ♀	1296	79	1.78	21.5	17.7	12.4	707	20 (♂)	4.5
Coturnix ♂	841	70	—	84.1	17	5.95 (♀)	2290	28.3 (♀)	3.2
Pigeons ♂ + ♀	4000	27	2.5	50.1	26.9	2.52	1000–3000	60.4	6.5
Lesser sandhill cranes ♂ + ♀	1200	—	—	—	25–50	—	—	40–60	1.0–4.5
Bullfrogs ♂ + ♀	2000	—	—	2000	400	—	4000	595	283–800
House sparrows ♀	—	78	—	35.4	21.0 (♂)	3.36	—	12.8	50.4
Canada geese ♂ + ♀	—	50–150	—	—	80	—	1790	5.95[a]	2.64
Gray partridge ♀	—	9 (♀)	—	—	—	16 (♂)	—	—	—
Mule deer ♂	—	75–150	—	—	—	22–44	200–400	100–350	20–30
Rainbow trout[b]	0.007	0.04	0.0018	1.5[c]	0.05	2	4.38[a]	—	8
Bluegill[b]	0.008	0.007	0.0005[d]	—	—	0.047	6.76[a]	—	11.2[d]
Shrimp, sand[b]	0.003	0.06	0.0028	—	—	0.011	—	—	—

From Tucker and Crabtree (1970); 14–30 days toxicity data.

[a]Lesser Canada geese ♂ + ♀.
[b]From Pimentel (1970); all data expressed in ppm, 24–48 hr LC_{50} value.
[c]Brook trout, 48 hr.
[d]96 hr.

provide a general order of toxicity of pesticides and also reveal the relative susceptibilities of animals to pesticides.

The data summarized in Table 10-6, for instance, indicate that among organophosphates parathion is generally more toxic than Dursban® or Abate® and that DDT is the least toxic chlorinated hydrocarbon insecticide among the three shown. Table 10-6 also shows the general order of susceptibility of various animals. Bullfrogs are generally resistant to all these insecticides. Among the birds, mallards are generally more resistant to these insecticides than pheasants or Canada geese, except for parathion, which is very toxic to them.

Macek and McAllister (1970) found that among 12 species of fish representing five different families, salmonids were always the most susceptible to all insecticides tested (nine chlorinated hydrocarbon, organophosphate, and carbamate insecticides); ictalurids and cyprinids were the least susceptible. Fish are known to have very inefficient mixed-function oxidase systems to detoxify these insecticides, which makes them vulnerable to them as environmental contaminants.

A compilation of data on pesticide toxicity for various nontarget species is now available (Pimentel, 1971).

10.2.2. Chronic Toxicity

It has become increasingly apparent that acute toxicity data alone cannot totally represent the ecological hazard of any pesticide. From the data shown in Table 10-6, for instance, it might be wrongly concluded that DDT is the safest insecticide for wildlife and that Zectran® is the most hazardous one. However, the long-term toxicity (chronic) of an insecticide may be quite different from its short-term toxicity, as is the case for DDT and Zectran®. (There are many other reasons for this discrepancy between acute LD_{50}s and ecological vulnerability; see later sections.)

In nature, the sudden (usually accidental) poisoning of wildlife by a single exposure to an insecticide generally does not create long-lasting and widespread ecological problems since such exposures are generally limited in geographical area and duration. What generally causes ecological threat is the long-term buildup of pesticides by environmental concentration. The data shown in Table 10-7 indicate the ratio between acute LD_{50} and chronic minimum lethal dosage. A low ratio can be regarded as the sign of a noncumulative poison (e.g., 2.4 for Zectran®), since it means that the daily dose (about one-third of the total dose) is not much less than the amount of pesticide needed to kill the animal in one exposure. In such a case, the animal is effectively eliminating the poison through detoxification or excretion

TABLE 10-7. Ratio Between Acute Oral LD_{50} and Chronic Minimum Lethal Dosage of Various Insecticides for Mallards

	Chronic minimum lethal dosage $(EMLD)^a$ (mg/kg/day)	Ratio of cumulativeness (acute LD_{50}/EMLD)
DDT	50	44.8
Dieldrin	1.25	76^b
Endrin	0.125	45
Abate®	2.5	32–40
Dursban®	2.5	30
Parathion	$3–6^c$	$2.7–5.3^c$
Sevin® (carbaryl)	125	17.4
Baygon® (arprocarb)	2.0	2.0
Zectran®	1.25	2.4

^{a}From Tucker and Crabtree (1970). The lowest daily oral dosage that produced death to one or two animals (out of six animals) by the end of 30 days (EMLD stands for 30-day empirical minimum lethal dosage).
^{b}Similar value for lindane (γ-BHC) is 67.
^{c}The data used are for gray partridges instead of mallards.

mechanisms at the end of each day. In contrast, high ratios such as for DDT, dieldrin, and endrin indicate the cumulativeness of chlorinated hydrocarbon insecticides.

10.2.3. Factors Influencing Toxicity

10.2.3a. Age and Size

Age and size apparently are the two most important factors influencing susceptibility. This is very much evident in fish, for which insecticide toxicity has been assessed by means of the concentration changes in the ambient water (not the amount given per unit body weight, as with other animals). For example, the water in Lake Michigan contains around 1 ppt of DDT, but even this low concentration is toxic enough to affect the hatchery operation for coho salmon. Bigger fish are not affected by it, so they have been imported into Lake Michigan for planting coho salmon colonies. Laboratory examinations of DDT toxicity to various stages and sizes of coho salmon (Buhler and Shanks, 1970) have confirmed that older fish and, within the same age group, bigger fish are more resistant to DDT (i.e., the median survival time is directly related to body weight). There are many other experimental data indicating the susceptibility of smaller fish (e.g., results of DDT spray on forest land in Canada). Older fish in turn accumulate more pesticides than the younger ones.

Reinert (1970) has proposed an interesting theory that the effects of age and size can be ascribed to the increase in fat content. The supporting evidence comes from the data he obtained from Lake Michigan and Lake Superior fish. A comparison of alewives, bloaters, lake trout, and yellow perch indicated that the ratio between DDT concentration and percent fat becomes very constant despite species and size differences (e.g., bloaters show three times as much DDT as yellow perch). It is not easy, however, to correlate the interspecific differences in pesticide accumulation with susceptibility variations. Some species can accumulate more than others without showing any poisoning effects.

Moreover, in fish, the role of fat mobilization with the resulting turnover of stored pesticides is not clearly understood. Bridges *et al.* (1963) observed that DDT residues in trout essentially remained the same for 15 months after treatment of a pond with 0.22 ppm, indicating a low or no turnover of DDT. Judging by the data obtained by Eberhardt *et al.* (1971), it is also evident that DDT levels decline very slowly in both small green sunfish and small carp. The obvious conclusion here is that susceptibility to a pesticide is species specific, but within a species toxicity is determined by the size and age of the animal as well as by the amount of pesticides it receives.

10.2.3b. Environmental Factors

Many environmental factors are known to influence the toxicity of pesticides. These include temperature, the pH of water and soil, the presence of other organisms, and the characteristics of the bottom in aquatic environments. The physiological state of the animal itself also influences the toxicity of pesticides. There are two different kinds of environmental factors to be considered: (1) factors influencing the availability of a pesticide through competitive absorption and degradation and (2) factors related to the physiological (or biological) and chemical state of the animal and the chemical nature of the pesticide itself.

The first factor is frequently observed in animals and plants in soil or water, since these two media can accommodate high levels of pesticides which can be competitively absorbed by different systems. For instance, pesticides are much less available in organic soils than in sandy (nonorganic) soils. Thus plants grown in sandy soil absorb higher amounts of pesticides (e.g., Lichtenstein and Schulz, 1960) than ones grown in soils with high organic matter. In aquatic fauna, the characteristics of the bottom greatly affect the availability of pesticides (Lee, 1970; Cope, 1966; Rudd, 1965).

Muddy sediments bind with pesticides and thereby reduce their availability to aquatic organisms.

The effects of temperature are less understood. For instance, DDT is known to be more toxic to many biological systems at lower temperatures. Does this mean that fish that live in cold temperatures are more severely affected? Or in cold climates could cold-blooded animals as a whole be at more of a disadvantage than warm-blooded animals?

There is evidence that a number of physiological factors influence the outcome of chronic toxicity experiments. Gish and Chura (1970) studied the effects of body weight, breeding condition, and sex on the susceptibility of Japanese quail to DDT in two sets of 21-day feeding experiments. They found that partially starved birds were most affected by DDT. The heavier birds survived longer, but weight loss was more evident in them. In addition to the sex differences (i.e., males were more susceptible than females), females in breeding condition showed some degree of resistance at the early stage of DDT poisoning.

Lincer *et al.* (1970) studied toxic effects of DDT and endrin on fathead minnows under static and dynamic (i.e., a continuous flow system to supply a constant test solution) assay conditions. They found that both insecticides (particularly DDT) were more toxic to the fish under static bioassay conditions. They reasoned that declining oxygen concentrations coupled with increasing metabolic wastes (ammonia, CO_2, etc.) enhanced the apparent toxicity of DDT during static exposure.

Schoettger (1970) examined the factors influencing the toxicity of Thiodan® to suckers and found that variation in calcium and magnesium salt concentrations did not significantly affect it.

10.2.4. Subtle Effects of Insecticides

10.2.4a. Birds

One of the most debated effects of sublethal dosages of insecticides has been their influence on eggshell thickness. The phenomenon was originally reported by Ratcliffe (1967) and by Hickey and Anderson (1968), who observed that the eggshell thickness of wild birds generally decreased as the levels of environmental contamination increased. Thinner eggshells, which result in egg breakage, have been suggested as the reason for the declining reproduction rate in several species of birds. Additional evidence (Anderson *et al.*, 1969) indicated that whenever the decreases were in excess of 10 % in raptor eggshells (museum and private collections were examined), the colony showed a general decline.

The validity of these correlative studies has been examined by several workers. Bitman *et al.* (1969) fed a low-calcium (0.56%) diet containing 100 ppm (or 0 for controls) o',p'-DDT and p,p'-DDT for 45 days to Japanese quail and found that eggshells were significantly thinner and contained less calcium than those of controls. The decline in eggshell thickness due to p,p'-DDT was of the order of 7%, which was significant at the 1% level. Although there was no significant difference in the total number of eggs laid, a lag period for the start of egg laying was observed in DDT-treated (particularly p,p'-DDT) groups. However, blood calcium concentration was not affected. Similar experiments have been conducted on mallards (Heath *et al.*, 1969). Eggshell thickness again declined about 10%, which is far less than the decline reported in wild birds (Jehl, 1969). Tucker and Haegele (1970) could not observe a significant decline in eggshell thickness in chronically DDT-fed (10 and 30 ppm) bobwhite quail and mallards at three different concentrations of calcium. In fact, the effects of calcium were more significant in this set of experiments. The effects of DDT became apparent only when these workers gave quite a high dose of DDT (1000 mg/kg) to fasting mallards, and then as much as a 25% reduction in thickness was observed. (DDT was given orally in the form of a gelatin capsule.) In more realistic studies, Porter and Wiemeyer (1969) examined American sparrow hawks kept on diets containing 0.28 and 0.84 ppm of dieldrin and 1.4 and 4.7 ppm of DDT. These birds produced eggshells roughly 10% thinner than normal ones. Other commonly observed patterns among raptorial populations, such as breakage and disappearance of eggs, were also noted. Later, Wiemeyer and Porter (1970) fed a diet containing 2.8 ppm of p,p'-DDE and in the second year again observed a 10% reduction in the thickness of eggshells. While there are other conflicting reports, it appears safe to state that some of the chlorinated hydrocarbon insecticides must cause a reduction in eggshell thickness in some bird species. There appear to be susceptible species and resistant species as far as eggshell thinning is concerned.

As for the mechanism of this eggshell thinning phenomenon, no adequate, or generally accepted, explanation exists. There are many reports that claim inhibition of carbonic anhydrase, generally acknowledged to play an important role in forming eggshells, by DDT and some other derivatives (e.g., Peakall, 1970; Bitman *et al.*, 1970), but there are just as many that deny it (e.g., Anderson and March, 1956; Dvorchik *et al.*, 1971). In brief, the carbonic anhydrase activity in the shell gland itself can be affected by DDT and DDE, as judged by an *in vitro* assay in experiments with birds fed a diet containing DDT or DDE. Degrees of inhibition found were 60% (Peakall, 1970) to 18% (Bitman *et al.*, 1970). On the other hand, when Dvorchik *et al.* (1971) directly tested DDT against semipurified bovine carbonic anhydrase *in vitro*, they observed no inhibition until the DDT concentration was raised above 500 ppm, and these workers argue that in the entire vertebrate kingdom there is no known carbonic anhydrase more susceptible to drug inhibition than the enzyme in red cells. In addition, these workers do not feel that 18 % inhibition of avian carbonic anhydrase *in vitro* can result in a 10–15 % decrease in eggshell thickness *in vivo* (i.e., in the experiment by Bitman *et al.*, 1970, 100 ppm of *p,p'*-DDT and *p,p'*-DDE was fed to Japanese quail and 18 % *in vitro* inhibition of carbonic anhydrase was observed; in Bitman *et al.*, 1969, 7–15 % thinner eggshells had been reported to result from identical experimental conditions).

Thus one is left with two alternatives to explain the phenomenon : (1) the carbonic anhydrase in the shell gland of avian species is unusually sensitive to chlorinated hydrocarbon insecticides, or (2) the *in vivo* inhibition observed is a result of some other effect, such as reduced enzyme production or the production of substances that suppress the activity of the enzyme.

There are a number of reports indicating more distinct biochemical and physiological effects of low doses of chlorinated hydrocarbon insecticides than their actions on carbonic anhydrases. The hepatic enzymes of avian species are known to be induced at relatively low levels of DDT (10 ppm) and dieldrin (2 ppm), and these two effects appear to be additive (Peakall, 1967, 1969). This is hardly surprising since avian systems (such as the chick embryo) have been utilized as a sensitive bioassay device to study induction phenomena; in addition, avian hepatic Δ-aminolevulenic acid synthetase (ALA synthetase) is more readily induced than that in mammalian liver systems.

Jefferies and French (1969) studied the effect of *p,p'*-DDT on the thyroid and the liver by feeding pigeons sublethal amounts of DDT (18, 36, and 72 mg/kg every second day). After 42 days the birds were examined, and a 24 % increase in liver weight was observed at the lowest dose as well as a distinct increase in thyroid weight at the two higher doses. The most striking effect was in the colloidal contents of the follicles, which showed a 60–94 % decrease at the lowest dose of DDT.

o,p'-DDT has been reported to show estrogenic effects at high doses, and both *o,p'*-DDT and *p,p'*-DDT have anti-estrogenic effects in the rat; the latter are probably due to induction of hepatic microsomal enzymes (Clement and Okey, 1972).

Because hepatic changes take place in avian species at relatively low doses of chlorinated hydrocarbons, it appears likely that the hormonal regulatory functions are most sensitive to these insecticides. Whether such changes can directly result in decrease of eggshell thickness cannot logically be discussed at present. However, avian reproductive systems are known to be highly sensitive to environmental changes which affect the hormonal regulatory mechanisms. For instance, decreased egg weight can be produced by either hyper- or hypothyroidism (Oloufa, 1953, Taylor and Burmeister,

1940). Both conditions have been reported (Jefferies, 1969; Glazener *et al.*, 1949) to appear in birds as the result of chlorinated hydrocarbon intake. Such changes in hormonal regulatory processes are expected to have multiple effects on the various steps of the reproductive process (Jefferies and French, 1969).

However, a number of environmental factors and chemicals other than chlorinated hydrocarbons are known to cause either hepatic induction or hormonal changes in avian species. As a consequence, no cause–effect relationship between environmental contamination by these insecticides and reproductive failure in some avian species can be established. Indeed, other types of pesticides, as well as PCBs, have been shown to cause eggshell thinning (Heath *et al.*, 1971; Tucker, 1973). Many more studies are needed to demonstrate causal relationships at ecosystem levels. It is also important to stress that the eggshell thinning cannot be the sole, or even a major, criterion of reproductive failure. Other criteria such as egg breakage, number of eggs laid, and number of eggs hatched (e.g., Heath *et al.*, 1969) are also important. Indeed, a report by Whitehead *et al.* (1972) indicates that BHC does not affect shell thickness in hens at doses equivalent to 10, 100, and 200 mg/kg (daily dose). Instead, the rate of egg production is significantly reduced.

10.2.4b. Fish and Other Aquatic Organisms

There are a number of reports of a relationship between sublethal treatment with chlorinated hydrocarbon insecticides and behavioral changes in fish. Warner *et al.* (1966), for instance, studied the effects of toxaphene (0.44–1.8 ppm for 96–264 hr at 25°C) on goldfish and found many dose-dependent behavioral changes.

Anderson and Peterson (1969) studied the effect of treatments with low DDT concentrations (20–60 ppb for 24 hr) on behavior directed by the central nervous system. They found the temperature at which a simple (propeller tail) reflex was blocked to be higher for treated brook trout; i.e., treated fish had less thermal acclimation ability. In trout treated with 60 ppb of DDT, cold blocking set in at 5°C, while control fish could tolerate 2.5°C (acclimation temperature was 18°C). In addition, during the course of establishing a conditional avoidance response, it was found that DDT-treated trout were almost incapable of learning. In this experiment, naive brook trout were acclimated at 9°C to 20 ppb of DDT and were given training by the application of electric shock to avoid whichever side of a two-chambered light–dark aquarium that they preferred (brook trout have individual preferences for either the light or the dark side of such an aquarium). Although untreated naive fish took only 30 trials to become conditioned, not

TABLE 10-8. Effect of 20 ppb of DDT on the Learning of a Conditional Avoidance Response in Brook Trout

Days after treatment	Average trials (No.)	Failure (misses) (%)	After shock escape (%)	Before shock avoidance (%)
			Success	
			After shock escape (%)	Before shock avoidance (%)
Control				
	30.3	13.0	52.0	35.0
Treated				
1	25	94.0	6	0
4	28.3	52	45.1	2.9
7	29.5	43.0	49.2	7.8

From Anderson and Peterson (1969).

one of the DDT-treated naive fish became conditioned (Table 10-8). However, Jackson *et al.* (1970) were able to modify the chamber conditions to facilitate learning, so that DDT-treated Atlantic salmon and speckled trout were able to learn to the same degree as nontreated fish. They concluded that sublethal doses of DDT do not *per se* affect "learning" but rather can alter the ability of fish to perform the tasks involved.

While this effect of DDT (Table 10-8) seems to be at least partially reversible, there appear to be other nonreversible effects. Davy *et al.* (1972), for instance, studied the effects of 10 ppb of DDT on goldfish locomotor behavior and found it to significantly affect a locomotor pattern related to a "memory" process within 4 days. Returning the fish to clean water and maintaining them for 130–139 days did not result in recovery to normal (positive correlation).

There are only a few reports on how pesticidal compounds affect biochemical systems and what the relationship is between such biochemical effects and the *in vivo* expression of toxicity in aquatic animals. Janicki and Kinter (1971) studied the effect of DDT (5.0 ppm or 1.4×10^{-5} M *in vitro*) on the $(Na^+ + K^+)$-ATPase of the eel (*Anguilla rostrata*) and found that 43% of the total enzyme activity was inhibited. When the intestinal sac was filled with Ringer solution containing 50 ppm of DDT, there was a 47% inhibition of water absorption. Since $(Na^+ + K^+)$-ATPase is generally acknowledged to be the machinery of osmoregulation in the teleost, they concluded that the inhibitory activities of DDT on the ATPase eventually result in impairment of fluid absorption in the intestinal sac. Indeed, there are reports indicating that various fish ATPases are sensitive to organochlorine insecticides (e.g., Koch, 1969; Cutkomp *et al.*, 1971, 1972; Janicki and Kinter, 1970). Such observations are also in agreement with *in vivo* data on disrupted osmoregulation, as judged by changes in ion concentrations in endrin-poisoned northern puffer and goldfish (Eisler and Edmunds, 1966; Eisler, 1970; Grant and Mehrle, 1970). In the northern puffer, the effect of 1 ppb of endrin (96 hr) was reduction of Na, K, and Ca in the liver and increase in Na, K, Ca, and cholesterol in serum (Eisler, 1970). Liver calcium contents, in particular, decreased 92% by this treatment.

Along somewhat similar lines, Matsumura (1972) demonstrated that the toxicities of DDT and BHC to the brine shrimp, *Artemia salina*, varied as the NaCl concentration changed, implying that the chlorinated hydrocarbon insecticides affect the NaCl tolerance mechanisms of this vertebrate species. This type of influence could be widespread among aquatic organisms. Batterton *et al.* (1972) were able to show that the mechanism of NaCl tolerance in a freshwater blue-green alga species, *Anacystis nidulans*, is greatly impaired by DDT. The loss of tolerance can be partially reversed by increasing Ca^{2+}, in agreement with the generally acknowledged theory of the Na transport mechanism (Table 10-9). In addition, the ouabain-sensitive portion of $(Na^+ + K^+)$-ATPase of this species was also sensitive to DDT.

TABLE 10-9. DDT Inhibition of NaCl Tolerance by the Blue-Green Alga *Anacystis nidulans*[a]

Culture	Normal medium		Normal + 125 mg $Ca(NO_2)_2$/liter	
	No NaCl	+1% NaCl	No NaCl	+1% NaCl
Control	2.01	1.41	1.89	1.84
Plus ethanol	2.13	1.40	2.14	1.76
Plus DDT in ethanol	1.95	0	1.80	1.53

From Batterton *et al.* (1972).

[a]The data are expressed as growth rates (i.e., large numbers for high growth; 0 for no growth). Note that the alga can tolerate either NaCl (1%) or DDT (800 ppb) but not both at the same time. Ca^{2+} can reverse this effect.

It is known that aquatic phytoplankton vary in their response to chlorinated hydrocarbons. Menzel *et al.* (1970) were able to show that *Dunaliella tertiolecta*, a high-salt-resistant species, was much more DDT resistant than other coastal and open-sea species as judged by response to DDT in terms of reduction in photosynthetic activities. Batterton *et al.* (1971) demonstrated that the freshwater *A. nidulans* was always more sensitive to eight cyclodiene analogues than the marine blue-green alga *Agmenellum quadruplicatum*. It remains to be seen whether such species differences in susceptibility to chlorinated hydrocarbon insecticides among aquatic phytoplankton are related to the abovementioned ATPase-regulated mechanisms.

10.3. BIOLOGICAL TRANSFER AND BIOACCUMULATION

There is no question that some persistent pesticides accumulate in various biological systems at levels much higher than those in their surroundings. Since the major concern about pesticide pollution is its effect on biological systems, this phenomenon has attracted the attention of many scientists. While there is unanimous agreement on the fact of biomagnification, there is much disagreement about its meaning, causes, and mechanisms. This is because of the difficulty of studying the ecology of any dynamic biological system. For instance, in order to understand the movement of pesticide residues, one has to know the energy cycle (i.e., budgeting of the food–energy relationship in an ecosystem) in relation to population changes, as well as take into account the "concentration factor" and metabolic changes of the particular pesticide.

Nevertheless, in studying the mechanism of biological transport, it is possible to apply a "black box" technique, where the input and the ultimate output are known but not the intrinsic individual processes. In other words, the DDT case, for instance, could be thought of as a gigantic ecological experiment on a continental or global scale. From the experiment, certain "outputs" have become apparent; i.e., a number of biological systems have been found to accumulate DDT. The magnitude of total DDT "input" and the timing are also known, as well as the fact that a rapid withdrawal period began after 1972. All of this provides an excellent opportunity to study the rate of DDT reduction in various ecosystems after many years of steady-state status. One could say that such an experiment resembles "flash incorporation" or pulse-labeling techniques in biochemistry, which are generally used to study the rate of biochemical reactions.

10.3.1. Route of Biological Transport

The total pesticide concentration in biological systems perhaps is not much compared to the amounts in soils, aquatic systems, and the air. However, because of the importance of biological systems in relation to pesticide pollution, the route of biological transport becomes the focus for environmental studies. Except for people who work in the pesticide industry or in agriculture, the bulk of human intake of pesticides is via biological transport routes.

10.3.1a. General Pattern

For the sake of clarity, let us assume that there are three major routes of pesticide transport within biological systems as shown in Fig. 10-3. Of course, the actual relationships are more complicated, but this simple classification of animals according to food sources will serve to illustrate general concepts.

There are some indications that the animals dependent on the route III system for their food source tend to accumulate more persistent pesticides.

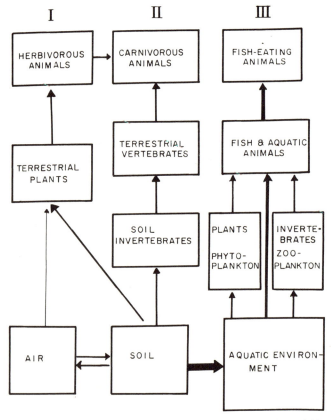

Fig. 10-3. Schematic diagram of biological transfer of pesticide residues.

TABLE 10-10. Typical DDT-R Levels Found in Bird and Mammalian Species with Different Food Habits

	Residues (ppm)	
	Typical relative value	Maximum value
Bird species		
Fish-eating birds	10.0	7.1–194
Predatory birds (owls, hawks)	3	1.42–175.8
(Bald eagle[a])	(8–20)	(0.3–100)
Herbivorous and insectivorous birds	2.0	0.16–55.6
Mammalian species		
Fish-eating mammals (seals)	10	8.3–16.3
Predatory mammals	1.0	0.13–0.24
Herbivorous mammals	0.1	0.02–12.6

From Edwards (1970).

[a]From Belisle *et al.* (1972), 1969–1970 data.

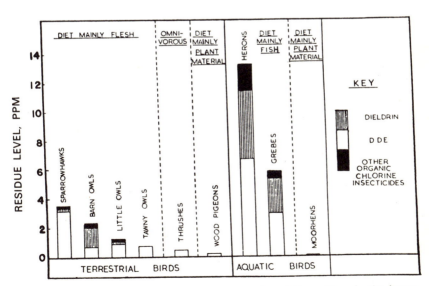

Fig. 10-4. Average concentration of organochlorine insecticide residues in the breast muscle of different types of birds. From Moore and Walker (1964).

According to Edwards (1970), fish-eating and predacious birds accumulate more DDT than other birds (Table 10-10). A similar conclusion was reached by Moore and Walker (1964), who studied dieldrin (HEOD), DDE, and other pesticide residues in several species of birds (Fig. 10-4). Therefore, the source of food apparently plays a very important role in determining residue levels. Studies of mammalian species are scarce, but general tendencies can be observed. Terrestrial mammalian species in particular do not tend to accumulate chlorinated hydrocarbon insecticides to the extent that avian species do, especially the fish-eating and, to a lesser extent, predacious (terrestrial) birds. This general trend also indicates the importance of food habits.

10.3.1b. Aquatic Environments

As mentioned above, the rate of bioaccumulation in aquatic environments generally appears to be higher than that in terrestrial environments. The most important reason for this appears to be the lipophilic nature of the

TABLE 10-11. Illustration of the Difference Between the Rate of Concentration of DDT-R in Terrestrial Environments and That in Aquatic Environments

System observed		Concentration factor[a]	
		Minimum	Maximum
Plants			
Soil	Plant foliage	—	0.08
	Plant roots	0.04	0.13
Water	Plankton	250	16,666
	Algae	0.34	33
	Higher plants	0.45	100,000
Invertebrates			
Soil	Earthworms	0.67	73
	Beetles	0.31	2.8
Water	Snails	144	1,480
	Oysters	15,000	70,000

From Edwards (1970) and Kenaga (1972).

[a]The concentration factors were obtained by dividing the DDT-R concentration found in the plants by the concentration found in either soil or water nearby.

persistent insecticides. In aquatic environments, compounds with low water solubility and high liposolubility are forced to seek organic-lipid containing material. In addition, water provides an excellent medium for transport of these insecticides. For example, terrestrial plants are not known to play very important roles in bioconcentration, while phytoplankton and aquatic plants do concentrate persistent insecticides from water. The magnitude of concentration is clearly of a different order from that of terrestrial plants (Table 10-11). The same tendency is also clear in aquatic vs. terrestrial soil particle-feeders (i.e., invertebrates). Another reason for the higher degree of biomagnification in aquatic environments is the availability of two separate routes of entry for pesticidal compounds (in higher organisms), i.e., direct and through the food chain. In the experiments summarized in Table 10-12,

TABLE 10-12. Uptake of Four Pesticidal Compounds by the Northern Silversides in the Presence (food chain plus direct pickup) and Absence (direct pickup only) of Mosquito Larvae

		Accumulation in fish (ppb)	
	Amount of pesticide on sand[a] (μg)	In the absence of mosquito larvae	In the presence of mosquito larvae
TCDD	1.62	2	708
DDT	1.79	458	337
γ-BHC	1.47	2904	1080
Zectran®	1.11	213	76

From Matsumura and Benezet (1973).

[a]The pesticides were put in the aquaria in the form of precoated dry sand.

the rates of bioaccumulation of four different insecticides in the presence and absence of food and mosquito larvae were compared. The results clearly indicate the importance of direct pickup of insecticides from the water. In comparison, it is almost inconceivable that terrestrial animals could pick up such quantities of insecticides from the air alone unless they were directly exposed to insecticidal spray. In a model ecosystem, Kapoor *et al.* (1970) found the accumulation of DDT-R (as measured by the total radioactivity recovered from the original ^{14}C-DDT used in the system) in snails (*Physa*), mosquito larvae (*Culex*), mosquito fish (*Gambusia*), and water to be 22.9, 8.9, 54.2, and 0.004 ppm, respectively. The final concentration in the mosquito

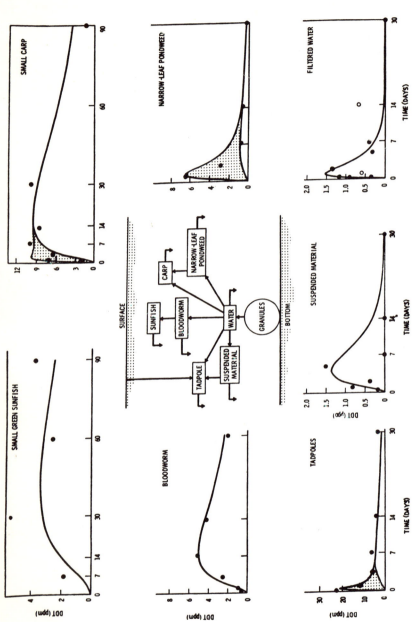

Fig. 10-5. Food chain model for selected items of a freshwater marsh. Plotted points represent observed concentrations of DDT at various times after application. Curves represent concentrations computed from the model equations. Shaded areas (carp, tadpoles, and narrow-leaf pondweed) represent the contribution of a "fast" compartment. From Eberhardt *et al.* (1971).

fish represents a 217,000-fold concentration factor, indicating that the aquatic organism at the top of this model ecosystem accumulates the most DDT-R.

As for direct absorption and pickup, Eberhardt *et al.* (1971) showed some species differences in the rates of quick pickup and slow accumulation of DDT-R. In their experiment, DDT was applied to a marshland in one dose (0.2 lb/acre) and many aquatic organisms were examined for their rates of DDT pickup and accumulation. Eberhardt *et al.* found (Fig. 10-5) that some species, aquatic plants and tadpoles, typically showed a biphasic curve: an initial quick buildup (0–14 days) followed by a slow decline in the level of DDT. Since the initial quick buildup curve closely resembles that for filtered water and suspended matter, it is likely that this portion of DDT-R pickup is attributable to direct surface absorption by the organisms. Thus it is clear that species differences are important. It is noteworthy, however, that the quick portion of DDT-R buildup does decline fast. Although there is no way of knowing the significance of the surface pickup in the total picture of long-term accumulation (more than 1 year) of pesticides, it is certain that both the food-chain and direct pickup routes are important in determining residue levels in aquatic organisms.

Perhaps the above examples, which have been taken from experiments where the insecticide was suddenly introduced to the natural ecosystem and to a model ecosystem, may not reveal the pattern of steady-state dynamics of pesticide accumulation, but they certainly represent situations immediately after spraying. Thus the "black box" approach to checking residue levels ("output") in various aquatic organisms is important.

In a long-term study on the fate of DDT-R, Dimond *et al.* (1971) followed the changes in its distribution for 10 years after an initial single spray at a rate of 1 lb/acre. It is apparent from their data (Table 10-13) that the general level of DDT-R declined sharply in the first 2–3 years (according to expectation) and the remainder persisted thereafter. Persistence was noticeable in insects, trout, and chubs. On the other hand, the residues in plants declined rapidly. However, the relative distribution of DDT-R was not much different (plants and kingfishers being exceptions) for the initial and the 10-year levels. That is, the organisms which showed a high rate of initial pickup also retained more DDT-R. On the whole, approximately 5–10% of the original levels of DDT-R were retained by all organisms after 10 years: thus the ratio vs. DDT-R in mud samples did not significantly change during 10 years for all animal samples from the streams. But it is apparent from the data that kingfishers accumulated the highest amount of DDT-R. While the birds apparently ranged over a wider area, and the data cannot be regarded in the same category as the other, they show that, in terms of the maximum levels of DDT-R attainable in an ecosystem, the fish-eaters do accumulate high levels of pesticide residues.

TABLE 10-13. Changes in DDT-R Distribution (ppm) in 10 Years Following a Single Spraying of DDT in a Lotic Ecosystem (watersheds in Maine)

Years after application	Mud	Aquatic plants[a]	Insects	Mussels	Trout (brook)	Chubs	King-fishers
0	0.83	0.74	2.41	0.20	8.21	9.84	—
1	1.08	0.06	—	—	3.51	5.46	—
2	—	—	0.33	0.06	—	—	9.44
3	0.21	—	0.21	—	2.10	1.55	—
4	0.21	0.02	0.28	—	0.99	1.34	—
5	0.59	0.04	0.43	0.03	0.79	0.76	—
6	0.06	0.01	0.06	—	0.68	—	4.21
7	0.21	0.05	0.22	—	0.36	—	—
8	0.13	—	0.27	—	0.27	0.21	—
9	0.03	0.01	0.44	—	0.44	—	—
10	0.07	0.02	0.27	—	0.35	0.36	—
Control[b]	0.006	0.007	0.05	0.01	0.07	0.04	0.39
Heavily sprayed area[c] (4–5 years)	0.35	0.09	0.58	—	2.11	3.48	10.06

From Dimond et al. (1971).

[a]Algae and higher plants mixed.
[b]Never been treated by DDT.
[c]Sprayed three times.

In aquatic environments, the specific process of pesticide accumulation by various species is certainly an important consideration. In this connection, both particle-feeders and bottom-feeders play an important role in accumulating pesticides, inasmuch as the bulk of the lipophilic pesticides are expected to be bound to the particles and to the bottom sediment. In Table 10-14 are shown the results of a study attempting to determine the pickup of DDT and TCDD (2, 3, 7, 8-tetrachlorodibenzo-p-dioxin) by aquatic organisms after application of the pesticides in the form of pre-coated dry sand particles. Since these compounds, particularly TCDD, have low water solubility, the pickup by the organisms is considered a dynamic biological process. It is obvious that the rate of TCDD pickup by mosquito larvae is extraordinary. They are bottom-feeders, and site observations confirmed that they actively pick up each sand particle and skim its surface with their mouth parts. The importance of particle-feeders has already been discussed in Section 10.1 with respect to their use as indicator organisms. Various shellfish, crabs, etc., are known to accumulate pesticides, and Odum et al. (1969) studied accumulation of DDT-R in fiddler crabs, which often

TABLE 10-14. Uptake of DDT and TCDD (2,3,7,8-tetrachlorodibenzo-*p*-dioxin) from Precoated Sand or Pretreated Algae by Several Aquatic Organisms in 7 Days

	TCDD (ppb)	DDT (ppb)	γ-BHC (ppb)	Zectran® (ppb)
Daphnia[a]	879	43,123	—	—
Ostracods[a]	279	36,391	—	—
Brine shrimp[b]	157	3,092	495	89
Mosquito larvae[b]	4,150	14,250	1,450	0
Fish[b]	2	458	2,904	213

From Matsumura and Benezet (1973).

[a]In these two experiments, algae were allowed to pick up pesticides from the aquarium surface. They were then added to another aquarium with the test organisms.
[b]Known amounts (1.62, 1.79, 1.47, and 1.11 μg for TCDD, DDT, γ-BHC, and Zectran®, respectively) were separately added to sand using a solvent. The solvent was evaporated and the sand was transferred to the aquarium with test organisms.

feed on particles derived from organic plant detritus. The crabs, which were fed detritus particles (greater than 500 μm) containing an average of 10 ppm DDT-R, tripled their DDT-R concentration in the large claw in 11 days.

10.3.1c. Terrestrial Environments

The initial rate-limiting factor in the biological transport system in terrestrial organisms is the process by which plants and soil invertebrates pick up residues from the soil, and these systems are relatively less efficient than the ones found in aquatic environments. On the whole, plant-mediated transport is more important simply because of the much larger biomass of plants in the biota. However, in terms of the specific rate of accumulation, the soil invertebrate–mediated system is more efficient.

Brown and Brown (1970) observed that in a subarctic environment herbivorous vertebrates accumulated less DDT-R than carnivorous ones (fish, mollusks, and insects). In this study (Table 10-15), samples were collected in 1967 in an area where DDT had not been used for 4 years (there had been a continuous DDT spray program from 1947–1963 at 0.22–0.25 lb/acre). Thus the air equilibrium for DDT-R distribution had been established in that area. The order of accumulation in terms of food habits of the

TABLE 10-15. DDT-R Levels in the Vertebrates in a Subarctic Environment[a]

Food habit	Species	Total DDT-R (ppm)
Vegetation	Red squirrel	1.97
	Willow ptarmigan	3.40
Seeds and insects	Tree sparrow	11.1
	White-crowned sparrow	17.4
Insects and mollusks	Northern phalarope	41.9
	Semipalmated plover	52.0
	Green-winged teal	36.0
	Shoveller	38.4
	Bonaparte gull	56.5
Fish and insects	Arctic tern	63.5

From Brown and Brown (1970).

[a]Vicinity of Fort Churchill, sprayed in 1947–1963 at 0.22–0.25 lb/acre DDT (twice annually from 1955 to 1963). Sampled in 1967 (4 years after spraying was discontinued) Data are expressed as ppm of DDT and its metabolites found in fat of various vertebrate species.

animals was herbivores < seed- and insect-eaters < insect- and mollusk-eaters < fish-eaters. Moore and Walker (1964) reached a similar conclusion that the order of accumulation was herbivores (terrestrial) < omnivores < raptors < fish-eaters among the bird species they studied in England (Fig. 10-4).

In large wild mammals, the level of accumulation appears to be relatively low (Walker *et al.*, 1965). In addition, differences in food habits (e.g., bears vs. other herbivorous animals) are apparently not important (Table 10-16). Similar residue levels were observed by Moore *et al.* (1968) for pronghorn antelope in South Dakota, where the value for DDT-R was 0.06–0.17 ppm. Other residues observed were lindane (0.04–0.05 ppm) and heptachlor epoxide (0.04–0.12 ppm). Unfortunately, there are not sufficient data on predatory mammals to conclude whether the residue levels in them are high, and the data on domestic animals are not too useful in assessing the magnitude of bioaccumulation. From the studies on birds, however, it can be concluded that bioaccumulation of pesticides via the plant-mediated route is not excessive, despite the large quantities of plants that herbivorous animals ingest, unless unusual localized plant contamination takes place (e.g., feeding of insecticide-treated seeds).

TABLE 10-16. Average DDT-R Levels in Large Wild Mammals

Animal	Location	Residues (ppm)		
		TDE	DDE	DDT[a]
Antelope	Idaho	<0.01	<0.01	0.098
Bear	Idaho	<0.01	<0.01	0.032
	Washington	<0.01	<0.01	0.045
Deer	Idaho	<0.01	0.01	0.109
	Washington	<0.01	0.01	0.122
Elk	Idaho	<0.01	0.03	0.071
	Washington	<0.01	0.04	0.056
Goat	Idaho	<0.01	<0.01	0.050
	Washington	<0.01	<0.01	0.023
Moose	Idaho	<0.01	0.01	0.087

From Walker *et al.* (1965).

[a] *p, p'*-DDT and *o, p'*-DDT.

As for the soil invertebrate–mediated route, the general indication is that high degrees of insecticide accumulation can be observed under certain circumstances. The infamous case of American robins feeding on earthworms containing DDT (e.g., Barker, 1958), for instance, clearly demonstrates the importance of this particular route. Here, the source of DDT was mainly from its use to control Dutch elm disease. Edwards (1970) compiled a comprehensive summary of accumulation of chlorinated hydrocarbon insecticides by soil invertebrates, and it is apparent from his figures that chlorinated hydrocarbon insecticides do accumulate in earthworms and other soil invertebrates. For earthworms, the range of accumulation was 10–680 ppm of DDT, 0.006–0.1 ppm of γ-BHC, 0.05–0.98 ppm of aldrin, and 0.3–4.6 ppm of dieldrin. The data for slugs and beetles appear to be of the same order of magnitude. Other examples of chlorinated hydrocarbon accumulation by soil invertebrates were presented in Table 10-5. The concentration factors appear to be much lower than those observed for aquatic organisms (0.06–73.1 for DDT in earthworms), but the comparison here is rather irrelevant since there is no upper limit, such as water solubility, in the soil. Rather, it should be noted that the general level of accumulation by soil invertebrates (typical value 1–10 ppm) is of the same order of magnitude as that by aquatic invertebrates.

The majority of feeders on soil invertebrates are bird species, and thus it is likely that several are affected by this route of pesticide transport. (Other animals that feed on earthworms are fish, moles, salamanders, shrews, and snakes. There are no available data to indicate the effects on these organisms.)

Stickel *et al.* (1965) fed heptachlor-containing earthworms (2.86 ppm) to woodcocks, and found that ten out of 12 birds died within 53 days (LT_{50} 35 days). These workers concluded that since earthworms from the area where 2 lb/acre of heptachlor was applied contained more than 3 ppm of heptachlor, this result signals an immediate danger to woodcocks. Considering the fact that woodcocks consume 77% of their body weight per day on the average, it is reasonable to assume that relatively large amounts of heptachlor must have been ingested. The average level of heptachlor epoxide in the dead birds was 13 ppm, apparently enough to kill them. Nevertheless, the degree of magnification here (from earthworms to woodcocks) does not appear to be great.

Jefferies and Davis (1968) studied the dynamics of dieldrin transfer in soil, earthworms, and song thrushes. They observed a decrease in dieldrin concentration from soil to song thrushes, in contrast to some other food chain studies. Exposure of earthworms to soil containing 25 ppm of dieldrin resulted in 18.4–24.9 ppm accumulation after 20 days of exposure. After 6 weeks of eating diets containing from 0.32 to 5.69 ppm of dieldrin, thrushes showed total body residues of 0.09–4.03 ppm. At least part of the reason why the birds do not concentrate dieldrin is that they excrete the bulk of it (at the highest dose, 381 μg/day, 8.5% of the dieldrin consumed was retained by the birds). These workers considered that birds which carry approximately 3 ppm dieldrin in the liver and 1 ppm in the brain are in danger.

Thus in wild birds one of the important limiting factors for high accumulation of chlorinated hydrocarbons appears to be their susceptibility to the insecticides; i.e., it is not possible for them to accumulate more insecticides than the lethal amount. This matter will be raised later when factors affecting the levels of bioaccumulation are discussed.

10.3.2. Bioaccumulation Through the Food Chain

Most people concerned with environmental pollution by pesticides are familiar with the terms *biomagnification, bioconcentration, food chain,* etc. While it is true that persistent, lipophilic pesticides do accumulate in the animals at the top of the food chain, the concept of biomagnification through the food chain has been oversimplified to the point where more harm might be expected through misunderstanding than good. The core of the problem is that this simplistic approach does not accommodate the idea that the levels of pesticides in any organism on top or in the middle of the food chain are determined by the rate of uptake and elimination. Thus it is entirely possible that the organisms in the middle of the food chain (i.e., low to intermediate trophic levels) may accumulate more pesticides than the ones above them.

10.3.2a. Aquatic Environments

Robinson *et al.* (1967) made a comprehensive study of the distribution of *p,p'*-DDE and dieldrin (HEOD) in a marine environment on the eastern coast of Britain (near Farne Islands, Northumberland). They compiled residue data on 22 marine species and on micro- and macrozooplankton and plotted the data against their trophic levels (Fig. 10-6). The overall picture roughly agrees with the concept of higher pesticide levels in higher trophic levels, although variation among the individual species within a trophic level was great. The variation was particularly great in trophic level 4, where several bird species (shags, ducks, gulls) are grouped together with some fish (e.g., cod and whiting). Shags, in particular, showed very high dieldrin (highest) and DDE (second highest) residues, thereby contributing to the wide variation in this trophic level. The data are summarized in Table 10-17.

As already discussed in Section 10.3.1b, Dimond *et al.* (1971) found in a freshwater ecosystem that the equilibrium values for DDT-R reached in 10 years (after aerial spray of DDT) were 0.07 ppm for mud, 0.02 ppm for aquatic plants, 0.27 ppm for insects, and about 0.36 ppm for two species of fish (Table 10-13). In more heavily treated areas, the corresponding values were 0.35 ppm, 0.09 ppm, 0.58 ppm, and 2.8 ppm (average), respectively. In

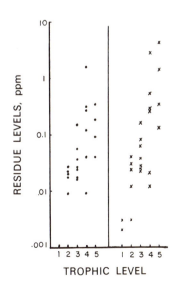

Fig. 10-6. Concentrations of dieldrin (left) and DDE (right) found in organisms of different trophic levels in an aquatic marine ecosystem. From Robinson *et al.* (1967).

TABLE 10-17. Typical Residue Levels Found in Marine Organisms in a Marine Ecosystem

Trophic level	Groups	Dieldrin (ppm)	DDE (ppm)
1	Seaweed	0.001	0.002
2,3	Microzooplankton and invertebrates	0.02	0.04
3	Macrozooplankton	0.16	0.16
3,4	Fish	0.04	0.02
4,5	Sea birds	0.5[a]	0.5[b]
5	Sea mammals (liver)	0.05	0.25

From Robinson *et al.* (1967).

[a]Shags alone showed an outstanding value at 1.6.
[b]Both shags (2.87 ppm) and cormorants (4.14 ppm) come outside of the general figures.

addition, the average residue level for the fish-eating kingfishers was on the order of 10 ppm. Thus a clear trend of increase was observed in this ecosystem having a history of prior pesticide application (i.e., the trend appeared to be more pronounced in the heavily sprayed area).

The data of Brown and Brown (1970) for a subarctic mixed aquatic and terrestrial environment (described in Section 10.3.1c) showed DDT-R residues in soil, aquatic sediments, plant leaves, fruits, and insects to be 0.088, 0.37, 0.1, 0.03, and 0.5 ppm, respectively. The corresponding values for vertebrates (Table 10-15) were considerably higher according to their food habits, showing increase in bioaccumulation in the order of trophic levels.

10.3.2b. Terrestrial Environments

For nonagricultural terrestrial areas, there is only a limited amount of information on food chain accumulation. (Note that examples cited so far are from data obtained from relatively confined ecosystems.) Lincer *et al.* (1970) studied DDE residue levels in two species of raptors in Alaska (along the Colville River, where there is no record of DDT spray) compared to those in their prey (Table 10-18). Their data indicate a modest biomagnification in the rough-legged hawk and a high concentration factor in the peregrine falcon. They reasoned that while the former preys exclusively on the native species (small ground mammals), the latter feeds on both native and migrant bird species. Another explanation of the difference might be that the rough-legged hawk feeds exclusively on ground mammals and the peregrine falcon on various bird species known (Brown and Brown, 1970) to be capable of accumulating DDT-R.

More relevant to bioaccumulation in agricultural areas, pesticide residues in bald eagles are now well described (Table 10-19). The total burden

TABLE 10-18. Residue Levels of DDE in Two Species of Alaskan Raptors and Their Prey[a]

Predator and prey	DDE in prey (ppm)	DDE in predator eggs (ppm)
Peregrine falcon		
Native birds		
Lagopus lagopus	0.19[b]	114
Migrant birds		
Anas carolinensis	0.15[b]	(131)
Spatula clypeata	0.21[b]	
Rough-legged hawk		
Native shrews		
Sorex cinereus[c]	0.24	1.21
S. arcticus[c]	0.48	(7.07)

From Lincer *et al.* (1970).

[a]Data are expressed as ppm in terms of the oven-dried weight in the muscle.
[b]There was a difference in DDE content in fat. The residue levels found in one each of *L. lagopus, A. carolinensis,* and *S. clypeata* were 0.89, 4.55, and 1.75 ppm, respectively.
[c]Whole skinned carcass instead of muscle. Common shrew and arctic shrew.

of pesticides, PCBs, and mercury in the bald eagle is heavy. Since bald eagles are at the top of the food chain, this circumstantial evidence indicates the possible role of bioaccumulation through the food chain. There are a number of reports on bioaccumulation of chlorinated hydrocarbon insecticides in

TABLE 10-19. Residues of Pesticides (ppm) in Carcass of Bald Eagles[a]

Compound	1966[b]	1967[b]	1968[b]	1969[c]	1970[c]
p, p'-DDE	11.8	16.55	4.92	6.9	18
p, p'-DDD	1.1	1.09	0.85	1.0	1.5
p,p'-DDT	0.2	0.2	0.15	0.22	0.10
Dieldrin	0.59	0.60	0.47	0.41	0.74
Heptachlor epoxide	0.07	0.08	0.08	0.06	0.10
Dichlorobenzophenone	1.20	0.71	0.53	0.42	trace
PCBs	—	—	—	10	20
Mercury	—	—	—	1.5	2.5

[a]Altogether 108 eagles found dead or moribund from 27 states.
[b]From Mulhearn *et al.* (1970).
[c]From Belisle *et al.* (1972).

agricultural areas. Korschgen (1970) studied the distribution of aldrin/ dieldrin residues (mostly found in the form of dieldrin) among several invertebrates and vertebrates in two adjacent cornfields in central Missouri. The fields had been treated with 1 lb/acre of aldrin for 16 and 15 of the previous 17 days. From the comparative food-habit data (Table 10-20), it is

TABLE 10-20. Soil–Food-Chain Relationship in Aldrin-Treated Cornfields in Central Missouri

	Aldrin/dieldrin residue (ppm wet weight)	Food habit (remarks)
Soil	0.31	—
Plant seeds	<0.02	—
Earthworms	1.49	Soil organic matter
Crickets	0.23	Mostly plant matter, some cannibalistic
Harpulus ground beetles	1.10	Larvae insectivorous, adults herbivorous
Peocilus ground beetles	9.67[a]	Highly predacious insectivorous
White-footed mice	0.98	Omnivorous
Toads	3.53	Insectivorous (and other invertebrates)
Bull snakes	1.25	Small rodents, birds, eggs
Garter snakes	12.35	Earthworms, salamanders, toads, birds, and small mammals

From Korschgen (1970).

[a]Much higher values (average 37.48 ppm) were recorded in June 1967.

apparent that the food chain through soil invertebrates (route II in Fig. 10-3) is more important in this ecosystem than the plant-mediated pathway (route I in Fig. 10-3). Important routes appear to be earthworm–(bird, salamander)– garter snake and soil insect–predacious insect–toad–garter snake. The predacious *Peocilus* beetles represent the case where organisms situated in the medium trophic level can accumulate higher residue levels than the ones above them (e.g., toads). The selected examples clearly indicate that generally the animals which are at the top of the food chain in an ecosystem because of their predacious food habits tend to accumulate more pesticide residues than the herbivorous animals or plants. At the same time, it is obvious that individual food habits may be most important in determining the final level of residues (e.g., Table 10–18) rather than simply position in the trophic level (Table 10-17).

One note of caution must be added here: there are a number of review articles which cite some astronomical biomagnification figures that greatly exceed the values found in the studies cited here. There are several reasons why such discrepancies can occur, the most obvious being that some reviewers collect data from many unrelated ecosystems in order to present a generalized or more dramatic picture. Thus the fish caught in Santa Monica Bay (which contain generally high levels of DDT-R) end up being compared to the phytoplankton in the northern Pacific Ocean. Another problem is the method of calculation of biomagnification with respect to water. Inasmuch as the water solubility of the chlorinated hydrocarbons is extremely low and the bulk of their reservoir is not in the water itself at any time after the initial peak resulting from spray has passed, the figures cited are often misleading. In considering the food chain accumulation (not direct pickup), calculation with respect to the original level in water should not be used in the strictest sense (only the primary producers such as phytoplankton accumulate residues solely from water).

Obviously, the important thing is to understand the processes that cause changes in the level of pesticides (upward or downward) and to analyze the degree of contribution of the food source to the total rate of pesticide intake and elimination.

10.3.3. Factors Affecting the Degree of Bioaccumulation

The complexity of the processes involved in bioaccumulation of pesticides in various ecological systems is overwhelming. It is difficult enough to understand the basis of population fluctuation as related to the food web relationship. Any attempt to mathematically predict the level of pesticide accumulation in an ecosystem must be considered a very gallant effort. Nevertheless, there are a number of influencing factors which can be clearly distinguished from others by virtue of their independent operation.

10.3.3a. Physicochemical Characteristics of Insecticides

The most obvious parameters for assessment of possible bioaccumulation are the chemical characteristics of the pesticides themselves. The characteristics most important are degree of persistence in the environment (including metabolic or conversion products), biological affinity (in most cases lipophilicity, but in some cases such as organomercury pesticides affinity for —SH protein), water solubility, and toxicity. Thus many bioconcentration problems are compound specific.

TABLE 10-21. Physicochemical Characteristics of Four Pesticidal Chemicals

	Water solubility	Benzene solubility (g/100)	Solvent solubility/ water solubility	Partition coefficiency (water vs. hexane)	Accumulation in brine shrimp[a] (ppb)
TCDD	0.2 ppb	0.047	10^6	1,000	157
DDT	1.2 ppb	80	10^{10}	100,000	3,092
γ-BHC	10 ppm	80	10^5	1,700	495
Zectran®	100 ppm	—	10^4	100	85

From Matsumura and Benezet (1973).

[a] Each compound (5 pmoles or approximately 1.5 μg) was deposited on 1 g of dry sand. The sand was added to the test aquarium containing the brine shrimp in 200 ml of water. The shrimp were collected after 24 hr, washed, and analyzed for residues.

Perhaps the first chemical characteristics that should be determined are partition coefficiency (between water and solvent or oil), water solubility, and solvent (lipid) solubility, in addition to chemical stability. Table 10-21 presents a summary of these parameters for four pesticidal compounds compared to bioaccumulation levels in shrimp. It is apparent from the data that the parameters interact. For instance, despite the low water solubility of DDT, bioaccumulation is high because of its high partition coefficiency. TCDD, on the other hand, does not accumulate in biological systems (under these experimental conditions) because of its low solubility in water and solvents. Zectran,® on the other hand, shows high water solubility coupled with a low partition coefficiency.

Because lipophilicity is a factor of unquestioned importance in bio-accumulation of pesticidal compounds, the second important step is to determine the lipid content of the animals and plants. The data obtained for Great Lake fish by Reinert (1970) have already been cited (Table 10-3). In essence, he showed that there is indeed a good correlation between fat content and the degree of accumulation of DDT-R by the fish (regardless of age). Earnest and Benville (1971) studied the relationship between lipid concentrations and the levels of DDT-R accumulated in eight species of fish and one crab species in San Francisco Bay and also found a positive correlation (Table 10-22).

The third factor to be considered probably is the concentration of insecticides in the surrounding area. This is particularly important for determining short-term uptake of residues. It is reasonable to assume that a law of diffusion should be applicable. Thus the rate of uptake is expected to be

TABLE 10-22. Correlation Between Lipid Concentrations and DDT-R Levels[a] in Eight Species of Fish and One Species of Crab from San Francisco Bay

Species	Total DDT-R[b] (ppb)	Lipid composition (%)	Correlation coefficient for lipid within each species
Dwarf perch	94–366	6.41	−0.53
Pile perch	33–410	4.36	0.79[c]
Shiner perch	126–281	3.44	0.36
Starry flounder	73–127	2.48	0.46
White perch	32–161	2.83	0.53[c]
Speckled sand dab	29–90	2.74	0.17
Staghorn sculpin	27–111	1.87	0.41
English sole	28–124	2.03	0.59[c]
Market crab	11–124	1.28	0.03

From Earnest and Benville (1971).

[a]Water concentration 4 ppt.
[b]From the data from Paradise Beach alone.
[c]Positive correlation at the 0.05 level.

high when the concentration gradient is steep. For instance, Gish (1970) was able to obtain good correlation between the concentrations of DDT-R found in earthworms and those in the soil, regardless of crop or soil type. Here, the concentration of the pesticide and its chemical nature appeared to be important determinants (correlations were poor for other pesticides examined). Such a concentration-dependent relationship was also observed by Wheatley and Hardman (1968) (Fig. 10-1), who noted a species difference in bioaccumulation. Even in aquatic environments, where biological systems generally pick up pesticides against the gradient of concentration, the dose–accumulation relationship appears to hold for many lipophilic insecticides. This is probably because of the quick establishment of partitioning equilibrium for the insecticides between water and the organisms, so that the rate-limiting factor in aquatic environments is not the speed of transfer. Thus accumulation in the primary producer (non-food-chain accumulation) is clearly influenced by the external concentration. However, in aquatic systems it is not easy to distinguish direct uptake from food chain pickup. In one experiment, Chadwick and Brocksen (1969) fed dieldrin-containing tubificid worms to fish (sculpins) and observed a positive correlation (Fig. 10-7). It is reasonable to assume that any biological system has a certain saturation level for a given compound. Such a saturation level could be

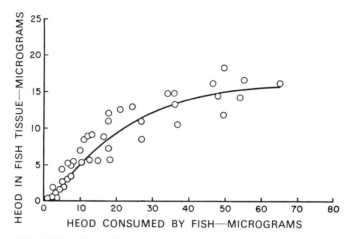

Fig. 10-7. Relationship between total HEOD in fish (sculpins) and total HEOD present in tubificid worms consumed by the fish. From Chadwick and Brocksen (1969).

determined by the susceptibility of the organism (scientists agree that in most cases dead organisms pick up less pesticide residues than live ones) to saturation of all available components due to solubility limits or even physical limitations. Whatever the cause, the relative effect of a dose is expected to be greater at concentrations far below saturation levels, and these situations are likely to occur in nature.

10.3.3b. Competition: "Biological Concentration" vs. "Biological Dilution"

Competition could play a very significant role in certain cases. Accumulation of a given pesticide in an organism within any ecosystem is determined by the degree of affinity of the organism vs. the availability of other competing matter which also picks up the same pesticide. Introduction of 1 μg of dieldrin to 1 g of algae in varying volumes of water would result in a tenfold accumulation (1 ppm) over another system involving 10 g of algal matter (0.1 ppm). Thus it is usually true that the levels of pesticide residues are generally lower in lakes with higher degrees of eutrophication than in the ones with lower degrees (e.g., Lake Michigan vs. Lake Erie; Reinert, 1970).

More drastic effects of biological competition have been observed by Rudd and Herman (1971). In brief, these workers have followed the effects of heavy DDD (TDE) application (120,726 lb in three installments between 1949 and 1957) on the ecosystem of Clear Lake, California. The most significant original effect of this contamination was total cessation of breeding in a colony of fish-eating birds, the western grebe (*Aechmophorus occidentalis*).

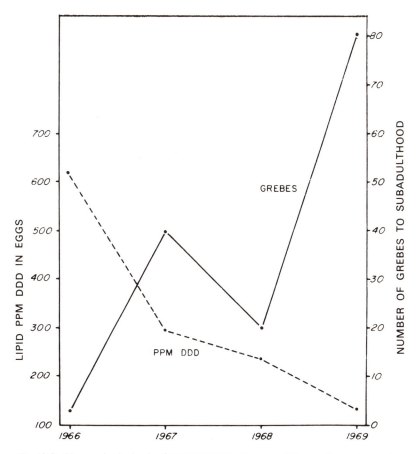

Fig. 10-8. Changes in the levels of DDD (TDE) in the eggs of Clear Lake western grebes during 1966–1969 compared to changes in the population of grebes. Note that the level of DDD sharply declined in 1967 when a new species of fish (silversides) was introduced into the ecosystem. From Rudd and Herman (1972).

These workers noted, after a 17-year period of life-history observation, that suddenly in 1969 the grebe populations recovered, with near-normal reproductive success (Fig. 10-8). They believe this turn of events was most likely caused by the successful introduction of a small atherinid fish ("silversides" or *Menidia audens*) in the fall of 1967. They increased in numbers unprecedented in the history of fish introduction, and this huge mass of "lipid pool" in the enclosed ecosystem certainly must have had the effect of diluting the level of DDD per gram of biological material. Assuming that the grebes ate on the average the same amount of fish as before, the total DDD intake should have decreased. (In reality, there is at least one more level to the food chain, since centrarchid fishes are normally eaten by grebes. The silversides serve as a readily available food source for the centrarchid fish.) Rudd and Herman use the term "nutrient shunt" to explain the phenomenon.

In an enclosed aquatic environment, where the major force of the movement of lipophilic pesticides is partitioned among various components, changes in the biomass of one species inevitably affect the pesticide levels in others, whether or not they are any way related via the food web. The phenomenon itself represents "biological dilution." Knowing, however, that such fluctuations in pesticide levels, upward or downward, merely reflect the result of "competition" for a fixed amount of pesticidal compound, a term such as "biological partitioning" might be more appropriate. The former nevertheless does facilitate understanding of the dynamics of competitive partitioning of pesticide residues in enclosed ecosystems by pointing out the possibility of dilution as opposed to "biological concentration," which is so often mentioned.

Live biological materials are not the only ones giving "competition" effects. Organic matter in the sediments and soil, soil particles and constituents (e.g., fulvic acid), and oil slicks and deposits are factors to consider. The fact that organic matter contents affect the rate of insecticide pickup by plants has already been discussed. The relationship in the earthworm–soil system is not readily apparent, as Gish (1970) could not demonstrate profound effects of soil type on the rate of pickup of various insecticides by earthworms in his exhaustive survey. In aquatic environments, the relationship is understandably more apparent. It has long been known that pesticides can be partitioned into suspended material and sediments (e.g., Green *et al.*, 1966; Keith, 1966; Keith and Hunt, 1966; and many others). Thus in an enclosed system the amount of pesticide available to the biological system is influenced by the texture of the aquatic environment. Earnest and Benville (1971) feel that the reason why the fish in San Francisco Bay (Table 10-22), where the water showed similar levels of pesticide pollution as Lake Michigan, had only one-tenth the pesticide level as the fish from the latter ecosystem is that the bottom of San Francisco Bay is muddy and that of Lake Michigan is sandy. Partitioning of lipophilic pesticides into oil slicks and sediments is also expected. Hartung and Klingler (1970) studied the oil sediments in the Detroit River, Michigan, a known location of oil pollution, and found that the partition coefficient for p,p'-DDD between water and sediment oil was on the order of 1.45×10^6 (i.e., DDT level in unfiltered water 40 ppt and that in oil sediment about 58 ppm). In contrast, according to the data obtained by Keith and Hunt (1966) for a lake system, the partition coefficients for DDT and related compounds between water and ordinary lake sediments and between water and suspended particles were on the order of 7×10^3 and 2.4×10^4.

10.3.3c. Rate of Food Consumption and Body Size

Other factors which affect the rate of bioaccumulation are the amount of food intake and the speed of elimination. The dynamics of accumulation and elimination will be discussed in the next section. Kenaga (1972) summarized the general relationship between the body weight of an animal and the amount of food consumption. The general tendency appears to be that the smaller the animal the higher the ratio of food consumption per unit weight, regardless of food habit (Table 10-23). Thus within the same food-

TABLE 10-23. Examples of the Relationship Between Body Size and Amount of Food Consumption

Animal	Body weight (g)	Food eaten per day (g)	Percent of body weight eaten per day
Dairy cattle	900,000	12,600	1.4
Swine	227,000	3,410	1.5
Man	65,000	715	1.1
Fox	5,700	143	2.5
Domestic fowl	1,800	61	3.4[a]
White rat	300	15	5.0
White owl	164–172	9.1–9.4	5.5
Kestrel	200	15.4	7.7
Song thrush	89	8.8	9.8[b]
White mouse	25	3.0	12.0
Chaffinch	22	2.9	13.2[a]
Robin (European)	16	2.35	14.7
Hamster	25	3.8	15.0
Great tit	18	4.4	26.0[b]
Blue tit	11	3.3	30.0[b]

From Kenaga (1972).

[a]Seeds and grains.
[b]Dry weight.

habit group, smaller individuals are likely to ingest larger amounts of pesticides. Since the small individuals also have a higher surface to body weight ratio (which is important in aquatic systems), they are even more likely to be subjected to residues than the large animals. Despite the obvious importance of the size factor, however, the residue levels actually observed in the environment show the reverse relationship; that is, levels in large animals are higher than those in small organisms (e.g., fish vs. plankton). The explanation can be sought in such phenomena as food chain accumulation, lipid content, longer life expectancy for larger animals, high susceptibility to pesticides for small organisms, and food habits. One factor which should not be excluded is differences in the rate of metabolism. The smaller animals have higher rates of metabolism, and so their rate of excretion–elimination should be higher. Whatever the reasons, in terms only of the final level of pesticide accumulation, the rate of food consumption by organisms of different sizes does not seem to play as important a role as might be expected.

10.3.3d. Dynamics of Pesticide Bioaccumulation

Perhaps the most important concept governing the translocation–accumulation behavior of pesticide residues among biological systems is that the pesticide level in any organism at any point in time is determined by the balance of two opposing reactions, uptake and elimination. It is important to stress that all three mathematical models cited in the previous chapter adopt this concept (Eberhardt *et al.*, 1971; Robinson, 1967; Woodwell *et al.*, 1971). The unifying theme is a first-order kinetic equation

$$\frac{dX}{dt} = K_1 X_{-1} - K_2 X$$

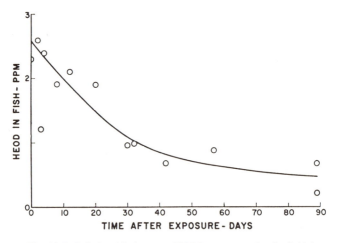

Fig. 10-9. Relationship between HEOD concentration in dieldrin-exposed fish (sculpins) and exposure time in water without HEOD. From Chadwick and Brocksen (1969).

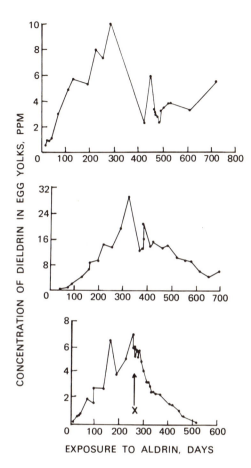

Fig. 10-10. Top: Concentration of dieldrin (ppm) in the yolks of eggs of a chicken fed 1 ppm aldrin in the diet for 700 days. Middle: The same except the chicken was fed 1 ppm dieldrin for 700 days. Bottom: Concentration of dieldrin (ppm) in the yolks of a chicken fed 1 ppm aldrin for 250 days. X marks the start of return to a normal diet for 250 days. In all cases, the dieldrin level starts declining around day 300 after the start of the experiment. From Brown *et al.* (1965).

where X and X_{-1} are concentrations of a persistent insecticide in an organism (or component within a body) and K_1 and K_2 are the rate of uptake and the rate of loss as illustrated below:

$$X_{-1} \xrightarrow[\substack{\text{rate of} \\ \text{uptake}}]{K_1} X \xrightarrow[\substack{\text{rate of} \\ \text{loss}}]{K_2} X_{+1}$$

Thus accumulation could take place whenever the amount of uptake is larger than the amount of loss. Other than being concentration dependent, these constants (K_1, K_2) are system (e.g., species, ecosystem, tissue, insecticide) specific. Despite the generalized nature of the above equation, there are a number of indications that it describes the tendencies of pesticide movements, at least in selected cases.

Eberhardt *et al.* (1971), for instance, could calculate the rate of pickup (transfer coefficient) and the rate of loss in 16 and 18 species of aquatic organisms in an enclosed 4-acre marsh following a DDT spray. Butler (1967*a, b*) demonstrated that the average loss of seven organochlorine pesticides in 7 days in four species of clams and one species of oyster was between 30 and 90%. The soft clam, which accumulated the highest level of pesticides among the five species, also lost the highest percentage (90%). Chadwick and Brocksen (1969) showed that the level of dieldrin in fish exposed to 1.28 ppb for 12 days declined once they were returned to water without dieldrin. The shape of the curve in Fig. 10-9 (loss vs. time) indicates the relationship, which followed first-order kinetics. Brown *et al.* (1965) kept chickens on a constant dieldrin and aldrin diet at a 1 ppm level for approximately 2 years and examined the dieldrin content in the yolks of their eggs. The results shown in Fig. 10-10 clearly indicate that the dieldrin levels increased to maximum in all cases in around 300 days and then declined (whether the birds were returned to a normal diet or not). Although such a result is most likely due to more long-term effects (e.g., liver "induction"), the fact that the insecticide level did not simply keep increasing as a result of continuous feeding clearly indicates that attainment of an equilibrium between intake and elimination in the chicken. The reason for shifts in this equilibrium state must be sought in physiological changes in the animal.

10.4. REFERENCES

Anderson, A. D., and R. B. March (1956). *Can. J. Zool.* **34**:68.

Anderson, D. W., J. J. Hickey, R. W. Risebrough, D. R. Hughes, and R. E. Christensen (1969). *Can. Field Naturalist* **83**:91.

Anderson, J. M., and M. R. Peterson (1969). *Science* **164**:440.

Barker, R. J. (1958). *J. Wildlife Management* **22**:269.

Batterton, J. C., G. M. Boush, and F. Matsumura (1971). *Bull. Environ. Contam. Toxicol.* **6**:589.

Batterton, J. C., G. M. Boush, and F. Matsumura (1972). *Science* **176**:1141.

Belisle, A. A., W. L. Reichel, L. N. Locke, T. G. Lamont, B. M. Mulhern, R. M. Prouty, R. B. DeWolf, and E. Cromartie (1972). *Pesticides Monitoring J.* **6**:133.

Bitman, J., H. C. Cecil, S. J. Harris, and G. F. Fries (1969). *Nature* **224**:44.

Bitman, J., H. C. Cecil, and G. F. Fries (1970). *Science* **168**:594.

Breidenbach, W. W., C. G. Gunnerson, F. K. Kawahara, J. J. Lichtenberg, and R. S. Green (1967). *Publ. Health Rep. Wash.* **82**:139.

Bridges, W. R., D. J. Kullman, and A. K. Andrews (1963). *Trans. Am. Fishery Soc.* **92**:421.

Brown, E., and Y. A. Nishioka (1967). *Pesticides Monitoring J.* **1**:38.

Brown, N. J., and A. W. A. Brown (1970). *J. Wildlife Management* **34**:929.

Brown, V. K., A. Richardson, J. Robinson, and D. E. Stevenson (1965). *Food Cosmet. Toxicol.* 3:675.

Buhler, D. R., and W. E. Shanks (1970). *J. Fish. Res. Board Can.* 27:347.

Burnett, T. (1971). *Science* 174:608.

Butler, P. A. (1967a). Bureau of Commercial Fisheries pesticide monitoring program. In *Proceedings of the Gulf and South Atlantic Shellfish Sanitation Research Conference.* U.S. Public Health Service, Bureau of Water Hygiene, Cincinnati, O., Publication No. 999-UIH-A.

Butler, P. A. (1967b). In *National Symposium on Estuarine Pollution*, Stanford University, p. 107. Stanford University Press, Palo Alto, California.

Chadwick, G. G., and R. W. Brocksen (1969). *J. Wildlife Management* 33:693.

Clement, J. G., and A. B. Okey (1972). *Can. J. Physiol. Pharmacol.* 50:971.

Cohen, J. M., and C. Pinkerton (1966). Organic Pesticides in the Environment, *Adv. Chem. Ser.* 60:163. American Chemical Society, Washington, D.C.

Cope, O. B. (1966). *J. Appl. Ecol.* 3:33 (Suppl.).

Cope, O. B. (1971). *Ann. Rev. Entomol.* 16:325.

Cutkomp, L. K., H. H. Yap, E. Y. Cheng, and R. B. Koch (1971). *Chem. Biol. Interact.* 3:439.

Cutkomp, L. K., D. Desaiah, and R. B. Koch (1972). *Life Sci.* 2:1123 (Part II).

Davis, B. N. K. (1968). *Ann. Appl. Biol.* 61:29.

Davy, F. B., H. Kleerekoper, and P. Gensler (1972). *J. Fish. Res. Board Can.* 29:1333.

Dimond, J. B., A. S. Getchell, and J. A. Blease (1971). *J. Fish. Res. Board Can.* 28:12.

Dvorchik, B. H., M. Istin, and T. H. Maren (1971). *Science* 172:728.

Earnest, R. D., and P. E. Benville, Jr. (1971). *Pesticides Monitotoring J.* 5:235.

Eberhardt, L. L., R. L. Meeks, and T. J. Peterle (1971). *Nature* 230:60.

Edwards, C. A. (1970). *Crit. Rev. Environ. Control* 1:7. Chemical Rubber Co., Cleveland, O.

Eisler, R. (1970). Pesticide induced stress profiles. In *FAO Technical Conference on Marine Pollution and Its Effects on Living Resources and Fishing*, Food and Agriculture Organization, Rome, December 9–18.

Eisler, R., and P. H. Edmunds (1966). *Trans. Am. Fish. Soc.* 95:153.

Gish, C. D. (1970). *Pesticides Monitoring J.* 3:241.

Gish, C. D., and N. J. Chura (1970). *Toxicol. Appl. Pharmacol.* 17:740.

Glazener, E. W., C. S. Shaftner, and M. A. Jull (1949). *Poultry Sci.* 28:834.

Grant, B. F., and P. M. Mehrle (1970). *J. Fish. Res. Board Can.* 27:2225.

Green, R. S., G. C. Gunnerson, and J. J. Lichtenberg (1966). In *Agriculture and the Quality of Our Environment*. American Association for the Advancement of Science, Washington, D.C., p. 137.

Hammerstrom, R. J., V. L. Casper, E. A. Robertson, Jr., J. C. Buggs, Jr., and J. L. Gaines (1967). Studies of pesticides in shellfish and estuarine areas of Louisiana and Alabama. In *Proceedings of the Gulf and South Atlantic Shellfish Sanitation Research Conference.* U.S. Public Health Service, Bureau of Water Hygiene, Cincinnati, Publication No. 999-UIH-9.

Hartung, R., and G. W. Klingler (1970). *Environ. Sci. Technol.* 4:407.

Heath, R. G., J. W. Spann, and J. F. Kreitzer (1969). *Nature* 224:47.

Heath, R. G., J. W. Spann, J. F. Kreitzer, and C. Vance (1971). In *Proceedings of the XVth International Ornithological Congress*. p. 20, ed. K. H. Voous, and E. J. Brill Publishing Co., Oude Rijw 39 Leiden, Netherlands.

Hickey, J. J., and D. W. Anderson (1968). *Science* 162:271.

Hylin, J. W. (1971). Personal communication.

Jackson, D. A., J. M. Anderson, and D. R. Gardner (1970). *Can. J. Zool.* 48:577.

Janicki, R. H., and W. B. Kinter (1970). *Am. Zoologist* 10:540.

Janicki, R. H., and W. B. Kinter (1971). *Science* 173:1146.

Jefferies, D. J. (1969). *Nature* **222**:578.

Jefferies, D. J., and N. K. Davis (1968). *J. Wildlife Management* **32**:441.

Jefferies, D. J., and M. C. French (1969). *Science* **166**:1278.

Jehl, J. R., Jr. (1969). *Environment Southwest* **418**:4 (June).

Kapoor, I. P., R. L. Metcalf, R. F. Nystrom, and G. K. Sangha (1970). *J. Agr. Food Chem.* **18**:1145.

Keith, J. O. (1966). *J. Appl. Ecol.* **3**:71 (Suppl.).

Keith, J. O., and E. G. Hunt (1966). In *Transactions of the 31st North American Wildlife Natural Resources Conference* p. 150. Wildlife Management Institute, Washington, D.C.

Kenaga, E. E. (1972). In *Environmental Toxicology of Pesticides*. F. Matsumura, G. M. Boush, and T. Misato, eds. Academic Press, New York, p. 193.

Koch, R. B. (1969). *Chem.-Biol. Interact.* **1**:199.

Korschgen, L. J., (1970). *J. Wildlife Management* **34**:1.

Lamar, W. L., D. F. Goerlitz, and L. M. Law (1966). Organic Pesticides in the Environment, *Adv. Chem. Ser.* **60**:187. American Chemical Society, Washington, D.C.

Lee, G. F. (1970). Factors affecting the transfer of materials between water and sediments. In *Eutrophication Information Program*. University of Wisconsin, Madison, July.

Lichtenstein, E. P., and K. R. Schulz (1960). *J. Agr. Food Chem.* **8**:452.

Lincer, J. L., T. J. Cade, and J. M. Devine (1970). *Can. Field-Naturalist* **84**:255.

Lowden, G. F., C. L. Saunders, and R. W. Edwards (1969). Organochlorine insecticides in water (Part II). In *Proceedings of the Society for Water Treatment and Examination.* The Water Research Assoc., Bucks., England.

Macek, K. J., and W. A. McAllister (1970). *Trans. Am. Fish. Soc.* **99**:20.

Matsumura, F. (1972). In *Environmental Toxicology of Pesticides*. F. Matsumura, G. M. Boush, and T. Misato, eds. Academic Press, New York, p. 525.

Matsumura, F., and H. Benezet (1973). *Envir. Health Perspect.*

Menzel, D. W., J. Anderson, and A. Randtke (1970). *Science* **167**:1724.

Metcalf, R. L. (1964). Report on National Academy of Science Traveling Symposium on Pesticides. Nov. 15–21, Washington, D.C.

Moore, G. L., Y. A. Greichus, and E. J. Hugghins (1968). *Bull. Environ. Contam. Toxicol.* **2**:269.

Moore, N. W., and C. H. Walker (1964). *Nature* **201**:1072.

Mulhern, B. M., W. L. Reichel, L. N. Locke, T. G. Lamont, A. A. Belisle, E. Cromartie, G. E Bagley, and R. M. Prouty (1970). *Pesticides Monitoring J.* **4**:141.

Nicolson, H. P. (1969). Occurrence and significance of pesticide residues in water, *Proc. Wash. Acad. Sci.* **59**:77.

Odum, W. E., G. M. Woodwell, and C. F. Wurster (1969). *Science* **164**:576.

Oloufa, M. M. (1953). *Poultry Sci.* **32**:391.

Peakall, D. B. (1967). *Nature* **216**:505.

Peakall, D. B. (1969). *Nature* **224**:1219.

Peakall, D. B. (1970). *Science* **168**:592.

Pillmore, R., and R. B. Finley, Jr. (1963). Residues in game animals resulting from forest and range insecticide applications. In *Transactions of the 28th North American Wildlife Natural Resources Conference*, p. 409. Wildlife Management Institute, Washington, D.C.

Pimentel, D. (1971). *Ecological Effects of Pesticides on Non-target Species*. Executive Office of the Presidency, Office of Science and Technology, Government Printing Office, Washington, D.C., No. 4106-0029.

Porter, R. D., and S. N. Wiemeyer (1969). *Science* **165**:199.

Ratcliffe, D. A. (1967). *Nature* **215**:208.

Reinert, R. E. (1970). *Pesticides Monitoring J.* **3**:233.

Robinson, J. (1967). *Nature* **215**:33.

Robinson, J., A. Richardson, A. N. Crabtree, J. C. Coulson, and G. R. Potts (1967). *Nature* **214**:1307.

Rudd, R. L. (1965). *Pesticides and Landscape*. University of Wisconsin Press, Madison.

Rudd, R. L., and S. G. Herman (1972). In *Environmental Toxicology of Pesticides*. F. Matsumura, G. M. Boush, and T. Misato, eds. Academic Press, New York, p. 471.

Schoettger, R. A. (1970). *Toxicology of Thiodan in Several Fish and Aquatic Invertebrates*. U.S. Department of the Interior, Washington, D.C., p. 35.

Stickel, W. H., D. W. Hayne, and L. F. Stickel (1965). *J. Wildlife Management* **29**:132.

Tarrant, K. R., and J. O'G. Tatton (1968). *Nature* **219**:725.

Taylor, L. W., and B. R. Burmeister (1940). *Poultry Sci.* **19**:326.

Tucker, R. K. (1973). Personal communication.

Tucker, R. K., and D. G. Crabtree (1970). *Handbook of Toxicity of Pesticides to Wildlife*. Bureau of Sports Fisheries and Wildlife, Denver, Colo., Resource Publication No. 84.

Tucker, R. K., and H. A. Haegele (1970). *Bull. Environ. Contam. Toxicol.* **5**:191.

USDA (1968). *The Pesticide Review*. Agricultural Stabilization and Conservation Service, Washington, D.C.

USDI (1964). *Pesticide-Wildlife Studies—1963*. Fish and Wildlife Service Circular, Washington, D.C., No. 199, p. 129.

Walker, K. C., D. A. George, and J. C. Maitlen (1965). *Residues of DDT in Fatty Tissues of Big Game Animals in the States of Idaho and Washington in 1962*. U.S. Department of Agriculture, Agricultural Research Service, Washington, D.C., ARS 33-105, 21 pp.

Warner, R. E., K. K. Peterson, and L. Borgman (1966). *J. Appl. Ecol.* **3**:223 (Suppl.).

Weaver, L., C. G. Gunnerson, A. W. Breidenbach, and J. J. Lichtenberg (1965). *Publ. Health Rep. Wash.* **80**:481.

West, I. (1964). *Arch. Environ. Health* **9**:626.

Wheatley, G. A., and J. A. Hardman (1965). *Nature* **207**:486.

Wheatley, G. A., and J. A. Hardman (1968). *J. Sci. Food Agr.* **19**:219.

Whitehead, C. C., J. N. Downie, and J. A. Phillips (1972). *Nature* **239**:411.

Wiemeyer, S. N., and R. D. Porter (1970). *Nature* **227**:737.

Witt, J. M., F. M. Whiting, and W. H. Brown (1966). Respiratory exposure of dairy animals to pesticides. In Organic Pesticides in the Environment, *Adv. Chem. Ser.* **60**:99. American Chemical Society, Washington, D.C.

Woodwell, G. M., P. P. Craig, and H. A. Johnson (1971). *Science* **174**:1101.

Zabik, M. J., B. E. Pape, and J. W. Bedord (1971). *Pesticides Monitoring J.* **5**:301.

Chapter 11

Hazards to Man and Domestic Animals

11.1. INTRODUCTION

Pesticides can come in contact with man and domestic animals by accidental encounter or as residues in food, water, and air. The former route often causes acute poisoning, which has been studied by medical and veterinary toxicologists. Occupational exposure (e.g., pesticide factory workers and pest control operators) usually results in chronic poisoning, which provides rough estimates of toxicity to humans for some of the pesticides. However, pesticide intake through food, water, and air has been neglected, since the amount of intake was thought to be insignificant. Studies on the subtle effects of this chronic exposure to persistent pesticides have begun only recently.

Because of the difficulties in using human subjects, toxicologists have traditionally relied on three experimental approaches: (1) case studies of accidental poisoning, (2) occupational exposure studies along with epidemiological studies on various segments of the population, and (3) controlled feeding tests. Evaluation of safety is always a difficult matter. According to the report of the Ad Hoc Subcommittee on Use of Human Subjects and Safety Evaluation (National Research Council, 1965), the kinds of studies in man that are useful are as follows:

A. Planned exposure studies.
  1. Metabolism, mode of action, enzymatic effects (followup of preliminary animal experiments).

2. Response to "safe" levels in simple feeding tests.
3. Absorption and storage.
4. Skin sensitization.
   B. Environmental and epidemiological studies.
   1. Chronic occupational exposure.
   2. Acute poisoning (accidental, suicidal, and occupational).
   3. Population surveys: tissue storage, placental transmission, and mammary transmission.

In all cases, experiments on human subjects are preceded by similar studies on experimental animals. Thus species differences in responses to various insecticides become very important. Some efforts have been made (e.g., Baker, 1971) in the field of pharmacology to correlate animal results with human effects, but such studies are not common in the field of insecticide toxicology.

The discussion in this chapter is divided into three separate parts: acute poisoning, chronic poisoning and subtle effects, and hazards of exposure to residues.

11.2. ACUTE POISONING BY INSECTICIDAL CHEMICALS

Studies on acute poisoning have a long history, for it is customary that the toxicity of a chemical is determined in experimental animals as soon as it begins to show any promise as an insecticide. Generally this phase of toxicology, followed by symptomatological observation and gross pathological examination, has been well worked out. There are a number of books and manuals which describe the toxicities of insecticides and the symptoms and treatments of accidental poisoning. Some examples are listed below:

Anonymous. *Clinical Handbook on Economic Poisons.* Atlanta, Ga.: Communicable Disease Center, U.S. Department of Health, Education and Welfare, 1963. Bulletin No. 476.

R. L. Brown. *Pesticides in Clinical Practice.* Springfield, Ill.: Charles C Thomas, Publisher, 1966. 492 pp.

R. H. Dreisbach. *Poisoning: Diagnosis and Treatment,* 5th ed. Los Altos, Calif.: Lange Medical Publications, 1966. 483 pp. (pp. 85–104 for insecticides.)

L. T. Fairhall. *Industrial Toxicology,* 2nd ed. New York: Hafner Publishing Co., 1969. 376 pp.

M. N. Gleason, R. E. Gosselin, H. C. Hodge, and R. R. Smith. *Clinical Toxicology and Commercial Products,* 3rd ed. Baltimore: Williams and Wilkins Co., 1969.

S. Moeschlin. *Poisoning.* English translation of 4th ed. New York: Grune & Stratton, 1965. 707 pp. (pp. 467–486 for insecticides.)

It is not within the scope of this book to describe the clinical details and treatments of poisoning for each insecticide. Instead, efforts will be made to explain the general principles involved by citing a few examples. The descriptions herein of toxic symptoms, methods of treatment, common routes of entry, etc., have been taken from the books just listed. Other specialized references of special interest are as follows:

D. F. Heath. *Organophosphorus Poisons.* New York: Pergamon Press, 1961. 403 pp.

K. W. Jager. *Aldrin, dieldrin, endrin and Telodrin.* Amsterdam: Elsevier Publishing Co., 1970. 234 pp.

R. D. O'Brien. *Toxic Phosphorus Esters.* New York: Academic Press, 1960.

While detailed explanations of toxic symptoms and effects are excluded from this book, such studies, particularly on human subjects, are extremely important. Simple statistics indicate that insecticides are responsible for 15,000–60,000 accidental poisonings among American children (Anonymous, 1967a). As shown in Table 11-1, insecticides ranked fifth among the

TABLE 11-1. Frequency of Ingestion Accidents Among American Children in 1966[a]

		Percent of total[b]
1.	Aspirin	24.9
2.	Soaps, detergents, cleansers	4.0
3.	Vitamins and iron	3.8
4.	Bleach	3.5
5.	Insecticides (excluding mothballs)	3.0
6.	Plants	2.6
7.	Analgesics and antipyretics	2.3
8.	Disinfectants and deodorizers	2.2
9.	Hormones	2.1
10.	Polishes and waxes	2.1
11.	Other	49.5

[a]See Anonymous (1967a) for reference.
[b]The total ranges from 500,000 to 2 million cases each year. About 500 of those under 5 years of age die.

major causes of accidental poisoning in 1966, and in terms of acute toxicity insecticides are the most poisonous of the substances listed. Therefore, pesticide scientists are much concerned with safety estimates and control, diagnosis, and treatment.

11.2.1. Chlorinated Hydrocarbon Insecticides: General Group Characteristics and Method of Treatment

As we discussed in Chapter 4, the precise mode of action of the chlorinated hydrocarbon insecticides is not known. From a toxicological point of view, they are mainly central nervous system poisons, eliciting a variety of CNS symptoms, among them neuromuscular and behavioral symptoms. Their effects on other vital organs cannot be neglected. A number of degenerative alterations are caused by chlorinated hydrocarbon insecticides, particularly in the liver and the kidney.

On the whole, however, this group of compounds is not as acutely toxic as the organophosphates, carbamates, and other poisons such as nicotine and veratrine. The number of accidental deaths caused by chlorinated hydrocarbon insecticides is far less than that caused by organophosphates.

As for general methods of treatment, the most successful approach has been the use of narcotic or anesthetic agents, such as chloral hydrate, or barbiturates, such as phenobarbital and pentobarbital. The toxicant should be removed from the animal or patient if possible. If poisoning was by topical contact, washing with water and detergent is important. For removal of poisons taken internally, syrup of ipecac, gastric lavage, and saline laxatives may be used. Oil laxatives should not be used since they promote absorption of these insecticides and of many organic solvents. When convulsions have occurred with regularity, the body temperature becomes high and washing with cold water helps bring it down. Tranquilizers have been used successfully by veterinarians to control the more violent forms of activity in animals. The object of sedation is not to induce sleep but to restore a relative calm. However, the proper dosage may be so high that it would induce anesthesia if poisoning were not present. If the animal is dull, listless, depressed, and unreactive, stimulants should be used.

11.2.2. DDT Poisoning

11.2.2a. Animals

The acute oral toxicity of DDT to mammals has been established by the general LD_{50} figure of 250 mg/kg for rats. However, rats are more susceptible to DDT than other mammals. The effect of DDT is greatly enhanced when it is dissolved in dietary fat rather than given as bulk crystals in a fat-free meal. DDT is moderately dangerous when applied in an oil solution or in other organic solvents by reason of its absorption through the skin. Whereas DDT solutions on the skin may be toxic, suspensions and powders are not. DDT-impregnated clothing may be worn without hazard. Open cuts in the skin are not special entrance sites for DDT. The eye is not affected by DDT applied in ointments or emulsions and is not a point of entry for general poisoning.

DDT is not irritating to the skin. Cattle, sheep, goats, pigs, and horses are unaffected by a single application of 8 % DDT emulsion. This is a level at which other chlorinated hydrocarbons (chlordane, toxaphene, lindane) induce symptoms and even death. Application to the skin of animals does introduce the hazard that they may poison themselves orally by licking.

DDT aerosols and mists that are used to control household and livestock insects are of negligible toxicity. For most experimental animals, toxicity does not appear until a concentration of 20 mg/liter (20 ppm) of DDT is reached, which is 4000 times the concentration required for insect control. Inhalation of some of the finer aerosol droplets is possible and absorption of DDT may occur in the lungs. Inhalation of DDT dusts or aerosols is a minor hazard to the smallest and/or youngest animals, which are more susceptible.

DDT is a cumulative poison. There have been many feeding experiments showing the effects of DDT at various dietary levels. When DDT is ingested, between 50 and 95 % of it is absorbed. Actually, acute poisoning by DDT is unlikely to occur in animals; the more common situation is chronic poisoning at low levels over long periods of time. When administered in low chronic doses, it is absorbed as readily in solid form as in solution. The result is gradual appearance of symptoms, especially the first ones. DDT is secreted in the milk of mammals and can be dangerous to the young.

11.2.2b. Symptoms of DDT Poisoning in Animals

Animals poisoned by DDT first become nervous and hyperexcitable, with excessive blinking, cold skin, ruffled fur, lack of appetite, and muscular weakness. Then follows the onset of fine tremors due to muscular fibrillation, particularly in the heart muscle, the hind legs, and the back. The tremors appear more rapidly in younger, faster-growing animals, in females rather than in males, and in starved rather than fully fed individuals. Continual shaking of the body muscles, coupled with the anorexia, causes a rapid loss of weight. There may be anemia and leukocytosis of the blood. The advanced stages of poisoning culminate in paralysis, clonic convulsions, and death.

11.2.2c. Human Toxicology

Mackerras and West (1946) report an instance in which 25 men ingested various amounts of DDT by mistake (for baking powder). Within $2\frac{1}{2}$ hr, all the men became weak and giddy. Four vomited, but all recovered within 48 hr. There have been numerous case reports of human DDT poisoning where fairly low dosages caused severe symptoms and even death. However, such results are unusual, and there is a large body of evidence which testifies to the general safety in the use of DDT as far as acute poisoning is concerned. When allowances are made for the toxicity of the solvent and the psychic and environmental conditions of the human subject, it is clear that the inhalation or skin absorption hazard of DDT is low indeed. However, there are some people with special susceptibility to DDT, and allergic reactions may be involved. DDT can be toxic by the oral route. Even if a person experiences no symptoms after ingesting DDT, there still could be liver damage. But this type of pathology, caused by low levels of dietary DDT, is rapidly rectified when the compound is no longer present.

11.2.3. BHC Poisoning

The γ-isomer of BHC, lindane, stimulates the mammalian nervous system, resulting in a rise in blood pressure, a fall in the rate of heartbeat (bradycardia), and an irregular encephalogram. The β- and δ-isomers of BHC are depressants of the mammalian nervous system and may partially eliminate, by antagonism, the effects of lindane. The symptoms of BHC poisoning are typical of nervous derangement: excitation, tremors, ataxia, convulsions, paralysis, and death from respiratory failure. Lindane does not accumulate to any great extent in the body, but is found in adipose tissue and in the kidney. Unlike DDT, lindane is taken up by the brain. BHC is secreted in cow's milk, where its characteristic musty odor may be detected.

Lindane is more toxic than DDT, but the other isomers of BHC are less toxic than DDT. The acute oral LD_{50} of lindane in rats is 190 mg/kg and that of crude BHC is 1250 mg/kg. The γ-isomer is not an irritant, but the β-, α-, and δ-isomers are strong irritants to the eye, nose, throat, and skin. BHC is absorbed through the skin, and lindane is also more toxic than the other BHC isomers by this route of entry. The single dangerous dose of BHC to man has been estimated as about 30 g and that of lindane 7–15 g. However, these figures may be too high, since individuals have suffered severe illness and convulsions at much lower doses. On the other hand, some people have withstood larger doses, especially of undissolved or poorly dispersed material, without serious effect. Children are more susceptible to poisoning by these compounds than adults, and much smaller amounts (even mg/kg) are often fatal to them. Similarly, young animals and emaciated or lactating animals are also more susceptible.

The histological changes caused by BHC and/or lindane include enlargement of the liver; necrosis, congestion, and fatty degeneration of the liver and kidney; congestion of the bladder with desquamation of its epithelium; hemorrhages in the gastrointestinal tract, heart, and lungs; and edema of the brain and spinal cord.

The barbiturates are used as antidotes to lindane and BHC poisoning. Atropine prevents the bradycardia, but phenobarbital is required to counteract the rise in blood pressure and convulsions.

11.2.4. Aldrin and Dieldrin Poisoning

11.2.4a. Symptoms and Toxicity

Generally speaking, the symptoms of dieldrin poisoning are similar to those described earlier. The peculiarity is that sometimes an apparently fully recovered animal will develop full-blown symptoms of poisoning as long as a month after the initial illness. The single dangerous oral dose of dieldrin to man is set at 10 mg/kg. Symptoms of dieldrin poisoning may appear as soon as 20 min after exposure, and in no instance has a latent period of more than 12 hr been confirmed in connection with a single exposure. In animals, the acute dermal toxicity of dieldrin in xylene is about 40 times that of DDT. With other solvents, the factor is about 6. Whereas undissolved dieldrin is not absorbed through the skin, dissolved dieldrin is readily absorbed.

Dieldrin is not readily metabolized and hence is stored in the body fat of animals. Residues once established in the tissues disappear very slowly; sometimes 6 months is required for reduction of dieldrin residues to 1 ppm. Dieldrin is excreted in the milk of lactating animals. It also appears to cross the placental barrier in horses. Animals have convulsed as much as 120 days following the last dermal dose of dieldrin. Recurrent illness has also been observed in humans poisoned by dieldrin. Treon (1954) studied dieldrin poisoning in mammals, and some of his findings were convulsions, weight loss, and anorexia. He found that in acute poisoning convulsions may bring death before much weight is lost. When convulsions do occur, they are followed at once by coma which at first is brief but progressively lengthens as they recur. Weight loss, a common sign of dieldrin poisoning in many species, is due to starvation resulting from anorexia and not to other metabolic disturbances. It is the long-continued coma and the weight loss that are fatal. Other symptoms Treon (1954) observed include hyperexcitation, poor coordination, weakness, excess salivation, jaw champing, muscle twitches, and personality changes. Nakamura (1960) found that chlorinated hydrocarbons sprayed over rats for 3 months induced a high frequency of anemia in addition to the symptoms of nervous excitation and weight loss.

In man, the signs of dieldrin poisoning are similar. Specifically, they include headache, nausea, vomiting, dizziness, and general malaise (Anonymous, 1963). In severe cases, convulsions follow the early symptoms, and sometimes convulsions are the only symptom observed (Anonymous, 1963). A coma may or may not follow a convulsion. Hyperexcitation and irritability are common. However, all these symptoms do not always appear in human poisoning. In some spray operators with repeated exposure, a condition similar to epilepsy has resulted. Laboratory tests have shown the presence of dieldrin in the tissues and urine after poisoning, but this finding is not proof of poisoning for dieldrin has been found in the blood and urine of spraymen who showed no symptoms. Workers who had had convulsions and other signs of poisoning tended to show a high concentration of dieldrin in the blood. In one case of acute aldrin poisoning with complete recovery, 40 ppm of dieldrin was found in the body fat.

11.2.4b. Autopsy Results on Dieldrin-Poisoned Animals

Treon (1954, 1955) found no specific lesions sufficient to account for death in autopsies of animals killed by dieldrin. In acute poisoning, the lesions they found were nonspecific, and small hemorrhages occurred at random throughout the body but were more consistently found at the heart. There was a cloudy swelling of most of the viscera and a blanching of the intestines. Generally the lungs were heavily congested, were dark in color, and showed some hemorrhages. There were also degenerative changes in the liver and kidney. In experiments on feeding various amounts of dieldrin to rats, Treon (1954) found the only characteristic histopathology to be in the liver, with hepatic cell swelling, cytoplasmic vacuolation, and homogeneity and peripheral grouping of cytoplasmic granules. Altered cells appeared sporadically distributed in the liver or else were concentrated in the lobular mid-zone and central zone. Treon and Cleveland (1955) found that dogs which died from dieldrin ingestion showed diffuse degenerative changes in the brain, liver, and kidneys. Fitzhugh *et al.* (1964) found microscopic lesions in the liver of rats fed all concentrations of chlorinated hydrocarbon insecticides. He also found fatty degenerative changes in the liver, kidney, and bone marrow in dogs fed 1 mg/kg dieldrin. Histological examination by Nakamura (1960) showed changes in the tissues of the respiratory organs and kidney and a cloudiness and swelling in the heart.

11.2.5. Organophosphate Poisoning

11.2.5a. Symptoms

O'Brien (1960) summarizes the usual symptoms of organophosphate poisoning in mammals as defecation, urination, lacrimation, muscular twitching and fibrillation, and convulsions, generally clonic (i.e., rapid repetitive movements) and less often tonic (i.e., limbs stretched and rigid). Usually death follows a clonic convulsion. Smaller doses lead to corresponding but less severe symptoms and occasionally to local paralysis. Chronic poisoning usually produces protracted and milder forms of these symptoms. Salivation, diarrhea, and muscular weakness are common. Animals can become adapted to chronic poisoning.

These insecticides can be absorbed through the skin as well as through the respiratory system and gastrointestinal tract, although skin absorption tends to be slow. However, these compounds are difficult to remove, so skin absorption may be prolonged. High temperature increases skin absorption, as does the presence of dermatitis.

Radeleff (1964) states that it is common practice to divide the signs and symptoms of organophosphate poisoning in both man and animals into (1) muscarinic, (2) nicotinic, and (3) central nervous:

1. Muscarinic effects are symptoms due to actions on the postganglionic nerve elements and to excessive stimulation of autonomic effector cells. They can include nausea and vomiting, anorexia, abdominal pain, cramps, gastrointestinal hypermotility, sweating, increased bronchial secretion, excessive lacrimation and salivation, diarrhea, respiratory difficulty, heart spasms, dyspnea, pallor, miosis, bradycardia, lung edema, fall in blood pressure, cyanosis, and incontinence of feces and urine. Atropine can antagonize these effects.

2. Nicotinic effects are the result of actions on somatic nerve elements which result in stimulation, followed by paralysis, of voluntary muscle. Twitching or spasms occur in the muscles of the tongue, eyelids, face, and, ultimately, general musculature, followed by weakness, diminished or absent tendon reflexes, flaccidity, rise in blood pressure, and paralysis. Atropine is not considered to be effective in combating these symptoms, except in large domestic animals.

3. Central nervous system effects are due to direct action on its elements. First there is a stimulation of action and then a depression. Symptoms may include headache, irritability, giddiness, tension, restlessness, apprehension, foreboding, ataxia, deep general tremor, drowsiness, stiffness of the neck, mental confusion, fever, convulsions,

loss of reflexes, flaccid paralysis, and coma. Heart block and arrest may occur. Atropine controls these symptoms.

11.2.5b. Cause of Death

It is generally agreed (O'Brien, 1960) that asphyxiation is the ultimate cause of death in mammals poisoned by organophosphates, and artificial respiration is known to enable animals to survive otherwise fatal doses. O'Brien (1960) lists the mechanisms involved in death as bronchoconstriction, lowered blood pressure, neuromuscular block of the respiratory muscles, and failure of the respiratory center. The sequence of events is (1) inhibition of cholinesterase, (2) acetylcholine accumulation, (3) disruption of nerve function, either centrally or peripherally, (4) respiratory failure, and (5) death by asphyxia.

11.2.5c. Example: Description of Malathion Poisoning

Malathion is one of the safest of the organophosphates, however, human poisonings still occur. Namba et al. (1970) have summarized the general trend of malathion poisoning, and the following is based on their description.

In Japan (population 99 million), there were 63 poisonings, resulting in ten deaths, from occupational handling or accident and 480 poisonings, including 404 deaths, from suicidal or homicidal attempts during 1957–1961 and 1965–1966 (Anonymous, 1967b). In Guyana (population 662,000), there were three deaths by accident and 43 deaths by suicide during 1959–1964 (Mootoo and Singh, 1966). Man may be more susceptible to malathion poisoning than experimental animals. There are various reports of death in adults after ingestion of about 5, 25, 35, or 70 g (Faraga, 1967). The amount of malathion absorbed was usually less, since in most patients some was vomited or removed by gastric lavage. Poisoning due to dermal absorption has been reported to occur, although the quantity absorbed does not appear to be sufficient to cause manifestations. It is not clear whether the estimated oral lethal dose of malathion for man is lower than for experimental animals because of greater human susceptibility or because of the presence of organic solvents that accelerate absorption (particularly xylene) in commercial preparations.

The onset of signs and symptoms of malathion poisoning has varied from 5 min to 3 hr after ingestion. They are similar to manifestations of poisoning due to other organophosphates and include muscarinic effects, nicotinic effects, and central nervous system effects. Malathion has a garlic-like odor resembling that of parathion, and almost all patients have this odor for days after poisoning has occurred.

11.2.6. Carbamate Poisoning

11.2.6a. Symptoms and Treatment

Symptoms of poisoning by carbamate insecticides are rather similar to those caused by organophosphates. The major ones are related to the accumulation of acetylcholine (a parasympathetic preponderance) and include bradycardia and decreased stroke volume accompanied by diarrhea and vomiting due to gastrointestinal hypermotility. Other symptoms include tremors, convulsive seizures of muscles, and increased secretion of bronchial, lacrimal, pancreatic islet, salivary, and sweat glands.

Atropine is the preferred antagonist, because it blocks the postsynaptic depolarization. The major difference in treatment for carbamate poisoning from treatment for organophosphate poisoning is that 2-PAM or other oximes are not used since they may cause further inhibition of cholinesterase. As explained in Chapter 4, the decarbamylation process is relatively fast and the use of oximes does not facilitate recovery from carbamate poisoning.

11.2.6b. Example: Description of Carbofuran Poisoning

The general description of Furadan® (carbofuran) poisoning that follows has been taken from a handbook prepared by the Niagara Chemical Division, FMC Corporation. As with organophosphates, the signs and symptoms of poisoning by carbofuran are those of cholinesterase inhibition. No study to date has shown any other mechanism of toxicity.

The early symptoms are headache, light-headedness, weakness, and nausea. Later, constriction of the pupils (may be preceded by a transient dilation), blurred vision, abdominal cramps, excessive salivation and perspiration, and vomiting may occur. Later symptoms probably resemble those caused by organophosphate poisoning. The length of the interval between exposure and the onset of signs and symptoms is related to the size of the dose and may vary from a few minutes to an hour or so. This period is shorter than for corresponding doses of organophosphate insecticides of the phosphorothioate type (such as parathion). The duration of symptoms is also related to dosage. In mild to moderate intoxication, spontaneous recovery may be anticipated in 1–4 hr; removal from exposure and administration of 2 mg atropine sulfate, intramuscularly or orally, are usually sufficient treatment. The reversal of the enzyme inhibition is rapid. The patient should be observed for several hours following exposure and atropine sulfate may be administered if symptoms persist or recur.

In more serious cases, treatment consists of artificial respiration and oxygen if signs of cyanosis (blueness of the lips and skin) or respiratory distress appear. Atropine sulfate, 4 mg, preferably intravenously, should be given as soon as the cyanosis disappears. The atropine should be repeated at 10-min intervals until signs and symptoms are controlled or the patient is fully atropinized.

11.2.7. Botanical Insecticides

11.2.7a. Nicotine

The initial reaction to nicotine is marked stimulation, followed by depression. Early manifestations are dizziness, diarrhea, headache, nausea, vomiting, elevated blood pressure, and excess salivation and sweating. Subsequent severe symptoms include convulsions, prostration, cardiac irregularity, respiratory depression, and coma.

Nicotine is an acute poison : death can occur within a few minutes or up to 5 hr later. The LD_{50} in mammals is cited as about 1–10 mg/kg. Nicotine penetrates the skin and absorption through inhalation is also rapid, while absorption through the gastrointestinal tract is relatively poor.

No antidotes are available. Treatment mainly consists of catharsis, emesis, and gastric lavage. Other treatments include absorption by activated charcoal and use of anticonvulsants such as barbiturates.

11.2.7b. Rotenone

Respiratory depression is the most marked symptom of poisoning by rotenone. Death occurs from respiratory paralysis, but kidney and liver damage occur from chronic poisoning. In severe poisoning, convulsions and coma may occur. The lethal dose for man is 100–200 g/kg orally.

11.3. CHRONIC TOXICITY AND STUDIES ON SUBTLE EFFECTS

11.3.1. Studies on Occupational Exposure and Human Feeding Tests

Studies on people subjected to pesticide exposure through their occupations, such as pesticide factory workers and pest control operators, provide a wealth of information about toxicity. General data on occupational diseases caused by pesticides indicate the source of the hazards. For instance, Table 11-2 shows that in California in 1957 the major problem was in the

TABLE 11-2. Number of Cases of Occupational Disease Attributed to Pesticides and Agricultural Chemicals in California, 1957

	Farm		
	Farms	Service	Manufacturing
Organophosphates[a]	109	67	23
Chlorinated hydrocarbons[b]	27	14	7
Cyanamide	2	2	1
Lead and/or arsenic	3	9	3
Herbicides	19	16	3
Fertilizers	19	4	2

From Kleinman *et al.* (1960).

[a]One death by demeton poisoning.
[b]Includes some fumigants such as methyl bromide and carbon tetrachloride.

manufacture and use of organophosphorus insecticides, pointing out the necessity of improving handling processes. Such studies do not necessarily provide the means to prevent or control occupational health problems (West, 1969), but they certainly give scientists a clue as to the possible safety levels of individual pesticide formulations.

11.3.1a. Chlorinated Hydrocarbon Insecticides

Laws *et al.* (1967) studied the workers in the Montrose Chemical Corporation, which has been manufacturing DDT since 1947. Altogether 35 workers (20, 12, and 3 people in high-, medium-, and low-exposure groups, respectively) were examined for their health, and DDT residues were determined. These people had been exposed to DDT for an average of 15 years (11–19 years) at estimated daily doses of about 18, 7, and 5 mg/man for high-, medium-, and low-exposure groups, respectively, as compared to an average dose of 0.04 mg/man/day for the general population. The overall range of storage of DDT-R in the men's fat was 38–647 ppm, compared to 4–8 ppm for the general population.

Despite such high levels of exposure (100- to 500-fold more than the general public), these workers showed no ill effects as judged by routine clinical laboratory tests and chest X-ray examinations. The occurrence of a lymphocyte/granulocyte ratio greater than 1:0 has been reported by Ortelee (1958) and also by Laws *et al.* (1967). Laws *et al.* consider the percentages of occurrence (5% in the former study and 14% in the latter) well within the normal variation for adults.

There are two interesting sidelights to the work by Laws *et al.* (1967): (1) correlation of fat and serum levels ($r = +0.64$) shows that the average concentration of DDT-R in fat is 338 times greater than in serum, and (2) the proportion of DDE compared to DDT is less in heavily exposed workers (41%) than in the general population (81%). These scientists conclude that the difference is due to intensity of DDT exposure. Thus at high doses detoxification of DDT via the DDE route becomes relatively less important than the route via DDA.

In controlled feeding tests, Hayes *et al.* (1971) studied the effects of long-term oral administration of pure and technical DDT to man. Twenty-four volunteers ingested technical *p,p'*-DDT at rates up to 35 mg/man/day for 21.5 months. They were then observed for an additional 21.5 months to 5 years (16 people). The levels of DDT accumulated in their fats (Table 11-3)

TABLE 11-3. Accumulation of DDT in People Who Have Taken Long-Term Oral Doses[a]

	DDT dose (mg/day/man)			
	Control	3.5-tech.	35-tech.	35-pure
Number of men	4	6	6	8
Before exposure	4.3	3.4	4.1	9.0
12.2 months exposure	10.3	32.2	201.2	211.0
18.8 months exposure	16.3	49.2	205.0	208.6
21.5 months exposure	22.0	50.2	280.5	325.0
(Feeding stopped)	—	—	—	—
4.9 months recovery	13.3	31.4	124.2	160.8
11.5 months recovery	19.0	26.3	139.8	235.5
18.0 months recovery	18.0	36.8	126.7	156.4
25.5 months recovery	20.5	33.2	99.8	105.1

From Hayes *et al.* (1971).

[a] The data are expressed as mean concentrations of DDT in ppm found in the fat of volunteers.

were comparable to the storage data obtained from DDT factory workers by Laws *et al.* (1967). Again, Hayes *et al.* were unable to find any ill effects of DDT. The tests conducted were general health, blood cells, cardiovascular, liver function, carbonic anhydrase, hearing, and gait.

Experiments on feeding dieldrin to human subjects were carried out by Shell scientists (Hunter and Robinson, 1967; Hunter *et al.*, 1969). Four groups of volunteers (three or four persons per group) ingested doses of 14, 24, 64, and 225 μg/day. No clinical abnormalities were found.

A few cases of occupational intoxication have been reported (Jager, 1970). A slow buildup can occur in workers who handle aldrin and dieldrin, resulting in initial symptoms of headache, lassitude, fatigue, loss of appetite, weight loss, insomnia, frequent nightmares, inability to concentrate, loss of memory, and hyperirritability. Hoogendam *et al.* (1965) reported 17 nonfatal cases of intoxication with convulsions in workers in an aldrin, dieldrin, and endrin manufacturing plant at Pernis, Netherlands; in all, the EEG tracings showed pathological abnormalities. The average recovery time to normal EEG pattern was 0.5 month for endrin and 1-4 months for aldrin/dieldrin. The most apparent EEG signs of cyclodiene intoxication are (1) a spike and wave complex, (2) a gradual increase in pathological changes, (3) too frequent shifting of tracings from normal to abnormal, and (4) other qualitative characteristic changes; myoclonic jerks are a related symptom. In some cases, liver function, as determined by thymol turbidity and SGOT and SGPT (serum glutamic oxaloacetic transaminase and pyruvic transaminase) tests, was altered, but recovery appeared to be quick. There are no separate endrin studies to show the effect of feeding or occupational hazards.

So far, the results of all feeding tests and most of the occupational exposure tests have not revealed the critical level at which these pesticides cause poisoning symptoms or pathological changes (with the exception of the EEG data for dieldrin). However, some attempts have been made to correlate residue levels in certain human tissues with critical levels *in vivo*. First it became apparent that adipose tissue is a nonvital storage site for chlorinated hydrocarbon insecticides and that the portion present in or subsequently released into the bloodstream actually represents the important factor in poisoning. Kazantzis *et al.* (1964) reported some evidence that there is a dynamic equilibrium between the dieldrin concentrated in the blood and that in the adipose tissue. Dale *et al.* (1962) similarly concluded that a continual equilibrium between the blood and the fat exists in the case of DDT and DDE. However, the results of various animal experiments indicate that the correlation between pathological effects and pesticide levels is highest in the brain (Dale *et al.*, 1963; Barnes and Heath, 1964; Harrison *et al.*, 1963). The data obtained by Dale *et al.* (1963), for instance, indicate that the slight tremors which appear to be the most sensitive criterion for DDT poisoning (Table 11-4) in the rat (male, 100 days old) begin 4 hr after administration of DDT, when the brain concentration of DDT reaches 242-371 (average 287) ppm. It is not really feasible, however, to obtain brain data for human subjects, and so data on the plasma level become extremely important. The plasma level found by Laws *et al.* (1967) in the factory workers with high exposure was 0.1387-2.2017 (average 0.7371) ppm, which is far below the concentration observed in the rat.

**TABLE 11-4. Relationship Between the Levels of
DDT and DDE in the Rat and Symptoms of
Poisoning[a]**

Time after DDT administration (ave. hr)	DDT (ppm)			DDE in plasma (ppm)
	Brain	Plasma	Fat	
Group 1: No clinical signs of poisoning				
2	90–168	178–416	29–84	<0.5
	(119)	(268)	(58)	
Group 2: Slight tremor				
4.75	242–371	178–1190	54–99	<0.5
	(287)	(833)	(68)	
Group 3: Severe tremor				
9.17	386–433	215–1086	118–266	99–193
	(404)	(645)	(213)	(129)
Group 4: Convulsions				
7.05	289–606	643–1348	173–512	<0.5–145
	(468)	(1007)	(293)	(98)
Group 5: Convulsion and death				
9.06	524–848	417–891	309–424	298–476
	(737)	(685)	(361)	(372)
Group 6: Recovery after severe tremor or convulsion or both				
26	138–213	119–178	377–810	298–476
	(176)	(131)	(598)	(405)

From Dale *et al.* (1963).

[a]The average concentration is indicated in parentheses.

Brown *et al.* (1964) studied the amount of dieldrin circulating in the bloodstream in relation to the clinical signs of poisoning in man and animals. In the experiments with dogs, doses ranging from 0.2 to 0.8 mg/kg body weight of dieldrin in capsules were fed to the point of intoxication. The concentrations of dieldrin in eight intoxicated dogs ranged from 27 to 127 μg/100g of whole blood (0.27–1.27 ppm). They concluded from these data, along with the data from poisoned human subjects, that 0.15–0.20 ppm of dieldrin in the whole blood is likely to be the critical concentration for intoxication (convulsion). These experiments were repeated by Keane and Zavon (1969), who adopted more subtle criteria for poisoning in addition to using the definite (convulsion) intoxication symptom. Their schedule of feeding was 2.0 mg/kg/day (up to 63 days) for the acute poisoning group and 0.2–2.0 mg/kg/day (up to 150 days) for the subacute group. Blood samples of the animals showing subsequent symptoms were analyzed for dieldrin. They found a direct relationship between the concentration of dieldrin in the blood and the severity of the clinical signs as follows: When a 40% decrease in food consumption was used as a criterion, the blood concentration was 0.37–0.39 ppm on the average. When a 10% decrease in body weight was observed, the average was 0.38–0.5 ppm. The levels for muscular spasm and convulsion were 0.51–0.58 ppm and 0.74–0.84 ppm, respectively. A similar experiment by Mount *et al.* (1966) indicated that the critical concentration of endrin in whole blood of channel catfish (*Ictalurus punctatus*) was 0.3 ppm. It is interesting to note that most of these workers have observed a relatively constant blood/fat ratio for the dieldrin concentration, in contrast to the data on DDT (see also Hunter *et al.*, 1969).

Probably the amount of endrin which causes the onset of poisoning symptoms is 0.2–0.25 mg/kg (Hayes, 1963). Coble *et al.* (1967) found 53 ppb of endrin 30 min after convulsion in a person poisoned by ingestion of contaminated bread. The serum endrin level dropped to 0.038 μg/ml (38 ppb) 20 hr after the onset of the convulsion. This patient recovered from poisoning. Weeks (1967) described four major outbreaks of acute endrin poisoning in Quatar and Saudi Arabia in 1967. Altogether 874 persons were hospitalized and 26 died after eating bread made from flour contaminated with 2–4 g/kg endrin. Blood from the patients contained 7–23 ppb endrin.

Studies on other chlorinated hydrocarbon insecticides are less complete. West (1967) described a few cases of possible lindane poisoning, and concluded that lindane is capable of causing serious blood dyscrasias, primarily aplastic anemia, in addition to some allergic reactions. There are also several reports claiming a similar cause–effect relationship between lindane and bone marrow damage (Best, 1963; Sanchez-Medal *et al.*, 1963; Lodge, 1965). Samuels and Milby (1971), however, could not find any clinical indication of abnormality, particularly in hematopoietic depression and renal or hepatic dysfunction, among 79 human subjects known to have been exposed to lindane for several weeks to years. In this particular instance, the mean blood lindane level in the highest-exposure group was 30.6 ppb (Milby *et al.*, 1968). This level is only one-tenth that found in a man showing clinical signs of lindane intoxication (290 ppb in plasma) 6 hr after ingestion (Dale *et al.*, 1966).

11.3.1b. Organophosphates and Carbamates

As shown in Table 11-2, organophosphorus compounds rank first as the cause of occupational poisoning. Furthermore, Metcalf (1957) found that in

California alone there had been 14,188 cases of accidental organophosphate poisoning. A more recent survey (Whitlock *et al.*, 1972) has shown that there were 627 poisonings in a 12-month period in southern California. Thus regions of occurrence have not changed, and there appears to be very little change in the occupational hazards. In Japan, there were about 6000 cases of parathion poisoning in the years 1953–1959 (Namba and Hiraki, 1958). Silverio (1969) estimates that the most commonly seen organophosphate poisoning cases are those of children poisoned by malathion, dichlorvos, and parathion. Heath (1961) considers that the high-risk groups can be classified as (1) manufacturers and spray operators, (2) children, particularly those of the preceding users, (3) suicide-prone people, (4) researchers.

The majority of the poisonings caused by spray occur in hot and humid field conditions, where it is difficult to force operators and farmers to wear heavy clothing for their own protection. Physiological factors, such as rapid sweating, affecting the susceptibility of workers cannot be overlooked. Durham (1967), in discussing the effects of high temperature on the toxicity of various insecticides, noted that many physiological factors promote toxicity. For instance, the *p*-nitrophenol level in the urine tends to go up at high temperatures, indicating high rates of skin absorption of parathion. Thus poisoning incidents are remarkably numerous in the South, in California, and in the rice fields of Asia, etc. In the Indian states of Andhra Pradesh, Madras, Mysore, and Kerala in 1961–1964, 668, 3582, 1091, and 405 poisoning cases, respectively, were reported. Organophosphates (parathion and to a lesser extent diazinon) were the most common causes of poisoning, followed by endrin (Indian Council of Agricultural Research, 1967).

Thienes and Haley (1972) summarized general toxicities of various organophosphates to humans (Table 11-5). While there are other similar data, individual susceptibility differences in man make all these estimates only approximations in any case. However, there is one great advantage in assessing the effects of organophosphates on man: their precise mode of action is well known. As far as their acute actions are concerned, these compounds owe their toxicities to their ability to inhibit cholinesterase. Although other side effects are known for certain compounds, these are neither abrupt nor fatal. Thus the methods for estimating the degree of poisoning are well enough developed so that chronically or mildly poisoned individuals can be detected before external symptoms begin to appear.

The most widely adopted approach is to measure the cholinesterase levels in the blood of the patient known or suspected to be poisoned. There are two distinct types of cholinesterases in the blood, erythrocyte cholinesterase and plasma cholinesterase. There are a number of different names for

TABLE 11-5. Estimated Fatal Toxicities of Organophosphorus Insecticides to Humans

	Estimated fatal dose (g)	Tolerance (ppm)
Chlorthion®	60	0.8
Coumaphos	10	0.1
Delnav®	5	2.1
Diazinon®	25	0.75
Dipterex®	25	0.1
EPN	0.3	0.5–3.0
Guthion®	0.2	2.0
Malathion[a]	60	0 (dairy products) 2–8 (others)
Methylparathion	0.15	1.0
Parathion	0.015–0.030	1.0
Pestox®	0.2	0.75
Phosdrin®	0.15	0.25–1.0
Systox®	0.02	0.3–12.5
TEPP	5.0	0
Trithion®	0.6	0.8

From Thienes and Haley (1972).

[a]For children, the estimated fatal dose is 0.1 mg/kg.

these substances, which are listed below:

1. Red cell cholinesterase.
   a. Other names: erythrocyte cholinesterase, AChE, true acetyl-cholinesterase type I.
   b. Characteristics: acetyl-β-methylcholine is a preferred substrate; excess substrate inhibition (optimum ACh concentration 10^{-3}M).
2. Plasma cholinesterase.
   a. Other names: ChE, pseudocholinesterase type II.
   b. Characteristics: hydrolyzes butyrylcholine faster than erythrocyte cholinesterase; no excess substrate inhibition.

There are basically three different techniques for cholinesterase assay in human blood: (1) potentiometric (i.e., pH changes), (2) radiometric, and (3) colorimetric determination of either the reaction products (acetic acid for pH metry) or the remaining unhydrolyzed substrate. The pH metry technique has been used most commonly in the past (e.g., Wolfsie, 1957; Augustinsson, 1957; Witter, 1963). In a typical case, blood samples are taken from the fingertip or ear and put into a heparinized capillary tube, which is sealed and centrifuged. The assay is done within 24 hr (if stored at room temperature) or within a few days (if stored at 0–4°C). The drop in pH, which is the result of hydrolysis of acetylcholine into acetic acid, is measured.

The radiometric technique (Winteringham and Disney, 1962) has been employed less frequently but has the advantage that it can be used on a microscale sample of blood (1 μl). The basic principle of the method is that ^{14}C-acetate-labeled acetylcholine, when hydrolyzed by cholinesterase, yields ^{14}C-acetic acid, which volatilizes on gentle warming with 0.5 N HCl on a glass plate, leaving the remaining acetylcholine. This method requires skill and a scintillation counter; however, some efforts are being made to expand its applicability to field scale.

Although there are a number of colorimetric methods available for cholinesterase assay (notably Hestrin's method, 1949), the most successful one for health hazard monitoring purposes is that developed by Voss and Sachsse (1970) requiring a sample of 10 μl of whole blood, taken from the fingertip. It is based on the very sensitive color development reaction between acetylthiocholine and 5,5-dithio-*bis*-2-nitrobenzoic acid (DTNB) (Ellman *et al.*, 1961; Voss and Schuler, 1967).

Despite all these developments, the data relating blood cholinesterase levels to specific toxic symptoms of organophosphate and carbamate poisoning in man are rather limited. Basically there are two groups of organophosphates whose differences in mechanism of action result in qualitatively different effects on blood cholinesterases. These are the quick-acting true phosphates and the thiophosphates. The latter require metabolic conversion in the body, so that there is a lag period before their effect on cholinesterases occurs. With respect to the former, Lebrun (1960) observed that after two oral doses of Dipterex® the plasma cholinesterase could be reduced to 10% and that of the red cells to 50% of normal value without the appearance of toxic symptoms. Edson (1955) conducted chronic feeding tests on schradan and observed that in some of his experiments blood cholinesterases were depressed (to 23% in red cells and 50% in plasma) in individuals who showed no apparent ill effects. Thus in this mode of administration the organophosphate level in the blood apparently rises rapidly before that in the central nervous system goes up. On the other hand, if the contact is through vapor, mist, application on skin, etc., it is possible that the local symptom (not the central effects) can precede the fall in cholinesterase activities in the blood. Such local effects include irritation of the skin, contraction of the pupil, and asthma-like respiratory difficulty.

In the case of thiophosphates, transport is mainly accomplished in the form of the original thiophosphate rather than phosphate. Thus the local poisoning signs do not precede the blood cholinesterase inhibition. Administration of a large single dose can cause central toxic symptoms even before the drop in blood cholinesterases takes place (in contrast to oral ingestion of Dipterex®).

Blood cholinesterase testing would therefore be useful in examining slow, chronic poisoning cases, particularly with organophosphates, which are known to form a much more stable complex with cholinesterases than carbamates do. However, Barnes *et al.* (1957) observed that in slow poisoning with parathion and Systox® the blood cholinesterase level could fall as low as 20% of normal before any toxic signs began to appear. Thus the monitoring of blood cholinesterases would be most applicable for diagnosing the long-term effects of thiophosphates or the effects of oral ingestion of true phosphates and carbamates.

In reviewing the significance of blood cholinesterase inhibition, Gage (1967) suggests that toxic effects are not likely to be encountered if the red cell and plasma activities remain above 50 and 25%, respectively, of their normal values. His suggestion of a threshold limit for blood cholinesterases is 30% inhibition, above which the worker involved must avoid further contact with the insecticide. This conclusion is based on the fact that blood cholinesterase levels vary to the extent of 15–25% for plasma and 10–15% for red cells.

Aside from establishing threshold inhibition levels, there are other useful purposes for checking blood cholinesterases. First, recovery from poisoning can be effectively monitored by this method. Leach (1953), for instance, studied recovery from parathion poisoning and concluded that although in general symptomatic recovery is quick, in some cases recovery in cholinesterase levels takes as long as 10 weeks. During such a recovery period, patients are advised to avoid further exposure.

Second, cholinesterase inhibition measurements can assist in establishing safe doses for various organophosphate and carbamate insecticides. Brown and Bush (1950) measured the cholinesterase activities of people who had been exposed to parathion ranging from 1 to 8 mg/10 m³ and concluded that 2–8 mg/10 m³ is potentially dangerous. When there are ample data on *in vivo* inhibition of blood cholinesterases in human poisoning cases, *in vitro* comparisons against animals could give valuable information about species differences in susceptibility (e.g., Sachsse and Voss, 1971).

Other methods for detection of organophosphates and carbamates in humans are not widely used. However, in some instances (e.g., legal cases) the detection of certain residues of insecticides is so important that these other techniques are employed. For instance, Jain *et al.* (1965) have analyzed ten organophosphorus compounds by gas chromatography. In specialized cases, gas chromotographic detection of insecticide metabolites offers a convenient means for analysis. Lieban *et al.* (1953) examined people poisoned by parathion and found *p*-nitrophenol, a rather easily detected hydrolysis product, in their urine. Davis *et al.* (1967) and Wyckoff *et al.* (1968) decided that urinary *p*-nitrophenol is an excellent indicator of parathion poisoning. Thus when the source of poisoning is nearly certain, the same technique can be applied for, e.g., methylparathion, dicapthon, and Chlorthion.® The elevated level of urinary *p*-nitrophenol continues throughout the poisoning period and correlates well with symptoms such as miosis and hypertension. In studying the detection of metabolites in the urine, Mattson and Sedlak (1960) conducted controlled feeding tests on a volunteer. The man was fed 0.16 mg/kg and 0.84 mg/kg of technical malathion, but showed no ill effects or blood cholinesterase inhibition. The level of the ether-extractable phosphates in the urine increased quickly, and a total of 19–23 % of the original doses were found in the urine in two experiments. Knaak *et al.* (1968) fed 2.0 mg/kg of carbaryl to two volunteers and in both cases found 25 % of the administered dose in the urine in the form of 1-naphthol derivatives, as judged by a colorimetric method (Best and Murray, 1962). The two men showed no apparent ill effects. Knaak *et al.* (1968) also studied the appearance of carbaryl metabolites in the urine by using a fluorescence detection technique on samples separated on a DEAE-cellulose column. The majority of the components which showed fluorescence were glucuronides, and some were sulfates of naphthol derivatives. Both of these techniques offer very sensitive means of analysis.

Two other reports on detection methods are worthy of attention. Jager *et al.* (1970) found that 17 workers out of 36 exposed to dimethylvinylphosphates showed electromyographic signs of impaired nerve and muscle function, while the blood cholinesterases did not show definite changes, possibly indicating that electromyographic detection is more sensitive. Roan *et al.* (1969) believe that direct determination of serum parathion and methylparathion give a much more accurate picture of poisoning in man than blood cholinesterase or urinary *p*-nitrophenol assays do.

In summary, the blood cholinesterase tests are valuable tools for determining the danger levels in human poisoning. Their usefulness is particularly apparent for people occupationally liable to constant insecticide exposure. Direct tests for blood insecticide levels and urinalysis for specific insecticide residues and/or specific metabolites offer a great promise in detecting early or mild poisoning when the source is known.

11.3.2. Nonfatal and Subtle Effects of Insecticides

Insecticides have other effects besides acute and chronic toxicity to their direct targets (e.g., cholinesterases of various tissues, the nervous system as a whole). These are (1) side- and after-effects, which generally occur at doses close to the fatal amount, and (2) subtle, often unexpected, effects occurring at much lower doses. With public concern about pesticide pollution increasing, it is these subtle effects that are drawing much of the attention. The problems discussed in the previous section are mainly those of people who professionally or intentionally come in contact with pesticides primarily in their original forms, whereas the problems associated with the subtle effects of pesticides are those of the general public, who may be affected in ways previously not suspected. There has been a recent outpouring of new information on the latter subject, and the following is a general summary of the situation.

11.3.2a. Pathological and Histological Changes

It is now clear that chlorinated hydrocarbon insecticides cause liver damage at high concentrations (e.g., at about 1000 ppm). The nature of damage ranges from increase in liver weight and fat content to cell necrosis (Hayes, 1959; Sarett and Jandorg, 1947; Durham, 1963). It is understandable that there are great variations in the doses which induce such effects, even among individual animals belonging to the same species, for general health and nutritional conditions are known to greatly affect the susceptibility of the liver. In the rat at least, sex also plays an important role in the manifestation of these histopathological changes. Cytological effects, representing "induction" phenomena, can take place at very low doses.

Ortega (1962) examined liver samples of DDT-poisoned rats by both light and electron microscopy. Some cytological changes were observed in the liver of male rats fed 5–15 ppm DDT in the diet, while over 200 ppm was necessary to cause similar effects in the liver of females. He noted that the cytological changes are readily reversible, indicating that they are "induction" related. (Liver "induction" is discussed in Section 11.3.2*b*). In man, the damaging effects of chlorinated hydrocarbon insecticides have not been well documented. Jager (1970) has summarized the results of liver function tests on both aldrin/dieldrin and endrin factory workers and states that there is a slight increase in liver function (as judged by SGPT and SGOT tests) among the exposed groups compared to those not exposed. The increases suggest liver enzyme "induction." A similar effect was observed by Kolmodin *et al.* (1969), who found that there is a significant reduction in the plasma half-life of antipyrine in workers exposed occupationally to chlorinated hydrocarbon insecticides. Thus it is likely that the effects of chlorinated hydrocarbons in the human liver at low (e.g., 1–10 ppm) daily dietary levels would be associated with "induction" rather than more permanent damage such as necrosis.

West (1967) believes that lindane causes hematological reactions, although the findings have not been confirmed in the case of chronically exposed human subjects (Samuels and Milby, 1971).

Nelson and Woodard (1949) fed TDE (DDD) to dogs and observed cytotoxic effects on both the liver and the adrenal cortex. There are a number of reports confirming this observation (Nichols and Henninger, 1957; Vilar and Tullner, 1959; Kaminsky *et al.*, 1962). The result of such atrophy is reduced secretion of glucocorticoids in response to ACTH, although the adrenal cortex seems to retain its ability to conserve sodium even at lowered blood sodium concentrations (Nichols and Richardson, 1960; Cazorla and Moncloa, 1962). According to Kaminsky *et al.* (1962), mitochondria of the adrenal cortex begin to swell 12 hr after the oral administration of 200 mg/kg of technical grade DDD to the dog. Hart *et al.* (1973) studied this phenomenon in detail by using three purified isomers of DDD. According to them, an intravenous dose of 60 mg/kg markedly inhibited ACTH-induced steroid production. IT_{50} (time to inhibit 50%) values were 27 min for m,p'-DDD, 87 min for o,p'-DDD, and 4–18 hr for p,p'-DDD. The mitochondrial effects were most marked at the zona fasciculata reticularis (Fig. 11-1), but not in the zona glomerulosa. An interesting observation is that the sequence of events follows the order of inhibition of steroid ultrastructural changes, and histopathological lesions.

Other histopathological lesions have not been clearly established. Ivanova (1972) claims that DDT and polychlorinated pinene cause histochemical changes in the myocardium at daily doses of 0.01, 3.5, and 4.25 mg/kg in rats. Structural alterations reported are granular changes and pyknotic nuclei in isolated fibers, eventually causing myolysis and karyolysis. Fowler (1972) studied the ultrastructural alterations in the proximal tubules of the kidney in rats fed diets containing dieldrin (5.0 ppm). The most marked morphological effects were observed in the pars recta. Interestingly, the response was greater in the female rats. An increase in smooth endoplasmic reticulum (SER) was also observed.

The effects of chlorinated hydrocarbon insecticides on other tissues of the body are not well studied. Uzoukwu and Sleight (1972) found that dieldrin at a dose of 15 mg/kg or 30 mg/kg (and subsequent doses of 0 or 15 mg/kg at 5-day intervals) caused histopathological damage to the lungs of the guinea pig.

The two main pathological or histological changes caused by some of the organophosphorus insecticides are neuropathological effects and dermatitic effects. With respect to the neuropathological effects, it is known that TOCP (triorthocresylphosphate) causes "delayed ataxia" due to a demyelination of the nerve sheath. The symptom is often called "ginger paralysis" (or "ginger Jake-paralysis") owing to the historical incident that the symptom

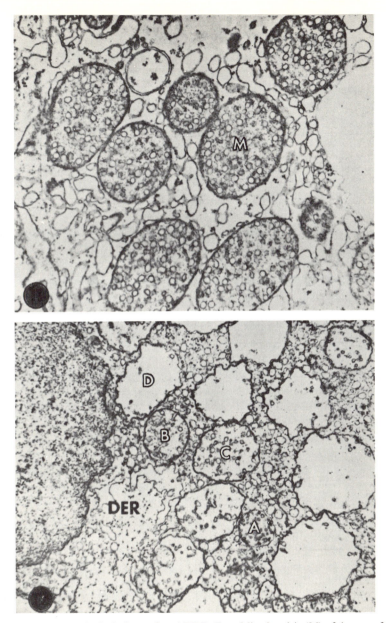

Fig. 11-1. Histopathological effects of *m,p'*-DDD. Top: Mitochondria (M) of the zona fasciculata from the adrenal cortex of an untreated dog. Bottom: The same region 2 hr after the
m,p'-DDD and ACTH. Mitochondria are seen in various stages of dissolution. A few appear
nearly normal (A) with many vesicular cristae and a small round shape. Most mitochondria
are in stages of swelling (B), internal dissolution (C), and collapse (D). The honeycombed
appearance of the endoplasmic reticulum is intact in many places, but occasional foci of disruption appear (DER). Nuclei are relatively intact. × 17,500. From Hart *et al.* (1973).

was first observed among the people who consumed adulterated Jamaica ginger during prohibition days in the United States. In addition, a number of poisonings have been caused by contamination of cooking oil with various petroleum oil products containing TOCP (Durham, 1963). The species specificity of this phenomenon is important: it is said that humans, chickens, cats, and calves are particularly susceptible.

The most-cited example of delayed demyelination effects is the mipafox poisoning of three workers in the manufacturing plant (Bidstrup *et al.*, 1953). They initially developed acute cholinergic symptoms which were cured by atropine treatments. However, two of the patients developed weakness in the legs in 10 days, one of whom experienced complete paralysis there. The process of recovery was slow (2–10 years). This "delayed ataxia" syndrome is not to be confused with the acute anticholinergic actions of organophosphates.

A crucial question is how many of the compounds having anticholinesterase activity can cause "delayed ataxia" in the human nervous system. The compounds known to have distinct demyelination effects are DFP (diisopropylphosphorofluoridate), diethylphosphorofluoridate, mipafox, TOCP, and tri(ethylphenyl)phosphate (O'Brien, 1960). In addition, Durham *et al.* (1956) found that in chickens malathion and EPN cause paralytic effects somewhat different from "ginger paralysis" that take place immediately. The characteristics of the immediate paralytic effects are localization of the symptoms and unusual persistence of the general muscle weakness commonly caused by other organophosphates. Frawley *et al.* (1956) found that myelin degeneration could occur in EPN-dosed chickens which developed muscle weakness, but no such lesion could be observed in malathion-poisoned chickens. The compounds tested so far which show no paralytic actions are Chlorthion,[®] dichlorvos, demeton, diazinon, sarin, soman, schradan, and tabun (O'Brien, 1960; Durham *et al.*, 1956).

There have been a number of efforts to explain why some organophosphates cause paralysis and others do not (including all carbamate insecticides tested so far). The theories of the mechanisms of such specific nerve reactions have been reviewed by O'Brien (1967). In brief, none of the theories up to 1967 was adequate to explain the phenomenon. Later developments, however, have apparently helped to clarify the subject. Johnson (1969a) found that an enzyme in the chicken nerve was phosphorylated by paralysis-causing organophosphates, but not by other cholinesterase inhibitors. From examination of its specificity, Johnson (1969b) concluded that this enzyme, which hydrolyzes phenyl phenylacetate (i.e., phenyl ester of 1-phenylacetic acid), is the action site at which DFP and mipafox cause the delayed paralytic effects.

The finding is further supported by the phenomenon that several carbamates which structurally resemble phenyl phenylacetate offer *in vivo* protection from DFP effects (Johnson and Lauwerys, 1969). The reason behind the protection appears to be that esterase inhibition by the carbamates is reversible. Thus these structurally similar carbamates (particularly phenyl benzylcarbamate and phenyl phenylcarbamate) protect the enzyme from being permanently phosphorylated by DFP by temporary competitive inhibition (as shown *in vitro*). Although this particular esterase represents only 3–5% of the total phenyl phenylacetate hydrolyzing activities of the chicken nervous system, it is technically possible to distinguish it from other esterases.

There are other less well-defined neuropathological effects of organophosphates. There have been a number of reports in Japan relating organophosphate poisoning to eye damage, particularly an apparently reversible "tunnel-vision" phenomenon. Otsuka (1971) conducted a controlled experiment on the effect of an organophosphate insecticide, Ekatin® (thiometon), using beagles. Various kinds of optic neuritis were apparent in the beagles given doses of 5, 10, and 15 mg/dog/day. One of the most persistent and marked pathological changes observed was elongation of the optic axis, which should cause "nearsightedness." More data are needed to confirm these effects of organophosphates, particularly information on the chemical characteristics of the organophosphates which allegedly cause eye damage.

While there have been many irate complaints of "skin irritation" from the users of insecticides, actual proof was not found until fairly recently. Dikshith and Datta (1968), for instance, examined the effects of cutaneously administered parathion on the skin of a female guinea pig. They topically applied 1 ml of 1 ppm parathion (98.7% technical in 50% ethanol) to an area 4 by 4 cm daily for 15 days and observed mild proliferation of connective tissue around hair follicles and glands within a few days, and hyperkeratinization and thickening of the stratum corneum in 5–10 days.

Naled (Dibrom®) is also known to cause dermatitis. Edmundson and Davies (1967) observed it in nine out of 12 female workers at a chrysanthemum grower's farm. Their symptoms ranged from mild irritation to maculopapular patchy eruption. Sensitization tests clearly indicated that the original naled itself was responsible for the skin trouble; other suspects (xylene, KBr, dicofol, captan, and DDT) did not show any sign of sensitization. These effects were confirmed by Phillips *et al.* (1972), who also found that Hercules 9007® (a carbamate, ENT-27334-Gb) causes skin irritation in man. It is interesting that two additional compounds, formaldehyde and Dowco 214® (*O,O*-dimethyl-*O*-3,5,6-trichloro-2-pyridylphosphorothioate), were mild irritants to rabbit skin (but not to human skin) as judged by the Draize rabbit skin irritancy test.

Pyrethrum is also known to cause dermatitis (Rickett *et al.*, 1972), although it is not the active principles that cause skin problems. Rickett *et al.* extracted pyrethrum flowers and found that the active dermatitis-causing substances have high molecular weight and are not dialyzable.

11.3.2b. Biochemical Changes

By far the most important biochemical change caused by insecticides is the "induction" phenomenon. The mechanisms of "induction" of liver microsomal enzymes have been discussed in general in Chapter 5. Several pioneering works have firmly established in experimental animals (mainly rats) that hepatic enzymes are induced by various chlorinated hydrocarbon insecticides such as DDT analogues, chlordane, toxaphene, lindane, and dieldrin (e.g., Hart *et al.*, 1963; Hart and Fouts, 1963, 1965; Kinoshita *et al.*, 1966). Now there are many other reports demonstrating these effects. Therefore, efforts will be made here only to mention the data on human subjects or those from the animal experiments designed for extrapolation. Stimulation of microsomal enzyme production along with proliferation of smooth endoplasmic reticulum (SER) by these apolar substances appears to be an adaptive phenomenon. In general, xenobiotics (foreign chemicals) are metabolized by microsomal enzymes to more polar compounds. Polar metabolites can be either directly excreted by the urinary system or further conjugated to form excretable material. In addition to chlorinated hydrocarbon insecticides, several organophosphates (parathion, OMPA, EPN, and malathion) and synergists (SKF-525A, piperonylbutoxide, and sesamex) are known to cause induction phenomena in mammals (Durham, 1967). Others are paraoxon, disulfoton, and carbaryl (Stevens *et al.*, 1972) and pyrethrum (Springfield, 1972).

The importance of the hepatic microsomal system as a sensitive biochemical indicator can be illustrated in the following example. Cranmer *et al.* (1972) studied the effects of *p,p'*-DDT on the squirrel monkey by administering 0.05, 0.5, 5, or 50 mg/kg/day of *p,p'*-DDT in peanut oil and then monitoring several blood enzyme activities and examining the hepatic enzyme activities at the end of the experiments. They could not demonstrate any hematological abnormality in any of the monkeys tested. Enzyme tests included those on plasma amylase, aldolase, glutamic-pyruvic transaminase, and isocitric dehydrogenase. However, two tests for mixed-function oxidase activity in the hepatic microsomal system showed a positive change in DDT-treated monkeys: by the criterion of EPN degradation the threshold level for induction was 5 mg/kg/day, and by the *p*-nitroanisole (PNA)

TABLE 11-6. Effect of *p,p'*-DDT on Hepatic Microsomal Enzyme Activity of Squirrel Monkeys[a]

p, p'-DDT (mg/kg/day)	μg EPN hydrolyzed/50 mg liver			μg PNA hydrolyzed/50 mg liver		
	2 months	4 months	6 months	2 months	4 months	6 months
Control	6, 7	6, 7	6, 6	13, 16	14, 15	8, 12
0.05	7, 4	5, 5	4, 8	14, 17	9, 16	10, 15
0.5	6, 8	4, 5	4, 6	28, 34[b]	26, 32[b]	28, 30[b]
5.0	14, 18[c]	15, 19[c]	15, 17[c]	37, 48[c]	45, 58[c]	50, 62[c]

From Cranmer *et al.* (1972).

[a]Each figure represents the average of duplicate results. For EPN the duplicates varied 0–66% from the average, and the PNA determination varied 0–11.1% from the average.
[b]Significantly more activity than controls and 0.5 mg/kg group ($p < 0.05$).
[c]Significantly more activity than all other groups ($p < 0.05$).

O-desmethylation criterion the level was lower at 0.5 mg/kg/day (Table 11-6). Moreover, induction appears to occur in less than 2 months (the shortest time period) and then remain relatively constant despite the intake of additional DDT by the monkeys. It is understood that at high doses other enzymes are also affected. Krampl and Grigel (1972) showed that liver glutamic-pyruvic transaminase and aldolase activities increased after a single dose of aldrin or dieldrin (40 mg/kg) in the female rat. After 72 hr, all these enzyme activities were depressed in the liver and increased in the serum. Chung *et al.* (1967, 1968) have demonstrated that both DDT and dieldrin cause increased protein synthesis in Hela S cell culture (human epidermoid carcinoma of the cervix in continuous cell culture for 8–9 years) at 0.5 ppm in the medium. At higher concentrations, the rates of synthesis, as measured by ^{14}C-leucine incorporation per milligram of cell, are inhibited. While their effects on DNA metabolism are not certain, the rate of RNA synthesis is inhibited at 0.5–10 ppm levels by both insecticides as assayed by the amount of ^{14}C-uridine incorporated per milligram of cell. In all cases, the effects are not significant according to the amount of ^{14}C-uridine or ^{14}C-leucine incorporated per culture.

Fisher and Mueller (1971) tested the effects of γ-BHC (lindane) on the turnover of phosphorus in phosphatidylinositol of human lymphocyte cultures. In this culture, the growth of cells is stimulated by phytohemagglutinin (PHA), which greatly enhances P_i turnover. At 10^{-3} M (final concentration), γ-BHC inhibited this PHA stimulation by 85%. Similarly, it inhibited RNA and DNA syntheses which were initiated by PHA. The concentration used by these workers (approximately 300 ppm) was rather high, and the significance of the inhibition must be further examined at lower doses to be relevant to human health problems.

Gertig *et al.* (1971*a,b*) made efforts to measure biochemical changes induced by various chlorinated hydrocarbons in the serum enzymes by using human blood *in vitro*. The effects of aldrin, lindane, DDT, DDE, and TDE (DDD) on both acid and alkaline phosphatases were slight. Their effects on human aminotransferases *in vitro* were varied, but on the whole they activated both L-aspartate:2-oxoglutarate aminotransferase (SGOT) and L-alanine:2-oxoglutarate aminotransferase (SGPT) activities. However, the solvent used appeared to have a marked effect on the enzymes. Thus the lack of sensitivity would probably make it difficult to use these enzymes as a diagnostic tool for assessing chronic poisoning in man. Kacew *et al.* (1972) injected 100 mg/kg of *o,p'*-DDT (intramuscularly) in the rat and observed increases in various renal gluconeogenic enzymes—pyruvate carboxylase, phosphoenolpyruvate carboxykinase, fructose-1,6-diphosphatase, and glucose-6-phosphatase—to the levels of 298, 273, 300, and 298 % of the control values, respectively. This effect on the kidney appears to be a direct one, since neither adrenalectomy nor concurrent administration of triamcinolone affected the increase. DDT is also known to inhibit carbonic anhydrase in man (Enns, 1972). The concentration required to inhibit the external, membrane-bound red cell enzyme was 1–2 moles of DDT per mole of the enzyme.

While there are a number of claims (and counterclaims) concerning the possible effects of some of the organophosphates on vitamin E utilization, confirmatory evidence is lacking. Goyer *et al.* (1970), for instance, could not observe such an effect in rats treated with 30 ppm (in the diet) of paraoxon and parathion for 8 weeks.

One of the most sensitive indicators of poisoning by some of the anticholinergic agents is the serum β-glucuronidase activity level. Within a few hours of administration of parathion, paraoxon, or Banol® to rats (12.5–50 % of the LD_{50}), Williams (1969) found up to an 800-fold (parathion) increase in this enzyme in serum. When sublethal doses of paraoxon were given, the serum β-glucuronidase activity increased up to 400-fold in 2, 4, and 6 hr without affecting erythrocyte, serum, and brain cholinesterases or causing any symptoms of poisoning. The blood β-glucuronidase activity returned to normal within 18 hr. The effects of Banol® were less drastic. The effects must be specific, since administration of other insecticides such as aldrin, chlordane, and Mobam® at doses causing other biochemical lesions did not influence the β-glucuronidase levels. The interesting observation is that the liver β-glucuronidase level is depressed while the level in blood remains high (Williams, 1970).

As for the paralysis-related biochemical changes, Faina *et al.* (1971) reported that acid peptidase and neural peptidase activities in peripheral nerves were stimulated by DFP. Such stimulatory activities were observed in the heterogeneous fraction of the nerve homogenates (not in the myelin fraction). The significance of such a biochemical change is not known at this stage.

The effects of pesticides on endocrine systems are now well established. DDT is known to have estrogenic activity, and some people consider this action to be related to its structural similarity to diethylstilbestrol, a synthetic estrogen; however, the validity of such an argument has not been critically examined. After some earlier confusion about the estrogenic activity of DDT, Welch *et al.* (1969) clarified the situation by finding that *o,p'*-DDT, a contaminant of technical-grade DDT formulations, is superior in estrogenic activity to pure *p,p'*-DDT. The data shown in Table 11-7 clearly

TABLE 11-7. Effect of Various Analogues of DDT on Uterine Wet Weight in the Rat[a]

Treatment	Uterine wet weight (mg $\pm$ SE)	Percent increase in uterine wet weight
Control	20.8 ± 0.7	—
o, p'-DDT	31.0 ± 0.6[b]	49
DDT (tech.)	29.8 ± 0.8[b]	43
Methoxychlor	28.5 ± 0.8[b]	37
p, p'-DDT	26.6 ± 1.2[b]	28
o, p'-DDD	24.3 ± 1.5[b]	17
m, p'-DDD	22.7 ± 1.2[c]	9
p, p'-DDE	21.1 ± 0.7[c]	1
p, p'-DDD	20.8 ± 0.7[c]	0

From Welch *et al.* (1969).

[a]Immature female rats were killed 6 hr after intraperitoneal injection of 50 mg/kg of DDT or one of its analogues. Six animals were used in each group.
[b]Significantly different from controls ($P < 0.05$).
[c]Not significantly different from controls ($P > 0.05$).

indicate that *o,p'*-DDT outstandingly increases uterine weight in the rat. (In 90 hr, the difference between *o,p'*- and *p,p'*-DDT becomes even greater: 151 % and 28 % increase, respectively). Also, Bitman *et al.* (1968) have shown that *o,p'*-DDT is estrogenic in rats, chickens, and quail, while *p,p'*-DDT has almost no effect. The difference is not due to selective accumulation in the reproductive organ, nor is it due to the ovarian activities.

In addition to direct estrogenic activities, DDT analogues are also known to act on the adrenal cortex to cause antisteroid effects. The histopathological aspects of such actions have already been described. o,p'-DDT causes a marked increase in the rate of secretion of 17-hydroxycorticosteroids in dogs (Vilar and Tullner, 1959), and o,p'-DDD induces a decrease in the response of the adrenal cortex to ACTH (1962) within 2 hr of administration (60 mg/kg). The mechanism of suppression of steroids could be twofold: a direct inhibitory activity and an extra-adrenal metabolic factor that reduces the levels of steroids. Another possible source of interference is the change in the binding pattern of steroids by the insecticide. Tronko and Kravchenko (1971) found that the activity of the corticosteroid-binding globulin (transcortin) was stimulated by 100 mg/kg of o',p'-DDD. DDT at high doses also appears to stimulate pituitary MSH (melanocyte-stimulating hormone) activity (Peaslee et al., 1972). Rats fed 500 ppm DDT in the diet showed a significant increase in pituitary MSH activity in 1 month; however, the level returned to normal after 2 months.

It is more than likely that other insecticidal compounds also affect endocrine secretion and metabolism. Wakeling et al. (1972) studied the binding behavior of 5a-dihydrotestosterone with specific receptor proteins of the rat ventral prostate in vitro. They found that the rate of binding in the presence of 10 ppm of dieldrin was significantly lower (approximately 33 % inhibition) than the control value. Kusevitskiy et al. (1970) state that carbaryl affects the function of thyroid gland. In rats treated with 0.7 or 2 mg/kg of carbaryl, the amounts of iodine bound to serum protein increased by 59 and 96 %, respectively. In the absence of added iodine, the figure rose to 250 % in 3.5 months after the treatment. Much more work is needed to understand the scope of insecticide influence on hormone secretion and metabolism.

There is some evidence that vitamin storage is affected by pesticides. Ferrando (1971) found that DDT administration for 5 days (5 mg/kg/day) had a significant effect on the storage of vitamin A, which was given at 50 μg/day for 8 successive days following the DDT administration. The percentage of vitamin A stored was 2.63 % for the DDT-treated rats and 4.92 % for the controls. Phillips and Hatina (1972) studied similar effects by using chlordane, lindane, toxaphene, methoxychlor, dieldrin, and DDT against weanling rats reared on a 100 ppm diet (except dieldrin: 100 ppm diet for 16 days and 10 ppm diet thereafter) until breeding age (about 100 days). Dieldrin reduced vitamin A in the liver of dams and progeny (27 % reduction), while other insecticides affected the progeny only (10 % reduction). The intake of these insecticides was calculated to be approximately 5 mg/kg/day, and much higher than normal residue levels were found.

There are some signs that pesticidal compounds can suppress immunological responses. Wassermann *et al.* (1971) approached the problem by using two different antigens, washed and heat-killed *Salmonella typhi* and washed sheep red blood cells, against rabbits treated with 200 ppm of *p,p'*-DDT in their drinking water for 38 days, beginning 18 days prior to the antigen injection. They found that the formation of antibodies as judged by γ-globulins and the antibody titer was markedly reduced in the DDT-treated rabbits, especially those tested against *Salmonella*, which causes a high level of antibody formation in controls. The major decrease was due to 7S globulins. Perelygin *et al.* (1971) studied both phagocytic activity of leukocytes and antibody formation and observed declines in both at doses of 0.5 mg/kg for DDT and 20 mg/kg for carbaryl.

Since insecticides are neuropoisons, it is natural to expect that they would change the levels of biogenic amines in body tissues. Also, insecticides known to induce hydroxylase activities should stimulate local production of some of the amines which are synthesized via hydroxylation routes. Indeed, Cloutier and Gascon (1971), from a comparison of DDT with the known inducers phenobarbital and 3-methylcholanthrene, concluded that all three stimulate the production of noradrenaline in the accessory sexual tissues. DDT also causes an increase in serotonin in the brain stem of the rat (44 % increase at 600 mg/kg; Peters *et al.*, 1972). Since the rate of 5-hydroxyindoleacetic acid production does not increase, this effect is not the result of monoamine oxidase inhibition by DDT. Other enzyme systems not affected by DDT are tryptophan hydroxylase and 5-hydroxytryptophan decarboxylase, so that in this case serotonin increase is not linked to increase in hydroxylase activities.

More direct insecticidal stimulation of specific storage sites (organ) or at synaptic levels can be suspected, particularly with anticholinergic agents which can easily act on cholinergic trigger mechanisms to release biogenic amines. For instance, the adrenal gland is known to be operated by an acetylcholine-sensitive system. Brzezinski and Rusiecki (1970) studied the effects of Disyston® (disulfoton) on urinary adrenaline and noradrenaline. At doses of 5 % (repeated) and 10 % (single) of the LD_{50}, the levels of these catecholamines in the urine increased beyond those attained by a single administration of Disyston® at 20 and 30 % of the LD_{50} dose. These authors suggest that release from readily accessible tissues, such as the heart and the kidney, accounts for the increase. Brzezinski (1972) later studied the changes in the levels of these catecholamines in the rat plasma, brain, and adrenals using Disyston,® dichlorvos, and fenitrothion and observed that Disyston® had the greatest effect. As expected, the plasma levels increased to 250 % of normal within 1 hr after the administration of Disyston® at a dose 40 % of the LD_{50}. Adrenal adrenaline declined to one-half in 15 min and brain adrenaline to one-third in 2 hr. Hassan (1971) similarly studied the effect of carbaryl on catecholamine metabolism in the rat. He found over 300% increase in urinary 3-methoxy-4-hydroxymandelic acid (a catecholamine degradation product) following the application of 700 mg/kg of carbaryl. This effect lasted for 195 days, with a peak of 30 days. Since there was a concurrent increase in the synthesis of norepinephrine (noradrenaline), and since essentially the same tendency was observed in adrenalectomized and hypophysectomized rats, Hassan concluded that the increased turnover phenomenon was due to the general increase in sympathoadrenergic activities in the poisoned animals.

There are indications that at least some of the insecticides can react directly with the enzyme systems directly involved in the metabolism of biogenic amines. Matsumura and Beeman (1972) and Beeman and Matsumura (1973) found that chlordimeform greatly increases brain serotonin and noradrenaline in the rat at 200 mg/kg (slightly lower than the LD_{50} dose). Chlordimeform and its analogues inhibit monoamine oxidase *in vitro*.

The levels of other neurotransmitters are also influenced by some insecticides. O'Brien (1967) summarized the research work up to 1964–1965 concerning the influence of cyclodiene insecticides on γ-aminobutyric acid and related compounds. In essence, dieldrin (or in some cases telodrin) increases the levels of several amino acids and γ-aminobutyric acid analogues. The problem is that elevation of γ-aminobutyric acid should have anti-convulsive effects, which is contradictory to the symptomatic observations.

Waseda (1971) recently reexamined the effects of endrin, aldrin and dieldrin on brain aspartic acid, glutamic acid, glutamine, and γ-amino-butyric acid by comparing the *in vivo* effects with *in vitro* reactions. *In vitro* (slices of cerebral cortex), the insecticides uniformly increased the levels of these substances at low concentrations and uniformly decreased them at high concentrations. *In vivo*, however, only aspartic acid and glutamine were increased; glutamic and γ-aminobutyric acids were decreased. The most interesting aspect of this finding is that these cyclodiene insecticides could act as stimulators of γ-aminobutyric acid at low doses and as inhibitors at high doses. Although it does not appear that interactions with γ-aminobutyric acid and the other amino acids are in any way related to the primary action mechanisms of the cyclodienes, it is nevertheless interesting to know the biochemical consequences of poisoning, for there is a wealth of information on various drugs which specifically elevate or lower the level of these neuro-transmitters. Such knowledge concerning poisoning thus facilitates under-standing of the relationship between symptomatological expression and change in transmitter levels.

In studying biochemical lesions caused by pesticidal compounds, it appears necessary to distinguish the lesions resulting from general stress and insult to the body systems from those which represent specific biochemical reactions. The specific reactions include inhibition of cholinesterases and other esterases, organophosphates and carbamates, ATPase inhibition by chlorinated hydrocarbon insecticides, action of DDD analogues on the adrenal cortex, and aconitase inhibition by fluoroacetate-generating agents. Some of the nonspecific responses are definitely related to the hepatic changes which can be induced by many pesticides, including induction of serum aminotransferases (e.g., SGOT and SGPT), lactic dehydrogenase, and alkaline phosphatase. Changes in levels of hormones and various biogenic amines are indirectly related to these hepatic changes. Since the hepatic changes take place at relatively low pesticide concentrations, and since the hormones and biogenic amines do act on vital biological sites at very low doses, it is likely that subtle biochemical changes can be produced by relatively low doses of pesticides.

As for the detection of these subtle effects of pesticides in man, there is no convenient monitoring method sensitive enough to be useful at this time (exclusive of blood cholinesterase tests). Naturally it is not feasible to collect liver samples from patients, and biochemical changes in the blood, feces, urine, etc., cannot be detected at low pesticide levels. Perhaps, as Cornish (1971) argues, it is necessary to develop reliable assay methods for changes in isoenzyme patterns or immunological testing methods (Wassermann *et al.*, 1971). Other alternatives would be to conduct a complete diagnostic investigation such as a complete blood count, serum bilirubin, thymol turbidity, gold sol reaction, leucine aminopeptidase, plasma sodium potassium, serum carotene, and serum protein hemoglobin in addition to the other routine tests already discussed. Such complete tests have been said to be effective in detecting prepoisoning cases among spray operators in Hungary (Rosner *et al.*, 1971).

11.3.2c. Carcinogenicity–Tumorigenicity

Public concern about the possible carcinogenicity of pesticidal compounds has created a great demand for research in this area. However, it is very difficult to prove that any chemical present in the environment is in fact carcinogenic. Some of the reasons for the difficulty in determining safety with respect to carcinogenic potential will be explained here.

First of all, the distinction between carcinogenesis and tumorigenesis is not always clear. It should be based on the potential to cause malignancy; that is, uncontrolled somatic cell division including formation of malignant tumors is carcinogenesis and formation of benign tumors is tumorigenesis. For instance, "liver carcinoma" is reserved for the former case, while "hepatoma" refers only to the latter. Certain researchers propose that the transplantability of a tumor be the criterion for malignancy (Andervont and

Dunn, 1952), although such efforts do not always give consistent results. Lemon (1967) believes that it is actually impossible to distinguish between benign and malignant liver tumors in the mouse. Thus in many experimental animal tests the enigma is simply ignored—no attempts at distinction are made. To be blunt, when researchers talk about carcinogenicity sometimes tumorigenicity is meant and sometimes both carcinogenicity and tumorigenicity.

Second, dose–effect relationships are not clearly established. Weil (1972) cites examples of many types of dose–effect relationships that are mainly determined by the nature of the inducer and the tissue but are also susceptible to other environmental and artificial factors. Moreover, it is not usually possible to establish a threshold value or minimum dose for induction of carcinogenesis.

Third, species specificity makes it difficult to relate the effects (or the lack of effects) of new chemicals on certain experimental animals to human safety. Species specificity involves differences in lifespan. Some scientists stress the importance of nonhuman primate tests, while others advocate the entire-life tests that are possible with mice and other animals having short life spans.

Fourth, there is a lack of adequate epidemiological data on human cases that could serve to establish species guidelines relative to other experimental animals. While many chemicals are epidemiologically implicated to be carcinogens, not all of them can cause carcinogenesis in animals under experimental conditions simulating the epidemiological ones (National Research Council, Food Protection Committee, 1960).

Despite these difficulties, the usefulness of animal experiments and epidemiological studies is apparent. A few pesticides have indeed been implicated as carcinogens. While it is surprising that there is still no established or universally accepted standard method for screening and testing, several proposals have been made to calculate the safety factor as 100-fold (Druckrey, 1967) to 5000-fold (Weil, 1972) of the minimum effect level, depending on the nature of the biological effects.

The question of the carcinogenicity of chlorinated hydrocarbon insecticides is not quite settled. Kemeny and Tarjan (1966) and Tarjan and Kemeny (1969) conducted a long-term multigeneration feeding study on the effects of DDT in mice maintained on a diet containing 2.8–3.0 ppm of p,p'-DDT, which corresponds to 0.4–0.7 mg/kg/day. They observed that the increased incidence of leukemia and tumors was statistically significant with respect to controls in the second and third generations. By the fifth generation, the incidence of pulmonary carcinoma had increased 25-fold. However, no effects on reproduction were found. The meaning of such a multigeneration study is not certain; the fetus is subjected to both direct exposure to DDT and indirect stress through poisoning of the mother. Innes *et al.* (1969) screened a number of pesticides for one generation and listed chemicals that gave statistically significant increases in mouse hepatic tumor incidence (Table 11-8). Although the dose employed for DDT was rather high (i.e., lower than

TABLE 11-8. Pesticidal Compounds Causing Increase in Hepatic Tumors in Mice

Chemical	Use[a]	Daily dosage[b]		Vehicle
		mg/kg	ppm[c]	
PCNB	F	464	1206	0.5% gelatin
p,p'-DDT	I	46.4	140	0.5% gelatin
Mirex	I	10	26	0.5% gelatin
Avadex®	F	215	560	0.5% gelatin
Ethylselenac	F	10	26	0.5% gelatin
Ethylene thiourea	F	215	646	0.5% gelatin
Chlorobenzilate®	I	215	603	0.5% gelatin
Strobane®	I	4.64	11	0.5% gelatin
Bis(2-chloroethyl)ether	I	100	300	Distilled water
N-(2-Hydroxyethyl)hydrazine	H	2.15	5	Distilled water
Bis(2-hydroxyethyl)dithio-carbamic acid, potassium salt	F	464	1112	0.5% gelatin

From Innes *et al.* (1969). Highest tolerable doses of pesticides were used. Altogether 120 compounds were tested. Twenty other compounds which showed some tumorigenicity, indicating the need for further testing, were piperonylbutoxide, piperonylsulfoxide, SDDC, *o,p'*-DDD, monuron, *p,p'*-DDD, ethyltellurac, Perthane,® Chloranil,® Vancide PB,® Redax,® Ledate,® Omal® (Dowside 25®), Agerite® powder, azobenzene, cyanamide, Vancide BL,® Ethyltuads,® Zectran,® and CCC.

[a]F, fungicide; I, insecticide; H, herbicide.
[b]Administered via stomach-tubing method for days 7–28 of age.
[c]Dosage in diet, given *ad libitum* (after 28 days of age) until necropsy at 18 months.

the LD_{50} but much higher than the calculated daily intake in man, which is approximately 0.5 $\mu g/kg/day$) this study at least established that DDT could cause hepatoma in the mouse. However, there is a large gap between such a finding and the establishment of carcinogenicity.

But even though the carcinogenicity of DDT has not been satisfactorily proven, it is clear that tumorigenic—whether malignant or benign—effects intensify during continuous exposure of generations of mice to DDT. Tomatis (1970) found that tumors occurred in the second generation of mice (BALB/c strain) administered 2.8–3.0 ppm of DDT. Tomatis *et al.* (1972)

found that in a two-generation experiment with the CF-1 minimum-inbred strain of mice the incidence of liver tumors, but not of lymphomas, osteomas, or lung tumors, increased at all levels (i.e., 2, 10, 50, and 250 ppm in the diet) of DDT and also appeared at an earlier age than in controls. According to Tomatis *et al.*, the tumors are "well-differentiated nodular growths, compressing, but not infiltrating, the surrounding parenchyma, or nodular growths in which the architecture of the liver is obliterated, often showing trabecular or glandular patterns."

That DDT causes more profound tumorigenic effects on subsequent generations than on the parental generation has also been shown by Shabad *et al.* (1972). In this experiment, mice were administered a daily dose of 0.1 ml of 10 ppm DDT in sunflower seed oil via stomach tube (this dose amounts to 1 μg/mouse/day or, for a 50-g mouse, 20 μg/kg/day, which is the equivalent of 1.4 mg DDT daily intake for a 70-kg man). A few female mice were killed at 19–20 days of pregnancy in each generation, and the fetal lungs were removed and minced. The fragments of tissue were then explanted to start tissue culturing. The effect of transplacentally administered DDT was thus studied for five successive generations (F_5) involving explant examination of 1076 treated and 712 control fetuses. The incidence of hyperplasia

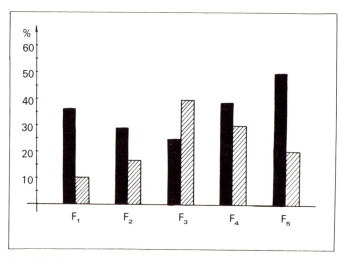

Fig. 11-2. Histogram showing hyperplasia and dystrophic lesions in organ cultures of embryonic tissues in five generations of mice treated with 1 μg/mouse/day of DDT via stomach tube. The incidence of hyperplastic changes (striped bars) is statistically significant; dystrophic lesions (solid bars) were not statistically analyzed. Hyperplasia is considered the first stage in the development of malignant tumors. From Shabad *et al.* (1972).

clearly increased until the third generation and then showed a modest decrease in F_4 and F_5 (Fig. 11-2). The increase is statistically significant. Both types of hyperplasia, diffuse and focal, were observed, sometimes including focal adenomatous growth.

Since diffuse and focal hyperplasia have been considered as the first two stages in the development of malignant lung tumors, they reveal DDT to be an oncogenic agent. This blastomogenic effect is apparently dose related, since a parallel experiment with 50 ppm (5 μg/mouse/day) showed an even more definite increase in the incidence of hyperplasia.

Two comments can be made about these experiments. First, the dose employed was somewhat high with respect to daily intake figures for general human populations. It is on the order of a forty-fold increase, although such exposure can be found in persons engaged in occupational handling of DDT or in human infants nursed on milk from mothers who have a fourfold increase in residue level over that of the general public (i.e., when mothers with a fourfold increase in DDT residue level nurse their babies, the infants show a fortyfold increase). However, considering the safety factor of 100 (or preferably more) this level probably represents a rather realistic hazard. Second, by the experimenters' own assertion, DDT does not rank high with respect to known carcinogenic agents which readily cause hyperplastic changes including the appearance of true adenoma, which DDT failed to cause. Thus DDT should be considered a relatively weak oncogenic agent. Nevertheless, these two considerations do not give full assurance that present general levels of DDT are safe, because human beings are exposed for longer periods of time (i.e., per-generation exposure for mice is only 9–12 weeks) and also the lung epithelium does not appear to be the most susceptible tissue to DDT (e.g., Tomatis *et al.*, 1972).

It is unfortunate that all these studies were made on only one species, but considering the short time that scientists have had to work on the question of the carcinogenicity of DDT it is probably all we can ask for at present. The only report indicating tumorigenic properties of DDT in other species is that of Halver (1967), who found that DDT causes hepatic cell tumor in trout at relatively low doses.

It would be unfair not to mention the studies indicating that DDT and its analogues are not carcinogenic and those showing that these compounds have antitumorigenic properties in certain cases. Ortega *et al.* (1956) and Ottoboni (1969) were not able to show that DDT has carcinogenic effects in the rat. Unfortunately, because these studies were conducted for a relatively short period, they do not resolve the extremely important question of whether this difference is species specific. Studies comparable to the mice experiments just discussed (including multigeneration studies) would have answered the question. Although there is some evidence that DDT might increase the incidence of hepatic tumors in the rat (Fitzhugh and Nelson, 1947) at very high doses, the issue of the effect of DDT in the rat is not settled.

Laws (1971), on the other hand, found that DDT had an antitumorigenic effect on the rate of success ("take") in transplanting an experimental ependymoma in the mouse. These animals were exposed to a 5.5 mg/kg/day dose of technical DDT (given in the diet at 33.3 mg/kg), and an ependymoma (the Zimerman ependymoma) known to have a 100% "take" rate in mice was transplanted 1 week after the initial feeding of DDT. The incidence of subcutaneous tumors was 92.1% in the DDT-treated mice and 100% in controls. Of 89 animals receiving DDT and a transplant, seven never developed any tumors throughout the experiments. Moreover, the DDT-fed mice lived longer. There are a few other studies indicating that DDT-treated animals are less likely to develop cancer in response to experimentally introduced chemical carcinogenic agents, probably because DDT has the ability to induce hepatic microsomal detoxification mechanisms. Okey (1972), for instance, studied this effect by using dimethylbenzanthracene, known to cause mammary tumor, against rats. p,p'-DDT concentrations in the diet as low as 10 ppm caused a significant reduction in tumor incidence. It must be remembered, however, that the effects of induction can be multi-fold. Induction clearly decreases the danger of carcinogenic chemicals that act directly, because metabolic transformation detoxifies them, but there are other compounds whose metabolic products are the carcinogenic agents (e.g., tryptophan metabolites) so that induction could increase the danger. Various induction aspects of chlorinated hydrocarbons in relation to carcinogenicity have been discussed by Falk (1971).

Some of the DDT analogues are already in practical, medical use as anticarcinogenic agents in conditions related to functional alteration of the adrenal cortex or as therapeutic agents for patients with adrenal carcinoma. The antisteroid action of o,p'-DDD has already been mentioned. In essence, o,p'-DDD suppresses the secretion of some of the corticosteroids and reduces the hormonal effects of the adrenal cortex by reducing its ACTH-induced stimulatory function. Administration of high doses of o,p'-DDD thus creates a practical state of "chemical adrenalectomy." Such a treatment can have therapeutic effects on many side-effects of adrenal dysfunction including pulmonary metastases (Perevodchikova *et al.*, 1972) and other secondary syndromes (e.g., "an abdominal mass, virilism, advanced bone age and elevated urinary ketosteroids"; Helson *et al.*, 1971). o,p'-DDD also appears to act as a purely anticarcinogenic agent; Kravtsova *et al.* (1971) found that o,p'-DDT had a direct cytostatic action on mammary gland tumors produced by 9,10-dimethyl-1,2-benzanthracene. It is suspected that o,p'-DDD must have additional therapeutic effects by stimulating the degradative metabolism of hormones.

Whatever its mechanism of action, DDD is well established as a useful chemotherapeutic agent for patients with "inoperable" adrenocortical tumors and as a hormonal anticarcinogenic agent, and case reports are now accumulating that attest to the apparent success of this approach. However, the doses used for such purposes are excessive—e.g., 1–12 g of o,p'-DDD for 1.5 months to 4 years (Perevodchikova *et al.*, 1972). These patients definitely exhibit DDD intoxication symptoms, which usually disappear on cessation of treatment. Long-term observations of the end results of DDD high-dose treatments are badly needed. In the rat at least, o,p'-DDD (at a dose of 0.6 mg/kg) has already been shown to cause testicular interstitial tumors after 10 months of treatment (Lacassagne, 1971).

Studies on other chlorinated hydrocarbon insecticides are less extensive. BHC isomers have been examined by Japanese scientists. Nagasaki *et al.* (1971) fed mice 6, 66, and 660 ppm of technical BHC and found that hepatomas developed in 24 weeks in all 20 of those fed 660 ppm. Later, Nagasaki *et al.* (1972) compared the effects of four isomers of BHC and found that only α-BHC induced hepatoma in mice at 250–500 ppm levels in the diet after 24 weeks. Yellowish nodules up to 0.3–2.0 cm in diameter appeared at these doses of α-BHC; no carcinogenic effects were observed with the β-, γ- (lindane), and δ-isomers. Independently, Goto *et al.* (1972) studied the effects of α-, β-, and γ-BHC and their possible metabolites, 1,2,4-trichlorobenzene, 2,3,5-trichlorophenol, and 2,4,5-trichlorophenol, on male mice. Their finding was essentially the same, that α-BHC is the most active analogue in causing hepatoma in mice. They observed tumors of 0.5–1.5 cm diameter.

Studies on the carcinogenicity of cyclodiene insecticides are also limited. According to Innes *et al.* (1969), mirex and Strobane® are definitely tumorigenic, while Telodrin® and Thiodan® (endosulfan) are not at doses of 0.215 mg/kg (0.646 ppm) and 1.0 mg/kg (3 ppm), respectively, fed by stomach tube for up to 28 days (or in the diet after 28 days).

As for dieldrin and aldrin, Fitzhugh *et al.* (1964) studied their effects on rats for 2 years. They found 18 tumors (in 41 rats) at 0.5 ppm, 15 tumors (in 41 rats) at 2 ppm, and 16 tumors (in 40 rats) at 10 ppm of aldrin or dieldrin in the diet. In contrast, Deichmann *et al.* (1967) and Walker *et al.* (1968) could not observe any increase in tumor incidence in rats fed aldrin (5 ppm) or dieldrin (0.1, 1.0, and 10 ppm) for 2 years. Aldrin and dieldrin are known to increase the occurrence of hepatic tumor in mice. Davis and Fitzhugh (1962) found increased hepatic tumors in mice fed 10 ppm of aldrin/dieldrin for 2 years. The experiments in the Tunstall Laboratory (Shell Chemicals in Britain) confirmed this observation. At 10 ppm, hepatic tumors appeared in about 9 months; however, at 1.0 ppm tumors were rare (Jager, 1970). Assuming that 1.0 ppm in the diet is the minimum effect level, aldrin/dieldrin concentrations in food commodities (0.01–0.1 ppm) are in excess of that allowed by the 100-fold safety factor.

In the rat, there is evidence that aldrin, dieldrin, and endrin have mild antitumorigenic activities. Deichmann *et al.* (1970) fed diets containing 20, 30, or 50 ppm of aldrin or dieldrin or 2, 6, or 12 ppm of endrin to 900 albino rats for their lifetime. The mean life spans of females fed 50 ppm aldrin and dieldrin were 13.0 and 16.6 months, respectively, compared to the control life span of 19.5 months. Altogether 257 tumors of all types were observed in 793 treated rats and 79 tumors in 163 controls. This reduction in tumor incidence was attributed to increased hepatic activity in the treated rats. The most frequent tumors were in the lungs, mammary tissues, lymph nodes, liver, and kidneys. Treon (1956) could not observe any sign of carcinogenicity in rats fed 1, 5, 25, 50, and 100 ppm of endrin in their diet for 2 years. The frequency of tumor incidence in the treated animals was identical to that in controls. Cabral *et al.* (1972) gave five doses of 10 mg/kg heptachlor in 2 days to 95 suckling rats and compared tumor incidence after 106–110 weeks with that in control rats (which received only corn oil). They concluded that heptachlor is not carcinogenic, as their studies revealed no statistically significant differences.

Studies on the carcinogenicity of organophosphates and carbamates are almost nonexistent. Andrianova and Alekseev (1970) have concluded that carbaryl is a carcinogenic agent in the rat. They either fed 30 mg/kg of carbaryl or administered 20 mg carbaryl via paraffin pellets implanted in the subcutaneous cellular tissue. Six tumors developed in 22 rats after 22 months as compared to only one tumor per 46 control rats. They state that the tumors observed had malignant characteristics. Much more evidence would be needed to confirm or deny such an allegation, however. The lack of data on organophosphates is astonishing considering the importance of this group of compounds as replacements for the more persistent chlorinated hydrocarbon insecticides.

Innes *et al.* (1969) surveyed the carcinogenicity of many pesticidal compounds by daily administration at maximum tolerated levels to mice and examination for the appearance of hepatic tumors after 18 months. Important insecticidal compounds which did not exhibit any tumorigenic tendency in these tests were carbaryl, isopropyl-*N*-methylcarbamate, piperonylbutoxide (in solvent, Butacide®), rotenone, phenothiazine, Isolan,® Telodrin,® and Thiodan.® Thus it is certainly not the lack of possible carcinogenicity that is the reason for the conspicuous absence of organophosphate data. The organophosphates have not been examined so far.

Perhaps in one compound the question of carcinogenicity has been established. Aramite,® 2-(*p*-tertiary-butylphenoxy)-isopropyl-2-chloroethyl sulfite, is an acaricidal compound for fruit and food crops. Oser and Oser

(1960) observed a single focus of nodular hyperplasia in 20 rats fed 500 ppm of Aramite® in the diet for 2 years. Similarly, Popper *et al.* (1960) found hepatic lesions in rats fed 1580 ppm of Aramite®; in two out of 21 rats, the pathological changes were classified as malignant. Sternberg *et al.* (1960) examined the effects of Aramite® in dogs fed 500–1429 ppm for over 3 years and found carcinomas in the biliary tract. The bile-concentrating areas such as the intrahepatic bile ducts, the ampulla of Vater, the gall bladder, and the common ducts were affected.

Finally, the overall question of the carcinogenicity potential of insecticides must be reconsidered here. The data are confusing even to specialists. All assortments of methods and doses have been employed, often leading to conflicting results, and so it is very difficult to present a definite concept of carcinogenic potential in these pages. One of the first things that needs to be done to remedy this situation is the preliminary screening of a great many insecticidal chemicals for carcinogenic tendencies in experimental animals even at high doses for at least their entire life span. This is because none of the epidemiological studies done so far has produced reliable carcinogenicity data. For instance, Durham (1963) states that "In a very extensive study carried out in the Wenatchee, Washington, area between 1937 and 1940 . . . for 1231 persons who lived and worked in this apple growing region, where large quantities of lead arsenate had been used since about 1900, . . . no increase in the incidence of any other [than acute lead arsenate intoxication] disease, including cancer, was noted." This is surprising in view of consistent rumors that arsenicals are carcinogenic (e.g., there are several reports on liver and skin cancer in vineyard workers in Germany). There are numerous other examples indicating the difficulties in obtaining epidemiological evidence for the regional effects of environmentally present carcinogens. This includes even the most clearly established carcinogens, such as aflatoxins (Wogan, 1969).

Second, despite some arguments against it, the establishment of dose–effect relationships is important. As Weil (1972) points out, this is the only proper way to estimate the safety of any chemical, and it requires a correct experimental design. Reliable judgments on safety factors are dependent on the extrapolation of such relationships.

Third, it is advisable to use more than one species of mammals in large numbers (e.g., more than 100 per dose group for any compound, as the FDA recommends). Species specificity is a rather unpredictable item at present. Epstein (1970) cites the example of thalidomide, which manifests teratogenic effects in man at 0.5 mg/kg/day (lowest dose known to induce such an effect), whereas the corresponding doses for mice, rats, dogs, and hamsters are 30, 50, 100, and 350 mg/kg/day, respectively. Thus man is the most sensitive species (along with some primate and avian species) to thalidomide. Also,

2-naphthylamine, known to cause bladder cancer, has been shown to act only in man and dogs, exhibiting no carcinogenic properties in rats, mice, guinea pigs, and rabbits. With respect to insecticides, mice, for instance, are apparently more sensitive to chlorinated hydrocarbons than rats.

Fourth, the assessment of the potential hazard of any pesticide must include metabolic studies to determine the nature of the actual *in vivo* carcinogen. The possibility of toxic contaminants, such as polychloro-*p*-benzodioxins and benzofurans, or of contaminants that could become carcinogenic *in vivo*, such as ethylene thiourea, must be examined.

It is clear that all these requirements for ascertaining the safety of insecticides are complicated and time consuming. Thus examination for carcinogenicity must follow an orderly sequence: screening tests with experimental animals of short life span, such as mice, at relatively high doses (like the ones employed by the Bionertic laboratory groups; Innes *et al.*, 1969), followed by a more careful examination of the dose–effect relationships using a large number of animals whenever the screening tests indicate carcinogenic potential, and then final tests on other experimental animals, preferably primates, pigs, etc. (Epstein, 1970), to relate the findings to human hazard. A rational decision concerning the safety factor must be based on sound judgment of the seriousness of the particular type of cancer, the frequency of malignancy, the slope of the dose–effect relationship, the stability of the chemical, and potential retention and concentration in specific organs.

In conclusion, it is clear that certain insecticides can induce tumors in some experimental animals (mostly rodents) even at relatively low concentrations; the lowest figure for such effects is approximately 1 ppm. While such effects have not been carefully assessed in terms of human hazard, they certainly represent potential danger, close enough to warrant further investigation.

11.3.2d. Mutagenicity and Teratogenicity

Mutagenic effects represent the end result of genetic impairments caused by introduced chemicals. Since actions that are mutagenic occur at the cellular level (particularly in chromosomal DNA and other cellular and nuclear constituents related to the function of cell division), they are also potentially carcinogenic with respect to somatic processes. However, the science of mutation or mutagenicity limits itself to detection of increases (or in limited instances "decreases") in the rate of genetic mutation above normal: the field is limited by methodological approaches which will be explained shortly. Therefore, the mutagenicity of pesticides is studied independently from carcinogenicity.

What are mutations? In a broad sense, mutations include all abnormal processes of genetic inheritance including abnormal somatic cell division. In more specific terms, mutations range from single gene changes (point mutations) to large chromosomal aberrations. Somatic mutations usually result in the death of nonvital cells and, unless related to cancer, are of minor importance, and quite often unnoticeable. However, an exception is somatic mutation occurring at the embryonic stage, which can result in "teratogenic" expression.

Teratogenic effects in higher animals can be produced either by embryonic somatic mutation or by mutation of genetic material (i.e., sperms and eggs). These effects can be observed only when (1) they are not extensive enough to cause death and early abortion and (2) they are significant enough to be morphologically (not necessarily externally) noticeable. There are naturally a number of mutations which are too mild in their phenotypic expression to be noticed, or the disorders are biochemical and functional rather than morphological. There are many known human genetic diseases originated from mutations; accordingly, teratogenicity is only a small part of the manifestation of mutagenic actions.

A few comments must be added to explain the scope and magnitude of mutagenic hazards. As with other toxicological effects, the severity of mutations is dose related; however, the major danger of mutation lies in the "mild" expressions that occur at low concentrations of inducers. The reason is that severe mutations are generally lethal and therefore are not transmitted to future generations. There are many more "mild" mutations than severe ones at low doses of mutagenic substances, and environmental contaminants are more likely to be present at low concentrations than at high ones. Furthermore, not all mutations are genetically "dominant" and immediately detectable. The effects of a mutation are likely to be spread over generations and some unlucky child whose parents happen to have the same mutant gene will be the one affected. Finally, there is no way to repair genetic damage. Once done, it is irreversible and likely to stay in the population unless it is either lethal or causes disadvantages in genetic propagation (e.g., physical difficulty in finding a mate). In the case of man, medical science has developed to the point that the probability of eliminating unwanted genes from the general gene pool is greatly reduced. Modern medicine can prolong the life of individuals with genetic defects, often allowing them to reproduce.

Mutagenicity is measured by statistical analyses of the effects of a particular chemical introduced into a population. According to Epstein and Legator (1971), the methods available for mutagenicity testing are as follows:

1. Ancillary submammalian systems.
    a. Bacterial methods (for "point" mutation).

    b. Neurospora method.
    c. Phage transformation method.
    d. Plant methods.
        i. Seed treatment.
        ii. Specific-locus method in pollen.
        iii. Root-tip method for chromosomal aberrations.
        iv. Somatic mutation method.
    e. Drosophila methods.
        i. Sex-linked recessive lethal tests.
        ii. Two-generation reciprocal translocation test.
        iii. One-generation sex chromosome loss test.
        iv. Bithorax test.
2. Mammalian systems.
    a. Cytogenetics and somatic cell genetics (visible chromosomal changes only).
    b. Host-mediated assay (indirect method: injection of microorganisms and mutagens into mammalian host to determine mutation in microorganisms).
    c. Specific-locus test (usually mice; coat-color, morphology changes).
    d. Dominant lethal test.

As one can easily see, the nonmammalian tests offer easy and rather statistically accurate accounts of the mutagenicity of compounds, including information on point mutations, but their usefulness in assessing potential mammalian hazards is less than ideal. The mammalian tests are more relevant, but the criteria for detection are limited to visible chromosomal changes and lethal mutations only, which severely affects statistical applicability and accuracy.

Based on the available methodology, Epstein and Legator (1971) suggest that all pesticides in use be tested with

> (1) Three mammalian systems, the dominant-lethal, host-mediated and *in vivo* cytogenetic, by appropriate routes of administration which reflect human exposure, and also parentally at high-dose levels, such as maximal tolerated doses,
> (2) ancillary microbial systems, preferably those detecting both single nucleotide changes and effects involving more than one gene. The precision of testing, both in mammalian and ancillary systems, should be such that a doubling of the control level of mutation would be statistically significant at the 5 percent level.

They point out that such a program takes only a year to complete. Lederberg (1971) suggests that the standard for mutagenicity should be set at approximately 1% above the spontaneous rate, or one recessive mutation per 1000 gametes per generation of typical exposure. A similar view was expressed in *Lancet* (Anonymous, 1971*b*) in a recommendation that all pesticides be tested for mutagenic potential.

The actual scientific data on mutagenicity and teratogenicity of insecticides are not numerous. Epstein and Legator's (1971) review of the literature on the subject was compiled from a survey by the Environmental Mutagen Information Center (EMIC) registry. Table 11-9 summarizes this and includes some more recent data. It is obvious that the information on hand is meager at best. The methods employed are diverse, and in most cases the doses applied are too large to be significant from the viewpoint of environmental toxicology.

It is not surprising at all to find that insect chemosterilants such as tepa, hempa, and metepa, are mutagenic, for these compounds are alkylating agents known to have mutagenic or radiation-like effects on biological tissues. The lack of suitable data for other insecticidal chemicals is rather distressing, although it is certain that some of them are mutagenic. A number of insecticides are known to interact with RNA, DNA, and histone (e.g., Meisner and Sorensen, 1966; Anonymous, 1971*a*; Isenberg and Small, 1972; Rosenkranz and Rosenkranz, 1973) at relatively low doses, and it is logical to assume that such effects could, in some instances, result in mutagenic manifestations. For example, Meisner and Sorenson (1966) observed an arrest of chromosomes at metaphase in the presence of rotenone, and Vacquier and Brachet (1969) found ethidium bromide to cause progressive clumping of chromosomes at interphase.

However, from the literature review on other mutagens studied it is evident that the commonly adopted criteria, such as chromosomal aberration and dominant lethal tests, can detect only the extreme mutagenic manifestations of any chemical. And it is quite probable that actual harmful mutations take place at much lower doses than the ones which induce gross chromosomal aberrations. In the final analysis, though, Epstein and Legator (1971) are correct in stating that the general assessment of mutagenicity is relatively easy. Comparing the mutagenicity of pesticidal chemicals to that of substances already known to be mutagens should give at least some indication of potential danger. For mammalian studies, particular emphasis must be placed on long-term, chronic administration experiments rather than the single high-dose experiments that have been so frequently employed by researchers in the past (e.g., Table 11-8).

11.4. INSECTICIDE RESIDUES IN MAN

It has long been known that stable pesticidal chemicals, including their conversion products and impurities, can accumulate in man (e.g., Howell,

TABLE 11-9. Survey of Available Literature on Mutagenicity of Insecticidal Chemicals

Insecticides	Assay method and effects	Dose	Reference
Carbaryl	Plant root tips	0.5 and 0.25 saturation	Epstein and Legator (1971)
DDT	Rat, three-generation teratogenic	100–500 mg/kg daily	Weil et al. (1972)
	Mice sperm	105 mg/kg	Epstein and Legator (1971)
	Plant root tips	saturated solution	Epstein and Legator (1971)
	Rat sperm	50–70 mg/kg	Epstein and Legator (1971)
	Marsupial somatic cell, chromosome aberration		Epstein and Legator (1971)
	Rat, dominant lethal	10–50 ppm	Legator (1970)
	Drosophila	80 mg/kg (δ)	Vogel (1972)
	Salmonella and Serratia, dominant lethal	no effect below 0.14 mmole/liter	Buselmaier et al. (1972)
DDA	Drosophila, recessive lethal	—	Vogel (1972)
	Salmonella and Serratia, dominant lethal	0.14 mmole/liter	Buselmaier et al. (1972)
DDE, DDD, DDOM	Drosophila	—	Vogel (1972)
	Salmonella and Serratia (DDD)	no effect	Buselmaier et al. (1972)
Dichlorvos	Onion root tips	—	Epstein and Legator (1971)
	Bacterial sp. including Salmonella	3.2–6.5 mmoles/liter	Voogd et al. (1972)
	Escherichia coli	Vapona® strip	Ashwood-Smith et al. (1972)
Dieldrin	Sprouts, Crepis capillaris	10% solution	Epstein and Legator (1971)
Endrin	Albino rat, chromosomal changes	0.25 mg/testis	Dikshith and Datta (1973)
	Barley meiosis, no effect	1000 ppm soaked	Epstein and Legator (1971)
DFP	Chick embryo, teratogenic	—	Flockhart and Casida (1972)
Ethylene dibromide	Salmonella and Serratia	—	Buselmaier et al. (1972)
Ethylene oxide	Fungi, point mutation, reverse mutation		Epstein and Legator (1971)
	Neurospora crassa	0.14 M	Epstein and Legator (1971)
	Maize cells, chromosome break	1 part per 20 parts air	Epstein and Legator (1971)
Fenthion	Mouse embryo, mild teratogenic	40, 80 mg/kg	Budreau and Singh (1973)
Hempa	Mice, dominant lethal	25, 50, 125 mg/kg 4 times/week (δ)	Sram (1971)

Compound	Effect	Dose	Reference
Lindane	Onion, root tips, chromosome break	12.5–50 ppm	Epstein and Legator (1971)
	Allium cepa root tips	0.00125–2%	Epstein and Legator (1971)
	Root tips of several plants	solid particles	Epstein and Legator (1971)
	Pisum sativum root cells	250 ppm	Baquar and Khan (1970)
Malathion (tech. 95%)	Leghorn chick embryo, teratogenic	3.99, 6.42 mg/egg	Greenberg and LaHam (1969)
Malathion	Leghorn chick embryo, teratogenic	5.7 mg/egg	Gill and LaHam (1972)
	Human hematopoietic cells, no effect on chromosomes	23.50 μl/ml	Huang (1973)
Methylparathion	Human hematopoietic cells and mice bone marrow cells, no effect on chromosomes	5–100 mg/kg	Huang (1973)
Metepa	Male mice, first 3 weeks mutagenic at low doses	0.782–100 mg/kg	Epstein *et al.* (1970)
Parathion	*Allium cepa* root tips, mitosis	0.01, 0.005, 0.0075%	Epstein and Legator (1971)
Phosphamidon	Barley, meiosis slight effect	1000 ppm soaked, 500 ppm spray	Epstein and Legator (1971)
Sodium arsenate	Mouse female, teratogenic	45 mg/kg, i.p. injection	Hood and Bishop (1972)
Systox® (demeton)	Mouse embryo, mildly teratogenic	7 or 10 mg/kg	Budreau and Singh (1973)
Trichlorfon	Calf thymus DNA, reaction	100 μg	Rosenkranz and Rosenkranz (1973)
Tepa	Male mice, first 3 weeks mutagenic at low doses	0.156–20 mg/kg	Epstein *et al.* (1970)
	Male mice, dominant lethal and translocation	2.5 mg/kg, i.p. injection	Sram (1971)

1948). Many people find the thought that pesticide residues are present in their own bodies rather disturbing, and naturally the first question that anybody raises is about the probability of bodily harm. Scientists as well are concerned with such issues as the origins of contamination, epidemiological signs, threats of future buildup, and kinetics of elimination vs. intake.

There have been many excellent surveys made in various countries on chlorinated hydrocarbon insecticide levels in human adipose tissue. Concurrent studies on pesticide residues in food and feed intended to monitor daily intake are also numerous. Accordingly, today a fairly accurate picture of the pesticide intake and residue situation is available for most populations. Studies on other aspects of pesticide residues in man are lagging far behind. The major problem is that epidemiological studies of pesticide effects on such a scale are very difficult. Also, it has been difficult to regulate pesticide intake in man because of the minute quantities concerned; that is, most people in large populations have been taking microgram quantities of pesticides daily through food commodities, which uniformly contain rather low levels of these substances. Only when a country withdraws a particular pesticide completely from the market can the dynamics of elimination be studied, and even in these cases the effects may take years to become apparent.

Despite the problems, the data available indicate interesting trends with respect to the movement of these chemicals through human populations. Following are some key points that have become apparent from the extensive surveys on pesticide residues in man:

1. The chemicals found to accumulate in humans are mostly chlorinated hydrocarbons. These are stable and lipophilic and are detectable at very low concentrations.
2. The levels of any given pesticide vary geographically and among various segments of the population.
3. Changes in the levels of stable pesticides over time are rather slight.
4. The major factor determining the distribution of pesticides in the body is fat content; however, there are a number of other factors influencing the final distribution of residues.

11.4.1. Geographical Variation in Residues Found in Man

Residents of different countries have been found to contain different levels of pesticides (Table 11-10). These data for human subjects are trustworthy because of the ease of collecting case histories as well as information on distribution, age, etc. The efficiency of governmental regulatory organizations in controlling the total pesticide intake by citizens is revealed by such

TABLE 11-10. Levels of Chlorinated Hydrocarbon Insecticides Stored in Human[a] Fat (mg/kg, or ppm mean values)

Survey	Year surveyed	Country	DDT and related compounds	DDE	Aldrin/ dieldrin	γ-BHC
Dale and Quinby (1963)	1961–62	U.S.A.	4.9	3.8	0.15	0.20
Hoffman et al. (1964)	1962–63	U.S.A.	7.6	7.0	0.14	0.48
Hayes et al. (1965)	1964	U.S.A.	10.0	6.9	0.29	0.06
Morgan and Roan (1970)	1970	U.S.A. (Arizona)	6.12	4.58	0.14	—
Wyllie et al. (1972)	1970	U.S.A. (Idaho)	9.3	7.2	0.2	0.3
Dale et al. (1965)	1964	India (vegetarians)	28	11.6	0.03	1.7
	1964	India (meat-eaters)	12	6.4	0.06	0.9
Egan et al. (1965)	1963–64	Britain	3.1	2.0	0.21	0.34
Bick (1965)	1967	Australia	1.7	0.93	0.05	—
de Vlieger et al. (1968)	1968	Netherlands	2.0	1.7	0.17	0.1
Hayes et al. (1963)	1961	France	5.2	3.2	—	—
del Vecchio and Leoni (1968)	1967	Italy	8.2	7.5	0.45	0.06
Wassermann et al. (1967)	1965–66	Israel	18.1	9.9	—	—

[a]All public samples.

data, and they also indicate the general levels of contamination in food commodities, which are mainly obtained from pesticide-treated, cultivated lands. Robinson (1970) made a literature survey on this subject and came up with the following DDT-R figures (in ppm) for the various general populations: Germany (FRG) 2.3, Netherlands 2.0, Denmark 3.1, Great Britain 3.0, Czechoslovakia 9.2, Italy 10.1, Germany (DDB) 11.4, Hungary 12.4, and Poland 13.4. The cause of such a wide variation is not immediately apparent, but no doubt differences in agricultural practices, food habits, and government regulatory processes play very important roles.

The data on the geographical variation in residue levels in North America are summarized in Table 11-11. In addition, the data on the total DDT-type residues in Alaskan residents from an earlier study (Durham et al., 1961), indicating 3.0 ppm, are of interest, although a direct comparison with the data in the table is somewhat difficult. As evident from the table, variation in residue levels within the United States is remarkably small. Brown's (1967)

TABLE 11 11. Residues of Chlorinated Hydrocarbon Insecticides (ppm mean values) in Human Adipose Tissues from Several Geographical Locations in North America as Determined with the GLC-Electron Capture Detector

Region	p,p'-DDT	o,p'-DDT	DDE	Dieldrin	Heptachlor epoxide
U.S.A.					
Northeast[a]	1.59	0.11	3.89	0.38	0.07
Deep South[a]	2.57	0.27	4.12	0.27	0.13
Midwest[a]	1.88	0.12	4.17	0.29	0.08
Far west[a]	3.37	0.16	6.25	0.12	0.11
U.S. cities (4)[a]	2.35	0.16	4.63	0.31	0.10
Chicago[b]	2.4	—	6.4	0.14	—
Arizona[c]	1.54	—	4.58	0.14	—
Florida[d]	4.3	—	7.0	—	—
Honolulu, Hawaii[e]	1.3	—	4.5	0.04	—
Utah[f]	1.5	—	5.0	0.17	—
Idaho[g]	1.9	0.1	7.2	0.2	0.1
Miami, Florida[h]	2.81	—	6.67	0.215	—
Canada					
Maritime provinces[i]	0.79	0.05	3.15	0.11	0.01
Montreal[i]	1.08	0.07	3.36	0.11	0.03
Ottawa[i]	0.90	0.04	3.23	0.08	0.02
Toronto[i]	1.25	0.05	4.00	0.16	0.03
Winnipeg[i]	1.16	0.05	3.75	0.12	0.05
Edmonton[i]	0.65	0.03	2.13	0.09	0.06
Vancouver[i]	0.85	0.03	3.40	0.15	0.07

[a] Zavon et al. (1965).
[b] Hoffman et al. (1967)
[c] Morgan and Roan (1970).
[d] Davies et al. (1968).
[e] Casarett et al. (1968).
[f] Warnick (1972).
[g] Wyllie et al. (1972).
[h] Fiserova-Bergerova et al. (1967).
[i] Hurtig (1972).

data obtained in Canada indicate that the DDT/DDE residue level was 3.8 ppm in Toronto. Since it is not likely that there is any great difference in food habits between people in Toronto and, for instance, Chicago (8.8 ppm; Hoffman et al., 1967), the discrepancy, if true, must result from differences in the regulatory processes of Canada and the United States. It is noteworthy that the data obtained in Hawaii do not show marked differences in chlorinated hydrocarbon residues (although pentachlorophenol contamination is high among Hawaiian residents; Klemmer, 1972) despite its distance from the continental United States and differences in racial composition and some food habits. Thus in terms of large population pools variations in residue levels among countries are more pronounced than those among geographical areas within one country, regardless of distance from each other and some differences in agricultural practice. (There is a similar tendency toward uniform residues within Great Britain; see Robinson, 1970.)

11.4.2. Factors Influencing the Residue Levels in Man

Despite the apparent uniformity of residue levels in the general population within one country, there is enough evidence to show that residue levels are quite different among distinct segments of the general population. Although high levels of residues are to be expected in the people occupationally exposed to insecticidal chemicals, there are other rather striking differences. The most noticeable is that between males and females: males invariably accumulate higher levels of DDT-R (Table 11-12). Since it is not likely that food habits are a factor here, the explanation must be sought in differences in environmental exposure or in physiological processes. In this regard, it is interesting to note that the sex difference is not apparent before maturity (age 20–30), supporting the view that it, at least in part, is related to hormonal factors.

Also of importance is the race difference, which has been shown to be statistically significant in several instances. In addition to the data cited in Table 11-12, Hoffman *et al.* (1967) found statistically significant differences in BHC and DDT/DDE levels between autopsy samples of Negro and white populations: the DDT/DDE level for white men was 9.8 and that for Negro men 17.0 ppm; for white females it was 8.3 and for Negro females 12.6 ppm. This race difference is quite puzzling. In some instances these residue differences might be related to exposure, but in many cases the explanation must be sought elsewhere. In contrast to the sex difference, the race difference becomes apparent in the youngest age groups. The balance of evidence suggests that it must come from environmental factors (including social and food habits) rather than from a hormonal factor. In fact, Warnick (1972) found lower residue levels in blood samples of Indians (43 samples) in Utah than in Caucasian samples. The proportions of the difference due to race are rather constant; the Negro (nonwhite) population shows about 40–50% higher levels than those found in the white population residing in the same area. Considering that overall residue levels are rather uniform throughout the United States, such an enormous difference must come from some definite source. Further studies are needed to shed light on this point.

TABLE 11-12. Average or Median Chlorinated Hydrocarbon Residues Found in Adipose Tissues of Various Segments of the U.S. Population

Group	Number tested DDT-R, dieldrin	DDT	DDE	Dieldrin
Warnick (1972)				
Male	71, 71	1.68	5.36	0.19
Female	32, 32	1.20	4.31	0.14
Old (>21)	88, 88	1.58	5.19	0.19
Young (<21)	15, 15	1.25	4.11	0.09
Caucasian	96, 96	1.32	4.75	0.16
Negro	4, 4	3.70	7.90	0.20
Zavon *et al.* (1965)				
Male	41, 41	2.52	5.39	0.31
Female	23, 23	2.18	3.61	0.24
Davies *et al.* (1968), Edmundson *et al.* (1969)				
Male, Caucasian	52, 51	2.9	5.0	0.23
Female, Caucasian	38, 37	2.5	4.7	0.24
Male, nonwhite	19, 16	5.6	12.9	0.19
Female, nonwhite	16, 13	4.8	7.7	0.23
Male, Caucasian				
Age 0–5	11, 8	1.3	2.	0.19
Age 6–10	7, 6	3.5	4.6	0.29
Age 11–20	17, 16	3.1	4.7	0.14
Age 21–30	13, 13	3.3	5.0	0.21
Age >31	15, 16	1.9	6.4	0.37
Female, Caucasian				
Age 0–5	6, 6	2.6	2.4	0.18
Age 6–10	7, 7	4.1	5.3	0.18
Age 11–20	9, 8	2.9	5.6	0.18
Age 21–30	12, 13	2.4	5.4	0.19
Age >31	10, 9	2.0	3.4	0.21
Male, nonwhite				
Age 0–5	10, 8	2.2	2.4	0.12
Age 6–10	3, 2	8.0	8.7	0.09
Age 11–20	2, 2	8.1	9.1	0.02
Age 21–30	7, 5	5.6	11.7	0.12
Age >31	7, 7	6.1	14.6	0.08
Wyllie *et al.* (1972)				
Caucasian, both sexes				
Age 0–20	16, 16	1.01	3.47	0.14
Age 21–40	47, 47	1.91	6.73	0.13
Age 41–60	80, 80	1.67	7.06	0.17
Age 61–90	59, 59	2.08	8.72	0.16

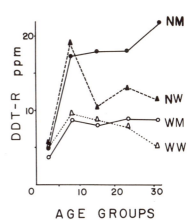

Fig. 11-3. Median levels of total DDT-derived residues in 159 necropsy tissues, Dade County, Florida, 1965–1967. NM, Negro men; NW, Negro women; WM, white men; WW, white women. From Davies *et al.* (1968).

The influence of age is less well defined, although it is generally acknowledged that residue levels rise with increasing age, in general agreement with the data obtained from animals. Nevertheless, the relationship is not always clear. Hoffman *et al.* (1967) found that the level of DDT/DDE rapidly increased from about 5.5 to 11.5 ppm up to the age group 35–44 years but became stabilized at 9.0 ppm in the four subsequent age groups (45–85 years old). Davies *et al.* (1968) found that the increase was apparent in both white and nonwhite males, but in females the residue levels actually declined after reaching a maximum in the 6–10 years age group (Fig. 11-3). It has been pointed out that obese persons do seem to accumulate less residues than the nonobese (Robinson, 1970), although such observations are largely based on human and animal feeding tests (Hunter and Robinson, 1968; Keane and Zavon, 1969) rather than on field observations. As women tend to accumulate more fat, the sex difference has been related to this factor. However, the feeding experiments were conducted with dieldrin (HEOD), which does not show clear sex differences. Accordingly, although there is good evidence to indicate the role of a hormonal factor in the sex difference in residue accumulation, any mention of fat metabolism as another must be regarded premature.

Naturally factors such as food habits, physiological state, and life style must play important roles in determining the residue levels in man. The problem is that currently there are no concrete data indicating the significance of any of them. For instance, the data from a residue survey of wild animals and some observations (Hayes *et al.*, 1958), along with simple logic, suggest that meat-eaters should accumulate more chlorinated hydrocarbon residues than vegetarians. The problem is that the food habits of human vegetarians are not strictly comparable with those of herbivorous animals because man's general life habits are much more complicated. For instance, vegetarians from India have more DDT residues in their fats than meat-eaters (Table 11-10). Also, many vegetarians consume relatively large quantities of oily foods (e.g., fried), which could contain higher levels of pesticide residues than lean meat. In short, only the three factors of sex, race, and age are conclusively known to be related to the final expression of residue levels among the non-occupational general population groups within one country.

11.4.3. The Meaning of Residues in Man

Since it has been firmly established that stable chlorinated hydrocarbon insecticides do accumulate in the human body and that probably most people in the world have been affected, the remaining central question is the implication of all this: How safe are these residues and what is going to take place? Obviously so far no catastrophic disaster has occurred, but on the other hand there are indications that some forms of wildlife are already suffering.

11.4.3a. Change in Residue Levels with Time

The data summarized in Table 11-10 indicate that the changes in DDT-R levels in the United States are not consistent; that is, no definite downward trend can be noted despite the general decline in the use of DDT since the early 1960s. Differences in sampling and analytical techniques probably account for the variation in data rather than actual changes in the residue levels. Accordingly, the only way to know how the residue levels are responding to the pesticide input into man's environment is for the same research team to examine the same population year after year using the same sampling and analytical techniques.

The data shown in Table 11-13 have been taken from Warnick (1972). It is apparent from his data that the decline, if there is one, is difficult to notice. However, a consistent small decline may be seen in the adipose tissue samples where the same sampling and analytical techniques were consistently employed. Moreover, data from blood samples in general are less reliable than those from adipose tissues; this point will be discussed further in Section 11.4.3c. The only other experiments which have been conducted on the same population by the same analytical techniques are controlled feeding experiments on volunteers. Hayes *et al.* (1971) fed 11 persons 3.5 mg/man/day (0.048–0.064 mg/kg/day) of technical DDT for 21.5 months and observed the changes in the residue levels in their fat for an additional 37.8 months, for a total of 59.3 months. It is apparent from Fig. 11-4 that the DDT residue level starts declining as soon as DDT is removed from the volunteer's diet, whereas the level of DDE residues continues to increase. The net result is the

TABLE 11-13. Changes in the Average Chlorinated Hydrocarbon Residue Levels in Human Serum and Adipose Tissue in the Residents of Utah From 1967 to 1971

| Year fiscal | Sample number | Adipose tissue[a] (ppb) | | | Sample number | Serum[b] (ppb) | | |
		p,p'-DDE	Total DDT-R	Dieldrin		p,p'-DDE	Total DDT-R	Dieldrin
1967	—	—	—	—	72	19.5	32.8	4.5
1968	48	5.95	9.01	0.20	237	15.4	24.7	2.1
1969	15	4.81	7.15	0.15	267	20.8[b]	31.2[b]	1.6
1970	40	4.02	5.33	0.15	439	18.7	25.1	1.6
1971	—	—	—	—	402	22.9	31.2	1.4

From Warnick (1972).

[a]No changes in methodology.
[b]There was a slight change in the methodology. The blood samples for 1967 and 1968 were extracted once with hexane, whereas those for 1969, 1970, and 1971 were extracted three times. The latter improvement(?) increased the average value for total DDT-R from 26.1 ppb to 31.2 ppb, and increased DDE recovery. This increase could have offset the general decline.

reversal of the order (from DDT > DDE to DDE > DDT) at the end of the 5-year experimental period. The causes of such a differential in the dynamics of accumulation–elimination could be both metabolic conversion of DDT to DDE (DDT is also further degraded but DDE probably is not) and preferential storage of DDE, which is the more lipophilic and stable of the two. Whatever the reasons, it is important to stress that it is DDE that eventually accumulates in man and other animals.

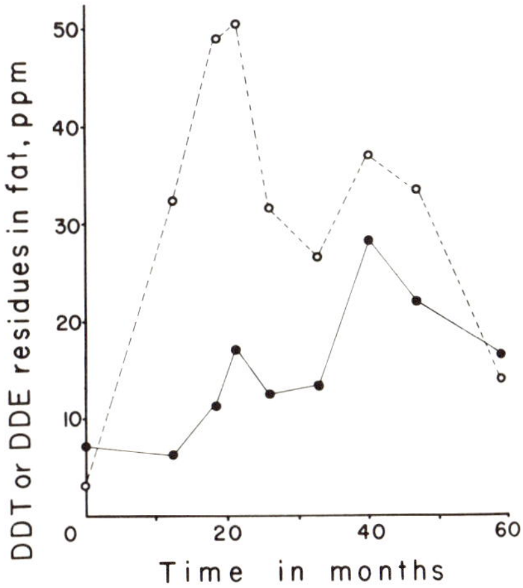

Fig. 11.4. Changes in residue levels of DDT (○) and DDE (●) in human volunteers fed 3.5 mg/man/day of technical DDT for 21.5 months. Drawn from the data of Hayes *et al.* (1971).

Although these experiments were conducted at a level about 100 times greater than the estimated average intake by the general population, the data supply valuable information inasmuch as experiments with animals have shown that general elimination dynamics in the body are dose dependent for each compound. Dose dependency suggests that the rate of elimination will be slowed down as the level of the residue becomes low. The balance of evidence available indicates that these pesticide residues do decline, but rather slowly.

O'Neill and Burke (1971) used various assumptions to develop an interesting model to predict the levels of DDT-R in human adipose tissue for several decades to come (Table 11-14). The first· assumption was that DDT usage would decrease at the rate of 5 million pounds per year. The second assumption was the complete cessation of DDT use in 1972 (close to reality in the United States). These two sets of projections gave almost insignificant differences in the predicted amount of DDT-R accumulation in man throughout the early part of the twenty-first century. The third assumption, that DDT usage would stabilize at 5 million pounds per year, resulted in slightly higher accumulations than for the first two predictions. The fourth,

TABLE 11-14. DDT-R Concentrations in Human Adipose Tissue Predicted by a Mathematical Model

Year	Continued reduction at 5 million lb/year	Zero future usage of DDT[a]	DDT usage maintained at 5 million lb/year	DDT usage maintained at 1966 level
1970	5.14	5.14	5.14	5.30
1974	4.18	4.13	4.20	5.08
1978	3.41	3.35	3.53	5.32
1982	2.78	2.73	3.02	5.60
1986	2.27	2.23	2.60	5.84
1990	1.86	1.82	2.25	6.03
1994	1.52	1.49	1.98	6.20
1998	1.24	1.21	1.75	6.32
2002	1.01	0.99	1.56	6.43
2006	0.82	0.81	1.41	6.52
2010	0.67	0.66	1.29	6.59
2014	0.55	0.54	1.18	6.65
2018	0.45	0.44	1.10	6.70
2022	0.37	0.36	1.03	6.73

From O'Neill and Burke (1971).

[a]Zero application in 1972.

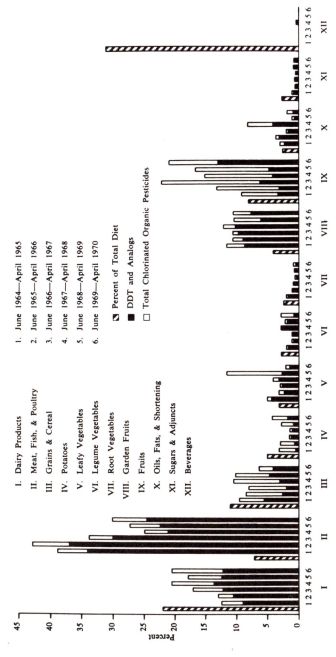

I. Dairy Products
II. Meat, Fish, & Poultry
III. Grains & Cereal
IV. Potatoes
V. Leafy Vegetables
VI. Legume Vegetables
VII. Root Vegetables
VIII. Garden Fruits
IX. Fruits
X. Oils, Fats, & Shortening
XI. Sugars & Adjuncts
XII. Beverages

1. June 1964—April 1965
2. June 1965—April 1966
3. June 1966—April 1967
4. June 1967—April 1968
5. June 1968—April 1969
6. June 1969—April 1970

▨ Percent of Total Diet
■ DDT and Analogs
☐ Total Chlorinated Organic Pesticides

Fig. 11-5. Distribution of total chlorinated hydrocarbon insecticides and DDT analogues among food classes compared to the percent of total diet represented by total food class. The data represent a summary of a residue survey on ready-to-eat foods collected from 30 markets in 28 different cities. From Duggan and Corneliussen (1972).

based on continued use at 1966 levels, naturally resulted in much higher DDT-R concentrations.

11.4.3b. Human Intake of Pesticides Through Food Residues

The rate at which the daily intake of pesticides changes in the general population can be estimated rather accurately, since it is not too difficult to analyze the residue levels in major food commodities, nor is it difficult to arrive at average percentages of intake for any particular food item in relation to the total daily food intake. The data shown in Fig. 11-5 have been taken directly from Duggan and Corneliussen (1972) to demonstrate the way the daily intake figures can be derived (in this case, from an extensive survey on ready-to-eat food commodities from 30 markets in the United States). In brief, the materials which show both high residue levels and high "percent

TABLE 11-15. Total Dietary Intake of Pesticidal Chemicals in the United States[a]

Compounds	FAO–WHO acceptable daily intake	Years surveyed (1964–1970)						Six-year average
		64–65	65–66	66–67	67–68	68–69	69–70	
Aldrin/dieldrin	0.1	0.09	0.1	0.06	0.06	0.07	0.07	0.08
Carbaryl	20	20	0.5	0.1	—	0.04	—	0.5
DDT-R[b]	5	0.9	1	0.8	0.7	0.5	0.4	0.7
γ-BHC	12.5	0.07	0.06	0.07	0.04	0.02	0.02	0.05
Bromide[c]	1,000	390	220	290	410	240	240	300
Heptachlor-R[d]	0.5	0.03	0.05	0.02	0.03	0.03	0.02	0.03
Malathion	20	—	0.1	0.2	0.04		0.2	0.2	0.1
Diazinon	2	—	0.02	0.001	0.001	0.004	0.001	0.01
Parathion	5	0.001	0.005	0.01	—	0.01	0.003	0.01
BHC	—	0.03	0.04	0.03	0.04	0.02	0.02	0.03
Dicofol (Kelthane®)	—	0.04	0.1	0.2	0.1	0.1	0.05	0.1
Endrin	—	0.009	0.004	0.004	0.01	0.004	0.005	0.005
Total chlorinated insecticides	—	1.2	1.6	1.2	1.0	0.8	0.6	1.1
Total organo-phosphates	—	—	0.14	0.25	0.07	0.23	0.26	0.19
Total herbicides	—	0.12	0.22	0.05	0.06	0.05	0.008	0.008

From Duggan and Corneliussen (1972).

[a]The data are expressed as μg/kg/day ingested in foods.
[b]DDT + DDE.
[c]Including naturally occurring bromides.
[d]Heptachlor + heptachlor epoxide.

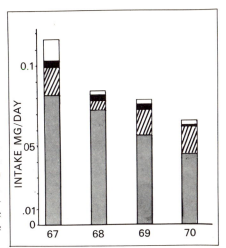

Fig. 11-6. Total daily intake of pesticide residues per man (assuming that the average weight of a 16- to 19-year-old male is 69 kg) according to chemical classes. From Duggan and Corneliussen (1972). Gray area: chlorinated organics. Diagonal line area: organic phosphates. Black area: herbicides. White area: carbamates.

of total diet" are the ones that affect the daily doses the most (e.g., dairy products, meat-fish-poultry, and fruits). Although it is not immediately apparent from this figure, the general levels of chlorinated hydrocarbon insecticides in the total diet of U.S. citizens are definitely going down. The data cited in Table 11-15 (Duggan and Corneliussen, 1972) show that a general decline occurred in the period 1964–1969 for most pesticide residues except organophosphate insecticides, whose usage steadily increased. Somewhat similar data were obtained by Corneliussen (1972), and Duggan and Corneliussen (1972) subsequently summarized these (Fig. 11-6). The only residues that increased in the period were those of organophosphates. In terms of location, residues in foods from Baltimore were consistently lower than those in foods from Boston, Los Angeles, Kansas City, and Minneapolis; Boston and Los Angeles levels were the highest of these. The average U.S. citizen consumes (1970 figure) about 70 μg of pesticide residues (not counting bromide residues) per day, of which 67 % is due to chlorinated hydrocarbons and 27 % to organophosphates. As a whole, these intake levels are far below the maximum levels recommended by the World Health Organization Expert Committee on Pesticide Residues (WHO, 1967) in conjunction with the Food and Agriculture Organization (FAO). The amount of bromides in the total diet samples equals less than one-half of the FAO-WHO acceptable level. Some of the bromide residues come from ethylene dibromide and methyl bromide, expected to occur in the form of a nonvolatile bromine complex. Unfortunately, it is not known how much is due to fumigation and how much is natural. Other compounds that come close to the FAO-WHO levels in the U.S. diet are aldrin and dieldrin. However, the FAO-WHO levels should not be construed to indicate "absolute" safety, for we must always remember that man has not yet experienced exposure for an entire life span, since use of these modern insecticides began only about 30 years ago.

11.4.3c. Pharmacokinetics of Pesticide Residues in Man

From the foregoing description of residue patterns and intake figures, it becomes apparent that an average man (70 kg) in the United States takes in 35 μg of DDT-R per day (1969 figure) and has about 5 ppm of DDT-R in his fat, of which 70–80% is in the form of DDE. The net daily intake amounts to only 0.5 ppb, assuming that the DDT-R distributes itself uniformly throughout the entire body. Even if the DDT-R were to localize in fat, which comprises about 10% of the entire body, this figure would be just 5 ppb, or 1/1000 of the level generally found in the adipose tissue. The question is how daily intake affects the level of DDT-R in the fat. In a controlled feeding test, Hayes *et al.* (1971) noted a tenfold increase (Fig. 11-4) in the DDT level in 12.2 months when volunteers were given 35 mg/day/man of DDT. Apparently, daily intake of DDT of this magnitude (a hundred times greater than normal) has a strong influence at least on the level of DDT. Assuming a linear relationship, a regular daily dose of 35 μg/man should influence the DDT-R in the fat by 10% in a year. In reality, however, the DDT-R level in the general public showed a slight decline in 1969–1971, indicating that the equilibrium point must be somewhat higher than this daily intake level as well as that there is no linear relationship between intake and rate of accumulation. In the same study, DDT residues in control volunteers who received 184 μg/day/man of DDT increased from 4.3 to 10.3 ppm in 12.2 months, to 16.3 ppm in 18.8 months, and to 22.0 in 21.5 months. No increase in DDE was observed in this period, but it started rising after 33 months and stayed at this level even at the end of the experiment after 59.3 months. Thus it is reasonable to assume that DDT intake at regular dietary levels must also affect DDT storage at an early stage and DDE storage after a considerable delay. Judging by the data for 1964–1965, when DDT-R levels in the fat remained constant in the public, this equilibrium point appears to be on the order of 70 μg/day/man, in agreement with the above observations, so the degree of turnover is 0.2% a day at equilibrium. Of course, at higher doses it is expected that both storage and excretion will increase. Hayes *et al.* (1971) found that volunteers fed 3.5 mg/day excreted on the average 0.46 mg or 13% of the daily dose in their urine in the form of DDA. The regular 35-μg daily dose constitutes 12.8 mg/man/year, which, if all stored in the fat, should amount to 1.5 ppm or 30% of the DDT-R level at 5 ppm. Since the net residue level in the fat is rather stable on the average, this must mean that the turnover rate is 30% per year at this dose. There is also other circumstantial evidence to indicate that storage is not static (i.e., dead storage) but rather is dynamic and based on an equilibrium between intake and discharge even at this low pesticide level. (Dynamics at higher concentrations have already been mentioned.)

Ever since pesticides were first discovered to be present in man, scientists have been trying to find a substitute for adipose tissue in the measurement of residues. Adipose tissues are difficult to obtain, particularly in suitable quantities, from healthy persons. In the past, they have been obtained at autopsy or from patients undergoing surgery, but sample numbers are thus severely limited. The most convenient materials would be blood and urine, and as accuracy in analytical methods improves, these materials could be used to determine residues if they accurately reflect the real levels in the body. The problem is that they do not. Hayes *et al.* (1971) found occasional sudden increases in urinary DDA levels which did not correlate at all with the general levels of DDT-R in fat. Wyllie *et al.* (1972) conducted an extensive survey of the relationship of serum DDT and DDE levels to adipose tissue levels and found rather poor correlations. Schafer and Campbell (1966) found a little better correlation between these two parameters among samples from Montgomery County, Ohio. Their data (Fig. 11-7) indicate a correlation index of $R^2 = 0.77$. However, considering that the number of samples was 18 and that one sample was not included because of a complete deviation (5.4 ppm in fat and 94 ppb in blood), this correlation must also be regarded as relatively poor. It should be pointed out that the correlation could be good at higher concentrations for other insecticidal compounds. For instance, Robinson and Hunter (1966) obtained a very good correlation (linear on log vs. log plotting) for the dieldrin levels in fat and blood samples from volunteers fed known doses. The levels of dieldrin in the fat in this case were from around 0.2 to almost 200 ppm. Although Keane and Zavon (1969) observed in dogs fed dieldrin a direct relationship between fat and blood levels, they found that the relationship between the total body burden and the fat level was indirect.

Other evidence indicating the dynamic state of residue storage is that residue levels can be altered by various inducing agents. The initial experiment by Street (1964) showing that the storage of dieldrin in the body fat of rats could be reduced by simultaneously administered DDT has been confirmed many times. The cause of such an interaction is not apparent, but it is not simple induction of microsomal enzyme activity, as shown by the speed at which dieldrin is discharged, the lack of inhibitory effects of antibiotics known to inhibit protein synthesis, and the occurrence of certain opposite effects such as the following: Deichmann *et al.* (1969) showed that the concentration of *p,p'*-DDT plus *p,p'*-DDE in the body fat of dogs could be increased by feeding DDT plus aldrin. In this experiment, one group of six

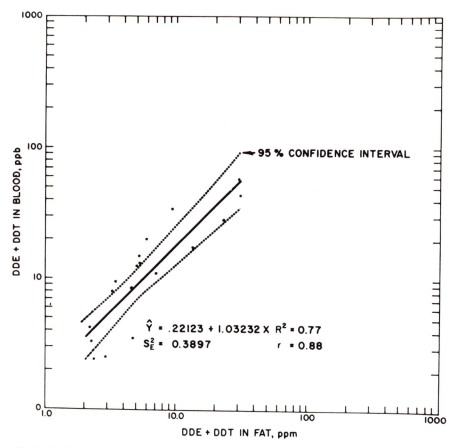

Fig. 11-7. Relationship between DDT/DDE levels in adipose tissue and those in serum among the residents of Montgomery County, Ohio. From Schafer and Campbell (1966).

dogs was fed 24 mg/kg of DDT alone and another group 12 mg/kg of DDT plus 0.3 mg/kg of aldrin. The retention of p,p'-DDT or p,p'-DDE (or both together) was greatly enhanced by the presence of aldrin; in fact, the dogs fed DDT plus aldrin showed 2.5–4 times more retention than the dogs fed DDT alone. DDE retention is shown in Fig. 11-8. An *in situ* experiment indicates that dieldrin release from liver tissue can be stimulated by phenobarbital, DDT, and even electric current (Matsumura and Wang, 1968). It is apparent that the rate of storage and release, which in turn determines the level of residues, is governed by various physiological states of the host animal. Furthermore, one cannot ignore the turning over of the adipose tissues themselves during a relatively long period of time, such as 5–10 years.

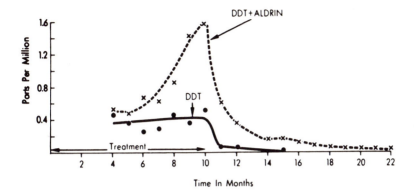

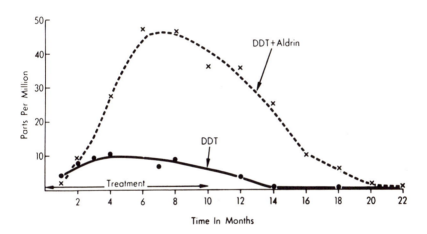

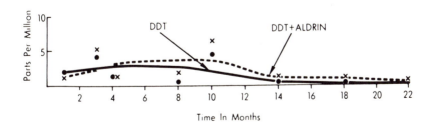

Fig. 11-8. Stimulation of DDE retention by aldrin. Solid line, DDT 24 mg/kg/day; broken line, DDT 12 mg/kg/day plus aldrin 0.3 mg/kg/day fed to six male beagle dogs. From Deichmann *et al.* (1969).

**TABLE 11-16. Increase in the Percentage of DDE
with Respect to Total DDT-R (including DDE) in
Human Fat**

References	Years surveyed	Percentage of DDE
Hayes *et al.* (1958)	1950s	58
Dale and Quinby (1963)	1956–63	56.5–67.0
	1963–65	67.3–73.9
Hoffman *et al.* (1967)	1962–63	71.9
	1964–66	73.9
Warnick (1972)	1968	66.0
	1969	67.2
	1970	75.4
Wyllie *et al.* (1972)	1970	78.5

Thus the relative stability of chlorinated hydrocarbon residues in the fat of the general population for many years does not indicate that residue storage is static. Such an end result most likely comes from dynamic equilibrium between intake and storage–discharge and from simple averaging of large numbers of individuals.

As mentioned previously, another phenomenon that has been noted by scientists is that the relative proportion of DDE with respect to other DDT-R complexes is increasing with time. Some of these data are summarized in Table 11-16 to illustrate the point. In particular, Hoffman *et al.* (1967) compared their own data for 1962–1963 and 1964–1966, collected by identical analytical methods and concluded that this difference is real. Examination of other sets of data obtained by the same research teams on the same populations also supports the existence of this phenomenon in that a small but definite tendency toward increase in DDE can be observed (but not as much as once suspected by Dale and Quinby, 1963). The causes could be severalfold: for instance, increase in DDE content in food, the time elapsed after high DDT exposures, dynamics of preferential storage of DDE, and concentration dependency (i.e., people with high DDT-R residues have lower percentages of DDE). The first factor is not as apparent as some scientists claim. Examination of the data of Duggan *et al.* (1971) and Duggan and Corneliussen (1972) reveals that between 1965 and 1970 the DDE content in food hardly changed. The actual ratios in the total daily diet for these years were 33.3, 20.0, 25.0, 28.8, 20.0, and 25.0% respectively, with the 6-year average being 28.8%. At present, therefore, it must be concluded that evidence for the other causes is better. The data shown in Fig. 11-4, for instance, clearly indicate that (1) there is a differential retention of DDT and DDE,

(2) at high DDT exposures the DDT ratio is likely to go up, and (3) relative DDE concentrations do increase as time from the sudden exposure to DDT elapses.

11.4.3d. Assessment of Safety

As we have learned (so far by mostly empirical approaches), present levels of insecticide residues in the general public have no apparent ill effects. Indeed, all human feeding experiments, such as the ones conducted by Hayes *et al.* (1971) and Robinson and Hunter (1966), were tried at hundredfold higher levels for DDT and ten- to fortyfold higher levels for dieldrin without any indication of ill effects. (In the latter experiment, the daily dose used was 50 μg or 211 μg per man for 3 or 6 months. The calculated daily aldrin/dieldrin intake for the U.S. general public is 5 μg for a 70-kg man.) It is therefore safe to say that the effects, if any, would be subtle, especially since in the United States at least daily pesticide intake is generally below the FAO-WHO recommended maximum allowable doses.

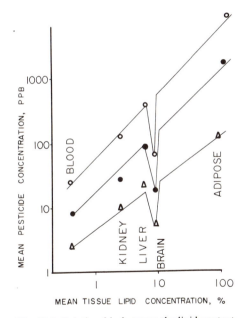

Fig. 11-9. Relationship between the lipid content of tissues and the levels of three chlorinated hydrocarbon insecticides in man. Note that the levels in the brain are much lower than those expected by its lipid content. Redrawn from Morgan and Roan (1970).

In considering the subtle effects, the first step is to look for any possible weakness in the human system. The most vulnerable site for insecticides in man would be the nervous system, for these chemicals are nerve poisons. The fact that EEG changes are the first to be observed in some of the occupationally exposed people (Jager *et al.*, 1970) attests to this.

Fortunately, there is some evidence that the pesticide residues in the brain are unusually low in comparison with those in other parts of the body. In Fig. 11-9 are summarized the data by Morgan and Roan (1970), who studied the relationship between the fat content of tissues and the level of chlorinated hydrocarbon residues. These as well as others support the observation. Casarett *et al.* (1968), for instance, found about one-tenth as much DDT-R and dieldrin in lipids extracted from the brain as in lipids from all other organs. With respect to the total quantity (rather than concentration) of residues in the brain, there is less than 0.1 % of the total body burden of DDT/DDE (Table 11-17). Indeed, the data reveal that altogether almost 97 % of DDT/DDE in the body is concentrated in the fat.

TABLE 11-17. Distribution of Chlorinated Hydrocarbon Insecticides Among Various Tissues and Organs

Tissues	Lipid content[a] (%)	Whole tissue basis 44 autopsies[a,b] (ppm)	Fat basis, one example[a,b] (ppm)	Total amount in each organ[c,d] (mg)
Fat[e]	55.7	6.03	6.95	43.7
Liver	2.1	0.285	17.20	0.47
Kidney	3.2	0.30	1.61	0.033
Brain	7.9	0.0989	0.25	0.040
Blood	2[f]	0.026[f]	—	0.048
Gonad	1.3	0.0875	5.35	—
Lung	0.7	0.0766	—	—
Spleen	0.6	0.0469	3.00	—
Adrenal	10.5	1.06	—	—
Protein (muscle)	—	—	—	1.03
Total	—	—	—	45.321

[a]Casarett *et al.* (1968).
[b]DDT + DDE + DDD + heptachlor + dieldrin.
[c]Schafer and Campbell (1966). DDT + DDE. The following assumption was made by these workers: blood 6% of total volume, protein 15% by weight, and fat 14% by weight.
[d]Average of three samples of low residue levels.
[e]Organ tissue for Casarett *et al.*, and total extracts from the various tissues for Schafer and Campbell.
[f]Morgan and Roan (1970).

Other susceptible sites for insecticides could be the reproductive organs and the liver. The level of residues in the gonads does not appear to be particularly high on a whole organ basis (Table 11-17), although on a fat basis it is. The highest value obtained by Casarett *et al.* (1968) for the gonads was 80.7 ppm on a fat basis, which amounts to about 1 ppm on a whole organ basis. This represents only 10 μg in a 10-g organ, still far from the value of 250 μg of endrin that Dikshith and Datta (1973) employed to induce mutagenic effects in the testis of the rat, especially considering the difference in size of man and rat. The available data will not be sufficient to prove the safety or danger of residues in the gonads until somebody actually investigates the rate of chromosomal aberration or performs some other mutagenicity test at 1 ppm levels of chlorinated hydrocarbon insecticides (or any other pesticides).

Another potentially susceptible reproduction-related system is the mother–child or mother–fetus transfer of pesticides via milk or via placental transport processes. Practically nothing is known about placental transfer of pesticides in humans, although several experiments indicate that it does occur in animals (e.g., Ackermann and Engst, 1970; Fish, 1966; Villeneuve *et al.*, 1972). With respect to human milk, Laug *et al.* (1951) reported that the mean concentration of DDT-R was 0.13 ppm in 32 samples from the Washington, D.C., area. This figure is much higher than that found in blood. Quinby *et al.* (1965) found a slightly lower value of 0.08 ppm on an individual basis and 0.07 ppm for pooled milk from various areas of the United States (Chicago, Wenatchee, Phoenix, and cities in Colorado). They also cite the data from Hungary suggesting that the levels there are on the order of 0.13–0.26 ppm in whole milk (Dénes, 1964). To quote Quinby *et al.* (1965), "If a 5 kg (11 lb) infant consumed 0.7 liter of milk daily containing DDT at an average concentration of 0.08 ppm, the resulting dosage would be 0.0112 mg/kg/day. This value may be compared with the average adult dosage (daily intake) of 0.0005 mg/kg/day. Also, there is a considerable amount of evidence that young animals, including foetal ones, are somewhat more susceptible than adults to many chemicals including DDT. However, the difference is only a few-fold." Assuming this "few-fold" factor means a two- to threefold increase in sensitivity, the babies have been subjected to an insult approximately 50 times greater than that normal adults endure; in terms of "acceptable daily intake," this amounts to eight times more than the FAO-WHO recommended maximum acceptable dose! With respect to other pesticides, Tanabe (1972) reports that although the average DDT-R level in human milk in Japan is only about 0.05 ppm on a whole milk basis, the value for total BHC comes to 0.1–0.2 ppm, which is similar to the value for cow's milk in that country. The dieldrin concentration is reported to be on the order of 0.005 ppm in human milk. This amounts to 0.022 mg/kg/day of

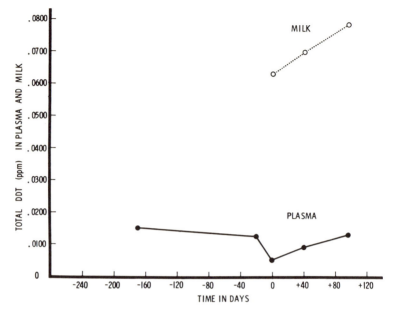

Fig. 11-10. Relationships of DDT-R levels in whole human milk and plasma of five women to the pregnancy and postdelivery time. From Curley and Kimbrough (1969).

BHC and 0.001 mg/kg/day of dieldrin, 1000 and 10 times more than normal U.S. adults receive, or, in the case of dieldrin, 10 times more than the FAO-WHO maximum acceptable level without even considering the extra susceptibility of children. The levels of total BHC, heptachlor epoxide, and dieldrin in human milk in the United States in 1968 were 0.0075, 0.0010, and 0.0069 ppm, respectively (Curley and Kimbrough, 1969). Note that the figure for dieldrin is even higher than that found in Japan. A similar survey in Great Britain (Egan *et al.*, 1965) put the DDT-R and dieldrin concentrations at 0.13 and 0.006 ppm. With the exception of Tanabe's survey, there is a uniform, puzzling tendency for residue levels of chlorinated hydrocarbon insecticides in human milk to be higher than those found in cow's milk. Furthermore, the residue level appears to increase after parturition (Fig. 11-10). Probably the factors affecting the levels of pesticides in both the blood and the milk are increases in serum lipids (46%) and decreases in serum albumin (25%), increases in progesterone and estrogen, and intake of various medications prior to and after delivery (Curley and Kimbrough, 1969). Whatever the causes, it is certain that newborn babies are subjected to much higher pesticide burdens than adults. The increases range from 10 to 1000 times more than adult levels, and in most cases exceed the FAO-WHO recommended maximum acceptable levels.

In adults, the most likely effects of low-level insecticide residues are the "induction" effects on the hepatic microsomal systems. Earlier, the data of the Tuskegee Institute group (Chung *et al.*, 1967, 1968) were cited that DDT and dieldrin can cause induction effects in cultured human cells (HeLa S cell culture) at a concentration of 0.5 ppm. This level is very close to what has actually been found in humans (see Table 11-17). Datta and Nelson (1968) fed DDT to rats at 0.04–0.10 mg/kg/day and observed statistically significant increases in hepatic enzymatic activity. The DDT-R concentration reached 11–21 ppm in their fat; indeed very close to that found in human adipose tissue.

Naturally, it can be argued that extrapolation of animal data, or even that from human cells in culture, is not possible. However, because these intake and residue levels are indeed very close to those of the general human population and no safety factor of 100 or 1000 enters in, as is usual in toxicological investigations, the balance of evidence and logic suggests that there is a good probability of these chlorinated hydrocarbon residues influencing human hepatic microsomal systems even at the current levels.

In summary, current pesticide residues do not appear to be harmful, and the most likely subtle effects are those resulting from induction of hepatic microsomal systems. These biochemical changes are reversible, and no permanent injury is expected from them. However, induction processes could lead to other changes, e.g., could affect hormone levels, and certainly more studies are needed to fathom the significance of such secondary influences. Effects on other vulnerable systems are either unlikely or very difficult to prove, although those on reproductive organs must be studied in the future. Perhaps the most concern should be directed toward breast-fed infants, for whom the bodily insult from chlorinated hydrocarbon residues is by far the greatest.

11.5. REFERENCES

Ackermann, H., and R. Engst (1970). *Arch. Toxicol.* **26**:17.
Andervont, H. B., and T. B. Dunn (1952). *J. Natl. Cancer Inst.* **13**:455.
Andrianova, M. M., and I. V. Alekseev (1970). *Vopr. Pitaniya* **29**:71. (Indirectly cited from *Health Aspects of Pesticides*, Abst. 71-2765.)
Anonymous (1963). *Clinical Handbook on Economic Poisons*. U.S. Department of Health, Education and Welfare, Communicable Disease Center, Atlanta, Ga., Bulletin No. 476.
Anonymous (1967a). Survey of products most frequently named in ingestion accidents, *Bull. Natl. Clearing House Poison Control Center*. September–October issue, Washington, D.C.

Anonymous (1967b). *Organophosphate Poisoning.* Tokyo Ministry of Health and Welfare, 1958–1962, 1966, 1967.

Anonymous (1968). Niagara Chemical Division Handbook. FMC Corp. Middleport, New York.

Anonymous (1971a). *Health Rep.* **86**:605.

Anonymous (1971b). *Lancet* **2(7733)**:1073.

Ashwood-Smith, M. J., J. Trevino, and R. Ring (1972). *Nature* **240**:418.

Augustinsson, K. B. (1957). In *Methods of Biochemical Analysis.* D. Glick, ed. Interscience, New York, p. 1.

Baker, S. B. D. (1971). *Proc. Europ. Soc. Stud. Drug Toxicity* **12**:381. Excerpta Medica, Amsterdam.

Barnes, J. M., and D. F. Heath (1964). *Brit. J. Ind. Med.* **21**:280.

Barnes, J. M., W. J. Hayes, Jr., and K. Kay (1957). *Bull. World Health Org.* **16**:41.

Baquar, S. R., and N. R. Khan (1970). *Rev. Biol.* **7**:195.

Beeman, R. W., and F. Matsumura (1973). *Nature* **242**:273.

Best, W. R. (1963). *J. Am. Med. Assoc.* **185**:286.

Best, E. M., Jr., and B. L. Murray (1962). *J. Occup. Med.* **4**:507.

Bick, M. (1965). *Med. J. Austral.* **1**:1127.

Bidstrup, P. L., J. A. Bonnell, and A. G. Beckett (1953). *Brit. Med. J.* **1**:1068.

Bitman, J., H. C. Cecil, S. J. Harris, and G. F. Fries (1968). *Science* **162**:371.

Brown, H. V., and A. F. Bush (1950). *Arch. Ind. Hyg. Occup. Med.* **1**:633.

Brown, J. R. (1967). *Can. Med. Assoc. J.* **97**:367.

Brown, V. K., C. G. Hunter, and A. Richardson (1964). *Brit. J. Ind. Med.* **21**:283.

Brzezinski, J. (1972). *Diss. Pharm. Pharmacol.* **24**:217.

Brzezinski, J., and W. Rusiecki (1970). *Diss. Pharm. Pharmacol.* **22**:507.

Budreau, C. H., and K. P. Singh (1973). *Toxicol. Appl. Pharmacol.* **24**:324.

Buselmaier, W., G. Roehrborn, and P. Propping (1972). *Biol. Zentralbl.* **91**:311.

Cabral, J. R., M. C. Testa, and B. Terracini (1972). *Tumori* **58**:49.

Casarett, L. J., G. C. Fryer, W. L. Yauger, and H. W. Klemmer (1968). *Arch. Environ. Health* **17**:306.

Cazorla, A., and F. Moncloa (1962). *Science* **136**:47.

Chung, R. A., I-L. Huang, and R. W. Brown (1967). *J. Agr. Food Chem.* **15**:497.

Chung, R. A., Y.-D. Lin, and R. W. Brown (1968). *J. Agr. Food Chem.* **16**:298.

Cloutier, G., and A. L. Gascon (1971). *Biochem. Pharmacol.* **20**:2319.

Coble, Y., P. Hildebrandt, J. Davis, F. Raasch, and A. Curley (1967). *J. Am. Med. Assoc.* **202**:153.

Corneliussen, P. E. (1972). *Pesticides Monitoring J.* **5**:313.

Cornish, H. H. (1971). *Crit. Rev. Toxicol.* **1**:1.

Cranmer, M., A. Peoples, and R. Chadwick (1972). *Toxicol. Appl. Pharmacol.* **21**:98.

Curley, A., and R. Kimbrough (1968). *Arch. Environ. Health* **18**:156.

Dale, W. E., and G. E. Quinby (1963). *Science* **142**:593.

Dale, W. E., T. B. Gaines, and W. J. Hayes, Jr. (1162). *Toxicol. Appl. Pharmacol.* **4**:89.

Dale, W. E., T. B. Gaines, and W. J. Hayes, Jr. (1963). *Science* **142**:1474.

Dale, W. E., T. B. Gaines, and W. J. Hayes, Jr. (1965). *Bull. World Health Org.* **33**:471.

Dale, W. E., A. Curley, and C. Cueto (1966). *Life Sci.* **5**:47.

Datta, P. R., and M. J. Nelson (1968). *Toxicol. Appl. Pharmacol.* **13**:346.

Davies, J. E., W. F. Edmundson, N. J. Schneider, and J. C. Cassady (1968). *Pesticides Monitoring J.* **2**:80.

Davis, J. H., J. E. Davies, and A. J. Fisk (1967). *Symposium on Biological Effects on Mammalian Systems.* New York Academy of Sciences, New York, May.

Davis, K. J., and O. G. Fitzhugh (1962). *Toxicol. Appl. Pharmacol.* **4**:187.

Deichmann, W. B., M. Keplinger, F. Sala, and E. Glass (1967). *Toxicol. Appl. Pharmacol.* **11**:88.

Deichmann, W. B., M. Keplinger, I. Dressler, and F. Sala (1969). *Toxicol. Appl. Pharmacol.* **14**:205.

Deichmann, W. B., W. E. MacDonald, E. Blum, M. Bevilacqua, J. Radomski, M. Keplinger, and M. Balkus (1970). *Ind. Med. Surg.* **39**:426.

del Vecchio, V., and V. Leoni (1969). *Nuovi Ann. Ig. Microbiol.* **28**:107.

de Vlieger, M., J. Robinson, M. K. Baldwin, A. N. Crabtree, and M. C. van Dijk (1968). *Arch. Environ. Health* **17**:759.

Dénes, A. (1974). *1963 Year-book of the Institute of Nutrition (Budapest)*, p. 47. (Indirectly cited from Quinby *et al.*, 1965.)

Dikshith, T. S. S., and K. K. Datta (1968). *Experientia* **28**:169.

Dikshith, T. S. S., and K. K. Datta (1973). *Bull. Environ. Contam. Toxicol.* **9**:65.

Druckrey, H. (1967). In *Potential Carcinogenic Hazards from Drugs: Evaluation of risks*. R. Truhaut, ed. UICC Monograph Series, Vol. 7, Springer-Verlag, Berlin and New York.

Duggan, R. E., and P. E. Corneliussen (1972). *Pesticides Monitoring J.* **5**:331.

Duggan, R. E., G. O. Lipscomb, E. L. Cox, R. E. Heatwole, and R. C. Kling (1971). *Pesticides Monitoring J.* **5**:73.

Durham, W. F. (1963). *Residue Rev.* **4**:33. F. A. Gunther, ed. Springer-Verlag, New York.

Durham, W. F. (1967). *Residue Rev.* **18**:21. F. A. Gunther, ed. Springer-Verlag, New York.

Durham, W. F., T. B. Gaines, and W. J. Hayes, Jr. (1956). *Arch. Environ. Health* **13**:326.

Durham, W. F., J. F. Armstrong, W. M. Upholt, and C. Heller (1961). *Science* **134**:1880.

Edmundson, W. F., and J. E. Davies (1967). *Arch. Environ. Health* **15**:89.

Edmundson, W. F., J. E. Davies, M. Cranmer, and G. A. Nachman (1969). *Ind. Med. Surg.* **38**:45.

Edmundson, W. F., J. E. Davies, and W. Hall (1972). *Pesticides Monitoring J.* **2**:86.

Edson, E. F. (1955). Mimeographed release. Fisons Pest Control, Ltd., Medical Department.

Egan, H. (1965). *Brit. Med. J.* **2**:66.

Egan, H., R. Goulding, J. Roburn, and J. O'G. Tatton (1965). *Brit. Med. J.* **11**:66.

Ellman, G. L., D. K. Courtney, V. A. Andres, and R. M. Featherstone (1961). *Biochem. Pharmacol.* **7**:88.

Enns, T. (1972). *Fed. Proc.* **31**:819.

Epstein, S. S. (1970). *Nature* **228**:816.

Epstein, S. S., and M. S. Legator (1971). In *The Mutagenicity of Pesticides: Concepts and Evaluation*. MIT Press, Cambridge, Mass., and London, p. 220.

Epstein, S. S., E. Arnold, K. Steinberg, D. Mackintosh, H. Shafner, and Y. Bishop (1970). *Toxicol. Appl. Pharmacol.* **17**:23.

Faina, L., G. Futtori, M. Pirotta, and G. Procellati (1971). *Acta Neurol.* **26**:243.

Falk, H. L. (1971). *Progr. Exptl. Tumor Res.* **14**:105.

Faraga, A. (1967). *Arch. Toxikol.* **23**:11.

Ferrando, R. (1971). *Bull. Acad. Natl. Med.* **155**:117.

Fiserova-Bergerova, V., J. L. Radomski, J. E. Davies, and J. H. Davies (1967). *Ind. Med. Surg.* **36**:65.

Fish, S. A. (1966). *Am. J. Obstet. Gynecol.* **96**:1148.

Fisher, D. B., and G. C. Mueller (1971). *Biochem. Pharmacol.* **20**:2515.

Fitzhugh, O. G., and A. A. Nelson (1947). *J. Pharmacol. Exptl. Therap.* **89**:18.

Fitzhugh, O. G., A. A. Nelson, and M. L. Quaife (1964). *Food Cosmet. Toxicol.* **2**:551.

Flockhart, I. R., and J. E. Casida (1972). *Biochem. Pharmacol.* **21**:2591.

Fowler, B. A. (1972). *Am. J. Pathol.* **69**:163.

Frawley, J. P., R. E. Zwickey, and H. N. Fuyat (1956). *Fed. Proc.* **15**:424.

Gage, J. C. (1967). *Residue Rev.* **18**:159. F. G. Gunther, ed. Springer-Verlag, New York.

Gertig, H., W. Nowaczyk, and B. Sawicki (1971*a*). *Diss. Pharm. Pharmacol.* **23**:541.

Gertig, H., W. Nowaczyk, and M. Sierzant (1971*b*). *Diss. Pharm. Pharmacol.* **23**:545.

Gill, G. R., and Q. N. LaHam (1972). *Can. J. Zool.* **50**:349.

Goto, M., M. Hattori, and T. Miyagawa (1972). *Chemosphere* **1**:153.

Goyer, G. R., E. A. Martin, P. Paganuzzi, and J. Brodeur (1970). *Can. J. Physiol. Pharmacol.* **48**:342.

Greenberg, J., and Q. N. LaHam (1969). *Can. J. Zool.* **47**:539.

Halver, J. E. (1967). *Bureau Sport Fish. Wildlife Res. Rep.* **70**:78.

Harrison, D. L., P. E. G. Maskell, and D. F. L. Money (1963). *Vet. Bull.* **33**:4097.

Hart, L. G., and J. R. Fouts (1963). *Proc. Soc. Exptl. Biol. Med.* **114**:388.

Hart, L. G., and J. R. Fouts (1965). *Arch. Exptl. Pathol. Pharmacol.* **249**:486.

Hart, L. G., R. W. Shultice, and J. R. Fouts (1963). *Toxicol. Appl. Pharmacol.* **5**:371.

Hart, M. M., R. L. Reagan, and R. H. Adamson (1973). *Toxicol. Appl. Pharmacol.* **24**:101.

Hassan, A. (1971). *Biochem. Pharmacol.* **20**:2299.

Hayes, W. J., Jr. (1959). In *DDT, the Insecticide Dichlorodiphenyltrichloroethane and Its Significance*, Vol. II. P. Muller, ed. Birkhäuser Verlag, Basel, p. 33.

Hayes, W. J., Jr. (1963). *Clinical Handbook of Economic Poisons*. U.S. Public Health Service, Department of Health, Education and Welfare, Atlanta, Ga.

Hayes, W. J., Jr., G. E. Quinby, K. L. Walker, J. W. Elliott, and W. M. Upholt (1958). *Am. Med. Assoc. Arch. Ind. Health* **18**:398.

Hayes, W. J., W. E. Dale, and R. LeBreton (1963). *Nature* **199**:1189.

Hayes, W. J., Jr., W. E. Dale, and V. W. Birse (1965). *Life Sci.* **4**:1611.

Hayes, W. J., W. E. Dale, and C. I. Pirkle (1971). *Arch. Environ. Health* **22**:119.

Heath, D. F. (1961). *Organophosphorus Poisons*. Pergamon Press, New York, 403 pp.

Heath, D. F., and M. Vanderkar (1964). *Brit. J. Ind. Med.* **21**:269.

Helson, L., N. Wollner, L. Murphy, and M. K. Schwartz (1971). *Clin. Chem.* **17**:1191.

Hestrin, S. (1949). *J. Biol. Chem.* **180**:249.

Hoffman, W. S., W. I. Fishbein, and M. B. Andelman (1964). *Arch. Environ. Health* **9**:387.

Hoffman, W., H. Adler, W. I. Fishbein, and F. C. Bauer (1967). *Arch. Environ. Health* **15**:758.

Hood, R. D., and S. L. Bishop (1972). *Arch. Environ. Health* **24**:62.

Hoogendam, I., J. P. J. Versteeg, and M. DeVlieger (1965). *Arch. Environ. Health* **10**:441.

Howell, D. E. (1948). *Proc. Okla. Acad. Sci.* **29**:1.

Huang, C. C. (1973). *Proc. Soc. Exptl. Biol. Med.* **142**:36.

Hunter, C. G., and J. Robinson (1967). *Arch. Environ. Health* **15**:614.

Hunter, C. G., and J. Robinson (1968). *J. Food Cosmet. Toxicol.* **6**:253.

Hunter, C. G., J. Robinson, and M. Roberts (1969). *Arch. Environ. Health* **18**:12.

Hurtig, H. (1972). In *Environmental Quality and Safety*. F. Coulston, and F. Korte, eds. Georg Thieme Verlag, Stuttgart, Academic Press, New York, p. 58.

Indian Council of Agricultural Research (1967). *Harmful Effects of Pesticides*. Report of the Special Committee, Delhi, 93 pp.

Innes, J. R. M., B. M. Ulland, M. G. Valerio, L. Petrucelli, L. Fishbein, E. R. Hart, A. J. Pallotta, R. R. Bates, H. L. Falk, J. J. Gart, M. Klein, I. Mitchell, and J. Peters (1969). *J. Natl. Cancer Inst.* **42**:1101.

Isenberg, I., and E. Small (1972). *Binding of Dieldrin and Hydrocarbons to Histones and Histone–DNA Complexes*. Annual Progress Report, Oregon State University Environmental Health Science Center, pp. 39–44.

Ivanova, S. I. (1972). *Fiziol. Zh.* (*Kiev*) **18**:391. (Indirectly cited from *Health Aspects of Pesticides*, Abst. 73-0902.)

Jager, K. W. (1970). *Aldrin, Dieldrin, Endrin and Telodrin.* Elsevier, Amsterdam, 234 pp.
Jager, K. W., D. V. Roberts, and A. Wilson (1970). *Brit. J. Ind. Med.* **27**:273.
Jain, N. C., C. R. Fontan, and P. L. Kirk (1965). *J. Pharm. Pharmacol.* **17**:362.
Johnson, D. W., and S. Lew (1970). *Pesticides Monitoring J.* **4**:57.
Johnson, M. K. (1969*a*). *Biochem. J.* **111**:487.
Johnson, M. K. (1969*b*). *Biochem. J.* **114**:711.
Johnson, M. K., and R. Lauwerys (1969). *Nature* **222**:1066.
Kacew, S., R. L. Singhal, P. D. Hrdina, and G. M. Ling (1972). *J. Pharmacol. Exptl. Therap.* **181**:234.
Kaminsky, N., S. Luse, and P. Hartroft (1962). *J. Natl. Cancer Inst.* **29**:127.
Kazantzis, G., A. I. G. McLaughlin, and P. F. Prior (1964). *Brit. J. Ind. Med.* **21**:46.
Keane, W. T. and M. R. Mitchell (1969). *Arch. Environ. Health* **19**:36.
Keane, W. T., and M. R. Zavon (1969). *Bull. Environ. Contam. Toxicol.* **4**:1.
Kemeny, T., and R. Tarjan (1966). *Experientia* **22**:748.
Kinoshita, F. K., J. P. Frawley, and K. P. DuBois (1966). *Toxicol. Appl. Pharmacol.* **9**:505.
Kleinman, G. D., I. West, and M. S. Augustine (1960). *Arch. Environ. Health* **1**:118.
Klemmer, H. W. (1972). *Residue Rev.* **41**:55. F. A. Gunther and J. D. Gunther, eds. Springer-Verlag, New York.
Knaak, J. B., M. J. Tallant, S. J. Kozbelt, and L. J. Sullivan (1968). *J. Agr. Food Chem.* **16**:465.
Kolmodin, B., D. L. Azarnoff, and F. S. Öqvist (1969). *Clin. Pharmacol. Therap.* **10**:638.
Krampl, V., and M. Grigel (1972). *Prac. Lek.* **24**:121. (Indirectly cited from *Health Aspects of Pesticides.*)
Kravtsova, O. L., L. I. Korenevskyy, and O. H. Reznikov (1971). *Dopov. Akad. Nauk Ukr. RSR Ser. B* **30(10)**:943. (Indirectly cited from *Health Aspects of Pesticides*, Abst. 72-1291.)
Kusevitskiy, I. A., A. Y. Kirlich, and L. A. Khovayeva (1970). *Veterinariya* **46**:73.
Lacassagne, A. (1971). *Bull. Cancer* **58**:235.
Lagator, M. S. (1970). Quoted in *Chem. Eng. News* **48**:51.
Laug, E. P., F. M. Kunze, and C. S. Prickett (1951). *Arch. Ind. Hyg.* **3**:245.
Laws, E. R. Jr. (1971). *Arch. Environ. Health* **23**:181.
Laws, E. R., Jr., F. M. Morales, W. J. Hayes, Jr., and C. R. Joseph (1967). *Arch. Environ. Health* **15**:766.
Leach, P. H. (1953). *Calif. Med.* **78**:491.
Lebrun, A. (1960). *Bull. World Health Org.* **22**:579.
Lederberg, J. (1971). In *The Mutagenicity of Pesticides: Concepts and Evaluation.* S. S. Epstein, and M. L. Legator, eds. MIT Press, Cambridge and London, p. viii.
Lemon, P. G. (1967). In *Pathology and Laboratory Rats and Mice.* E. Cotchin, and F. J. C. Roe, eds. F. A. Davis Co., Philadelphia, p. 25.
Lieban, J., R. K. Waldman, and L. Krause (1953). *Arch. Ind. Hyg. Occup. Med.* **7**:93.
Lodge, J. P. (1965). *J. Am. Med. Assoc.* **193**:110.
Mackerras, I. M., and R. F. K. West (1946). *Med. J. Austral.* **1**:400.
Matsumura, F., and R. W. Beeman (1972). *Report on the Mode of Action of chlordimeform.* CIBA-Geigy Ltd., Basel. August.
Matsumura, F., and C. M. Wang (1968). *Bull. Environ. Contam. Toxicol.* **3**:203.
Mattson, A. M., and V. A. Sedlak (1960). *J. Agr. Food Chem.* **8**:107.
Meisner, H. M., and L. Sorensen (1966). *Exptl. Cell Res.* **42**:291.
Metcalf, R. L. (1957). *Arch. Ind. Health* **16**:337.
Milby, T. H., A. J. Samuels, and F. Ottoboni (1968). *J. Occup. Med.* **10**:584.
Mootoo, C. L., and B. Singh (1966). *West Indian Med. J.* **15**:11.
Morgan, D. P., and C. C. Roan (1970). *Arch. Environ. Health* **20**:452.
Mount, D. I., L. W. Vigor, and M. L. Slafer (1966). *Science* **152**:1388.

Nagasaki, H., S. Tomii, T. Mega, M. Murakami, and N. Ito (1971). *Gann (Cancer)* **62**:431.
Nagasaki, H., S. Tomii, T. Mega, M. Murakami, and N. Ito (1972). *Gann (Cancer)* **63**:393.
Nakamura, N. (1960). Chronic poisoning by various insecticides in rats, *Nippon Yakurigaku Zasshi* **56**:829.
Namba, T., and K. Hiraki (1958). *J. Am. Med. Assoc.* **166**:1834.
Namba, T., M. Greenfield, and D. Grob (1970). *Arch. Environ. Health* **21**:533.
National Research Council (1965). *Some Considerations in the Use of Human Subjects in Safety Evaluation of Pesticides and Food Chemicals.* Report of the Ad Hoc Subcommittee on Use of Human Subjects in Safety Evaluation, National Academy of Sciences, National Research Council, Washington, D.C., Publ. No. 1270, 22 pp.
National Research Council, Food Protection Committee, Food and Nutrition Board (1960). *Problems in the Evaluation of Carcinogenic Hazard from Use of Food Additives.* National Academy of Sciences, National Research Council, Washington, D.C., Publ. No. 749.
Nelson, A. A., and G. Woodard (1949). *Arch. Pathol (Chicago)* **48**:387.
Nichols, J., and G. Henninger (1957). *Exptl. Med. Surg.* **15**:310.
Nichols, J., and A. W. Richardson (1960). *Proc. Soc. Exptl. Biol. Med.* **104**:539.
O'Brien, R. D. (1960). *Toxic Phosphorus Esters.* Academic Press, New York.
O'Brien, R. D. (1967). *Insecticides: Action and Metabolism.* Academic Press, New York.
Okey, A. B. (1972). *Life Sci.* **11**:833.
O'Neill, R. V., and O. W. Burke (1971). *A Simple Systems Model for DDT and DDE Movement in the Human Food Chain.* (Indirectly cited from *Report of the DDT Advisory Committee,* Environmental Protection Agency, Washington, D.C., September, 1971.)
Ortega, P. (1962). *Fed. Proc.* **21**:306.
Ortega, P., W. J. Hayes, Jr., W. F. Durham, and A. Mattson (1956). *Public Health Monogr.* No. 43. U.S. Public Health Service, Washington, D.C., Publ. No. 484.
Ortelee, M. F. (1958). *Am. Med. Assoc. Arch. Ind. Health* **18**:433.
Oser, B. L., and M. Oser (1960). *Toxicol. Appl. Pharmacol.* **2**:441.
Otsuka, J. (1971). *Ganka (Ophthalmology)* **13**:715.
Ottoboni, A. (1969). *Toxicol. Appl. Pharmacol.* **14**:74.
Peaslee, M. H., M. Goldman, and S. E. Milburn (1972). *Comp. Gen. Pharmacol.* **3**:191.
Perelygin, V. M., M. B. Shpirt, O. A. Aripov, and V. I. Ershova (1971). *Gig. Sanit.* **36**:29. (Indirectly cited from *Health Aspects of Pesticides,* Abst. 72-1024.)
Perevodchikova, N. I., L. V. Platinskiy, and V. I. Kertsman (1972). *Vop. Onkol.* **18**(11):24. (Indirectly cited from *Health Aspects of Pesticides,* Abst. 73-1227.)
Peters, D. A. V., P. D. Hardina, R. L. Singhal, and G. M. Ling (1972). *J. Neurochem.* **19**:1131.
Phillips, L., Jr., M. Steinberg, H. I. Maibach, and W. A. Akers (1972). *Toxicol. Appl. Pharmacol.* **21**:369.
Phillips, W. E. J., and G. Hatina (1972). *Nutr. Rep. Internat.* **5**:357.
Popper, H., S. S. Sternberg, B. L. Osher, and M. Oser (1960). *Cancer* **13**:1035.
Quinby, G. E., J. E. Armstrong, and W. F. Durham (1965). *Nature* **207**:726.
Radeleff, R. D. (1964). *Veterinary Toxicology.* Lea and Febiger, Philadelphia.
Rickett, F. E., K. Tyszkiewicz, and N. C. Brown (1972). *Pyrethrum Post* **11**:85.
Roan, C. C., D. P. Morgan, N. Cook, and E. H. Paschal (1969). *Bull. Environ. Contam. Toxicol.* **4**:362.
Robinson, J. (1970). *Ann. Rev. Pharmacol.* **10**:353.
Robinson, J., and C. G. Hunter (1966). *Arch. Environ. Health* **13**:558.
Rosenkranz, H. S., and S. Rosenkranz (1973). *Experientia* **28**:386.
Rosner, E., G. Pansztor, and A. Sawinsky (1971). *Egeszsegtudomany* **15**:195. (Indirectly cited from *Health Aspects of Pesticides,* Abst. 71-2406.)

Sachsse, K. R., and G. Voss (1971). *Residue Rev.* **37**:61. F. G. Gunther, and J. D. Gunther, eds. Springer-Verlag, New York.

Samuels, A. J., and T. H. Milby (1971). *J. Occup. Med.* **13**:147.

Sanchez-Medal, L., J. P. Castanedo, and F. Garcia-Rojas (1963). *New Engl. J. Med.* **269**:1365.

Sarett, H. P., and B. J. Jandorg (1947). *J. Pharmacol. Exptl. Therap.* **91**:340.

Shabad, L. M., T. S. Kolesnichenko, and T. V. Nikonova (1972). *Internat. J. Cancer* **9**:365.

Schafer, M. L., and J. E. Campbell (1966). *Pesticides in the Environment.* American Chemical Society Series 60, Washington, D.C.

Silverio, J. (1969). *J. School Health* **39**:607.

Springfield, A. C. (1972). *Diss. Abst. Internat.* **32**:5960B.

Sram, R. J. (1971). *Cesk. Hyg.* **6**(7/8):262. (Indirectly cited from *Health Aspects of Pesticides*, Abst. 72-1370.)

Sternberg, S. S., H. Popper, B. L. Oser, and M. Oser (1960). *Cancer* **13**:780.

Stevens, J. T., R. E. Stitzel, and J. J. McPhillips (1972). *J. Pharmacol. Exptl. Therap.* **181**:576.

Street, J. C. (1964). *Science* **146**:1580.

Tanabe, H. (1972). In *Environmental Toxicology of Pesticides.* F. Matsumura, G. M. Boush, and T. Misato, eds. Academic Press, New York, p. 239.

Tarjan, R., and T. Kemeny (1969). *Food Cosmet. Toxicol.* **7**:215.

Thienes, C. H., and T. J. Haley (1972). *Clinical Toxicology*, 5th ed. Lea and Febiger, Philadelphia, p. 110.

Tomatis, L. (1970). In *Proceedings of the 4th International Congress of Rural Medicine.* H. Kuroiwa, ed. Japanese Association for Rural Medicine, Tokyo, Japan.

Tomatis, L., V. Turusov, N. Day, and R. T. Charles (1972). *Internat. J. Cancer* **10**:489.

Treon, J. F. (1954). Report, Kettering Laboratory, Department of Preventive Medicine and Industrial Health, University of Cincinnati, College of Medicine, December.

Treon, J. F. (1955). Report, Kettering Laboratory, Department of Preventive Medicine and Industrial Health, University of Cincinnati, College of Medicine, February.

Treon, J. F. (1956). *The Toxicology and Pharmacology of Endrin.* Kettering Laboratory Report, University of Cincinnati.

Treon, J. F., and F. P. Cleveland (1955). *J. Agr. Food Chem.* **3**:402.

Tronko, M. D., and V. I. Kravchenko (1971). *Fiziol. Zh. (Kiev)* **17**:245. (Indirectly cited from *Health Aspects of Pesticides*, Abst. 72-1295.)

Turner, N. (1954). *Conn. Agr. Expt. Sta. New Haven Bull.*, No. 594.

Uzoukwu, M., and S. D. Sleight (1972). *Am. J. Vet. Res.* **33**:579.

Vacquier, V. D., and J. Brachet (1969). *Nature* **222**:193.

Vilar, O., and W. W. Tullner (1959). *Endocrinology* **65**:80.

Villeneuve, D. C., R. F. Willes, J. B. Lacroix, and W. E. J. Phillips (1972). *Toxicol. Appl. Pharmacol.* **21**:542.

Vogel, E. (1972). *Mutation Res.* **16**:157.

Voogd, C. E., J. J. J. A. A. Jacobs, and J. J. Van der Stel (1972). *Mutation Res.* **16**:413.

Voss, G., and K. Sachsse (1970). *Toxicol. Appl. Pharmacol.* **16**:764.

Voss, G., and J. Schuler (1967). *Bull. Environ. Contam. Toxicol.* **2**:357.

Wakeling, A. E., T. J. Schmidt, and W. J. Visek (1972). *Fed. Proc.* **31**:725.

Walker, A. I. T., D. E. Stevenson, J. Robinson, L. W. Ferrigan, and M. Roberts (1968). Tunstall Laboratory Report TL/23/68, Shell Chemical Co., Sittingbourne, Kent, Great Britain.

Warnick, S. L. (1972). *Pesticides Monitoring J.* **6**:9.

Waseda, Y. (1971). *Nippon Hoigakuzasshi (Japan. J. Legal Med.)* **25**:64.

Wassermann, M., D. Wassermann, L. Zellermayer, and M. Gon (1967). *Pesticides Monitoring J.* **1**:15.

Wassermann, M., D. Wassermann, E. Kedar, and M. Djavaherian (1971). *Bull. Environ. Contam. Toxicol.* **6**:426.

Weeks, D. E. (1967). *Bull. World Health Org.* **37**:499.

Weil, C. S. (1972). *Toxicol. Appl. Pharmacol.* **21**:454.

Weil, C. S., M. D. Woodside, and C. P. Carpenter (1972). *Toxicol. Appl. Pharmacol.* **21**:390.

Welch, R. M., W. Levin, and A. H. Conney (1969). *Toxicol. Appl. Pharmacol.* **14**:358.

West, I. (1967). *Arch. Environ Health* **15**:97.

West, I. (1969). In *Chemical Fallout*. M. W. Miller, and G. G. Berg, eds. Charles C. Thomas, Publisher, Springfield, Ill., p. 447.

Whitlock, N. W., J. E. Keil, and S. H. Sandifer (1972). *J. South Calif. Med. Assoc.* **69**:109.

WHO (1967). *Evaluation of Some Pesticide Residues in Food.* Report of a Joint Meeting of the FAO Working Party and WHO Expert Committee on Pesticide Residues, 1966, World Health Organization, Geneva, Switzerland, 32 pp.

Williams, C. H. (1969). *Toxicol. Appl. Pharmacol.* **14**:283.

Williams, C. H. (1970). *Toxicol. Appl. Pharmacol.* **16**:533.

Wills, J. H. (1969). In *Chemical Fallout*. M. W. Miller, and G. G. Berg, eds. Charles C. Thomas, Publisher, Springfield, Ill., p. 461.

Winteringham, F. P. W., and R. W. Disney (1952). *Nature* **195**:1303.

Witter, R. F. (1963). *Arch. Environ. Health* **6**:537.

Wogan, G. N. (1969). *Progr. Exptl. Tumor Res.* **11**:134.

Wolfsie, J. H. (1957). *Am. Med. Assoc. Arch. Ind. Health* **16**:403.

Wyckoff, D. W., J. E. Davies, A. Barquet, and J. H. Davis (1968). *Ann. Int. Med.* **68**:878.

Wyllie, J., J. Gabica, and W. W. Benson (1972). *Pesticides Monitoring J.* **6**:84.

Zavon, M. R., C. H. Hine, and K. D. Parker (1965). *J. Am. Med. Assoc.* **193**:837.

Author Index

Subject Index